W9-ANN-513

ADVANCES IN CHEMICAL PHYSICS

VOLUME LXXV

EDITORIAL BOARD

Advances in
CHEMICAL PHYSICS

EDITED BY

I. PRIGOGINE

University of Brussels
Brussels, Belgium
and
University of Texas
Austin, Texas

AND

STUART A. RICE

Department of Chemistry
and
The James Franck Institute
The University of Chicago
Chicago, Illinois

VOLUME LXXV

AN INTERSCIENCE® PUBLICATION
JOHN WILEY & SONS
NEW YORK • CHICHESTER • BRISBANE • TORONTO • SINGAPORE

An Interscience® Publication

Library of Congress Catalog Number: 58-9935

ISBN 0-471-62219-2

Printed in the United States of America

10 9 8 7 6 5 4 3 2 1

CONTRIBUTORS TO VOLUME LXXV

GIUSEPPE ALLEGRA, Dipartimento di Chimica, Politecnico di Milano, Milan, Italy

V. A. BENDERSKII, Academy of Sciences of the USSR, Institute of Chemical Physics, Moscow, USSR

*ALEKSANDRA BORYSOW, Physics Department, University of Texas at Austin, Austin, Texas

CLIFFORD E. DYKSTRA, Department of Chemistry, University of Illinois, Urbana, Illinois

MANFRED EIGEN, Max Planck Institut für Biophysikalische Chemie, Göttingen, Federal Republic of Germany

LOTHAR FROMMHOLD, Physics Department, University of Texas, Austin, Texas

FABIO GANAZZOLI, Dipartimento di Chimica, Politecnico di Milano, Milan, Italy

O. I. GERASIMOV, Katholieke Universitet Leuven, Laboratorium Voor Malekuulfysika, Leuven, Belgium; Odessa State University, Department of Theoretical Physics, Odessa, USSR

V. I. GOL'DANSKII, Academy of Sciences of the USSR, Institute of Chemical Physics, Moscow, USSR

ERICH, P. IPPEN, Department of Electrical Engineering and Computer Science, Massachusetts Institute of Technology, Cambridge, Massachusetts

PAOLO LAZZERETTI, Dipartimento di Chimica dell'Universitā degli Studi di Modena, Modena, Italy

*Present address: Joint Institute for Laboratory Astrophysics, University of Colorado, Boulder, Colorado.

*Shi-yi Liu, Department of Chemistry, University of Illinois, Urbana, Illinois

John McCaskill, Max Planck Institut für Biophysikalische Chemie, Göttingen, Federal Republic of Germany

David J. Malik, Department of Chemistry, Purdue University at Indianapolis, Indianapolis, Indiana

Keith A. Nelson, Department of Chemistry, Massachusetts Institute of Technology, Cambridge, Massachusetts

Peter Schuster, Institut für Theoretische Chemie und Strahlenchemie der Universität Wien, Wien, Austria

L. I. Trakhtenberg, Academy of Sciences of the USSR, Institute of Chemical Physics, Moscow, USSR

*Present address: Allied Signal Corporation, Des Plaines, Illinois.

INTRODUCTION

Few of us can any longer keep up with the flood of scientific literature, even in specialized subfields. Any attempt to do more and be broadly educated with respect to a large domain of science has the appearance of tilting at windmills. Yet the synthesis of ideas drawn from different subjects into new, powerful, general concepts is as valuable as ever, and the desire to remain educated persists in all scientists. This series, *Advances in Chemical Physics*, is devoted to helping the reader obtain general information about a wide variety of topics in chemical physics, which field we interpret very broadly. Our intent is to have experts present comprehensive analyses of subjects of interest and to encourage the expression of individual points of view. We hope that this approach to the presentation of an overview of a subject will both stimulate new research and serve as a personalized learning text for beginners in a field.

ILYA PRIGOGINE
STUART A. RICE

CONTENTS

ADVANCES IN CHEMICAL PHYSICS

VOLUME LXXV

FEMTOSECOND COHERENT SPECTROSCOPY

KEITH A. NELSON*

Department of Chemistry
Massachusetts Institute of Technology
Cambridge, Massachusetts 02139

ERICH P. IPPEN

Department of Electrical Engineering and Computer Science
Massachusetts Institute of Technology
Cambridge, Massachusetts 02139

CONTENTS

I. INTRODUCTION

The development of laser sources that can produce pulses of duration under 100 femtoseconds (fs) began in 1981 [1] and has progressed rapidly since [2]. Clearly this has great significance for many areas of physical chemistry. It is

*Alfred P. Sloan Fellow and Presidential Young Investigator Awardee.

now possible, in principle at least, to time-resolve almost any nuclear motion including collective vibrations in condensed media, molecular orientational motion and intermolecular collisions in gases and liquids, and even many intramolecular vibrations. Since these elementary motions of atoms and molecules are involved in most events of chemical interest (e.g., chemical bond formation and breakage in any phase of matter and local and collective structural rearrangement in condensed phases), the prospect exists for direct time-resolved observation of such events. It is also possible to time-resolve electronic excited-state relaxation and dephasing in many previously inaccessible cases, permitting mechanistic questions concerned with electron–electron and electron–nuclear interactions to be addressed in detail.

Although speculation about femtosecond time-resolved physical chemistry experiments has far outpaced their execution, it is fair to say that the outlook of experimental physical chemists has been altered significantly by the possibilities. In molecular spectroscopy, time-resolved observation of nonequilibrium molecular species including transition states is now discussed routinely. Molecular dynamics in liquids and other condensed phases are discussed in the context of experimental observation on the time scale of intermolecular collisions and comparison to corresponding computer simulations. The focus of study of electronic excited-state dynamics in semiconductors and metals, molecular crystals, light-sensitive biological species, and other condensed materials is shifting increasingly toward the femtosecond regime.

Several recent reviews have presented broad overviews of ultrafast time-resolved spectroscopy [3–6]. We shall concentrate instead on a selected, rather small subset of femtosecond time-resolved experiments carried out (and to a very limited extent, proposed) to date. In particular, we shall review experiments in which phase-coherent electronic or, more often, nuclear motion is induced and monitored with time resolution of less than 100 fs. The main reason for selectivity on this basis is the rather ubiquitous appearance of phase-coherent effects (especially vibrational phase coherence) in femtosecond spectroscopy. As will be discussed, nearly any spectroscopy experiment on molecular or condensed-phase systems is likely to involve phase-coherent vibrational motion if the time scale becomes short enough. Since the coherent spectral bandwidth of a femtosecond pulse often exceeds collective or molecular vibrational frequencies, such a pulse may perturb and be perturbed by a medium in a qualitatively different manner than a longer pulse of comparable peak power. The resulting spectroscopic possibilities are of special interest to these reviewers.

Measurements of electronic phase coherence and its decay are discussed in the next section. Pump–probe experiments with two incident laser pulses and polarization grating experiments with three incident pulses are described. Determination of homogeneous and inhomogeneous dephasing dynamics in organic dye molecules is reviewed.

The last section begins by showing that coherent vibrational (in general, nuclear) motion on ground- or excited-state potential surfaces can be initiated by a femtosecond pulse through either "impulsive" stimulated Raman scattering or optical absorption. Subsequent ultrashort pulses can be used to probe the coherently vibrating species at various stages of vibrational distortion, that is, to time-resolve individual cycles of vibrational oscillation. Experiments of this type on phonon and polariton dynamics in crystalline solids, molecular librational motion in liquids, and intramolecular vibrations are reviewed. Femtosecond time-resolved photochemistry experiments in which bond breakage or formation is monitored are also discussed. Finally, anticipated progress in optical control of molecular motion on ground- and excited-state potential surfaces is summarized, with attention focused on the prospects for "phase-coherent chemistry" and time-resolved observation of transition states.

II. ELECTRONIC PHASE COHERENCE

A. Measurement of Electronic Energy Relaxation and Dephasing

In the first reported measurements made with picosecond pulses, an optical beam splitter was used to pick off a portion of the pulse train and a variable optical delay path was introduced between the two beams [7]. The main beam was used to excite (pump) a dye sample, and the weak (probe) beam was used to monitor the recovery of dye transmission as a function of delay. Over the past two decades, this pump–probe method has been extended to a variety of measurement geometries and used to measure electronic polarization dephasing times as well as population lifetimes.

Interactions of pump and probe pulses with a material absorption are usually described by density matrix equations for a distribution of two-level or three-level systems. The formulation of these equations can be found in textbooks and other reviews [8, 9]. We try here simply to describe in physical terms the ways in which the different parameters of the equations manifest themselves experimentally. The simplest theoretical case is the two-level system, in which T_2 is the dephasing time of the coherently induced electronic polarization and T_1 is the energy relaxation time. We begin our discussion by simplifying even further and assuming that T_2 is very short compared to the optical pulse durations. Then, the coherent polarization follows the optical field, and the induced change in absorption (due to excitation of population) has a temporal behavior given by

$$\Delta\alpha \sim \int_{-\infty}^{t} e^{-(t-t')/T_1} E_1(t')E_1^*(t')\,dt', \tag{1}$$

where $E_1(t)$ is the electric field of the incident pump pulse. The detected signal in a pump–probe measurement is the change in the energy of the transmitted probe pulse. It has the form

$$\gamma(\tau) \sim \int_{-\infty}^{\infty} E_2(t-\tau)E_2^*(t-\tau) \int_{-\infty}^{t} e^{-(t-t')/T_1} E_1(t')E_1^*(t')\,dt'\,dt, \tag{2}$$

where $E_2(t)$ is the electric field of the probe pulse and τ is the temporal delay between the pump and probe pulses. For a more complicated interaction in which the two fields may have different polarizations and the energy relaxation is nonexponential, Eq. (2) is easily extended to

$$\begin{aligned}\gamma(\tau) &\sim \int_{-\infty}^{\infty} E_2(t-\tau)E_2^*(t-\tau) \int_{-\infty}^{t} R_{2211}(t-t')E_1(t')E_1(t')\,dt'\,dt \\ &= \int_{-\infty}^{\tau} R_{2211}(\tau-\tau') \int_{-\infty}^{\infty} I_2(t-\tau')I_1(t)\,dt\,d\tau',\end{aligned} \tag{3}$$

which is the convolution of the energy (in general, susceptibility) relaxation function R_{2211} with the (separately measureable) intensity correlation function of the optical pulses. This description, however, ignores the optical interferences of the two pulses when they overlap in time. That interference gives rise to an additional detected signal of the form

$$\beta(\tau) \approx \int_{-\infty}^{\infty} E_2^*(t-\tau)E_1(t) \int_{-\infty}^{\infty} R_{2121}(t-t')E_2(t'-\tau)E_1^*(t')\,dt'\,dt. \tag{4}$$

This signal, often referred to as the "coherent artifact", has been discussed by many authors [10–14]. It may be thought of as a scattering of the pump beam into the probe beam (and vice versa) by the periodic interference they produce in the medium. For the case $E_1(t) = E_2(t)$, $\gamma(0) = \beta(0)$. If the pump and probe have different polarizations, these magnitudes may not be equal. Reorientation (or loss of polarization memory within the medium) during the pulse can reduce $\beta(\tau)$. If pump and probe have different frequencies, $\beta(\tau)$ depends upon the ability of the material response to follow the beat. Thus, this coherent coupling term may also be used to obtain information about both the medium and the optical pulses. Recent experiments have shown that it can have dramatic spectral as well as temporal manifestations [15, 16].

Pump–probe experiments of the type described in the preceding can be used to study energy relaxation in both excited and ground states. Even with pump and probe pulses of the same wavelength, vibrational relaxation in the electronic excited state, solvent reorganization, and other processes may be distinguished from T_1 recovery to ground state by their different temporal signatures [17]. With a broadband femtosecond continuum probe, excited-state spectral dynamics can be studied in detail [18]. In these experiments, the coupling to vibrational modes may appear as transient sidebands of a spectral hole burned into the absorption spectrum by the pump pulse.

B. Polarization Dephasing

Different pump–probe coupling phenomena arise if T_2 is not negligibly small compared to the pulse duration. In this domain, studies of coherent transients [19] [the photon echo [20, 21] is a notable example] make it possible to distinguish between different, homogeneous and inhomogeneous, polarization dephasing mechanisms. Consider first the effects of extended polarization coherence on our two-pulse pump–probe experiment. One direct consequence is a distortion of the pump–probe signal described previously in the region of pulse overlap [14]. This distortion does not, however, depend strongly upon either T_2 or the dephasing mechanism. A more clear signature is given by light scattered ("self-diffracted") into a different direction. See Fig. 1. When the noncollinear pump and probe pulses overlap in the absorber, they produce a periodic (gratinglike) excitation that couples them together. It also diffracts the probe beam into the direction defined by $2\mathbf{k}_2 - \mathbf{k}_1$, where $\mathbf{k}_2$ and $\mathbf{k}_1$ are the propagation vectors of the probe and pump beams, respectively. Even when the probe pulse arrives delayed with respect to the pump, it can still produce a grating and scatter into $2\mathbf{k}_2 - \mathbf{k}_1$ if the coherent polarization created by the pump pulse has not completely died out (i.e., been dephased). In the limit of T_2 much greater than the pulse durations, the scattered energy depends upon delay τ between pump and probe as

$$S(\tau > 0) \sim e^{-2\tau/T_2} \tag{5}$$

if the system is homogeneously broadened. If the system is very inhomogeneous, this dependence changes to

$$S(\tau > 0) \sim e^{-4\tau/T_2} \tag{6}$$

and the signal is an "echo" that appears after a delay τ following the probe pulse. When the approximation of long T_2 is not valid, the situation is more complicated, but the results are easily calculated from third-order density matrix perturbation theory. The symmetry of the situation implies that when the pump is delayed with respect to the probe, a similar scattering signature should be observed in the direction $2\mathbf{k}_1 - \mathbf{k}_2$. Thus, when T_2 effects are present, there will be a temporal asymmetry between scattering in the two directions [22, 23]. Conversely, such an asymmetry might be taken as an indication of polarization coherence. Unfortunately, this latter assumption is not always true. Asymmetry in the scattering depends not only on T_2 but also on T_1 and the pulse shape [24]. Figure 2 illustrates this asymmetry for two different pulseshapes in the limit $T_2 \ll T_p$, the pulsewidth FWHM (full width at half maximum), and $T_1 \gg T_p$. The experimental points were taken using a methanol solution of the dye malachite green as the sample. These data, along with intensity autocorrelation measurements of the scattered pulse durations, have been interpreted to mean that $T_2 < 40$ fs in this dye [24].

With a three-pulse geometry it is possible to remove the ambiguity caused by T_1 and pulseshape [25, 26]. Such a geometry is given in Figure 3. A third pulse is used to monitor the grating established by the first two pulses. We emphasize, however, that in contrast to most transient grating experiments, including those described in the next section, the important parameter here is the delay between the first two pulses. The third pulse scatters ("diffracts") into the background-free directions $\mathbf{k}_4 = \mathbf{k}_3 + (\mathbf{k}_1 - \mathbf{k}_2)$ and $\mathbf{k}_5 = \mathbf{k}_3 - (\mathbf{k}_1 - \mathbf{k}_2)$. For pulses much shorter that the inverse absorption spectral width of the sample, the scattered energies exhibit unique signatures of dephasing. If the absorbing system is homogeneously broadened, scattering into the two directions is always symmetric with regard to the delay between pulses 1 and 2:

$$S_4(\tau) = S_5(\tau) \sim e^{-2|\tau|/T_2}. \tag{7}$$

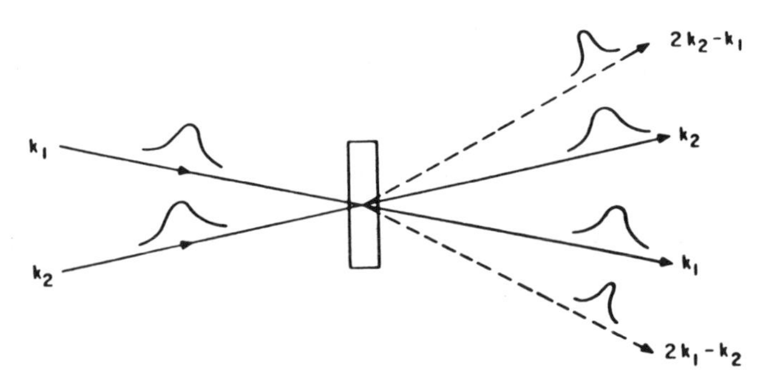

Figure 1. Interaction geometry for two-pulse studies of polarization coherence.

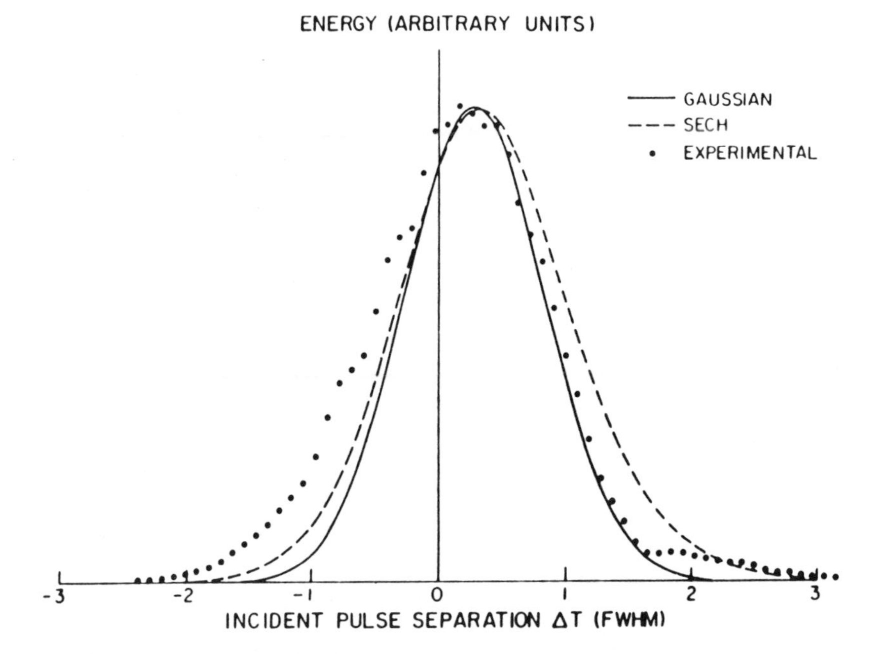

Figure 2. Total energy (integrated intensity) of scattered signal versus incident pulse separation ΔT in units of incident pulse FWHM. The experimental data were plotted assuming Gaussian incident pulses.

On the other hand, if the system is inhomogeneously broadened, scattering is no longer symmetric. In the limit of predominantly inhomogeneous broadening,

$$S_4(\tau) = 0, \qquad S_5(\tau) \sim e^{-4\tau/T_2} \tag{8}$$

for $\tau > 0$, and

$$S_4(\tau) \sim e^{4\tau/T_2}, \qquad S_5(\tau) = 0 \tag{9}$$

for $\tau < 0$. Here positive τ indicates that pulse 1 precedes pulse 2. In this case, the diffracted pulse may be described as a stimulated echo [27] delayed by τ relative to pulse 3.

This asymmetry provides a clear and simple criterion for differentiating between the two types of line broadening. It can be explained in the following way. Following excitation by the first pulse, the coherent polarizations of the subsystems (of an inhomogeneous system) assume their natural frequencies

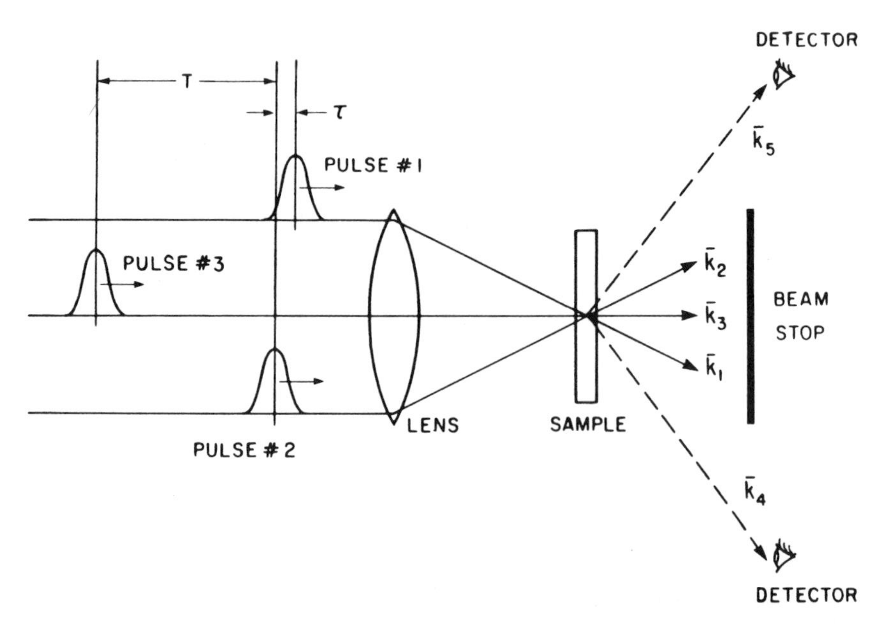

Figure 3. Experimental geometry for dephasing measurements by three-pulse scattering technique. Pump pulses (1 and 2) produce a grating response that diffracts probe pulse (3) into two orders. In grating experiments described in Section III (nuclear phase coherence), $\tau = 0$ and probe pulse is incident at the phase-matching angle for Bragg diffraction into only one order.

and develop relative phase shifts with time. The second pulse then interacts with these polarizations to produce a set of subsystem population gratings shifted spatially with respect to each other. Arrival of pulse 3 then generates for each grating a third-order polarization whose initial phase is determined by its spatial shift. At time $T + \tau$, all of the polarization components interfere constructively to form a phased array for radiation in direction $\mathbf{k}_4$ for $\tau < 0$ and $\mathbf{k}_5$ for $\tau > 0$. Thus, for a fixed T, scattering occurs preferentially in a single direction.

For pulses much longer than T_2, the expression for the scattered energy is simply the envelope of the electric field autocorrelation squared for both the homogeneous and inhomogeneous cases:

$$S_4(\tau) = S_5(\tau) = \left| \int_{-\infty}^{\infty} E_2(t-\tau) E_1^*(\tau)\, dt \right|^2. \tag{10}$$

This property has been used to measure the coherence properties of laser pulses [28]. Because it is readily obtainable by Fourier transform from the pulse spectrum, the $T_2 = 0$ limit can be determined experimentally [25]. Therefore, fast dephasing times can be resolved by looking for small differences between the scattering data and the transform-determined instantaneous response. The results of two experimental comparisons are shown in Figure 4. Here, T_2 is limited by the inverse dye absorption width, so dashed curves are calculated using the Fourier transform of the pulse spectrum multiplied by the dye absorption spectrum. Since the absorption curves of the two dyes overlap the pulse spectrum differently, different scattering curves are expected. The solid lines are thermal grating scattering data obtained to confirm this prediction. Information about system inhomogeneity was then obtained by comparing these curves to data taken using orthogonal polarizations for pulses 1 and 2 to eliminate the thermal effect. Data for a variety of dyes in solution at room temperature indicated homogeneous dephasing on a time scale of less than 20 fs [26].

True evidence of inhomogeneous broadening has been obtained from three-pulse studies of dye molecules in thin films of polymethyl methacrylate (PMMA) [29]. Figure 5 shows the experimental results. The curves of Figure 5*a* demonstrate that the system is homogeneous at room temperature. Scattering in both directions is symmetric about $T = 0$ and is determined by pulse coherence. The data of Figure 5*b* clearly show asymmetry indicating that inhomogeneous broadening is present. It is interesting that most of the

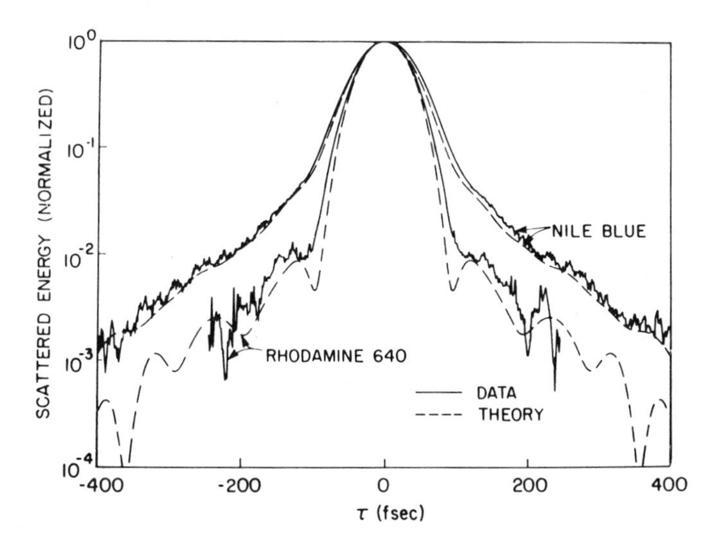

Figure 4. Three-pulse scattering data for Rhodamine 640 and Nile Blue in methanol using parallel polarizations. Solid lines, integrated scattered intensity; dashed lines, obtained by Fourier transform from pulse spectrum modified by dye absorption.

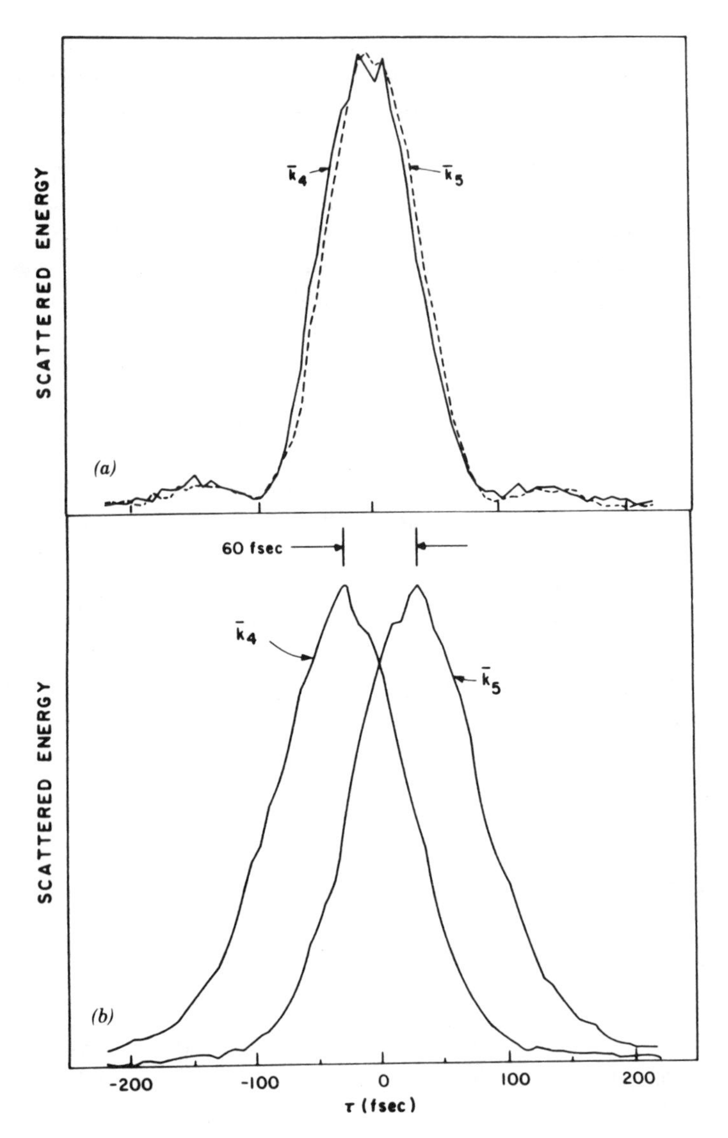

Figure 5. Measured scattering energy for cresyl violet in PMMA as function of delay between pulses 1 and 2. The delay of the third pulse was set to 1.3 ps. Temperatures: (*a*) 290 K; (*b*) 15 K.

asymmetry is in the peak shift; there is no corresponding $\exp(-4|\tau|/T_2)$ behavior in the wings of the curves. The explanation apparently lies in the fact that the molecular absorption must be described by a multilevel system rather than two-level structures. Unlike inhomogeneous dephasing, beating between several lines of a single molecule leads to irreversible dephasing of the coherent polarization associated with it. The asymmetry of the data in Figure 5*b* indicates that the complicated spectra of individual molecules are indeed resolvable at low temperatures but that the irregularity of their level spacings suppresses the tail of the time domain data. As the temperature increases, the individual lines of a molecule become unresolvable. Each multilevel molecule behaves as a homogeneously broadened two-level system with a linewidth equal to the total absorption width. In this context, it should be pointed out that it is possible to burn holes in the absorption spectra of multilevel molecules even though they are homogeneously broadened from a dephasing part of view. The difference between the multilevel case and that of an inhomogeneous distribution of two-level systems is that in the former each molecule has the same absorption spectrum.

The three-pulse scattering geometry offers the possibility of extracting still more information about the sample under investigation by variation of T, the arrival time of the third pulse. As time progresses after interference of pulses 1 and 2, the spatially chirped population gratings of an inhomogeneous system can lose their inhomogeneous character through spectral cross-relaxation. If this spectral cross-relaxation proceeds more rapidly than energy relaxation through T_1, the asymmetry of the scattering will also change as a function of T. Observation of this change would provide information about configurational or environmental fluctuation in the medium. With dye molecules in the polymer, no spectral cross-relaxation could be detected [24], implying that for the temperature at which the medium became inhomogeneous very little environmental change occurs within T_1. There should, however, be other molecular systems for which this probe of spectral cross-relaxation would be especially useful.

Finally, it should be noted that dephasing dynamics may also be investigated with temporally incoherent pulses [30–32]. To the extent that the two- and three-pulse T_2 measurements described in the preceding depend upon the electric field correlation and not upon the intensity autocorrelations, femtosecond temporal resolution may be obtained with any broadband source. Spectral cross-relaxation and a variety of T_1 effects, however, can reduce the effectiveness of this technique.

III. NUCLEAR PHASE COHERENCE

When a femtosecond laser pulse passes through *nearly any medium*, coherent vibrational excitation (in general, initiation of coherent wavepacket propagation) is likely [33, 34]. One- or two-photon absorption of a visible or ultraviolet pulse into an electronic excited state can result in phase-coherent motion in the excited-state potential [35]. Impulsive stimulated Raman scattering can initiate phase-coherent vibrational motion in the electronic ground state [33, 34]. In noncentrosymmetric materials, phonon–polariton modes can be excited coherently through the inverse electro-optic effect [36]. The main criterion for any of these coherent excitation mechanisms is that the laser pulse duration be short compared to a single oscillation cycle of the vibrational mode that is to be excited. Equivalently, the transform-limited bandwidth of the pulse must exceed the vibrational frequency.

The coherent motion initiated by an excitation pulse can be monitored by variably delayed, ultrashort probe pulses. Since these pulses may also be shorter in duration than the vibrational period, individual cycles of vibrational oscillation can be time resolved and spectroscopy of vibrationally distorted species (and other unstable species) can be carried out. In the first part of this section, the mechanisms through which femtosecond pulses may initiate and probe coherent lattice and molecular vibrational motion are discussed and illustrated with selected experimental results. Next, experiments in the areas of liquid state molecular dynamics and chemical reaction dynamics are reviewed. These important areas can be addressed incisively by coherent spectroscopy on the time scale of individual molecular collisions or half-collisions.

A. Phase-Coherent Excitation Mechanisms

1. *Impulsive Stimulated Scattering*

Most stimulated scattering experiments are carried out by overlapping spatially and temporally two laser outputs of appropriately tuned frequencies and wave vectors $(\omega_1, \mathbf{k}_1)$ and $(\omega_2, \mathbf{k}_2)$ to excite coherent phonons (or other Raman-active modes) of the difference frequency and wave vector $(\Omega = \omega_1 - \omega_2, \mathbf{q}_0 = \mathbf{k}_1 - \mathbf{k}_2)$ [37]. The excitation process can be described by the stimulated scattering equation of motion [9] for a damped, nondispersive vibrational mode,

$$\rho\left(\frac{\partial^2 Q}{\partial t^2} + 2\gamma\frac{\partial Q}{\partial t} + \omega_0^2 Q\right) = \sum_{i,j} \frac{a_{ij} E_i E_j}{8\pi}, \tag{11}$$

where Q is the normal vibrational coordinate, ρ is the reduced mass density or moment of inertia, ω_0 is the natural undamped frequency, and γ is the dephasing rate of the mode. The right-hand side of Eq. (11) is the driving term in which $\mathbf{E}$ is the net electric field and $a_{ij} = (\partial\varepsilon_{ij}/\partial Q)_0$, where ε is the dielectric tensor, is the light-scattering coupling constant for incident and scattered light of polarizations i and j. For long pulses or continuous-wave (cw) lasers, the two excitation fields take the form $e^{i(\omega_1 t - \mathbf{k}_1 \cdot \mathbf{r})}$ + c.c. (complex conjugate) and $e^{i(\omega_2 t - \mathbf{k}_2 \cdot \mathbf{r})}$ + c.c., and the product contains difference-frequency terms that drive a coherent traveling-wave vibrational response, that is, $E_i E_j \sim \cos(\Omega t - \mathbf{q}_0 \cdot \mathbf{r}) \sim Q(\mathbf{r}, t)$.

For pulses whose duration τ_L is shorter than the vibrational period, $\tau_{\rm osc} = 2\pi/\Omega$, the transform-limited spectral bandwidth exceeds the vibrational frequency $\Omega = (\omega_0^2 - \gamma^2)^{1/2}$. In this case, it is not necessary to overlap two pulses of different, discrete frequencies ω_1 and ω_2. Two short pulses with the *same* central frequency ω_L can be crossed, and stimulated Raman scattering will occur through mixing among their Fourier components [33]. The excitation fields (assuming Gaussian temporal profiles) take the form

$$e^{-t^2/2\tau_L^2}[e^{i(\omega_L t - \mathbf{k}_1 \cdot \mathbf{r})} + \text{c.c.}] \quad \text{and} \quad e^{-t^2/2\tau_L^2}[e^{i(\omega_L t - \mathbf{k}_2 \cdot \mathbf{r})} + \text{c.c.}].$$

The driving term in Eq. (11) contains terms of the form

$$E_i E_j \sim \delta(t) \cos \mathbf{q}_0 \cdot \mathbf{r} \sim \delta(t)[\delta(\mathbf{q} + \mathbf{q}_0) + \delta(\mathbf{q} - \mathbf{q}_0)], \tag{12}$$

where the approximation of an "impulse" (i.e., delta function) driving force holds if $\tau_L \ll \tau_{\rm osc}$. Equation (12) gives expressions for $E_i E_j$ as a function of $(\mathbf{r}, t)$ or $(\mathbf{q}, t)$, the latter of which will be used in what follows. Equations (11) and (12) show that the crossed excitation pulses exert a spatially periodic, temporally impulsive driving force on a vibrational mode. This produces a standing-wave vibrational response:

$$Q(\mathbf{q}, t) \propto a_{ij}[\delta(\mathbf{q} + \mathbf{q}_0) + \delta(\mathbf{q} - \mathbf{q}_0)] G^Q(t), \tag{13}$$

where $G^Q(t)$ is the impulse response function (Green's function) given for underdamped modes by

$$G^Q(t) = e^{-\gamma t} \sin \Omega t / \rho \Omega. \tag{14}$$

The standing-wave vibrational oscillations are most easily detected by coherent scattering of a variably delayed probe pulse phase matched for "diffraction" with scattering wave vector $\mathbf{q}_0$. The oscillations give rise to spatial modulation of the dielectric tensor:

$$\delta\varepsilon_{kl}(\mathbf{q}, t) = \left(\frac{\partial \varepsilon_{kl}}{\partial Q}\right)_0 Q(\mathbf{q}, t) = a_{kl} Q(\mathbf{q}, t). \tag{15}$$

For incident and diffracted probe polarizations k and l and excitation pulse polarizations i and j, the intensity of the diffracted signal is

$$I(\mathbf{q}, t) \propto |\delta\varepsilon_{kl}(\mathbf{q}, t)|^2 \propto |a_{ij} a_{kl} G^Q(\mathbf{q}, t)|^2 = |G_{ijkl}^{\varepsilon\varepsilon}(\mathbf{q}, t)|^2, \tag{16}$$

where $\mathbf{G}^{\varepsilon\varepsilon}$ is the dielectric tensor impulse response function (often called the nonlinear optical susceptibility and labeled χ). For an underdamped vibrational mode, Eqs. (14) and (16) show that

$$I(t) \propto e^{-2\gamma t} \sin^2 \Omega t \propto e^{-2\gamma t}(1 - \cos \Omega t). \tag{17}$$

Impulsive stimulated Raman scattering (ISRS) signal shows oscillations at twice the vibrational frequency and decay at twice the vibrational dephasing rate.

In general, several Raman-active modes may be excited simultaneously by the ISRS excitation pulses and may contribute to coherent scattering of the probe pulse. The dielectric response tensor $\mathbf{G}^{\varepsilon\varepsilon}(\mathbf{q}, t)$ may be written as a sum of terms representing different material modes. In addition, depending on the polarizations used, several tensor components of $\mathbf{G}^{\varepsilon\varepsilon}(\mathbf{q}, t)$ may be sampled in a given experiment. For simplicity, we write

$$I(\mathbf{q}, t) \propto |G^{\varepsilon\varepsilon}(\mathbf{q}, t)|^2, \tag{18}$$

where the scalar $G^{\varepsilon\varepsilon}$ is taken to mean the projection of $\mathbf{G}^{\varepsilon\varepsilon}$ that is sampled by the polarizations chosen [33, 38].

Using picosecond laser pulses, acoustic phonons in liquids and solids have been characterized through impulsive stimulated Brillouin scattering [39–43]. With femtosecond pulses, higher frequency excitations including optic phonons [44–46] and molecular vibrations [34, 47–49] can be studied through ISRS. Figure 6 shows temperature-dependent ISRS data from optic phonons in the organic molecular crystal α-perylene, whose excimer formation reaction will be discussed further in what follows. The data show a "beating" pattern because two phonon modes of energies 80 and 104 cm^{-1} are excited. In terms of Eq. (16), the signal takes the form $I(t) \propto (a_1^2 e^{-\gamma_1 t} \sin \Omega_1 t + a_2^2 e^{-\gamma_2 t} \sin \Omega_2 t)^2$, where the subscripts label the modes. Sum and difference frequency terms contribute to the data. From this type of data, temperature-dependent phonon dephasing rates were determined. Similar data have been recorded from the isomorphous excimer-forming crystal pyrene [45] and other crystals [33].

Figure 7 shows ISRS data from dibromomethane liquid. The oscillations in the data correspond to oscillations in the 173-cm^{-1} bromine "bending" mode. The data also contain a nonoscillatory contribution due to molecular orientational motion. Both vibrational and orientational dynamics are accounted for in the fit [34, 47].

The impulsive stimulated scattering (ISS) experiment is a stimulated, time domain analog of spontaneous, frequency domain light-scattering (LS) spec-

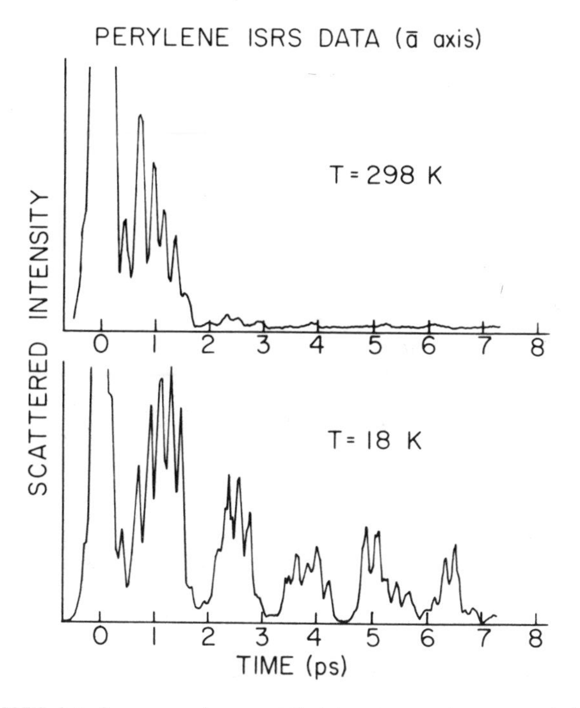

Figure 6. ISRS data from α-perylene crystal at two temperatures, recorded with transient grating experimental arrangement. Oscillations in each sweep due to 80- and 104-cm^{-1} optic phonons. Data contain sum and difference frequencies that produce "beating" pattern.

troscopy. The connection between the two is clear from the classical limit expressions for the LS spectrum [50, 51],

$$I(\mathbf{q}, \omega) \propto (k_B T/\omega)\mathrm{Im}[G^{\varepsilon\varepsilon}(\mathbf{q}, \omega)] \propto \int_{-\infty}^{\infty} dt\, e^{i\omega t} C^{\varepsilon\varepsilon}(\mathbf{q}, t), \tag{19}$$

where the frequency-dependent response function $G^{\varepsilon\varepsilon}(\omega)$ is the Fourier transform of $G^{\varepsilon\varepsilon}(t)$. The second part of Eq. (19) relates the LS spectrum to the dielectric tensor time correlation function $\mathbf{C}^{\varepsilon\varepsilon}(t)$, whose elements are $C^{\varepsilon\varepsilon}_{ijkl}(\mathbf{q}, t) = \langle \varepsilon^*_{ij}(\mathbf{q}, 0)\varepsilon_{kl}(\mathbf{q}, t)\rangle$. The connection between ISS data and $\mathbf{C}^{\varepsilon\varepsilon}$ is made through the relation

$$\mathbf{G}^{\varepsilon\varepsilon}(t > 0) = -(k_B T)^{-1}\frac{\partial}{\partial t}[\mathbf{C}^{\varepsilon\varepsilon}(t)]. \tag{20}$$

A detailed comparison of ISS and LS techniques, including simulated data

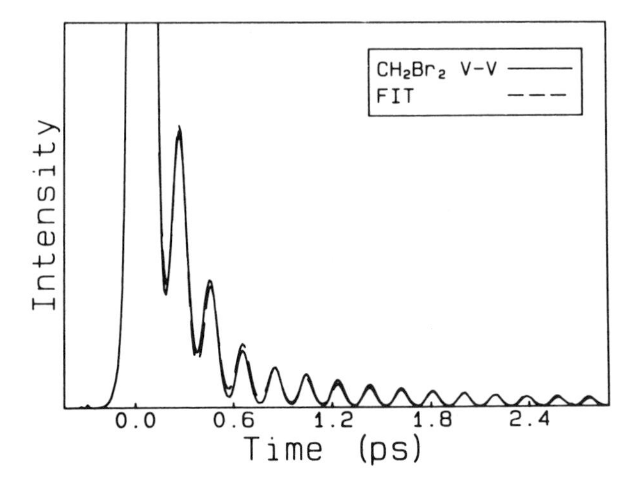

Figure 7. Time-resolved observation of molecular vibrations in CH_2Br_2 liquid recorded with a transient grating experimental arrangement using V-polarized pulses. The 5.2-THz oscillations correspond to oscillations of 173-cm^{-1} bromine "bending" mode which was excited coherently through ISRS.

and a discussion of their relative advantages in different cases, has been presented [33, 52].

The data shown in the preceding were collected using spatially and temporally overlapped femtosecond excitation pulses as shown in Figure 1 (with $\tau = 0$). However, it was predicted theoretically [53] and later shown experimentally [47] that a *single* ultrashort laser pulse will also excite a coherent vibrational response through ISRS. For an i-polarized, z-propagating excitation pulse the driving force in Eq. (11) contains terms of the form $E_i^2 \sim \delta(t')$, where the finite speed c/n of the pulse in the sample (neglected earlier) is accounted for by defining a "local time" variable $t' = t - zn/c$. The single pulse exerts a temporally impulsive, *spatially uniform* force on a Raman-active mode. The vibrational response is therefore also spatially uniform:

$$Q(z, t) \propto a_{ii} G^Q(t') \propto e^{-\gamma(t - zn/c)} \sin(\Omega t - qz), \tag{21}$$

where $q = \Omega n/c$ is the vibrational wave vector that is collinear with that of the pulse (i.e., along the z axis). Physically, the excitation pulse first drives a vibrational response in the front of the sample, then the middle, and then the back, so the vibrational phase varies linearly with z. The vibrational wavelength is equal to the distance light travels in the sample during one vibrational period; that is, the optical and vibrational phase velocities are equal ($\Omega/q = c/n$). The ISRS excitation process with one pulse can be understood as forward Raman scattering that is stimulated because the fundamental and Stokes-shifted frequencies are already present in the incident pulse [53]. *Forward-ISRS* excitation occurs *whenever* a sufficiently short pulse passes through a Raman-active medium.

The traveling-wave excitation described by Eq. (21) affects the dielectric tensor, as described by Eq. (15). The effects can be detected by a variably delayed probe pulse that is phase matched for coherent scattering, that is, collinear (in practice, nearly collinear) with the excitation pulse and the vibrational wave vector. Since the probe pulse follows the excitation pulse through the sample at the same speed c/n (neglecting dispersion), it "surfs" along a crest or null of the vibrational wave. The probe pulse therefore encounters each region of the sample with identical coherent vibrational distortion.

The probe pulse is altered in several ways as it propagates through the medium. Its transit time through the sample, its polarization properties, and its spectral content can all be measured to reveal the presence of the traveling wave.

For an i-polarized probe pulse, the speed of propagation through the sample oscillates as a function of delay relative to the excitation pulse since at any point in the sample $\delta\varepsilon_{ii}(q, t) \sim G^{\varepsilon\varepsilon}_{iiii}(q, t) \sim (a_{ii})^2 Q(q, t)$ oscillates at the vibrational frequency. The transit time of a variably delayed probe pulse can be measured by autocorrelation or other techniques [54]. Similar effects can be measured for a j-polarized probe pulse since $\delta\varepsilon_{jj}(q, t) \sim G^{\varepsilon\varepsilon}_{iijj}(q, t) \sim a_{ii}a_{jj}Q(q, t)$ oscillates as well. Since in general $a_{ii} \neq a_{jj}$, the sample birefringence, given by $\delta\varepsilon_{ii} - \delta\varepsilon_{jj}$, also shows an oscillatory time dependence that can be determined in an optical Kerr effect (or other) experimental configuration. Such experiments have been carried out in CH_2Br_2 and other molecular liquids [34, 48, 49].

The probe pulse may also undergo oscillatory changes in its *spectral content* due to coherent scattering by the vibrational wave. The unique effects that occur can be understood by consideration of the impulsive forces exerted by the excitation *and probe* pulses on the vibrational mode. Just like successive "pushes" of a pendulum, the two pulses exert driving forces on the vibrational mode that may be in or out of phase. If the probe pulse arrives in phase, that is, delayed by an integral multiple of vibrational periods, it amplifies the vibrational motion, gives up energy to the vibrational wave, and emerges red shifted. If the probe pulse arrives out of phase, it *opposes* the coherent vibrational motion, takes up energy from the vibrational wave, and emerges blue shifted. The spectrum of the transmitted probe pulse therefore "wags" back and forth at the vibrational frequency. Observation of this effect in dibromomethane, as shown in Figure 8, confirmed the general occurrence

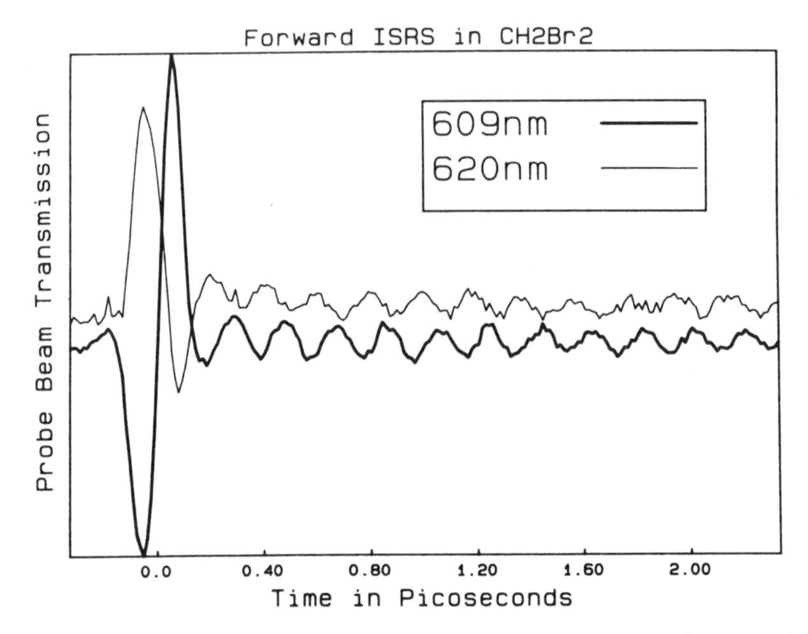

Figure 8. Molecular vibrations in CH_2Br_2 excited by single pulse and monitored by measuring intensity of different spectral components of transmitted probe pulse centered at 615 nm. Probe pulse spectrum alternately red and blue shifts, leading to antiphased oscillations in intensities of red (620-nm) and blue (609-nm) components.

of ISRS excitation with a single ultrashort pulse [47]. Similar observations from molecular crystals [46] are discussed subsequently.

Finally, forward ISRS excitation can be followed by measurement of time-dependent absorption or Raman spectra, second harmonic generation efficiency, or any other optical property that may be affected by vibrational distortion. Preliminary time-resolved absorption spectroscopy of nonequilibrium, vibrationally distorted species is discussed further in the next section [46].

2. *Inverse Electro-optic Effect*

In noncentrosymmetric crystals, difference frequency mixing of two optical frequency inputs can occur to generate far-IR radiation [9]. This can be thought of as the inverse electro-optic effect (or optical rectification) since two optical fields are mixed to produce a low-frequency or dc field. If a single ultrashort pulse is used, far-IR radiation can be generated through mixing among its Fourier components. This was first carried out with picosecond pulses, producing far-IR output with frequencies up to 15 cm^{-1} [55]. More recently, 50-fs pulses were used and the IR output spanned the frequency range 0.1–5 THz [56]. The IR output was an electromagnetic pulse whose duration was approximately 300 fs, or one cycle of $\sim$ 5-THz radiation. In these experiments the speed of the optical pulse that generates the far-IR output exceeds the IR phase velocity due to dispersion in the dielectric constant. The condition for Cherenkov radiation is therefore met, and the far-IR output radiates outward from the optical beam path in a conical "shock wave" [57, 58].

Unlike ISS, the electro-optic effect (or its inverse) can occur only in noncentrosymmetric media and in general does not lead to any real material excitation. However, if there are low-frequency IR-active modes in the crystal, they may be excited impulsively [36, 59]. Such phonons couple strongly to IR radiation to form mixed modes called polaritons. Impulsive stimulated polariton scattering can be described approximately by coupled equations of motion for the polarization contributions P_i and P_e due to ionic motions (i.e., phonons) and electronic motions, respectively [9, 60]:

$$\frac{\partial^2 P_i}{\partial t^2} + \frac{\partial P_i}{\partial t} + \omega_i^2 P_i = \alpha_i E + \beta_i P_e^2, \tag{22}$$

$$\frac{\partial^2 P_e}{\partial t^2} + \gamma_e \frac{\partial P_e}{\partial t} + \omega_e^2 P_e = \alpha_e E + 2\beta_e P_e P_i, \tag{23}$$

where γ_i and γ_e are ionic and electronic damping factors, ω_i and ω_e are the ionic and electronic resonant frequencies, α and β are linear and nonlinear coupling factors, and E is the total electric field. The high-frequency optical field drives the ionic response through the P_e^2 term in Eq. (22), which is analogous to the driving term of Eq. (11). The resulting oscillations couple linearly to the IR frequency electric field and electric polarization.

The coupled phonon–polariton oscillations can be detected by measurement of oscillatory birefringence with a variably delayed probe pulse. (The transit time and the spectral content of the probe pulse also should show oscillatory time dependences.) As in forward ISRS, this pulse "surfs" along a crest or null of the polariton wave. Since the polariton radiates outward from the excitation beam, the probe pulse need not be overlapped spatially with the excitation pulse. By varying the spatial separation between the two parallel-propagating beams, the polariton group velocity and dispersion can be determined. Phonon–polariton dynamics in $LiTaO_3$ crystals were determined in this manner [36, 59]. An example of data is shown in Figure 9.

3. *Optical Absorption*

Optical absorption of an ultrashort pulse can lead to coherent vibrational oscillations in electronic excited states if the shapes or minima of the ground

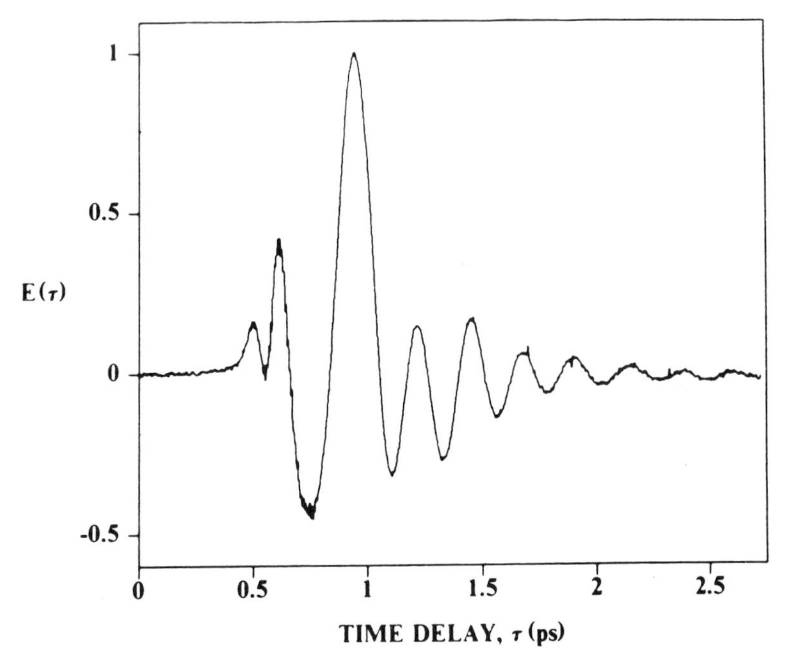

Figure 9. Oscillations of phonon–polariton mode in lithium tantalate crystal excited through inverse electro-optic effect and impulsive stimulated polariton scattering. Time-dependent birefringence measured with probe pulse, which propagated parallel to but not collinear with excitation pulse. (Reprinted with permission from ref. 36.)

and excited-state potentials differ. As illustrated in Figure 10, the situation may be viewed as "vertical" excitation (by single-photon or multiphoton absorption) from the bottom of the ground-state potential to the side of an excited-state potential. If the excitation pulse duration is short compared to the excited-state (e.g., S_1) vibrational frequency, then the excited species in the sample will undergo phase-coherent oscillation. This has been observed in several molecular liquids through time-dependent absorption measurements [35, 61–64]. Data from ethyl violet in ethanol is shown in Figure 11. In general, time-resolved fluorescence, birefringence [65], or other properties may also be monitored.

The time-dependent oscillations in absorption or other observables can be thought of as quantum beats resulting from coherent excitation of several vibronic levels contained within the bandwidth of the ultrashort excitation pulse. In a formal sense, the experiment is the same as other "quantum beat" experiments carried out on femtosecond or longer time scales. However, in most such experiments *different molecular vibrational degrees of freedom* that

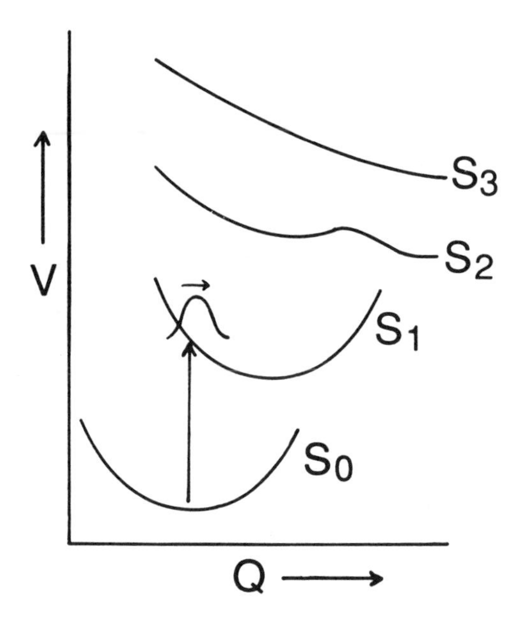

Figure 10. Coherent wavepacket propagation initiated by absorption of ultrashort pulse. Absorption into S_1 leads to coherent excited-state oscillations. Absorption into S_2 or S_3 may lead to synchronized (i.e., coherent) photodissociation.

coincidentally have near-degenerate energy levels (or more precisely, energy levels closely spaced relative to the laser bandwidth) are excited coherently. The quantum beat frequency is then given by the difference between the two fundamental vibrational frequencies. In the case under discussion here, the coherently excited vibronic levels represent different quanta of the same molecular vibrational mode, and the quantum beat frequency is just the fundamental vibrational frequency in the S_1 state.

A quantum-mechanical treatment has been given for the coherent excitation and detection of excited-state molecular vibrations by optical absorption of ultrashort excitation and probe pulses [66]. Here we present a simplified classical-mechanical treatment that is sufficient to explain the central experimental observations. The excited-state vibrations are described as damped harmonic oscillations [i.e., by Eq. (11) with no driving term but with initial condition $Q(0) < 0$.] We consider the effects of coherent vibrational oscillations in S_1 on the optical density OD_λ at a single wavelength λ within the $S_0 \rightarrow S_1$ absorption spectrum. Due to absorption from S_0 to S_1 and stimulated emission from S_1 and S_0,

$$OD_\lambda = \varepsilon_0 b c_0(t) + \varepsilon_1(t) b c_1(t), \tag{24}$$

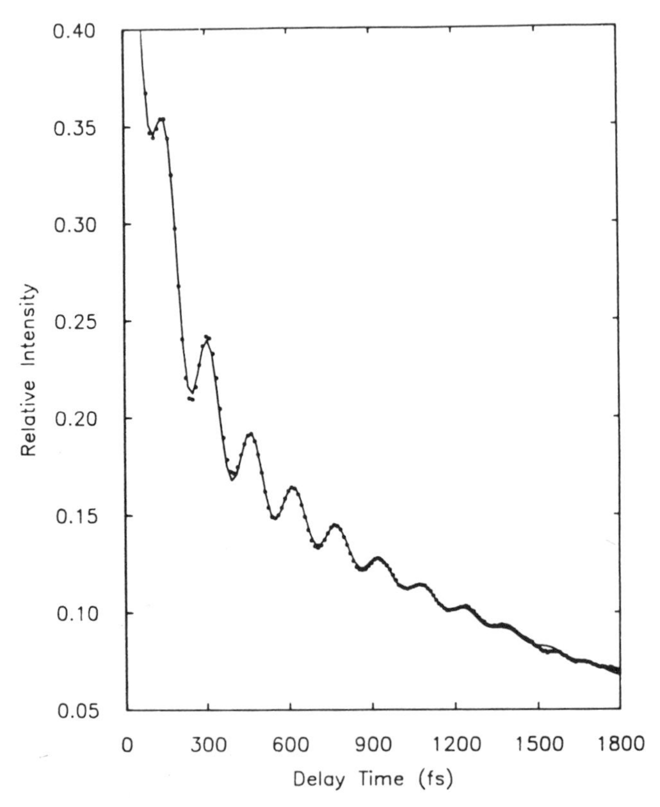

Figure 11. Molecular vibrational oscillations of ethyl violet in ethylene glycol solution superimposed with electronic excited-state decay. (Reprinted with permission from ref. 35.)

where c_0 and c_1 are S_0 and S_1 concentrations, respectively, ε_0 and ε_1 are the corresponding extinction coefficients ($\varepsilon_0 > 0$ and $\varepsilon_1 < 0$), and b is the sample length. The oscillations in stimulated emission probability [i.e., in $\varepsilon_1(t)$] give rise to oscillations in OD_λ. In addition, excited-state relaxation is accounted for in Eq. (24). The ISRS initiation of coherent oscillations in S_0, which would cause ε_0 to oscillate at the ground-state vibrational frequency, has been neglected. We note, however, that resonance-enhanced ISRS effects may in fact be observable [65].

To simplify the analysis substantially, we assume that $\varepsilon_1(t)$ varies linearly with $Q(t)$, the vibrational displacement. (The actual variation, which is not necessarily monotonic, depends on the shapes and positions of S_0 and S_1 potentials, the probe central wavelength and spectral bandwidth, the width of the ground-state wavepacket, excited-state absorption into higher lying levels, and other factors.) In this case,

$$\varepsilon_1(t) = -\varepsilon_0 + a(1 - \cos\omega t)e^{-\gamma t}, \tag{25}$$

where ω and γ are the excited-state vibrational frequency and dephasing rate, respectively, and a is a constant. The S_1 relaxation rate is τ; that is, $c_1(t) = c_1(0)e^{-t/\tau}$ and $c_0(t) = c - c_1(t)e^{-t/\tau}$, where c is the total concentration. The time-dependent change in optical density following excitation at $t = 0$ is

$$\begin{aligned}\Delta\mathrm{OD}_\lambda &= \mathrm{OD}_\lambda(t) - \mathrm{OD}_\lambda(t < 0)\\ &= bc_1(0)e^{-t/\tau}[2\varepsilon_0 - a(1 - \cos\omega t)e^{-\gamma t}].\end{aligned} \tag{26}$$

Thus $\Delta\mathrm{OD}_\lambda$ shows exponential decay with the S_1 lifetime τ, modulated by damped oscillations that give the S_1 vibrational frequency and dephasing rate.

An interesting result of this or a more detailed analysis [66] is that the excited-state vibrational parameters can be extracted from this type of time domain experiment even when the corresponding frequency domain observation of vibrational progressions in the $S_0 \rightarrow S_1$ absorption spectrum is impossible due to inhomogeneous broadening of the electronic transition. In malachite green, the vibrational dephasing rate is about twice as rapid in S_1 as in S_0 [63].

Depending on the nature of the excited-state potential, wavepacket propagation following excitation may not be oscillatory. For example, in unimolecular photodissociation from an unstable excited state (such as S_2 or S_3 in Figure 10), the wavepacket may simply move in one direction and never return. If the excitation pulse duration is short compared to the time required for photodissociation, then the photodissociation events in different excited molecules will be synchronized and wavepacket propagation can still be considered coherent. Experiments of this type will be discussed in the next section.

4. *Other Excitation Schemes: Phase-Coherent Chemistry*

The mechanisms described in the preceding for imparting phase coherence to ground-state or excited-state nuclear motion are easily carried out. In many cases, coherent excitation through one or another such mechanism is unavoidable. Here we discuss straightforward variations on the foregoing themes designed not just to impart phase coherence but to exercise experimental control over coherent wavepacket propagation. There are various motivations for trying these schemes, but perhaps most intriguing are the prospects for elucidating and controlling wavepacket propagation along chemically reactive potential surfaces.

One approach is to reach the potential surface of interest (e.g., S_2 in Figure 10) indirectly by first exciting to a different level (e.g., S_1) and then pumping $S_1 \rightarrow S_2$ absorption. The simplest possibility would be to use a long pulse for $S_0 \rightarrow S_1$ excitation and wait for molecules to equilibrate in the S_1 potential minimum. An ultrashort pulse could then undergo $S_1 \rightarrow S_2$ absorption, initiating coherent motion on the S_2 potential with a different starting point than would be reached from direct $S_0 \rightarrow S_2$ excitation. A more powerful approach would be to use two ultrashort pulses [67, 68]. The first initiates coherent motion along S_1, and the second is absorbed into S_2. Depending upon the delay between the two pulses, an entire set of initial conditions (i.e., initial positions and momenta) for propagation along S_2 could be selected. In *stimulated emission pumping* schemes the intermediate level could be S_1 and the final level could be S_0, on which controlled coherent propagation would take place. In exceptional cases in which the final potential had several possible reaction pathways, this approach could permit control over the ratio of reaction products formed [67, 68].

Other schemes may become possible if *large-amplitude* ISRS excitation can be achieved. Vibrational amplitudes of less than 10^{-3} Å have been reached to date, but through resonance enhancement of ISRS and other techniques, amplitudes in excess of 0.1 Å may be possible [53]. Following initiation of large-amplitude motion in S_0, a second ultrashort pulse could be used for absorption into S_2 or other excited states. Phase-coherent motion in those states would then begin. Depending on the amplitude of ISRS excitation and on the timing between the two pulses, the initial position and momentum on the excited-state potential could be independently varied. This would be of interest, for example, in exploring propagation over potential barriers in chemically reactive excited states. Note that both the magnitude and the direction of the initial propagation velocity could be varied.

More variations on these themes could be discussed, but the basic idea should be clear. Through absorption, stimulated emission, and stimulated scattering, it should become possible to exert considerable influence on coherent molecular motion in ground and excited electronic states.

B. Liquid State Molecular Dynamics

The dynamics of molecular motion in simple (i.e., nonviscous) liquids have long been of interest in their own right and because of their importance in mediating liquid state chemical reactions. "Collective" orientational relaxation times, which measure the return of partially aligned liquids to their isotropic equilibrium states and are usually in the 5–100-ps range, have been determined in many fluids from Rayleigh linewidths or optical Kerr relaxation as well as other methods [4, 69]. Individual molecular collisions and "local" equilibration between near neighbors occur on subpicosecond time scales, as indicated mainly by broad (non-Lorentzian) Rayleigh-wing spectra [69–71]. The first femtosecond time-resolved optical Kerr effect (OKE) experiments on carbon disulfide (CS_2) at 300 K clearly resolved two relaxation times [54, 72, 73], of 1.6 ps and $\sim$ 240 fs, which were consistent with the widths of the depolarized Rayleigh-scattering features. Experiments on argon, neon, and oxygen liquids yielded similar results [74].

Impulsive stimulated scattering experiments on CS_2 with 65-fs pulses revealed additional short-time dynamics [33, 34, 75–77]. In particular, the *vibrational* character of orientational motion was clearly seen in data such as that shown in Figure 12. In these experiments, the excitation pulses exert impulsive torques on CS_2 molecules through stimulated rotational Raman scattering, that is, through the CS_2 single-molecule polarizability anisotropy. The coherent scattering of the probe pulse measures the net orientational alignment (also through the single-molecule polarizability anisotropy). The inertial rise and rapid fall of signal in room temperature CS_2 were originally fit with an overdamped oscillator model for librational motion [33, 76, 77]. In data at lower temperatures (Figure 12) a weakly oscillatory response indicating underdamped librational motion is observed [34, 78, 79].

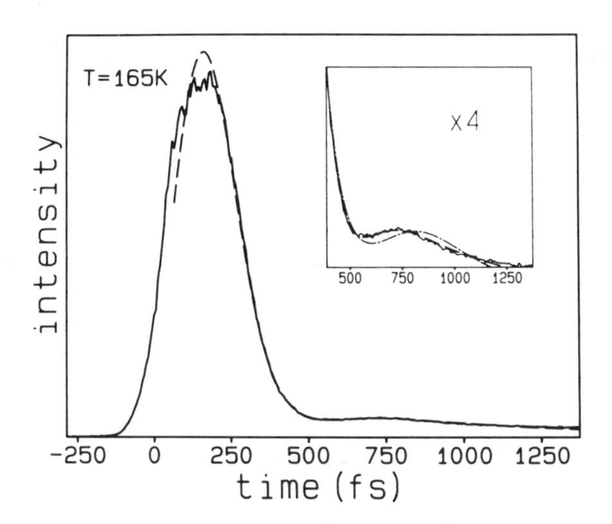

Figure 12. ISS data from CS_2 liquid at 165 K recorded in transient grating experimental configuration with V- and H-polarized excitation pulses. Weakly oscillatory signal indicates that orientational motion is librational in character. Inset: fits to data based on models of homogeneous dephasing (broken curve) or inhomogeneous dephased (dashed curve, under data).

That the short-time orientational motion in molecular liquids shows vibrational character is expected based on the physical picture of a molecule in a liquid, surrounded by a "cage" of neighbors, undergoing vibrational oscillations about its (transient) local potential minimum. In most liquids, this motion is extremely difficult to characterize or even discern by conventional methods such as Rayleigh scattering [69–71, 80]. Time domain vibrational spectroscopy, yielding data such as that in Figure 12, permits determination of librational frequencies and also dephasing dynamics with reasonable accuracy. The dephasing in most liquids appears to be predominantly inhomogeneous [34, 75, 79]. The data in Figure 12 are described approximately by

$$G^{\varepsilon\varepsilon}_{1212}(t) = Ae^{-\Delta^2 t^2/2} \sin \omega_A t + Be^{-\Gamma t}, \tag{27}$$

where ω_A is the configuration-averaged librational frequency, Δ is the extent of inhomogeneity in the frequency, Γ is the collective orientational relaxation rate, and A and B are constants. In CS_2, ω_A increases (from $\sim$ 3.4 to 7.2 ps^{-1}) and Δ decreases (from $\sim$ 5.7 to 4.0 ps^{-1}) upon cooling from 300 to 165 K. These trends are reasonable in view of the thermal contraction of the local "cages" (which increases ω_A) and reduced thermal access to differing local configurations (which reduces Δ).

Data similar to that shown in Figure 12 were recorded with an optical Kerr effect (OKE) configuration [34, 77, 78] which also measures $G^{\varepsilon\varepsilon}_{1122}(t)$ and with other ISS polarization combinations which measure $G^{\varepsilon\varepsilon}_{1111}(t)$, $G^{\varepsilon\varepsilon}_{1122}(t)$, and various linear combinations [34, 77, 81]. The $C^{\varepsilon\varepsilon}(t)$ and simulated LS spectra have been determined from Eqs. (20) and (19), respectively [82]. As has been discussed extensively in connection with interaction-induced light scattering [69–71, 74, 80], $C^{\varepsilon\varepsilon}(t)$ may have contributions not only from single-molecule polarizability anisotropies but also from polarizabilities of molecule pairs, triplets, and so on. For example, the ISS excitation pulses may exert forces on neighboring molecule pairs inducing relative translational motion which is detected by the probe pulse. Arguments have been presented [75, 79] suggesting that single-molecule polarizabilities and single-molecule orientational correlations may dominate the signal in CS_2. Neglecting other contributions to ISS data, angular velocity correlation functions have been extracted [82]. The measured values of ω_A and Δ can be used to determine the configuration-averaged restoring force (i.e., torque) against librational motion and the extent of inhomogeneity in the force. Thus information about intermolecular vibrational frequencies and "force constants" can be extracted from time domain vibrational spectroscopy of the liquid.

ISS data have been recorded in many pure and mixed molecular liquids [34, 49, 75, 83, 83–85]. In most cases, the data are not described precisely by Eq. (27). Rather, an additional decay component appears at intermediate times (decay times $\sim$ 500 fs). This has been interpreted [49, 84] in terms of higher order polarizability contributions to $C^{\varepsilon\varepsilon}(t)$ which represent translational motions, an interpretation supported by observations in CCl_4 (whose single-molecule polarizability anisotropy vanishes by symmetry). This interpretation is not consistent with several molecular dynamics simulations of CS_2 [71, 86]. An alternative analysis has been presented [82] that incorporates theoretical results showing that even the single-molecule orientational correlation function $C_i^{or}(t)$ should in fact show decay on the 0.5-ps time scale of "cage" fluctuations [87, 88].

While there remain many open questions about the detailed interpretation of ISS data from molecular liquids, the ability to characterize the vibrational nature of short-time intermolecular motion in many samples is a significant step forward. ISS experiments have recently been carried out in diamond anvil cells, and pressure-dependent experiments on molecular liquids should help clarify many of the outstanding issues. Also useful are experiments on dilute solutions and ISRS experiments on intramolecular vibrations, such as those discussed earlier on CH_2Br_2. Since only single-molecule polarizabilities contribute to the oscillatory signals in Figures 7 and 8, independent determination of $C_i^{or}(t)$ may be possible [34, 48]. It is clear that femtosecond spectroscopy [including that carried out with incoherent pulses of longer duration [89]] will yield further significant advances in our understanding of liquid state molecular dynamics.

C. Chemical Reaction Dynamics

Several observations of coherent wavepacket dynamics on reactive potential surfaces have been reported. The results provide unmistakable indications of the possibilities and of the focus of intense future activity.

In the gas phase, results on the photodissociation of ICN to yield I + CN were recently reported [90, 91]. Optical absorption of a 120-fs, 307-nm pulse produced an excited state that is unstable with respect to dissociation. The dynamics were probed by determining the time-resolved excited-state absorption strength (measured by laser-induced fluorescence) at various wavelengths. Data are shown in Figure 13. The appearance and decay of unstable intermediates along the dissociation pathway was noted. The clarity of this experiment lies in the simplicity of the sample, a small molecule in the gas phase. In conjunction with resonance Raman spectra [92, 93] and other results, these results may make possible a rather complete determination of dissociation dynamics and of the reactive potential surface.

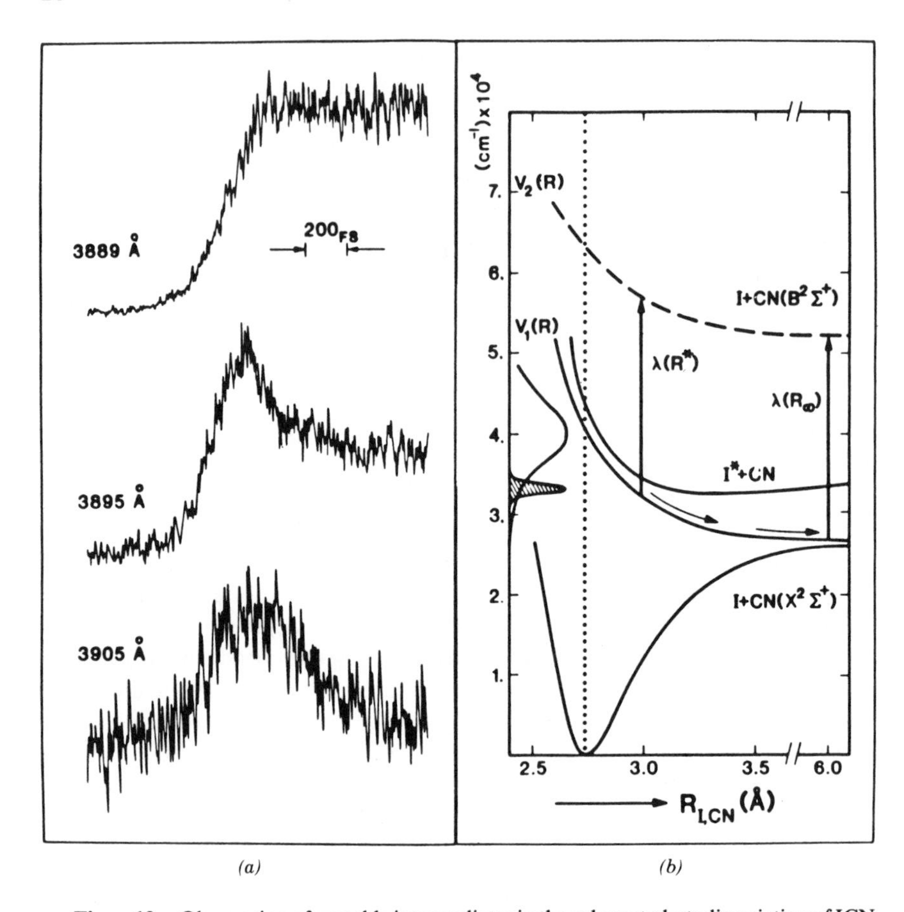

(a) (b)

Figure 13. Observation of unstable intermediates in the coherent photodissociation of ICN. The excitation pulse is absorbed into the V_1 potential which is unstable with respect to dissociation as indicated on the right. Coherent motion along $V_1(R)$ is monitored by absorption of variably delayed, tunable probe pulses into $V_2(R)$. The excited-state absorption progressively red shifts as indicated on the right and by the data on the left. The data show that absorption intensity shifts from 3905Å to 3895Å to 3889Å as dissociation occurs. Reprinted with permission from reference 90.

In condensed phases, dissipation and other effects destroy the one-to-one correspondence between reactive excited-state potential and reaction dynamics. However, semiquantitative determination of reactive potential surfaces should be possible through measurements of reaction dynamics. Equally important, the effects of interactions between the reacting species and its environment can be studied.

Transient absorption measurements have recently been recorded from the organometallic species chromium hexacarbonyl in ethanol solution [94]. Absorption of a 65-fs, 310-nm excitation pulse was followed by measurement of excited-state absorption of a 65-fs, 480-nm probe pulse. The data shown in Figure 14 indicate a rapid nonexponential decay at short times followed by a gradual exponential rise. The slower feature was observed previously [95] and is known to correspond to the solvent complexation of $Cr(CO)_5$ to yield $Cr(CO)_5(MeOH)$. The initial feature, which is observed at other probe wavelengths as well, is believed to correspond to the initial ligand loss reaction. Note that this case is different from ICN in that the initially excited wavepacket is not on the side of the S_1 potential but rather (as is clear from the molecular symmetry) on a local S_1 potential maximum. The wavepacket must then "spread"; that is, dissociation along either direction is equally likely. The rapid nonexponential decay was analyzed in terms of classical kinematics along a dissociative potential.

Transient absorption measurements on dimanganese decacarbonyl, $(CO)_5Mn\text{–}Mn(CO)_5$, which undergoes CO ligand loss or metal–metal (M–M) bond cleavage in methanol were also reported [94]. Oscillations in the excited-state absorption suggested that the projection of the reactive potential along the M–M "stretch" coordinate may be weakly bound (similar

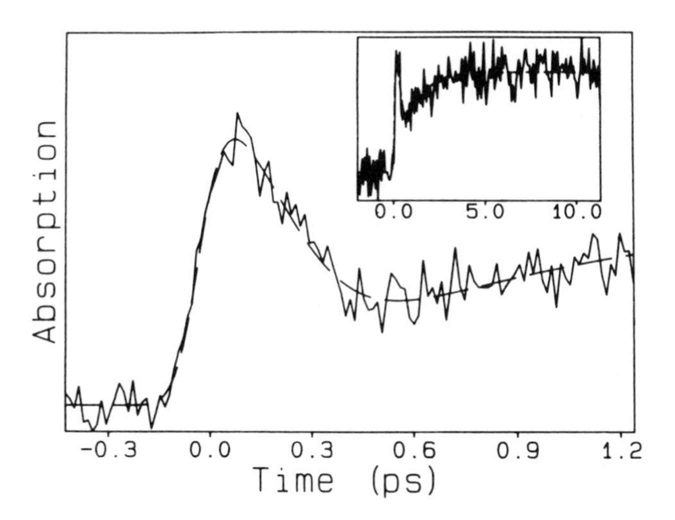

Figure 14. Transient 480-nm absorption of $Cr(CO)_6$ following 310-nm excitation into excited state that is unstable with respect to CO ligand loss. Nonexponential decay of excited-state absorption at short times is believed to reflect the process of bond breakage. This is followed by solvent complexation at longer times (see inset).

to S_2 in Figure 10) by solvent molecules that oppose separation of the large photofragments. If this interpretation proves correct, then solvent effects on reaction dynamics and yields are observed in a very direct way.

Experiments on bacteriorhodopsin (BR), which is the basis for a light-driven proton pump in halobacteria, were recently reported [96]. The primary photoreaction is believed to be a trans-to-cis isomerization. Absorption of a 620-nm pulse by BR in membranes was followed by measurements of stimulated emission at various probe wavelengths between 695 and 930 nm. The rapid ($\tau_1 \sim 200$ fs) decay of stimulated emission itensity at the bluer wavelengths, slower decay ($\tau_2 \sim 500$ fs) at redder wavelengths, and biexponential decay at intermediate wavelengths were interpreted in terms of partially coherent rotational motion along the S_1 potential surface.

In crystalline solids, highly oriented bimolecular reactions may be observed in the time domain. Transient grating and pump–probe experiments on excimer formation in the pyrene and α-perylene crystals discussed earlier have been reported [44–46]. In these crystals, pairs of planar molecules align in "sandwich" fashion. Excimer formation, which involves motion of neighboring molecules in a pair toward each other [97], was complete in about 150 fs following photoexcitation (into high-lying levels above S_1) in each crystal at room temperature. The result was interpreted in terms of initially synchronized intermolecular approach following photoexcitation. The absence of oscillations in the signal following excimer absorption indicates that phase coherence is lost during the reaction. It is possible that coherent oscillations about the excimer geometry would be observed following excitation directly into the S_1 origins.

The 620-nm excitation pulses in these experiments also excite coherent optic phonons, through ISRS, including one mode whose motion is a vibration of neighboring molecules against each other [44–46]. In other words, the phonon motion is similar to that which occurs in the excited-state reaction. Figure 15 shows data from α-perylene at 180 K in which the probe wavelength, 480 nm, is weakly absorbed by the ground state but is not absorbed by the excimer. Thus there is no excited-state signal at $t > 150$ fs. However, the phonon oscillations are observed as oscillations in the intensity of the transmitted probe pulse. There are two contributions to this signal. The first and most interesting is that due to $S_0 \rightarrow S_1$ absorption spectral shifts induced by the phonons. These shifts could provide a direct measure of the reactive excited-state potential as a function of phonon displacement, that is, of the electron–phonon interactions that drive the reaction. The other contribution to the signal is coherent scattering of the (nearly collinear) probe pulse. As discussed earlier, this leads to alternating blue and red shifts of the probe pulse. Ordinarily this would not significantly affect total transmitted intensity, but since the probe central wavelength lies on the red edge of the

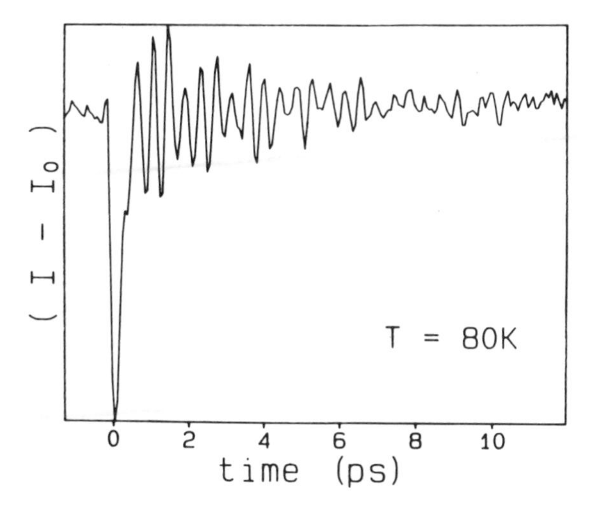

Figure 15. Transient 480-nm absorption in α-perylene crystal at 80 K following ISRS excitation of 80- and 104-cm^{-1} coherent optic phonons. (compare Figure 6.) Oscillations can arise from coherent scattering of the probe pulse, as in Figure 8, and from phonon-induced shifts of crystalline absorption spectrum.

$S_0 \rightarrow S_1$ absorption band, shifting of the probe wavelength affects probe transmission just as does shifting of the crystalline absorption spectrum. In Figure 15, the major contribution is that due to coherent scattering. Quantitative determination of the phonon-induced spectral shift is impossible, but an approximate upper limit can be set. Its value ($\sim 1\ cm^{-1}$ for a 10^{-4}-Å phonon amplitude) indicates that the *initial slope* of the S_1 potential with respect to intermolecular separation is rather gradual and that the slope must increase as displacement increases [98]. Thus, in addition to information about excimer formation dynamics, partial (and at this point, qualitative) information about the reactive potential surface has been elucidated.

These examples, although few and preliminary, nonetheless indicate the direction in which time-resolved spectroscopy of reactive species is headed. More detailed examinations of unstable structures between chemical reactants and products will certainly follow. A major goal in this area will be direct observation of coherent wavepacket propagation through local potential maxima (i.e., transition states). Experimental control over wavepacket momentum through potential maxima will be especially important in evaluating solvent effects, barrier recrossing probabilities, and so on. Methods that permit observation and control of transition state production may be anticipated.

Acknowledgments

E. P. Ippen acknowledges collaboration with J. G. Fujimoto, A. M. Weiner, and S. De Silvestri. K. A. Nelson gratefully acknowledges collaboration with S. Ruhman, Y.-X. Yan, L. R. Williams, B. Kohler, and A. G. Joly. Collaboration between the two groups in experiments on molecular crystals is also mutually acknowledged. Work in both groups was supported by JSEP Grant No. DAAL03-86-K-0002. K. A. Nelson acknowledges support from NSF Grant No. DMR-8704352 and NSF and private contributions to a Presidential Young Investigator Award.

References

1. R. L. Fork, B. I. Greene, and C. V. Shank, *Appl. Phys. Lett. 38*:671 (1981).
2. W. H. Knox, *IEEE J. Quantum Electron. 24*:388 (1988).
3. G. R. Fleming, *Ann. Rev. Phys. Chem. 37*:81–104 (1986).
4. G. R. Fleming, *Chemical Applications of Ultrafast Spectroscopy*. New York: Oxford, 1986.
5. *Rev. Phys. Appl. 22* (1987) (special issue, ultrafast lasers and spectroscopy).
6. *IEEE J. Quant. Electron. 24* (1988) (special issue, ultrafast lasers and spectroscopy).
7. J. A. Armstrong, *Appl. Phys. Lett. 10*:16 (1967).
8. N. Bloembergen, *Nonlinear Optics*. Reading, MA: Benjamin, 1965.
9. Y. R. Shen, *The Principles of Nonlinear Optics*. New York: Wiley, 1983.
10. E. P. Ippen and C. V. Shank, *Ultrashort Light Pulses*, S. L. Shapiro (ed.), New York: Springer-Verlag, 1983.
11. A. von Jena and H. E. Lessing, *Appl. Phys. 19*:131 (1979).
12. Z. Vardeny and J. Tauc, *Opt. Commun. 39*:396 (1981).
13. B. S. Wherrett, A. L. Smirl, and T. F. Boggess, *IEEE J. Quant. Electron. QE-19*:680 (1983).
14. M. W. Balk and G. R. Fleming, *J. Chem. Phys. 83*:4300 (1985).
15. B. Fleugel, N. Peyghambarian, G. Olbright, M. Lindberg, S. W. Koch, M. Joffre, D. Hulin, A. Migus, and A. Antonetti, *Phys. Rev. Lett. 59*:2588 (1987).
16. C. H. Brito Cruz, J. P. Gordon, P. C. Becker, R. L. Fork, and C. V. Shank, *IEEE J. Quant. Electron. 24*:261 (1988).
17. A. M. Weiner and E. P. Ippen, *Chem. Phys. Lett. 114*:456 (1985).
18. C. H. Brito Cruz, R. L. Fork, W. H. Knox, and C. V. Shank, *Chem. Phys. Lett. 132*:341 (1986).
19. M. D. Levenson, *Introduction to Nonlinear Laser Spectroscopy*, New York: Academic Press, 1982, Chapter 6.
20. I. D. Abella, N. A. Kurnit, and S. R. Hartmann, *Phys. Rev. A 141*:391 (1966).
21. W. H. Hesselink and D. A. Wiersma, *Phys. Rev. Lett. 43*:1991 (1979).
22. T. Yajima and Y. Taira, *J. Phys. Soc. Japan 47*:1620 (1979).
23. T. Yajima, Y. Ishida, and Y. Taira, *Picosecond Phenomena II*. R. M . Hochstrasser, W. Kaiser and C. V. Shank (eds.), New York: Springer-Verlag, 1980, p. 190.
24. J. G. Fujimoto and E. P. Ippen, *Opt. Lett. 8*:446 (1983).
25. A. M. Weiner and E. P. Ippen, *Opt. Lett. 9*:53 (1984).
26. A. M. Weiner, S. deSilvestri, and E. P. Ippen, *J. Opt. Soc. Am. B 2*:654 (1985).
27. T. Mossberg, A. Flusberg, R. Kochru, and S. R. Hartmann, *Phys. Rev. Lett. 42*:1665 (1979).
28. H. J. Eichler, U. Klein, and D. Langhans, *Appl. Phys. 21*:215 (1980).

29. S. deSilvestri, A. M. Weiner, J. G. Fujimoto, and E. P. Ippen, *Chem. Phys. Lett. 112*:195 (1984).
30. M. Morets and T. Yajima, *Phys. Rev. A 30*:2525 (1984).
31. S. Asaka, H. Nakatsuka, M. Fujiwara, and M. Matsuoka, *Phys. Rev. A 29*:2286 (1984).
32. K. Kurokawa, T. Hattori, and T. Kobayashi, *Phys. Rev. A 36*:1298 (1987).
33. Y.-X. Yan, L.-T. Cheng, and K. A. Nelson, *Advances in Nonlinear Spectroscopy*, R. J. H. Clark and R. E. Hester (eds.), Chichester: Wiley, 1988, pp. 299–355.
34. S. Ruhman, A. G. Joly, B. Kohler, L. R. Williams, and K. A. Nelson, *Revue Phys. Appl. 22*:1717–1734 (1987).
35. F. W. Wise, M. J. Rosker, and C. L. Tang, *J. Chem. Phys. 86*:2827–2832 (1987).
36. K. P. Cheung and D. H. Auston, *Phys. Rev. Lett. 20*:2152–2155 (1985).
37. A. Lauberau and W. Kaiser, *Rev. Mod. Phys. 50*:607 (1978).
38. Y.-X. Yan and K. A. Nelson, *J. Chem. Phys. 87*:6240–6256 (1987).
39. K. A. Nelson, R. J. P. Miller, and M. D. Fayer, *J. Appl. Phys. 77*:1144–1152 (1982).
40. M. M. Robinson, Y.-X. Yan, E. B. Gamble, L. R. Williams, J. S. Meth, and K. A. Nelson, *Chem. Phys. Lett. 112*:491–996 (1984).
41. M. R. Farrar, L.-T. Cheng, Y.-X. Yan, and K. A. Nelson, *IEEE J. Quant. Electron. QE-22*:1453–1456 (1986).
42. L.-T. Cheng and K. A. Nelson, *Phys. Rev. B 37*:3603–3610 (1988).
43. Y.-X. Yan, L.-T. Cheng, and K. A. Nelson, *J. Chem. Phys. 88*:6477–6486 (1988).
44. S. De Silvestri, J. G. Fujimoto, E. P. Ippen, E. B. Gamble, L. R. Williams, and K. A. Nelson, *Chem. Phys. Lett. 116*:146–152 (1985).
45. L. R. Williams, E. B. Gamble, K. A. Nelson, S. De Silvestri, A. M. Weiner, and E. P. Ippen, *Chem. Phys. Lett. 139*:244–249 (1987).
46. L. R. Williams and K. A. Nelson, *J. Chem. Phys. 87*:7346–7347 (1987).
47. S. Ruhman, A. G. Joly, and K. A. Nelson, *J. Chem. Phys. 86*:6563–6565 (1987).
48. S. Ruhman, A. G. Joly, and K. A. Nelson, *IEEE, J. Quant. Electron. 24*:460–469 (1988).
49. D. McMorrow, W. T. Lotshaw, and G. A. Kenney-Wallace, *IEEE J. Quant. Electron. 29*: 443–454 (1988).
50. C. H. Wang, *Spectroscopy of Condensed Media*, Orlando: Academic, 1985.
51. B. J. Berne and R. Pecora, *Dynamic Light Scattering*, New York: Wiley, 1976.
52. Y.-X. Yan and K. A. Nelson, *J. Chem. Phys. 87*:6257–6265 (1987).
53. Y.-X. Yan, E. B. Gamble, and K. A. Nelson, *J. Chem. Phys. 83*:5391–5399 (1985).
54. J.-M. Halbout and C. L. Tang, *Appl. Phys. Lett. 40*:765–767 (1982).
55. K. H. Yang, P. L. Richards, and Y. R. Shen, *Appl. Phys. Lett. 19*:320–323 (1971).
56. D. H. Auston, K. P. Cheung, J. A. Valdmanis, and D. A. Kleinman, *Phys. Rev. Lett. 53*:1555–1558 (1984).
57. D. H. Auston, *Appl. Phys. Lett. 43*:713 (1983).
58. D. A. Kleinman and D. H. Auston, *IEEE J. Quant. Electron. QE-20*:964–970 (1984).
59. M. C. Nuss and D. H. Auston, *Ultrafast Phenomena V*, G. R. Fleming and A. E. Siegman (eds.), Berlin: Springer-Verlag, 1986, pp. 284–286.
60. A. S. Barker and R. Loudon, *Rev. Mod. Phys. 44*:18–47 (1972).
61. M. J. Rosker, F. W. Wise, and C. L. Tang, *Phys. Rev. Lett. 57*:321–324 (1986).

62. J. M. Y. Ha, H. J. Maris, W. M. Risen, J. Tauc, C. Thomson, and Z. Vardeny, *Phys. Rev. Lett.* *57*:3302 (1986).
63. K. A. Nelson and L. R. Williams, *Phys. Rev. Lett.* *58*:745 (1987).
64. C. L. Tang, F. W. Wise, and I. A. Walmsley, *Rev. Phys. Appl.* *22*:1695–1703 (1987).
65. J. Chesnoy and A. Mokhtari, *Rev. Phys. Appl.* *22*:1743–1747 (1987).
66. M. Mitsunaga and C. L. Tang, *Phys. Rev. A* *35*:1720–1728 (1987).
67. D. J. Tannor and S. A. Rice, *J. Chem. Phys.* *83*:5013–5018 (1985).
68. D. J. Tannor, R. Koslaff, and S. A. Rice, *J. Chem. Phys.* *85*:5805–5820 (1986).
69. W. G. Rothschild, *Dynamics of Molecular Liquids*, New York: Wiley, 1984.
70. D. Kivelson and P. A. Madden, *Ann. Rev. Phys. Chem.* *31*:523–558 (1980).
71. P. A. Madden, *Ultrafast Phenomena IV*. D. H. Auston and K. B. Eisenthal, Berlin: Springer-Verlag, 1984, pp. 244–251.
72. B. I. Greene and R. C. Farrow, *J. Chem. Phys.* *77*:4779–4780 (1982).
73. B. I. Greene and R. C. Farrow, *Chem. Phys. Lett.* *98*:273–276 (1983).
74. B. I. Greene, P. A. Fleury, H. L. Carter, and R. C. Farrow, *Phys. Rev. A* *29*:271–274 (1984).
75. S. Ruhman, B. Kohler, A. G. Joly, and K. A. Nelson, *IEEE J. Quant. Electron.* *24*:470–481 (1988).
76. L. R. Williams, S. Ruhman, A. G. Joly, B. Kohler, and K. A. Nelson, *Advances in Laser Science, II*, M. Lapp, W. C. Stwalley, and G. A. Kenney-Wallace (eds.), New York: American Institute of Physics, 1987, pp. 408–416.
77. S. Ruhman, L. R. Williams, A. G. Joly, B. Kohler, and K. A. Nelson, *J. Phys. Chem.* *91*:2237–2240 (1987).
78. C. Kalpouzos, W. T. Lotshaw, D. McMorrow, and G. A. Kenney-Wallace, *J. Phys. Chem.* *91*:2029–2030 (1987).
79. S. Ruhman, B. Kohler, A. G. Joly, and K. A. Nelson, *Chem. Phys. Lett.* *141*:16–24 (1987).
80. P. A. Madden and T. I. Cox, *Mol. Phys.* *43*:287–305 (1981).
81. J. Etchepare, G. Grillon, J. P. Chambaret, G. Hamoniaux, and A. Orszag, *Opt. Commun.* *63*:329–334 (1987).
82. S. Ruhman and K. A. Nelson (to be published).
83. W. T. Lotshaw, D. McMorrow, C. Kalpouzos, and G. A. Kenney-Wallace, *Chem. Phys. Lett.* *136*:323–328 (1987).
84. C. Kalpouzos, D. McMorrow, W. T. Lotshaw, and G. A. Kenney-Wallace, *Chem. Phys. Lett.* *150*:138–146 (1988).
85. J. Etchepare, G. Grillon, G. Hamoniaux, A. Antonetti, and A. Orszag, *Rev. Phys. Appl.* *22*:1749–1753 (1987).
86. L. C. Geiger and B. M. Ladanyi, *J. Chem. Phys.* *87*:191–202 (1987).
87. R. J. Lynden-Bell and W. A. Steele, *J. Phys. Chem.* *88*:6514–6518 (1984).
88. W. T. Coffee, G. J. Evans, M. Evans, and G. H. Wegdam, *Mol. Phys.* *29*:310–323 (1977).
89. T. Kobayashi, T. Hattori, A. Terasaki, and K. Kurokawa, *Rev. Phys. Appl.* *22*:1773–1785 (1987).
90. M. Dantus, M. J. Rosker, and A. H. Zewail, *J. Chem. Phys.* *87*:2395–2397 (1987).
91. A. H. Zewail, *Ann. Rev. Phys. Chem.* *39* (in press).
92. E. J. Heller, *Acc. Chem. Res.* *14*:368 (1981).
93. D. Imre, J. L. Kinsey, A. Sinha, and J. Krenos, *J. Phys. Chem.* *88*:3956 (1984).

94. A. G. Joly and K. A. Nelson, *J. Phys. Chem.* (submitted).

95. J. D. Simon and X. Xie, *J. Phys. Chem. 90*:6751 (1986).

96. R. A. Mathies, C. H. Brito Cruz, W. T. Pollard, and C. V. Shank, *Science 240*:777 (1988); J. Dobler, W. Zinth, W. Kaiser, and D. Oesterhelt, *Chem. Phys. Lett. 144*:215 (1988).

97. A. Warshel and E. Huler, *Chem. Phys. 6*:463 (1974).

98. L. R. Williams and K. A. Nelson (to be published).

AB INITIO DETERMINATION OF MOLECULAR ELECTRICAL PROPERTIES

CLIFFORD E. DYKSTRA AND SHI-YI LIU*

Department of Chemistry
University of Illinois
Urbana, Illinois 61801

DAVID J. MALIK

Department of Chemistry
Indiana University—Purdue University at Indianapolis
Indianapolis, Indiana 46223

CONTENTS

*Current Address: Allied Signal Corporation, 50 E. Algonquin Road, Des Plaines, IL 60017-5016.

I. INTRODUCTION

Theoretical knowledge of molecular electrical moments and polarizabilities has reached an important threshold. Methods have advanced considerably in the last several years and have reached the point that elusive electrical properties of molecules are obtainable from calculation. At the same time, there has been significant advancement in understanding the role of these properties in certain chemical phenomena, and application of electrical interaction ideas is likely to become even more widespread. A survey of calculated properties is presented here.

An important concern with calculated molecular electrical properties is the assessment of accuracy. In part, accuracy is a "stand-alone" question; in principle, it can be analyzed entirely with theoretical methods, if sufficiently exhaustive. Short of that, comparison with laboratory measurements is in order, though often what is calculated and what is measured are different in subtle ways. Thus, part of this review is intended to provide a surer connection between ab initio calculation of electrical properties and laboratory measurement. In doing this, one cannot help but think about how electrical properties can be utilized, through the application of fields, to alter or control certain spectroscopic phenomena. Then a little more is learned about how to understand and analyze experiments that have been done and maybe something about experiments that might be done.

Definitions of molecular electrical multipoles and polarizabilities are available at the textbook level [1], but we find that expressing the properties in a certain manner, to be described, offers convenience and facilitates a number of applications. Classical electrostatics dictates that the energy of interaction for placing a charge q_i in an electrical potential $V(x, y, z)$ is the product of the charge and the potential at the position of the charge, $r_i = (x_i, y_i, z_i)$. For a distribution of N point charges in an electrical potential, the classical interaction energy is a sum:

$$E_{\text{int}} = \sum_{i}^{N} q_i V(x_i, y_i, z_i). \tag{1}$$

The potential function V can be expressed as a power series expansion about some point, which is most conveniently taken to be the origin 0:

$$\begin{aligned}
V(x, y, z) = V_0 &+ x\frac{\partial V}{\partial x}\bigg|_0 + y\frac{\partial V}{\partial y}\bigg|_0 + z\frac{\partial V}{\partial z}\bigg|_0 \\
&+ \tfrac{1}{2}x^2\frac{\partial^2 V}{\partial x^2}\bigg|_0 + \tfrac{1}{2}y^2\frac{\partial^2 V}{\partial y^2}\bigg|_0 + \tfrac{1}{2}z^2\frac{\partial^2 V}{\partial z^2}\bigg|_0 \\
&+ xy\frac{\partial^2 V}{\partial x\,\partial y}\bigg|_0 + xz\frac{\partial^2 V}{\partial x\,\partial z}\bigg|_0 + yz\frac{\partial^2 V}{\partial y\,\partial z}\bigg|_0 \\
&+ \tfrac{1}{3}\cdot\tfrac{1}{2}x^3\frac{\partial^3 V}{\partial x^3}\bigg|_0 + \tfrac{1}{3}\cdot\tfrac{1}{2}y^3\frac{\partial^3 V}{\partial y^3}\bigg|_0 + \tfrac{1}{3}\cdot\tfrac{1}{2}z^3\frac{\partial^3 V}{\partial z^3}\bigg|_0 \\
&+ \tfrac{1}{2}x^2y\frac{\partial^3 V}{\partial x^2\,\partial y}\bigg|_0 + \tfrac{1}{2}x^2z\frac{\partial^3 V}{\partial x^2\,\partial z}\bigg|_0 + xyz\frac{\partial^3 V}{\partial x\,\partial y\,\partial z}\bigg|_0 \\
&+ \cdots .
\end{aligned} \tag{2}$$

Using this expansion for the $V(x_i, y_i, z_i)$ values in Eqn. (1) and the subscript notation where

$$V_{xy} = \frac{\partial^2 V}{\partial x\,\partial y}\bigg|_0$$

yields the usual multipole expansion:

$$\begin{aligned}
E_{\text{int}} = {} & V_0\sum_i q_i \\
&+ V_x\sum_i q_i x_i + V_y\sum_i q_i y_i + V_z\sum_i q_i z_i \\
&+ \tfrac{1}{2}V_{xx}\sum_i q_i x_i^2 + \tfrac{1}{2}V_{yy}\sum_i q_i y_i^2 + \tfrac{1}{2}V_{zz}\sum_i q_i z_i^2 \\
&+ V_{xy}\sum_i q_i x_i y_i + V_{xz}\sum_i q_i x_i z_i + V_{yz}\sum_i q_i y_i z_i \\
&+ \tfrac{1}{6}V_{xxx}\sum_i q_i x_i^3 + \cdots .
\end{aligned} \tag{3}$$

The sums appearing in Eqn. (3) correspond to moments of the charge distribution: $Q = \Sigma_i q_i$ is the total charge or zero moment; $\mu_x = \Sigma_i q_x x_i$ is the x component of the first or dipole moment. Next are second-moment terms, then third-moment terms, and so on. In general, the elements of each moment depend on the specific directionality of the coordinate axes. Any moment that has an odd-order dependence on x_i, y_i, or z_i will change sign upon reversing the direction of the corresponding axis, while rotating the coordinate axes will mix moment elements.

The choice of the evaluation center affects moments beyond the first nonvanishing moment. With a neutral charge distribution ($\Sigma_i q_i = 0$), a dipole component evaluated relative to a point other than the origin would be unchanged in value:

$$\sum_i q_i(x_i - a) = \sum_i q_i x_i - a \sum_i q_i = \sum_i q_i x_i. \tag{4}$$

In the same way, the sum $\Sigma_i q_i x_i^2$ would be unchanged only if $\Sigma_i q_i = 0$ *and* $\Sigma_i q_i x_i = 0$. For an off-diagonal tensor component such as $\Sigma_i q_i x_i y_i^2$, the value of the moment element would be invariant to repositioning the origin along the x axis if the system were neutral. It would be invariant to repositioning along the y axis if the system were neutral and the y component of the dipole were zero. Clearly, the nth-moment elements can be evaluated at a spatially translated origin if the set of moments through the nth moment is known at the original origin. This implies that with a truncated moment expansion the interaction energy depends on the evaluation center, and so, the validity of a *truncated* moment expansion depends on there being a physically reasonable choice of the evaluation center and/or truncation [2]. A spatially distributed set of moments [3, 4] is a corresponding idea for minimizing truncation error.

Laplace's equation, $\nabla^2 V = 0$, means that the number of unique elements needed to evaluate an interaction energy can be reduced. For the second moment this amounts to a transformation into a traceless tensor form, a form usually referred to as the quadrupole moment [5]. Transformations for higher moments can be accomplished with the conditions that develop from further differentiation of Laplace's equation. With modern computation machinery, such reduction tends to be of less benefit, and on vector machines, it may be less efficient in certain steps. We shall not make that transformation and instead will use *traced Cartesian moments.* It is still appropriate, however, to refer to *quadrupoles* or *octupoles* rather than to *second* or *third moments* since for interaction energies there is no difference. Logan has pointed out the convenience and utility of a Cartesian form of the multipole polarizabilities [6], and in most cases, that is how the properties are expressed here.

Applequist has set forth a useful organization for electrical properties [7], which we follow here. It avoids the letter convention of Buckingham [5] (e.g., α, β, . . . , A, B, C, . . .), which is exhausted by the properties that are nowadays calculable. Applequist defines a Cartesian polytensor of the first degree as a sequence of Cartesian tensors, each arranged in a column. We take the column ordering of the Cartesian tensors to be anticanonical, which means the ordering of the moment elements is as follows:

$$\begin{aligned}&\text{First: } x, y, z;\\&\text{Second: } xx, xy, xz, yx, yy, yz, zx, zy, zz;\\&\text{Third: } xxx, xxy, xxz, xyx, xyy, \ldots, zzz.\end{aligned} \tag{5}$$

The anticanonical order has the first index changing slowest. The first-degree polytensor is a stacked list of the (column) moment tensors in increasing rank. This will be designated **M**, with M_x being the x component of the first moment, M_{xy} being the xy component of the second moment, and so on. Notice that in this list, equivalent values such as M_{xy} and M_{yx} are all included:

$$\mathbf{M} = \begin{bmatrix} M_0 \\ M_x \\ M_y \\ M_z \\ M_{xx} \\ M_{xy} \\ \cdot \\ \cdot \\ \cdot \\ M_{zz} \\ M_{xxx} \\ M_{xxy} \\ \cdot \\ \cdot \end{bmatrix} = \begin{bmatrix} M_0 \\ M_1 \\ M_2 \\ M_3 \\ \cdot \\ \cdot \\ \cdot \\ \\ \\ \\ \\ \\ \\ \\ \end{bmatrix} \tag{6}$$

The alternate labeling of the elements in the rightmost part of Eqn. (6) is convenient for computational purposes. An integer i designates a row in **M** and thereby specifies a particular moment component M_i. It will be helpful to consider the moment components as abstract independent directions in the "space of multipoles" since this makes their organization into a column vector quite appropriate.

The power series in Eqn. (2) has as many partial derivative values as there are multipole moments. They can also be arranged in a first-degree Cartesian polytensor **V** with row indices in the same anticanonical order. The energy of interaction can now be expressed as

$$E_{\text{int}} \equiv \mathbf{V}^{\mathrm{T}} \cdot \mathbf{M}, \tag{7}$$

which is consistent with the following definitions:

$$\begin{aligned}
M_0 &\equiv \sum_i q_i = \text{total charge}, \\
M_x &\equiv \sum_i q_i x_i = \mu_x, \\
&\vdots \\
M_{xx} &\equiv \frac{1}{2} \sum_i q_i x_i^2 = \frac{1}{2} Q_{xx}, \\
M_{xy} &\equiv \frac{1}{2} \sum_i q_i x_i y_i = \frac{1}{2} Q_{xy}, \\
&\vdots \\
M_{xxx} &\equiv \frac{1}{6} \sum_i q_i x_i^3 = \frac{1}{6} R_{xxx}, \\
M_{xxy} &\equiv \frac{1}{6} \sum_i q_i x_i^2 y_i = \frac{1}{6} R_{xxy}. \\
&\vdots
\end{aligned} \tag{8}$$

The factors of $\frac{1}{2}$ for each of the second moments and $\frac{1}{6}$ for the third moments come about because of the equivalent terms (e.g., $V_{xy} M_{xy}$ and $V_{yx} M_{yx}$) in Eqn. (7). More frequently used moment forms (e.g., μ, Q, R, . . .) are related to the M elements by these factors as indicated.

For quantum-mechanical systems, the moment definitions in Eqn. (8) can be made into definitions of moment operators by replacing position coordinates with their corresponding position operators. The molecular Hamiltonian for a molecule experiencing an external electrical potential following the convention of Eqn. (7) is

$$\hat{H} = \hat{H}_0 + \mathbf{V}^{\mathrm{T}} \cdot \hat{\mathbf{M}}, \tag{9}$$

where $\hat{H}_0$ is the free molecule Hamiltonian and **V** is now a set of parameters in the Schrödinger equation. That is, for any particular choice of **V** one has a new Schrödinger equation to solve. This is exactly what is done in finite-field calculations where the molecular energy is determined at a specified field strength.

The Hamiltonian does not include polarization terms, but the quantum-mechanical charge distributions are certainly polarizable. Polarization develops because of how the wavefunction adjusts to a potential **V**. In other words, the wavefunction and likewise the energy eigenvalues, E, can (and do) have nonlinear dependencies on the elements of **V** even though the Hamiltonian is only linear in **V**.

The classical interaction energy expression of Eqn. (3) tells us that the value of a dipole moment component is the derivative of the energy with respect to either V_x, V_y, or V_z evaluated at $\mathbf{V} = 0$. Higher moments are first derivatives as well. By correspondence, the quantum-mechanical expression for any moment is the first derivative of the energy with respect to the appropriate element of **V**, that derivative evaluated at $\mathbf{V} = 0$.

It is possible to differentiate the quantum-mechanical electronic energy beyond first order, and means for doing this are discussed in Section III. The second derivatives are the usual polarizabilities, the third derivatives are the hyperpolarizabilities, and so on. These properties are associated with a power series expansion of the energy in terms of the elements of **V**. A second-degree polytensor $\mathbf{P}^{(2)}$ is introduced for handling all the polarizabilities [7]. It is a square matrix whose rows and columns are labeled, in anticanonical order, by the same indices that label the elements of the column array **M**. For example,

$$P^{(2)}_{x,x} = \left.\frac{\partial^2 E}{\partial V_x^2}\right|_{\mathbf{V}=0}$$

is the dipole polarizability. The general definition with the convention used herein is

$$P^{(2)}_{i,j} \equiv \left.\frac{\partial^2 E}{\partial V_i \partial V_j}\right|_{\mathbf{V}=0}, \tag{10}$$

where i and j are the directions in the "space of multipoles" [Eqn. (6)]. For an external potential represented by **V**, the interaction energy is

$$E^{(2)}_{\text{int}} = \tfrac{1}{2}\mathbf{V}^{\mathrm{T}}\cdot\mathbf{P}^{(2)}\cdot\mathbf{V}. \tag{11}$$

The hyperpolarizabilities are the third derivatives:

$$P^{(3)}_{i,j,k} \equiv \left.\frac{\partial^3 E}{\partial V_i \partial V_j \partial V_k}\right|_{\mathbf{V}=0}. \tag{12}$$

Fourth derivatives and beyond come from continuing the differentiation. Equations (10) and (12) use a convention for the elements of the polarizability polytensor $\mathbf{P}^{(2)}$ and the hyperpolarizability polytensor $\mathbf{P}^{(3)}$. It is that they are strictly the energy derivatives with respect to the elements of $\mathbf{V}$ that multiply the moment operators in Eqn. (7). This is opposite the most common sign convention [8]. Furthermore, for quadrupole or higher moments, the factors in Eqn. (8) show up in relating the $\mathbf{P}^{(n)}$ elements to the more traditional tensor forms. For example, the usual Cartesian dipole and quadrupole polarizabilities enter into the $\mathbf{P}^{(2)}$ polytensor as

$$\mathbf{P}^{(2)} = -\left(\begin{array}{c|c} \alpha & \frac{1}{2}A \\ \hline \frac{1}{2}A^{\mathrm{T}} & \frac{1}{2}C \ldots \end{array}\right). \tag{13}$$

To keep the conventions clear, we use $\mathbf{P}$ and $\mathbf{M}$ to give values that are defined by Eqns. (8), (10), and (12). Any values given as A, B, C, . . . adhere to the traditional conventions, though using traced Cartesian forms, not traceless forms [6]. The polytensor and moment definitions are helpful for computation because the interaction formulas are then uniform for *all* multipoles. Also, our means for obtaining the polarizabilities [9] generates derivatives. Using the operator definitions of Eqn. (8) will directly give elements of $\mathbf{P}^{(2)}$, $\mathbf{P}^{(3)}$, . . . in exactly the right form.

For neutral molecules, the dipole polarizabilities and hyperpolarizabilities are invariant to the choice of the moment center. Other multipole polarizabilities may be invariant in certain cases of high molecular symmetry. The changes that may occur in $\mathbf{P}^{(2)}$, $\mathbf{P}^{(3)}$, . . . upon shifting an evaluation center are determined by the changes in the moments or moment operators. If a particular origin translation leads to

$$\hat{M}_i' = \hat{M}_i + a\hat{M}_j, \tag{14}$$

then the corresponding energy differentiation is

$$\frac{\partial}{\partial V_i'} = \frac{\partial}{\partial V_i} + a\frac{\partial}{\partial V_j}. \tag{15}$$

The translation $x' = x - b$ for a neutral molecule gives $M_{xx}' = M_{xx} - 2bM_x$. The second-moment polarizability evaluated with the origin shifted to $x = b$ is obtained as

$$\frac{\partial^2}{\partial V_{xx}'^2}E = \begin{cases} \left(\dfrac{\partial}{\partial V_{xx}} - 2b\dfrac{\partial}{\partial V_x}\right)\left(\dfrac{\partial}{\partial V_{xx}} - 2b\dfrac{\partial}{\partial V_x}\right)E, & (16) \\[2ex] \dfrac{\partial^2}{\partial V_{xx}}E - 4b\dfrac{\partial^2}{\partial V_{xx}\partial V_x}E + 4b^2\dfrac{\partial^2}{\partial V_x^2}E. & (17) \end{cases}$$

That is, the "translated" polarizability is related to the original quadrupole–quadrupole polarizability, the dipole–quadrupole polarizability, and the dipole–dipole polarizability.

The moments of a molecular system change in the presence of a field. This change is obtained by integration and for the effect of $\mathbf{P}^{(2)}$ it is

$$\mathbf{M}_{\text{ind}}(\text{via } \mathbf{P}^{(2)}) = \mathbf{P}^{(2)} \cdot \mathbf{V}. \tag{18}$$

In a like manner, polarizabilities are induced via higher polarizabilities; for example, $\mathbf{P}^{(n)}_{\text{ind}}$ arises, in part, from $\mathbf{P}^{(n+1)} \cdot \mathbf{V}$.

Atomic units (a.u.) are obviously convenient when finding electrical properties by ab initio calculations. The conversion factor to SI units (Coulombs, meters, and joules) is

$$A^{n+2m+3k+\cdots}(Q/U)^{n+m+k+\cdots}U, \tag{19}$$

where $A = 0.52917706 \times 10^{-10}$ m (1.0 a.u. of length), $Q = 1.6021892 \times 10^{-19}$ C (1.0 a.u. of charge), and $U = 4.3598124 \times 10^{-18}$ J (1.0 a.u. of energy) [10]. The set of integers n, m, k, . . . are the orders of dependence of the moments. For instance, for the dipole–dipole–quadrupole hyperpolarizability, $n = 2$ for the dipole dependence and $m = 1$ for the quadrupole dependence; the other integers are zero. Table I lists certain of these conversion factors.

II. MEASUREMENT OF PERMANENT MOMENTS AND POLARIZABILITIES

There has been dramatic evolution in the experimental means for the determination of electrical moments, polarizabilities, and hyperpolarizabilities [5, 11–13]. Very recent results of Muenter and co-workers [14] obtained from molecular beam electrical resonance (MBER) spectroscopy have yielded electrical moments to remarkably high precision for ground and first excited vibrational states. Dudley and Ward [15] have employed dc electric field induced second harmonic generation to determine first and second hyperpolarizabilities of HF and HCl. Stark effect, pulsed molecular beam microwave spectroscopy, as first carried out by Flygare and co-workers [16], has provided the capability for precise dipole moment measurements for transient species and species with very small dipoles. There are a host of sophisticated experimental designs and instrumentation available today that make it possible to measure those subtle features and properties of molecules that describe a response to electric and magnetic fields. In this chapter, we are interested in mentioning certain of the techniques for determining the re-

TABLE I

Conversion Factors for Molecular Electrical Properties from Atomic Units

Electric charge (zero moment):
1.602189×10^{-19} C
4.803242×10^{-10} cm$^{3/2}$ g$^{1/2}$ s^{-1}
Dipole (first) moment:
8.478418×10^{-30} C m
2.541765×10^{-18} cm$^{5/2}$ g$^{1/2}$ s^{-1}
2.541765 D
Quadrupole (second) moment:
4.486584×10^{-40} C m^2
1.345044×10^{-26} cm$^{7/2}$ g$^{1/2}$ s^{-1}
1.345044 B
Octupole (third) moment:
2.374197×10^{-50} C m^3
7.117664×10^{-35} cm$^{9/2}$ g$^{1/2}$ s^{-1}
Dipole polarizability:
1.648778×10^{-41} C^2 m^2 J^{-1}
0.148185 Å^3
Dipole–quadrupole polarizability:
8.724947×10^{-52} C^2 m^3 J^{-1}
7.841588×10^{-2} Å^4
Quadrupole–quadrupole and dipole–octupole polarizability:
4.617042×10^{-62} C^2 m^4 J^{-1}
4.149589 10^{-2} Å^5
Dipole hyperpolarizability:
3.206334×10^{-53} C^3 m^3 J^{-2}
4.149589×10^{-2} Å^5
Second-dipole hyperpolarizability:
6.235278×10^{-65} C^4 m^4 J^{-3}
1.162003×10^{-2} Å^7
Electric field strength:
5.142250×10^{9} V cm^{-1}
Electric field gradient, 1 V = 1 J/C:
9.717447×10^{17} V cm^{-2}

sponse of molecules to applied electric fields and how static electrical properties are deduced. These techniques emphasize particular and sometimes different molecular features.

A. Spectroscopic Measurements

When an electric field is applied to a dipolar molecule in a Stark experiment, the magnetic quantum number degeneracy is at least partially removed and the molecule reorients itself in the field. The energy levels are shifted, and there is a low-order dependence on the dipole moment and the strength of the applied field [17]. Except for linear molecules, the lowest order energy correction is linear in the field strength [18]. By making variations in the field strength and measuring transition energies, the dipole moment can be obtained. With the lowest order dependence being quadratic for linear molecules, only the magnitude and not the sign of the dipole can be obtained. Isotopic substitution, though, may change the rotational magnetic moment and then the sign of the dipole may be deduced [19, 20]. Since the energy correction terms also contain the polarizability anisotropy, high-field experiments allow for the determination of static polarizabilities as well.

Pressure broadening of spectral lines is a manifestation of the nature of intermolecular forces between molecules. Important features evident from the spectra include energy shifts, linewidths, and lineshapes. These can be used to predict the tensor properties of molecules, and studies have been done on pure gases as well as mixtures. From vibrational–rotational spectra, the moments of the band can be used to find electrical properties indirectly. Gordon [21] has shown that the second and fourth moments of the band are related to the mean square torque in the radiating molecule. Once the torque is known, the hypervirial theorem connects the torque to the intermolecular potential energy function. If a multipole model is used for the potential energy function, the tensor properties can be found from the long-range anisotropic terms. Gordon applied the model to CO–Ar and CO–CO systems, and Armstrong et al. [22] applied it to rare gas–tetrahedral molecule systems. This same idea is often used in multiproperty analyses (described later) for determination of moments and polarizabilities.

Nuclear magnetic resonance (NMR) has been used to determine electrical properties in a conventional setup [23], but recently, applied electric fields have been incorporated for the determination of properties [24, 25]. Polar liquids and solutions of polar molecules align when a strong electric field (about 300 kV/cm) is applied. The anisotropic spin interactions essentially modify the NMR spectrum, and determinations of the lowest order dipole polarizability can be made. To low order, the interaction energy may be taken to be

$$W = -\tfrac{1}{2}\mathbf{F}\cdot\mathbf{S}\cdot\mathbf{F} = -\tfrac{1}{2}[\tfrac{1}{3}\Delta S(3\cos^2\theta - 1) + \tfrac{1}{2}\delta S(\sin^2\theta\cos 2\phi) + \tfrac{1}{3}\mathrm{Tr}(\mathbf{S})]\,|\mathbf{F}|^2, \tag{20}$$

where $\mathbf{F}$ is an arbitrary field (electric or magnetic) and $\mathbf{S}$ is a second-rank property (α, or the magnetic susceptibility χ). This puts the energy correction in terms of the anisotropy and asymmetry $\Delta S = S_{zz} - \frac{1}{2}(S_{xx} + S_{yy})$ and $\delta S = S_{xx} - S_{yy}$ of the tensor. A problem that remains is that the actual internal field strength must be known, and two models, due to Lorentz [26] and Onsager [27], have been used with comparable success. Both rely on the dielectric permittivity of the medium to predict the field. As in electro-optic experiments, the method does not directly provide the individual components of the polarizability tensor, and an independent measurement of the mean polarizability is required. A feature of electric field NMR is that a single measurement provides both the anisotropy and asymmetry of the static polarizability.

B. Molecular Beams

Molecular beam methods for property measurement have their origins in the magnetic field experiments of Stern and Gerlach. Use of electric fields has become widespread [28–30] for probing molecular details, with quadrupolar fields used for focusing or with hexapole plus dipole fields for orienting molecules [31]. In MBER, electromagnetic radiation resonant with a transition frequency can bring about selective deflection of molecules because of how the properties change by excitation to another quantum state. These molecular beam deflection techniques are important tools for investigating permanent moments, even very small ones, as Klemperer and co-workers have reported [32].

If an rf field is applied at right angles to the homogeneous fields, the electrical resonance techniques allow single rotational states to be studied and moments determined [33, 34]. If sufficiently strong fields are applied, line broadening, changes in the average dipole moment, and changes in the moment of inertia result from the coupling of vibrational states. Thus, Stark effects can be related to geometry differences among vibrational states. The significant advantage of MBER techniques over spectroscopic electrical resonance is linewidth. Since the lifetime of a rotational state is about the same as the time between collisions, the linewidth of a resonant absorption associated with reorientation in a strong electric field is very large. In a molecular beam experiment, this is not the case.

A demonstration of the efficacy of MBER spectroscopy is the recent experiments on HF carried out by Bass, DeLeon, and Muenter [14]. In an effort to obtain Stark, Zeeman, and hyperfine properties, measurements were made that gave accurate values for both the ground and first excited vibrational levels of HF. Conventional resonance experiments can be done if the $v = 1$ state can be sufficiently populated. Using a color center IR laser to excite HF to $v = 1, J = 1$ levels, all the properties measured for the $v = 0$ and $v = 1$ states had essentially identical precision. The results included dipole moments, magnetic shielding anisotropies, rotational magnetic moments, magnetic susceptibilities, transition moments, and first and second derivatives with respect to internuclear separation of the properties.

C. Light Scattering

When an electric field is applied to a material, the induced birefringence is the difference between the index of refraction for directions parallel and perpendicular to the applied field [35], as measured by the Kerr constant,

$$K_m = \begin{cases} \dfrac{6n}{(n^2+2)^2(\varepsilon+2)^2}\left[(n_{\parallel}-n_{\perp})\dfrac{V_m}{F^2}\right]_{F\to 0}, & (21) \\[2ex] A_k + \dfrac{B_k}{V_m} + \cdots, & (22) \end{cases}$$

where n is the isotropic index of refraction, ε is the dielectric constant, V_m is the molar volume, F is the uniform electric field strength, and A_k is a Kerr virial coefficient and depends on the dynamic α, β, and γ polarizabilities. Essentially, the absorption of light is monitored with fields applied perpendicular and parallel to polarization directions. Buckingham and Longuet-Higgins [36] showed that the presence of a nonuniform field leads to induced birefringence depending on field gradient components with an explicit dependence on the product $(\alpha_{\parallel}-\alpha_{\perp})\theta$, where θ is the quadrupole moment. Thus, the quadrupole moment is obtainable via the Kerr effect. In this case, the definition of coordinate center is important since θ refers to the quadrupole defined at the effective quadrupole center. If the center were the center of mass, then a Kerr measurement describes the anisotropy of the polarizability [37]. The quadrupole moment could be obtained from the measured product if the magnitude of the polarizability difference were obtained in a depolarization experiment and the sign were determined from bond polarizabilities. Kerr experiments are not adequate to define all of the tensor components of the polarizabilities. If the experiments use polar molecules, a mean polarizability and a depolarization ratio of scattered polarized light are required to uniquely define the components. If nonpolar molecules are investigated, however, the previous information is not independent and data from crystal refraction measurements or Cotton–Mouton effect measurements may be required.

D. Multiproperty Analysis

The moments and polarizabilities of molecules can be determined by indirect means. In collision experiments, the nature of the interaction is governed by the potential energy surface, itself a function of the molecular properties of the colliding partners. Usually the potential energy is written in a multipole expansion whereby the electrical properties are displayed in the long-range terms [38]. The potential that is generated must satisfy simultaneously molecular beam scattering data (total and state-to-state differential cross sections), virial coefficients, spectral moment data, and so on [39]. However, this experimental data may vary widely in sensitivity to the intermolecular surface. Close-in, virial coefficients tend to prescribe the strong repulsive wall. For properties incorporated into the long-range terms, experiments must reflect sensitivity in that region. Molecular beam data can be useful in equilibrium and long-range regions. If measurements are made of the total differential cross section, features such as rainbow maximum and glory oscillations can reliably predict the isotropic part of the potential though less reliably the anisotropic components of the potential [40]. If the state-to-state differential cross sections are accurately measured, then an accurate anisotropic potential can be given in conjunction with other data. Recent applications include He–CO [41] and Ar–CH_4 [42, 43], where from the derived potentials, values of the polarizabilities have been extracted. Because of the indirect nature of a multiproperty analysis, its results as well as the heuristic association of contributing pieces from the multiproperty potential with excitation energies, ionization energies, or polarizability elements must be taken carefully. The ab initio results given in the next section afford the opportunity to evaluate the veracity of these types of procedures.

III. CALCULATION OF STATIC MOLECULAR ELECTRICAL PROPERTIES

A. Analytical Differentiation and Derivative Hartree–Fock Theory

Many important molecular properties are directly defined as a derivative of an energy. The dipole moment, of course, is properly defined as the first derivative of the energy with respect to the strength of an applied, uniform field. The force constant associated with a bond is the second derivative of the molecular potential energy function with respect to a displacement coordinate along that bond. Derivative-related properties can be obtained in several ways. One way is by finite differences. For instance, an electronic wavefunction and energy of a molecule might be calculated with a uniform field applied along the x axis with a strength of 0.0005 a.u. This would be accomplished by adding a one-electron operator to the Hamiltonian that corresponds to the interaction of the electrons with such a field. It would be the x axis dipole moment operator multiplied by 0.0005 a.u. The difference between the "finite-field" energy and the energy obtained without the added operator divided by 0.0005 a.u. would be a finite-difference estimate of the x component of the molecular dipole moment. Possible difficulties in a finite-difference procedure are discussed later.

Calculation of a dipole might be accomplished by taking the expectation value of the dipole moment operator. That result will be equivalent to the result obtained from invoking the strict definition of the dipole moment as a derivative in the case where the wavefunction obeys the Hellman–Feynman theorem [1, 44] or, in general, where the wavefunction is completely variational.

Derivatives of the electronic energy of a molecule can be found analytically, and finding ways to do that has been the focus of many investigations in recent years. What may be regarded as the underlying idea behind analytical derivative methods is rigorously differentiating the true quantum-mechanical energy. Generally this involves solving a type of eigenfunction equation or a simultaneous system of linear equations. To illustrate how this is accomplished, a straightforward procedure, not specific to electronic wavefunctions, will be discussed.

1. *Differentiation of the Schrödinger Equation*

A molecular Hamiltonian usually has embedded parameters. Within the Born–Oppenheimer approximation, for instance, the nuclear positions are parameters. Without loss of generality, any parameters can be embedded in an arbitrary Hamiltonian, and these will be designated $a, b, c, \ldots,$

$$H = H(a, b, c, \ldots). \tag{23}$$

The parameter list may include geometric parameters that would typically represent displacements from some chosen position, or it might include field strengths or some type of influencing perturbation. It is convenient to have a zero-order choice or "equilibrium" specification of parameters in working out the derivative Schrödinger equations. We take as that zero-order specification the situation when all parameters are zero. This means using relative parameters in place of any that actually are nonzero at the equilibrium. The specific choice of equilibrium parameters amounts to defining the zero-order Hamiltonian as

$$H_0 = H(a=0, b=0, c=0, d=0, \ldots). \tag{24}$$

The Schrödinger equation is simply

$$H\Psi = E\Psi \quad \text{or} \quad (H-E)\Psi = 0. \tag{25}$$

At the equilibrium choice of the embedded parameters, the Schrödinger equation is written as

$$(H_0 - E_0)\Psi_0 = 0. \tag{26}$$

The task of differentiating Eqn. (25) begins with differentiating the Hamiltonian. The partial derivatives of the Hamiltonian are designated as

$$H^a = \frac{\partial H}{\partial a}. \tag{27}$$

At this point, the partial derivative of the Hamiltonian may or may not still have dependence on some of the parameters. Thus, the derivative of the Hamiltonian becomes a specific operator when the parameters are set to some particular values. The equilibrium choice is designated:

$$H_0^a = \left.\frac{\partial H}{\partial a}\right|_{a=0,\,b=0,\,c=0,\,\ldots}. \tag{28}$$

The first derivative of the Schrödinger equation with respect to one of the parameters, a, is

$$H^a\Psi + H\Psi^a = E^a\Psi + E\Psi^a. \tag{29}$$

Assuming that the zero-order Schrödinger equation has been solved or that its solutions are somehow known, each derivative expression can be made to take advantage of what is known by selecting the equilibrium parameter choice. Evaluated at equilibrium, the first-derivative equation is

$$H_0^a\Psi_0 + H_0\Psi_0^a = E_0^a\Psi_0 + E_0\Psi_0^a. \tag{30}$$

This equation includes the first derivative of the energy with respect to the parameter a, E_0^a. It is also an equation with a very real correspondence to first-order perturbation theory, and that suggests how best to use it. Indeed, the general procedure being outlined here differs from a perturbation expansion in only one minor way. A perturbation expansion is in terms of powers of one or more parameters. The derivative expansion is a Taylor-series-type expansion that has each nth power series term divided by $n!$. That factor converts perturbative energy corrections into energy derivatives. So, Eqn. (30) is conveniently rearranged, just as is usually done in an elementary introduction to perturbation theory:

$$E_0^a\Psi_0 = (H_0 - E_0)\Psi_0^a + H_0^a\Psi_0. \tag{31}$$

Integration of this equation using the zero-order wavefunction yields an expression for the first derivative of the energy:

$$E_0^a\langle\Psi_0|\Psi_0\rangle = \langle\Psi_0|H_0 - E_0|\Psi_0\rangle + \langle\Psi_0|H_0^a|\Psi_0\rangle. \tag{32}$$

If the zero-order wavefunction does, in fact, satisfy Eqn. (26), then the first term on the right in Eqn. (32) is identically zero. This means that the first derivatives are obtained as an expectation value of the derivative Hamiltonian. This last statement is the Hellmann–Feynman theorem.

Another useful way of rearranging Eqn. (29) is to collect terms involving the unknown derivative function. Again, this follows perturbation theory, and one obtains

$$(H-E)\Psi^a = -(H^a - E^a)\Psi. \tag{33}$$

This expression is an eigenfunction expression similar to Eqn. (25). Instead of the operator $(H-E)$ yielding zero, however, it yields another function, that which is on the right side of Eqn. (33). It will be seen that all derivative equations can be put into the same type of eigenfunction form, but with different right-hand sides. Evaluating Eqn. (33) at the equilibrium choice of parameters gives another way of writing Eqn. (31):

$$(H_0 - E_0)\Psi_0^a = -(H_0^a - E_0^a)\Psi_0. \tag{34}$$

With the result of Eqn. (32), the only unknown in Eqn. (34) is the derivative of the wavefunction, which is the solution to be garnered from Eqn. (34).

The Schrödinger equation can be formally differentiated to any order with respect to one parameter or any number of parameters. Continuing to second differentiation with respect to another parameter, b, Eqn. (29) leads to

$$H^{ab}\Psi + H^a\Psi^b + H^b\Psi^a + H\Psi^{ab} = E^{ab}\Psi + E^a\Psi^b + E^b\Psi^a + E\Psi^{ab}. \tag{35}$$

Rearranging Eqn. (35) and integrating with the zero-order wavefunction yields an expression for the second derivative of the energy:

$$\begin{aligned} E_0^{ab}\langle\Psi_0|\Psi_0\rangle &= \langle\Psi_0|H_0^{ab}|\psi_0\rangle + \langle\Psi_0|H_0^a - E_0^a|\psi_0^b\rangle \\ &\quad + \langle\Psi_0|H_0^b - E_0^b|\Psi_0^a\rangle + \langle\Psi_0|H_0 - E_0|\Psi_0^{ab}\rangle. \end{aligned} \tag{36}$$

The last term of the right side is zero if Eqn. (26) is satisfied. Rearrangement of Eqn. (35) into eigenfunction form gives

$$(H-E)\Psi^{ab} = -(H^a - E^a)\Psi^b - (H^b - E^b)\Psi^a - (H^{ab} - E^{ab})\Psi. \tag{37}$$

This equation can be evaluated at equilibrium and then solved to obtain the second derivative of the wavefunction. The uniformity of the derivative eigenfunction equations is illustrated by giving the third-derivative equation and comparing it with Eqns. (33) and (37):

$$(H-E)\Psi^{abc}=-(H^a-E^a)\Psi^{bc}-(H^b-E^b)\Psi^{ac}-(H^c-E^c)\Psi^{ab}-(H^{ab}-E^{ab})\Psi^c$$
$$-(H^{ac}-E^{ac})\Psi^b-(H^{bc}-E^{bc})\Psi^a-(H^{abc}-E^{abc})\Psi. \tag{38}$$

In all cases, the right side is entirely determined from lower order derivative results. Those terms can be collected and the equation to be solved has the same form, no matter what the level of differentiation.

As in perturbation theory, there is a $2n+1$ *rule* [45–47]. Using the nth-order wavefunction, energy derivatives up to $2n+1$ may be evaluated directly, and the derivative wavefunctions between the n and $2n+1$ orders are not required explicitly. For example, with the first-derivative wavefunctions known explicitly, the third energy derivatives may be calculated directly. To see this for the *abc* derivative of Eqn. (38), one evaluates the equation at the equilibrium choice of parameters and then integrates with the zero-order wavefunction. That is,

$$E^{abc}=\langle\psi_0|H_0-E_0|\Psi_0^{abc}\rangle+\langle\psi_0|H_0^{abc}|\psi_0\rangle+\langle\Psi|H_0^a-E_0^a|\Psi_0^{bc}\rangle$$
$$+\langle\Psi_0|H_0^b-E_0^b|\Psi_0^{ac}\rangle+\langle\Psi_0|H_0^c-E_0^c|\Psi_0^{ab}\rangle+\langle\psi_0|H_0^{bc}-E_0^{bc}|\Psi_0^a\rangle$$
$$+\langle\Psi_0|H_0^{ac}-E_0^{ac}|\Psi_0^b\rangle+\langle\Psi_0|H_0^{ab}-E_0^{ab}|\psi_0^c\rangle. \tag{39}$$

The first term is zero if Eqn. (26) is satisfied. This is an example of the simple $n+1$ rule where the nth-derivative wavefunctions are used to obtain the $n+1$ derivatives. Other terms in Eqn. (39) involve the second derivatives of the wavefunctions. However, Eqns. (33) and (37) allow for substitutions that eliminate the second-derivative wavefunctions explicitly. First, Eqn. (33) is used to "swap" differentiation of the operators with differentiation of the wavefunctions, as in

$$\langle\psi_0|H_0^a-E_0^a|\psi_0^{bc}\rangle=-\langle\Psi_0^a|H_0-E_0|\psi_0^{bc}\rangle. \tag{40}$$

This result follows from evaluating Eqn. (37) at the equilibrium and then integrating with ψ_0^{bc} and using the Hermitian properties of the operators. Equation (37) is used to swap second-order differentiation of the operators in the same way, this time integrating with Ψ_0^a:

$$\langle\Psi_0|H_0^{bc}-E_0^{bc}|\Psi_0^a\rangle=-\langle\Psi_0^{bc}|H_0-E_0|\psi_0^a\rangle$$
$$+\langle\Psi_0^b|H_0^c-E_0^c|\Psi_0^a\rangle+\langle\Psi_0^c|H_0^b-E_0^b|\Psi_0^a\rangle \tag{41}$$

Essentially, the third-derivative expression is recast in terms of the first derivatives of the wavefunction and the first derivatives of the Hamiltonian and the energy.

$$E^{abc}=\langle\Psi_0|H_0^{abc}|\Psi_0\rangle+2\langle\Psi_0^b|H_0^c-E_0^c|\Psi_0^a\rangle$$
$$+2\langle\Psi_0^c|H_0^b-E_0^b|\Psi_0^a\rangle+2\langle\Psi_0^c|H_0^a-E_0^a|\Psi_0^b\rangle. \quad (42)$$

In this manner, formal differentiation of the Schrödinger equation is easily accomplished.

2. *Basis Set Expansions of Derivative Equations*

Basis sets can be employed to solve derivative Schrödinger equations as naturally as employing them to solve the basic Schrödinger equations. An organized way of using basis sets, and a way that is quite suited for computational implementation, is to cast operators into their matrix representations in the given basis. This needs to be done for the zero-order Hamiltonian and for each derivative Hamiltonian operator. The zero-order Schrödinger equation for one state in matrix form is

$$(H_0-E_0)\mathbf{C}_0=0 \quad (43)$$

where $\mathbf{C}_0$ is the eigenvector, H_0 is the matrix representation of the "equilibrium" Hamiltonian, and E_0 is the eigenenergy. The first-derivative Schrödinger expression, when developed in a basis set expansion, becomes

$$(H_0-E_0)\mathbf{C}_0^a=-(H_0^a-E_0^a)\mathbf{C}_0. \quad (44)$$

Converting the equations in the first part of this section to their basis set forms involves little more than replacing a wavefunction (or derivative wavefunction) with a vector of coefficients, a column vector if to the right of an operator, and replacing an operator by its matrix representation.

Comparing Eqns. (43) and (44) reveals that the equations have the same form except that there is a nonzero column vector on the right side of Eqn. (44). All higher derivative expressions look the same, and so the matrix expression that needs to be solved in every case has the same inhomogeneous form:

$$(H_0-E_0)\mathbf{C}_0^\alpha=\mathbf{X}^\alpha. \quad (45)$$

A general algorithm for solving the coupled linear equations of Eqn. (45) is all that is necessary, and there are many standard algorithms for this purpose.

The column vector on the right side of Eqn. (45) will be determined in every case from lower order results.

The energy derivatives can be found from each equation by multiplying on the left with $\mathbf{C}_0^\dagger$. This corresponds to the $n+1$ rule as illustrated with the first-derivative equation

$$E_0^a = \mathbf{C}_0^\dagger H_0^a \mathbf{C}_0. \tag{46}$$

Using the $2n+1$ rule presents no added complication when used with a basis set expansion if the basis is not dependent on the differentiation parameters. If the basis set is dependent on the parameters, then derivatives to the $2n+1$ order of the integrals over the basis functions are required. King and Komornicki [47] have given a very general and well-organized development of the relationships leading to the $2n+1$ rule and at the same time have derived expressions for errors in derivatives from using imperfectly optimized wavefunctions.

In electronic structure calculations, it is not unlikely for a basis set to be dependent on the parameters. The most obvious case involves geometric parameters. The atomic orbital basis functions used to construct molecular orbitals are generally chosen to follow the atomic centers. This means that the functions are dependent on the molecular geometry, and so there will be nonzero derivatives of the usual one- and two-electron integrals. In the case of parameters such as an electric field strength, there is no functional dependence of the standard types of basis functions. The derivatives of all the basis functions with respect to this parameter are zero, and so all derivative integrals involving the zero-order Hamiltonian terms are zero as well.

Sadlej [48–51] has recommended use of basis sets that *are* dependent on the perturbation because this may afford using a smaller basis set. Sadlej and co-workers have done this for electric field properties by incorporating a specific field dependence into Gaussian basis functions. Hudis and Ditchfield [52] have more recently suggested a different way of selecting field-dependent bases, and both schemes seem quite promising. Ditchfield had much earlier accomplished the same thing for the challenging problem of solving for NMR chemical shifts [53].

3. *Derivative Hartree–Fock (DHF) Theory*

Dykstra and Jasien [9] used the general equations given in the preceding to implement an approach for the calculation of derivatives of the Hartree–Fock or SCF energy. Unique to the DHF approach is its open-endedness. The computer program that was written was immediately able to compute a tenth-dipole hyperpolarizability and beyond, if desired. Derivatives involving geometric parameters could be obtained, too, but then there exists the requirement for the corresponding derivatives of the two-electron integrals over basis functions.

In the application of DHF to multipole polarizabilities, the first task is assembling the derivative Fock operators because in SCF the Fock operator is the effective (one-electron) Hamiltonian for the system. The parameters of differentiation are the elements of the expansion of the electrical potential, $V_x, V_y, V_z, V_{xx}, V_{xy}, \ldots, V_{xxx}, \ldots$. From Eqn. (28), one may see that the first derivatives of the Fock operator with respect to these parameters are the component multipole moment operators. For example,

$$F_0^{V_x} = \mu_x = M_x, \tag{47a}$$

$$F_0^{V_{xy}} = M_{xy}. \tag{47b}$$

The basis set is the set of atomic one-electron functions, and we shall take, for example, $F_0^{V_x}$ to be matrix representations in that basis.

Differentiating the Hartree–Fock equation is more involved than differentiating the Schrödinger equation because the Hartree–Fock equation must yield several (one-electron) states or orbitals. We begin with the equation itself:

$$\mathbf{FC} = \mathbf{SCE}, \tag{48}$$

where $\mathbf{S}$ is the matrix of overlap of the atomic basis functions, $\mathbf{C}$ is the matrix of coefficients of the atomic orbitals making up the molecular orbitals, and $\mathbf{F}$ is the Fock operator matrix appropriate for the given electronic state. In general, $\mathbf{F}$ should be a matrix of single-substitution Hamiltonian elements that would be zero by Brillouin's theorem for a given wavefunction, as in a one-Fock-operator method for open-shell or multiconfigurational SCF. If set up that way, the development can be carried through for all types of states and all types of SCF wavefunctions. Here, $\mathbf{E}$ is the orbital energy matrix. Recall that the condition that it be block diagonal serves to determine the orbital set. Direct differentiation of Eqn. (48) with respect to one parameter yields

$$\mathbf{F}_0^a\mathbf{C}_0 + \mathbf{F}_0\mathbf{C}_0^a = \mathbf{S}_0^a\mathbf{C}_0\mathbf{E}_0 + \mathbf{S}_0\mathbf{C}_0^a\mathbf{E}_0 + \mathbf{S}_0\mathbf{C}_0\mathbf{E}_0^a. \tag{49}$$

As before, the zero subscript means that the parameters in the Hamiltonian, $V_x, V_y, V_z, V_{xx}, \ldots$, have been set to zero, the equilibrium. This is in contrast to Eqn. (48), which implicitly represents the set of all SCF equations for the infinite number of possible choices of the field, field gradient, and so forth. The symbol $\mathbf{C}_0$ is just the set of SCF orbitals, or orbital expansion coefficients, with $\mathbf{V} = 0$. A derivative orbital set such as $\mathbf{C}_0^a$, $\mathbf{C}_0^b$, or $\mathbf{C}_0^{ab}$ can be related to the original $\mathbf{C}_0$ matrix by a matrix multiplication involving a new set of matrices:

$$\mathbf{C}_0^a = \mathbf{C}_0\mathbf{U}_0^a \quad \text{and} \quad \mathbf{C}_0^{ab} = \mathbf{C}_0\mathbf{U}_0^{ab}. \tag{50}$$

Each $\mathbf{U}$ matrix is a representation of the derivative orbital functions in the original orbital basis. Finding the $\mathbf{U}$ matrices, then, is the same thing as finding the derivatives of the orbitals.

The first step in finding a $\mathbf{U}$ matrix is to multiply through by $\mathbf{C}_0^\dagger$ and substitute according to Eqn. (50). Doing this for Eqn. (49) yields

$$\mathbf{C}_0^\dagger\mathbf{F}_0^a\mathbf{C}_0 + \mathbf{C}_0^\dagger\mathbf{F}_0\mathbf{C}_0\mathbf{U}_0^a = \mathbf{C}_0^\dagger\mathbf{S}_0^a\mathbf{C}_0\mathbf{E}_0 + \mathbf{C}_0^\dagger\mathbf{S}_0\mathbf{C}_0\mathbf{U}_0^a\mathbf{E}_0 + \mathbf{C}_0^\dagger\mathbf{S}_0\mathbf{C}_0\mathbf{E}_0^a. \tag{51}$$

A simplification using the fact that $\mathbf{C}_0^\dagger\mathbf{S}_0\mathbf{C}_0 = 1$ and $\mathbf{C}_0^\dagger\mathbf{F}_0\mathbf{C}_0 = \mathbf{E}_0$ yields

$$\mathbf{E}_0\mathbf{U}_0^a - \mathbf{U}_0^a\mathbf{E}_0 = \mathbf{C}_0^\dagger\mathbf{S}_0^a\mathbf{C}_0\mathbf{E}_0 - \mathbf{C}_0^\dagger\mathbf{F}_0^a\mathbf{C}_0 + \mathbf{E}_0^a. \tag{52}$$

Equation (52) is the standard coupled-perturbed Hartree–Fock (CPHF) equation in matrix form [54, 55]. This equation is sufficient for finding the elements of $\mathbf{U}_0^a$ wherever the corresponding elements of $\mathbf{E}_0^a$ are zero. The matrix $\mathbf{E}_0$ is usually diagonalized completely, but diagonalization of $\mathbf{E}_0$ within the diagonal blocks corresponding to all occupied or to all virtual orbitals is an arbitrary condition for which the SCF energy is invariant. Thus, only off-diagonal blocks of $\mathbf{E}_0^a$ are necessarily zero, meaning that Eqn. (52) serves to determine only the corresponding off-diagonal block elements of $\mathbf{U}_0^a$. As King and Komornicki have discussed [47], the reason only part of any $\mathbf{U}_0$ is found from this type of procedure is that the entire set of molecular orbital coefficients are not independently variable due to orthonormality of the orbitals. Shortly, we shall discuss constraint equations that complete the determination of $\mathbf{U}_0$, though there is an alternative, which is using orbital rotation parameters that are independent [56–59].

It is useful to introduce two matrices with the following definitions:

$$\mathbf{R}_0^\alpha \equiv \mathbf{C}_0^\dagger\mathbf{S}_0^\alpha\mathbf{C}_0 \quad \text{and} \quad \mathbf{G}_0^\alpha \equiv \mathbf{C}_0^\dagger\mathbf{F}_0^\alpha\mathbf{C}_0, \tag{53}$$

where the superscript α designates an arbitrary derivative. These $\mathbf{R}$ and $\mathbf{G}$ matrices are simply derivative overlap and Fock operator matrices transformed to the molecular-orbital basis. Introducing them makes the subsequent derivations more compact. It also corresponds to an effective computational organization for using $\mathbf{S}^\alpha$ and $\mathbf{F}^\alpha$ since once these matrices are obtained for a given α, they can be used in the **R** and **G** form to find the next-higher derivatives.

In terms of **R** and **G** matrices, Eqn. (52) is

$$\mathbf{E}_0\mathbf{U}_0^a - \mathbf{U}_0^a\mathbf{E}_0 = \mathbf{R}_0^a\mathbf{E}_0 - \mathbf{G}_0^a + \mathbf{E}_0^a. \tag{54}$$

Equation (54) is used to find the elements of the **U** matrix as in conventional CPHF by rewriting it for a specific element such as that involving the m occupied orbital and the r virtual orbital:

$$U_{mr}^a(\varepsilon_m - \varepsilon_r) = R_{mr}^a\varepsilon_r - G_{mr}^a, \tag{55}$$

where ε_m and ε_r are orbital energies. (The zero subscript will be suppressed in expressions involving individual matrix elements.) Equation (55) represents an iterative process because the **U** matrix being determined defines the derivative orbitals, which in turn are needed to construct the derivative Fock operator. Thus, following some type of initial guess, a **U** matrix is found, then a new derivative Fock matrix, and then another **U** matrix. The process is converged when the **U** matrix changes negligibly between cycles. Alternatively, this equation can be solved as well by writing it as a set of coupled linear equations [60, 61].

A second-derivative equation takes the form

$$\mathbf{E}_0\mathbf{U}_0^{ab} - \mathbf{U}_0^{ab}\mathbf{E}_0 = \mathbf{R}_0^{ab}\mathbf{E}_0 - \mathbf{G}_0^{ab} + \mathbf{E}_0^{ab} + \mathbf{\Gamma}_0^{ab}. \tag{56}$$

The matrix $\mathbf{\Gamma}_0^{ab}$ is completely determined by the lower order derivatives, and for this case it is given by

$$\begin{aligned}\mathbf{\Gamma}_0^{ab} = {} & (\mathbf{R}_0^a\mathbf{U}_0^b + \mathbf{R}_0^b\mathbf{U}_0^a)\mathbf{E}_0 + \mathbf{R}_0^a\mathbf{E}_0^b + \mathbf{R}_0^b\mathbf{E}_0^a \\ & + \mathbf{U}_0^a\mathbf{E}_0^b + \mathbf{U}_0^b\mathbf{E}_0^a - \mathbf{G}_0^a\mathbf{U}_0^b - \mathbf{G}_0^b\mathbf{U}_0^a.\end{aligned} \tag{57}$$

The $\mathbf{\Gamma}$ matrix facilitates the general treatment for derivatives. It is fixed throughout the iterative cycle to find **U** and is easily constructed. In developing explicit equations for second derivatives of MC–SCF orbitals, Dupuis has shown [61] that the first- and second-derivative equations take on a general form, as in Eqn. (56).

The general equations for iteratively finding the off-diagonal blocks of elements of the **U** matrix are easily developed and implemented. Letting α collectively represent differentiation to the ith, jth, and so forth powers of the parameters,

$$\alpha = a^i b^j c^k \ldots. \tag{58}$$

Then for any α, one must solve

$$U^{\alpha}_{mr}(\varepsilon_m - \varepsilon_r) = R^{\alpha}_{mr}\varepsilon_r - G^{\alpha}_{mr} + \Gamma^{\alpha}_{mr}. \tag{59}$$

The general expression for the $\mathbf{\Gamma}^{\alpha}$ matrix is

$$\begin{aligned}\mathbf{\Gamma}^{\alpha} = &\sum_{\substack{i'+i''+i'''=i \\ i'\neq i, i''\neq i, i'''\neq i}} \sum_{\substack{j'+j''+j'''=j \\ j'\neq j, j''\neq j, j'''\neq j}} \cdots \quad \mathbf{R}^{a^{i'} b^{j'} c^{k'}} \cdots \mathbf{U}^{a^{i''} b^{j''} c^{j''}} \cdots \\ &\mathbf{x}\, \mathbf{E}^{a^{i'''} b^{j'''} c^{j'''}} \cdots f(i', i'', i''') f(j', j'', j''') \cdots \\ &- \sum_{\substack{i'+i''=i \\ i'\neq i, i''\neq i}} \sum_{\substack{j'+j''=1 \\ j'\neq j, j''\neq j}} \cdots \quad \mathbf{G}^{a^{i'} b^{j'} c^{k'}} \cdots \mathbf{U}^{a^{i''} b^{j''} c^{k''}} \cdots \\ &\times f(0, i', i'') f(0, j', j'') \cdots .\end{aligned} \tag{60}$$

Each summation involves three indices. For example, the first summation means sum over all positive or zero values of i', i'', and i''' such that their sum equals i but that neither equals i alone. The factor f means

$$f(i', i'', i''') = \binom{i+i''}{i''}\binom{i'+i''+i'''}{i'''}. \tag{61}$$

Again, the parameters $a, b, c \ldots$ can be of all types including field strengths and bond lengths.

One simple but very effective logic procedure developed for DHF keeps track of the derivatives. An integer list is constructed for each particular derivative where there is one integer in the list for each parameter. For first- and second-multipole polarizabilities, there will be nine parameters, three for the first moment and six for the second moment, after ignoring equivalent elements in $\mathbf{V}$. One way of ordering them is V_x, V_y, V_z, V_{xx}, V_{xy}, V_{xz}, V_{yy}, V_{yz}, and V_{zz}, and then nine integers, ordered the same, are associated with each α derivative:

0 0 0 0 0 0 0 0 0	Zero-order
1 0 0 0 0 0 0 0 0	x component of first moment
0 1 0 0 0 0 0 0 0	y component of first moment
0 0 1 0 0 0 0 0 0	z component of first moment
0 0 0 1 0 0 0 0 0	xx component of second moment
0 0 0 0 1 0 0 0 0	xy component of second moment

⋮

0 0 0 0 0 0 0 0 1	zz component of second moment
2 0 0 0 0 0 0 0 0	x, x dipole polarizability
1 1 0 0 0 0 0 0 0	x, y dipole polarizability
1 0 1 0 0 0 0 0 0	x, z dipole polarizability
1 0 0 1 0 0 0 0 0	x, xx dipole–quadrupole polarizability
⋮	
0 0 0 2 0 0 0 0 0	xx, xx quadrupole polarizability
0 0 0 1 1 0 0 0 0	xx, xy quadrupole polarizability
⋮	
3 0 0 0 0 0 0 0 0	x, x, x dipole hyperpolarizability
2 1 0 0 0 0 0 0 0	x, x, y dipole hyperpolarizability
⋮	

These integer lists are useful for computing the lower order pieces in the derivative equations. It is a general result that for a derivative given by a list $[i, i', i'', \ldots]$, the lists for the elements in any contributing low-order product terms must add to give this list. This facilitates searching for the low-order terms. A specific example is the calculation of the $\alpha = [2\ 0\ 0\ 0\ 1\ 0\ 0\ 0\ 0]$ derivative. To identify low-order terms that would be used in products to find this derivative, sums of lists may be checked:

$$\begin{array}{rl} \alpha_i = & [0\ 0\ 0\ 0\ 1\ 0\ 0\ 0\ 0] \\ +\alpha_j = & [2\ 0\ 0\ 0\ 0\ 0\ 0\ 0\ 0] \\ \hline \alpha = & [2\ 0\ 0\ 0\ 1\ 0\ 0\ 0\ 0] \end{array} \qquad \begin{array}{rl} \alpha_k = & [1\ 0\ 0\ 0\ 0\ 0\ 0\ 0\ 0] \\ +\alpha_l = & [1\ 0\ 0\ 0\ 1\ 0\ 0\ 0\ 0] \\ \hline \alpha = & [2\ 0\ 0\ 0\ 1\ 0\ 0\ 0\ 0] \end{array}$$

In this example, two combinations of two lists have been found that sum to the list of the particular derivative being sought. Each is a "match," and so in using Eqn. (60) to find $\mathbf{\Gamma}^{\alpha}$, the products $\mathbf{G}^{\alpha_i}\mathbf{U}^{\alpha_j}$ and $\mathbf{G}^{\alpha_k}\mathbf{U}^{\alpha_l}$ are identified as contributing.

The diagonal blocks of the $\mathbf{U}$ matrices are obtained from differentiation of the orbital orthogonormality condition:

$$\mathbf{C}^{\dagger}\mathbf{S}\mathbf{C} = \mathbf{1}. \tag{62}$$

The procedure corresponds to that of differentiating Eqn. (48):

$$\mathbf{C}_0^{\dagger a}\mathbf{S}_0\mathbf{C}_0 + \mathbf{C}_0^{\dagger}\mathbf{S}_0^a\mathbf{C}_0 + \mathbf{C}_0^{\dagger}\mathbf{S}_0\mathbf{C}_0^a = \mathbf{0}, \tag{63}$$

$$\mathbf{U}_0^{a\dagger} + \mathbf{U}_0^a + \mathbf{R}_0^a = \mathbf{0}. \tag{64}$$

Equation (64) follows from Eqn. (63) using the definitions of Eqns. (50) and (53). For an arbitrary derivative, the expression required is

$$\mathbf{U}_0^{\alpha\dagger} + \mathbf{U}_0^{\dagger} + \mathbf{R}_0^{\alpha} + \mathbf{\Delta}_0^{\alpha} = \mathbf{0}, \tag{65}$$

where $\mathbf{\Delta}$ is analogous to $\mathbf{\Gamma}$ in that it is completely determined by the lower order differentiation,

$$\mathbf{\Delta}_0^{\alpha} = \sum\sum \cdots \mathbf{U}_0^{a^{i'} b^{j'}\dagger} \cdots \mathbf{R}_0^{a^{i''} b^{j''}} \cdots \mathbf{U}_0^{a^{i'''} b^{j'''}} \cdots f(i', i'', i''') \cdots. \tag{66}$$

The restrictions on the summations in Eqn. (66) are the same as those in Eqn. (60). This result can also be obtained from the idempotency of the one-electron density, which is usable through all orders of differentiation too [62, 63]. Equation (65) is used for elements of $\mathbf{U}$ that are in diagonal blocks. One may freely choose those blocks to be symmetric and then the elements are given by

$$U_{mn}^{\alpha} = -\tfrac{1}{2}(R_{mn}^{\alpha} + \Delta_{mn}^{\alpha}). \tag{67}$$

The final energy derivative evaluation is most easily illustrated for a closed-shell wavefunction. The energy expression is

$$\varepsilon = \tfrac{1}{2}\langle \mathbf{D}(\mathbf{h} + \mathbf{F})\rangle. \tag{68}$$

All that is required is to directly differentiate Eqn. (68). The simplifications that make this practical require considering the forms of the operators in closer detail than done in the preceding. The derivative of the Fock operator matrix can be broken into derivatives of constituent matrices:

$$\mathbf{F}^a = \mathbf{h}^a + 2\mathbf{J}(\mathbf{D}^a) - \mathbf{K}(\mathbf{D}^a) + 2\mathbf{J}^a(\mathbf{D}) - \mathbf{K}^a(\mathbf{D}), \tag{69}$$

where the terms involving the derivatives of the two-electron integrals over the basis functions are designated $\mathbf{J}^a$ and $\mathbf{K}^a$. Specifically,

$$[\mathbf{J}^a(\mathbf{D})]_{st} = \sum_{uv} D_{uv}(st|uv)^a. \tag{70}$$

From the derivatives of the orbitals the derivative of the one-electron density matrix (or the corresponding nth iteration guess) can be found by simple matrix multiplication:

$$\mathbf{D}^a = \mathbf{U}^{a\dagger}\mathbf{D} + \mathbf{D}\mathbf{U}^a. \tag{71}$$

Finally, a standard operator identify (for general Coulomb or exchange operators) needs to be used:

$$\langle \mathbf{D}\mathbf{J}(\mathbf{D}^a) \rangle = \langle \mathbf{D}^a \mathbf{J}(\mathbf{D}) \rangle. \tag{72}$$

Then, direct differentiation of Eqn. (68) and simplification yields

$$\varepsilon_0^a = \langle \mathbf{D}_0^a \mathbf{F}_0 \rangle + \langle \mathbf{D}_0 \mathbf{h}_0^a \rangle + \tfrac{1}{2}\langle \mathbf{D}_0 [2\mathbf{J}_0^a(\mathbf{D}_0) - \mathbf{K}_0^a(\mathbf{D}_0)] \rangle. \tag{73}$$

For an arbitrary derivative one has

$$\varepsilon_0^\alpha = \langle \mathbf{D}_0^\alpha \mathbf{F}_0 \rangle + \langle \mathbf{D}_0 \mathbf{h}_0^\alpha \rangle + \tfrac{1}{2}\langle \mathbf{D}_0 [2\mathbf{J}_0^\alpha(\mathbf{D}_0) - \mathbf{K}_0^\alpha(\mathbf{D}_0)] \rangle + \mathbf{\Phi}^\alpha. \tag{74}$$

The quantity $\mathbf{\Phi}^\alpha$ is

$$\mathbf{\Phi}_0^\alpha = \sum \sum \ldots \mathbf{D}_0^{a^{i'} b^{j'}} \cdots (\mathbf{h}_0^{a^{i''} b^{j''}} \cdots + \mathbf{F}^{a^{i'''} b^{i'''} \cdots}) f(i', i'', i''') \cdots, \tag{75}$$

where Φ_0^α is obtained entirely from lower order derivative results just as was done for the $\mathbf{\Gamma}$ and $\mathbf{\Delta}$ matrices in earlier steps, and the sums in Eqn. (75) are restricted as in Eqn. (60). Equation (74) represents the $n+1$ rule expression. The $2n+1$ rule can be developed by the swapping of terms discussed earlier.

The DHF procedure has certain features that make it an attractive organization for the calculation of properties. The order of differentiation can be carried as far as desired. The only seeming limitation comes about in the special summation restrictions as in Eqns. (60), (66), and (75). Our practical solution, the integer lists for each derivative, is strictly open ended with respect to the order higher of differentiation of the parameters. When $\mathbf{\Gamma}^\alpha$, $\mathbf{\Delta}^\alpha$, or $\mathbf{\Phi}^\alpha$ are to be found, the integer lists for α is selected. Then, integer lists for $\alpha' < \alpha$ and $\alpha'' < \alpha$ are searched to find each (α', α'') pair such that their list elements summed give the list of α. That identifies all matrix pairs, such as $\mathbf{D}^{\alpha'} \mathbf{F}^{\alpha''}$, that are needed. A search is also done for the set $(\alpha', \alpha'', \alpha''')$ whose list elements combine to give the list of α. This identifies matrix terms such as $\mathbf{R}^{\alpha'} \mathbf{U}^{\alpha''} \mathbf{E}^{\alpha'''}$ that are needed. With this form of the logic, the process can truly be continued to any order.

The organization is transparent to the type of parameter provided that the derivative Fock operators are suitably constructed. The separation of terms in Eqn. (69) mentioned here corresponds to the idea of Takada, Dupuis, and King for "skeleton" Fock matrices [64]. A geometric derivative can be obtained with the same code as an electrical property, as can mixed derivatives. Furthermore, magnetic properties, which are unique because second derivatives of the Hamiltonian operator are not necessarily zero, fit into the DHF structure without modification. Open-shell wavefunctions can be employed, but the Fock operator needs to be more generally defined. A one-Fock-operator method, which employs projection operators, is formally quite workable for DHF treatment of open-shell states.

Schaefer and co-workers have presented several reports [65] about derivatives of SCF wavefunctions, including harmonic transition moments that are electrical property derivatives. They elect to give expressions directly in terms of one- and two-electron integrals. This alternative formulation of the general problem is the most immediate means of solution, but it must be done tediously, order by order. However, it has been successfully worked out to low order entirely for closed-shell, open-shell, and certain MC–SCF wavefunctions. Derivatives of correlated wavefunctions may be found by the general schemes discussed at the outset of this section.

B. Finite-Difference Methods

The ab initio calculation of moments and polarizabilities can be accomplished directly from solution of the Schrödinger equation with an augmented Hamiltonian. The finite-field approach incorporates the electric field terms directly into the Hamiltonian for a fixed field strength and the Schrödinger equation is solved at that strength. An alternate approach uses one or more fixed charges placed in the vicinity of the molecule, but this adds in contributions not only of a uniform field but also of higher gradient components.

The perturbed total energies or other properties of the system can be written as an expansion in terms of moment and polarizability components (see Section I). If different values of the field strength or charge positions are used, a system of simultaneous equations can be written from the truncated series, and these equations are solved to find the unknown polarizabilities. The system of equations must be chosen sufficiently large to ensure that the truncation error is minimized, but sometimes it is not practical to carry out the number of finite-field calculations that this might call for.

1. *Finite-Field Methods*

Finite-field methods were first used to calculate dipole polarizabilities by Cohen and Roothaan [66]. For a fixed field strength V_z, the Hamiltonian potential energy term for the interaction between the electric field and ith electron is just $V_z \mu_{z_i}$. The induced dipole moment with the applied field can be calculated from the Hartree–Fock wavefunction by integrating the dipole moment operator with the one-electron density since this satisfies the Hellmann–Feyman theorem. With the usual dipole moment expansion,

$$\mu_z = \mu_z^0 + \alpha_{z,z} V_z + \cdots + (\tfrac{1}{2}) A_{z,zz} V_{zz} + \cdots, \qquad (76)$$

where μ_z^0 is the permanent field-free moment. The polarizability may be obtained as a finite difference,

$$\alpha_{z,z}(V_z) = \frac{\mu_z - \mu_z^0}{V_z}. \tag{77}$$

If this is done for several choices of V_z, it may be possible to take the limit as $V_z \rightarrow 0$ and obtain the dipole polarizability as strictly defined:

$$\alpha_{i,j} = \left.\frac{\partial \mu_i}{\partial V_j}\right|_{\mathbf{V}=0}. \tag{78}$$

In practice, the fields used must be small to minimize the truncation error in the series and ensure that a numerical derivative is accurate. Contamination of the polarizabilities by the next higher order term can be removed if fields of the same magnitude but of opposite sign are used.

If only uniform fields are used in the finite-field method, then it is only possible to calculate the dipole polarizabilities and hyperpolarizabilities. In order to obtain any higher multipole contributions, field gradients and further higher order field gradients must be applied. This increases the computational effort noticeably, though the procedures are the same for all multipoles. A one-electron operator, which consists of a particular moment–element operator scaled by the choice of the finite field, or the finite field gradient, and so on, is added to the Hamiltonian. Typical values for finite-field and field gradient strengths employed in these types of calculations range from 0.0001 to 0.01 a.u. A nice feature of finite-field approaches is that they can be carried out at the SCF level, CI level, MBPT level, and so on, with equal ease. A large number of calculations using the finite-field method have been reported for both atoms and molecules. Selecting just a few representative examples, there are studies of Group I and II atoms [67], Li_2 [68], HF [69], H_2^+ (and other isotopically substituted hydrogen ions) [70], LiH [71], water and methane [72], nitrogen [73, 74], and small hydrocarbons [75].

2. *Fixed-Charge Method*

Moments and polarizabilities can also be obtained by the fixed-charge method [76]. This technique allows for the single-step incorporation of the nonuniform electric field contributions due to gradients and higher order field derivatives. One or more charges are placed around the molecule in regions where the molecular wavefunctions are negligible. It is important that the basis set used for the field-free molecule be the same as that used in the presence of the field and that the molecule basis be adequate to describe any polarization of the molecule in the presence of this field. Again, the fields applied should be small enough to let a truncated expansion accurately represent the moment or energy and strong enough to ensure an effect can be calculated with numerical significance. Generally charges are placed 20–25 Å from the center of the molecule, and the resulting uniform field components range from 0.002 to 0.05 a.u. If the charges are placed in special arrangements around the molecule, the contributions from the field gradients can be eliminated (recall that the field components and gradients are derivatives evaluated at the origin of the molecule).

A number of calculations allow for internal consistency checks. For example, if either $A_{x,xx}$ or $B_{x,x,xx}$ is desired, one can use either total dipole moment or quadrupole moment expansions. These properties could not be calculated by uniform field computations alone. The finite-charge approach is easily carried out at all electronic structure levels (e.g., SCF, CI, and MBPT) because it amounts to adding an extra "nucleus," one with a negative or positive charge. It does not seem to be as widely employed as finite-field calculations. Examples include calculations on methane [77], Li_2 and ions [78], and HF [79].

3. *Series Convergence Properties*

Both the finite-field and fixed-charge methods use the energy or total induced moment expansions to evaluate polarizabilities. A complication is that the energy of the system with an applied field really represents a pseudoeigenvalue or a resonant-state energy rather than a true bound-state energy. However, for suitably weak fields, the use of a finite basis restricts the description to the bound state. (Of course, the states must have complex energies; yet the perturbation series yields real ones [80].) A number of calculations have been done on hydrogen in intense fields to ascertain if perturbative series can converge to these energies. The Rayleigh–Schrödinger series actually diverges and has zero radius of convergence in terms of a field applied along a Cartesian axis [81]. In the case of finite-field and fixed-charge methods, the convergence behavior of the energy series in terms of moments and polarizabilities has not been the subject of extensive study, but it is important to consider its validity and reliability. It has been numerically established that the polarizabilities do not at all begin to vanish as their order (tensor rank) increases. In fact, second and third hyperpolarizabilities can be sizable [82]. Thus, the coefficients in the energy expansion may be increasing. While this alone is not sufficient to establish divergence of a power series, it must be realized that if the series were diverging, then any number of methods using the same series for determination of the moments and polarizabilities may very well calculate identical polarizabilities; however, they would all be erroneous.

Recently, Larter and Malik [83] have explored the properties of the energy expansion using N-variable rational approximants (analogous to single-variable Padé approximants). Using ab initio polarizabilities for LiH [84], estimates of the radius of convergence of the series were calculated as a function of internuclear geometry (Figure 1). Of particular interest was the effect of the nonuniformity of the electric field on the apparent convergence of the interaction energy. Their results showed a significant influence on the series behavior. If there were no gradient components in the field, the behavior was optimal. However, contributions to the first-order gradient along the molecular axis, both negative and positive, usually reduced the radius of convergence. For fixed gradient and values of the internal coordinate larger than the equilibrium distance, the radius of convergence decreased significantly, indicating that the expansion is less reliable at larger bond distances. The effects were strongest near and less than the field-free equilibrium distance. If one estimates polarizabilities from truncated expansions, these results suggest the more reliable results would be obtained for

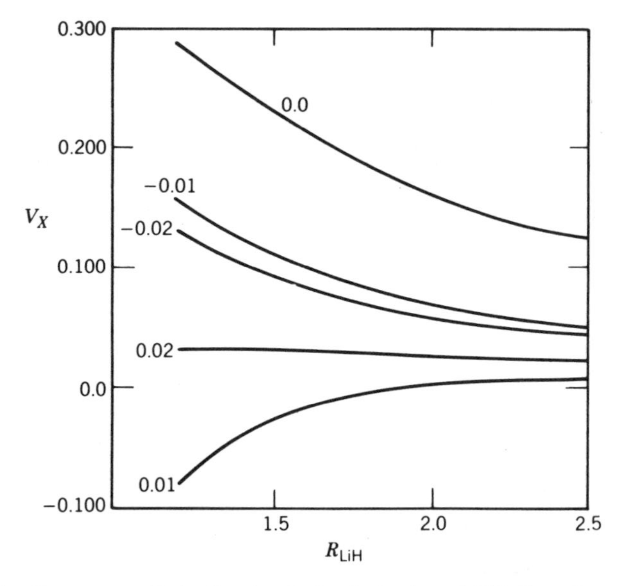

Figure 1. Convergence limits for electrical polarization of LiH (from Larter and Malik [83]). Each curve represents limit to which it is possible to extract polarizabilities from finite-field or fixed-charge calculations. Five curves drawn for different field gradients, V_{xx} (a.u.), with $V_{yy} = -V_{xx}$. Electrical field strength (a.u.) is on vertical axis. Five curves show narrowing of range of convergence with increasing magnitude of axial field gradient. They also reveal that there may be accidental points (i.e., specific field, field gradient, and bond length) where convergence is nearly impossible.

the shorter bond distances. Another consideration in their calculations was accuracy of the high-order polarizabilities. The radii of convergence were somewhat sensitive to the values of the highest order polarizabilities. As one might expect, since the sizes of the polarizabilities can be increasing, the estimation of series behavior is readily apparent and the higher-order terms play an important role.

C. Calculated Properties

In the last several years, there has been a sharp expansion in the number of reported high-level ab initio calculations of electrical properties. We focus here on a subset, a series of our calculations done at comparable levels on a variety of small molecules and complexes [9, 82, 84–90]. In the course of these calculations, a good number of basis set tests have been carried out. From these and other studies that have characterized basis set effects for electrical properties [79, 91, 92], we have selected bases that are reasonably good quality and probably as economical as possible for the quality obtained. We have designated these as ELP (electrical properties) bases [82]. They are standard triple-zeta core-valence sets [93, 94] supplemented with diffuse valence and polarization functions, as shown in Table II. One feature is uniformity from element to element; lack of uniformity in treating a heteronuclear molecule creates an imbalance in the basis that degrades the quality of the results. Also, basis sets should be balanced among the l-type subsets (e.g., $s, p, d, \ldots$) and they must include diffuse functions. This has guided the prescription for ELP bases.

Tables III–XVII give calculated permanent moments. Selected comparisons with experimental values or calculations of others are also listed. All values are in atomic units, and traceless rather than Cartesian forms are distinguished with Greek letters, θ (quadrupole) and Ω (octupole). Coordinates for the atomic centers are listed. These specify the geometry used, which were equilibrium geometries, and implicitly the multipole expansion center ($x=0$, $y=0$, $z=0$). The moments are given at both the SCF level and at the well-correlated level of coupled-cluster theory [95–102]. ACCD [103–106] was the particular coupled-cluster approach, and the moments were evaluated by expectation [102] with the cluster expansion truncated at single and double substitutions.

It is interesting that the correlation effects on the moments are generally small. Typically, correlation changes dipole moments and second-moment tensor elements by only a few percent. In many cases the third-moment tensor elements show the same insensitivity to correlation. For small molecules, the higher moments sample the charge distribution with greater weighting in the outer regions of the molecule. This is where correlation affects the electron density the least, and so sensitivity to correlation effects of

TABLE II
The ELP Basis Set Augmentation to Atomic Triple-Zeta Sets

Element	Exponents of Uncontracted Gaussians		
	s	*p*	*d*
H	0.06	0.9	
		0.1	
Li[a]	0.008	5.0	0.9
		1.0	0.13
		0.15	0.02
		0.02	
B	0.035	0.02	0.9
		0.0035	0.13
			0.02
C	0.05	0.03	0.9
		0.005	0.13
			0.02
N	0.06	0.04	0.9
		0.006	0.13
			0.02
O	0.06	0.05	0.9
		0.007	0.13
			0.02
F	0.06	0.06	0.9
		0.008	0.13
			0.02
Ne	0.07	0.07	0.9
		0.009	0.13
			0.02

[a] For lithium, augmentation was to 5*s* valence set without *p* functions.

TABLE III
Permanent Moments (a.u.) of H_2: ELP Basis[a]

Q_{xx}	SCF	−1.0671
	CISD	−1.0830
Q_{yy}	SCF	−1.5569
	CISD	−1.5342
θ_{xx}	SCF	0.4898
	CISD	0.4512 (0.4568[b])

[a] Coordinates (Å): (±0.3717, 0, 0).
[b] Adiabatic calculation of Poll et al. [121].

TABLE IV
Permanent Moments (a.u.) of LiH: ELP Basis[a]

μ_x	SCF	2.3664 (2.3625[b])
	ACCD	2.3084 (2.921[c])
Q_{xx}	SCF	−7.3287
	ACCD	−7.2201
Q_{yy}	SCF	−3.8978
	ACCD	−4.0281
θ_{xx}	SCF	−3.4309 (−3.3719[b])
	ACCD	−3.1920 (−3.0699[c])
R_{xxx}	SCF	28.9757
	ACCD	28.4709
R_{xyy}	SCF	7.4437
	ACCD	7.4720
Ω_{xxx}	SCF	6.6446
	ACCD	6.0549

[a] Atomic coordinates (Å): Li = (0.2003768, 0, 0), H = (−1.3949232, 0, 0).
[b] SCF results of Roos and Sadlej [122].
[c] CASSCF results of Roos and Sadlej [122].

TABLE V
Permanent Moments (a.u.) of CH_4: ELP Basis[a]

Q_{xx}	SCF	−6.2674
	ACCD	−6.309
R_{xyz}	SCF	1.0061
	ACCD	0.989
Ω_{xyz}	SCF	2.5153 (2.61[b], 3.23[c])
	ACCD	2.4725 (2.54[b])

[a] Atomic coordinates (Å): H_1 = (0.6296777, 0.6296777, 0.6296777), H_2 = (−0.6296777, 0.6296777, −0.6296777), H_3 = (−0.6296777, −0.6296777, 0.6296777), H_4 = (0.6296777, −0.6296777, −0.6296777).
[b] SCF and CI results of Amos [77].
[c] Experimental [42].

only a few percent is not unlikely. However, use of smaller bases, ones that would be poorer in the outer regions, could exaggerate the correlation effect because of basis set deficiency. This is probably why it is sometimes thought that correlation plays a more sizable role in determining electrical moments than that shown here.

Tables XVIII–XXXII give the DHF values of polarizabilities and hyperpolarizabilities obtained with the ELP bases. An interesting feature of these

TABLE VI
Permanent Moments (a.u.) of NH_3: ELP Basis[a]

μ_x	SCF	−0.6422 (−0.635[b])
	ACCD	−0.6075 (−0.595[b], −0.5785[c])
Q_{xx}	SCF	−6.7394
	ACCD	−6.9561
Q_{yy}	SCF	−4.5065
	ACCD	−4.7300
θ_{xx}	SCF	−2.2329
	ACCD	−2.2261
R_{xxx}	SCF	−0.0482
	ACCD	0.1588
R_{xyy}	SCF	−0.8706
	ACCD	−0.7879
R_{yyy}	SCF	1.3705
	ACCD	1.3105
Ω_{xxx}	SCF	2.5636
	ACCD	2.5225
Ω_{yyy}	SCF	3.4263
	ACCD	3.2763

[a] Atomic coordinates (Å): N = (0.0676610, 0.0, 0.0), H_1 = (−0.3133669, 0.9375296, 0.0), H_2 = (−0.3133669, −0.4687648, −0.8119244), H_3 = (−0.3133669, −0.4687648, 0.8119244).

[b] SCF and CEPA-1 results with basis B of Werner et al. [91].

[c] Experimental [123].

results is the regular decline [82] in the isotropic dipole polarizability of AH_n molecules with the atomic number of A. This is illustrated in Figure 2. The declining isotropic polarizability follows the filling of the $n = 2$ valence shell of the first-row atom until polarization has to be accomplished with $n = 3$ orbitals.

Dipole polarizabilities for hydrides containing two first-row atoms reveal an interesting pattern [82]. The $\bar{\alpha}$ values of these molecules tend to follow those of the AH_n molecules in the following sense. The $\bar{\alpha}$ value of a given ABH_n molecule is well estimated (to within about 15%) by the sum of the $\bar{\alpha}$ values for the A and B simple hydrides, diminished by about 6 a.u. if there is a double bond between A and B, or diminished by about 12 a.u. if there is a triple bond:

$$\bar{\alpha}(ABH_n) \approx \bar{\alpha}(AH_m) + \bar{\alpha}(BH_l) - 6.0(N_{AB} - 1), \tag{79}$$

where N_{AB} is the A–B bond order. Thus, for HCN, one has $\bar{\alpha}$ (estimated) = $\bar{\alpha}(CH_4) + \bar{\alpha}(NH_3) - 12 = 16.8$ compared to the calculated value of 16.6.

TABLE VII
Permanent Moments (a.u.) of H_2O: ELP Basis[a]

μ_x	SCF	−0.7921 (−0.782[b], 0.73[c], −0.81[d], 0.7296[e])
	ACCD	−0.743 (−0.723[b], 0.7268[f], −0.7848[g])
Q_{xx}	SCF	−4.4642
	ACCD	−4.621
Q_{yy}	SCF	−3.1306
	ACCD	−3.293
Q_{zz}	SCF	−5.6781
	ACCD	−5.837
θ_{xx}	SCF	−0.0599 (−0.15[d], −0.10[h])
	ACCD	−0.056 (−0.075[g])
θ_{yy}	SCF	1.9406 (1.81[d], 1.96[h])
θ_{zz}	SCF	−1.8807 (−1.66[d], −1.86[h])
	ACCD	−1.880 (−1.943[g])
R_{xxx}	SCF	−0.2364
	ACCD	−0.2063
R_{xyy}	SCF	−1.6486
	ACCD	−1.6544
R_{xzz}	SCF	0.2107
	ACCD	0.2100
Ω_{xxx}	SCF	1.9205
	ACCD	1.9603 (1.735[g])
Ω_{xyy}	SCF	−3.2844
	ACCD	−3.3107 (−3.032[g])
Ω_{xzz}	SCF	1.3639
	ACCD	1.3504 (1.297[g])

[a] Atomic coordinates (Å) H_1 = (−0.5103130, 0.7569503, 0.0), H_2 = (−0.5203130, −0.7569503, 0.0), O = (0.0655692, 0.0, 0.0).
[b] SCF and CEPA-1 results of basis B by Werner et al. [91].
[c] Experimental [124].
[d] Basis B results of Ahlström et al. [125].
[e] Experimental [127].
[f] Experimental [128].
[g] CI results of John et al. [72].
[h] Experimental [126].

Table XXXIII compares the estimated $\bar{\alpha}$ with those computed. The accuracy of this simple estimator suggests that the isotropic polarizability is significantly determined by the heavy atoms, not the hydrogens, and that multiple bonds are, as expected, less polarizable. The anisotropy of the dipole polarizability, of course, has very much to do with the hydrogens.

For certain of the molecules, the evaluation of the polarizabilities has been carried to high levels, such as fifth derivatives (e.g., δ), or to subtle properties such as the dipole–octupole polarizabilities. While in most situations of

TABLE VIII
Permanent Moments (a.u.) of HF: ELP Basis[a]

μ_x	SCF	−0.7665 (−0.737[b], −0.707[c], −0.757[d], −0.7571[e])
μ_x	ACCD	−0.7153 (−0.703[f], −0.7089[g])
Q_{xx}	SCF	−2.4684
	ACCD	−2.6393
Q_{yy}	SCF	−4.2471
	ACCD	−4.3892
θ_{xx}	SCF	1.7787 (1.714[h])
	ACCD	1.7499 (1.740[e], 1.643[i], 1.690[h])
R_{xxx}	SCF	−2.0724
	ACCD	−1.7948
R_{xyy}	SCF	0.1690
	ACCD	0.2211
Ω_{xxx}	SCF	−2.5794 (−2.508[h], −2.6265[e])
	ACCD	−2.4581 (−2.387[h])

[a] Atomic coordinates (Å): H = (−0.8706155, 0.0, 0.0), F = (0.0461845, 0.0, 0.0).
[b] Optimized EFV GTO basis set results of Sadlej [50].
[c] Experimental [129] .
[d] SCF results using basis B of Werner et al. [91].
[f] CEPA results using basis B of Werner et al. [91].
[g] SDQ-MBPT(4) results using $6s5p4d/5s3p$ basis of Bartlett et al. [107].
[h] SCF and CI results of Maillard et al. [130].
[i] Experimental [131].
[e] "B8" basis results of Bishop et al. [79].

current interest, these properties may be of minor importance, the paucity of such values encourages listing them. They serve to give guidance for future investigations of what these properties are like for small molecules. At the same time, it must be made clear that sizable refinements in the values of certain of the more exotic properties may come about through even larger basis set studies.

Accuracy in the DHF level polarizabilities is limited by the neglect of correlation effects as well as by basis set quality. For covalent species, the refinements in dipole polarizabilities from including electron correlation effects are often ~10–15% [74, 91, 107]. Of the species considered here, lithium hydride is probably furthest from being covalent, and so, the correlation effect is larger, about ~20–30% [68, 71, 108]. It is possible that the correlation effects could be greater for higher derivative properties, but a general assessment of that sort remains to be done. The need for high-quality basis sets can be understood by realizing that molecular charge polarization can be both intra-atomic and valence charge (interatomic) polarization. The

TABLE IX
Permanent Moments (a.u.) of CO: ELP Basis[a]

μ_x	SCF	−0.0911 (−0.122[b], 0.0481[c], −0.092[d])
	ACCD	0.0357 (0.122[d])
Q_{xx}	SCF	−9.1411
	ACCD	−9.1598
Q_{yy}	SCF	−7.5924
	ACCD	−7.6696
θ_{xx}	SCF	−1.5486 (−1.7184[e], −1.9[f])
	ACCD	−1.4902
R_{xxx}	SCF	8.7981
	ACCD	8.7708
R_{xyy}	SCF	1.4592
	ACCD	1.6504
Ω_{xxx}	SCF	4.4205 (4.5644[e])
	ACCD	3.8196

[a] Atomic coordinates (Å): C = (−0.6446681, 0.0, 0.0), O = (0.4836549, 0.0, 0.0).
[b] Optimized EFV GTO results of Sadlej [50].
[c] Experimental [132].
[d] SCF and CEPA-1 results of Werner et al. [91].
[e] SCF results of Gready et al. [133].
[f] Experimental [134].

TABLE X
Permanent Moments (a.u.) of N_2: ELP Basis[a]

Q_{xx}	SCF	−8.6661
	ACCD	−8.7798
Q_{yy}	SCF	−7.7803
	ACCD	−7.6952
θ_{xx}	SCF	−0.8858 (−1.1576[b], −1.09[c])
	ACCD	−1.0846 (−1.3455[b])

[a] Atomic coordinates (Å): (±0.557, 0.0, 0.0).
[b] SCF and correlated results of Gready et al. [133].
[c] Experimental [135].

TABLE XI
Permanent Moments (a.u.) of HCCH: ELP Basis[a]

Q_{xx}	SCF	−4.9989
	ACCD	−5.3255
Q_{yy}	SCF	−10.4245
	ACCD	−10.2504
θ_{xx}	SCF	5.4256
	ACCD	4.9249

[a] Atomic coordinates (Å): C = (±0.601, 0.0, 0.0), H = (±1.661, 0.0, 0.0).

TABLE XII
Permanent Moments (a.u.) of HCN: ELP Basis[a]

μ_x	SCF	-1.2997 (-1.308^b, 1.174^c, -1.295^d)
	ACCD	-1.1954
Q_{xx}	SCF	-6.8217
	ACCD	-7.0550
Q_{yy}	SCF	-8.9093
	ACCD	-8.8036
θ_{xx}	SCF	2.0876 (1.508^b, 2.3^e)
	ACCD	1.7486
R_{xxx}	SCF	-9.0499
	ACCD	-8.4691
R_{xyy}	SCF	0.3323
	ACCD	0.4525
Ω_{xxx}	SCF	-10.0468 (-8.944^b)
	ACCD	-9.8266

[a] Atomic coordinates (Å): H = (−1.6239227, 0.0, 0.0), C = (−0.5580227, 0.0, 0.0), N = (0.5950774, 0.0, 0.0).
[b] SCF results of Gready et al. [133].
[c] Experimental [135].
[d] SCF results of McLean et al. [136].
[e] Experimental [137].

TABLE XIII
Permanent Moments (a.u.) of F_2: ELP Basis[a]

Q_{xx}	SCF	-6.5128
	ACCD	-6.5547
Q_{yy}	SCF	-7.0575
	ACCD	-7.2427
θ_{xx}	SCF	0.5447
	ACCD	0.6880

[a] Atomic coordinates (Å): (±0.70595, 0.0, 0.0).

description of intra-atomic polarization has the most stringent basis set requirements, with diffuse and higher *l*-type functions being very essential, just as they would be in describing polarization of an isolated atom. Interatomic valence polarization is not as demanding in basis set flexibility and is described partly with usual valence basis functions [86]. Thus, small molecules require the most carefully chosen basis sets. The ELP bases are large sets, but refinements in the properties are likely to be found with even more complete sets. Quadrupole and octupole polarizabilities are probably less well determined with these sets than purely dipole polarizabilities.

TABLE XIV
Permanent Moments (a.u.) of *trans*-N_2H_2: ELP Basis[a]

Q_{xx}	SCF	−9.6245
	ACCD	−9.7968
Q_{yy}	SCF	−7.7583
	ACCD	−7.9180
Q_{zz}	SCF	−9.3533
	ACCD	−9.2798
Q_{xy}	SCF	2.7714
	ACCD	2.6960
θ_{xx}	SCF	−1.0687
	ACCD	−1.1979
θ_{yy}	SCF	1.7306
	ACCD	1.6203
θ_{zz}	SCF	−0.6619
	ACCD	−0.4224
θ_{xy}	SCF	4.1571
	ACCD	4.0425

[a] Atomic coordinates (Å): $N_1 = (0.626, 0.0, 0.0)$, $N_2 = (-0.626, 0.0, 0.0)$, $H_1 = (0.9239834, 0.9838648, 0.0)$, $H_2 = (-0.9239834, -0.9838648, 0.0)$.

The final list of calculated polarizabilities are DHF values for Be and Mg atoms, given in Table XXXIV. Sen and Schmidt have obtained values for these and other atoms from numerical CPHF [109], and Sen has suggested an interesting estimating formula for atomic polarizabilities [110]. The moderate-sized basis DHF values are within about 10% of the numerical CPHF results of Sen and Schmidt. The DHF calculations on Be_2 (at $R = 2.45$ Å) give the isotropic dipole polarizability to be −107.41 a.u. This is more than twice the polarizability of a Be atom and suggests an interesting enhancement for such a simple system. Small metal clusters are probably an important area for electrical property studies, and some surprising features are possible.

IV. ROLE OF VIBRATION IN MOLECULAR ELECTRICAL PROPERTIES

Molecular electrical properties change as a molecule stretches or bends, and one may consider there to be surfaces of electrical properties. Equilibrium properties such as those tabulated earlier, are but a "zero-order" picture of electrical response because vibration is continuously changing the properties.

TABLE XV
Permanent Moments (a.u.) of *cis*-N_2H_2: ELP Basis[a]

μ_y	SCF	1.2036
	ACCD	1.1528
Q_{xx}	SCF	−9.1804
	ACCD	−9.3751
Q_{yy}	SCF	−8.0912
	ACCD	−8.2578
Q_{zz}	SCF	−9.3384
	ACCD	−9.2614
θ_{xx}	SCF	−0.4656
	ACCD	−0.6155
θ_{yy}	SCF	1.1682
	ACCD	1.0605
θ_{zz}	SCF	−0.7026
	ACCD	−0.4450
R_{xxy}	SCF	4.2495
	ACCD	3.9503
R_{yyy}	SCF	−0.3909
	ACCD	−0.8383
R_{yzz}	SCF	−1.2976
	ACCD	−1.2865
θ_{yyy}	SCF	−4.8188
	ACCD	−4.8340

[a] Atomic coordinates (Å): $N_1 = (0.6255, 0.0, 0.0)$, $N_2 = (-0.6255, 0.0, 0.0)$, $H_1 = (1.0186166, 0.9585173, 0.0)$, $H_2 = (-1.0186166, 0.9585173, 0.0)$.

Thus, it is natural to expect a polarizability or a permanent moment to be subject to vibrational influence. Likewise, a laboratory measurement of a particular electrical property may not compare directly with a value calculated for a fixed structure.

A. Vibrational Contributions to Electrical Properties

The vibrational excursions of a molecule may cause it to have sharply changing electrical properties from state to state. This, of course, is essential for mechanisms of absorption and emission of radiation. How sharp these changes may be is illustrated for HF in Figure 3. The curves show the axial elements of α, A, and β in the vicinity of the equilibrium bond length as a function of the H–F distance. The types of changes that may be found in a polyatomic molecule are illustrated by Figures 4 and 5. They show contours of the dipole polarizability and hyperpolarizability elements over the two stretching coordinates of HCN. Both β_{xxx} and β_{xyy} have zero contours

TABLE XVI
Permanent Moments (a.u.) of H_2O_2: ELP Basis[a]

μ_z	SCF	0.6763 (0.99877[b], 0.8655[c])
Q_{xx}	SCF	−8.6302
Q_{yy}	SCF	−6.8129
Q_{zz}	SCF	−8.7150
Q_{xy}	SCF	−2.3329
θ_{xx}	SCF	−0.8663
θ_{yy}	SCF	1.8597
θ_{zz}	SCF	−0.9934
θ_{xy}	SCF	−3.4993
R_{xxz}	SCF	1.5758
R_{xyz}	SCF	−2.6813
R_{yyz}	SCF	1.5180
R_{zzz}	SCF	−1.3612
Ω_{xxz}	SCF	1.00626
Ω_{xyz}	SCF	0.2927
Ω_{yyz}	SCF	1.7341
Ω_{zzz}	SCF	8.0138

[a] Atomic coordinates (Å): O = (±0.7335, 0.0, 0.0), H = (±0.87613608, ∓0.82653491, 0.47720015).
[b] DZ basis results of Snyder [138].
[c] Experimental [139].

TABLE XVII
Permanent Moments (a.u.) of C_2H_4: ELP′ Basis[a]

Q_{xx}	SCF	−9.0712
	ACCD	−9.2460
Q_{yy}	SCF	−9.2176
	ACCD	−9.2819
Q_{zz}	SCF	−12.0134
	ACCD	−11.8405
θ_{xx}	SCF	1.5443
	ACCD	1.3152
θ_{yy}	SCF	1.3247
	ACCD	1.2614
θ_{zz}	SCF	−2.8690
	ACCD	−2.5766

[a] Moments obtained with basis smaller than complete ELP set. There were no diffuse *s* functions, only one augmented *p* function on carbon ($\alpha = 0.02$) and two *d* functions on carbon ($\alpha = 0.9, 0.1$). Atomic coordinates (Å): C = (±0.669500, 0.0, 0.0), H = (±1.2299387, ±0.9290498, 0.0).

TABLE XVIII
DHF/ELP Basis Multipole Polarizabilities and Hyperpolarizabilities of H_2

$P^{(2)}(\alpha)$	x, x	−6.443
	y, y	−4.440
$P^{(2)}(C/2)$	xx, xx	−3.6725
	xx, yy	−0.78025
	xy, xy	−1.3858
	yy, yy	−2.0318
$P^{(3)}(B/2)$	x, x, xx	57.15
	x, x, yy	13.07
	x, y, xy	20.955
	y, y, xx	13.625
	y, y, yy	40.46
	y, y, zz	26.29
	y, z, yz	8.585
$P^{(3)}$	xx, xx, xx	105.01
	xx, xx, yy	17.413
	xx, xy, xy	20.85
	xx, yy, yy	9.3638
	xy, xy, yy	14.75
	xy, xy, zz	9.9438
	xy, xz, yz	2.4038
	yy, yy, yy	28.663

TABLE XIX
DHF/ELP Basis Polarizabilities and Hyperpolarizabilities of LiH

$P^{(2)}(\alpha)$	x, x	−22.322	$P^{(3)}(B/2)$	x, x, xx	−622.45
	y, y	−24.018		x, x, yy	−420.38
$P^{(2)}(A/2)$	x, xx	−44.110		x, y, xy	−351.53
	x, yy	4.4187		y, y, xx	−136.53
	y, xy	−23.112		y, y, yy	−1,316.5
				y, y, zz	−364.84
$P^{(2)}(C/2)$	xx, xx	−119.86		y, z, yz	−475.85
	xx, yy	−1.2398	$P^{(3)}$	x, xx, xx	−2,971.8
	xy, xy	−33.816		x, xx, yy	−604.10
	yy, yy	−33.546		x, xy, xy	−450.76
	yy, zz	−12.532		x, yy, yy	540.34
	yz, yz	−10.507		x, xy, zz	−5.8048
$P^{(3)}(\beta)$	x, x, x	353.20		x, yz, yz	273.07
	x, y, y	137.64		y, xx, xy	−881.68

(Continued)

TABLE XIX (*Continued*)

	y, xy, yy	−923.28
	y, xy, zz	−335.26
	y, xz, yz	−294.01
$P^{(3)}$	*xx, xx, xx*	−12,124
	xx, xx, yy	−1,608.4
	xx, xy, xy	−1,945.5
	xx, yy, yy	−168.28
	xx, yy, zz	−246.91
	xx, yz, yz	39.317
	xy, xy, yy	−1,779.0
	xy, xy, zz	−522.08
	xy, xz, yz	−628.46
	yy, yy, yy	−5,953.8
	yy, yy, zz	−806.16
	yy, yz, yz	−1,286.9
$P^{(4)}(\gamma)$	*x, x, x, x*	−51,598
	x, x, y, y	−15,130
	y, y, y, y	−42,280
	y, y, z, z	−14,093
$P^{(4)}$	*x, x, x, xx*	1,734.8
	x, x, x, yy	16,355
	x, x, y, xy	11,098
	x, y, y, xx	−8,693.3
	x, y, y, yy	25,688
	x, y, y, zz	4,554.6
	x, y, z, yz	10,566
	y, y, y, xy	−6,301.5
	y, y, z, xz	−2,100.5
$P^{(4)}$	*x, x, xx, xx*	−70,444
	x, x, xx, yy	−16,192
	x, x, yy, yy	−35,893
	x, x, yy, zz	−10,397
	x, x, yz, yz	−12,748
	x, y, xx, xy	−17,946
	x, y, xy, yy	−24,590
	x, y, xy, zz	−11,203
	x, z, xy, yz	−6,693.5
	y, y, xx, xx	−13,424
	y, y, xx, yy	1,379.6
	y, y, xx, zz	−708.28
	y, y, xz, xz	−1,930.4
	y, y, yy, yy	-1.8265×10^5
	y, y, yy, zz	−40,883
	y, z, xx, yz	1,043.9
	y, z, xy, xz	−14,789
	y, z, yy, yz	−42,324
	z, z, yy, yy	−13,350
$P^{(4)}$	*x, xx, xx, xx*	-4.1981×10^5
	x, xx, xx, yy	−82,336
	x, xx, yy, yy	−19,812
	x, xx, yy, zz	−19,675
	x, xx, yz, yz	−68.586
	x, xy, xy, yy	−2,131.0
	x, xy, xy, zz	−9,964.7
	x, xy, xz, yz	3,916.9
	x, yy, yy, yy	2.0246×10^5
	x, yy, yy, zz	12,925
	x, yy, yz, yz	47,383
	y, xx, xx, xy	−94,227
	y, xx, xy, yy	−65,398
	y, xx, xy, zz	−26,774
	y, xx, xz, yz	−19,312
	y, xy, xy, xy	−51,307
	y, xy, xz, xz	−17,102
	y, xy, yy, yy	-1.0978×10^5
	y, xy, yy, zz	−31,347
	y, xy, yz, yz	−13,254
	y, xz, yy, yz	−25,964
	z, xz, yy, yy	−5,928.9
$P^{(4)}$	*xx, xx, xx, xx*	-2.9936×10^6
	xx, xx, xx, yy	-4.6668×10^5
	xx, xx, xy, xy	-4.0060×10^5
	xx, xx, yy, yy	-1.0638×10^5
	xx, xx, yy, zz	−57,420
	xx, xx, yz, yz	−24,481
	xx, xy, xz, yz	−82,629
	xx, yy, yy, yy	−78,386
	xx, yy, yy, zz	2,642.9
	xx, yz, yz, zz	−20,257
	xy, xy, xy, xy	-2.6416×10^5
	xy, xy, xz, xz	−88,054
	xy, xy, yy, yy	-4.4293×10^5
	xy, xy, yz, yz	−83,069
	xy, xy, zz, zz	−45,167
	xz, xz, yy, zz	−77,910
	yy, yy, yy, yy	-2.5792×10^6
	yy, yy, yy, zz	-1.8554×10^5
	yy, yy, yz, yz	-3.9894×10^5
	yy, yy, zz, zz	−82,254
	yz, yz, yz, yz	-2.6047×10^5
$P^{(5)}$	*x, x, x, x, x*	5.2253×10^6
	x, x, x, y, y	1.0927×10^6
	x, y, y, y, y	1.3627×10^6
	x, y, y, z, z	4.5422×10^5

TABLE XX
DHF/ELP Basis Multipole Polarizabilities and Hyperpolarizabilities of CH_4

$P^{(2)}(\alpha)$	x, x	−16.000
$P^{(2)}(A/2)$	x, yz	6.5618
$P^{(2)}(C/2)$	xx, xx	−18.509
	xx, yy	−2.2019
	xy, xy	−10.026
$P^{(3)}(\beta)$	x, y, z	−11.032
$P^{(3)}(B/2)$	x, x, xx	−160.96
	x, x, yy	−35.568
	x, y, xy	−70.330
$P^{(3)}$	x, xx, yz	−21.383
	x, xy, xz	−41.198
	x, yy, yz	−26.398
$P^{(3)}$	xx, xx, xx	−480.55
	xx, xx, yy	−71.009
	xx, xy, xy	−129.70
	xx, yy, zz	11.611
	xy, xy, zz	−22.803
	xy, xz, yz	−55.074

TABLE XXI
DHF/ELP Basis Multipole Polarizabilities and Hyperpolarizabilities of NH_3

$P^{(2)}(\alpha)$	x, x	−13.188	$P^{(3)}(B/2)$	x, x, xx	−470.56
	y, y	−12.645		x, x, yy	−99.50
$P^{(2)}(A/2)$	x, xx	0.47031		x, y, xy	−140.27
	x, yy	0.91261		x, y, yy	12.12
	y, xy	1.5144		y, y, xx	−97.15
	y, yy	−1.8825		y, y, xy	14.59
$P^{(2)}(C/2)$	xx, xx	−14.783		y, y, yy	−216.92
	xx, yy	−2.7849		y, y, zz	−40.84
	xy, xy	−5.8054		y, z, yz	−88.04
	xy, yy	0.8356	$P^{(3)}$	x, xx, xx	−20.115
	yy, yy	−11.702		x, xx, yy	11.243
	yy, zz	−0.9032		x, xy, xy	27.551
	yz, yz	−5.3992		x, xy, yy	−7.635
$P^{(3)}(\beta)$	x, x, x	9.1259		x, xy, yz	−3.0425
	x, y, y	7.8012		x, xz, yz	10.678
	y, y, y	−9.7979		x, yy, yy	22.258
				x, yy, zz	−7.8200

(Continued)

TABLE XXI (*Continued*)

	x, yz, yz	15.039	*xx, yy, yy*	−114.1
	y, xx, xy	16.472	*xx, yy, zz*	−28.8
	y, xx, yy	4.7781	*xx, yz, yz*	−42.6
	x, xy, yy	34.358	*xy, xy, xy*	22.5
	y, xy, zz	4.3440	*xy, xy, yy*	−204.9
	y, xy, xy	−16.016	*xy, xy, zz*	−55.1
	y, xz, yz	15.008	*xy, xz, yz*	−74.9
	y, yy, yy	−88.622	*xy, yy, yy*	66.4
	y, yy, zz	21.262	*xy, yy, zz*	−17.3
	y, zz, zz	46.098	*xy, zz, zz*	−31.7
	z, yy, yz	12.418	*xz, yy, yz*	−7.2
			xz, yz, zz	−41.8
$P^{(3)}$	*xx, xx, xx*	−1,487.0	*yy, yy, yy*	−486.1
	xx, xx, yy	−281.4	*yy, yy, zz*	−49.1
	xx, xy, xy	−272.4	*yy, yz, yz*	−113.2
	xx, xy, yy	12.6	*yy, zz, zz*	−43.8
	xx, xy, yz	5.4	*yz, yz, zz*	−107.9
	xx, xz, yz	−18.0	*zz, zz, zz*	−491.3

TABLE XXII
DHF/ELP Basis Multipole Polarizabilities and Hyperpolarizabilities of H_2O

$P^{(2)}(\alpha)$	*x, x*	−8.3128		*x, x, zz*	−22.945
	y, y	−9.0770		*x, y, xy*	−30.190
	z, z	−7.8385		*x, z, xz*	−27.189
$P^{(2)}(A/2)$	*x, xx*	1.3703		*y, y, xx*	−20.330
	x, yy	1.4598		*y, y, yy*	−57.049
	x, zz	−0.1377		*y, y, zz*	−17.239
	y, xy	2.0853		*y, z, yz*	−26.874
	z, xz	0.39355		*z, z, xx*	−24.381
				z, z, yy	−17.331
$P^{(2)}(C/2)$	*xx, xx*	−5.8830		*z, z, zz*	−92.055
	xx, yy	−1.6640			
	xx, zz	−1.8783	$P^{(3)}$	*x, xx, xx*	22.246
	xy, xy	−3.5520		*x, xx, yy*	7.2792
	xz, xz	−2.0501		*x, xx, zz*	0.42326
	yy, yy	−6.4197		*x, xy, xy*	14.0869
	yy, zz	−1.0679		*x, xz, xz*	2.7404
	yz, yz	−2.3880		*x, yy, yy*	15.2623
	zz, zz	−6.7654		*x, yy, zz*	−0.94360
				x, yz, yz	3.7414
$P^{(3)}(\beta)$	*x, x, x*	5.4715		*x, zz, zz*	−4.4189
	x, y, y	10.029		*y, xx, xy*	12.595
	x, z, z	0.5445		*y, xy, yy*	21.723
$P^{(3)}(B/2)$	*x, x, xx*	−68.174		*y, xy, zz*	−0.35763
	x, x, yy	−18.336		*y, xz, yz*	5.5899

TABLE XXII (*Continued*)

	z, xx, xz	5.2794		yy, yy, yy	−100.93
	z, xy, yz	5.3323		yy, yy, zz	−13.167
	z, xz, yy	4.7789		yy, yz, yz	−28.359
	z, xz, zz	4.5628		yy, zz, zz	−28.913
				yz, yz, zz	−40.256
$P^{(3)}$	xx, xx, xx	−125.81		zz, zz, zz	−225.30
	xx, xx, yy	−25.085	$P^{(4)}(\gamma)$	x, x, x, x	−794.01
	xx, xx, zz	−34.337		x, x, y, y	−288.58
	xx, xy, xy	−39.672		x, x, z, z	−343.37
	xx, xz, xz	−28.764		y, y, y, y	−475.24
	xx, yy, yy	−21.582		y, y, z, z	−300.11
	xx, yy, zz	−9.4406		z, z, z, z	−1,309.1
	xx, yz, yz	−13.954			
	xx, zz, zz	−45.083	$P^{(5)}(\delta)$	x, x, x, x, x	5,265.7
	xy, xy, yy	−39.090		x, x, x, y, y	2,325.1
	xy, xy, zz	−10.342		x, x, x, z, z	835.99
	xy, xz, yz	−13.158		x, y, y, y, y	3,405.0
	xz, xz, yy	−10.409		x, y, y, z, z	1,192.7
	xz, xz, zz	−35.349		x, z, z, z, z	519.69

TABLE XXIII
DHF/ELP Basis Multipole polarizabilities and Hyperpolarizabilities of HF

$P^{(2)}(\alpha)$	x, x	−5.6121	$P^{(3)}(B/2)$	x, x, xx	−31.634
	y, y	−4.3199		x, x, yy	−6.8266
$P^{(2)}(A/2)$	x, xx	1.9048		x, y, xy	−9.3130
	x, yy	−0.06775		y, y, xx	−6.7547
	y, xy	0.23011		y, y, yy	−28.939
				y, y, zz	−10.558
$P^{(2)}(C/2)$	xx, xx	−3.9657		y, z, yz	−9.1907
	xx, yy	−0.52648	$P^{(3)}$	x, xx, xx	22.795
	xy, xy	−0.93001		x, xx, yy	0.22997
	yy, yy	−2.5480		x, xy, xy	2.5607
	yy, zz	−1.1518		x, yy, yy	−0.25524
	yz, yz	−0.69811		x, yy, zz	0.19454
$P^{(2)}$	x, xxx	−4.6820		x, yz, yz	−0.22489
	x, xyy	−1.1956		y, xx, xy	2.9554
	y, xxy	−1.2032		y, xy, yy	1.6475
	y, yyy	−3.4955		y, xy, zz	0.65299
	y, yzz	−1.1652		y, xz, yz	0.49726
$P^{(3)}(\beta)$	x, x, x	8.2739	$P^{(3)}$	xx, xx, xx	−57.627
	x, y, y	−0.09965		xx, xx, yy	−5.1236

(*Continued*)

TABLE XXIII (*Continued*)

	xx, xy, xy	−8.2297
	xx, yy, yy	−4.9830
	xx, yy, zz	−2.6787
	xx, yz, yz	−1.1522
	xy, xy, yy	−7.9122
	xy, xy, zz	−3.1715
	xy, xz, yz	−2.3704
	yy, yy, yy	−45.705
	yy, yy, zz	−14.842
	yy, yz, yz	−7.7158
$P^{(4)}(\gamma)$	*x, x, x, x*	−262.92
	x, x, y, y	−80.769
	y, y, y, y	−312.78
	y, y, z, z	−104.26
$P^{(4)}$	*x, x, x, xx*	200.24
	x, x, x, yy	−1.7173
	x, x, y, xy	20.914
	x, y, y, xx	16.843
	x, y, y, yy	−3.0620
	x, yy, zz	0.56302
	x, y, z, yz	−1.8125
	y, y, y, xy	25.165
	y, y, z, xz	8.3885
$P^{(4)}$	*x, x, xx, xx*	−710.47
	x, x, xx, yy	−108.86
	x, x, yy, yy	−90.525
	x, x, yy, zz	−45.714
	x, x, yz, yz	−22.405
	x, y, xx, xy	−129.63
	x, y, xy, yy	−135.10
	x, y, xy, zz	−53.894
	x, z, xy, yz	−40.603
	y, y, xx, xx	−92.705
	y, y, xx, yy	−114.70
	y, y, xx, zz	−46.058
	y, y, xz, xz	−32.313
	y, y, yy, yy	−768.81
	y, y, yy, zz	−208.72
	y, z, xx, yz	−34.321
	y, z, xy, xz	−44.190
	y, z, yy, yz	−153.16
	z, z, yy, yy	−156.15
$P^{(4)}$	*x, xx, xx, xx*	749.25
	x, xx, xx, yy	25.349
	x, xx, yy, yy	−18.607
	x, xx, yy, zz	−5.8721
	x, xx, yz, yz	−6.3678
	x, xy, xy, yy	21.271
	x, xy, xy, zz	13.188
	x, xy, xz, yz	4.0415
	x, yy, yy, yy	−45.049
	x, yy, yy, zz	−9.5048
	x, yy, yz, yz	−8.8839
	y, xx, xx, xy	85.534
	y, xx, xy, yy	28.777
	y, xx, xy, zz	14.349
	y, xx, xz, yz	7.2142
	y, xy, xy, xy	30.736
	y, xy, xz, xz	10.245
	y, xy, yy, yy	11.328
	y, xy, yy, zz	11.814
	y, xy, yz, yz	−3.1013
	y, xz, yy, yz	2.8581
	z, xz, yy, yy	−0.10459
$P^{(4)}$	*xx, xx, xx, xx*	−2,271.1
	xx, xx, xx, yy	−221.02
	xx, xx, xy, xy	−269.75
	xx, xx, yy, yy	−25.165
	xx, xx, yy, zz	−20.760
	xx, xx, yz, yz	−2.2027
	xx, xy, xz, yz	−45.570
	xx, yy, yy, yy	−154.42
	xx, yy, yy, zz	−51.673
	xx, yz, yz, zz	−25.686
	xy, xy, xy, xy	−129.31
	xy, xy, xz, xz	−43.105
	xy, xy, yy, yy	−270.85
	xy, xy, yz, yz	−24.065
	xy, xy, zz, zz	−43.508
	xz, xz, yy, zz	−109.05
	yy, yy, yy, yy	−2,546.6
	yy, yy, yy, zz	−770.38
	yy, yy, yz, yz	−296.04
	yy, yy, zz, zz	−394.88
	yz, yz, yz, yz	−81.215
$P^{(5)}(\delta)$	*x, x, x, x, x*	2,309.4
	x, x, x, y, y	243.49
	x, y, y, y, y	91.056
	x, y, y, z, z	30.352

TABLE XXIV
DHF/ELP Basis Multipole Polarizabilities and Hyperpolarizabilities of CO

Property	Components	Value	Property	Components	Value
$P^{(2)}(\alpha)$	x, x	−14.211	$P^{(3)}$	x, xx, xx	217.56
	y, y	−11.088		x, xx, yy	24.586
$P^{(2)}(A/2)$	x, xx	−6.3212		x, xy, xy	39.023
	x, yy	−0.56669		x, yy, yy	9.4224
	y, xy	−4.6304		x, yy, zz	2.8260
				x, yz, yz	3.2981
$P^{(2)}(C/2)$	xx, xx	−26.427		y, xx, xy	46.191
	xx, yy	−0.71705		y, xy, yy	49.772
	xy, xy	−11.343		y, xy, zz	10.877
	yy, yy	−9.0173		y, xz, yz	19.447
	yy, zz	−2.5988	$P^{(3)}$	xx, xx, xx	−788.01
	yz, yz	−3.2092		xx, xx, yy	−88.855
$P^{(3)}(\beta)$	x, x, x	30.810		xx, xy, xy	−134.09
	x, y, y	5.0260		xx, yy, yy	−15.031
				xx, yy, zz	−10.428
$P^{(3)}(B/2)$	x, x, xx	138.34		xx, yz, yz	−2.3013
	x, x, yy	25.396		xy, xy, yy	−118.33
	x, y, xy	42.340		xy, xy, zz	−30.429
	y, y, xx	21.694		xy, xz, yz	−43.950
	y, y, yy	89.173		yy, yy, yy	−224.58
	y, y, zz	25.819		yy, yy, zz	−48.679
	y, z, yz	31.677		yy, yz, yz	−43.974

passing through regions not far from the equilibrium. This means they may change sign during some vibrational excursion. The contours in Figures 4 and 5 tend to be near vertical lines because of the greater importance of the CN bond than the HC bond in determining the overall polarizabilities.

Vibrational motion means that over long times, the electrical properties of a molecule are, in part, averages. One can appreciate what significance there is in this by carrying out the vibrational averaging of the electrical properties of the electronic wavefunctions, that is, averaging over curves and surfaces such as those in Figures 3–5. This has been done for the diatomics LiH [84] and HF with numerical integration using Numerov–Cooley vibrational wavefunctions [111]. Table XXXV compares equilibrium and averaged values for LiH. Differences are found between equilibrium and ground-state values of up to ~ 10%. The differences are more sizable for excited vibrational states. A contributing factor in the changes among vibrational levels for averaged polarizabilities of LiH is anharmonicity of the LiH potential. The vibrational averaging of a linearly changing property for a harmonic

TABLE XXV
DHF/ELP Basis Multipole Polarizabilities and Hyperpolarizabilities of N_2

$P^{(2)}(\alpha)$	x, x	−14.83
	y, y	−9.66
$P^{(2)}(C/2)$	xx, xx	−18.78
	xx, yy	−0.6275
	xy, xy	−7.8725
	yy, yy	−7.7475
	yy, zz	−2.7375
	yz, yz	−2.505
$P^{(3)}(B/2)$	x, x, xx	−106.31
	x, x, yy	−21.03
	x, y, xy	−35.115
	y, y, xx	−14.165
	y, y, yy	−67.16
	y, y, zz	−22.395
	y, z, yz	−22.38
$P^{(3)}$	xx, xx, xx	−425.13
	xx, xx, yy	−52.00
	xx, xy, xy	−75.288
	xx, yy, yy	−13.60
	xx, yy, zz	−9.173
	xx, yz, yz	−2.214
	xy, xy, yy	−62.225
	xy, xy, zz	−17.613
	xy, xz, yz	−22.30
	yy, yy, yy	−169.50
	yy, yy, xx	−46.30
	yy, yz, yz	−30.788

oscillator will show no state-to-state differences. It is the mechanical anharmonicity of the potential combined with a property's nonlinear dependence on bond length (electrical anharmonicity) that gives rise to the state differences. Clearly, vibrational averaging can lead to noticeable differences in state electrical properties, and the consequences are that interaction potentials, even at long range, can depend somewhat on the vibrational states.

As Werner and Meyer [91] and Adamowicz and Bartlett [70] have clearly explained, molecular electrical properties have another contribution from vibrational motion. Recall that each electrical property is strictly defined as a derivative of the molecular state energy with respect to elements of **V**.

TABLE XXVI
DHF/ELP Basis Multipole Polarizabilities and Hyperpolarizabilities of HCCH

$P^{(2)}(\alpha)$	x, x	−31.36
	y, y	−19.04
$P^{(2)}$	x, xxx	−59.810
	x, xyy	−12.390
	y, xxy	−12.217
	y, yyy	−26.055
	y, yzz	−8.6851
$P^{(2)}(C/2)$	xx, xx	−61.415
	xx, yy	0.595
	xy, xy	−18.093
	yy, yy	−20.678
	yy, zz	−9.0775
	yz, yz	−5.800
$P^{(3)}(B/2)$	x, x, xx	332.87
	x, x, yy	86.64
	x, y, xy	137.86
	y, y, xx	19.12
	y, y, yy	324.16
	y, y, zz	119.58
	y, z, yz	102.29
$P^{(3)}$	xx, xx, xx	1,328.8
	xx, xx, yy	119.69
	xx, xy, xy	263.63
	xx, yy, yy	41.575
	xx, yy, zz	23.275
	xx, yz, yz	9.1438
	xy, xy, yy	275.13
	xy, xy, zz	102.01
	xy, xz, yz	86.575
	yy, yy, yy	1,026.6
	yy, yy, zz	326.0
	yy, yz, yz	175.13

Vibrational motion is affected by the presence of an external field, field gradient, and so on. This introduces a response to an applied potential beyond that of the vibrationally averaged *electronic* state properties. Physically, it arises from the change in the stretching potential experienced when the molecule is in a field.

TABLE XXVII
DHF/ELP Basis Multipole Polarizabilities and Hyperpolarizabilities of HCN

$P^{(2)}(\alpha)$	x, x	−22.402
	y, y	−13.751
$P^{(2)}(A/2)$	x, xx	5.5316
	x, yy	−0.11445
	y, xy	0.56725
$P^{(2)}$	xx, xx	−38.946
	xx, yy	−0.20082
	xy, xy	−11.983
	yy, yy	−12.355
	yy, zz	−4.9475
	yz, yz	−3.7038
$P^{(3)}(\beta)$	x, x, x	7.3464
	x, y, y	2.2554
$P^{(3)}(B/2)$	x, x, xx	−201.09
	x, x, yy	−42.672
	x, y, xy	−67.814
	y, y, xx	−16.459
	y, y, yy	−143.40
	y, y, zz	−49.605
	y, z, yz	−46.900
$P^{(3)}$	x, xx, xx	60.321
	x, xx, yy	0.64890
	x, xy, xy	5.2001
	x, yy, yy	−2.7384
	x, yy, zz	−0.079055
	x, yz, yz	−1.3301
	y, xx, xy	2.6984
	y, xy, yy	4.1042
	y, xy, zz	2.8109
	y, xz, yz	0.64653
$P^{(3)}$	xx, xx, xx	−833.57
	xx, xx, yy	−82.993
	xx, xy, xy	−141.84
	xx, yy, yy	−34.111
	xx, yy, zz	−20.118
	xx, yz, yz	−6.9964
	xy, xy, yy	−126.32
	xy, xy, zz	−42.175
	xy, xz, yz	−42.075
	yy, yy, yy	−387.80
	yy, yy, zz	−113.94
	yy, yz, yz	−68.467
$P^{(4)}(\gamma)$	x, x, x, x	−1,561.1
	x, x, y, y	−581.49
	y, y, y, y	−1,647.0
	y, y, z, z	−549.01
$P^{(5)}(\delta)$	x, x, x, x, x	2,445.3
	x, x, x, y, y	−164.96
	x, y, y, y, y	2,146.5
	x, y, y, z, z	715.51

To understand the complete role of vibration in determining electrical properties, it is useful to consider a diatomic molecule in the harmonic oscillator approximation, where the stretching potential is taken to be quadratic in the displacement coordinate. The doubly harmonic model takes the various electrical properties to be linear functions of the coordinate. This turns out to be most reasonable in the vicinity of an equilibrium structure, but it breaks down at long separations. Letting x be a coordinate giving the displacement from equilibrium of a one-dimensional harmonic oscillator, the dipole moment, dipole polarizability, and dipole hyperpolarizability, within the doubly harmonic (dh) model, may be written in the following way:

$$\mu^{\mathrm{dh}}(x) = \mu_0 + mx, \qquad \alpha^{\mathrm{dh}}(x) = \alpha_0 + ax, \qquad \beta^{\mathrm{dh}}(x) = \beta_0 + bx. \quad (80)$$

TABLE XXVIII
DHF/ELP Basis Multipole Polarizabilities and Hyperpolarizabilities of F_2

$P^{(2)}(\alpha)$	x, x	−14.555
	y, y	−5.2766
$P^{(2)}(C/2)$	xx, xx	−11.393
	xx, yy	−0.91842
	xy, xy	−3.9807
	yy, yy	−2.9103
	yy, zz	−1.2025
	yz, yz	−0.85389
$P^{(3)}(B/2)$	x, x, xx	−57.034
	x, x, yy	−13.316
	x, y, xy	−20.630
	y, y, xx	−9.6420
	y, y, yy	−24.692
	y, y, zz	−9.0601
	y, z, yz	−7.8161
$P^{(3)}$	xx, xx, xx	−131.30
	xx, xx, yy	−19.447
	xx, xy, xy	−26.067
	xx, yy, yy	−7.0642
	xx, yy, zz	−2.9346
	xx, yz, yz	−2.0649
	xy, xy, yy	−24.364
	xy, xy, zz	−8.6699
	xy, xz, yz	−7.8472
	yy, yy, yy	−33.020
	yy, yy, zz	−9.8854
	yy, yz, yz	−5.7837

An external uniform field along the x axis and of strength $\mathbf{V}_x$ will alter the stretching potential with a potential that is linear in x:

$$E'(x) = a' + b'x, \tag{81}$$

$$a' = -(\mu_0 V_x + \tfrac{1}{2}\alpha_0 V_x^2 + \tfrac{1}{6}\beta_0 V_x^3 + \cdots), \tag{82}$$

$$b' = -(mV_x + \tfrac{1}{2}aV_x^2 + \tfrac{1}{6}bV_x^3 + \cdots). \tag{83}$$

The effect of $E'(x)$ when added to a potential $E(x)$ that is quadratic in x is to

TABLE XXIX

DHF/ELP Basis Multipole Polarizabilities and Hyperpolarizabilities of *trans*-N_2H_2

$P^{(2)}(\alpha)$	x, x	−23.908
	x, y	−2.9840
	y, y	−17.382
	z, z	−13.247
$P^{(2)}(C/2)$	xx, xx	−32.438
	xx, xy	0.36454
	xx, yy	−3.6876
	xx, zz	−1.3134
	xy, xy	−19.191
	xy, yy	−6.0924
	xy, zz	−1.6848
	xz, xz	−13.462
	xz, yz	0.92172
	yy, yy	−21.025
	yy, zz	−2.2177
	yz, yz	−6.8511
	zz, zz	−13.368
$P^{(3)}(B/2)$	x, x, xx	−245.06
	x, x, xy	34.203
	x, x, yy	−68.385
	x, x, zz	−52.120
	x, y, xx	27.856
	x, x, xy	−136.09
	x, y, yy	−34.660
	x, y, zz	12.677
	x, z, xz	−77.087
	x, z, yz	3.0424
	y, y, xx	−64.055
	y, y, xy	−40.148
	y, y, yy	−192.58
	y, y, zz	−34.087
	y, z, xz	7.6121
	y, z, yz	−53.816
	z, z, xx	−28.118
	z, z, xy	18.308
	z, z, yy	−30.552
	z, z, zz	−142.24
$P^{(3)}$	xx, xx, xx	−921.11
	xx, xx, xy	179.98
	xx, xx, yy	−138.88
	xx, xx, zz	−139.66
	xx, xy, xy	−295.09
	xx, xy, yy	−1.0470
	xx, xy, zz	30.729
	xx, xz, xz	−169.95
	xx, xz, yz	31.267
	xx, yy, yy	−159.20
	xx, yy, zz	−32.056
	xx, yz, yz	−32.624
	xx, zz, zz	−55.148
	xy, xy, xy	9.3793
	xy, xy, yy	−330.13
	xy, xy, zz	−62.985
	xy, xz, xz	52.978
	xy, xz, yz	−75.537
	xy, yy, yy	−162.22
	xy, yy, zz	20.308
	xy, yz, yz	−2.6818
	xy, zz, zz	42.755
	xz, xz, yy	−32.061
	xz, xz, zz	−174.34
	xz, yy, yz	−12.577
	xz, yz, zz	37.056
	yy, yy, yy	−601.74
	yy, yy, zz	−77.309
	yy, yz, yz	−80.381
	yy, zz, zz	−49.491
	yz, yz, zz	−99.177
	zz, zz, zz	−556.15

change the location of the minimum and the minimum energy. It will not change the curvature, and so it will not change the vibrational frequency or the spacing between levels. The actual change in the minimum energy is $a' - b'^2/4c$, where c is the quadratic coefficient in $E(x)$. This energy change is, of course, experienced by all the vibrational energy levels; all are shifted down or up by the same amount for a given field strength.

TABLE XXX
DHF/ELP Basis Multipole Polarizabilities and Hyperpolarizabilities of *cis*-N_2H_2

$P^{(2)}(\alpha)$	x, x	−25.028		x, xy, zz	9.2446
	y, y	−17.010		x, xz, yz	−14.753
	z, z	−13.564		y, xx, xx	43.201
$P^{(2)}(A/2)$	x, xy	−5.7239		y, xx, yy	−32.976
	y, xx	−1.3894		y, xx, zz	14.159
	y, yy	−6.6185		y, xy, xy	−109.27
	y, zz	0.60505		y, xz, xz	8.5067
	z, yz	0.23987		y, yy, yy	−134.27
				y, yy, zz	1.6906
$P^{(2)}(C/2)$	xx, xx	−32.915		y, yz, yz	−19.293
	xx, yy	−4.8736		y, zz, zz	17.493
	xx, zz	−1.1974		z, xx, yz	3.8608
	xy, xy	−22.231		z, xy, xz	2.9455
	xz, xz	−12.461		z, yy, yz	−22.285
	yy, yy	−21.350		z, yz, zz	−1.8218
	yy, zz	−2.6305	$P^{(3)}$	xx, xx, xx	−820.81
	yz, yz	−7.0336		xx, xx, yy	−169.73
	zz, zz	−12.684		xx, xx, zz	−125.61
$P^{(3)}(\beta)$	x, x, y	−11.615		xx, xy, xy	−343.80
	y, y, y	−17.948		xx, xz, xz	−138.73
	y, z, z	6.6520		xx, yy, yy	−189.13
				xx, yy, zz	−31.191
$P^{(3)}(B/2)$	x, x, xx	−238.52		xx, yz, yz	−40.739
	x, x, yy	−94.170		xx, zz, zz	−45.632
	x, x, zz	−48.319		xy, xy, yy	−430.04
	x, y, xy	−163.90		xy, xy, zz	−64.238
	x, z, xz	−73.664		xy, xz, yz	−89.798
	y, y, xx	−80.816		xz, xz, yy	−46.109
	y, y, yy	−195.15		xz, xz, zz	−143.67
	y, y, zz	−40.613		yy, yy, yy	−631.60
	y, z, yz	−55.687		yy, yy, zz	−91.884
	z, z, xx	−25.796		yy, yz, yz	−94.819
	z, z, yy	−30.832		yy, zz, zz	−70.596
	z, z, zz	−135.35		yz, yz, zz	−103.52
$P^{(3)}$	x, xx, xy	−10.540			
	x, xy, yy	−108.31			

Using the definition of electrical properties as derivatives of the molecular energy with the doubly harmonic model gives

$$\mu = \mu_0, \tag{84}$$

$$\alpha = \alpha_0 + m^2/2c, \tag{85}$$

TABLE XXXI
DHF/ELP Basis Multipole Polarizabilities of HOOH

$P^{(2)}(\alpha)$	x, x	−19.900
	x, y	1.9024
	y, y	−12.407
	z, z	−10.899
$P^{(2)}(A/2)$	x, xz	−1.6898
	x, yz	0.6041
	y, xz	1.1423
	y, yz	−2.0624
	z, xx	−0.13691
	z, xy	0.97710
	z, yy	−1.2612
	z, zz	−1.4930
$P^{(2)}(C/2)$	xx, xx	−25.175
	xx, xy	0.3658
	xx, yy	−1.8018
	xx, zz	−2.3781
	xy, xy	−12.102
	xy, yy	3.4774
	xy, zz	0.2935
	xz, xz	−10.983
	xz, yz	1.6694
	yy, yy	−10.712
	yy, zz	−2.2707
	yz, yz	−4.3990
	zz, zz	−8.8767
$P^{(3)}(\beta)$	x, x, x	−2.2013
	x, y, z	−0.0079
	y, y, z	−6.3731
	z, z, z	−1.7992

$$\beta = \beta_0 + 3ma/2c, \tag{86}$$

$$\gamma = \gamma_0 + (4mb + 3a^2)/c. \tag{87}$$

These properties, which are for the entire vibrational state manifold of the harmonic oscillator, are equilibrium values plus corrections that result from the shift in the equilibrium position from the application of a field. This simple model is useful in several ways . For example, if the dipole moment is not dependent on x (if $m = 0$), then the vibrational motion will not affect α or

TABLE XXXII
DHF/ELP Basis Polarizabilities and Hyperpolarizabilities of H_2CCH_2

$P^{(2)}(\alpha)$	x, x	-36.774
	y, y	-24.687
	z, z	-22.779
$P^{(2)}(C/2)$	xx, xx	-63.245
	xx, yy	-6.8588
	xx, zz	-0.63543
	xy, xy	-38.283
	xz, xz	-23.615
	yy, yy	-35.225
	yy, zz	-3.5240
	yz, yz	-14.563
	zz, zz	-31.605
$P^{(3)}(B/2)$	x, x, xx	-403.69
	x, x, yy	-96.248
	x, x, zz	-130.33
	x, y, xy	-200.06
	x, z, xz	-212.81
	y, y, xx	-83.172
	y, y, yy	-265.03
	y, y, zz	-98.045
	y, z, yz	-148.19
	z, z, xx	-34.201
	z, z, yy	-79.818
	z, z, zz	-532.07
$P^{(3)}$	xx, xx, xx	$-1,522.8$
	xx, xx, yy	-231.51
	xx, xx, zz	-222.78
	xx, xy, xy	-530.65
	xx, xz, xz	-375.48
	xx, yy, yy	-175.14
	xx, yy, zz	-19.706
	xx, yz, yz	-83.524
	xx, zz, zz	-86.029
	xy, xy, yy	-504.94
	xy, xy, zz	-85.367
	xy, xz, yz	-195.33
	xz, xz, yy	-125.11
	xz, xz, zz	-522.96
	yy, yy, yy	-836.93
	yy, yy, zz	-165.18
	yy, yz, yz	-292.63
	yy, zz, zz	-298.61
	yz, yz, zz	-351.55
	zz, zz, zz	$-2,132.78$

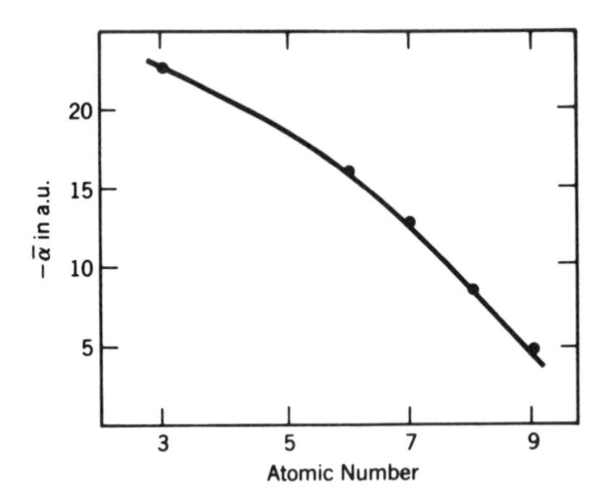

Figure 2. Isotropic dipole polarizability ($\bar{\alpha}$) (a.u.) of AH_n molecules versus the atomic number of A.

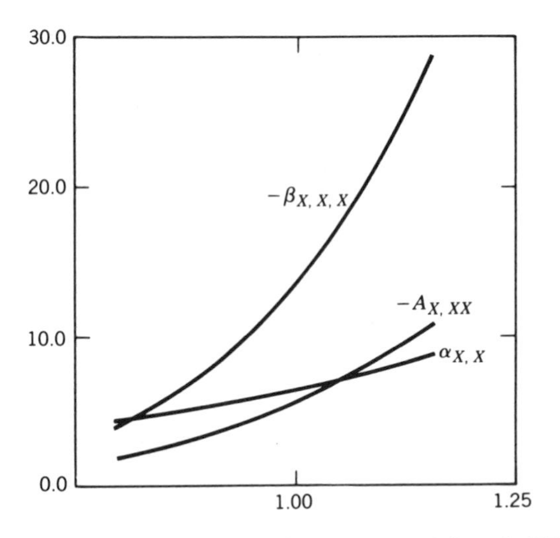

Figure 3. Bond length dependence of $\alpha_{x,x}$, $A_{x,xx}$, and $\beta_{x,x,x}$ in HF (a.u.).

β. So, within the doubly harmonic model, the vibrational state dipole polarizability of a molecule such as N_2, but not a molecule such as CO, will be identical with the equilibrium or pure electronic value.

There are two steps to take in going from the doubly harmonic model to a highly realistic treatment. One is to allow for the potential to be anharmonic, and the other is to allow the properties to be more complicated than just

TABLE XXXIII
Calculated and Estimated Isotropic Dipole Polarizabilities of ABH_n molecules (a.u.)

	Calculated	Estimated, Eqn. (79)	Difference (%)
CO	−12.13	−12.41	2
N_2	−11.38	−13.64	16
HCN	−16.63	−16.82	1
HCCH	−23.15	−20.00	−16
H_2CCH_2	−28.08	−26.00	−8
N_2H_2 (*cis*)	−18.18	−19.64	7
N_2H_2 (*trans*)	−18.53	−19.64	6
H_2O_2	−14.40	−16.82	14

TABLE XXXIV
Dipole and Quadrupole Polarizabilities[a] (a.u.) of Be and Mg Atoms

		Be	Mg
α	x, x	−44.139	−81.651
$P^{(2)}(C/2)$	xx, xx	−38.444	−107.42
	xx, yy	−13.517	−32.52
	xy, xy	−12.464	−37.45

[a] DHF results with $12s5p2d/7s3p2d$ Be basis and $14s9p3d/9s6p3d$ Mg basis.

being linear in the vibrational coordinate. As soon as the property functions become nonlinear, the effect of an external field will be more than just to change the equilibrium position. It will also change the shape of curvature at the minimum. This means the spacing between vibrational levels will be affected, and so the properties will not be alike for the vibrational states. The effects from the electrical properties being nonlinear are often referred to as *electrical anharmonicity* effects while the changes from using an anharmonic stretching potential are referred to as *mechanical anharmonicity* effects [112, 113].

For the HF molecule, we have rigorously evaluated the vibrational–electronic state electrical properties. For about a dozen bond length choices, electrical properties for HF were calculated. An ACCD potential energy curve, accurate enough to predict the $v = 0 \rightarrow v = 1$ field-free transition

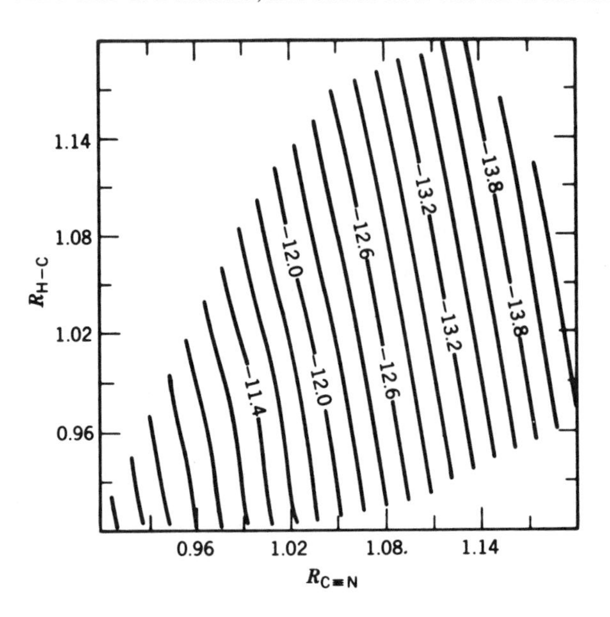

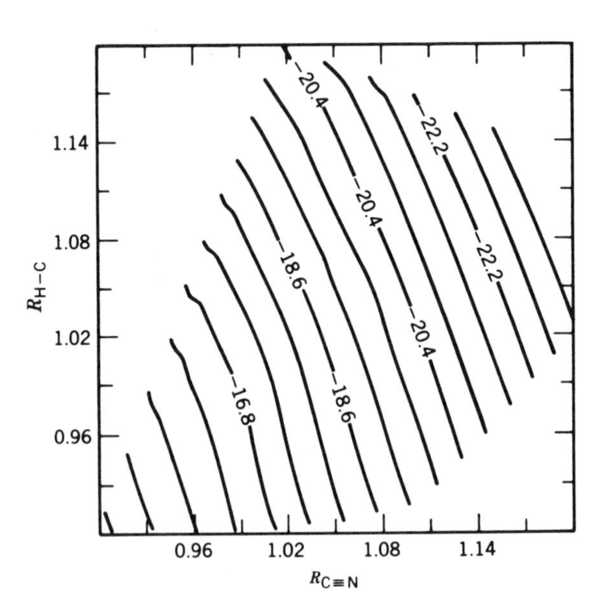

Figure 4. (*a*) α_{xx} contours (a.u.) for HCN. The H–C axis is vertical axis. The C–N axis is horizontal axis. (*b*) α_{yy} contours (a.u.).

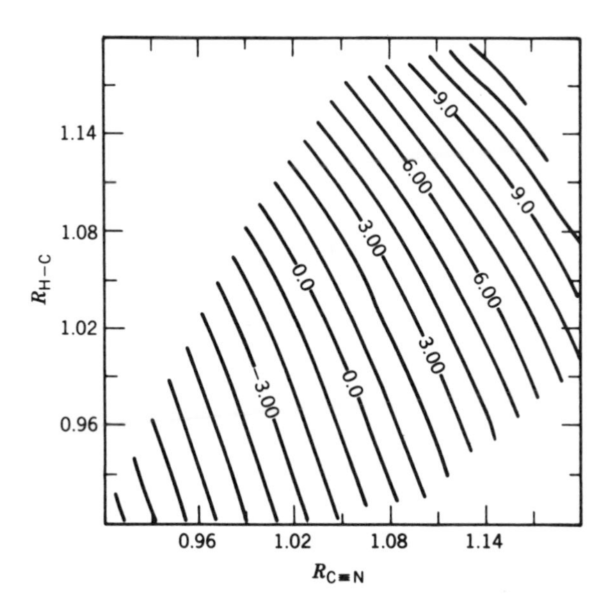

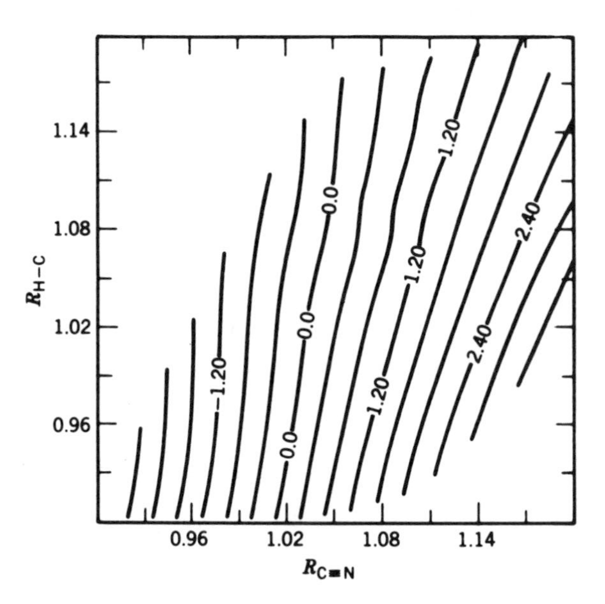

Figure 5. (a) β_{xxx} contours (a.u.) for HCN. Dashed lines are negative contours. (b) β_{xyy} contours for HCN.

TABLE XXXV
Vibrationally Averaged Polarizabilities (a.u.) of LiH[a]

	Property	Equilibrium Value	$v = 0$	$v = 1$	$v = 2$
α	x, x	−22.14	−23.01	−24.79	−26.73
	y, y	−22.68	−23.02	−23.69	−24.39
A	x, xx	−86.73	−91.12	−100.21	−110.19
	x, yy	8.494	8.901	9.734	10.636
	y, xy	−40.62	−41.75	−44.00	−46.36
β	x, x, x	382.5	421.0	503.5	598.8
	x, y, y	215.4	225.8	247.3	271.0

[a] From ref. 84. Basis was not ELP so equilibrium properties are slightly different than in Table XVIV. With very large basis and ACCD treatment of electron correlation, computed dipole moment values for three vibrational states are 5.943 ($v = 0$), 6.060 ($v = 1$), and 6.177 ($v = 2$) D [140]. Experimentally measured values obtained by Wharton et al. [141] are 5.882, 5.990, and 6.098 for these three states. Agreement is to 1%.

frequency to within 1.0 cm^{-1} of the measured value, was also calculated. Then, for 10×10 and 20×20 grids of field strengths and field gradient strengths, a correction was made to the stretching potential corresponding to the interaction of the molecule (at each of the dozen HF separations) with the field and field gradient. At each of the 100 or 400 grid spots, a Numerov–Cooley analysis was carried out to rigorously determine the vibrational state energies in the presence of the selected field and field gradients. Figure 6 gives energy contours of the $v = 0$ and $v = 1$ states over a very wide range of field and field gradient strengths. These energies are relative to the potential minima, so the contours represent changes due to how the *shape* of the potential is affected by fields and field gradients. Notice the sharper effect, and the more closely spaced contour lines, for the $v = 1$ state. Effects are more pronounced higher in the potential.

The data used to plot the contours in Figure 6 is sufficient to numerically compute the derivatives of the state energies. Let $E_{\min}(\mathbf{V})$ be the energy at the potential minimum, which is dependent on $\mathbf{V}$, and $E_{v=0}(\mathbf{V})$ be the energy of the $v = 0$ vibrational state relative to $E_{\min}$. Then $E(\mathbf{V}) = E_{\min}(\mathbf{V}) + E_{v=0}(\mathbf{V})$, and

$$\frac{\partial}{\partial V_x} E(\mathbf{V}) = \frac{\partial}{\partial V_x} E_{\min}(\mathbf{V}) + \frac{\partial}{\partial V_x} E_{v=0}(\mathbf{V}). \tag{88}$$

This is the breakdown into terms that reflect how the potential alone changes in the presence of a field and how the vibrational motions change. Table

XXXVI compares properties for hydrogen fluoride. Most of the contribution to the complete value for the vibrational–electronic ground state comes from the change in the potential, and this reinforces the validity of doubly harmonic analyses. The difference between the vibrationally averaged and complete values is small for α but quite significant for β, as Adamowicz and Bartlett first discovered [70].

In this analysis and in that of the next section, the vibrational motion effects presume a field source that is rotating with the molecule, such as when the electrical perturbation is due to a weakly complexed partner molecule. A freely rotating molecule in a laboratory-fixed field source, however, is different, and then evaluations of electrical properties should account for rotational state dependence as well [114, 115].

B. Derivative Numerov–Cooley Theory

The difference equation or numerical integration method for vibrational wavefunctions usually referred to as the Numerov–Cooley method [111] has been extended by Dykstra and Malik [116] to an open-ended method for the analytical differentiation of the vibrational Schrödinger equation of a diatomic. This is particularly important for high-order derivatives (i.e., hyperpolarizabilities) where numerical difficulties may limit the use of finite-field treatments. As in Numerov–Cooley, this is a procedure that invokes the Born–Oppenheimer approximation. The accuracy of the results are limited only by the quality of the electronic wavefunction's description of the stretching potential and of the electrical property functions and by the adequacy of the Born–Oppenheimer approximation.

The relevant equations for the derivative Numerov–Cooley (DNC) method closely follow Cooley's [111] presentation. Let R be the radial coordinate or bond displacement coordinate, $P(R)$ a radial eigenfunction, and $U(R)$ the potential function. The one-dimensional Schrödinger equation is then

$$\frac{1}{2\mu} P^{(2)}(R) = [U(R) - E] P(R), \tag{89}$$

TABLE XXXVI
Vibrational Effects in $v = 0$ State on Dipole Properties (a.u.) of HF

Property	Vibrationally Averaged Value	Complete value for Vibrational–Electronic State
μ_x	−0.7254	−0.7254
$\alpha_{x,x}$	−5.9456	−6.1154
$\beta_{x,x,x}$	10.5046	0.4639

where μ is the reduced mass and E is the energy. The superscript in $P^{(2)}(R)$ means that this is the second derivative with respect to R. There will be one $P(R)$ for each vibrational state, but for notational convenience we suppress denoting a particular vibrational state.

It will be assumed that there is a set of parameters of interest, $[a, b, c, \ldots]$, and that these are embedded in a known function of R. For convenience, this is taken to be part of the potential function $U(R)$. It is also possible to use model perturbations, such as the influence of a surrounding bath, to examine the response of the system, in which case the parameters may be any of those embedded in the model perturbation. As discussed in general terms previously, each derivative equation is solved with all parameters set to zero, and this will be presumed in the following discussion. A superscript placed outside the parentheses denoting differentiation with respect to R will serve to indicate differentiation with respect to a parameter from the set.

In succession, the equations generated by differentiating Eqn. (89) with respect to parameters a and then b are the following:

$$\frac{1}{2\mu} P^{(2)a}(R) = [U^a(R) - E^a]P(R) + [U(R) - E]P^a(R), \tag{90}$$

$$\frac{1}{2\mu} P^{(2)ab}(R) = [U^{ab}(R) - E^{ab}]P(R) + [U^a(R) - E^a]P^b(R) + [U^b(R) - E^b]P^a(R) + [U(R) - E]P^{ab}(R). \tag{91}$$

This process can be continued as far as desired with respect to any and all parameters, and for general problems, each equation has the form

$$\frac{1}{2\mu} P^{(2)\alpha}(R) = [U(R) - E]P^\alpha(R) - E^\alpha P(R) + X^\alpha(R), \tag{92}$$

where $X^\alpha(R)$ is a function that is fixed and obtained from lower order derivative energies and derivative wavefunctions. The $n+1$ energy derivatives are obtained from the nth-derivative wavefunctions developed by multiplying and integrating with the zero wavefunction. With a normalized $P(R)$, the following expression is obtained:

$$E^\alpha = \langle P(R) | X^\alpha(R) \rangle. \tag{93}$$

Orthonormality is a constraint that may be incorporated into the derivative Schrödinger equation or imposed separately [116].

A solution of the one-dimensional, vibrational state differential equation may be accomplished by converting it to a difference equation. This means partitioning or discretizing the coordinate R into uniform, small steps and finding the value of the wavefunction at each step. In practice, we use a step size that is smaller than 0.001 Å. Tests indicate that vibrational state energies for simple diatomics are unaffected to 0.01 cm^{-1} by further decreasing the step size. In fact, a step size of 0.01 Å can yield state energies to a numerical reliability of 0.1 cm^{-1} in many cases. Following Cooley, the step size is called h and there is a finite set of specific distances R_i measured relative to some initial or closest-in value:

$$R_i = R_0 + ih, \quad \text{where } i = 1, 2, \ldots, N. \tag{94}$$

Power series expansions for the function $P(R)$ at R_{i+1} and at R_{i-1} or for any of the derivatives of $P(R)$ with respect to R are summed:

$$P_{i+1}^{(l)\alpha} + P_{i-1}^{(l)\alpha} = \sum_{k=0}^{\infty} \frac{2h^{2k}}{(2k)!} P_i^{(2k+l)\alpha}. \tag{95}$$

Truncating at fourth power in h and using $l = 0$ yields

$$P_{i+1}^{\alpha} + P_{i-1}^{\alpha} = 2P_i^{\alpha} + h^2 P_i^{(2)\alpha} + \tfrac{1}{12} h^4 P_i^{(4)\alpha}. \tag{96}$$

Substitution using Eqn. (92) yields

$$P_{i+1}^{\alpha} + P_{i-1}^{\alpha} - 2P_i^{\alpha} = 2\mu h^2 [(U_i - E)P_i^{\alpha} - E^{\alpha} P_i + X_i^{\alpha}] + \tfrac{1}{12} h^4 P_i^{(4)\alpha}. \tag{97}$$

The power series expansion of the second derivative with respect to R [from Eqn. (95)] is truncated at the first term to give an expression for $P_i^{(4)\alpha}$:

$$P_{i+1}^{(2)\alpha} + P_{i-1}^{(2)\alpha} - 2P_i^{(2)\alpha} = h^2 P_i^{(4)\alpha}. \tag{98}$$

This is scaled by $\frac{1}{12}h^2$ and subtracted from Eqn. (97) to eliminate the h^4 term with the following definitions:

$$T_i^{\alpha} \equiv -E^{\alpha} P_i + X_i^{\alpha}, \tag{99}$$

$$Y_i^{\alpha} \equiv P_i^{\alpha} - \tfrac{1}{12} h^2 P_i^{(2)\alpha} = P_i^{\alpha} - \tfrac{1}{12} h^2 [(U_i - E)P_i^{\alpha} + T_i^{\alpha}]. \tag{100}$$

The following is obtained:

$$Y_{i+1}^{\alpha} + Y_{i-1}^{\alpha} - 2Y_i^{\alpha} = 2\mu h^2 [(U_i - E)P_i^{\alpha} + T_i^{\alpha}]. \tag{101}$$

If P^{α} and Y^{α} are known at two adjacent points, Y^{α} may be found at the next adjacent point using Eqn. (101). Then P_i^{α} is found from Y_i^{α} by Eqn. (100).

In standard Numerov–Cooley, integration or step-by-step use of Eqn. (101) begins both at close-in and at far-out extremes where the values of P^α and Y^α are near zero, assuming a bound state. These are guessed to be very small values. Then, the inward and outward functions that are obtained are matched in slope and value at some midway point by iterative adjustment of the energy. In Eqn. (101), the zero-order energy E is known already and so the process *requires no iteration.* This also means the integration needs to be done in only one direction. The P^α that is found will be a mixture of the true derivative wavefunction and the zero-order wavefunction, and so the last step is a projection step to ensure orthonormality.

C. Effects on Vibrational Spectra Arising from Electrical Interaction

Figure 6 gives contours of the $v = 0$ and $v = 1$ vibrational state energies for HF in a field–field gradient. The differences in those energies are transition energies ω_{01}. Figure 7 is a plot of how $\omega_{01}(V_x, V_{xx})$ differs from $\omega_{01}(V_x = 0, V_{xx} = 0)$. It is a contour plot of the transition frequency shifts due to applied fields and field gradients. The same type of result has also been obtained for LiH [84], and it shows that field gradients can augment or balance against an axial field in the effect on transition frequencies. The curvature and varying spacing of the contour lines result in part from the polarizabilities and hyperpolarizabilities. Figure 8, which gives contours for the $v = 0$ state, shows how the hyperpolarizabilities affect the state energies. Polarizabilities enter into the transition frequency shifts with about an order of magnitude greater importance than the hyperpolarizabilities.

Electrical interaction also affects transition moments because of the induced dipoles. As shown in Figure 9 for LiH, field gradients with fields may work differently in changing the transition moment of a fundamental versus an overtone transition.

A transition dipole is an off-diagonal element of the matrix representation of the dipole moment in the basis of molecular states:

$$D_{n,m} = \langle \psi_n | \mu | \psi_m \rangle. \tag{102}$$

With the approximation of being electrically harmonic (eh), the transition moment reduces to a matrix element of the vibrational coordinate r times the slope of the dipole moment function at equilibrium:

$$D^{\mathrm{eh}}_{n,m} = \langle \psi_n | r | \psi_m \rangle \frac{d\mu}{dr}. \tag{103}$$

Both expressions are easy to use for diatomics where numerical wavefunctions of the vibrational states make for a simple calculation of the

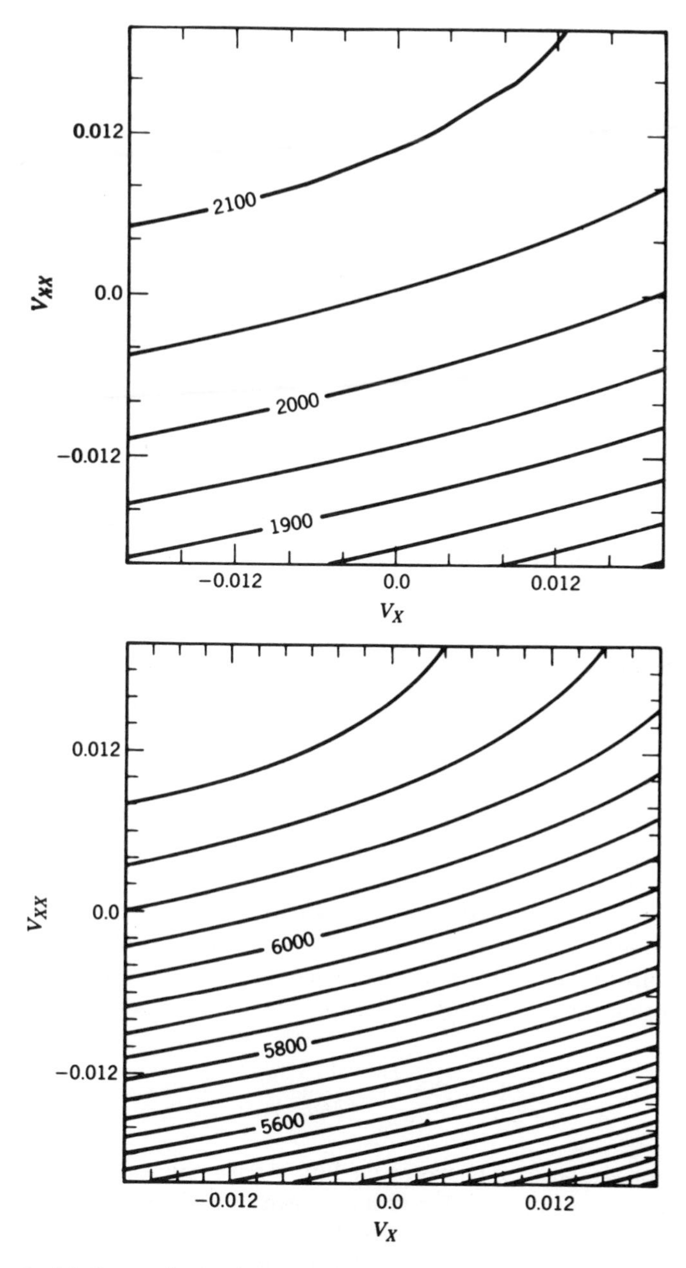

Figure 6. (*a*) Contour levels of ground vibrational state ($v = 0$) energy (cm^{-1}) of HF in presence of axial fields (horizontal axis) (a.u.) and field gradients (vertical axis) (a.u.). For field gradients $V_{yy} = -V_{xx}$. (*b*) Corresponding contour levels of $v = 1$ state energies.

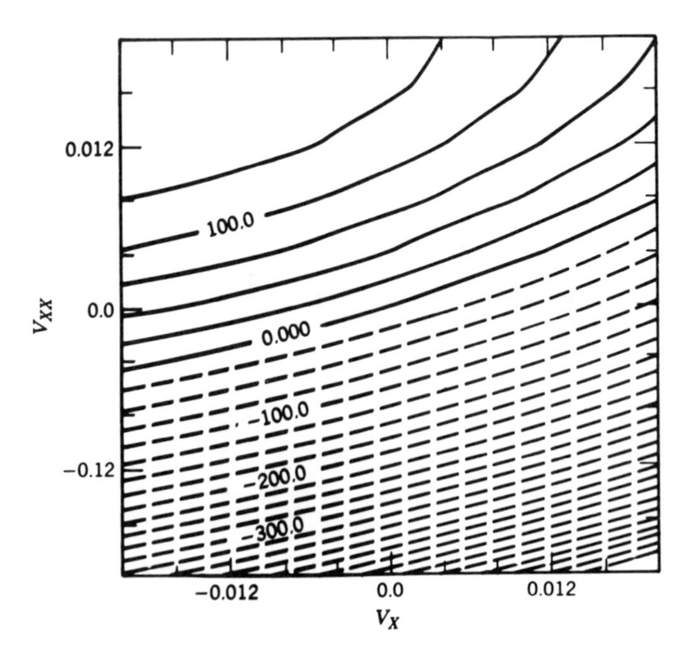

Figure 7. Contours of shift in $v = 0 \rightarrow v = 1$ transition frequency (cm^{-1}) of HF from applied fields and field gradients (a.u.).

integrals. The most approximate form of the transition moment is the doubly harmonic approximation. The integrals $\langle \psi_n | r | \psi_m \rangle$ are nonzero for a harmonic oscillator only if n and m are adjacent levels, and in those cases, the integrals are directly proportional to the square root of the quantum number. Thus, the doubly harmonic expression for the transition moments requires knowing only the slope of the dipole moment function at the equilibrium and the equilibrium force constant or else the harmonic vibrational frequency:

$$D^{\text{dh}}_{n,n+1} = \frac{(n+1)^{1/2}(d\mu/dr)}{(2m\omega_e)^{1/2}}. \tag{104}$$

An ab initio calculation [90] has given doubly harmonic and complete values for transition moments of HF and has shown how these change because of the presence of perturbing fields. The field-free $v = 0 \rightarrow v = 1$ moment that was obtained using Eqn. (102) was 0.09915 Debye, while a value of 0.09898 was obtained with the doubly harmonic approximation, Eqn. (104). The measured value is 0.0985 [117]. For the $v = 0 \rightarrow 2$ overtone transition

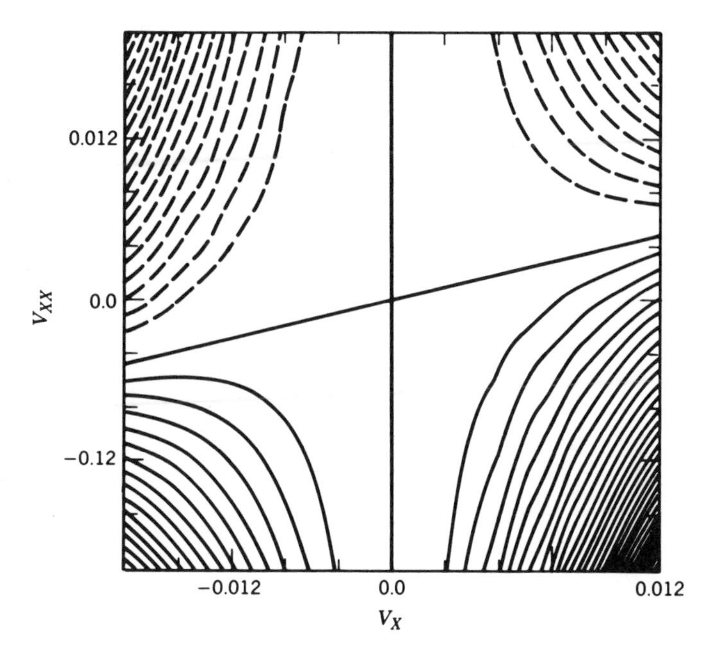

Figure 8. Contours of changes in $v = 0$ state energies of HF in applied fields and field gradients due to hyperpolarizabilities (dipole and quadrupole). Straight lines crossing in middle are zero contours. Each contour away is spaced at 0.1 cm^{-1}. Dashed lines are negative changes.

moment, the calculated value is − 0.01131 D. The doubly harmonic value is identically zero, and the experimental value is − 0.0127 D. A good determination of transition moments is clearly possible with ab initio techniques. For polyatomics, the harmonic model is computationally advantageous because the equilibrium frequencies and equilibrium slopes of the dipole moment functions can be obtained by differentiation methods.

The dipole polarizability can be used in place of the dipole moment function, and this will lead to Raman intensities. Likewise, one can compute electrical quadrupole and higher multipole transition moments if these are of interest.

Botschwina's calculations on HCN, HCP, and C_2N_2 [118] included IR (vibrational dipole) intensities. Compared against experimental values, the use of correlated wavefunctions seems to have given better results than the use of SCF wavefunctions. In some cases, SCF significantly overestimated the intensities relative to correlated wavefunction values.

Vibrational spectroscopic experiments that go beyond a Stark field and use field gradients in concert may uncover new phenomena while, according

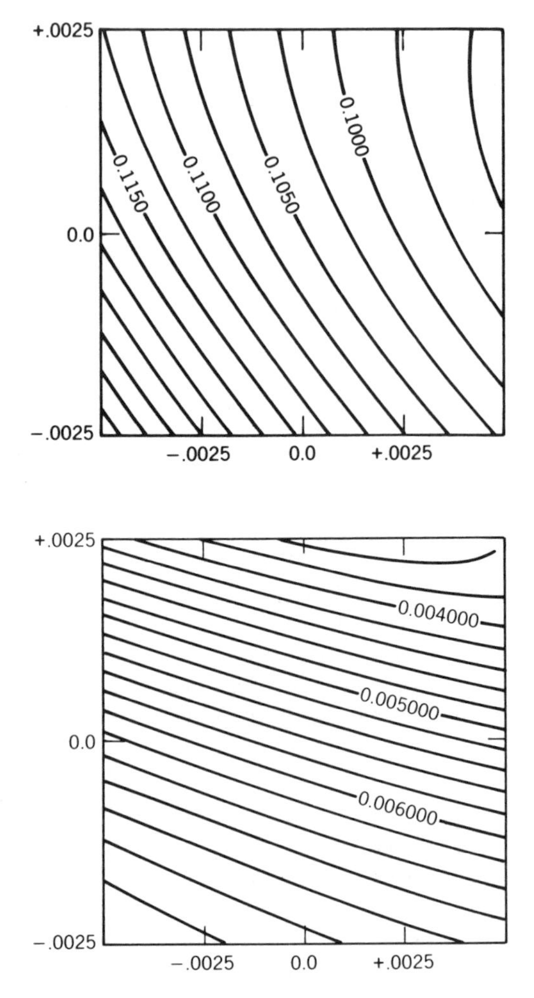

Figure 9. Contours of absolute dipole transition moments of LiH in presence of field (horizontal axis, a.u.) and field gradient (vertical axis, a.u.) for (*a*) $v = 0 \rightarrow v = 1$ transition and (*b*) $v = 0 \rightarrow v = 2$ transition.

to our analyses, they could provide a way of measuring subtle and elusive electrical properties. Particularly intriguing from the studies that we have carried out is that the range over which one might tune a vibrational frequency by applying a field is extended if a gradient is applied as well. Furthermore, the ratio of the fundamental to overtone transition probabilities can be tuned, probably by as much as two or three orders of magnitude.

Finally, one interesting way in which electrical effects are interlaced with vibrational effects is in hydrogen bonding. Intermolecular electrical interaction appears to be a dominant factor in understanding red shifts and transition moment enhancements in intramolecular vibrations of hydrogen-bonded molecules [90, 119, 120]. Undoubtedly, there will be other types of molecular problems that are revealed to be dependent on electrical features. The capability for ab initio study of electrical properties paves the way for this to happen.

Acknowledgments

The support of the National Science Foundation through a grant to C. E. D. from the Chemistry Division (Grant CHE 84-19496) has been most important in many of our studies discussed here. Acknowledgment is also made to the Donors of the Petroleum Research Fund, administered by the American Chemical Society, for partial support of this work (Grant 19786-AC6-C). We thank Ms. Sharilyn Loushin for carrying out certain of the DHF calculations on Be and Mg. An electrical property database has been set up at the University of Illinois, and new DHF calculated results and property values from other sources are being added. Information about this database or archive may be obtained from one of us (C. E. D.).

V. References and Notes

1. See for example, P. W. Atkins, *Molecular Quantum Mechanics*, 2nd ed., (Oxford University Press, Oxford, 1983).
2. D. E. Stogryn and A. P. Stogryn, *Molec. Phys. 11*:371 (1966).
3. A. J. Stone and M. Alderton, *Molec. Phys. 56*:1047 (1985).
4. A. D. Buckingham and P. W. Fowler, *Can. J. Chem. 63*:2018 (1985).
5. A. D. Buckingham, *Quart. Rev. (London) 13*:189 (1959); *Adv. Chem. Phys. 12*:107 (1967).
6. D. E. Logan, *Mol. Phys. 46*:271 (1982).
7. J. Applequist, *J. Math. Phys. 24*:736 (1983); *J. Chem. Phys. 83*:809 (1985); *Chem. Phys. 85*:279 (1984).
8. A common practice in electrical derivations [1, 5, 7] is to use fields and field gradients that differ in sign from the elements in the polytensor **V**. Applequist [7] constructs a field gradient rank-one polytensor, and if we designate it **F**, then $\mathbf{F} = -\mathbf{V}$. This does not imply a sign difference in the convention for moments, just a difference in the basic interaction expression:

$$E_{\text{int}} = \mathbf{V}^{\text{T}} \cdot \mathbf{M}^{\text{T}} = -\mathbf{F}^{\text{T}} \cdot \mathbf{M}.$$

For the $\mathbf{P}^{(n)}$ tensors, there are sign differences that may be clear from the equivalent expressions for the total interaction energy:

$$\begin{aligned} E_{\text{int}} &= \mathbf{V}^{\text{T}} \cdot (\mathbf{M} + \tfrac{1}{2}\mathbf{P}^{(2)}\mathbf{V} + \tfrac{1}{6}\mathbf{P}^{(3)}\mathbf{V}^2 + \cdots) \\ &= -\mathbf{F}^{\text{T}} \cdot (\mathbf{M} + \tfrac{1}{2}\tilde{\mathbf{P}}^{(2)}\mathbf{F} + \tfrac{1}{6}\tilde{\mathbf{P}}^{(3)}\mathbf{F}^2 + \cdots), \end{aligned}$$

where $\tilde{\mathbf{P}}^{(n)}$ designates polarizabilities using the traditional sign convention. By inspection, $\tilde{\mathbf{P}}^{(2n)} = -\mathbf{P}^{(2n)}$ and $\tilde{\mathbf{P}}^{(2n+1)} = \mathbf{P}^{(2n+1)}$.

9. C. E. Dykstra and P. G. Jasien, *Chem. Phys. Lett. 109*:388 (1984).
10. E. R. Cohen and B. N. Taylor, *J. Phys. Chem. Ref. Data 2*:663 (1973).
11. M. P. Bogaard and B. J. Orr, in *Molecular Structure and Properties*, A. D. Buckingham (ed.) (Butterworths, London, 1975).
12. T. M. Miller and B. Bederson, *Adv. At. Mol. Phys. 13*:1 (1977).
13. A. D. Buckingham and B. J. Orr, *Quart. Rev.* (*London*) *21*:195 (1967).
14. S. M. Bass, R. L. DeLeon, and J. S. Muenter, *J. Chem. Phys. 86*:4305 (1987).
15. J. W. Dudley II, and J. F. Ward, *J. Chem. Phys. 82*:4673 (1985).
16. T. J. Balle and W. H. Flygare, *Rev. Sci. Instrum. 52*:33 (1981); E. J. Campbell, W. G. Read and J. A. Shea, *Chem. Phys. Lett. 94*:69 (1983).
17. E. U. Condon and G. H. Shortley, *The Theory of Atomic Spectra* (Cambridge University Press, London, 1951).
18. W. H. Flygare, *Molecular Structure and Dynamics* (Prentice-Hall, Engelwood Cliffs, NJ, 1978).
19. C. H. Townes, G. C. Dousmanis, R. L. White, and R. F. Schwartz, *Disc. Far. Soc. 19*:56 (1955).
20. C. H. Townes and A. L. Schawlow, *Microwave Spectroscopy* (McGraw-Hill, New York, 1955).
21. R. G. Gordon, *J. Chem. Phys. 39*:2788 (1963); *40*:1973 (1964); *41*:1819 (1964).
22. R. L. Armstrong, S. H. Blumenfeld, and C. G. Gray, *Can. J. Phys. 46*:1331 (1968).
23. S. Rajan, K. Lalita, and S. Y. Babu, *J. Magnet. Reson. 16*:115 (1974).
24. B. H. Ruessink and C. MacLean, *J. Chem. Phys. 85*:93 (1986).
25. B. H. Ruessink and C. MacLean, *Molec. Phys. 60*:1059 (1987).
26. H. A. Lorentz, *The Theory of Electrons* (Dover, New York, 1952).
27. L. Onsager, *J. Am. Chem. Soc. 58*:1486 (1936).
28. N. Ramsay, *Molecular Beams* (Oxford, London, 1956).
29. C. R. Mueller, in *Physical Chemistry: An Advanced Treatise*, Douglas Henderson (ed.) (Academic Press, New York, 1970).
30. J. S. Muenter and T. R. Dyke, in *Molecular Structure and Properties*, Vol. 2, A. D. Buckingham (ed.) (Butterworths, London, 1975).
31. S. E. Choi and R. B. Bernstein, *J. Chem. Phys. 85*:150 (1986).
32. For example, L. Wharton, R. Berg, and W. Klemperer, *J. Chem. Phys. 39*:2023 (1963); R. Berg, L. Wharton, and W. Klemperer, *J. Chem. Phys. 43*:2416 (1965).
33. E. W. Kaiser, *J. Chem. Phys. 53*:1686 (1970).
34. B. Fabricant, D. Krieger, and J. S. Muenter, *J. Chem. Phys. 67*:1576 (1977); W. L. Ebenstein and J. S. Muenter, *J. Chem. Phys. 80*:3989 (1984).
35. C. J. F. Böttcher and P. Bordewijk, *Theory of Electric Polarization*, Vol. II, 2nd ed., (Elsevier, Amsterdam, 1978).
36. A. D. Buckingham and H. C. Longuet-Higgins, *Mol. Phys. 14*:63 (1968).
37. A. D. Buckingham, in *Physical Chemistry: An Advanced Treatise*, Vol. IV, Douglas Henderson (ed.) (Academic Press, New York, 1970).
38. See, for example, S. Stolte and J. Reuss in *Atom-Molecule Collision Theory: A Guide for the Experimentalist*, Richard B. Bernstein (ed.) (Plenum, New York, 1979).

39. G. C. Maitland, M. Rigby, E. B. Smith, and W. A. Wakeham, *Intermolecular Forces: Their Origin and Determination* (Clarendon Press, Oxford, 1981).
40. D. J. Malik, D. Secrest, and U. Buck, *Chem. Phys. Lett. 75*:965 (1980).
41. M. Keil and G. A. Parker, *J. Chem. Phys. 82*:1947 (1986).
42. P. Isnard, D. Robert, and L. Galatry, *Mol. Phys. 31*:1789 (1976).
43. U. Buck, J. Schleusener, D. J. Malik, and D. Secrest, *J. Chem. Phys. 74*:1707 (1981).
44. A. C. Hurley, *Introduction to the Electron Theory of Small Molecules* (Academic Press, New York, 1976).
45. N. C. Handy and H. F. Schaefer III, *J. Chem. Phys. 81*:5031 (1984).
46. T.-S. Nee, R. G. Parr, and R. J. Bartlett, *J. Chem. Phys. 64*:2216 (1976).
47. H. F. King and A. Komornicki, *J. Chem. Phys. 84*:5645 (1986).
48. A. J. Sadlej, *Chem. Phys. Lett. 47*:50 (1977).
49. A. J. Sadlej, *Molec. Phys. 34*:731 (1977).
50. A. J. Sadlej, *Theoret. Chim. Acta 47*:205 (1978).
51. K. Szalewicz, L. Adamowicz, and A. J. Sadlej, *Chem. Phys. Lett. 61*:548 (1979).
52. J. A. Hudis and R. Ditchfield, *Chem. Phys. 86*:455 (1984).
53. R. Ditchfield, *Chem. Phys. Lett. 15*:203 (1972).
54. R. M. Stevens, R. M. Pitzer, and W. N. Lipscomb, *J. Chem. Phys. 38*:550 (1963).
55. J. Gerratt and I. M. Mills, *J. Chem. Phys. 49*:1719 (1968).
56. P. Jorgensen and J. Simons, *J. Chem. Phys. 79*:334 (1983).
57. R. N. Camp, H. F. King, J. W. McIver, Jr., and D. Mullally, *J. Phys. 79*:1088 (1983); H. F. King, R. N. Camp, and J. W. McIver, Jr., *J. Chem. Phys. 80*:1171 (1984).
58. M. Page, P. Saxe, G. F. Adams, and B. H. Lengsfield III, *J. Chem. Phys. 81*:434 (1984).
59. J. Almlöf and P. R. Taylor, *Int. J. Quant. Chem. 27*:743 (1985).
60. J. A. Pople, R. Krishnan, H. B. Schlegel, and J. S. Binkley, *Int. J. Quant. Chem. 225*:S13 (1979).
61. M. Dupuis, *J. Chem. Phys. 74*:5758 (1981).
62. G. Diercksen and R. McWeeny, *J. Chem. Phys. 44*:3554 (1966).
63. J. L. Dodds, R. McWeeny, W. T. Raynes, and J. P. Riley, *Molec. Phys. 33*:611 (1977).
64. T. Takada, M. Dupuis, and H. F. King, *J. Comput. Chem. 4*:234 (1983).
65. Y. Osamura, Y. Yamaguchi, and H. F. Schaefer, *J. Chem. Phys. 75*:2919 (1981); Y. Yamaguchi, M. Frisch, J. Gaw, H. F. Schaefer, and J. S. Binkley, *J. Chem. Phys. 84*:2262 (1986); Y. Osamura, Y. Yamaguchi, and H. F. Schaefer, *Chem. Phys. 103*:227 (1986).
66. H. D. Cohen and C. C. J. Roothaan, *J. Chem. Phys. 43*:34 (1965).
67. W. Müller, J. Flesch, and W. Meyer, *J. Chem. Phys. 80*:3297 (1984).
68. J. E. Gready, G. B. Bacskay, and N. S. Hush, *Chem. Phys. 23*:9 (1977).
69. H. Sekino and R. J. Bartlett, *J. Chem. Phys. 84*:2726 (1986).
70. L. Adamowicz and R. J. Bartlett, *J. Chem. Phys. 83*:4988 (1986); *Phys. Rev. A37*:1 (1988).
71. J. E. Gready, G. B. Bacskay, and N. S. Hush, *Chem. Phys. 24*:333 (1977).
72. I. G. John, G. B. Bacskay, and N. S. Hush, *Chem. Phys. 38*:319 (1979); *51*:49 (1980).
73. R. D. Amos, *Molec. Phys. 39*:1 (1980).
74. M. A. Morrison and P. J. Hay, *J. Chem. Phys. 70*:4034 (1979).

75. R. D. Amos and J. H. Williams, *Chem. Phys. Lett. 66*:471 (1979).
76. A. D. McLean and M. Yoshimine, *J. Chem. Phys. 46*:3682 (1967).
77. R. D. Amos, *Mol. Phys. 38*:33 (1979).
78. D. M. Bishop and C. Pouchan, *J. Chem. Phys. 80*:789 (1984).
79. D. M. Bishop and G. Maroulis, *J. Chem. Phys. 82*:2380 (1985).
80. W. P. Reinhardt, *Int. J. Quant. Chem. 21*:133 (1982).
81. See, for example, E. J. Austin, *Chem. Phys. Lett. 66*:498 (1979); *Molec. Phys. 40*:893 (1980).
82. S.-Y. Liu and C. E. Dykstra, *J. Phys. Chem. 91*:1749 (1987). (*Note*: In this report polarizabilities were given in the $\tilde{\mathbf{P}}$ form [8], which means the signs were opposite for even-order polarizabilities.)
83. R. Larter and D. J. Malik, *Chem. Phys. 112*:301 (1987).
84. D. J. Malik and C. E. Dykstra, *J. Chem. Phys. 83*:6307 (1985).
85. C. E. Dykstra, *J. Chem. Phys. 82*:4120 (1985).
86. S.-Y. Liu and C. E. Dykstra, *Chem. Phys. Lett. 119*:407 (1985).
87. C. E. Dykstra, S.-Y. Liu, and D. J. Malik, *J. Molec. Struc.* (*Theochem*) *135*:357 (1986).
88. S.-Y. Liu, C. E. Dykstra, K. Kolenbrander, and J. M. Lisy, *J. Chem. Phys.,85*:2077 (1986).
89. D. E. Bernholdt, S.-Y. Liu, and C. E. Dykstra, *J. Chem. Phys. 85*:5120 (1986).
90. S.-Y. Liu, C. E. Dykstra, and D. J. Malik, *Chem. Phys. Lett. 130*:403 (1986).
91. H.-J. Werner and W. Meyer, *Mol. Phys. 31*:855 (1976).
92. P. A. Christiansen and E. A. McCullough, *Chem. Phys. Lett. 55*:439 (1978).
93. T. H. Dunning, *J. Chem. Phys. 55*:716 (1971).
94. S. Huzinaga, *J. Chem. Phys. 42*:1293 (1965).
95. J. Cizek, *J. Chem. Phys. 45*:4256 (1966); *Adv. Chem. Phys. 14*:25 (1969).
96. J. A. Pople, R. Krishnan, H. B. Schlegel, and J. S. Binkley, *Int. J. Quant. Chem. 14*:545 (1978).
97. R. J. Bartlett, *Annu. Rev. Phys. Chem. 32*:359 (1981); R. J. Bartlett and G. D. Purvis, *Int. J. Quant. Chem. 14*:561 (1978).
98. R. A. Chiles and C. E. Dykstra, *J. Chem. Phys. 74*:4544 (1981).
99. R. J. Bartlett, J. Paldus, and C. E. Dykstra, in *Advanced Theories and Computational Approaches to the Electronic Structure of Molecules*, C. E. Dykstra (ed.) (Reidel, Dordrecht, Holland, 1984).
100. J. Cizek and J. Paldus, *Int. J. Quant. Chem. 5*:359 (1971).
101. I. Lindgren, *Int. J. Quant. Chem. Symp. 12*:33 (1978).
102. P. G. Jasien and C. E. Dykstra, *Int. J. Quant. Symp. 17*:289 (1983).
103. K. Jankowski and J. Paldus, *Int. J. Quant. Chem. 18*:1243 (1980).
104. R. A. Chiles and C. E. Dykstra, *Chem. Phys. Lett. 80*:69 (1981).
105. S. Saebo and P. Pulay, *Chem. Phys. Lett. 131*:384 (1986).
106. C. E. Dykstra, S.-Y. Liu, M. F. Daskalakis, J. P. Lucia, and M. Takahashi, *Chem. Phys. Lett. 137*:266 (1987).
107. R. J. Bartlett and G. D. Purvis III, *Phys. Rev. A 20*:1313 (1979).
108. G. Karlström, B. O. Roos, and A. J. Sadlej, *Chem. Phys. Lett. 86*:374 (1982).
109. K. D. Sen and P. C. Schmidt, *Int. J. Quant. Chem. 19*:373 (1981).
110. K. D. Sen, *Int. J. Quant. Chem. 26*:1051 (1984).

111. J. W. Cooley, *Math. Comp. 15*:363 (1961).
112. T. DiPaolo, C. Bourdeson, and C. Sandorfy, *Can. J. Chem. 50*:3161 (1972).
113. C. Sandorfy, *Topics Curr. Chem. 120*:41 (1984).
114. M. Brieger, *Chem. Phys. 89*:275 (1984).
115. D. M. Bishop, B. Lam, and S. T. Epstein, *J. Chem. Phys. 88*:337 (1988).
116. C. E. Dykstra and D. J. Malik, *J. Chem. Phys. 87*:2806 (1987).
117. R. N. Sileo and T. A. Cool, *J. Chem. Phys. 65*:117 (1976).
118. P. Botschwina, *Chem. Phys. 81*:73 (1983).
119. S.-Y. Liu and C. E. Dykstra, *J. Phys. Chem. 90*:3097 (1986).
120. S.-Y. Liu and C. E. Dykstra, *Chem. Phys. Lett. 136*:22 (1987).
121. J. D. Poll and L. Wolniewicz, *J. Chem. Phys. 68*:3053 (1978).
122. B. O. Roos and A. J. Sadlej, *Chem. Phys. 94*:43 (1985).
123. M. D. Marshall and J. S. Muenter, *J. Molec. Spec. 85*:322 (1981).
124. W. S. Benedict and L. D. Kaplan, *J. Chem. Phys. 30*:388 (1959).
125. M. Ahlström, B. Jönsson and G. Karlström, *Molec. Phys. 38*:1051 (1979).
126. J. Verhoeven and A. Dymanus, *J. Chem. Phys. 52*:3222 (1970).
127. T. R. Dyke and J. S. Muenter, *J. Chem. Phys. 59*:3125 (1973).
128. S. A. Clough, Y. Beers, G. P. Klein, and L. S. Rothman, *J. Chem. Phys. 59*:2254 (1973).
129. J. S. Muenter and W. Klemperer, *J. Chem. Phys. 52*:6033 (1970).
130. D. Maillard and B. Silvi, *Molec. Phys. 40*:933 (1980).
131. F. H. DeLeeuw and A. Dymanus, *J. Molec. Spec. 48*:427 (1973).
132. J. S. Muenter, *J. Molec. Spec. 55*:490 (1975).
133. J. E. Gready, G. B. Bacskay, and N. S. Hush, *Chem. Phys. 31*:467 (1978).
134. A. D. Buckingham, R. L. Disch, and D. A. Dunmur, *J. Am. Chem. Soc. 90*:3014 (1968).
135. B. N. Battacharya and N. Gordy, *Phys. Rev. 119*:144 (1960).
136. A. D. McLean, M. Yoshimine, *IBM J. Res. Develop. Suppl. 11*:163 (1967).
137. S. L. Hartford, W. C. Allen, C. L. Norris, E. F. Pearson, and W. H. Flygare, *Chem. Phys. Lett. 18*:153 (1973).
138. L. C. Snyder, *J. Chem. Phys. 61*:747 (1974).
139. R. D. Nelson, D. R. Lide, and A. A. Maryott, *Natl. Stand. Ref. Data Ser.* 10 (1967).
140. S.-Y. Liu, M. F. Daskalakis, and C. E. Dykstra, *J. Chem. Phys. 85*:5877 (1986).
141. L. Wharton, L. P. Gold, and W. Klemperer, *J. Chem. Phys. 37*:2149 (1962).

LINESHAPE IN EELS OF SIMPLE DISORDERED SYSTEMS

O. I. GERASIMOV

Katholieke Universiteit Leuven,
Laboratorium voor Molekuulfysika,
Leuven, Belgium

and

Odessa State University,
Department of Theoretical Physics,
Odessa, USSR

CONTENTS

I. INTRODUCTION

Electron energy loss spectroscopy (EELS) has been a valuable technique for exploring electronic excitation in different systems [1]. In EELS an electron beam is passed through a cell containing the target system and the forward-scattered electrons are energy analyzed. The electron beam is produced with an electron monochromator. Those electron that scatter through a small angle are retarded to an energy of and then analyzed with an electron energy analyzer, a hemispherical analyzer being a typical example. The energy loss spectrum is obtained by providing a ramp between the target region and the local ground of the electron energy analyzer. The spectra produced are calibrated against well-resolved valence shell states that are recorded in the spectroscopic literature. For a detailed description of EELS see ref. 2. Double-differential inelastic cross sections have been studied extensively [1, 3], particularly in spectral line wings that correspond to electronic excitations within a single molecule.

The inelastic cross section in this region has usually been explained with impact or quasi-static theories of spectral line broadening, taking into account the electron–electron correlations but without considering the internuclear or intermolecular correlations. Because the wing region of a spectral line is formed primarily by the short-range interactions between an excited molecule with its environment, these theories were adequate for explaining dense systems where the intermolecular correlations are saturated. But in cases where the resonance is narrow, for example, in small-angle scattering in low-temperature systems with small Doppler width, the intermolecular contributions become nonnegligible.

Because EELS has become technologically advanced [1, 2], there is a possibility for studying energy loss near a narrow resonance with millivolt resolution. Therefore the first aim of this chapter is to develop a statistical approach to the description of narrow resonance lineshapes in disordered systems. We develop this treatment around a correlation–expansion method [4].

The direct reconstruction of a disordered system by EELS is an important technique by virtue of its wide applicability. The theory for this problem requires only simple expressions that connect the inelastic cross section with the structural characteristics of the system. The second goal of this chapter will be to develop this theory. Note that the modern source of information about distribution functions is the Fourier transform of the static structure factors for a system.

Another set of questions deals with the properties of single-particle localized excitations in disordered systems and its display in EELS. This problem is directly connected with the question of different mechanisms of energy transmission in disordered systems [5]. It is particularly interesting to investigate the EELS lineshape in a long-living single resonance near the critical point of a phase transition in a pure system or near the point of phase separation in a mixture [6].

It is possible to formulate a general theory of pressure broadening of line profiles, but because the problem is a many-body problem, the formal expressions require some additional approximations before they are useful for calculation. These approximations are determined by the molecular species and the deterministic conditions of the system. In connection with this, consider a monochromatic beam of fast electrons ($E \sim 10^3$ eV) that is inelastically scattered on a fixed angle $\theta \leqslant 1^\circ$ by energy losses (ε) in the narrow $\varepsilon \sim 10^{-2} \div 10^{-1}$ eV neighborhood of distinctive narrow resonance. This corresponds to excitation of a single molecule in one of its first excited states. In a simple disordered system of N symmetrical molecules with volume V and temperature $T \sim 10^2$ K, the scattering process may be described with Born's approximation [7, 8]. In this approximation, the excited electronic level contributes primarily to the sum under the electronic state.

Now, we shall suppose that the target molecules may be considered as being at rest during scattering and that the excited level degenerates. We want to emphasize that the broadening of the spectral resonance is due to the effects of molecular motion and backscattering $\Delta\,[\Delta \sim (mK_b\, TE\theta/M)^{1/2}]$, where m and M are the masses of electron and molecule, respectively, for the parameter values of our problem, $\Delta \lesssim 10^{-3}$ eV. In our problem the bound charge of the target molecules distributes not only in the region of localization of one single molecule but over the entire system volume. Therefore, the expression for inelastic cross section must be averaged over all the molecular configurations with the Gibbs distribution function. The resulting quantum-mechanical expression for the second-differential cross section of the electron scattering, $d^2\delta/d\varepsilon\, d\theta$ [or the cross section of energy losses $I_\theta(\varepsilon)$], with reference to the volume of system V is

$$I_\theta(\varepsilon) \equiv \frac{d^2\delta}{d\varepsilon\, d\theta} = \frac{1}{V}\left[\frac{e}{E\theta^2}\right]^2 \left\langle \sum_{(\alpha)} |\rho^{(N)}_{10,\alpha}(\mathbf{q};\{\mathbf{R}\})|^2 \times \delta[E_{1\alpha}(\{\mathbf{R}\}) - E_0(\{\mathbf{R}\}) - \varepsilon] \right\rangle, \tag{1.1}$$

where $\rho^{(N)}_{10,\alpha}(\mathbf{q};\{\mathbf{R}\})$ is the Fourier transformation of the matrix element from the charge density operator; $\mathbf{q}$ is the impulse of scattering; $\{\mathbf{R}\} = \mathbf{R}_1, \ldots, \mathbf{R}_N$ are the coordinates of the center of mass of N molecules; and $E_{1\alpha}(\{\mathbf{R}\})$ and $E_0(\{\mathbf{R}\})$ are, respectively, the energy of the final and initial states of the system in the adiabatic approximation. The sum from α goes over all the excited states of the system, which gives rise to the distinctive level of the isolated molecule; $\delta(\ \ )$ is the Dirac delta function and the angular brackets denotes a Gibbs distribution average.

Equation (1.1) is the generalization of the one-particle expressions for the cross section of inelastic electron scattering [7] in the case where we consider the disordered system as a single large molecule.

II. STATISTICAL THEORY OF EELS OF THE SIMPLE DISORDERED SYSTEMS

A. Correlational Expansion for the Cross Section of Energy Losses

Following ref. 8, let us expand the total energy loss scattering cross section $I_{\theta,N}(\varepsilon)$ in a correlational series. The terms of this series represent contributions to the cross section by isolated many-particle fluctuational complexes of different ranges. Using the identical representation for $I_{\theta,N}(\varepsilon)$,

$$\begin{aligned} I_{\theta,N}(\varepsilon;\{\mathbf{R}\}) = &\sum_{i=1}^{N} Q_{\theta,1}(\varepsilon;\mathbf{R}_i) + \sum_{i=1}^{N}\sum_{j=i+1}^{N} Q_{\theta,2}(\varepsilon;\mathbf{R}_i,\mathbf{R}_j) \\ &+ \sum_{i=1}^{N}\sum_{j=i+1}^{N}\sum_{k=j+1}^{N} Q_{\theta,3}(\varepsilon;\mathbf{R}_i,\mathbf{R}_j,\mathbf{R}_k) \\ &+ \cdots + Q_{\theta,N}(\varepsilon;\{\mathbf{R}\}), \end{aligned} \tag{2.1}$$

where

$$\begin{aligned} Q_{\theta,s}(\varepsilon;\mathbf{R}_1,\ldots,\mathbf{R}_s) = (-1)^s &\sum_{j=1}^{s} I_{\theta,1}(\varepsilon;\mathbf{R}_j) \\ &+ (-1)^{s-2}\sum_{i=1}^{s}\sum_{j=i+1}^{s} I_{\theta,2}(\varepsilon;\mathbf{R}_i,\mathbf{R}_j) \\ &+ \cdots + I_{\theta,s}(\varepsilon;\mathbf{R}_1,\ldots,\mathbf{R}_s) \end{aligned} \tag{2.2}$$

is the irreducible part of the scattering cross section by the s-particle complex $R_1,\ldots,R_s$; $I_\theta(\varepsilon) = \langle I_{\theta,N}(\varepsilon)\rangle$.

In the limit of a thermodynamic average, Eqn. (2.1) becomes

$$\begin{aligned} I_\theta(\varepsilon) = a'(\theta)\sum_{s=1}^{\infty}\frac{n^s}{s!}\int\cdots\int\Bigg[&\sum_{\alpha=1}^{\nu_s} Z_{s,\alpha}(\mathbf{q};\mathbf{R}_1,\ldots,\mathbf{R}_s)\delta(\varepsilon+\mu_{s,\alpha}(\mathbf{R}_1,\ldots,\mathbf{R}_s)) \\ &- \frac{s!}{(s-1)!}\sum_{\alpha=1}^{\nu_{s-1}} Z_{s-1,\alpha}(\mathbf{q};\mathbf{R}_1,\ldots,\mathbf{R}_{s-1}) \\ &\times\delta(\varepsilon+\mu_{s-1,\alpha}(\mathbf{R}_1,\ldots,\mathbf{R}_{s-1})) \\ &+ \frac{s!}{(s-2)!\,2!}\sum_{\alpha=1}^{\nu_{s-2}} Z_{s-2,\alpha}(\mathbf{q};\mathbf{R}_1,\ldots,\mathbf{R}_{s-2}) \\ &\times\delta(\varepsilon+\mu_{s-2,\alpha}(\mathbf{R}_1,\ldots,\mathbf{R}_{s-2})) \\ &- (-1)^s s\, Z_1(\mathbf{q};\mathbf{R}_s)\Bigg] g_s(\mathbf{R}_1,\ldots,\mathbf{R}_s)\,d\mathbf{R}_1\cdots d\mathbf{R}_s. \end{aligned} \tag{2.3}$$

Here $n = N/V$ is the particle density, $\{g_s(\mathbf{R}_1, \ldots, \mathbf{R}_s)\}$ are s-particle distribution functions [9], and $\mu_{s,\alpha}(\mathbf{R}_1, \ldots, \mathbf{R}_s)$ are the single excitation energy shifts due to intermolecular interactions in s-particle complexes. Also,

$$a'(\theta) = \left[\frac{e}{E\theta^2}\right]^2,$$

$$Z_{s,\alpha}(\mathbf{q}; \mathbf{R}_1, \ldots, \mathbf{R}_s) \equiv |\rho^{(s)}_{10,\alpha}(\mathbf{q}; \mathbf{R}_1, \ldots, \mathbf{R}_s)|^2.$$

Note that the functions $\{I_{\theta,s}(\varepsilon; \mathbf{R}_1, \ldots, \mathbf{R}_s)\}$ [or $\{Q_{\theta,s}(\varepsilon; \mathbf{R}_1, \ldots, \mathbf{R}_s)\}$] are symmetric and

$$I_{\theta,1}(\varepsilon; \mathbf{R}) = Q_{\theta,1}(\varepsilon; \mathbf{R})$$

is independent of R. In Eqn. (2.3), the sum from 1 to v_s covers all states of s-particle complexes.

The expression in square brackets in Eqn. (2.3) describes irreducible s-particle contributions to the inelastic scattering cross section. The parameters are fixed and adopted vide supra for disordered systems. Each term of Eqn. (2.3) is a functional of density and temperature as the functions $\{g_s\}$ are functionals of these parameters.

B. Binary Approximation in the Theory of EELS

In moderate density systems we can terminate the series (2.3) after a finite number of terms. For example, if we take only the first two terms into account, we obtain an expression for the inelastic cross section of two-atom quasi-molecules. The regular part is equal to

$$I_\theta(\varepsilon) = \tfrac{1}{2} a'(\theta)\, n^2 \int g_2(R) \sum_{(\alpha)} Z_{2,\alpha}(\mathbf{q}, \mathbf{R})\, \delta(\varepsilon + \mu_{2,\alpha}(\mathbf{R}))\, d\mathbf{R} \tag{2.4}$$

The quasi-molecule complexes consist of two atoms of the same element, one of which is in an excited state. The electronic states are divided into two groups, even (g) and odd (u), in accordance with the property of wavefunctions. Even states conserve sign under inversion in the plane of symmetry; odd states change sign. In Eqn. (2.4) α may be equal to g or u. Using zero-order perturbation theory and neglecting overlap interactions, the wavefunctions of the ground state $\Psi_0(\{\mathbf{r}\}, \mathbf{R})$ and the excited states $\Psi_{1,\alpha}(\{\mathbf{r}\}, \mathbf{R})$ may be written:

$$\begin{aligned} \Psi_0(\{\mathbf{r}\}, \mathbf{R}) &= \psi_0([\mathbf{r}_1])\psi_0([\mathbf{r}_2 - \mathbf{R}]), \\ \Psi_{1,\alpha}(\{\mathbf{r}\}, \mathbf{R}) &= 2^{-1/2}\left\{\psi_0([\mathbf{r}_1])\psi_1([\mathbf{r}_2 - \mathbf{R}]) \pm \psi_1([\mathbf{r}_1])\,\psi_0([\mathbf{r}_2 - \mathbf{R}])\right\}. \end{aligned} \tag{2.5}$$

Here $\psi_0([\mathbf{r}_1])$ and $\psi_1([\mathbf{r}_2 - \mathbf{R}])$ are wavefunctions of isolated atoms. Also, $\{\mathbf{r}\}$ is the set of electronic variables of the pair of atoms and $[\mathbf{r}_1]$ and $[\mathbf{r}_2 - \mathbf{R}]$ are, respectively, the set of electronic coordinates of the first and second atom; the $\pm$ sign corresponds to $\alpha = g$ and $\alpha = u$.

Taking the Fourier transform of the charge density operator

$$\rho^{(2)}_{10,\alpha}(\mathbf{q}, \mathbf{R}) = \sum_e \int \cdots \int e^{-i\mathbf{q}\mathbf{r}_e} \Psi^*_{1,\alpha}(\{\mathbf{r}\}, \mathbf{R}) \Psi_0(\{\mathbf{r}\}, \mathbf{R}) \{d\mathbf{r}\} \tag{2.6}$$

and expression (2.5), we obtain

$$Z_{2,\alpha}(\mathbf{q}, \mathbf{R}) = 2Z_1(\mathbf{q}) \begin{cases} \cos^2(\frac{1}{2}\mathbf{q}\,\mathbf{R}), & \alpha = g, \\ \sin^2(\frac{1}{2}\mathbf{q}\,\mathbf{R}), & \alpha = u. \end{cases} \tag{2.7}$$

Now we determine $\mu_{2,\alpha}(R)$, which is equal to $E^{(\alpha)}_{10}(R) - E_{00}(R)$, where $E^{(\alpha)}_{10}(R)$ and $E_{00}(R)$ are the ground- and excited-state interatomic interaction energies of the two-particle fluctuational complexes in the adiabatic approximation.

Now we need to determine which excited states are dipole allowed. To this end, we want to find the value for the crossing transition probability. In the dipole approximation the interatomic interaction is described with an operator $\hat{V}$:

$$\hat{V} = \frac{(\mathbf{d}_1 \cdot \mathbf{d}_2) - 3(\mathbf{d}_1 \cdot \mathbf{n})(\mathbf{d}_2 \cdot \mathbf{n})}{R^3}, \tag{2.8}$$

which is the dipole–dipole interaction operator. Here $\mathbf{d}_1$ and $\mathbf{d}_2$ are the dipole moments of both atoms and $\mathbf{n}$ is the unit direction vector between both atomic charge centers. For example, consider a $1S$ ground state and a $1P$ excited state for which the transition is dipole permitted. The matrix elements of the operator $\hat{V}$, which are given by

$$V^{(\alpha)}_{1i}(R) \equiv \langle \Psi_{1i,\alpha} | \hat{V} | \Psi_{1i,\alpha} \rangle = \int \cdots \int \Psi^*_{1i,\alpha}(\{\mathbf{r}_j\}, \mathbf{R}) \hat{V} \Psi_{1i,\alpha}(\{\mathbf{r}_j\}, \mathbf{R}) \{d\mathbf{r}_j\}, \tag{2.9}$$

may be calculated by using the wavefunctions (2.5), namely,

$$V^{(\alpha)}_{1i}(R) = \pi_{i\alpha} \frac{d^2_{1\alpha}}{R^3}, \tag{2.10}$$

where

$$\pi_{1\alpha} = \begin{cases} -2, & \alpha = g, \quad i = z, \\ 1, & \alpha = g, \quad i = x, y, \\ 2, & \alpha = u, \quad i = z, \\ -1, & \alpha = u, \quad i = x, y. \end{cases}$$

Here $i = x, y, z$ are the three perpendicular orientations of angular momentum. If $i = z$, the angular momentum and the axis of the quasi-molecule are collinear. Finally, using Eqs. (2.4) and (2.10), we obtain

$$I_\theta(\varepsilon) = \tfrac{2}{3}\pi a'(\theta)\frac{d_{1\alpha}^2}{\varepsilon^2}Z_1(q)n^2\sum_{s=-2}^{2}\left[1 + \mathrm{sgn}(s)\frac{\sin qR_s}{qR_s}\right]\theta(R_s)\,g_2(R_s), \quad (2.11)$$

where

$$\theta(x) = \begin{cases} 1, & x > 0, \\ 0, & x < 0, \end{cases} \qquad R_s = \left[-s\frac{d_{1\alpha}^2}{\varepsilon}\right]^{1/3}. \quad (2.12)$$

Let us consider a dipole forbidden transition, for example, the $1S$ to $2S$ transition. This transition has a probability that is zero to first order and is dependent only upon the interatomic separation. In this case $\mu_{2,\alpha}(R) \sim -(C_\alpha/R^6)$, where C_α is the atomic constant. We obtain

$$I_\theta(\varepsilon) = \frac{1}{8\pi}a'(\theta)Z_1(q)n^2\left[\sum_{v}\frac{R_v^2[1 + \sin qR_v/qR_v]}{|d\mu_{2,g}(R)/dR|_{R=R_v(\varepsilon)}|}g_2(R_v(\varepsilon))\right.$$
$$\left. + \sum_{v'}\frac{R_{v'}^2[1 - \sin qR_{v'}/qR_{v'}]}{|d\mu_{2,u}(R)/dR|_{R=R_{v'}(\varepsilon)}|}g_2(R_{v'}(\varepsilon))\right]. \quad (2.13)$$

where R_v and $R_{v'}$ are the solutions of the equations

$$\begin{aligned} \mu_{2,g}(R) &= \hbar(\omega - \omega_1) = -\varepsilon, \\ \mu_{2,u}(R) &= \hbar(\omega - \omega_1) = -\varepsilon, \\ \mu_{2,\alpha}(R) &= -C_\alpha/R^6. \end{aligned} \quad (2.14)$$

In general, if one of the terms in parentheses has a divisible root $\tilde{\omega}$, then in the limit $\omega \to \tilde{\omega}$ will be sharply peaked. Corresponding sharp peaks will appear in the electron energy loss spectrum. These singular points are analogous to Van Hove singularities for the spectral density of states. Suppose that for such a frequency interval there exists a solution for only one of the equations in (2.14). Then, for the corresponding interval of the distances we can directly determine the value of $g_2(R(\omega))$ by using Eqs. (2.13) and (2.14). If the splitting of the distinguishing level is negligible,

$$|\mu_{2,g}(R) - \mu_{2,u}(R)| \sim \Delta \langle\langle \tfrac{1}{2}|\mu_{2,g}(R) + \mu_{2,u}(R)|, \quad (2.15)$$

we obtain a simplified form of Eqn. (2.13):

$$I_\theta(\varepsilon) = \frac{1}{4\pi} a'(\theta) n^2 Z_1(\mathbf{q}) R^2 \left| \frac{dR}{d\varepsilon} \right| g_2(R(\varepsilon)),$$
$$\mu_{2,\alpha}(R) = \mu(R(\varepsilon)) = -\varepsilon = -C/R^6. \tag{2.16}$$

Expression (2.16) can be rewritten in the form

$$I_\theta(\varepsilon) = \frac{C^{1/2}}{24\pi} a'(\theta) Z_1(\mathbf{q})\, \varepsilon^{-3/2} n^2 g_2\left[\left[\frac{C}{\varepsilon}\right]^{1/2}\right]. \tag{2.17}$$

In the ideal-gas limit, when $g_2 = 1$, $I_\theta(\varepsilon) \sim \varepsilon^{-3/2}$ [7]. Expressions (2.11)–(2.16) allow $g_2(R)$ to be directly determined for the individual points in simple disordered systems. This supplants the usual procedure that involves taking the inverse Fourier transform of the scattering structure factors. Due to the irregular convergence of the Fourier series, this usual procedure does not readily produce an easy determination of $g_2(R)$ [10].

C. Statistical Model of Localized Excitations in Disordered Systems

Consider a model in which an excitation is localized on a single molecule and the probability of excitation exchange between molecules is negligible. This situation is a good model for a dipole-forbidden excitation, a situation where the resonance interaction between an excited molecule and a ground-state molecule decreases with interatomic distance more rapidly than R^{-3}. We shall suppose that the excited level of a single molecule is degenerated. Then, using zero-order perturbation theory. We obtain

$$Z_{s,\alpha}(\mathbf{q}; \mathbf{R}_1, \ldots, \mathbf{R}_s) \simeq Z_1(\mathbf{q}) \equiv |\rho_{1\alpha}^{(1)}(\mathbf{q})|^2, \qquad \alpha = s,$$
$$\mu_{s,\alpha}(\mathbf{R}_1, \ldots, \mathbf{R}_s) = \sum_{j=1}^{s-1}{}' \mu(|\mathbf{R}_j - \mathbf{R}_\alpha|), \tag{2.18}$$

where $\mathbf{R}_s = 0$; $\mu(R)$ is the exchange of energy of the single excitation in a pair of molecules that shift the one from the other on vector $\mathbf{R}$; and α, in this case, means that the excitation is localized at the molecule with number s. Due to the symmetry of the functions $\{g_s\}$, and the Fourier-reducing properties of the Dirac delta function, we can rewrite expression (2.3) as

$$I_\theta(\varepsilon) = a(\theta) Z_1(\mathbf{q}) \sum_{s=1}^{\infty} \frac{n^s}{(s-1)!} \int_{-\infty}^{\infty} d\tau \exp(-i\varepsilon\tau)$$
$$\times \int \cdots \int \prod_{j=1}^{s-1} [\exp(-i\mu(\mathbf{R}_j)\tau) - 1]$$
$$\times g_s(\mathbf{R}_1, \ldots, \mathbf{R}_{s-1}) d\mathbf{R}_1, \ldots, d\mathbf{R}_{s-1}, \qquad (2.19)$$

$$g_s(\mathbf{R}_1, \ldots, \mathbf{R}_{s-1}) \equiv g_s(\mathbf{R}_1, \ldots, \mathbf{R}_{s-1}, 0), \qquad a(\theta) \equiv \frac{a'(\theta)}{2\pi}.$$

Thus, if $\mu(R)$ is a finite function [i.e., $\mu(R) = 0$ outside a sphere with radius r_0], integration in Eqn. (2.19) goes over the volume of the sphere $V_0 = (4\pi/3)r_0^3$ with its center at the origin of the coordinate system. The series (4) for the finite function $\mu(R)$ is generated in the finite sum (we can only put a finite number of molecules in a sphere with radius $r_0 < \infty$). This means that, starting with some number s_0, all the functions $\{g_s\}$ $(s > s_0)$ become zero for all $R_s \geqslant r_0$. To prove that the correlation series converges when $\mu(R)$ is not finite but only a rapidly decreasing function, it is sufficient to introduce the multiplier $\exp(-\Delta^2\tau^2)$ under the integral sign in Eqn. (2.19). This is necessary for including spectral line broadening due to molecular motion and backscattering.

If all the functions $\{g_s\}(|g_s| < C < \infty)$ are uniformly limited, the convergence of the series (2.19) follows from

$$|I_\theta(\varepsilon)| \leqslant a(\theta)|Z_1(\mathbf{q})| nC \sum_{s=1}^{\infty} \frac{n^{s-1}}{(s-1)!} \int_{-\infty}^{\infty} d\tau \exp(-\Delta^2\tau^2)$$
$$\times \int \cdots \int \prod_{j=1}^{s-1} |\exp(-i\mu(R_j)\tau - 1| \, d\mathbf{R}_1 \cdots d\mathbf{R}_{s-1} = a(\theta)|Z_1(\mathbf{q})| Cn$$
$$\times \int_{-\infty}^{\infty} d\tau \exp\left[-\Delta^2\tau^2 + n \int d\mathbf{R} |\exp(-i\mu(R)\tau) - 1| \right]. \qquad (2.20)$$

Because all of the functions under the integral sign in Eqn. (2.19) are infinitely differentiable with respect to ε, every part of (2.19) decreases more rapidly than any power of ε when $\varepsilon \to \pm\infty$. This is in accordance with the usual properties of Fourier integrals.

For distinguishing the correlational contribution in the inelastic cross section, we shall write the s-particle distribution function in the form of a series of irreducible contributions:

$$\begin{aligned} g_s(\mathbf{R}_1, \ldots, \mathbf{R}_{s-1}) = 1 &+ \sum_{i=1}^{s} \sum_{j=i+1}^{s} h_2(\mathbf{R}_i - \mathbf{R}_j) \\ &+ \sum_{i=1}^{s} \sum_{j=i+1}^{s} \sum_{k=j+1}^{s} h_3(\mathbf{R}_i - \mathbf{R}_j; \mathbf{R}_i - \mathbf{R}_k) \\ &+ \cdots + h_s(\mathbf{R}_1, \ldots, \mathbf{R}_{s-1}), \end{aligned} \tag{2.21}$$

where $\{h_s(\mathbf{R}_1, \ldots, \mathbf{R}_{s-1})\}$ is the set of irreducible correlation functions:

$$\begin{aligned} h_2(\mathbf{R}) &= g_2(\mathbf{R}) - 1, \\ h_3(\mathbf{R}, \mathbf{R}') &= g_3(\mathbf{R}, \mathbf{R}') - g_2(\mathbf{R}) - g_2(\mathbf{R}') - g_2(\mathbf{R} - \mathbf{R}') + 2, \\ &\vdots \\ h_s(\mathbf{R}_1, \mathbf{R}_2, \ldots, \mathbf{R}_{s-1}) &= g_s(\mathbf{R}_1, \ldots, \mathbf{R}_{s-1}) \\ &\quad - g_{s-1}(\mathbf{R}_1, \ldots, \mathbf{R}_{s-2}) - \cdots - (-1)^s(s-1). \end{aligned} \tag{2.22}$$

Let us rewrite expression (2.19) with the help of (2.21) and distinguish noncorrelation contributions and the corrections to them due to two-, three-, . . ., s-particle correlations. We obtain the noncorrelational term $I_\theta^{(0)}(\varepsilon)$, which describes the shift and the broadening of the spectral lines as a function of pressure in a gas approximation if we put all $\{g_s\} = 1$ in Eqn. (2.19). After summing the power series we obtain in this case

$$\begin{aligned} I_\theta^{(0)}(\varepsilon) &= a(\theta) Z_1(\mathbf{q}) n \int_{-\infty}^{\infty} d\tau \exp\left[- i\varepsilon\tau + 4\pi n \int_0^{\infty} (e^{-i\mu(R)\tau} - 1) R^2 \, dR \right] \\ &\equiv a(\theta) Z_1(\mathbf{q}) n M_\theta^{(0)}(\varepsilon). \end{aligned} \tag{2.23}$$

If, for example we used in $\mu(R)$ only the long-range asymptote of this function, that is, $\mu(R) = -C/R^6$ (if we consider dipole-forbidden excitation), we immediately obtain the expression derived by Margenau [11] and others [12], which describes the behavior of the spectral line of a molecule dissolved in an ideal gas:

$$\begin{aligned} I_\theta^{(0,M)}(\varepsilon) &= a(\theta) Z_1(\mathbf{q}) n \lambda \varepsilon^{-3/2} \exp\left[-\frac{\pi\lambda^2}{\varepsilon} \right] = a(\theta) Z_1(\mathbf{q}) n M_\theta^{(0,M)}(\varepsilon), \\ \lambda &\equiv \tfrac{2}{3}\pi (C)^{1/2} n. \end{aligned} \tag{2.24}$$

In general, if we examine only the first two terms in Eqn. (2.21), after summing the corresponding power series that generates, Eqn. (2.19), we obtain

$$I_\theta(\varepsilon) = a(\theta) Z_1(\mathbf{q}) n \left\{ M_\theta^{(0)}(\varepsilon) + n \int \Delta M_\theta^{(0)}(\varepsilon + \mu(\mathbf{R})) h_2(\mathbf{R})\, d\mathbf{R} \right.$$
$$+ \tfrac{1}{2} n^2 \iint \left[\Delta M_\theta^{(0)}(\varepsilon + \mu(\mathbf{R}) + \mu(\mathbf{R}')) - \Delta M_\theta^{(0)}(\varepsilon + \mu(\mathbf{R})) \right.$$
$$\left.\left. - \Delta M_\theta^{(0)}(\varepsilon + \mu(\mathbf{R}')) \right] h_2(|\mathbf{R} - \mathbf{R}'|)\, d\mathbf{R}\, d\mathbf{R}' \right\} \tag{2.25}$$
$$\equiv a(\theta) Z_1(\mathbf{q}) n M_\theta^{(1)}(\varepsilon), \; \Delta M_\theta^{(0)}(\varepsilon + \mu(R)) \equiv M_\theta^{(0)}(\varepsilon + \mu(\mathbf{R})) - M_\theta^{(0)}(\varepsilon).$$

In the next approximation

$$I_\theta(\varepsilon) = a(\theta) Z_1(\mathbf{q}) n \left\{ M_\theta^{(1)}(\varepsilon) + \tfrac{1}{2} n^2 \iint \left[\Delta M_\theta^{(0)}(\varepsilon + \mu(\mathbf{R}) + \mu(\mathbf{R}')) \right.\right.$$
$$\left.\left. - \Delta M_\theta^{(0)}(\varepsilon + \mu(\mathbf{R})) - \Delta M_\theta^{(0)}(\varepsilon + \mu(\mathbf{R}')) \right] h_3(\mathbf{R}, \mathbf{R}')\, d\mathbf{R}\, d\mathbf{R}' \right\}, \tag{2.26}$$

and so on. In Eqs. (2.25) and (2.26), $M_\theta^{(0)}(\varepsilon)$ is formally determined by Eqn. (2.23). Thus, expressions such as (2.25) and (2.26) give the possibility for analytical (or numerical) calculation of corrections to expressions (2.23) and (2.24) in any order of correlation approximation.

D. Integral Connections between EELS and Static Structure Factors

Following Parceval's theorem [13], Eqs. (2.25) and (2.26) may be written as a relation between the second-differential inelastic scattering cross section $I_\theta(\varepsilon)$ and the structure factors of the medium $\{S_i\}$. For example, using (2.25) and (2.26), we obtain

$$I_\theta(\varepsilon) = a(\theta) Z_1(\mathbf{q}) n \left\{ M_\theta^{(0)}(\varepsilon) + \int K_2(\varepsilon, \mathbf{k}) h_2(\mathbf{k})\, d\mathbf{k} \right.$$
$$\left. + \iint K_3(\varepsilon, \mathbf{k}, \mathbf{k}') h_3(\mathbf{k}, \mathbf{k}')\, d\mathbf{k}\, d\mathbf{k}' \right\}, \tag{2.27}$$

where

$$K_2(\varepsilon, \mathbf{k}) = \frac{1}{(2\pi)^3} \int e^{i\mathbf{k}\mathbf{R}} \left[\Delta M_\theta^{(0)}(\varepsilon + \mu(\mathbf{R})) + \tfrac{1}{2} n \int e^{-i\mathbf{k}\mathbf{R}'} (\Delta M_\theta^{(0)}(\varepsilon + \mu(\mathbf{R}) \right.$$
$$\left. + \mu(\mathbf{R}')) - \Delta M_\theta^{(0)}(\varepsilon + \mu(\mathbf{R})) - \Delta M_\theta^{(0)}(\varepsilon + \mu(\mathbf{R}'))) \right] d\mathbf{R}\, d\mathbf{R}',$$
$$\tag{2.28}$$
$$K_3(\varepsilon, \mathbf{k}, \mathbf{k}') = \frac{n^2}{2(2\pi)^3} \iint e^{i\mathbf{k}\mathbf{R} + i\mathbf{k}'\mathbf{R}'} \left[\Delta M_\theta^{(0)}(\varepsilon + \mu(\mathbf{R}) + \mu(\mathbf{R}')) \right.$$
$$\left. - \Delta M_\theta^{(0)}(\varepsilon + \mu(\mathbf{R})) - \Delta M_\theta^{(0)}[\varepsilon + \mu(\mathbf{R}')) \right] d\mathbf{R}\, d\mathbf{R}'.$$

$$S_2(\mathbf{k}) = 1 + nh_2(\mathbf{k}),$$
$$S_3(\mathbf{k},\mathbf{k}') = 1 + n(h_2(\mathbf{k}) + h_2(\mathbf{k}') + h_2(\mathbf{k}-\mathbf{k}')) + n^2 h_3(\mathbf{k},\mathbf{k}'), \tag{2.29}$$

where $h_2(\mathbf{k}), h_3(\mathbf{k},\mathbf{k}'), \ldots$ are the Fourier transforms of $\{h_s\}$ [see Eqn. (2.22)], and $S_2(\mathbf{k}), S_3(\mathbf{k},\mathbf{k}'), \ldots$ the static structure factors of the medium, and so on. The static structure factors are the usual means of expressing the results of elastic X-ray and neutron-scattering experiments. These relations therefore allow us to treat alternate experiments with this same formalism.

E. Momentum Analysis of EELS

With Eqs. (2.3) and (2.9) we may obtain the total inelastic scattering cross section

$$P_0(\theta) = \int_{-\infty}^{\infty} I_\theta(\varepsilon)\, d\varepsilon = a'(\theta) Z_1(\mathbf{q}) n. \tag{2.30}$$

Expression (2.30) follows from the initial approximation regarding excitation localization. According to this approximation in the limit when $\Delta \to 0$, the inelastic cross section coincides with the thermodynamic limit of

$$I_\theta(\varepsilon) = \frac{N}{V} a'(\theta) Z_1(\mathbf{q}) \langle \delta(\varepsilon - \sum_j \mu(R_j)) \rangle. \tag{2.31}$$

If we integrate Eqn. (2.19) term by term using Eqn. (2.31) for the inelastic cross-sectional moments, we obtain

$$P_j(\theta) = \int_{-\infty}^{\infty} \varepsilon^j I_\theta(\varepsilon)\, d\varepsilon, \qquad j = 1, 2, \ldots . \tag{2.32}$$

In the limit $\Delta \to 0$, we obtain

$$P_1(\theta) = 4\pi a'(\theta) Z_1(\mathbf{q}) n^2 \int_{-\infty}^{\infty} R^2 g_2(R)\, \mu(R)\, dR,$$
$$P_2(\theta) = a'(\theta) Z_1(\mathbf{q}) \Big\{ 4\pi n^2 \int_{-\infty}^{\infty} R^2 g_2(R) \mu^2(R)\, dR$$
$$+ n^3 \int\!\!\int \mu(R_1) \mu(\mathbf{R}_2) g_3(\mathbf{R}_1, \mathbf{R}_2)\, d\mathbf{R}_1\, d\mathbf{R}_2 \Big\}, \ldots \tag{2.33}$$

and so on. Let us normalize the moments $P_1, P_2, \ldots, P_0$ and consider the series $\{P_j(\theta)\} = \{P_j(\theta)/P_0(\theta)\}, j = 1, 2, \ldots$. The first member of this series represents the shift of the centrum of the energy-loss line; the dispersion $[P_2(\theta) - P_1^2(\theta)]^{1/2}$ describes the effective halth width of the line, the third normalized moment may be connected with the asymmetry of the line, and so on. Knowing the normalized moment, 1, $P_1(\theta)$, $P_2(\theta)$, $P_3(\theta)$, . . . , we are able to narrow the class of functions that includes the inelastic cross section. Following the theorem by Nevanlinna [14], every nonnegative function $W_n(\varepsilon)$ that has the first $2n$ moments in common with the function $I_\theta(\varepsilon)$ may be represented as

$$W_n(\varepsilon) = \frac{P_0 \Delta_n^2}{\pi} \frac{v(\varepsilon)}{[\hat{Q}_{n+1}(\varepsilon) - P_{2n+1}(\varepsilon)\hat{Q}_n(\varepsilon) + U(\varepsilon)\hat{Q}_n(\varepsilon]^2 + v^2(\varepsilon)\hat{Q}_n^2(\varepsilon)}, \tag{2.34}$$

where

$$\hat{Q}_0 = 1, \qquad \hat{Q}_n(\varepsilon) = \left\| \begin{matrix} 1 & P_1 & \cdots & P_n \\ P_1 & P_2 & \cdots & P_{n+1} \\ \vdots & \vdots & & \vdots \\ P_{n-1} & P_n & \cdots & P_{2n} \\ 1 & \varepsilon & \cdots & \varepsilon^{2n} \end{matrix} \right\|,$$

$$\tag{2.35}$$

$$\Delta_n = \left\| \begin{matrix} 1 & P_1 & \cdots & P_n \\ P_1 & P_2 & \cdots & P_{n+1} \\ \vdots & \vdots & & \vdots \\ P_{n-1} & P_n & \cdots & P_{2n} \end{matrix} \right\|,$$

and $v(\varepsilon)$ and $U(\varepsilon)$ are, respectively, the imaginary and real parts of the boundary value $q(\varepsilon + i0)$ of some function $q(Z) = U(Z) + i\,v(Z)$. As an example,

$$q(Z) = \hat{a} + \int_{-\infty}^{\infty} \left[\frac{1}{t - Z} - \frac{t}{1 + t^2} \right] d\hat{\delta}(t), \tag{2.36}$$

where Im $a \geqslant 0$ and $\hat{\delta}(t)$ is a nondecreasing function that satisfies the next condition:

$$\int_{-\infty}^{\infty} \frac{\hat{\delta}(t)}{1 + t^2}\, dt < \infty.$$

In particular, when $n = 1$, the function $q(Z)$, as Eqn. (2.36), is connected with $I_\theta(\varepsilon)$ by the correlation

$$I_\theta(\varepsilon) = \frac{(P_0 E_2^4/\pi)\, \nu(\varepsilon)}{\{[(E_2^2\varepsilon^2 + (E_2^2 + E_1^2)\varepsilon E_1 - (E_1^2 + E_2^2)^2 + U(\varepsilon)(\varepsilon - E_1)]^2 + \nu^2(\varepsilon)(\varepsilon - E_1)^2\}}, \tag{2.37}$$

$$E_1 = P_1, \qquad E_2 = \sqrt{P_2 - P_1^2}.$$

With the known $2n$ moments, the singularities of $I_\theta(\varepsilon)$, in accordance with Eqn. (2.34), may be examined with a respective function $q(Z)$. Standard relations from the classical problem of moments and formulas from the Kristoffel–Darbu theory of orthogonal polynomials [14] allow us to establish the functions

$$q_n(Z) = \frac{\Delta_{n+1}}{\Delta_{n-1}}\left[P_{2n+1} - \frac{1}{q_{n+1}(Z) + Z}\right]. \tag{2.38}$$

To determine the $2n$ moments by a finite part of a spectrum, it is possible to approximate the inelastic cross section using (2.34). The function $q(Z)$ changes on a constant $i\tilde{h}$ ($\tilde{h} > 0$). This constant and the moments may be determined with coincidence between the experimental curve and the constructed approximation. If the function $\mu(r)$ conserves sign up to distances that are comparable with the molecular diameter, the line of losses will appear only in the Stokes (anti-Stokes) neighborhood of the respective molecular term. In this case $I_\theta(\varepsilon)$ is nonzero only on the negative (positive) half-axis, and the function $q(Z) = i\tilde{h}/\sqrt{-Z}$ ($= i\tilde{h}/\sqrt{Z}$), $\tilde{h} > 0$, is an appropriate choice.

F. EELS of Binary Mixtures

In this section we develop a statistical theory of EELS lineshapes for binary systems. This treatment will be developed around the correlation expansion.

Again consider a model where the excitation is localized on a single molecule and the probability of excitation exchange is negligible. This model is adequate for treating the situation where the excited state of one molecule type and the bound states of both molecular types have equal symmetry. A dipole-forbidden transition in the excited state is in example of this. In this case the resonance interaction between an excited molecule and one in the ground state depends only on the interatomic distance R in both types of two-particle fluctuation clusters in the first approximation. This resonance interaction decreases more rapidly than R^{-3}. For simplicity, we shall suppose again that the excited level of a single molecule is degenerated and that the shift of the energy of excitation in s-particle complexes is described with the help of the pair-additive model [see Eqn. (2.3)]. Following Eqn. (2.1), let us expand $I_\theta(\varepsilon)$ in this case in a correlational series. Each part of this series describes the contribution to the total inelastic cross section due to scattering by isolated many-particle fluctuational complexes. Using the identical representation,

$$
\begin{aligned}
I_{\theta,N}(\varepsilon;\{\mathbf{R}\}) = {} & \sum_{i=1}^{N_1} Q_{\theta,1}^{(1)}(\varepsilon;\mathbf{R}_1) + \left[\sum_{i=1}^{N_1}\sum_{j=i+1}^{N_2} Q_{\theta,2}^{(1,1)}(\varepsilon;\mathbf{R}_i,\mathbf{R}_j)\right. \\
& \left. + \sum_{i=1}^{N_1}\sum_{j=1}^{N_2} Q_{\theta,2}^{(1,2)}(\varepsilon;\mathbf{R}_i,\mathbf{R}_j)\right] + \left[\sum_{i=1}^{N_1}\sum_{j=i+1}^{N_1}\sum_{k=j+1}^{N_1} Q_{\theta,3}^{(1,1,1)}\right. \\
& \times(\varepsilon;\mathbf{R}_i,\mathbf{R}_j,\mathbf{R}_k) + \sum_{i=1}^{N_1}\sum_{j=1}^{N_2}\sum_{k=i+1}^{N_1} Q_{\theta,3}^{(1,2,1)}(\varepsilon;\mathbf{R}_i,\mathbf{R}_j,\mathbf{R}_k) \\
& + \sum_{i=1}^{N_1}\sum_{j=i+1}^{N_1}\sum_{k=1}^{N_2} Q_{\theta,3}^{(1,1,2)}(\varepsilon;\mathbf{R}_i,\mathbf{R}_j,\mathbf{R}_k) \\
& \left. + \sum_{i=1}^{N_1}\sum_{j=1}^{N_2}\sum_{k=j+1}^{N_2} Q_{\theta,3}^{(1,2,2)}(\varepsilon;\mathbf{R}_i,\mathbf{R}_j,\mathbf{R}_k)\right] + \cdots \\
& + \sum_{(\{N\})}\sum_{(P)} \hat{P} Q_{\theta,N}^{(\{(N\})}(\varepsilon;\{\mathbf{R}\}), \qquad (2.39)
\end{aligned}
$$

where

$$
\begin{aligned}
Q_{\theta,1}^{(1)}(\varepsilon;\mathbf{R}_i) &\equiv I_{\theta,1}^{(1)}(\varepsilon;\mathbf{R}_i), \\
Q_{\theta,2}^{(1,1)}(\varepsilon;\mathbf{R}_i,\mathbf{R}_j) &= \sum_\alpha I_{\theta,2,\alpha}^{(1,1)}(\varepsilon;\mathbf{R}_i,\mathbf{R}_j) - I_{\theta,1}^{(1)}(\varepsilon;\mathbf{R}_i) - I_{\theta,1}^{(1)}(\varepsilon;\mathbf{R}_j), \\
Q_{\theta,2}^{(1,2)}(\varepsilon;\mathbf{R}_i,\mathbf{R}_j) &= \sum_\alpha I_{\theta,2,\alpha}^{(1,2)}(\varepsilon;\mathbf{R}_i,\mathbf{R}_j) - I_{\theta,1}^{(1)}(\varepsilon;\mathbf{R}_i), \\
Q_{\theta,3}^{(1,1,1)}(\varepsilon;\mathbf{R}_i,\mathbf{R}_j) &= \sum_\alpha I_{\theta,3,\alpha}^{(1,1,1)}(\varepsilon;\mathbf{R}_i,\mathbf{R}_j,\mathbf{R}_k) - \sum_\alpha I_{\theta,2,\alpha}^{(1,1)}(\varepsilon;\mathbf{R}_i,\mathbf{R}_j) \\
&\quad - \sum_\alpha I_{\theta,2,\alpha}^{(1,1)}(\varepsilon;\mathbf{R}_i,\mathbf{R}_k) + I_{\theta,1}^{(1)}(\varepsilon;\mathbf{R}_i), \qquad (2.40) \\
Q_{\theta,3}^{(1,1,2)}(\varepsilon;\mathbf{R}_i,\mathbf{R}_j,\mathbf{R}_k) &= \sum_\alpha I_{\theta,3,\alpha}^{(1,1,2)}(\varepsilon;\mathbf{R}_i,\mathbf{R}_j,\mathbf{R}_k) - \sum_\alpha I_{\theta,2,\alpha}^{(1,1)}(\varepsilon;\mathbf{R}_i,\mathbf{R}_j) \\
&\quad - \sum_\alpha I_{\theta,2,\alpha}^{(1,2)}(\varepsilon;\mathbf{R}_i,\mathbf{R}_k) + I_{\theta,1}^{(1)}(\varepsilon;\mathbf{R}_i),
\end{aligned}
$$

$$Q_{\theta,3}^{(1,2,1)}(\varepsilon;\mathbf{R}_i,\mathbf{R}_j,\mathbf{R}_k)=\sum_{\alpha} I_{\theta,3,\alpha}^{(1,2,1)}(\varepsilon;\mathbf{R}_i,\mathbf{R}_j,\mathbf{R}_k)-\sum_{\alpha} I_{\theta,2,\alpha}^{(1,2)}(\varepsilon;\mathbf{R}_i,\mathbf{R}_j)$$
$$-I_{\theta,2}^{(1,1)}(\varepsilon;\mathbf{R}_i,\mathbf{R}_k)+I_{\theta,1}^{(1)}(\varepsilon;\mathbf{R}_i),$$
$$Q_{\theta,3}^{(1,2,2)}(\varepsilon;\mathbf{R}_i,\mathbf{R}_j,\mathbf{R}_k)=\sum_{\alpha} I_{\theta,3,\alpha}^{(1,2,2)}(\varepsilon;\mathbf{R}_i,\mathbf{R}_j,\mathbf{R}_k)-\sum_{\alpha} I_{\theta,2,\alpha}^{(1,2)}(\varepsilon;\mathbf{R}_i)$$
$$-\sum_{\alpha} I_{\theta,2,\alpha}^{(1,2)}(\varepsilon;\mathbf{R}_i,\mathbf{R}_k)+I_{\theta,1}^{(1)}(\varepsilon;\mathbf{R}_i),$$

and so on, where the subscripts indicate the molecular type:

$$\begin{aligned}
I_{\theta,1}^{(1)}(\varepsilon)&=a'(\theta)Z_1^{(1)}(\mathbf{q})\,\delta(\varepsilon),\\
I_{\theta,2,\alpha}^{(1,1)}(\varepsilon;\mathbf{R})&=a'(\theta)Z_{2,\alpha}^{(1,1)}(\mathbf{q};\mathbf{R})\,\delta(\varepsilon+\mu_{2,\alpha}^{(1,1)}(\mathbf{R})),\\
I_{\theta,2,\alpha}^{(1,2)}(\varepsilon;\mathbf{R})&=a'(\theta)Z_{2,\alpha}^{(1,2)}(\mathbf{q};\mathbf{R})\,\delta(\varepsilon+\mu_{2,\alpha}^{(1,2)}(\mathbf{R})),\\
I_{\theta,3,\alpha}^{(1,i,j)}(\varepsilon;\mathbf{R},\mathbf{R}')&=a'(\theta)Z_{3,\alpha}^{(1,i,j)}(\mathbf{q};\mathbf{R},\mathbf{R}')\delta(\varepsilon+\mu_{3,\alpha}^{(1,i,j)}(\mathbf{R},\mathbf{R}')),\qquad i,j=1,2,\ldots.
\end{aligned}\tag{2.41}$$

After averaging Eqn. (2.39), in the thermodynamic limit ($N, V\to\infty$, $n=N/V=\text{const}$) and taking into account the molecular symmetry, we obtain

$$\begin{aligned}
I_\theta(\varepsilon)={}&n_1 Q_{\theta,1}^{(1)}(\varepsilon)+\left[\frac{n_1^2}{2!}\int Q_{\theta,2}^{(1,1)}(\varepsilon;\mathbf{R})g_2^{(1,1)}(R)\,d\mathbf{R}\right.\\
&\left.+n_1n_2\int Q_{\theta,2}^{(1,2)}(\varepsilon;\mathbf{R})g_2^{(1,2)}(R)\,d\mathbf{R}\right]\\
&+\left[\frac{n_1^3}{3!}\iint Q_{\theta,3}^{(1,1,1)}(\varepsilon;\mathbf{R},\mathbf{R}')g_3^{(1,1,1)}(\mathbf{R},\mathbf{R}')\,d\mathbf{R}\,d\mathbf{R}'\right.\\
&+\frac{n_1^2n_2}{2!}\iint Q_{\theta,3}^{(1,2,1)}(\varepsilon;\mathbf{R},\mathbf{R}')g_3^{(1,2,1)}(\mathbf{R},\mathbf{R}')\,d\\
&+\frac{n_1^2n_2}{2}\iint Q_{\theta,3}^{(1,1,2)}(\varepsilon;\mathbf{R},\mathbf{R}')g_3^{(1,1,2)}(\mathbf{R},\mathbf{R}')\,d\mathbf{R}\,d\mathbf{R}'\\
&\left.+\frac{n_1n_2^2}{2!}\iint Q_{\theta,3}^{(1,2,2)}(\varepsilon;\mathbf{R},\mathbf{R}')g_3^{(1,2,2)}(\mathbf{R},\mathbf{R}')\,d\mathbf{R}\,d\mathbf{R}'\right]+\ldots,
\end{aligned}\tag{2.42}$$

where n_1 and n_2 are partial densities of both species of particles and

$$\{g_s^{\overbrace{(1,i,j,\ldots)}^{s}}(\mathbf{R}_1,\ldots,\mathbf{R}_{s-1})\}\equiv\{g_s^{\overbrace{(1,i,j,k,\ldots)}^{s}}(\mathbf{R}_1,\ldots,\mathbf{R}_{s-1},\mathbf{O})\}$$

is the full set of s-particle partial distribution functions [9] (the upper indices refer to the molecular type).

Following Section II.C we can rewrite Eqn. (2.42) as

$$I_\theta(\varepsilon) = a(\theta) Z_1^{(1)}(\mathbf{q}) \int_{-\infty}^{\infty} d\tau e^{-i\varepsilon\tau}\Bigg[n_1 + \frac{n_1^2}{2!} 2 \int \alpha^{(1,1)}(\tau, R) g_2^{(1,1)}(R)\, d\mathbf{R}$$

$$+ n_1 n_2 \int \alpha^{(1,2)}(\tau, R) g_2^{(1,2)}(R)\, d\mathbf{R}$$

$$+ \frac{n_1^3}{3!} 3 \iint \alpha^{(1,1)}(\tau, R)\alpha^{(1,1)}(\tau, R') g_3^{(1,1,1)}(\mathbf{R}, \mathbf{R}')\, d\mathbf{R}\, d\mathbf{R}'$$

$$+ \frac{n_1^2 n_2}{2!} \iint \alpha^{(1,1)}(\tau, R)\alpha^{(1,2)}(\tau, R') g_3^{(1,1,2)}(\mathbf{R}, \mathbf{R}'),$$

$$+ \frac{n_1 n_2^2}{2!} \iint \alpha^{(1,2)}(\tau, R)\alpha^{(1,2)}(\tau, R') g_3^{(1,2,2)}(\mathbf{R}, \mathbf{R}')\, d\mathbf{R}\, d\mathbf{R}'$$

$$+ \frac{n_1^4}{4!} 4 \iiint \alpha_1(\tau, R)\alpha_1(\tau, R')\alpha_1(\tau, R'') g_4^{(1,1,1,1)}(\mathbf{R}, \mathbf{R}', \mathbf{R}'')\, d\mathbf{R}\, d\mathbf{R}'\, d\mathbf{R}''$$

$$+ \frac{n_1^3 n_2}{3!} 3 \iiint \alpha_1(\tau, R)\alpha_1(\tau, R')\alpha_2(\tau, R'') g_4^{(1,1,1,2)}(\mathbf{R}, \mathbf{R}', \mathbf{R}'')\, d\mathbf{R}, d\mathbf{R}'\, d\mathbf{R}''$$

$$+ \frac{n_1^2 n_2^2}{[2!]^2} 2 \iiint \alpha_1(\tau, R)\alpha_2(\tau, R')\alpha_2(\tau, R'') g_4^{(1,1,1,2,2)}(\mathbf{R}, \mathbf{R}', \mathbf{R}'')\, d\mathbf{R}\, d\mathbf{R}'\, d\mathbf{R}''$$

$$+ \frac{n_1 n_2^3}{3!} \iiint \alpha_2(\tau, R)\alpha_2(\tau, R')\alpha_2(\tau, R'') g_4^{(1,2,2,2)}(\mathbf{R}, \mathbf{R}', \mathbf{R}'')\, d\mathbf{R}\, d\mathbf{R}'\, d\mathbf{R}''$$

$$+ \cdots \Bigg], \alpha^{(1,j)}(\tau; R) = e^{-i\mu^{(1,j)}(R)\tau} - 1, \quad j = 1, 2. \tag{2.43}$$

Here we are taking into account the symmetry of the functions using the Fourier reducing property of the Dirac delta functions and are calculating the matrix elements using zero-order perturbation theory.

Next we write the partial s-particle distribution function in the form of a series of irreducible contributions:

$$g_s^{\overbrace{(1,i,j,k,\ldots)}^{s}}(\mathbf{R}_1, \ldots, \mathbf{R}_{s-1}) = 1 + \sum_{i=1}^{s} \sum_{j=i+1}^{s} \sum_{i'=1}^{i} h_2^{(1,i')}(|\mathbf{R}_i - \mathbf{R}_j|)$$

$$+ \sum_{i=1}^{s} \sum_{j=i+1}^{s} \sum_{k=j+1}^{s} \sum_{i'=1}^{i} \sum_{j'=1}^{j} h_3^{(1,i',j')}(\mathbf{R}_i, \mathbf{R}_j, \mathbf{R}_i - \mathbf{R}_k)$$

$$+ \cdots + h_s^{\overbrace{(1,i,j,k,\ldots)}^{s}}(\mathbf{R}_1, \ldots, \mathbf{R}_{s-1}), \tag{2.44}$$

where $\{h_s^{\overbrace{(1,i,j,\ldots)}^{s}}\}$ is the full set of irreducible correlational functions:

$$h_2^{(1,j)}(R) = g_2^{(1,j)} - 1,$$
$$h_3^{(1,j,k)}(\mathbf{R},\mathbf{R}') = g_3^{(1,j,k)}(\mathbf{R},\mathbf{R}') - g_2^{(1,j)}(R) - g_2^{(1,k)}(R') - g_2^{(j,k)}(|\mathbf{R}-\mathbf{R}'|) + \cdots. \tag{2.45}$$

Now with Eqs. (2.43) and (2.44) we can distinguish the noncorrelational contributions to $I_\theta(\varepsilon)$ and the corrections due to their inclusion. We obtain the noncorrelational term $I_\theta^{(0)}(\varepsilon)$ that describes the shift and broadening of the EELS spectral line as a function of partial densities n_1 and n_2 in the gas approximation. If we put all

$$\{g_s^{\overbrace{(1,i,j,k,\ldots)}^{s}}\} \simeq 1,$$

and sum up the power series we obtain in this case,

$$I_\theta^{(0)} = a(\theta) Z_1^{(1)}(\mathbf{q}) n_1 \int_{-\infty}^{\infty} d\tau \exp[-i\varepsilon\tau + n_1 \alpha_1(\tau) + n_2 \alpha_2(\tau)], \tag{2.46}$$

where

$$\alpha_j(\tau) = \int (e^{-i\mu^{(1,j)}(R)\tau} - 1)\, d\mathbf{R}, \qquad j = 1, 2.$$

For example, if we consider a dipole-forbidden excitation $\mu^{(1,j)}(R)$ and include only the long-range portion, $\mu^{(1,j)}(R) \sim -C_j/R^6$, we obtain

$$I_\theta^{(0,g)}(\varepsilon) = a(\theta) Z_1^{(1)}(q) n_1 \lambda \varepsilon^{-3/2} \exp\left[-\frac{\pi\lambda^2}{\varepsilon}\right],$$
$$\lambda = \frac{2\pi}{3}(\sqrt{C_1 n_1} + \sqrt{C_2 n_2}). \tag{2.47}$$

which in the limit, when $n_2 \to 0$, tends to $I_\theta^{(0,M)}(\varepsilon)$:

$$I_\theta^{(0,g)}(\varepsilon) \xrightarrow[n_2 \to 0]{} I_\theta^{(0,M)}(\varepsilon),$$

described by Margenau and others (see Section II.C). In general, if we examine only the first two terms in Eqn. (2.44) for the approximation $\{g_s^{\overbrace{(1,1,j,k,\ldots)}^{s}}\}$

with the help of Eqn. (2.43), we obtain

$$I_\theta(\varepsilon) = I_\theta^{(0,g)}(\varepsilon) + \int \Delta I_\theta^{(0,g)}[\varepsilon + \mu^{(1,1)}(R)]\, h_2^{(1,1)}(R)\, d\mathbf{R}$$

$$+ \frac{n_2}{n_1} \int \Delta I_\theta^{(0,g)}\{\varepsilon + \mu^{(1,2)}(R)]\, h_2^{(1,2)}(R)\, d\mathbf{R}$$

$$+ \frac{n_1}{2} \iint \Big\{ \Delta I_\theta^{(0,g)}[\varepsilon + \mu^{(1,1)}(R) + \mu^{(1,1)}(R')]$$

$$- \Delta I_\theta^{(0,g)}[\varepsilon + \mu^{(1,1)}(R)] - \Delta I_\theta^{(0,g)}[\varepsilon + \mu^{(1,1)}(R')] \Big\}$$

$$\times\, h_2^{(1,1)}(|\mathbf{R} - \mathbf{R}'|)\, d\mathbf{R}\, d\mathbf{R}' + n_2 \iint \Big\{ \Delta I_\theta^{(0,g)}\ [\varepsilon + \mu^{(1,1)}(R) + \mu^{(1,2)}(R')]$$

$$- \Delta I_\theta^{(0,g)}[\varepsilon + \mu^{(1,1)}(R)] - \Delta I_\theta^{(0,g)}[\varepsilon + \mu^{(1,2)}(R')] \Big\}$$

$$\times\, h_2^{(1,2)}(|\mathbf{R} - \mathbf{R}'|)\, d\mathbf{R}\, d\mathbf{R}' + \frac{n_2^2}{2n_1} \iint \Big\{ \Delta I_\theta^{(0,g)}[\varepsilon + \mu^{(1,2)}(R) + \mu^{(1,2)}(R')]$$

$$- \Delta I_\theta^{(0,g)}[\varepsilon + \mu^{(1,2)}(R)] - \Delta I_\theta^{(0,g)}[\varepsilon + \mu^{(1,2)}(R')] \Big\} h_2^{(2,2)}(|\mathbf{R} - \mathbf{R}'|)\, d\mathbf{R}\, d\mathbf{R}',$$

$$\Delta I_\theta^{(0,g)}(\varepsilon + \xi) \equiv I_\theta^{(0,g)}(\varepsilon + \xi) - I_\theta^{(0,g)}(\varepsilon). \tag{2.48}$$

In the next approximation for $\{g_s^{\overbrace{(1,i,j,k,\ldots)}^{s}}\}$ we obtain, using Eqn. (2.43), expressions such as (2.48), which enable us to calculate corrections to Eqn. (2.46) to any order of approximation. Now in the limit $n_2 \to 0$, we obtain an expression that describes EELS lineshapes for a homogeneous disordered system. Following Parceval's theorem, expression (2.48) and similar ones may be written in the form of a relation between the second-differential cross section of inelastic electron scattering $I_\theta(\varepsilon)$ and the full set of the static structure factors of the medium $\{S_s^{\overbrace{(1,i,\ldots)}^{s}}\}$. For example, using Eqn. (2.48) and Parceval's theorem, one obtains

$$I_\theta(\varepsilon) = I_\theta^{(0,g)}(\varepsilon) + \sum_{i,j} \int M^{(i,j)}(\mathbf{K}) \Big[S_2^{(i,j)}(\mathbf{K}) - 1 \Big] d\mathbf{K}, \tag{2.49}$$

where

$$
\begin{aligned}
n_1 M^{(1,1)}(\mathbf{K}) &\equiv K_1^{(1,1)}(\mathbf{K}) + \tfrac{1}{2} n_1 K_2^{(1,1,1)}(\mathbf{K}, -\mathbf{K}), \\
\frac{n_1+n_2}{n_1 n_2} M^{(1,2)}(\mathbf{K}) &\equiv \frac{1}{n_1} K_1^{(1,2)}(\mathbf{K}) + K_2^{(1,1,2)}(\mathbf{K}, -\mathbf{K}), \\
M^{(2,2)}(\mathbf{K}) &= \frac{n_2}{n_1} K_2^{(1,2,2)}(\mathbf{K}, -\mathbf{K}), \\
K_1^{(i,j)}(\mathbf{K}) &\equiv \int \Delta I_\theta^{(0,g)}[\varepsilon + \mu^{(i,j)}(R)]\, e^{i\mathbf{KR}}\, d\mathbf{R}, \\
K_2^{(1,i,j)}(\mathbf{K},\mathbf{K}') &\equiv \iint \Big\{ \Delta I_\theta^{(0,g)}[\varepsilon + \mu^{(1,i)}(R) + \mu^{(1,j)}(R')] \\
&\quad - \Delta I_\theta^{(0,g)}[\varepsilon + \mu^{(1,i)}(R)] - \Delta I_\theta^{(0,g)}[\varepsilon + \mu^{(1,j)}(R')] \Big\} \\
&\quad \times e^{i\mathbf{KR} + i\mathbf{K}'\mathbf{R}'}\, d\mathbf{R}\, d\mathbf{R},
\end{aligned}
\tag{2.50}
$$

$$
S_2^{(i,j)}(\mathbf{K}) = 1 + n_{ij} h_2^{(i,j)}(\mathbf{K}), \qquad n_{11} \equiv n_1, \qquad n_{12} \equiv \frac{n_1 n_2}{n_1 + n_2}, \qquad n_{22} = n_2,
$$

and so on. Generalizing Section II.B about the binary approximation in EELS in the case of two-component binary mixtures leads to

$$
\begin{aligned}
I_\theta(\varepsilon) = a'(\theta) \Bigg\{ &\frac{n_1^2}{2} \int \sum_\alpha Z_{2,\alpha}^{(1,1)}(\mathbf{q},\mathbf{R})\, \delta\Big(\varepsilon + \mu_{2,\alpha}^{(1,1)}(\mathbf{R}) \Big) g_2^{(1,1)}(\mathbf{R})\, d\mathbf{R} \\
&+ n_1 n_2 \int Z_2^{(1,2)}(\mathbf{q},\mathbf{R})\, \delta\Big(\varepsilon + \mu_2^{(1,2)}(\mathbf{R}) \Big) g_2^{(1,2)}(\mathbf{R})\, d\mathbf{R} \Bigg\},
\end{aligned}
\tag{2.51}
$$

where $n_1, Z_{2,\alpha}^{(i,j)}(\mathbf{q},\mathbf{R}), \mu_{2,\alpha}^{(i,j)}(\mathbf{R})$, and $g^{(i,j)}(R)$ are partial characteristic parameters and functions determined in Section II.B but in two-component systems (indexes $i,j = 1,2$ indicate the type of component). Integration in Eqn. (2.51) (using the properties of the Dirac delta function) gives

$$
\begin{aligned}
I_\theta(\varepsilon) = a'(\theta) n_1 \Bigg\{ &\frac{1}{2} n_1 \sum_{(\nu,\alpha)} \overline{Z_{2,\alpha}^{(1,1)}(q, R_{\nu_\alpha}(\varepsilon))}^{(\Omega)} \frac{R_{\nu_\alpha}^2(\varepsilon)}{|d\mu_{\alpha,2}^{(1,1)}/dR|}\Bigg|_{R_{\nu_\alpha}(\varepsilon)} g_2^{(1,1)}(R_{\nu_\alpha}(\varepsilon)) \\
&+ n_2 \sum_{\nu'} \overline{Z_2^{(1,2)}(q, R_{\nu'}(\varepsilon))}^{(\Omega)} \frac{R_{\nu'}^2(\varepsilon)}{|d\mu_2^{(1,2)}/dR|}\Bigg|_{R_{\nu'}(\varepsilon)} g_2^{(1,2)}(R_{\nu'}(\varepsilon)) \Bigg\}.
\end{aligned}
\tag{2.52}
$$

Analysis of this expression for the realistic model of the $\mu^{(i,j)}$ and $g_2^{(i,j)}$ will be carried out in Section III.B.

III. EELS LINESHAPES NEAR CRITICAL POINTS

A. EELS Near Phase Transition Critical Points

The correlational series (2.19) allows us to examine localized excitations to varying orders of approximation. In this section we are concerned with single resonances in the energy loss spectrum near the critical point of a phase transition. We develop the formalism for this problem in the context of a model that has a resonance lifetime comparable with the typical relaxation time of the order parameter. Generally, the existence question of these resonances is connected with the problem of variable energy transmission modes in disordered systems [5]. For our treatment we shall realize this relaxation time condition artificially. As an example, we can quench a flashing dipoleforbidden resonance in the neighborhood of the critical point. In this region the correlational radius tends to infinity, which probably leads to stabilization of the localized excitation [6]. For our treatment we use the simplest Ornstein–Zernike correlation function $h_2(R)$:

$$h_2(R) = A\frac{e^{-xR}}{R}, \tag{3.1}$$

where A is a temperature-dependent coefficient with the dimension of a distance and x is the inverse correlational radius.

Taking Eqs. (2.25) and (3.1) with an assumption of moderate density, we may approximate $I_\theta(\varepsilon)$ as

$$\begin{aligned} I_\theta(\varepsilon) = a(\theta)Z_1(q)n\Bigg\{ M_{\theta(\varepsilon)}^{(0,M)} + 4\pi A n\lambda\Bigg[\int_{(C/\varepsilon)^{1/6}}^{\infty} (\varepsilon - C/R^6)^{-3/2} \\ \times \exp\left[-\frac{\pi\lambda^2}{\varepsilon - \dfrac{C}{R^6}} - xR \right] R\,dR - \varepsilon^{-3/2} \\ \times \exp\left[-\frac{\pi\lambda^2}{\varepsilon} \right] \int_{(C/\varepsilon)^{1/6}}^{\infty} \exp(-xR)R\,dR \Bigg]\Bigg\}. \end{aligned} \tag{3.2}$$

This expression includes terms up to order n. In previous work [8], we demonstrated that the contribution in Eqn. (2.25) due to correlational series that are nonsingular will have another order of density. These contributions may be separated. After changing the variable in the first integral and calculating the second integral, we can rewrite Eqn. (3.2) as

$$I_\theta(\varepsilon) = I_\theta^{(0,M)}(\varepsilon)\left(1 + 4\pi An\left\{\frac{1}{6}C^{1/3}\varepsilon^{1/6}\int_0^\infty y^{-4/3}\left[y + \frac{1}{\varepsilon}\right]^{3/6}\right.\right.$$
$$\times \exp\left[-\pi\lambda^2 y - \frac{xC^{1/6}}{\varepsilon^{1/3}}\left(\frac{1}{y} + \varepsilon\right)^{1/6}\right]dy$$
$$\left.\left. - x^2\Gamma\left(2, x\left(\frac{C}{\varepsilon}\right)^{1/6}\right)\right\}\right), \tag{3.3}$$

where $\Gamma(2, x(C/\varepsilon)^{1/6})$ is the incomplete Γ function [13],

$$\Gamma\left[2, x\left(\frac{C}{\varepsilon}\right)^{1/6}\right] = \left[1 + x\left(\frac{C}{\varepsilon}\right)^{1/6}\right]\exp\left[-x\left(\frac{C}{\varepsilon}\right)^{1/6}\right].$$

Now we examine the behavior of $I_\theta(\varepsilon)$ for limiting values of ε, with the assistance of Eqn. (3.3).

We define the integral $\Xi(\varepsilon, n, x)$:

$$\Xi(\varepsilon, n, x) \equiv \int_0^\infty y^{-4/3}\left(y + \frac{1}{\varepsilon}\right)^{5/6}\exp\left[-\pi\lambda^2 y - \frac{xC^{1/6}}{\varepsilon^{1/3}}\left(\frac{1}{y} + \varepsilon\right)^{1/6}\right]dy. \tag{3.4}$$

As $\varepsilon \to 0$, we can approximate the value of Ξ by the method of steepest descent [14]. We obtain

$$\Xi(\varepsilon, n, x)_{\varepsilon \to 0 + \Delta} \simeq \frac{6.3^{3/14}(2\pi)^{5/7}}{7^{1/2}}\lambda^{3/7}(xC^{1/6})^{-5/7}\varepsilon^{-25/42}$$
$$\times \exp\left[-\frac{7}{6}(6\pi C)^{1/7}\cdot\left(\frac{\lambda x^3}{\varepsilon}\right)^{2/7}\right], \tag{3.5}$$

combining Eqns. (3.5) and (3.3) yields

$$I_\theta(\varepsilon)_{\varepsilon \to 0 + \Delta} \simeq I_\theta^{(0,M)}(\varepsilon)\left(1 + 4\pi Anx^{-5/7}\left\{\frac{(2\pi)^{4/14}}{7^{1/2}3^{3/14}}C^{3/14}(\lambda\varepsilon)^{-3/7}\right.\right.$$
$$\times \exp\left[-\frac{7}{6}(6\pi C)^{1/7}\left(\frac{\lambda x^3}{\varepsilon}\right)^{2/7}\right] - x^{-2/7}\left(\frac{C}{\varepsilon}\right)^{1/6} \tag{3.6}$$
$$\left.\left.\times \exp\left[-x\left(\frac{C}{\varepsilon}\right)^{1/6}\right]\right\}\right)_{\varepsilon \to 0} \to I_\theta^{(0,M)}(\varepsilon), \left(\frac{I_\theta(\varepsilon)}{I_\theta^{(0,M)}(\varepsilon)}\right)_{\varepsilon \to 0} \to 1.$$

In the limit, when $\varepsilon \to \infty$, we obtain

$$\begin{aligned} I_\theta(\varepsilon)_{\varepsilon \to \infty} &\simeq I_\theta^{(0,M)}(\varepsilon)\left(1 + 4\pi An\left\{\frac{1}{6}\frac{C^{1/3}}{\lambda}\varepsilon^{1/6} - x^{-2}\left[1 + x\left(\frac{C}{\varepsilon}\right)^{1/6}\right]\right\}\right. \\ &\left.\times \exp\left[-x\left(\frac{C}{\varepsilon}\right)^{1/6}\right]\right)_{\varepsilon \to \infty} \to A\left(\frac{\varepsilon}{C}\right)^{1/6} I_\theta^{(0,M)}, \\ &\left(\frac{I_\theta(\varepsilon)}{I_\theta^{(0,M)}(\varepsilon)}\right)_{\varepsilon \to \infty} \to A\left(\frac{\varepsilon}{C}\right)^{1/6}. \end{aligned} \tag{3.7}$$

If we use Eqns. (3.6) and (3.7) to account for correlational corrections to the ideal-gas expression for $I_\theta(\varepsilon) = I_\theta^{(0,M)}(\varepsilon)$, the energy loss spectrum becomes complicated by the interatomic correlations.

It is particularly interesting to note that the relation of $I_\theta(\varepsilon)$ to $I_\theta^{(0)}(\varepsilon)$ asymptotically increases with increasing ε, as a power law $\sim \varepsilon^{1/6}$ [see Eqn. (3.7)] and exponentially tends to 1 rapidly with decreasing ε in accordance with Eqn. (3.6). Let us look now at the spectral lineshape, particularly in the critical point of the phase transition. For this aim, we shall formally put the inverse correlational radius equal to zero. In this case, the first integral in square brackets in Eqn. (3.2) can be calculated with the help of a change of variables and integration by parts; we obtain as an exact result

$$\begin{aligned} I_\theta(\varepsilon) = I_\theta^{(0,M)}(\varepsilon)&\left(1 + 2\pi An\left\{\left(\frac{C}{\varepsilon}\right)^{1/3} + \Gamma\left(\frac{2}{3}\right)C^{1/3}(\pi\lambda^2)^{-3/4}\varepsilon^{5/12}\right.\right. \\ &\times \exp\left(\frac{\pi\lambda^2}{2\varepsilon}\right)\left[\frac{5}{6}W_{1/12,-1/4}\left(\frac{\pi\lambda^2}{\varepsilon}\right) - \left(\frac{\pi\lambda^2}{\varepsilon}\right)^{1/2}\right. \\ &\left.\left.\left.\times W_{7/12,-3/4}\left(\frac{\pi\lambda^2}{\varepsilon}\right)\right]\right\}\right), \end{aligned} \tag{3.8}$$

where $W_{\nu,\mu}(z)$ are the functions of Whittaker [6] and $\Gamma\left(\frac{2}{3}\right)$ is the Γ function. For the asymptotic value of $I_\theta(\varepsilon)$ in the limit $\varepsilon \to 0$, taking into account the properties of $W_{\nu,\mu}(z)$ [13],

$$W_{\nu,\mu}(z)|_{z \to \infty} \simeq e^{-z/2}z^\nu,$$

we obtain

$$\begin{aligned} I_\theta(\varepsilon)_{\varepsilon \to 0 + \Delta} \simeq I_\theta^{(0,M)}(\varepsilon)&\left(1 + 2\pi An\left(\frac{C}{\varepsilon}\right)^{1/3}\left\{1 + \Gamma\left(\frac{2}{3}\right)\right.\right. \\ &\left.\left.\times\left[\frac{5}{6}\left(\frac{\pi\lambda^2}{\varepsilon}\right)^{-2/3} - \left(\frac{\pi\lambda^2}{\varepsilon}\right)^{1/3}\right]\right\}\right). \end{aligned} \tag{3.9}$$

In the limit $\varepsilon \to \infty$, using Eqn. (3.8) and

$$W_{\nu,\mu}(z)|_{z\to 0} \simeq \frac{\Gamma(-2\mu)}{\Gamma(\frac{1}{2}-\mu-\nu)} z^{\mu+1/2} e^{-z/2}, \tag{3.10}$$

we obtain

$$\underset{\varepsilon\to\infty}{I_\theta(\varepsilon)} \simeq I_\theta^{(0,M)}(\varepsilon)\left[1 + 2\pi An\left(\frac{C}{\varepsilon}\right)^{1/3} + A\left(\frac{\varepsilon}{C}\right)^{1/6}\right]_{\varepsilon\to\infty}$$

$$\to A\left(\frac{\varepsilon}{C}\right)^{1/6} I_\theta^{(0,M)}(\varepsilon), \tag{3.11}$$

This result is in exact accordance with Eqn. (3.7), which was obtained outside the critical point. Thus, from Eqns. (3.7) and (3.11), it follows that the asymptotic behavior of the spectral line ($\varepsilon \to \infty$) is modified due to interatomic correlations but is little influenced by the onset of critical behavior. However, the central part of the line, comparable to the Doppler width, is strongly affected near the critical point [see Eqs. (3.6) and (3.9)]. The relation of the cross section of energy loss $I_\theta(\varepsilon)$ to the ideal-gas approximation of the function $I_\theta^{(0)}(\varepsilon)$ in the critical region changes in the limit $\varepsilon \to 0$ as a power law, instead of exponentially [Eqn. (3.6)], as it does outside the critical region.

Because the parameters ε, n, and c belong (see Section II) to fixed intervals of values, ($\varepsilon \in [10^{-2}, 10^{-1}]$ eV, $n \in [10^{-2}, 10^{-3}]$ Å^{-3}, and $c \in [10, 10^2]$ eV Å^6, in practice, it is necessary to examine the estimating behavior of $I_\theta(\varepsilon)$, taking into account these conditions. With the parameters ε, n, and c already adopted, the dimensionless argument of Whittaker's functions in Eqn. (3.8), $\pi\lambda^2/\varepsilon$, belong to the interval $\pi\lambda^2/\varepsilon \in [10^{-3}, 1]$. Thus, in Eqn. (3.8), outside of resonances with width $\sim 10^{-2}$ eV, we can use series expansion (3.10) for Whittaker's functions $W_{\nu,\mu}(\pi\lambda^2/\varepsilon)$ instead of the general expressions. It means that expression (3.11) in our model approximately describes the behavior of $I_\theta(\varepsilon)$ in the neighborhood of (distinctive resonances $\varepsilon \in [10^{-2}, \infty]$ eV), particularly in the critical region. The accuracy of this expression increases with increasing ε.

Thus, as follows from Eqs. (3.6)–(3.11), in the wide interval of thermodynamic parameters, the inelastic cross section scales in relation to the ideal-gas limit of $I_\theta^{(0)}(\varepsilon)$ near the resonances. The value $\pi\lambda^2/\varepsilon$ plays a role in the scaling argument.

Expressions (3.6)–(3.11) would be useful for investigating the influence of many-particle correlational effects in atomic lineshapes in dense disordered systems. In addition, this theory can be used to describe atomic lineshapes in optical absorption and radiation emission in a similar system

B. Lineshape in Binary Approximation

In this section we establish simple model expressions that approximately describe narrow EELS resonances near a critical point in the binary approximation [8]. In this analysis we terminate the correlational series, taking into account only the pairwise contributions (the accuracy of this approximation is increased from the band of moderate densities). We use a Lennard–Jones potential

$$\mu_{2,\alpha}(R) = \frac{C_{12}^{(\alpha)}}{R^{12}} - \frac{C_{6}^{(\alpha)}}{R^{6}} \tag{3.12}$$

to model the pair-additional shift of the excitation energy. This potential includes both long-range and short-range contributions.

Using Eqs. (2.13), (2.14), and (3.12), we get an expression for $I_\theta(\varepsilon)$ that determines the inelastic cross section of a dipole-forbidden resonance near the point of a phase transition:

$$\begin{aligned} I_\theta(\varepsilon) = \frac{\pi}{3} a'(\theta) n^2 \sum_{\nu_\alpha} \overline{Z_{2,\alpha}(q, R_{\nu_\alpha})}^{(\Omega)} \frac{|R_{\nu_\alpha}(\varepsilon)|^3}{|\varepsilon - C_{12}^{(\alpha)} R_{\nu_\alpha}^{-12}(\varepsilon)|}. \\ \times \{1 + \exp[-x R_{\nu_\alpha}(\varepsilon)]/R_{\nu_\alpha}(\varepsilon)\}, \end{aligned} \tag{3.13}$$

where

$$R_{\nu_\alpha} = \left(\frac{C_6^{(\alpha)}}{2\varepsilon}\right)^{1/6} \left[1 + \bar{\bar{\delta}}\left(1 - \frac{4C_{12}^{(\alpha)}}{[C_6^{(\alpha)}]^2}\varepsilon\right)^{1/2}\right]^{1/6} \tag{3.14}$$

are the solutions of the equations

$$\varepsilon + \mu_{2,\alpha}(R) = 0, \tag{3.15}$$

$\bar{\bar{\delta}} = \pm 1$, and the solid line with index Ω means integration over the orientational variables. On the other hand, expressions (3.13) and (3.14) allow the direct determination of individual points in simple disordered systems, particularly near a critical point. These expressions allow us to determine the charge density distribution in the neighborhood of a localized excitation. Near the points $\varepsilon_{\mathrm{VH}} = [C_6^{(\alpha)}]^2/4C_{12}^{(\alpha)}$ sharp peaks may appear in the EELS spectrum. These singularities, as we stressed in Section II.B are analogous to Van Hove singularities for the corresponding spectral density of states. Let us consider several limits of Eqs. (3.13) and (3.14) when

$\varepsilon \in (\infty, ([C_{12}^{(\alpha)}]^2/4C_{12}^{(\alpha)} \equiv \varepsilon_{VH})$. With the help of Eqs. (3.13)–(3.15), one can obtain

$$I_\theta(\varepsilon) \simeq \frac{\pi}{3} a'(\theta) n^2$$

$$\sum_\alpha \overline{Z_{2,\alpha}\left(q, \left(\frac{C_6^{(\alpha)}}{\varepsilon}\right)^{1/6}\right)}^{(\Omega)} [C_6^{(\alpha)}]^{1/2} \frac{1 + A(\varepsilon/C_6^{(\alpha)})^{1/6} \exp[-x(C_6^{(\alpha)}/\varepsilon)^{1/6}]}{|1 - (C_{12}^{(\alpha)}/[C_6^{(\alpha)}]^2)\varepsilon|}$$

$$+ \sum_\alpha \overline{Z_{2,\alpha}\left(q, \left(\frac{C_{12}^{(\alpha)}}{C_6^{(\alpha)}}\right)^{1/6}\right)}^{(\Omega)} \left(\frac{C_{12}^{(\alpha)}}{C_6^{(\alpha)}}\right)^{1/2} \frac{1 + A(C_6^{(\alpha)}/C_{12}^{(\alpha)})^{1/6} \exp[-x(C_{12}^{(\alpha)}/C_6^{(\alpha)})^{1/6}]}{|\ \varepsilon - ([C_6^{(\alpha)}]^2/C_{12}^{(\alpha)})\ |},$$

$$\varepsilon > 0, \qquad \varepsilon \to 0 + \Delta;$$

$$\sum_\alpha \overline{Z_{2,\alpha}\left(q, \left(\frac{C_6^{(\alpha)}}{2\varepsilon}\right)^{1/6}\right)}^{(\Omega)} \left(\frac{C_6^{(\alpha)}}{2}\right)^{1/2} \frac{1 + A(2\varepsilon/C_6^{(\alpha)})^{1/6} \exp[-x(C_6^{(\alpha)}/2\varepsilon)^{1/6}]}{|\ 1 - (4C_{12}^{(\alpha)}/[C_6^{(\alpha)}]^2 \varepsilon\ |}$$

$$\varepsilon^{-3/2}, \varepsilon > 0, \qquad \varepsilon \to \varepsilon_{VH}; \tag{3.16}$$

$$\sum_\alpha \overline{Z_{2,\alpha}\left(q, \left(\frac{C_{12}^{(\alpha)}}{C_6^{(\alpha)}}\right)^{1/6}\right)}^{(\Omega)} \left(\frac{C_{12}^{(\alpha)}}{C_6^{(\alpha)}}\right)^{1/2} \frac{1 + A(C_6^{(\alpha)}/C_{12}^{(\alpha)})^{1/6} \exp[-x(C_{12}^{(\alpha)}/C_6^{(\alpha)})^{1/6}]}{|\ |\varepsilon| + ([C_6^{(\alpha)}]^2/C_{12}^{(\alpha)})\ |},$$

$$\varepsilon < 0, \qquad \varepsilon \to 0 + \Delta;$$

$$\frac{1}{2} \sum_\alpha \overline{Z_{2,\alpha}\left(q, \left(\frac{C_{12}^{(\alpha)}}{|\varepsilon|}\right)^{1/12}\right)}^{(\Omega)} (C_{12}^{(\alpha)})^{1/4} \left[1 + A\left(\frac{|\varepsilon|}{C_{12}^{(\alpha)}}\right)^{1/12} e^{-x\left(\frac{C_{12}^{(\alpha)}}{|\varepsilon|}\right)^{1/12}}\right]$$

$$\times |\varepsilon|^{-13/12}, \varepsilon < 0, \qquad \varepsilon \to \infty.$$

According to Eqn. (3.16), the spectral line is strongly asymmetric and the rate of change in the Stokes and anti-Stokes neighborhoods are different. Van Hove singularities occur at the points $\varepsilon_{VH} = [C_6^{(\alpha)}]^2/4C_{12}^{(\alpha)}$. Analysis of Eqn. (3.16) shows that energy loss spectra near single resonances behave non-monotonically. We stress that if $C_{12}^{(\alpha)} \to 0$, the anti-Stokes resonance disappears and the spectral intensity changes as $I_\theta^{(0)}(\varepsilon) \sim \varepsilon^{-3/2}$, assuming that we neglect correlational effects. Note that as $\varepsilon \to 0 + \Delta$, $\overline{Z_{2,\alpha}(q, (C_6^{(\alpha)}/\varepsilon)^{1/6})}^{(\Omega)}$ tends to the limiting value $\overline{Z_{2,\alpha}(q, \infty)}^{(\Omega)}$, which may be calculated with zero-order perturbation theory for interparticle interactions.

To simplify the following argument, we shall assume that the difference between even (g) and odd (u) terms is negligible:

$$C_6^{(g)} \sim C_6^{(u)} \equiv C_6, \qquad C_{12}^{(g)} \sim C_{12}^{(u)} \equiv C_{12}.$$

In addition, we introduce the spectral correlation density of states,

$$N(\varepsilon) = I_\theta(\varepsilon)/I_\theta^{(0)}(\varepsilon), \qquad N_{\text{corr}}(\varepsilon) = I_\theta^{\text{corr}}(\varepsilon)/I_\theta^{(0)}(\varepsilon), \qquad N(\varepsilon) = 1 + N_{\text{corr}}(\varepsilon).$$

Here $I_\theta^{(0)}(\varepsilon)$ is the inelastic cross section for the gas approximation without interparticle correlations and with a van der Waals energy shift for a single excitation. Within our framework of approximations we obtain

$$N(\varepsilon) \simeq 1 + \begin{cases} A\left(\dfrac{\varepsilon}{C_6}\right)^{1/6} \exp\left[-x\left(\dfrac{C_6}{\varepsilon}\right)^{1/6}\right] + \left\{1 + A\left(\dfrac{C_6}{C_{12}}\right)^{1/6} \right. \\ \left. \times \exp\left[-x\left(\dfrac{C_{12}}{C_6}\right)^{1/6}\right]\right\}\varepsilon^{3/2}, \ \varepsilon > 0, \qquad \varepsilon \to 0 + \Delta; \\ A\left(\dfrac{2\varepsilon}{C_6}\right)^{1/6} \exp\left[-x\left(\dfrac{C_6}{\varepsilon}\right)^{1/6}\right], \qquad \varepsilon > 0, \qquad \varepsilon \to \varepsilon_{\text{VH}}; \\ A\left(\dfrac{C_6}{C_{12}}\right)^{1/6} \exp\left[-x\left(\dfrac{C_{12}}{C_6}\right)^{1/6}\right], \qquad \varepsilon < 0, \qquad |\varepsilon| \to 0 + \Delta; \\ A\left(\dfrac{|\varepsilon|}{C_{12}}\right)^{1/12} \exp\left[-x\left(\dfrac{C_{12}}{|\varepsilon|}\right)^{1/12}\right], \qquad \varepsilon < 0, \qquad |\varepsilon| \to \infty. \end{cases} \tag{3.17}$$

Note that in Eqn. (3.17) we have carried out the corresponding limit transitions for $I_{(\theta)}^0$.

From (3.17), it follows that in the neighborhood of a critical point ($x = 0$), the correlational contribution scales in relation to the gas limit, near the impact cordinate Δ in the Stokes neighborhood and also in the asymptotic region of the anti-Stokes neighborhood of a single resonance:

$$N(\varepsilon) \approx 1 + \begin{cases} \left\{1 + A\left(\dfrac{C_6}{C_{12}}\right)^{1/6} \exp\left[-x\left(\dfrac{C_{12}}{C_6}\right)^{1/6}\right]\right\}\varepsilon^{3/2}, \\ \varepsilon > 0, \quad \varepsilon \to 0 + \Delta; \\ A\left(\dfrac{|\varepsilon|}{C_{12}}\right)^{1/12}, \qquad \varepsilon < 0, \qquad |\varepsilon| \to \infty. \end{cases} \tag{3.18}$$

In the neighborhood of the Van Hove singularities ($\varepsilon > 0$, $\varepsilon \to \varepsilon_{\text{VH}}$) this scaling behavior disappears. Finally, the anti-Stokes impact core neighborhood is practically constant. The limit transition $x \to 0 (T \to T_c)$

induces rigorous scaling behaviour in all the neighborhoods of resonance, that is,

$$
N(\varepsilon) \simeq 1 + \begin{cases} A\left(\dfrac{\varepsilon}{C_6}\right)^{1/6} + \left\{1 + A\left(\dfrac{C_6}{C_{12}}\right)^{1/6} \exp\left[-x\left(\dfrac{C_{12}}{C_6}\right)^{1/6}\right]\right\} \\ \times \varepsilon^{3/2}, \quad \varepsilon > 0, \quad \varepsilon \to 0 + \Delta; \\ A\left(\dfrac{2\varepsilon}{C_6}\right)^{1/6}, \quad \varepsilon > 0. \quad \varepsilon \to \varepsilon_{VH}; \\ A\left(\dfrac{C_6}{C_{12}}\right)^{1/6}, \quad \varepsilon < 0, \quad |\varepsilon| \to 0; \ . \\ A\left(\dfrac{|\varepsilon|}{C_{12}}\right)^{1/12}, \quad \varepsilon < 0, \quad |\varepsilon| \to \infty . \end{cases} \tag{3.19}
$$

Following Eqn. (3.19) the correlational contribution is practically independent from the critical behavior of the system in the asymptotic anti-Stokes neighborhood of a resonance. This is reasonable as the short-range part of the interparticle interactions dominate this neighborhood. In the anti-Stokes neighborhood, when $x \to 0$ the correlational contributions only renormalize the constant value. Note that the index of the power law, which describes the scaling behavior of the correlational structure of EELS, changes in the Stokes neighborhood ($\varepsilon > 0$, $\varepsilon \to 0 + \Delta$) near the impact core of a single resonance, according to the following rule:

$$
N(\varepsilon) \sim 1 + \begin{cases} \varepsilon^{3/2}, & x \neq 0; \\ A\left(\dfrac{\varepsilon}{C_6}\right)^{1/6}, & x \to 0. \end{cases} \tag{3.20}
$$

The Stokes region of a dipole-forbidden resonance is formed by long-range interparticle interactions. According to our model, an essential change of EELS correlational structure takes place close to the critical point where the correlational radius tends to infinity. This effects corresponds qualitatively to those of section III. A.

C. Behavior of EELS near Point of Phase Separation in Binary Mixtures

It is interesting to investigate the lineshape behavior of a long-living resonance near the phase separation critical point in a binary mixture. In binary mixtures, the differing molecules push out the coordinate sphere of the locally excited molecule. For simplicity we treat use a binary correlational approximation and a pair-additional Lennard–Jones model for the energy shift of a single excitation:

$$\mu_{2,\alpha}^{(1,1)}(R) = \frac{C_{12,\alpha}^{(1,1)}}{R^{12}} - \frac{C_{6,\alpha}^{(1,1)}}{R^6}, \qquad \mu_2^{(1,2)}(R) = \frac{C_{12}^{(1,2)}}{R^{12}} - \frac{C_6^{(1,2)}}{R^6},$$

$$C_{12,\alpha}^{(1,1)}, C_{6,\alpha}^{(1,1)}, C_{12}^{(1,2)}, C_6^{(1,2)} > 0, \tag{3.21}$$

where the superscripts in parentheses refer to the sort of molecules and $C_{12,\alpha}^{(1,1)}$, $C_{6,\alpha}^{(1,1)}$, $C_{12}^{(1,2)}$, $C_6^{(1,2)}$ are constants that can be calculated by quantum-mechanical methods [7].

Again as in Sections III. A and III. B we use the simplest Ornstein–Zernike correlation function

$$h_2^{(i,j)}(R) = g_2^{(i,j)}(R) - 1 = A^{(i,j)} \frac{e^{-x^{(i,j)}R}}{R}, \tag{3.22}$$

which describes order parameter fluctuations near the critical point of phase separation or liquid–vapor phase transitions. Here, $A^{(i,j)}$ is a proportionality constant that varies only slowly with temperature and density and $x^{(i,j)}$ is the inverse correlational radius between molecules of different or equal species and tends to zero for corresponding correlations as a power law near the critical point of phase separation or a gas–liquid phase transition ($i, j = 1, 2$). Now we investigate the inelastic cross section near the phase separation critical point. At this point the radius of correlation between the molecules of species 1 and 2 tends to infinity, but the correlation radius between identical molecules stays finite.

Thus, near the point of the phase separation we obtain

$$\begin{aligned} I_\theta(\varepsilon) = a'(\theta) n_1 \Bigg[\frac{1}{2} n_1 \sum_{v_\alpha} \overline{Z_{2,\alpha}^{(1,1)}(q, R_{v_\alpha}(\varepsilon))}^{(\Omega)} \frac{|R_{v_\alpha}^{(1,1)}(\varepsilon)|^3}{|\varepsilon - C_{12,\alpha}^{(1,1)}[R_{v_\alpha}^{(1,1)}(\varepsilon)]^{-12}|} \\ \times (g_2^{(1,1)}(R_{v_\alpha}^{(1,1)})) + n_2 \sum_{v'} \overline{Z_2^{(1,2)}(q, R_{v'}^{(1,2)}(\varepsilon))}^{(\Omega)} \\ \times \frac{|R_{v'}^{(1,2)}(\varepsilon)|^3}{|\varepsilon - C_{12}^{(1,2)}[R_{v'}^{(1,2)}(\varepsilon)]^{-12}|} \left(1 + A^{(1,2)} \frac{\exp(-x^{(1,2)} R_{v'_\alpha}^{(1,2)})}{R_{v'_\alpha}^{(1,2)}} \right) \Bigg] \end{aligned}$$

[compared with (3.13)]. (3.23)

Expression (3.23) allows us to determine the partial pair distribution functions at a point by measurement of the intensity of the energy losses in a simple two-component system, particularly across the gas–liquid critical lines. Also for a statistically determined system, (3.23) allows us to determine the charge density distribution in the neighborhood of a localized excitation in a binary complex.

Now we consider several limiting forms of Eqn. (3.23):

$$I_\theta(\varepsilon) \simeq a'(\theta) n_1 \begin{cases} \frac{1}{2} n_1 \sum_\alpha \overline{Z_{2,\alpha}^{(1,1)}(q,\infty)}^{(\Omega)} [C_{6,\alpha}^{(1,1)}]^{1/2} g_2^{(1,1)}\left[\left(\frac{C_{6,\alpha}^{(1,1)}}{\varepsilon}\right)^{1/6}\right] \varepsilon^{-3/2} \\ + \frac{1}{2} n_1 \sum_\alpha \overline{Z_{2,\alpha}^{(1,1)}\left(q, \left(\frac{C_{12,\alpha}^{(1,1)}}{C_{6,\alpha}^{(1,1)}}\right)^{1/6}\right)}^{(\Omega)} \frac{(C_{12,\alpha}^{(1,1)})^{3/2}}{(C_{6,\alpha}^{(1,1)})^{5/2}} \\ \times g_2^{(1,1)}\left[\left(\frac{C_{12,\alpha}^{(1,1)}}{C_{6,\alpha}^{(1,1)}}\right)^{1/6}\right] + n_2 \overline{Z_2^{(1,2)}(q,\infty)}^{(\Omega)} (C_6^{(1,2)})^{1/2} \\ \times \left\{1 + A^{(1,2)}\left(\frac{\varepsilon}{C_6^{(1,2)}}\right)^{1/6} \exp\left[-x^{(1,2)}\left(\frac{C_6^{(1,2)}}{\varepsilon}\right)^{1/6}\right]\right\} \varepsilon^{-3/2} \\ + n_2 \overline{Z_2^{(1,2)}\left(q, \left(\frac{C_{12}^{(1,2)}}{C_6^{(1,2)}}\right)^{1/6}\right)}^{(\Omega)} \left(\frac{C_{12}^{(1,2)}}{C_6^{(1,2)}}\right)^{1/2} \\ \times \left\{1 + A^{(1,2)}\left(\frac{C_6^{(1,2)}}{C_{12}^{(1,2)}}\right)^{1/6} \exp\left[-x^{(1,2)}\left(\frac{C_{12}^{(1,2)}}{C_6^{(1,2)}}\right)^{1/6}\right]\right\}; \\ \varepsilon > 0, \quad \varepsilon \to 0 + \Delta; \\ \frac{1}{2} n_1 \sum_\alpha \overline{Z_{2,\alpha}^{(1,1)}\left(q, \left(\frac{C_{6,\alpha}^{(1,1)}}{2\varepsilon}\right)^{1/6}\right)}^{(\Omega)} \left(\frac{C_{6,\alpha}^{(1,1)}}{2}\right)^{1/2} \\ \times \frac{g_2^{(1,1)}[(C_{6,\alpha}^{(1,1)}/2)^{1/6}]}{|1 - (4C_{12,\alpha}^{(1,1)}/[C_{6,\alpha}^{(1,1)}]^2)\varepsilon|} \varepsilon^{-3/2} \\ + n_2 \overline{Z_2^{(1,2)}\left(q, \left(\frac{C_6^{(1,2)}}{2\varepsilon}\right)^{1/6}\right)}^{(\Omega)} \left(\frac{C_6^{(1,2)}}{2}\right)^{1/2} \\ \times \frac{1 + A^{(1,2)}(2\varepsilon/C_6^{(1,2)})^{1/6}]\exp[-x^{(1,2)}(C_6^{(1,2)}/2\varepsilon^{1/6})]}{|1 - (4C_{12}^{(1,2)}/[C_6^{(1,2)}]^2)\varepsilon|} \\ \times \varepsilon^{-3/2}; \varepsilon > 0, \quad \varepsilon \to \{\varepsilon_{VH}\}; \end{cases} \tag{3.24}$$

$$
\begin{cases}
\frac{1}{2} n_1 \sum_{\alpha} \overline{Z_{2,\alpha}^{(1,1)}\left(q, \left(\frac{C_{12,\alpha}^{(1,1)}}{C_{6,\alpha}^{(1,1)}}\right)^{1/6}\right)}^{(\Omega)} \left(\frac{C_{12,\alpha}^{(1,1)}}{C_{6,\alpha}^{(1,1)}}\right)^{3/2} \frac{1}{C_{6,\alpha}^{(1,1)}} \\
g_2^{(1,1)}\left[\left(\frac{C_{12,\alpha}^{(1,1)}}{C_6^{(\alpha)}}\right)\right] + n_2 \overline{Z_{2,}^{(1,2)}\left(q, \left(\frac{C_{12}^{(1,2)}}{C_6^{(1,2)}}\right)^{1/6}\right)}^{(\Omega)} \frac{(C_{12}^{(1,2)})^{3/2}}{(C_6^{(1,2)})^{5/2}} \\
\times \left\{1 + A^{(1,2)} \left(\frac{C_6^{(1,2)}}{C_{12}^{(1,2)}}\right)^{1/6} \exp\left[-x^{(1,2)} \left(\frac{C_{12}^{(1,2)}}{C_6^{(1,2)}}\right)^{1/6}\right]\right\}, \\
\varepsilon < 0, \quad |\varepsilon| \to 0 + \Delta; \\
\frac{1}{2} n_1 \sum_{\alpha} \overline{Z_{2,\alpha}^{(1,1)}\left[q, \left(\frac{C_{12,\alpha}^{(1,1)}}{|\varepsilon|}\right)^{1/12}\right]}^{(\Omega)} (C_{12,\alpha}^{(1,1)})^{1/4} \\
\times g_2^{(1,1)}\left[\left(\frac{C_{12,\alpha}^{(1,1)}}{|\varepsilon|}\right)^{1/12}\right] |\varepsilon|^{-13/12} \\
+ n_2 \overline{Z_2^{(1,2)}\left[q, \left(\frac{C_{12}^{(1,2)}}{|\varepsilon|}\right)^{1/12}\right]}^{(\Omega)} (C_{12}^{(1,2)})^{1/4} \\
\left\{1 + A^{(1,2)} \left(\frac{|\varepsilon|}{C_{12}^{(1,2)}}\right)^{1/12} \exp\left[-x^{(1,2)} \left(\frac{C_{12}^{(1,2)}}{|\varepsilon|}\right)^{1/2}\right]\right\} \\
\times |\varepsilon|^{-13/12}, \ \varepsilon < 0, \quad |\varepsilon| \to \infty.
\end{cases}
$$

Following Eqn. (3.24) the spectral line is similar to that of a one-component system in that it is strongly asymmetric and the rate of change of $I_\theta(\varepsilon)$ in the Stokes and anti-Stokes neighborhoods is different. Analysis of (3.24) shows that EELS near the resonances has a complex behavior and depends strongly on the distance from the critical point, the values of the spectral intervals of energy loss close to the impact core Δ, and concentrations of the molecules of species 1 and 2.

Note that if all $\{C_{12,\alpha}^{(i,j)}\} \to 0$, the anti-Stokes neighborhood disappears and, in neglecting correlational effects ($g_2^{(i,j)} \sim 1$), we obtain $I_\theta^{(0)}(\varepsilon) \sim \varepsilon^{-3/2} +$ const. The numerical value of the constant in this expression [and in Eqn. (3.24)] depends on the relative concentrations of the molecules of both species. Again we introduce the spectral density of states; $N(\varepsilon) = I_\theta(\varepsilon)/I_\theta^{(0)}(\varepsilon)$, $N^{(\mathrm{corr})}(\varepsilon) = I_\theta^{(\mathrm{corr})}(\varepsilon)/I_\theta^{(0)}(\varepsilon)$ [$N(\varepsilon) = 1 + N^{(\mathrm{corr})}(\varepsilon)$]. Here $I_\theta^{(0)}(\varepsilon)$ means the inelastic cross section in the gas limit of the two-component mixture, neglecting the interparticle correlations (within the framework of a model with van der Waals shift of energy of the single excitation). By definition, $N(\varepsilon)[N^{(\mathrm{corr})}(\varepsilon)]$ describes the difference in behavior of $I_\theta(\varepsilon)$ [$I_\theta^{(\mathrm{corr})}(\varepsilon)$] in relation to $I_\theta^{(0)}(\varepsilon)$. For simplicity we suppose that the difference between even (g) and odd (u) terms is negligible (particulalry, it means that $C_{6,g}^{(i,j)} \simeq C_{6,u}^{(i,j)} \equiv C_6^{(i,j)}$, $C_{12,g}^{(i,j)} \simeq C_{12,u}^{(i,j)} \equiv C_{12}^{i,j}$). Using Eqn. (3.24), we obtain

$$N(\varepsilon) = 1 + \begin{cases} a(n_1,n_2,T)\varepsilon\cos\left(\dfrac{\tilde{\beta}^{(1,1)}}{\varepsilon^{1/6}}\right)\exp\left[-\dfrac{\tilde{x}^{(1,1)}}{\varepsilon^{1/6}}\right] + b(n_1,n_2,T)\varepsilon^{1/6} \\ \times\exp\left[-\dfrac{\tilde{x}^{(1,2)}}{\varepsilon^{1/6}}\right] + C(n_1,n_2,T)\varepsilon^{3/2},\ \varepsilon > 0, \qquad \varepsilon \to 0 + \Delta, \\ \left\{n_1\Gamma_1^{(1,1)}(2\varepsilon)h_2^{(1,1)}\left[\left(\dfrac{C_6^{(1,1)}}{2\varepsilon}\right)^{1/6}\right] \Big/ \left|1 - \dfrac{\varepsilon}{\varepsilon_{\mathrm{VH}}^{(1)}}\right|\right. \\ + n_2\Gamma_1^{(1,2)}(2\varepsilon)\left(\dfrac{2\varepsilon}{C_6^{(1,2)}}\right)^{1/6} A^{(1,2)}(n_1,n_2,T) \\ \left.\times\exp\left(-\dfrac{x^{(1,2)}}{(2\varepsilon)^{1/6}}\right)\Big/\left|1-\dfrac{\varepsilon}{\varepsilon_{\mathrm{VH}}^{(2)}}\right|\right\}\Big/\left(n_1\Gamma_1^{(1,1)}(2\varepsilon)\Big/\left|1-\dfrac{\varepsilon}{\varepsilon_{\mathrm{VH}}^{(1)}}\right|\right. \\ \left.+ n_2\Gamma_1^{(1,2)}(2\varepsilon)\Big/\left|1-\dfrac{\varepsilon}{\varepsilon_{\mathrm{VH}}^{(2)}}\right|\right), \\ \varepsilon > 0, \qquad \varepsilon \to \{\varepsilon_{\mathrm{VH}}^{(i)}\}, \qquad i = 1, 2, \\ \left\{n_1\Gamma_2^{(1,1)}h_2^{(1,1)}\left[\left(\dfrac{C_{12}^{(1,1)}}{C_6^{(1,1)}}\right)^{1/6}\right] + n_2\Gamma_2^{(1,2)}\left(\dfrac{C_6^{(1,2)}}{C_{12}^{(1,2)}}\right)^{1/6}\right. \\ \left.\times A^{(1,2)}(n_1,n_2,T)\exp\left[-x^{(1,2)}\left[\left(\dfrac{C_{12}^{(1,2)}}{C_6^{(1,2)}}\right)^{1/6}\right]\right]\right\}\Big/ \\ \left[n_1\Gamma_2^{(1,1)} + n_2\Gamma_2^{(1,2)}\right],\ \varepsilon < 0, \qquad |\varepsilon| \to 0 + \Delta, \\ \left\{n_1\Gamma_3^{(1,1)}(|\varepsilon|)h_2^{(1,1)}\left[\left(\dfrac{C_{12}^{(1,1)}}{|\varepsilon|}\right)^{1/12}\right] + n_2\Gamma_3^{(1,2)}(|\varepsilon|)\left(\dfrac{|\varepsilon|}{C_{12}^{(1,2)}}\right)^{1/2}\right. \\ \left.\times A^{(1,2)}(n_1,n_2,T)\exp\left[-x^{(1,2)}\left(\dfrac{C_{12}^{(1,2)}}{|\varepsilon|}\right)^{1/12}\right]\right\}\Big/ \\ (n_1\Gamma_3^{(1,1)}(|\varepsilon|) + n_2\Gamma_3^{(1,2)}(|\varepsilon|)),\ \varepsilon < 0, \qquad |\varepsilon| \to \infty, \end{cases} \tag{3.25}$$

where

$$a(n_1,n_2,T) = [C_6^{(1,1)}]^{-1/6}\frac{n_1\Gamma_1^{(1,1)}(\Delta)}{n_1\Gamma_1^{(1,1)}(\Delta) + n_2\Gamma_1^{(1,2)}(\Delta)}A^{(1,1)}(n_1,T),$$

$$b(n_1,n_2,T)=[C_6^{(1,2)}]^{-1/6}\frac{n_2\Gamma_1^{(1,2)}(\Delta)}{n_1\Gamma_1^{(1,1)}(\Delta)+n_2\Gamma_1^{(1,2)}(\Delta)}A^{(1,2)}(n_1,n_2,T),$$

$$C(n_1,n_2,T)=\left\{n_1\Gamma_2^{(1,1)}h_2^{(1,1)}\left[\left(\frac{C_{12}^{(1,1)}}{C_6^{(1,1)}}\right)^{1/6}\right]\right.$$

$$+n_2\Gamma^{(1,2)}\left(\frac{C_6^{(1,2)}}{C_{12}^{(1,2)}}\right)^{7/6}C_6^{(1,2)}$$

$$\left.\times A^{(1,2)}(n_1,n_2,T)\exp\left[x^{(1,2)}\left(\frac{C_{12}^{(1,2)}}{C_6^{(1,1)}}\right)^{1/6}\right]\right\}\Big/$$

$$[n_1\Gamma^{(1,1)}(\Delta)+n_2\Gamma^{(1,2)}(\Delta)],$$

$$\tilde{\beta}^{(1,1)}=\beta^{(1,1)}(C_6^{(1,1)})^{1/6},\ \tilde{x}^{(i,j)}=x^{(i,j)}(C_6^{(i,j)})^{1/6},\ i,j=1,2,$$

(3.26)

$$\Gamma_1^{(i,j)}(\Delta)=\tau_{ij}\,\overline{Z_2^{(i,j)}\left(q,\left(\frac{C_6^{(i,j)}}{\Delta}\right)^{1/6}\right)}^{(\Omega)},\qquad \tau_{11}=\tfrac{1}{2},\qquad \tau_{12}=1,$$

$$\Gamma_2^{(i,j)}=\tau_{ij}\,\overline{Z_2^{(i,j)}\left(q,\left(\frac{C_{12}^{(i,j)}}{C_6^{(i,j)}}\right)^{1/6}\right)}^{(\Omega)},$$

$$\{\varepsilon_{VH}^{(i)}\}=\frac{[C_6^{(1,i)}]^2}{4C_{12}^{(1,i)}},\qquad i=1,2,$$

$$\Gamma_3^{(i,j)}(|\varepsilon|)=\tau_{ij}\,\overline{Z_2^{(i,j)}\left(q,\left(\frac{C_{12}^{(i,j)}}{|\varepsilon|}\right)^{1/12}\right)}^{(\Omega)}(C_{12}^{(i,j)})^{1/4}.$$

Note that in Eqn. (3.25) we have carried out the corresponding limits for $I^{(0)}(\varepsilon)$. Also note that in (3.25) we use the next asymptotic form for $h_2^{(1,1)}(Z)$ when $Z\to\infty$ [15]:

$$h_2^{(1,1)}(Z)=\frac{\mathrm{A}^{(1,1)}(n_1,T)}{Z}\exp(-x^{(1,1)}Z)\cos(\beta^{(1,1)}Z+\delta^{(1,1)}),\quad (3.27)$$

where $A^{(1,1)}$, $x^{(1,1)}$, $\beta^{(1,1)}$, and $\delta^{(1,1)}$ are the parameters of the first component.

From Eqs. (3.25)–(3.27)) it follows that in the fixed neighborhood of the critical point of phase separation ($x^{(1,2)} \neq 0$) the scaling behavior of the correlational contribution in relation to the gas limit, takes place near the impact coordinate Δ in the Stokes neighborhood and also in the asymptotic region of the anti-Stokes neighborhood of a single resonance, that is,

$$N(\varepsilon) \simeq 1 + \begin{cases} C(n_1, n_2, T)\varepsilon^{3/2}, \qquad \varepsilon > 0, \qquad \varepsilon \to 0 + \Delta; \\ \dfrac{n_1 \Gamma_3^{(1,1)}(\infty)}{n_1 \Gamma_3^{(1,1)}(\infty) + n_2 \Gamma_3^{(1,2)}(\infty)} h_2^{(1,1)}(0) \\ + \dfrac{n_2 \Gamma_3^{(1,2)}(\infty)}{n_1 \Gamma_3^{(1,1)}(\infty) + n_2 \Gamma_3^{(1,2)}(\infty)} \\ \times A^{(1,2)}(n_1, n_2, T)|\varepsilon|^{1/12}, \\ \varepsilon < 0, \qquad |\varepsilon| \to \infty. \end{cases} \tag{3.28}$$

Note that with Eqn. (3.28) we may obtain $h_2^{(1,1)}(0)$ [or $g_2^{(1,1)}(0)$] in principle and the behavior of $h_2^{(1,1)}(R)$ [$g_2^{(1,1)}(R)$], respectively, for a small distance R. Then in the neighborhood of a Van Hove singularity, both types $\{\varepsilon_{\mathrm{VH}}^{(i)}\}$ of scaling behaviors disappear:

$$N(\varepsilon) \simeq 1 + \begin{cases} h_2^{(1,1)}\left[\left(\dfrac{C_6^{(1,1)}}{2\varepsilon}\right)^{1/6}\right], \qquad \varepsilon \to \varepsilon_{\mathrm{VH}}^{(1)}, \\ \left(\dfrac{2\varepsilon}{C_6^{(1,2)}}\right)^{1/6} A^{(1,2)}(n_1, n_2, T) \exp\left(-\dfrac{x^{(1,2)}}{(2\varepsilon)^{1/6}}\right), \qquad \varepsilon \to \varepsilon_{\mathrm{VH}}^{(2)}. \end{cases} \tag{3.29}$$

Following Eqn. (3.29), we find that the density of states in the Stokes neighborhood behaves differently for the various Van Hove points (in the case of the first type, the behavior is nonmonotonic).

Finally, the anti-Stokes impact core neighborhood is practically constant and depends only on the thermodynamic state of the system. In the limit $x^{(1,2)} \to 0$ the scaling behavior of $N(\varepsilon)$ is close to that of the second type of Van Hove points $N(\varepsilon) \simeq 1 + (2\varepsilon/C_6^{(1,2)})^{1/6}$, $A^{(1,2)}(n_1, n_2, T_s)$. Note that the correlational structure of EELS in the Stokes neighborhood near the impact core of a single resonance has a complex nonmonotonic behavior near the critical point. In the double limit $x^{(1,2)} \to 0$, $\varepsilon \to 0 + \Delta$ changes the index of the power law that describes scaling, in accordance with

$$
N(\varepsilon) \simeq 1 + \begin{cases} a(n_1, n_2, T)\varepsilon \cos \dfrac{\tilde{\beta}^{(1,1)}}{\varepsilon} \exp\left(-\dfrac{\tilde{x}^{(1,1)}}{\varepsilon^{1/6}}\right) + b(n_1, n_2, T)\varepsilon^{1/6} \\ \times \exp\left(-\dfrac{\tilde{x}^{(1,2)}}{\varepsilon^{1/6}}\right) + C(n_1, n_2, T)\varepsilon^{3/2} \to C(n_1, n_2, T)\varepsilon^{3/2}, \\ \varepsilon > 0, \quad \varepsilon \to 0 + \Delta, \quad x^{(1,2)} \neq 0; \\ \\ a(n_1, n_2, T)\varepsilon \cos \dfrac{\tilde{\beta}^{(1,1)}}{\varepsilon} \exp\left(-\dfrac{\tilde{x}^{(1,1)}}{\varepsilon^{1/6}}\right) + b(n_1, n_2, T)\varepsilon^{1/6} \\ + C(n_1, n_2, T)\varepsilon^{3/2} \to b(n_1, n_2, T_c)\varepsilon^{1/6} \\ + C(n_1, n_2, T_c), \varepsilon^{3/2} \to b(n_1, n_2, T_c)\varepsilon^{1/6}, \\ \varepsilon > 0, \quad \varepsilon \to 0 + \Delta, \quad x^{(1,2)} \to 0 \quad \text{[compare with (3.2)].} \end{cases} \tag{3.30}
$$

Thus close to the critical point of phase separation the behavior is first nonmonotonic and then the scaling behavior of $I_\theta(\varepsilon)$ becomes a two-index and then a one-index. Similar expressions can be derived for gas–liquid critical point behavior.

Thus in the Stokes neighborhood of a dipole-forbidden resonance formed by long-range interparticle correlations, the correlational structure of EELS takes changes close to the phase separation critical point. In the asymptotic anti-Stokes neighborhood, the correlational behavior is independent of the critical point behavior.

Similar analysis may be attempted by summing more terms of Eqn. (2.43).

Acknowledgements

The author acknowledges Professors I. Prigogine, G. Nicolis, R. Balescu, J.-P. Boon, and all the members of the seminar under the guidance of I. Prigogine at Université Libre Bruxelles (Belgium) for helpful discussions of the present work. He also expresses gratitude to the Laboratory of Molecular Physics of the Katholieke Universiteit Leuven (Belgium) and Professor Wim Van Dael for his hospitality and to Professor Yu. Kilmontovich, V. Adamian, and I. Fisher for constant attention paid and support of the present research. At last, he acknowledges Mrs. Anita Raets for help in the preparation of the manuscript.

References

1. G. J. Schulz, *Rev. Mod. Phys. 45*:378 (1973). N. F. Lane, *Rev. Mod. Phys. 52*:29 (1980). H. Raether, in *Springer Tracts in Modern Physics*, Springer-Verlag, New York, 1980. K. Sturm, *Adv. Phys. 31*:1 (1982)

2. T. D. Mark and G. H. Dunn (eds.) Electron Impact Ionisation, Springer-Verlag Wien, New York. 1985. A. A. Lucas et al., *Int. J. Quant. Chem. 19*:687 (1986). M. Liehz, *Phys. Rev. B 33*:5682 (1986).
3. V. V. Afrosimov, *Soviet Phys. JETP 55*:821 (1969).
4. R. Kubo, *J. Phys. Soc. Japan 17*:1100 (1962).
5. S. A. Rice and J. Jortner, *J. Chem. Phys. 12*:4250 (1965); *44*:4470 (1966).
6. K. Binder and D. Stauffer, *Adv. Phys. 25*:343 (1976). K. Binder and D. W. Heerman, "Growth of Domains and Scaling in the Late Stage of Phase Separation," in *Scaling Phenomenon in Disordered Systems* R. Pynn and A. Skjeltorp (eds.), Plenum Press, New York, 1985. S. Watanabe, *J. Phys. Soc. Japan. 54*:1665 (1985).
7. L. D. Landau and E. M. Lifshitz, *Quantum Mechanics*, Nauka, Moscow, 1973. L. A. Vainstein, I. I. Sobelman, E. A. Vkov, *Vosbushdenie atomov in ushirenie spectralnych liniy*, Nauka, Moscow, 1979.
8. V. M. Adamian and O. I. Gerasimov, *Ukrainsky Fisichesky J.* (*USSR*), *27*:935 (1982); in *Spectrum of Electron Energy Loss in Disordered Systems* (ITF, Kiev, 1985); *Teoreticheskaia l Matematicheskaia fisika* (USSR), *74*:412 1987. O. I. Gerasimov, and V. M. Adamian, *Phys. Rev. A*, (1988), Preprint KUL-MF-87/03 (1987). O. I. Gerasimov, *Europhys. Lett.*, (1988), Preprint KUL-MF-87/05 (1987).
9. N. N. Bogoliubov, *Selected Works*, Vol. 2, Naukova dymka, Kiev, 1970.
10. H. N. V. Temperly, J. S. Rowlinson, and C. S. Rushbrooke (eds.) *Physics of Simple Liquids, North-Holland, Amsterdam*, 1968. O. I. Gerasimov and V. Lisy, Phys. Lett. *98A*:60 (1983).
11. H. Margenau and W. W. Watson, *Rev. Mod. Phys. 8*:22 (1936).
12. J. Szudy and W. E. Baylis, *J. Quant. Spectr. Radiat. Transfer. 15*:641 (1975). C. Peach. *Adv. Phys. 30*:367 (1981). N. Allard and J. Kielkopf, *Rev. Mod. Phys. 54*:1103 (1982). C. Peach, *J. Phys. B: At. Mol. Phys. 17*:2599 (1984); *20*:1175 (1987).
13. M. Abramowitz and I. A. Stegun (eds.), *Handbook of Mathematical Functions*, Dover, New York, 1968.
14. N. I. Achiezer, *Classical Problem of Moments*, Nauka, Moscow, 1961.
15. N. H. March and M. Parinello, *Collective Effects in Solids and Liquids*, Adam Hilger, Bristol, 1982.

THE MOLECULAR QUASI-SPECIES

MANFRED EIGEN AND JOHN M[C]CASKILL

Max Planck Institut für Biophysikalische Chemie
D-3400 Göttingen, FRG

PETER SCHUSTER

Institut für Theoretische Chemie und Strahlenchemie der Universität Wien
A-1090 Wien Austria

CONTENTS

I. INTRODUCTION

1. Model and Reality

Physical chemists are well aware of the usefulness of models. An understanding of the fundamental properties of matter can hardly be gained from watching reality, requiring instead the posing of if–then questions that can be answered only by models. The nature of pressure or temperature of a gas as a collective property of its individual atomic or molecular constituents became obvious only through the billiard ball models of Clausius, Maxwell, and Boltzmann, despite our later insights that true atoms or molecules have quantized motion.

In trying to understand the physical behavior of *viable* matter, we are facing a quite similar situation today as did our forerunners the physical chemists in the late-nineteenth century, when they started to look into the nature of inanimate matter. What is the kind of physical problem that we are really concerned with? Analyzing life at the molecular level, we realize the fundamental dichotomy of a genotypic legislative and phenotypic executive represented by particular forms of macromolecular organization. Asking how such an order became established, we are referred to Darwin's principle of natural selection. There is general consensus among biologists that this principle not only has guided the evolution of species but must also have been as instrumental in the evolution of the molecular forms of organization, up to a state that might be called "the first living species." As we have learned more about the details of this organization, it has seemed more difficult to provide any answer to the question of the physical nature of life. Indeed, what stands out from our increasingly detailed perspective is the tremendous complexity of molecular organization in even the most primitive viable system. We are facing questions such as: why did that particular organization proving viable come about while myriads of alternative states of the system, structurally just as stable, do not share the property of being alive and is there a guiding principle that narrows down the number of alternatives so as to render the appearance of life on our planet possible with meaningful physical expectation? Another, not less stringent, question then immediately comes up: Why is this molecular organization of life, which historically evolved, so perfect? Why, for instance, are enzymes optimal catalysts? The number of structurally stable alternatives that could have been assembled from the same monomeric material is so large that one could not accommodate it within the spatial and temporal limits of our universe. If, as a consequence, we must conclude that all these alternatives never could be tested, why is it that the particular choice of nature turns out to be optimal in its dynamical performance? Half a century of enzyme kinetics has clearly demonstrated this fact. The performance of enzymes can hardly be improved. Nature's solution for the enzymes represents the optimum of what can be achieved with this type of macromolecular organization.

In this chapter we shall try to give answers to these questions concerning an understanding of the physics basic to life. We shall use defined models in order to derive quantitative results, and we shall discuss the relevance of those models for realistic scenarios that can be simulated in laboratory experiments. We shall *not*, however, present hypotheses regarding the historical origin of life. The historical process of evolution depends not only on physical principles but also on historical boundary conditions. These render the historical process unique. Physics, as Eugene Wigner [1] once stated clearly, cannot deal with such unique events. Physics can deal only with regularities among events. The regularity "life," not the historical reality "life," will be the subject of this Chapter.

2. Darwinian Systems

The logic of life may be condensed into four statements:

1. Life came about through evolution.
2. Evolution is the result of variation and natural selection under conditions far from thermodynamic equilibrium.
3. a) Natural selection is a consequence of self-reproduction under conditions far from thermodynamic equilibrium.
   b) Variation is due to imprecise reproduction or other modifications involved in the processing.
4. Self-reproduction is based on structural complementarity of a particular class of molecules.

Systems adhering to this scheme may be called Darwinian because Charles Darwin was the first to reason along this line. Statements such as the preceding are too general to be of more than heuristic value. Let us therefore apply this logic to a defined model and then analyze its consequences in more depth.

The central issue will be Darwin's principle of natural selection. In its original formulation as "survival of the fittest," it does not provide much new insight and it was promptly misunderstood. We really have to define the terms *survival* and *fittest* independently of this principle in order not to get caught in the tautological loop of "survival of the survivor."

Survival certainly is to be measured in terms of population numbers or, to use the language of the physical chemist, in terms of concentrations. A surviving type has to be present with a nonzero number of copies. Fittest, on the other hand, refers to a value parameter that characterizes the surviving type, for example, the one most efficient in producing offspring. For complex living beings with all their mutual interferences, it may be quite difficult to express it through numbers. For a self-reproductive molecular entity I_i, such as an RNA or DNA molecule possibly in combination with enzymic machinery, a selective value W_i can be precisely defined given that the reaction mechanism has been established. Then the quantitative correlation between W_i and the population number N_i is the essence of the selection problem. Survival of the fittest means that all types but the one with maximum selective value are bound to die out. In this form the selection principle is certainly only a zero-order approximation. In other words: there is no real correlation between population numbers and selective values except for the "fittest," which represents the entire system.

It is obvious that such a ruthless "all-or-none" decision could neither be a consequence of random production nor result from interactions as they are responsible for chemical equilibrium, which always settles on finite concentration ratios. It is indeed the peculiar mechanism of the reproduction process far from equilibrium that accounts for the fact of survival, and this mechanism is even active when the competitors are degenerate in their selective values, that is, if they are neutral competitors. In this limiting case, considered to be very important for the evolution of species, Darwin's principle indeed reduces to the mere tautology: survival of the survivor. Nevertheless, there are, even here, systematic quantitative regularities in the way that macroscopic populations of wild types rise and fall in a deterministic manner (as far as the process, not the particular copy choice, is concerned), which make it anything but a *trivial* correlation. This case of neutral selection has been called non-Darwinian. It should be emphasized, however, that Darwin was well aware of this possibility and described it verbally in a quite adequate way. The precise formulation of a theory of neutral selection, which then allows us to draw quantitative conclusions on the evolution of species is an achievement of the second half of this century. Kimura [2] has pioneered this new branch of population genetics.

We have called this all-or-none type of selection a *zero-order approximation*. As such, and wherever the boundary conditions allow for selection rather than for coexistence, it may hold only for competitors that do not have close kinship relations in their genetic inheritance. In fact, even with strong differences in selective value, evolution would not have proceeded very far if it were based on such a correlation for natural selection. For instance, we could not explain why the enzymes are catalysts of optimal efficiency. Evolution would soon have come to a standstill on a minor hill in the value landscape, waiting for an advantageous mutation that would appear much too rarely considering the huge number of possible alternatives. It is common to rely on Darwin's principle as a deterministic tool for selection of the advantageous mutant, while at the same time assuming that this advantageous mutant could come about by nothing but pure chance. One could shock today's biologists by saying that there *must* exist also some guidance, assuring that the advantageous mutants will appear with much higher rates than the disadvantageous ones, and indeed it is considered a heresy to say that evolution can be guided other than by chance.

We shall show that in a Darwinian system evolution indeed is guided to the peaks in the value landscape through biased mutations. To be sure, there is no correlation between the (intrinsically stochastic) act of mutation and the fitness of the product; yet there is bias provided by the fitness-dependent population distribution of mutants—so to speak a mass action guidance of probabilities of mutations. To appreciate this, we have to look at the higher approximations of the selection principle. The target of selection is not a singular wild-type sequence. The fact that sequence analysis as carried out for complete viral genomes yields unique primary structures is no proof that the individual sequence as such is really present to a major extent. It only confirms that at each position the found symbol (nucleotide or amino acid residue) is the prevailing one. A typical RNA virus has 10^4 different one-error copies. If each of them would appear with equal frequency, a population consisting solely of mutants would yield the wild-type sequence at each position with an accuracy of 0.9999 despite the fact that this particular sequence as an individual may not be present at all. It has been shown by Weissmann [3] and co-workers through cloning and rapid amplification of single mutants that wild-type distributions consist predominantly of mutants, the wild type as an individual remaining below the limit of detectability. These mutants—even a diluted test tube fraction usually contains some 10^9–10^{12} such individuals—must be included if we ask *how* are population numbers related to selective values. Each mutant in the wild-type population, though inferior to the wild type, must be assigned a selective value too. Hence mutants are produced not only as error copies of the wild type but also through self-replication, and this will bias the distribution of mutants according to their selective values. Value peaks will be clustered as mountains are in any landscape on earth. Evolutionary optimization then may be viewed as a hill-climbing process that proceeds to the mountain areas, eventually finding the highest peaks. How much this process is biased by the detailed structure of the mountain distribution came out as a true surprise from the rigorous treatment of the selection problem. In this chapter we shall start with the formulation and detailed analysis of a particular deterministic model, the quasi-species [4–7].

II. DETERMINISTIC APPROACH TO SELECTION

In conventional chemical kinetics, time changes of concentrations are described deterministically by differential equations. Strictly, this approach applies to infinite populations only. It is justified, nevertheless, for most chemical systems of finite population size since uncertainties are limited according to some $\sqrt{N}$ law, where N is the number of molecules involved. In a typical experiment in chemical kinetics N is in the range of 10^{18} or larger, and hence fluctuations are hardly detectable. Moreover, ordinary chemical reactions involve but a few molecular species, each of which is present in a very large number of copies. The converse situation is the rule in molecular evolution: the numbers of different polynucleotide sequences that may be interconverted through replication and mutation exceed by far the number of molecules present in any experiment or even the total number of molecules available on earth or in the entire universe. Hence the applicability of conventional kinetics to problems of evolution is a subtle question that has to be considered carefully wherever a deterministic approach is used. We postpone this discussion and study those aspects for which the description by differential equations can be well justified.

1. The Sequence Space

We consider a set of sequences of uniform length comprising ν monomeric subunits of which κ classes exist. If the model is applied to single-stranded RNA or double-stranded DNA sequences—as throughout this chapter—κ assumes the value 2 or 4, depending on whether only purines (R = G, A) and pyrimidines (Y = C, U, or T) or the four individual bases (G, A, C, U, or T, respectively) are specified. The total number of different sequences in such a set amounts to κ^{ν}. To give some examples: a particular tRNA ($\nu=76$) represents one choice out of 10^{46}, a particular ribosomal 5S mRNA ($\nu=120$) one out of 10^{72}, and a particular viral genome such as Qβ ($\nu = 4200$) one out of 10^{2529} alternative sequences of the given length. We specify now two sequences I_i and I_k in such a set. The Hamming distance $d(i, k)$ counts how many positions in these sequences are occupied by different monomers (symbols: R and Y or G, A, C and U/T). For any given reference sequence I_m (which later will usually be the master sequence of the wild type) the Hamming distance $d = d(m, k) \in (1, 2, \ldots, \nu)$ partitions the mutants into classes, each comprising N_d different sequences [cf. eqn. (A1.1) in Appendix 1]. A correct ordering of mutants according to their mutual Hamming distances requires a ν-dimensional space, in which each dimension consists of κ equivalent points (Figure 1). *Equivalent* here means that one can jump to any of the κ points within a given dimension according to a one-error mutation. (It does not necessarily mean that all such jumps occur with equal probabilities.) In nucleic acids, transitions, which are mutations that conserve the base class R or Y, respectively, indeed do occur more frequently than transversions, which are mutations that change those classes. Considering only transversions leads to the simpler limiting case of binary sequences ($\kappa = 2$).

An important feature of this high-dimensional sequence space, as is obvious from Figure 1, is the enormous increase in the number of shortest mutational routes between two given mutant sequences with increasing Hamming distance d. There are $d!$ such connecting 2^d states. Another related effect of high dimensionality is that many states become confined to a close neighborhood. In other words all distances among such an enormous number of states remain small so that a target state can be reached in relatively few steps provided a guiding gradient exists. Moreover, all states are connected through many alternative routes, which may pass through saddle points of higher order, that is, points where the selective value may increase in k and decrease in $\nu - k$ directions. In fact, each point is linked to ν neighboring points, thereby yielding a multiply looped network. The value function, or landscape, of such a space looks quite different from what we are used to with the topographical landscape over the two-dimensional surface of earth. Such a landscape would appear very bizarre, typically showing drastic changes on short distances. This will be of importance if we look at the (selective) value and population landscapes over such a sequence space.

Now consider the sequences to be self-reproductive. Mutants will appear through copying errors. We introduce a fidelity q_i as the probability that a symbol i is copied correctly ($0 < q_i < 1$). Correspondingly, the probability of producing an error at that position is $1 - q_i$. The probability of copying a complete sequence I correctly then is $q_1 \cdot q_2 \cdots q_\nu = \bar{q}^\nu$, where $\bar{q}$ is the geometric mean taken over the fidelities for all positions of the sequence and $\bar{q}^\nu$ defines a quality of sequence copying and is identical with the Q_d value for $d = 0$, as listed in Appendix 1.

In order to characterize the average mutational behavior, we now assume uniform fidelities q for all positions and obtain for the probabilities of

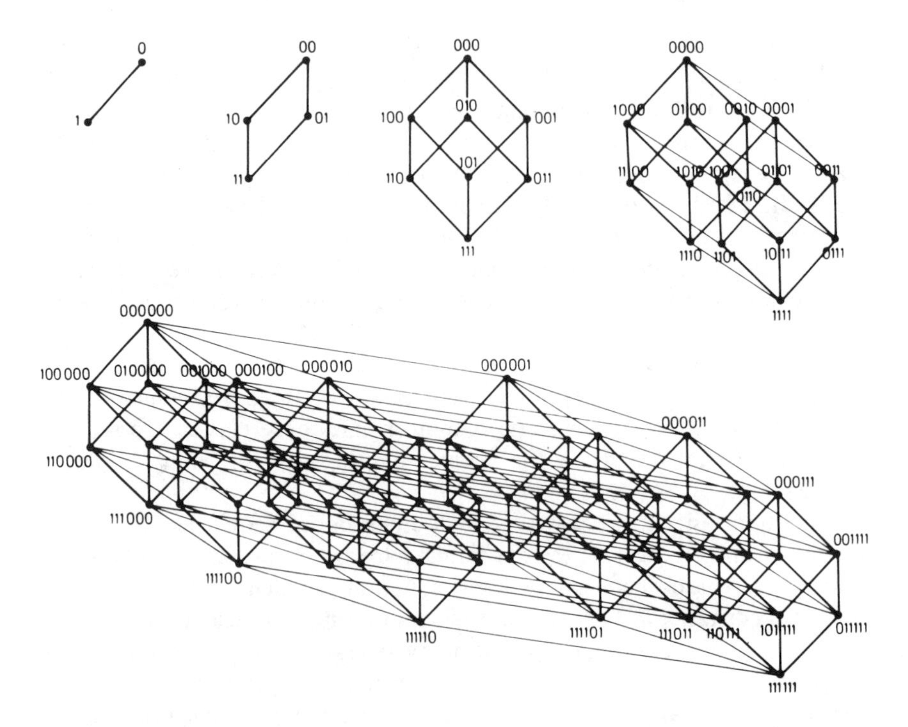

Figure 1. "Lexicographic" ordering of sequences through successive duplications of sequence space. As shown, for binary sequences the sequence space of dimension n can be constructed by duplication of sequence space of dimension $n - 1$. This iterative procedure is used in Appendix 2 to construct mutation matrix in such a way that eigenvalues and eigenvectors can be computed easily [8]. Each of 2^ν points specifies binary (R, Y) sequence. If, in addition, two alternative base classes (R = G or A, Y = C or U) are specified, then to each of points in binary sequence space another subspace of binary specification is added, yielding total of 4^ν points or dimension of hypercube of 2ν.

producing mutants with Hamming distances d the expressions in Appendix 1. (By this procedure we neglect, of course, inhomogeneities such as hot or cold spots.) Note that the probability of producing a particular mutant I_i from a reference copy I_k strongly decreases with increasing Hamming distance $d(i, k)$ [Eqn. (A1.5) in Appendix 1)] since q is usually close to 1.

The mutation frequencies between all pairs (I_i, I_k) of the κ^ν sequences of chain length ν are best given a matrix representation

$$Q = \{Q_{ik};\ i, k = 1, 2, \ldots, \kappa^\nu\}$$

Q being called the *mutation matrix*.

For the uniform error model it is suggestive to order the individual elements according to classes defined by their Hamming distances. Rumschitzki [8], however, has proposed a different procedure, which constructs the mutation matrix for chain length ν from that for $\nu - 1$ recursively (cf. Appendix 2). In this way it is possible to calculate the eigenvalues and eigenvectors of Q for arbitrary chain lengths ν, which in this particular model are all real and positive.

Physically, the elements Q_{ik} are associated with the production rates of the mutants (i.e., producing I_i by miscopying I_k). They do not resemble the frequencies by which these mutants appear in the population. First, these frequencies depend also strongly on the value topology of the mutant space, that is, on the individual efficiencies of reproduction. Second, in order to arrive at population numbers, we have to solve the kinetic equations.

2. The Kinetic Equations

The model is based on the following assumptions [9]:

(i) Sequences as defined in Section II.1 form and decompose steadily. Any individual type I_i of sequence is present with $n_i(t)$ copies per unit volume. This concentration may vary with time t, the rate being $\dot{n}_i(t) \equiv dn_i(t)/dt$.

(ii) Sequences form exclusively through either faithful or erroneous copying of sequences already present. The rates are first order in the concentrations of those sequences that act as templates.

(iii) The substrates of the formation reaction, that is, the energy-rich monomers from which new sequences are to be assembled, are assumed to be present in large excess and hence are buffered. Their constant concentration terms then can be included in the rate coefficients of the formation reactions. (This assumption may be relaxed later on if systems under constant-flow conditions are considered. This assumption turned out not to be decisive for the main behavior of the model.)

(iv) Sequences decompose according to a first-order rate law. The corresponding constant half-lives refer to a constant environment.

Conditions (ii)–(iv) allow the definition of two types of rate terms of a linear system: a diagonal term $W_{ii}n_i(t)$ and a nondiagonal term $W_{ik}n_k(t)$, referring to a value matrix $W = \{W_{ik};\ i, k = 1, 2, \ldots, \kappa^\nu\}$. The diagonal term comprises the effects of faithful replication and of decomposition, both being proportional to $n_i(t)$: $W_{ii} = A_iQ_{ii} - D_i$. Here, A_i describes autocatalytic amplification, that is, replication catalyzed by template I_i, of which only the fraction Q_{ii} (cf. Section III.1) leads to replica that are identical with the template i. The decomposition term D_i counts the average number of decays of template i per unit time. The inverse of D_i is the average lifetime of i (cf. Figure 2).

The nondiagonal terms refer to the fraction of erroneous copying processes for which the conservation relation must hold

$$\sum_l A_l(1 - Q_{ll})n_l(t) = \sum_l \sum_k W_{lk} n_k(t). \tag{II.1}$$

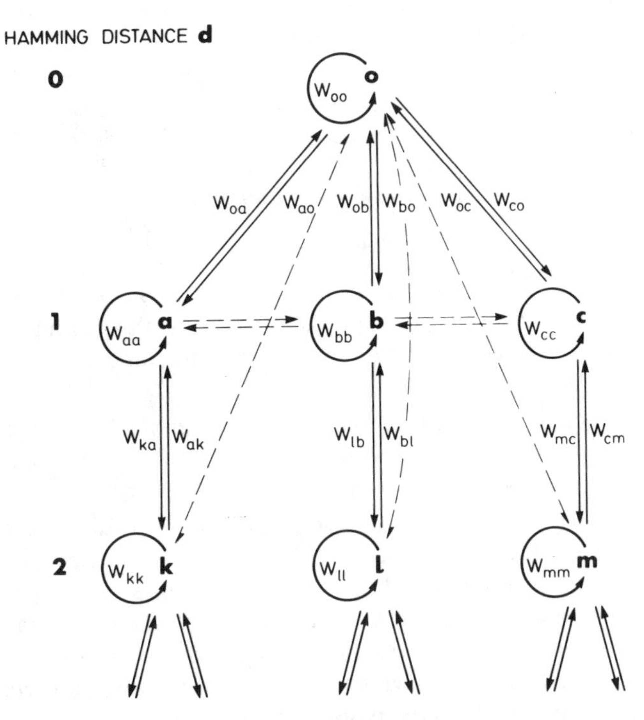

Figure 2. Reaction scheme of quasi-species. Index zero is assigned to the master copy, W_{00} being rate coefficient for correct (excess) production of master copy. Likewise W_{ii} refers to correct (excess) production of mutant copy i. Off-diagonal rate coefficients refer to mutation rates. W_{ij} represents the production of mutant i by miscopying template j. Rate coefficients are always associated with relative concentration terms x_j that refer to second of two subscripts (in agreement with conventional matrix notation). Hamming distances in this scheme are relative to master copy. It is important to note that diagonal coefficients are by an order of magnitude $(k-1)/(q^{-1}-1)$ larger than off-diagonal coefficients that refer to one-error miscopying, and that these in turn are correspondingly larger than coefficients for two-error miscopying, and so on. Error matrix may be ordered recursively in way similar to that demonstrated for buildup of sequence space (Figure 1) such that the antidiagonal represents mutation terms with extreme errors. Following this procedure explicit solutions could be obtained for certain rate coefficient schemes in uniform error model [8].

This conservation is a consequence of assumption (ii), namely, that mutants originate exclusively through erroneous replication and not through external interferences such as radiation or chemical attack. (If this assumption is relaxed, destruction terms must be subtracted in the conservation law to balance the additional first-order off-diagonal mutation terms.) The non-diagonal elements of the value matrix W_{ik} depend strongly on the Hamming distance $d(i, k)$ between template i and erroneous replica k. For the uniform error rate model the expression reads

$$W_{ik} = \left(\frac{q^{-1} - 1}{\kappa - 1} \right)^{d(i,k)} A_k Q_{kk} \equiv \varepsilon^{d(i,k)} A_k Q_{kk}. \tag{II.2}$$

The conservation relation (II.1) allows for a simplification of the summed rate terms, in terms of the total excess production rates E_k, such that the quality factors Q_{ii} no longer appear: as average excess production $\bar{E}(t)$ we define $\sum E_k n_k(t) / \sum n_k(t)$.

In addition to changes due to chemical reactions we must account also for changes caused by transport processes. Since we are not interested in spatially inhomogeneous distributions (cf. the example of a stirred flow reactor), we introduce a general dilution flux term $\Phi = \bar{E}(t)$ that removes material in proportion to the amount produced.

With these assumptions we can write the kinetic ansatz as shown in Appendix 3. Selection represents a kinetic evaluation of sequences relative to one another. It is therefore appropriate to introduce relative population variables

$$x_i(t) = n_i(t) \Big/ \sum_k n_k(t). \tag{II.3}$$

As is seen from Appendix 2, the flux terms, in the form introduced in the preceding, do not appear in the rate equations referring to relative population variables. Instead the average excess production enters as a threshold of selection. This form of rate equations is not limited to a steady state but rather holds also for the relative growth behavior in time-variable systems [4].

3. How Realistic Is the Kinetic Ansatz?

Replication is a multiple-step reaction that usually involves sophisticated enzymic machinery. Is such a process adequately described by a straightforward linear autocatalytic model?

The kinetics of RNA replication by virus replicases have been studied recently in some more detail [10–12]. Rates have been measured under a variety of conditions. These include variable concentrations of substrate, enzyme, and template; asymmetries of plus and minus strand concentrations; annealing processes; and the presence of competitor strands. The mechanisms have been elucidated analytically and reproduced through computer simulation. Although the reaction mechanism (Figure 3) turned out to be quite involved, the phenomenological description is largely in agreement with the linear autocatalytic ansatz used in this chapter. This holds generally for low concentrations of template where the linear dependence of rates on template concentrations applies throughout. At larger template concentrations saturation effects typical for enzymic reactions are observed. The upgrowth of new templates, even if the enzyme is saturated by other templates (selected in previous steps), is exponential until the new template has become dominant, that is, until it is selected. Neither the fact that the replication process involves many consecutive steps, of which one or a few may or may not be rate limiting, nor the cross-catalytic nature of the plus–minus strand instruction

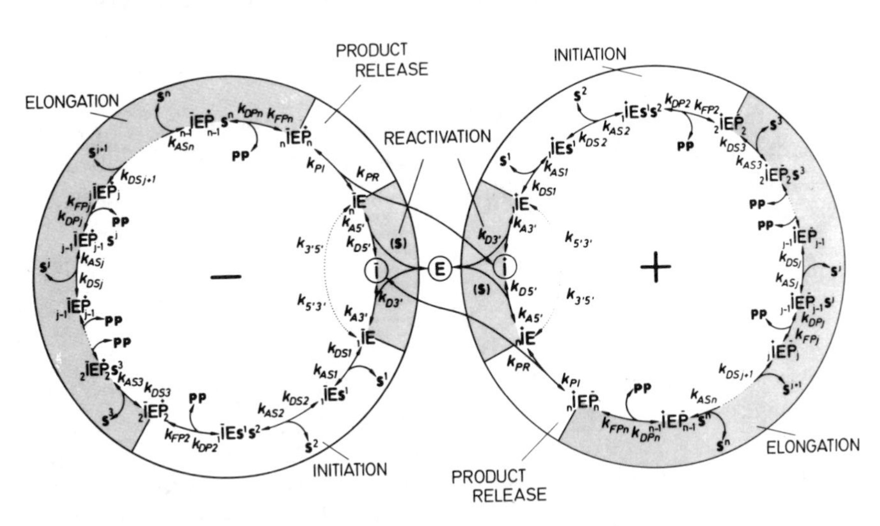

Figure 3. Reaction scheme of complementary replication of single-stranded RNA. Reaction consists of four phases: initiation, elongation, product release, and template reactivation. Reaction product (replica) is complementary to template. Substrates are four nucleoside triphosphates: ATP, GTP, UTP, and CTP. Pyrophosphate (pp) is waste product at each step of incorporation. Symbols: *I*, RNA template chain; E, enzyme (replicase); P, growing RNA replica chain. Indexes: A, association; D, dissociaton; S, substrate; F, phosphate diester bond formation; PR, product release; the numbers 3′, or 5′, refer to end of the RNA chain to which the enzyme binds or from which it dissociates (cf. ref. 10).

alters the overall dependence of the growth rate on template concentration from the model used here. All it does is to superimpose a spectrum of relaxation processes during an induction period of the reaction. During such an induction period the concentrations of the various intermediate states and also their sums for each of the plus and the minus strands assume fixed ratios. All states then grow concomitantly according to the linear autocatalytic rate law. This was verified with natural and artificial RNA sequences. It holds as well for complete virus particles.

DNA replication, although a process that usually depends on much more sophisticated enzymic machinery [13], phenomenologically adheres to the same type of model. The various steps involved contribute to the magnitude of the overall rate, yielding under suitable reaction conditions a defined doubling time of the double-stranded molecule. Since cell division is linked with this replication mechanism, the linear autocatalytic mechanism even holds for autonomous organisms and thus found its way into population biology.

There are, however, reproduction mechanisms that cannot be represented by the linear model. We have mentioned the case of saturation, which in many ways can be materialized, for example, through limited supply, inhibitory interference, or blocking of machinery. The loss of autocatalytic amplification is accompanied by a (partial or total) loss of competition and selection, allowing different sequences to coexist. On the other hand, there are reproductive processes the rates of which depend stronger than linearly on the growing concentration. A virus, for instance, encodes important parts of its reproduction machinery, which is exclusively adapted to the virus template. Hence in the early phase of infection the buildup of the virus population depends on both the virus template and the virus replicase, which is encoded in the template. The growth rate thus depends stronger than linearly on the virus template concentration. We have classified a whole category of such hypercyclic reaction networks that under certain conditions may exhibit hyperbolic rather than exponential growth. The effect is a once-and-for-all type of selection of that compound which managed first to populate the reaction space [4].

Apart from the linear autocatalytic ansatz, the rate equation involves some further assumptions the relevance of which should be discussed. First, the neglect of production terms other than those due to template-instructed reproduction seems straightforward. Whenever template instruction can become effective, it soon will outgrow any other type of production as far as the formation of specific sequences is concerned. Second, the assumption of a buffered level of substrates may seem unnatural. We have studied the effect of exhaustion of substrates by the upgrowth of a more efficiently reproducing mutant in a medium for which the influx of substrate is kept constant. The more efficient consumption of substrate causes a lowering of its concentration level. Selection under such conditions of constant flows remains qualitatively the same but requires quantitative modifications of the equations involved [14]. The same is true for the assumptions made with respect to fluxes. The "proportional" dilution is most easily realized under experimental conditions. Under this condition the flux terms do not influence at all the relative population number and hence need not be specified further.

4. Solutions of the Rate Equations

In section II.2 we arrived at the following form of the rate equations in relative (i.e., fractional) population variables:

$$\dot{x}_i(t) = [W_{ii} - \bar{E}(t)]x_i(t) + \sum_{k \neq i} W_{ik} x_k(t). \tag{II.4}$$

Due to the fact that the term $\bar{E}(t)$ is inherently nonlinear, these equations have been solved by approximations corresponding to second-order perturbation theory [9]. The solutions reflect the threshold nature of selection and the consequences on length limitations for sequences that can be selected and resist an accumulation of errors (i.e., an error catastrophy).

Exact solutions can be derived for these equations [15, 16]. We follow here essentially the method of Jones et al. [16].

The nonlinear term $\bar{E}(t)$ can be removed through the following substitution:

$$x_i(t) = z_i(t) f(t) \quad \text{with} \quad f(t) = \exp\left(\int_0^t \bar{E}(\tau)\, d\tau \right), \quad i = 1, 2, \ldots, \kappa^\nu, \tag{II.5}$$

yielding the differential equations in the coordinates $z(t)$:

$$\dot{z}_i(t) = \sum_k W_{ik} z_k(t), \tag{II.6}$$

where the sum now includes the diagonal terms too.

We realize from Eqn. (II.4) that with $\sum x_k(t) = 1$, $f(t)$ can be expressed as

$$f(t) = \left(\sum_k z_k(t) \right)^{-1}. \tag{II.7}$$

The solution of Eqn. (II.6) follows the standard procedure of linear algebra.

This procedure and the solutions obtained are presented in Appendix 4.

The solutions $z_i(t)$ have the form shown in Eqn. (A3.10) in Appendix 3. Using Eqs. (II.5) and (II.7), they can be transformed back to the $x_i(t)$ [cf. Eqn. (A4.12) in Appendix 4]. The nature of these solutions becomes more obvious if we express them through the variables $y_i(t)$ as introduced in Appendix 3. The differential equations written in these variables assume the form

$$\dot{y}_i(t) = [\lambda_i - \bar{\lambda}(t)]\, y_i(t), \qquad \text{(II.8)}$$

where the mean eigenvalue equals the mean productivity: $\bar{\lambda}(t) = \Sigma_k \lambda_k y_k(t) = \Sigma_k E_k x_k(t) = \bar{E}(t)$. The variables $y_i(t)$ are directly related to the normal modes of the relative population variables $x_i(t)$. They are, like the $x_i(t)$, normalized to 1, that is, $\Sigma_k y_k(t) = 1$.

From the form of Eqn. (II.8) of the differential equations the nature of selection as a consequence of a self-organizing process becomes obvious. The target of selection is a quasi-species defined by the maximum or dominant eigenvalue $\lambda_{\text{max}} \equiv \lambda_0$ and its attributed normal mode, the dominant eigenvector, related to the composite population variable y_0. It is characterized by exclusively positive components, and therefore the quasi-species is not the equivalent of any single sequence. Nevertheless, the quasi-species may be dominated by a single sequence; we call it the master copy. The transformation from x to y coordinates may be physically viewed in the following way: Instead of looking at n single sequences, we look at n different clans, that is, combinations of sequences in which kinship relations are taken into consideration. These clans compete for selection, and the one with the largest eigenvalue λ_0 will win the competition. Equation (II.7) shows that the average of all eigenvalues [being equal to the average net productivity $\bar{E}(t)$] acts as a threshold. All eigenvalues $\lambda_i < \bar{\lambda}(t)$ yield a negative sign for y and hence cause that particular combination to die out. Likewise all eigenvalues $\lambda_i > \bar{\lambda}(t)$ cause the corresponding populations to grow. Hence combinations with small eigenvalues disappear; those with large eigenvalues build up. What happens to the mean eigenvalue $\bar{\lambda}(t) = \bar{E}(t)$ during selection? Let us assume that we started with exclusively nonnegative variables $y_k(0) \geq 0$ and that all eigenvalues λ_k are real. Both conditions need not be fulfilled in realistic systems, but for the sake of coherence we postpone a discussion of the general case to the next section (see also Appendix 5). The increase of $\bar{\lambda}(t)$ will continue until it equals the maximum eigenvalue; the threshold increases until it reaches λ_0, where only one combination is left that can match it. We call this stable combination the quasi-species; it does not change with time any more: $\dot{y}_0(t) \to 0$. Hence selection is a self-organizing process inherently caused by the autocatalytic nature of the formation rates. It can be characterized — in analogy to other self-organizing processes, such as equilibration — by an extremum principle:

$$\bar{E}(t) = \bar{\lambda}(t) \rightarrow \lambda_0. \tag{II.9}$$

This principle holds as an absolute principle only for the quasi-linear case treated in the preceding, that is, for constant rate coefficients W_{ik}. As a local principle it covers also a larger group of nonlinear systems involving concentration- or time-dependent rate parameters.

Explicit expressions for the eigenvalues λ_i and for the components of the eigenvectors belonging to λ_i are to be obtained from Eqs. (A4.5)–(A4.7) in Appendix 4 for any specific set of rate parameters W_{ik}. If the nondiagonal elements are sufficiently small as compared to the diagonal terms and if the diagonal elements can be brought into some hierarchical order, where $W_{mm} > W_{kk}$ for any k, second-order perturbation theory yields the following approximations for the largest eigenvalue λ_0 and the corresponding eigenvector $\mathbf{l}_0$ (see also Section III.1),

$$\lambda_0 = W_{mm} + \sum_{k \neq m} \frac{W_{mk} W_{km}}{W_{mm} - W_{kk}}, \tag{II.10}$$

and the dominant eigenvector $\mathbf{l}_0$,

$$\frac{l_{k0}}{l_{m0}} = \frac{W_{km}}{W_{mm} - W_{kk}} + \sum_{j \neq k, m} \frac{W_{ki} W_{jm}}{(W_{mm} - W_{kk})(W_{mm} - W_{jj})}. \tag{II.11}$$

While in general these expressions are of only limited realistic value — since many mutants indeed are (nearly) degenerate — they show very clearly one fact. The surviving quasi-species for which $\lambda_i = \lambda_0$ is dominated by the sequence with the largest diagonal rate coefficient W_{mm}. We call it the master species. It is dominant as long as the sum term of Eqn. (II.10) is negligible, that is, $W_{mm} > W_{kk}$ for any $k \neq m$. Since the products $W_{mk} W_{km}$ are usually very small, the differences between master I_m and mutants I_k need not be large. Taking the known mutation rates of RNA viruses and their genome size of a few thousand nucleotides, an advantage of a tenth of a percent in W_{mm} values is sufficient to clearly define the master. The mutants then are grouped around the master in such way that often their average sequence equals that of the master, which though being the most abundant individual sequence in the distribution may be present as only a very minor fraction of the total set of all mutants.

5. Potential Functions, Optimization, and Guided Evolution

The results presented in the preceding section suggest the need for a fundamental reinterpretation of Darwinian behavior for self-replicating macromolecules. First, selection can be shown as a mere (physical) consequence of constrained self-replication. It is a characteristic of biological systems since there is no other way for living entities to come about than through self-reproduction or recombinative reproduction, both being based on replication of DNA. Conversely, however, selection is not inherently linked to life and may occur wherever self-reproduction or complementary-reproduction or autocatalysis in general is involved. One well-studied example from physics is selection in laser modes [17].

In order to visualize the molecular selection process in the more general context of optimization of replication rates, we consider the simple case of replication with ultimate accuracy first. In this case we have $Q_{ik} = \delta_{ik}$, the value matrix W is diagonal ($W_{ik} = W_{kk}\delta_{ik} = W_k = A_k - D_k$) and the corresponding system of differential equations is weakly coupled by the $\bar{E}(t)$ term only:

$$\dot{x}_k = x_k(W_k - \bar{E}), \quad k = 1, 2, \ldots, n. \tag{II.12}$$

The solutions of Eqs. (II.12) are of the form

$$x_k(t) = x_k(0)\exp\left(W_k t - \int_0^t \bar{E}(\tau)\,d\tau\right), \quad k = 1, 2, \ldots, n. \tag{II.13}$$

The system converges asymptotically to a homogeneous state in which the most efficiently replicating species is present exclusively:

$$\lim_{t\to\infty} x_0(t) = 1, \quad \lim_{t\to\infty} x_k(t) = 0 \; \forall k \neq m \quad \text{and} \quad W_0 = \max[W_j; j = 1, 2, \ldots, n].$$

We calculate the time dependence of the average excess production and find

$$\frac{d\bar{E}}{dt} = \frac{d}{dt}\left(\sum_k W_k x_k\right) = \sum_k x_k\left((W_k - \bar{E})^2 + \frac{dW_k}{dt}\right). \tag{II.14a}$$

In the case of time-independent rate constants the second term vanishes and

$$\frac{d\bar{E}}{dt} = \sum_k x_k(W_k - \bar{E})^2 \geqslant 0, \tag{II.14b}$$

which is a nonnegative expression since $x_k \geqslant 0$ and W_k, a combination of rate constants, is always real by definition. Thus, the average excess production $\bar{E}$ is a nondecreasing function of time. It represents a quantity that is optimized during selection.

A function $V(x)$, algebraically identical to the average excess production, can be shown to represent a potential for the selection process (Appendix 5):

$$V(x_1, x_2, \ldots, x_n) = \sum_{k=1}^{n} W_{kk} x_k. \tag{II.15}$$

Now we can visualize evolutionary optimization as a "hill-climbing" process on a "landscape" that is given by an extremely simple potential [Eqn. (II.15)]. This potential, an $(n-1)$-dimensional hyperplane in n-dimensional space, seems to be a trivial function at first glance. It is linear and hence has no maxima, minima, or saddle points. However, as with every chemical reaction, evolutionary optimization is confined to the cone of nonnegative concentration restricts the physically accessible domain of relative concentrations to the unit simplex $(x_1 > 0, x_2 > 0, \ldots, x_n > 0;\ \Sigma x_k = 1)$. The unit simplex intersects the $(n-1)$-dimensional hyperplane of the potential on a simplex (a three-dimensional example is shown in Figure 4). Selection in the error-free scenario approaches a corner of this simplex, and the stationary state corresponds to a "corner equilibrium," as such an optimum on the intersection of a restricted domain with a potential surface is commonly called in theoretical economics.

Let us now consider replication with errors. For sufficiently accurate replication the system approaches a stationary mutant distribution. We do not observe selection of a single species. The target of the selection process is a unique combination of species determined by the dominant eigenvector of the value matrix W. We perform a linear transformation of variables and choose the eigenvectors of the matrix W as the new basis of the coordinate system:

$$\mathbf{x}(t) = (x_1(t), x_2(t), \ldots, x_n(t)) = \sum_{k=1}^{n} x_k(t) \cdot \mathbf{e}_k = \sum_{k=1}^{n} y_k(t) \cdot \mathbf{l}_k. \tag{II.16}$$

Herein the vectors $\mathbf{e}_k$ are the unit eigenvectors of the Cartesian coordinate system and $\mathbf{l}_k$ the right-hand eigenvectors of the value matrix W as discussed in Appendix 4. Formally the transformed differential equation is of the same general type as (II.12):

$$\dot{y}_k = y_k(\lambda_k - \bar{E}), \quad k = 1, 2, \ldots, n,$$

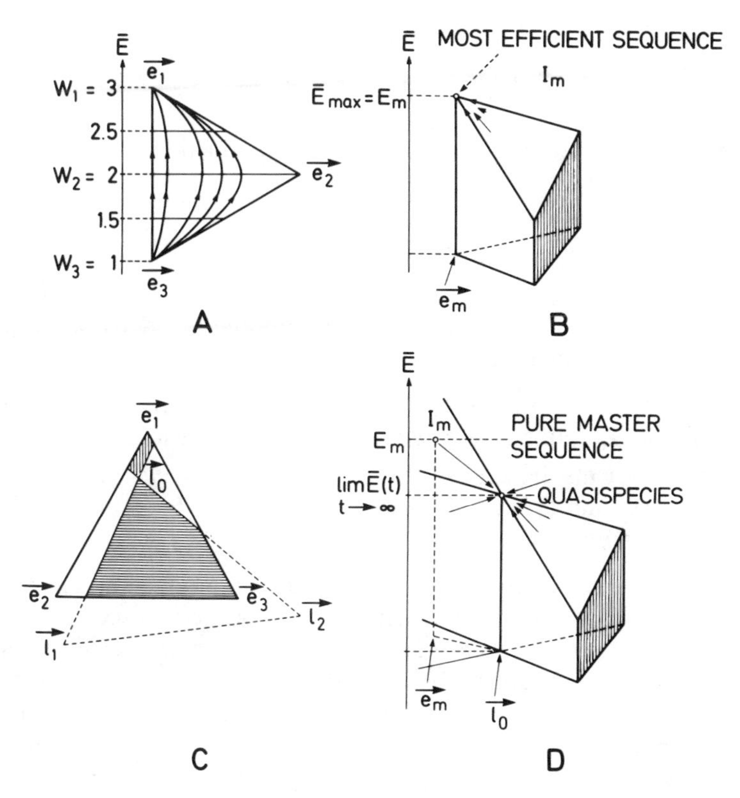

Figure 4. Potential functions and "corner equilibria" in evolutionary optimization. (A) Trajectories and constant level sets of potential V [Eqn. (II.15)] for error-free replication in system with three molecular species. Constants $W_1 = 3$, $W_2 = 2$, and $W_3 = 1$ were chosen. In general, trajectories do not cross constant level sets at right angles. Generalization of definition of gradients based on Riemannian metric allows one to rescale angles such that all trajectories cross constant level sets at right angles (Appendix 5). (B) Demonstrates meaning of "corner equilibrium". "Hill-climbing" process ends at corner of simplex since trajectories cannot cross boundaries. (C) Concentration simplex and triangle spanned by three eigenvectors of value matrix W for one particular choice of constants: $n = 3$, $A_1 = 10$, $A_2 = 3$, and $A_3 = 1$; $Q_{kk} = 0.6$ and $Q_{ik} = 0.2$ for $i \neq k$ and $i, k = 1, 2, 3$. Only dominant eigenvector $\mathbf{l}_0$ lies inside concentration simplex. Entire domain of physically accessible values of relative concentrations is split into four zones. Two are characterized by monotonous behavior of average excess production: in the large, horizontally hatched area $E(t)$ is nondecreasing function. In top zone, hatched vertically, $E(t)$ is nonincreasing. No general predictions can be made for other two areas. (D) Behavior of average excess production $E(t)$ illustrated by means of potential V. Now corner equilibrium lies inside concentration simplex since invariant lines that are not to be crossed by trajectories are given by boundaries of triangle spanned by eigenvectors of value matrix W: $\mathbf{l}_0$, $\mathbf{l}_1$, and $\mathbf{l}_2$.

and the same is true for the solutions,

$$y_k(t) = y_k(0)\exp\left(\lambda_k t - \int_0^t \bar{E}(\tau)\,d\tau\right), \quad k = 1, 2, \ldots, n,$$

and the time dependence of the average excess production,

$$\frac{d\bar{E}}{dt} = \sum_k y_k(\lambda_k - \bar{E})^2.$$

When it comes to details, however, there are substantial differences between the two differential equations and their solutions (see e.g. ref. 18):

1. The variables x_k are relative concentrations and are nonnegative by definition: $x_k \geq 0$. No such relation holds for the variables y_k.
2. The coefficients W_k are obtained from rate constants and mutation frequencies and hence represent real quantities. The λ_k are the eigenvalues of a nonsymmetric matrix and need not be real.

As a consequence of 1 and/or 2, the average excess production is no longer a nondecreasing function of time. Optimization of the average excess production may still occur, but then it is restricted to certain choices of initial conditions (Appendix 5). Jones [19] derived a more complicated function $\bar{\varepsilon}(t)$ shown in Appendix 5 that represents a universal optimization criterion in the replication–mutation system, but the physical meaning of this Lyapunov function is unclear.

The existence of complex eigenvalues of the value matrix W implies that the coefficients in Eqn. (II.15) are complex and rules out the existence of a real-valued potential function. Transient oscillations in the concentrations may occur, but in the limit of long times the system nevertheless converges toward the dominant eigenvector. The corresponding largest eigenvalue is real and positive, and hence all oscillations in concentrations have to fade out inevitably.

The fact that the average excess production may decrease during a selection process can be illustrated by means of a simple example. Consider a homogeneous population consisting exclusively of the master sequence I_m at time $t = 0$. Clearly the excess production is largest since the master sequence is characterized by the maximum selective value. During selection, mutants are formed that have smaller selective values, and finally when the system approaches the stationary distribution, which is the quasi-species, the average excess production reaches the stationary value from above, that is, from higher, total efficiency of replication.

For exclusively real eigenvalues of W the time dependence of the average excess production is determined by the choice of initial conditions. As shown in Appendix 5, optimization of $\bar{E}(t)$ is restricted to initial conditions in the positive orthant $[y_k(0) > 0;\ k = 0, 1, \ldots, n]$. These initial conditions are not difficult to fulfil, and they will apply to many cases in reality. We should keep in mind, nevertheless, that there are other choices of initial conditions, such as the start with a pure master sequence, for which the simple principle does not hold. For one particular type of choice, $y_1(0) > 1$ and $y_k(0) \leqslant 0$ for all $k \neq 1$, the average excess production decreases monotonically.

How likely is it to have a value matrix W with pairs of conjugate complex eigenvalues? Rumschitzki [8] showed that the value matrix W can be converted to a symmetric matrix W' by means of a similarity transformation provided the corresponding mutation matrix Q is symmetric $(Q_{ij} = Q_{ji})$. Then all eigenvalue of W are real. Equal frequency of mutations in both directions, $I_j \rightarrow I_i$ and $I_i \rightarrow I_j$, is a realistic assumption unless the polynucleotide sequences under consideration contain so-called hot spots. These are positions at which point mutations are particularly frequent. It is unlikely that the reverse mutation leading to the sequence with the original hot spot is also an unusually frequent event. Therefore we expect a mutation matrix Q lacking symmetry in these cases.

6. Population Structures

The fundamental reinterpretation of Darwinian behavior to which we referred in the preceding is a consequence of one central fact; namely, that not a single sequence but rather a master sequence and its entire quasi-species distribution appears to be the target of selection. Natural selection being the consequence of self-reproduction has been stressed already in population biology (mainly through the work of Fisher [20], Haldane [21], and Wright [22]). Hence an advantageous mutant appearing to any stochastically significant extent is bound to increase, making selection finally a deterministic event. On the other hand, the appearance of the particular mutant that proves to be selectively advantageous was assumed to be a stochastic event, having no target- or goal-directed bias. This view cannot be sustained any more as a result of the following argument.

The chance of finding an advantageous mutant will increase with increasing Hamming distance (i.e., the mutation distance from the wild type), the main reason being the large increase of the number of mutants and hence possible candidates with increasing distance. Whether a distant mutant can appear will depend on how its precursors are populated. If the precursors of a desired mutant are more populated than those of a nondesired one, there will be guidance of the evolutionary process along preferred routes. The theory states that it is not the single master sequence representing the wild-type but rather the localized distribution in the sequence space that is the target of selection. Then it is very important to know *how* the various mutants are populated. The equations quoted in the preceding correlate population numbers with kinetic parameters called *selective values*. Our problem now has two aspects: How are selective values distributed in sequence space and how does the value distribution map into a population distribution.

It is immediately seen that guidance to optimal performance requires first that mutant population numbers critically depend on their own selective values (and not only on wild-type selective value) and that the value distribution, as with the altitude distribution on earth, is not entirely random but rather clustered along more or less cohesive routes. The kinetic theory, as previously presented, makes quantitative assertions about the correlations between selective values and population numbers.

According to Eqn. (A1.3) Q_d describes (for a uniform error model) the probability of producing a mutant with Hamming distance d in terms of the normalized error rate ε [Eqn. (A2.1)]. This probability decreases exponentially with increasing Hamming distance. The example of a virus with $\nu = 4200$ (discussed in Section V) yields an ε value of about 10^{-4}, hence a 10-error mutant would occur only with a probability of 10^{-40}. A mutant, once it has been produced by miscopying of the wild type or of some intermediate precursor, may also replicate itself, possibly with a rate not too much smaller than that of the master copy. Hence the probability of making a given error copy is not identical with the stationary rate of appearance of that copy.

The appearance of a mutant copy I_i of the master copy I_m in a stationary quasi-species can be calculated by the system of rate equations setting $\dot{x}_i = 0$. In the realm of second-order perturbation theory the explicit expressions can be obtained through recursion. The procedure is best explained with the help of Figure 1. Starting with the master copy, we assume that error copies form only along downhill routes including jumps of any length but neglecting any looped routes (i.e., routes in which Hamming distances do not monotonically increase). This entails the neglect of two kinds of processes:

(i) consecutive change of symbols at a given position (e.g., A → G → U instead of A → U) and

(ii) reversion of a change at a given position.

Since any mutational process costs at least a factor ε, those looped routes are negligible unless mutants with increasing distance substantially increase in their population numbers (which would violate the validity of second-order perturbation expressions, for which W_{ii} must not be too close to W_{mm}).

The stationary solutions ($\dot{x} = 0$) then yield for the relative population numbers

$$\frac{\bar{x}_{di}}{\bar{x}_m} = \varepsilon^d \frac{W_{mm}}{W_{ii}} f_{di}, \tag{II.17}$$

where ε is the quotient introduced with Eqn. (A2.1). The index d_i [abbreviated di in Eqn. (II.17)] refers to the ith sequence in the error class defined by Hamming distance d, and f_{di} is obtained through recursion:

$$\begin{aligned}
f_{1i} &= \frac{W_{ii}}{W_{mm} - W_{ii}}, \\
f_{2i} &= \frac{W_{ii}}{W_{mm} - W_{ii}} \left(1 + \sum_{j=1}^{\binom{2}{1}} f_{1j} \right), \\
f_{3i} &= \frac{W_{ii}}{W_{mm} - W_{ii}} \left(1 + \sum_{j=1}^{\binom{2}{1}} f_{1j} + \sum_{j=1}^{\binom{3}{2}} f_{2j} \right), \\
&\vdots \\
f_{ki} &= \frac{W_{ii}}{W_{mm} - W_{ii}} \left(1 + \sum_{j=1}^{\binom{k}{1}} f_{1j} + \ldots + \sum_{j=1}^{\binom{k}{k-1}} f_{k-1,j} \right).
\end{aligned} \tag{II.18}$$

In these expressions W_{ii} always refers to the individual I_i in the corresponding error class under consideration, while the j's in the sums refer to the corresponding precursors of I_i, for example, in f_{1i} to all d one-error precursors or in $f_{d-1,i}$ to all d of the $(d-1)$-error precursors of I_i. Note that due to the iterative nature, the first sum term in f_{di} contains $d!$ d-fold products of $W_{kk}/(W_{mm} - W_{kk})$ terms.

In this approximation neutral mutants (i.e., $W_{ii} = W_{mm}$) and hence singularities in the f_i terms are precluded. Most mutants will have selective values W_{ii} that are small compared to W_{mm} or that may even be zero (nonviable mutants). If all mutants in fact were of this kind (i.e., $W_{ii} \ll W_{mm}$), the mutant appearance would resemble the Poissonian-type distribution, yielding for any individual in error class d the probability ε^d. Expression (II.18) comprises all contributions of mutant states (classes d' ranging from 1 to ν, including stepwise mutations as well as any jumps up to length d (occurring in direction $0 \rightarrow d$). Contributions from reverse mutations are neglected in this approximation. They could occur if truly neutral or nearly neutral mutants were present that reversibly are populated with numbers that increase with increasing d. This approximation still allows for quite large terms $W_{ii}/(W_{mm} - W_{ii})$ extending to several orders of magnitude. In fact, all that is required is that these terms do not become as large as ε^{-1}. Then the x_{di} decrease with increasing Hamming distance d, yet much less drastically than ε^d does (cf. Figure 5 and the more detailed discussion in Section IV.4). In the previously mentioned example of a virus, where for the 10-error mutant, ε^d drops to 10^{-40}, such an individual mutant, if it were in a domain of precursors with W_{ii} values within 1— of W_{mm}, would be raised to a realistic probability of appearance, in contrast to corresponding mutants in low-value domains. This requires not only singular (nearly) neutral mutants in the distribution but also a clustering of such mutants in more or less cohesive domains, as in landscapes on earth where planar and mountainous regions are also clustered.

On the other hand, a comparison with landscapes on earth may be misleading since the sequence space is of high dimension. All distances remain relatively small, while, as Figure 1 demonstrates, all states are highly interconnected. A mountain trip in such a space may proceed along ridges that are multiply interconnected and thereby link up the peaks. Those ridges and peaks are the points in sequence space that are most heavily populated by mutants, which thereby provide guidance to the evolutionary process. Here we come back to the other still unanswered question: how are values distributed in sequence space? If this distribution were entirely random, it could not guide the evolutionary process over any substantial length of path.

Mandelbrot [23] has shown that the "most random" type of height distribution to be expected on earth is of a fractal type. The same should be true for a value distribution in the ν-dimensional sequence space. Such a fractal distribution is highly connective, that is, anything but uncorrelated. Moreover, we know that functional efficiency is clustered around certain sequences. The functional efficiency of an enzyme depends on the correct spatial arrangement of certain amino acid residues that comprise the active center: this is achieved by three-dimensional folding of the polypeptide chain [24]. Hence there exists a correlation:

$$\begin{array}{c} \text{Primary sequence} \\ \text{Protein folding} \;\downarrow \\ \text{Tertiary structures} \\ \downarrow \\ \text{Functional efficiency} \end{array}$$

Similarities in primary sequences in general will map into similarities of folding, which in turn will lead to similarities in tertiary structures and functional efficiencies.

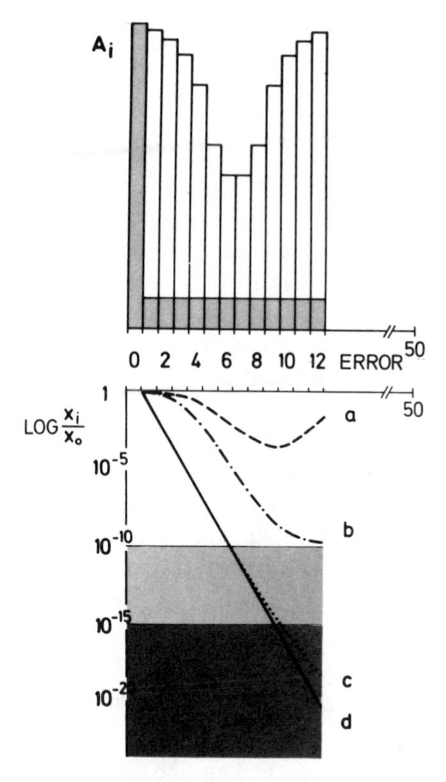

Figure 5. Importance of mutant bias in evolution. Simulation demonstrates role of ridges in selection value landscape in directing route of evolution. Binary sequence of complexity 50 is assumed. In upper part of picture, two value profiles are shown as functions of Hamming distances from reference master 0. Selective value maximum connected to particular mutant at $d = 13$ (excluded from stationary state calculation) by three different landscapes: (i) lowland where all intermediate mutants have inferiority of 0; (ii) plateau where all mutants have inferiority of 0.2 (open bars); and (iii) smooth mountain ridge with monotonous decrease of inferiority to 0.5: at $d = \{6, 7\}$ followed by rise to 0.98 at $d = 12$ (filled bars). Adding sequence at $d = 13$ that is neutral with master (inferiority 1.0) would produce symmetric distribution between two maxima. For each profile two limiting scenarios have been calculated, and results for mutant populations plotted beneath. Curves b and d: elevated inferiorities (W_{ii}/W_{00}) are found for only one particular route comprising 12 subsequent defined mutation steps. All other mutants have negligible inferiority values. Curves a and c: All mutants with same Hamming distance to nearest maximum have degenerate inferiority values (W_{dd}/W_{00}); i.e., all direct routes are possible because succession of mutation steps is arbitrary. (Consult Figure 1 in order to correlate with routes. Note that routes for $d = 12$ comprise only minute fraction of possible connections in 50-dimensional space, i.e., $2^{12}/2^{50} = 10^{-15}$.) Solid curve d coincides with relative mutant populations according to Poissonian distribution ($W_{dd}/W_{00} = 0$). Curves c, d and a, b refer to inferiority value distributions of plateau and smooth ridge profiles, respectively. Following features are seen: Relative populations, around a peak in selection value landscape that rises abruptly from lowland, drop rapidly with increasing Hamming distance from master. Probabil-

Evidence for such a similarity transformation has been provided by experiments on site-directed mutagenesis. Changes in the primary sequence of enzyme molecules produced through site-specific genetic manipulation in many cases turned out to be of only minor consequence to the catalytic performance. Those changes, of course, must not occur at certain strategic positions that are determinants for the structure of the active site. Otherwise, there are many positions that are relatively insensitive to substitution, especially by related amino acids. The same conclusion can be reached by considering phylogenetic sequence relations. A well-studied case is the enzyme cytochrome-*c*. Sequences in different species may be dishomologous in more than 70% of the positions, and yet the same reaction is catalyzed (almost) equally well. Hence there is no doubt that a similarity relation corresponding to a clustered-value distribution in sequence space does exist. However, the relation itself is anything but simple, and no straightforward method of calculating those relations seems to be at hand. For the quasi-species model the existence of a clustered-value distribution as such is of paramount importance. Since the theory shows how value distributions map into population distributions, a quantitative analysis of mutant distributions in selection experiments could provide the evidence and allow for quantitatively estimating the effect of mutational guidance in evolution.

III. ERROR THRESHOLD FOR QUASI-SPECIES LOCALIZATION

The organizing principle embodied in the kinetic equations presented in Section II is given a global character by the appearance of the off-diagonal mutational terms, which provide the variation necessary to the Darwinian logic. Thus, under the physical conditions, where every sequence may be reached by a succession of mutations, the model describes an evolutionary process capable of finding the fittest from arbitrary initial conditions and of returning to it after perturbation. The special feature of the organization

---

ity of reaching another peak with Hamming distance of 12 is very low for typical laboratory populations. In case where sharp crest with low selection values is attached to peak, probability of finding 12-error mutant is only slightly higher. If peak is surrounded by low plateau (between $d = 0$ and $d = 12$), this probability rises by nearly seven orders of magnitude. Effect of higher selective values is drastic: Just one sharp crest with profile depicted above brings probability of occurrence of 12-error mutant up to such order of magnitude that population necessary to find one strand can be realized readily in laboratory. If all routes leading away from summit have same profile, large multiplicity of 12-error mutants will occur (deterministically) in laboratory population. Bias of selective values guiding evolutionary path along ridges in selective value landscape is clearly demonstrated. Note also cumulative effect of connected high-value regions in ν-dimensional landscape.

allowing the selection of a particular sequence type is the relative weakness of mutational interactions. The nonequilibrium Darwinian organizing principle has a parallel in the existence of particular ordered states at equilibrium, such as the low-temperature phases of ferro- and antiferromagnets and ferro- and antiferroelectrics or the condensed states of nucleoside bases in stacks. There the temperature represents the equivalent of the mutation rate. At certain critical temperatures we observe order–disorder equilibrium transitions. Likewise, there is a threshold for selection of a particular wild-type sequence in the nonequilibrium Darwinian organizing principle.

In this section we demonstrate the existence of a sharp error threshold in the kinetic model developed in the preceding and explore its dependence on the value topology of Darwinian fitnesses W_i and the mean fidelity $q = \bar{q}$ of single-digit replication. The error threshold result delineates the domain where evolution proceeds to the survival of a stably localized quasi-species. With higher error rates, the class of sequences distinguished as fittest, by the stationary solution of the deterministic kinetic equations, becomes so large that it cannot be sampled by any biological population. A stochastic interpretation of the dynamics as a random drift covering this class is then called for. The character of the present discussion is then intermediate between a deterministic and stochastic account, and although we shall refer to statistical distributions of replication rates, the dynamics will be that of the deterministic kinetics.

1. Error Threshold and Selective Advantage

For a particular sequence I_m to compete successfully with all the other mutants, it has to be produced at a (net) rate W_{mm} faster than the excess production rate $\bar{E}_{k \neq m}$ of these other mutants. The latter rate is enhanced by the inclusion of inexact replication (leading mostly to further mutants and only rarely back to the wild type). Thus, not only does a sequence I_m generally need to have the maximum net rate W_m for exact replication in order to be conserved indefinitely in a population, but it also needs to satisfy the more stringent requirement

$$W_{mm} > \bar{E}_{k \neq m} = \frac{\sum_{k \neq m} E_k \bar{x}_k}{\sum_{k \neq m} \bar{x}_k}, \tag{III.1}$$

where the $\bar{x}_k$ are the stationary relative population numbers.

Indeed, the neglect of backflow of mutants to the wild type, that is, identification of W_{mm} with the diagonal rate coefficient of the master sequence, is equivalent to a second-order perturbation result, and to this order a calculation of the wild-type population yields

$$\frac{\bar{x}_m}{\sum_k \bar{x}_k} = \frac{W_{mm} - \bar{E}_{k \neq m}}{E_m - \bar{E}_{k \neq m}}, \tag{III.2}$$

where $E_m = A_m - D_m$ is the excess productivity of the wild type.

According to Eqn. (III.2), the relative equilibrium concentration of the wild type $\bar{x}_m / \Sigma \bar{x}_k = \bar{x}_m / c_0$ vanishes for some critical error rate at which $W_{mm} = \bar{E}_{k \neq m}$. This, of course, is merely a consequence of the perturbation approach that neglects backflow of mutations to the wild type. The phenomenological calculation of $\bar{x}_m$, based on the use of $\bar{E}_{k \neq m}$, can be carried out exactly if we introduce a mean backflow rate:

$$W_M = \frac{\sum_{k \neq m} W_{mk} \bar{x}_k}{\sum_{k \neq m} \bar{x}_k} = \frac{1}{c_0 - \bar{x}_m} \sum_{k \neq m} W_{mk} \bar{x}_k.$$

Then we obtain the relative equilibrium concentration of the wild type as the positive root of a quadratic equation:

$$\frac{\bar{x}_m}{\sum_k \bar{x}_k} = \tfrac{1}{2}\left(\mathscr{A} - \mathscr{B} + \sqrt{(\mathscr{A} - \mathscr{B})^2 + 4\mathscr{B}}\right) \tag{III.3}$$

with

$$\mathscr{A} = \frac{W_{mm} - \bar{E}_{k \neq m}}{E_m - \bar{E}_{k \neq m}} \quad \text{and} \quad \mathscr{B} = \frac{W_M}{E_m - \bar{E}_{k \neq m}}.$$

It is readily seen that $\bar{x}_m$ does not vanish at the critical point $\mathscr{A} = 0$, or $W_{mm} = \bar{E}_{k \neq m}$, but is given by

$$\frac{\bar{x}_m}{\sum_k \bar{x}_k} = \frac{\mathscr{B}}{2}\left(\sqrt{1 + \frac{4}{\mathscr{B}}} - 1\right) \quad \text{for} \quad \mathscr{A} = 0,$$

which is small, on the order of

$$\sqrt{\mathscr{B}} = \sqrt{\frac{W_M}{E_m - \bar{E}_{k \neq m}}}.$$

The exact equation (III.3) may be solved if the phenomenological parameters $\mathscr{A}$ and $\mathscr{B}$ can be computed. For the special test case of a single advantageous master with homogeneously poorer competitors, $\mathscr{A}$ may be computed exactly and $\mathscr{B}$ computed as a function of $\bar{x}_m/\Sigma_k \bar{x}_k$ under the assumption that all other mutants are equally populated. The solutions for $\bar{x}_m$ given by (i) second-order perturbation theory Eqn. (III.2), (ii) the phenomenological Eqn. (III.3) with the above assumption for $\mathscr{B}$, and (iii) the exact solution are presented in Figure 6 for sequence lengths 7 and 20. For large sequence lengths the results are not significantly different from the solution of Eqn. (III.3) with $\mathscr{B} = \varepsilon(1 - \bar{x}_m/\Sigma_k \bar{x}_k)$, $\varepsilon \to 0$ (two straight lines).

Equation (III.2) may be rewritten to isolate the dependence on the copying fidelity q in order to demonstrate that for a given set of replication parameters there is an error-rate-dependent threshold sequence length for quasi-species instability. To this end the selective advantage or superiority parameter σ was introduced:

$$\sigma = \frac{A_m}{D_m + \bar{E}_{k \neq m}}, \qquad \text{(III.4)}$$

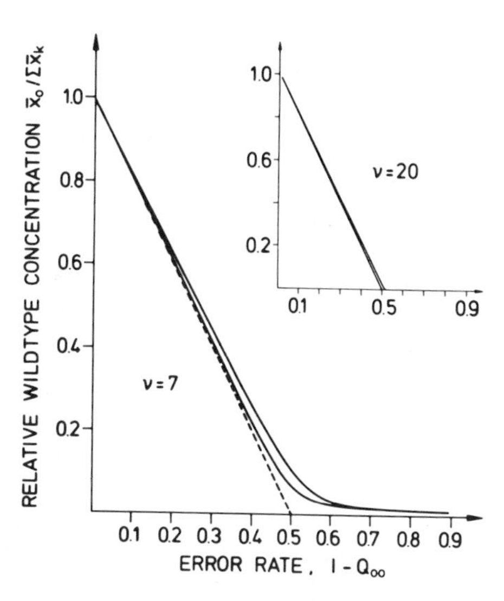

Figure 6. Relative wild-type concentrations as function of error rate $1 - Q_{00}$. We show exact solution curve (upper full line) and compare it with result of perturbation theory [Eqn. (III.2); broken line) and exact solution of Eqn. (III.3) with B calculated as in text (lower full line). Following parameters and rate constants applied: $A_0 = 10$, $\bar{A}_{k \neq 0} = 5$, $D_0 = D_1 = \cdots = D_n$ and $\nu = 7$, $n = 2^\nu - 1$.

which is not explicitly dependent on q and in terms of which the inequality (III.1) becomes $Q_m > \sigma^{-1}$, or using Eqs. (A1.1–A1.4) in Appendix 1 to relate the quality factor to the sequence length:

$$\nu < \nu_{max} \doteq -\frac{\ln \sigma}{\ln q} \simeq \frac{\ln \sigma}{1-q}. \qquad \text{(III.5)}$$

This relationship shows that the amount of information that can be maintained in the form of a specific stable sequence is limited by the copying fidelity q. Thus the price that is paid for the global nature of the organization via variation is a limitation on the discrimination of fitness. Typical variations in fitness can only be selected for up to the critical sequence length ν_{max} since the fidelity of copying the whole sequence decreases exponentially with increasing sequence length.

It should be briefly pointed out that simply decreasing the replication rates of other sequences to increase the superiority of the master does not indefinitely postpone the error threshold. Once the replication rates for mutant sequences fall below the destruction rate for the master sequence, the domain of viable existence for the quasi-species is limited by $A_m Q_m > D_m$. In Section III.3 a more fundamental statistical argument restricts the effective superiority to a finite value of order 1.

The dependence of the threshold on the replication rate parameters A_k, D_k expressed through Eqn. (III.1) or (III.4) is dependent on the populations of each of the mutants that appear to weight the average excess productivity $\bar{E}_{k \neq m}$ [cf. Eqn. (III.1)]. In particular, mutants that are distant from the wild type may require a net exact replication rate much closer to that of the wild type than does a nearby mutant in order to reach a substantial fraction in the population and increase $\bar{E}_{k \neq m}$ above W_{mm}. On the other hand, a more detailed analysis of the error threshold in terms of the many rate parameters characterizing the different mutant sequences appears unwieldy, as it requires full solution of an extremely high-dimensional eigenvalue problem.

One possible option is to adopt a statistical description of the kinetic parameters and to ask how likely it is for the quasi-species to be localized about the wild type. This undertaking requires an analysis beyond the second order in perturbation theory since a distant mutant with a selective value very close to that of the wild type may jeopardize the stability of the latter in the population. We were however encouraged by the progress that had been made with a problem of similar difficulty in the very different area of electron or spin localization in disordered solids. Indeed, it turns out that an expression of the form of Eqn. (III.5) may be obtained, with an explicit expression for the superiority parameter σ_0, dependent on the distribution of replication rates but not on any average involving population variables.

More exciting than this, the selective advantage of the wild type turned out to be nonzero even with a continuous distribution of selective values (for the different mutants) up to that of the wild type. These developments are reviewed in the following section.

2. Localization Threshold for Statistically Distributed Replication Rates

We consider once again the deterministic kinetic Eqn. (II.4) with the uniform error rate expression Eqn. (A1.5) relating mutation rates to replication rates:

$$W_{ik} = \left(\frac{q^{-1} - 1}{\kappa - 1} \right)^{d(i,k)} (W_k + D_k) = V_{ik} W_k. \tag{III.6}$$

If we ignore the destruction terms D_k, the different mutants are fully characterized through their rates for exact replication W_k. This assumption simplifies the following discussion by allowing a single distribution of value parameters $f(W)$ to completely characterize the mutant spectrum, but this is not essential to the argument. Accordingly, we write

$$V_{ik} = \left(\frac{q^{-1} - 1}{\kappa - 1} \right)^{d(i,k)}, \tag{III.7}$$

expressing the sharp dependence of mutation rate on the Hamming distance in the mutant space.

In Appendix 6 we summarize the necessary modification of the Rayleigh–Schrödinger perturbation theory, which was applied to replication dynamics [15, 16], in order to obtain explicit expressions for higher order terms. This Brillouin–Wigner approach yields a self-consistent expression for the dominant eigenvalue, Eqn. (A6.8), but one involving a perturbation series that may be immediately written down to arbitrarily high order [Eqn. (A6.3) evaluated at $s = \lambda$]. In particular, the wild-type fraction in the population is given by the reciprocal of Eqn. (A6.9) for $k = m$, the result being not restricted to second order. Localization of the quasi-species about the wild type I_m is then a question [29] of the convergence of Eqn. (A6.9), or equivalently of the series

$$T_m = \sum_{i \neq m} T_{im} = \sum_{i \neq m} \frac{{}^{c}W_{im}(\lambda)}{W_{mm}} \tag{III.8}$$

determined by Eqn. (A6.3), or more conveniently by Eqn. (A7.1). Here, λ denotes the dominant eigenvalue of the matrix W ($\lambda > W_{ii}$ for all i).

The question of convergence of this series in turn depends on the exact replication rates W_{ii} of all the different mutants. If we describe these statistically in the absence of more detailed information, then convergence must be

TABLE I

Effective Superiority of Wild Type for Various Replication Rate Distributions[a]

Distribution	Description	Probability Density	$\langle (W_0 - W)/W \rangle_{\ln}$	$\ln \sigma_{\text{eff}}$
Delta function	Constant value $W = \bar{W} < W_0$	$f(W) = \delta(W - \bar{W})$	$(W_0/\bar{W}) - 1$	$\ln(W_0/\bar{W})$
Rectangular	1. All values $0 \leqslant W \leqslant W_0$ equiprobable	$f(W) = \dfrac{1}{W_0}$, $0 \leqslant W \leqslant W_0$ (0 otherwise)	1	$\ln 2$
	2. All values, $W_L \leqslant W \leqslant W_0$ equiprobable "near neutral"	$f(W) = \dfrac{1}{W_0 - W_L}$, $W_L \leqslant W \leqslant W_0$ (0 otherwise)	$1 - (W_L/W_0)$ for $W_L \to W_0$	$1 - (W_L/W_0)$ for $W_L \to W$
Exponential (*see also gamma distribution*, $\eta = 1$)	Mean $\bar{W}$ truncated at $W_0 \gg \bar{W}$, c_1 is Euler's constant (0.577 . . .)	$f(W) = \dfrac{1}{\bar{W}} e^{-W/\bar{W}}$, $0 \leqslant W \leqslant W_0$ (0 otherwise)	$(W_0/\bar{W})e^{c_1}$	$c_1 + \ln(W_0/\bar{W})$
Gamma, truncated at $W_0 \gg \bar{W}$	Parameters β, η, mean $\bar{W} = \beta\eta$, relative variance $\sigma/\bar{W} = \eta^{-1/2}$	$f(W) = \dfrac{W^{\eta-1}}{\beta^{\eta}\Gamma(\eta)} e^{-W/\beta}$, $0 \leqslant W \leqslant W_0$ (0 otherwise)	$(W_0/\bar{W})e^{c_\eta}$, $c_\eta = c_1 + \ln(\eta - 1) - \sum_{r=1}^{\eta-1} 1/r$	$c_\eta + \ln(W_0/\bar{W})$
		$\eta = 1$, $\sigma/\bar{W} = 1$	$c_1 = 0.58$	
		$\eta = 2$, $\sigma/\bar{W} = 0.71$	$c_2 = 0.27$	
		$\eta = 3$, $\sigma/\bar{W} = 0.58$	$c_3 = 0.18$	
		$\eta = 4$, $\sigma/\bar{W} = 0.50$	$c_4 = 0.13$	$\ln(W_0/\bar{W})$
		$\eta = 9$, $\sigma/\bar{W} = 0.33$	$c_9 = 0.036$	
		$\eta = \infty$, $\sigma/\bar{W} = 0$	$c_\infty = 0$	

Normal, truncated at 0 and at $W_0 \gg \bar{W}$	Mean $\bar{W}$, variance σ^2	$f(W) = (2\pi\sigma^2)^{-1/2} e^{-1/2[(W-\bar{W})/\sigma]^2}$, $0 \leqslant W \leqslant W_0$ (0 otherwise)	$(W_0/\bar{W})e^{\sigma^2/\bar{W}^2}$	$\frac{1}{2}\frac{\sigma^2}{\bar{W}^2} + \ln\frac{W_0}{\bar{W}}$
		$\sigma/\bar{W} = 0.71$	$\sigma^2/2\bar{W}^2 = 0.25$	$\ln(W_0/\bar{W})$
		$\sigma/\bar{W} = 0.50$	$\sigma^2/2\bar{W}^2 = 0.125$	
		$\sigma/\bar{W} = 0.33$	$\sigma^2/2\bar{W}^2 = 0.055$	
		$\sigma/\bar{W} = 0$	$\sigma^2/2\bar{W}^2 = 0$	

[a] Here $f(W)$ represents an underlying probability distribution from which (in a possibly correlated manner) replication rates of different mutants are sampled. $\bar{W}$ is average of this distribution and W_0 the additionally specified maximum value (which truncates distribution in latter cases). This maximum value is assigned to master sequence m, so we may write $W_{mm} = W_0$.

discussed probabilistically. The question is then how probable localization of the quasi-species is for a given distribution of replication rates.

In order to answer this question, a significant source of statistical correlation arising from mutation paths that visit a particularly advantageous mutant more than once must be considered. In the perturbation theory these paths are represented by products of factors involving the mutant replication rates, and it is necessary to remove the strong correlation that arises between these factors where repeated indices are present in order to obtain a tractable statistical analysis of convergence. The Watson renormalization procedure [29], the application of which to the steady-state quasi-species is summarized in Appendix 7, accomplishes just this [30]. The cost is a consecutive modification of the denominator, which may however be simplified to good approximation, as in Eqn. (A7.5).

The individual contributions to the series T_m may then be shown, as outlined in Appendix 8, to have a narrow probability distribution about their logarithmic mean value provided the replication rates have no long-range correlations with respect to the Hamming distance. The contributions of paths involving products of N mutations may then be summed to an expression that lies with a probability approaching 1 in a narrow range of values about L^N as N becomes large. Thus the convergence of the series T_m, or the nonvanishing of the wild-type population, depends on whether the parameter L is greater than or less than 1. It is further possible to demonstrate that λ is not significantly different from W_{mm} even at threshold, which is then characterized by the equation

$$1 = L(W_{mm}) \simeq (q^{-\nu} - 1)\left\langle \frac{W}{W_{mm} - W} \right\rangle_{\ln}, \tag{III.9}$$

where the logarithmic average defined in Eqn. (A8.9) is over a probability distribution $f(W)$ of replication rates taking values up to that of the wild type W_{mm}. The maximum sequence length corresponding to the preceding threshold is

$$\nu_{\max} = \frac{\ln[1 + \langle (W_{mm} - W)/W \rangle_{\ln}]}{-\ln q} = \frac{\ln \sigma_{\text{eff}}}{-\ln q}. \tag{III.10}$$

In comparison with Eqn. (III.4) the result of the statistical analysis is to provide an expression for the effective superiority parameter σ_{eff} of the wild type in terms of the distribution of replication rates of its mutants.

Some values of the logarithmic average appearing in Eqn. (III.9) and of $\ln \sigma_{\text{eff}}$ are recorded for various replication rate distributions in Table I. For the special case where all the mutants have identical replication rate $\bar{W}$, the superiority parameter of Section III.1 is independent of population variables and may be explicitly evaluated, agreeing with the effective superiority defined by Eqn. (III.10). At the opposite extreme, where all values less than or equal to the wild-type replication rate W_{mm} are equally probable, a finite answer $\ln \sigma_{\mathrm{eff}} = \ln 2 = 0.69\ldots$ is obtained. As the replication rate distribution becomes more weighted toward the wild-type value, a neutral mutant extreme is approached for which $\ln \sigma_{\mathrm{eff}}$ steadily decreased toward zero. These last two results are at first surprising because the continuous distributions imply the existence of mutants arbitrarily close in selective value to the wild type. That the effective superiority of the wild type is nonzero is a reflection of the crucial effect of mutational distance and the structured population of the quasi-species.

The remaining three distributions in Table I are appropriate when the wild-type replication rate is regarded as the largest value of many sequences sampled. We shall focus on this view in the following section, which allows rather general conclusions to be made with the aid of extreme-value theory.

It will be apparent that the probability distributions, upon which the localization threshold result of Eqn. (III.9) depend, contain no reference to possible correlations between the selective values of neighboring sequences in the mutant space. A consideration of the statistical argument summarized in Appendix 8 shows that the localization threshold depends only on the absence of long-range correlations between the selective values. We illustrate *short-* and *long-range* correlation by means of examples. Short-range correlation occurs in systems where the selective value of the mutant has always a fixed probability of a value near that of each of its neighbors. This would lead, for example, to those *direct-track* mutants, between the wild-type and an unusually good competitor, having a higher probability of a high selective value than others. A long-range correlation, which would jeopardize the result, occurs when the replication rate is dependent upon two properties determined by disjoint subsets of the sequence. Such correlations occur, for example, when a sequence codes for several proteins or when only a fraction of the sequence determines the fitness (cf. Section III.4).

3. Extreme-Value Theory for Effective Superiority

The evaluations of the error threshold and effective superiority discussed so far [Eqn. (III.10)] assume specific knowledge of the wild-type net rate of exact replication W_{mm}. It is clear, however, that the wild type is distinguished from the other mutants only insofar as it has the maximum selective value, and so it is sensible to regard this value as the random extreme of n trials from the probability distribution $f(W)$ [29]. We may then speak of the general localization threshold properties of a quasi-species belonging to a particular distribution dependent upon n instead of W_{mm}. The results are of more than academic interest in their remarkable insensitivity to n for large populations. Clearly, n is the number of different mutant sequences sampled in the history of the quasi-species.

The extreme value $W_0^{(n)}$ of n trials from a probability density $f(W)$ with the cumulant distribution $F(W)$ satisfying the large W behavior

$$\lim_{W \to \infty} \frac{d}{dW}\left(\frac{1 - F(W)}{f(W)}\right) = 0 \qquad \text{(III.11)}$$

has, asymptotically, the extreme-value distribution of exponential type in that

$$\lim_{n \to \infty} \text{Prob}\{(W_0 - l_n)nf(l_n) \leqslant u\} = \exp[-\exp(-u)], \qquad \text{(III.12)}$$

where l_n depends on the underlying distribution $F(l_n) = 1 - 1/n$. The preceding result is one form (due to van Mise, see ref. 32) of the extreme-value theorem, the (cumulant) distribution on the right of Eqn. (III.12) being known as the limiting extreme-value distribution of exponential type.

The physically reasonable distributions in the absence of a specified maximum are not truncated but have diminishing probability for very high values satisfying Eqn. (III.11). The extreme-value parameter l_n, which is the mode of the extreme, and $nf(l_n)$, which is related to its variance, are displayed for exponential, gamma, and normal distributions for large n in Table II. The modal estimate becomes reliable for large n, and the error threshold may be estimated with $W_0 = l_n$, as the table indicates. The n dependence of the threshold through $\ln(\ln n)$ is so extremely weak that rather precise estimates can be made. Thus, in the $Q\beta$ replicase in vitro evolution experiments of the Spiegelman type, the number of molecules limits n to $10^{12 \pm 3}$ even taking into account consecutive sampling of sequences (on a laboratory time scale of 10^2 generations), and $\ln(\ln n)$ is then 3.3 ± 0.3. At the other extreme of the maximum possible number of sequences sampled in the course of the evolutionary history of the earth, $n < 10^{50}$, $\ln(\ln n)$ is only 4.75. The numbers in the table are for $n = 10^{12}$. The results completely confirm the original view that $\ln \sigma$ should be a number of order unity. In comparison with experiments, it should be emphasized that the threshold sequence length must be greater than the observed sequence length for a stably localized quasi-species. For example, a single independent measurement of v and q for a $Q\beta$ viral RNA [34, 35] indicates $\ln \sigma_{\text{eff}} > 1.4$ for this virus.

TABLE II
Extreme-Value Parameters and Corresponding Effective Superiorities for Various Underlying Distributions[a]

Underlying Distribution	Probability Density	l_n	$nf(l_n)$	$\ln \sigma_{\text{eff}}$
Exponential (see also gamma distribution, $\eta = 1$)	$f(W) = \frac{1}{\bar{W}} e^{-W/\bar{W}}$	$\bar{W} \ln n$	$1/\bar{W}$	$c_1 + \ln(\ln n)$
Gamma:parameters β, η (c.f. Table I) Numerical	$f(W) = \frac{W^{\eta-1}}{\beta_\eta \Gamma(\eta)} e^{-W/\beta}$	$\frac{\bar{W}}{\eta} \ln[n/\Gamma(\eta)]$	$\eta/\bar{W}$	$c_\eta - \ln \eta + \ln \ln[n/\Gamma(\eta)]$
	$\eta = 1$ $(\sigma/\bar{W} = 1)$	$27.6\,\bar{W}$		3.9
	$\eta = 2$ (0.71)	$13.8\,\bar{W}$		2.9
	$\eta = 3$ (0.50)	$8.9\,\bar{W}$		2.4
	$\eta = 4$ (0.50)	$6.5\,\bar{W}$		2.0
	$\eta = 9$ (0.33)	$1.9\,\bar{W}$[b]		1.2[b]
Normal $(\bar{W}, \sigma)$	$f(W) = (2\pi\sigma^2)^{-1/2} e^{-1/2[(W-\bar{W})/\sigma]^2}$	$\bar{W} + \sigma(2 \ln n)^{1/2}$	$\frac{1}{\sigma}(2 \ln n)^{1/2}$	$\frac{1}{2}\sigma^2/W^2 + \ln(\sigma/\bar{W}) + \frac{1}{2}\ln(\ln n)$
	$\sigma/\bar{W} = 0.71$	$6.3\,\bar{W}$		1.5
	$\sigma/\bar{W} = 0.50$	$4.7\,\bar{W}$		1.1
	$\sigma/\bar{W} = 0.33$	$3.5\,\bar{W}$		0.6

[a] Here n is number of trials from probability distribution $f(W)$ of rates for exact replication having mean W. l_n is value of W for which there is probability $1/n$ of sampling higher replication rate. Numerical values of l_n and $\ln \sigma_{\text{eff}}$ are for $n = 10^{12}$ (see text).

[b] Approximation ill founded.

4. Relaxed Error Threshold and Gene Duplication

So far we considered every part of the polynucleotide sequence as being equally important for the replication process. This very restrictive assumption will not be justified for most systems in reality. This is quite obvious for eukaryotes, where the larger fraction of the genome consists of noncoding intervening regions (introns). Though introns, depending on their functional role, may well require conserved sequences, the fidelity requirements will vary greatly and differ from those in expressed regions (exons). Even in the coding regions, mutations in some parts of the polymer have little or practically no influence on the "fitness" of the polynucleotide. In test tube replication experiments such unimportant parts are represented by bases outside the regions of enzymic recognition, when, in addition, they do not contribute to the secondary structure required for replication. In the genomes of organisms these mutations concern changes in the base sequence, which, when translated into protein, yield enzymes that differ very little in catalytic efficiency from those of the wild type. Other mutations are "silent" and lead to identical proteins upon translation. Others may concern parts of the DNA that are not translated at all and that have no influence on replication as well. Such mutations are commonly considered as being *selectively neutral.*

Selectively neutral mutants were observed in test tube evolution experiments [36]. They are much harder to detect with small viruses, which often have highly condensed genomes including overlapping genes, read-through positions, and so on. Neutral mutants were found in chemostat experiments with bacteria [37–39] and in higher organisms [2]. The latter have in addition large fractions of their genomes apparently unused for coding protein or regulation. Changes within these regions of the DNA will be selectively neutral provided they do not interfere with recombination.

Given that such regions of little or no influence on the fitness of a polynucleotide sequence do exist, we expect intuitively a relaxation of the error threshold. Typical situations in which this relaxation appears to be of major importance are gene duplication, multigene messengers, and segmented genes. In this section we present the results of a quantitative analysis of replication with partially relaxed constraints on accuracy [40]. The same approach can be applied also to related biological problems such as the occurrence of "selfish" RNA or DNA or the exon–intron structure of eukaryotic genes. We consider a polynucleotide of chain length ν that is built up from two segments A and B. The two segments are ν_A and ν_B bases long, respectively: $\nu = \nu_A + \nu_B$. Both segments exist in various mutant forms, $A_1, A_2, \ldots, A_r$ and $B_1, B_2, \ldots, B_s$ (Figure 7). We may visualize such a polynucleotide as a primitive genome consisting of two genes only. Both genes are present in several alleles. Segments A and B need not be joined

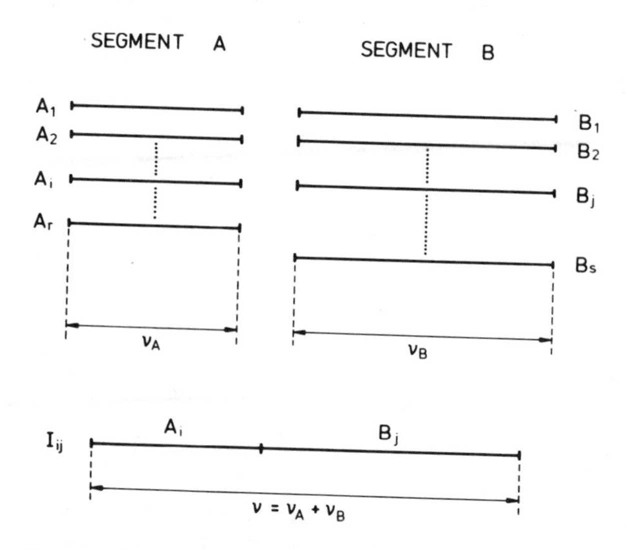

Figure 7. Structure of segmented polynucleotide sequences. Sequence is built from two segments of chain lengths ν_A and ν_B, respectively; rs different sequences can be formed by combination of r segments of type A and s segments of type B.

physically. For instance, segment B could be split into some or many parts that are inserted into A at different positions. We shall only assume that changes in all parts of A have a strong influence on the fitness of the polynucleotide, whereas those in the parts of B are of little or no influence.

An individual polynucleotide sequence is labeled by two indices, i and j: I_{ij} denotes the combination A_i–B_j. There are, of course, rs individual combinations of this kind, that is, rs polynucleotide sequences on which selection acts. We introduce relative concentrations (which carry double indices accordingly), $c_{ij}=[\mathrm{I}_{ij}]$:

$$x_{ij}=c_{ij}\Big/\sum_{kl=1}^{rs} c_{kl} \tag{III.13}$$

with $0\leqslant x_{ij}$ and $\Sigma x_{ij}=1$. The selection equation is of the form

$$\dot{x}_{ij}=\sum_{kl=1}^{rs} W_{ij,kl}x_{kl}-\bar{E}x_{ij}, \tag{III.14}$$

The selective values and the rate constants are defined as before; in particular, we have

$$W_{ij,kl}=A_{kl}Q_{ij,kl}-D_{ij}\delta_{ij,kl}.$$

In principle, Eqn. (III.14) does not differ from the cases treated above in the preceding.

Now, we shall focus upon individual regions of the polynucleotide sequence $A_{(i)}$ and $B_{(j)}$. We introduce new variables that represent summations over the distribution of the alleles in the other region:

$$x_{i0} = \sum_{j=1}^{s} x_{ij} \quad \text{and} \quad x_{0j} = \sum_{i=0}^{r} x_{ij}. \tag{III.15}$$

Usage of these variables requires definition of average selective values and average rate constants, which are summarized in Appendix 9.

The analysis of the selection equation is based on two assumptions:

1. We consider single-step mutations only. Precisely, we assume that either part A or part B may be changed during one act of replication. Simultaneous changes in both parts are neglected.
2. We shall assume that the quality factors for replication of A and of B do not depend on the particular base sequences. We denote these general factors by Q_A and Q_B.

The second assumption is consistent with the uniform mutation frequency model discussed in previous sections. It is, however, less restrictive than that: Error rates need not be uniformly distributed in A or B as long as the total quality factors remain the same.

With these two assumptions we derive separate selection equations for part A and part B (see Appendix 9 and ref. 39):

$$\dot{x}_{i0} = \left[\left(1 - \sum_{k \neq i} Q_{ik}^{A}\right)\bar{A}_{i0} - \bar{D}_{i0} - \bar{E}\right] x_{i0} + \sum_{k \neq i} Q_{ik}^{A} \bar{A}_{k0} x_{k0}, \tag{III.16}$$

$$\dot{x}_{0j} = \left[\left(1 - \sum_{l \neq j} Q_{jl}^{B}\right)\bar{A}_{0j} - \bar{D}_{0j} - \bar{E}\right] x_{0j} = \sum_{l \neq j} Q_{jl}^{B} \bar{A}_{0l} x_{0l}, \tag{III.17}$$

with $i, k = 1, \ldots, r$ and $j, l = 1, \ldots, s$.

These two differential equations describe how selection acts on parts of the replicating unit. The equations are equivalent to the selection Eqn. (III.15) in so far as no simplifying assumptions were made except the two concerning the structure of the mutation matrix Q. The interaction between the two parts of the polynucleotide occurs via the average rate constants $\bar{A}_{i0}$, $\bar{D}_{i0}$, $\bar{A}_{0j}$ and $\bar{D}_{0j}$ as well as implicitly through the common average excess production:

$$\bar{E} = \sum_{i=1}^{r} (\bar{A}_{i0} - \bar{D}_{i0}) x_{i0} = \sum_{j=1}^{s} (\bar{A}_{0j} - \bar{D}_{0j}) x_{0j}. \tag{III.18}$$

Let us now compare the mathematical structures of the selection Eqn. (III.15) and the coupled systems of Eq. (III.16) and (III.17): The original equation had rs variables and one conservation relation and was linear apart from the mild nonlinearity caused by $\bar{E}$. Equations (III.16) and (III.17) contain $r+s$ variables only; they fulfil two conservation relations but are highly nonlinear through the coupling terms. We recall from Appendix 9 that, For example,

$$\bar{A}_{i0}=\frac{1}{x_{i0}}\sum_{j=1}^{s} A_j x_{ij}.$$

Hence, solution of these equations will be rather involved in the general case, and one has to rely on numerical procedures. These are readily available because closely related problems are frequently encountered in numerical mathematics [41]. We dispense here with the details of the corresponding equation for the general case where the sequences in part B contribute differently to the fitness. As expected, it is not possible to remove the coupling terms completely.

Let us now consider an important special case in which the coupling terms have a particularly simple structure. We shall assume that the individual base sequences in part B have no specific influence on the replication rate of the whole polynucleotide A–B. They are selectively neutral, and we may write

$$A_{ij}=A_i f_B,$$

wherein A_i is the sequence specific effect of part A and f_B is the unspecific effect of part B on the rate of replication. This ansatz has the consequence that the two mean values are of different form:

$$\bar{A}_{i0}=A_1 f_B \quad \text{and} \quad \bar{A}_{0j}=f_B\frac{1}{x_{0j}}\sum_{j=1}^{r} A_i x_{ij},$$

which implies highly unsymmetric coupling of the two differential equations.

For the sake of simplicity we assume equal rates of degradation for all sequences ($D_{ij}=D$). Then the contributions of D to W and $\bar{E}$ cancel in both differential equations, and we may simply neglect them. The selection equation of part A can now be written in a form that closely resembles the original differential equation (Appendix 3):

$$\dot{x}_{i0}=f_B\left((A_i Q_{ii}^A-\bar{E}')x_{i0}+\sum_{k\neq i} A_k Q_{ik}^A x_{k0}\right) \qquad \text{(III.19)}$$

with

$$E' = \sum_i A_i x_{i0} \quad \text{and} \quad Q_{ii}^A = 1 - \sum_{k \neq i} Q_{ik}^A \qquad (i, k = 1, \ldots, r).$$

In principle, the common factor f_B can be absorbed into the time axis by means of a linear transformation: $\tau = t f_B$. Apart from this scaling the differential Eqs. (III.19) are completely independent of part B. The equations are now confined to part A of the polynucleotide, and the error threshold is determined by the quality factor Q_{ii}^A. This quality factor is substantially larger than the quality factor of the entire sequence,

$$Q_{ij,ij} = 1 - \left(\sum_{k \neq i} Q_{ik}^A + \sum_{l \neq j} Q_{jl}^B \right) \quad \text{and} \quad Q_{ii}^A - Q_{ij,ij} = \sum_{l \neq j} Q_{jl}^B.$$

In other words, the formula derived for the maximum number of bases in the polynucleotide sequence (ν_{max}) applies now to ν_A only and not to the total length.

In general, chain lengths ν are not conserved in polynucleotide replication. Insertions and deletions are common. Duplications of longer parts of sequences occur as well. Every polynucleotide thus has to compete not only with molecules of the same chain length but also with longer and shorter sequences.

Whether or not a given polynucleotide can compete successfully with competitors of different length is determined largely by the chain length dependence of the rate of replication. This dependence is rather complicated [10–12] and varies with environmental conditions. As an example, we consider the limit of long polynucleotide chains. If template concentrations are in vast excess of polymerase concentrations, a polynucleotide will bind one enzyme molecule or none. The mean time for the complete replication process increases linearly with the chain length, and the rates of replication are roughly proportional to ν^{-1}. If the enzyme is available in excess of templates, however, many polymerase molecules can bind to the same template. The number of polymerase molecules operating on one polynucleotide strand is proportional to its length, and then the rate for the replication of sufficiently long chains becomes independent of the chain length.

The factor f_B is an appropriate measure of the fitness of polynucleotides that carry unused or redundant parts and compete with shorter and longer sequences. In case the rate of replication does not depend on the chain length, f_B will be close to 1 and longer sequences may compete successfully with shorter ones. Rates of replication that are largely insensitive to chain lengths are particularly important for the process of gene duplication. A coding sequence may be duplicated accidentally by a replication error. The longer polynucleotide formed has to compete with shorter sequences for some intermediate period of time during which its fitness remains practically unchanged ($f_B \approx 1$). One part of the sequence is not required for replication and hence may vary unrestrictedly. If variation of the unused part leads to a sequence that has a positive effect on the fitness of the whole polynucleotide, it will proliferate and eventually become dominant in the population. An increase in fitness may be caused simply by changes in the secondary structure of the polynucleotide. More interesting is a case when the unused part is translated and the translation product gains catalytic activity. Successive gene duplications may lead to whole families of proteins with different substrate specificities.

The new catalytic function may eventually improve the accuracy of replication. Then gene duplication represents a powerful evolutionary mechanism to increase the catalytic capacity of a replicating system, which makes replication more precise whenever this is necessary and possible.

Present-day structures of proteins provide some hints on the role of gene duplications in evolution. Proteins commonly occur in families. These are structurally related enzymes catalyzing reactions of the same class with different substrate specificities. Examples are families of proteases or dehydrogenases. In addition to this, one observes interesting regularities in the structures of many globular proteins: Substructures (so-called motifs) are often repeated exactly or with minor modifications only. Such repetitions were found in the same protein molecule as well as in different protein molecules. Both the modular structure of polymers as well as the existence of protein families can be explained by a gene duplication mechanism.

Last but not least, we may mention the exon–intron structure of eukaryotic genomes. Though error rates are generally adapted to the full genome size and throughout found to be smaller than 10^{-10}, fidelity requirements may differ for exons and introns and even vary within both. For instance, part of the intron regions may have signal character and therefore ought to be highly conserved. The relations referred to in this section will allow a quantitative treatment of such cases.

5. Analogies to Phase Transitions

Numerical studies on quasi-species localization [42]—as they will be presented and discussed in Section IV—showed a sharpening of error thresholds with increasing chain lengths ν. *Sharpening* means here that the transition zone from localized quasi-species to delocalized sequence distributions becomes narrower. This phenomenon is reminiscent of cooperative transitions observed with conformational equilibria in biopolymers. We shall investigate here whether or not error thresholds show analogy to phase transitions in the limit $\nu \to \infty$. The equilibrium statistical mechanics of lattice models and the dynamics of Markov processes can be described within the same mathematical discipline of "*Markov random fields*" [43], and therefore, similarities in global behavior such as the existence of cooperative phenomena or phase transitions are not completely unexpected. Spin lattice models are particularly well-suited candidates in the search for analogies between equilibrium properties and replication dynamics because several of them are sufficiently simple to allow derivations of analytical expressions of thermodynamic functions [44].

Analogies between replication dynamics and spin lattice models were investigated in recent publications by Demetrius [45] and Leuthäusser [46, 47]. Both approaches are based on a common concept, and we shall discuss them here together. Replication dynamics is considered as a dynamical system in discrete time and modeled by the difference equation

$$x_i(t+\Delta t) - x_i(t) = \sum_{k=1}^{n} W_{ik} x_k(t) - [D_i + \Phi(t)] x_i(t), \quad i = 1, \ldots, n, \tag{III.20}$$

which is defined on a time axis $\ldots, t-\Delta t, t, t+\Delta t, \ldots$. The discrete dynamics of the replication–mutation process is, of course, not identical to the dynamics of the kinetic Eqn. (A4.1) since it assumes synchronized replications. In the limit of infinite time ($t \to \infty$), however, the distributions of polynucleotide sequences in the discrete system converge asymptotically toward the stationary solutions of the differential equation [48].

The discrete dynamical system can be modeled by a multitype branching process provided the flux term $\Phi(t)$ is neglected. Then the distribution of sequences is derived from proper statistics of *genealogies*. A genealogy is an individual time-ordered series of sequences that represents one particular recording of successive descendants: each member of a given genealogy was synthesized by correct or erroneous copying of its precursor as template (Figure 8). The individual genealogies are compared with one-dimensional lattices of generalized spins. Every spin has $n = \kappa^\nu$ different states, 2^ν for binary sequences and 4^ν for polynucleotides. Interactions on the lattice are restricted to nearest neighbors. This restriction follows naturally from replication dynamics or from the Markov property of the branching process: the probability distribution of copies—correct replicas and mutants—depends on the template to be copied and *not* on its precursors in the genealogy. For the spin lattices the nearest-neighbor approximation is a more serious restriction since magnetic interactions may reach much further. It is nevertheless the basis of a well-studied class of models known as Ising lattices.

Probability distributions on the genealogies correspond to probability distributions on the lattice of spins. The connection between the discrete

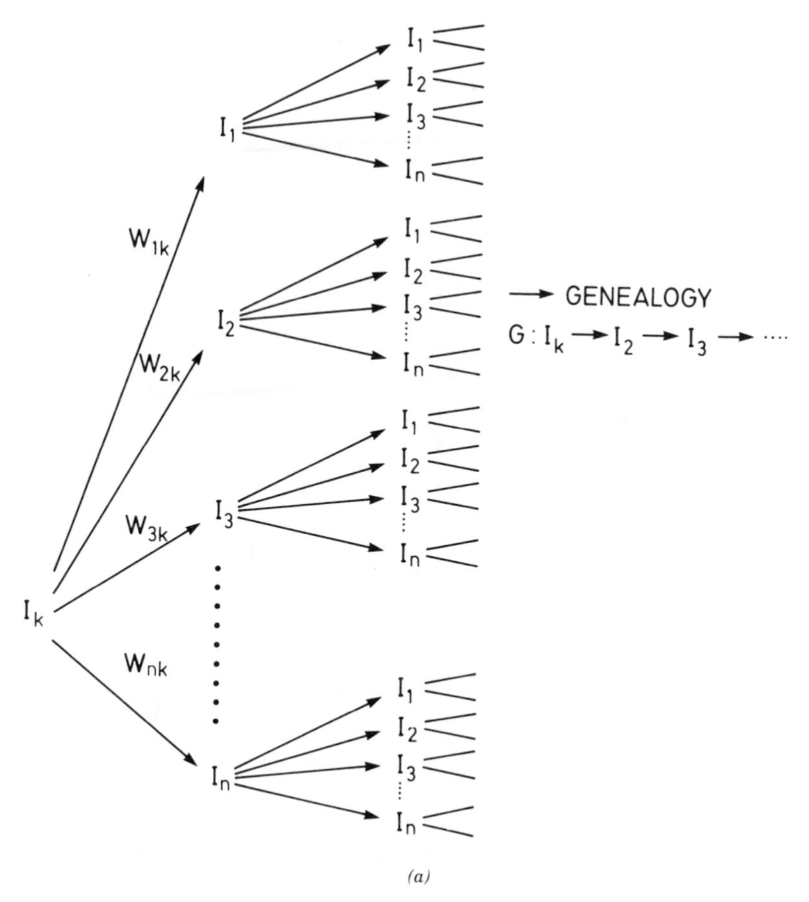

Figure 8. Polynucleotide replication as multitype branching process is compared with spin lattice models: (A) One-dimensional model is based on generalized one-dimensional Ising lattice: Every spin is assumed to exist in n different states corresponding to n different polynucleotide sequences. Genealogy of branching process is considered as analog of particular one-dimensional arrary of spins.

dynamics and the lattice model is based on the following fact: The dominant eigenvalue of the value matrix W, λ, satisfies a variational principle formally identical to the minimization of the free energy in statistical mechanics. As a consequence, the macroscopic quantities in replication dynamics can be compared with the thermodynamic variables in lattice models. The free energy, for example, corresponds to the logarithm of the dominant eigenvalue of the value matrix W: $F \propto \log \lambda$. The product of entropy S and temperature T is the negative analog of the *complexity parameter* H of the stationary

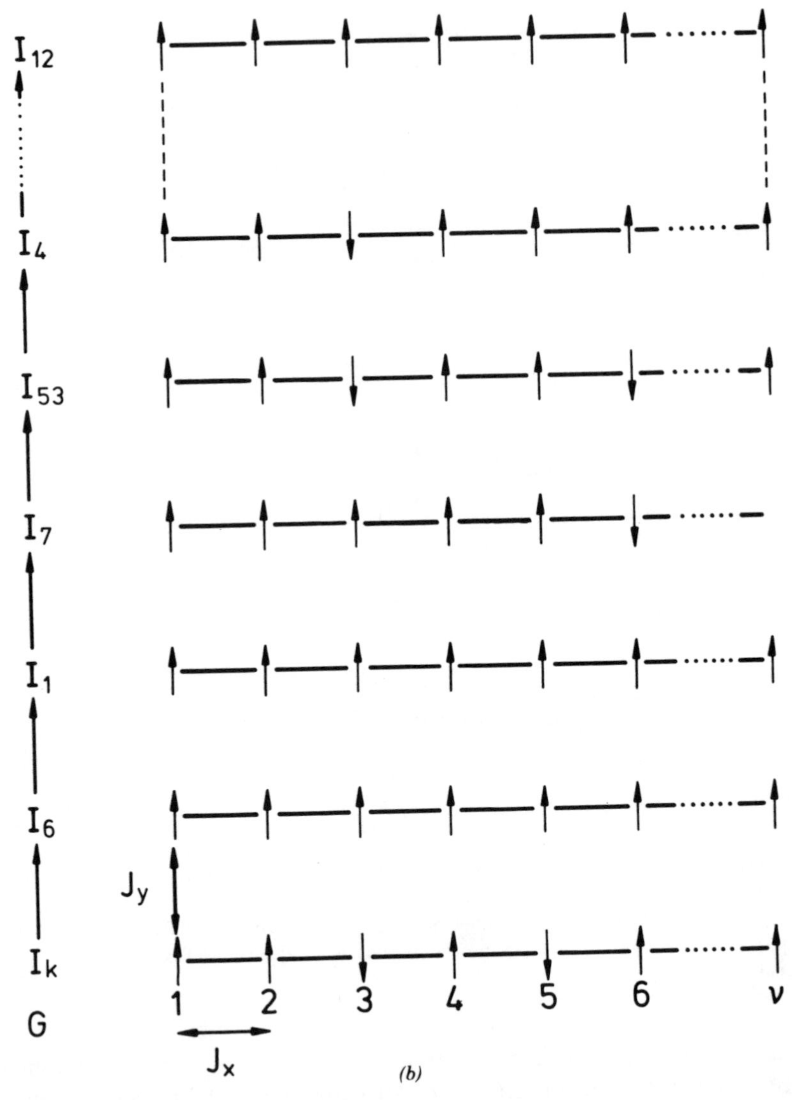

(b)

Figure 8. (B) In two-dimensional model genealogy is represented by two-dimensional spin lattice. Individual spins exist in two states (↑ and ↓) corresponding to two digits in binary sequences. Every row of lattice consists of ν digits and corresponds to polynucleotide sequence. "In-row" interaction, described by spin–spin coupling constant J_x, is property of individual sequence and contributes to rate constant of replication. "Vertical" coupling constant J_y, on the other hand, is measure of mutation frequency.

sequence distribution: $TS \propto -H$. This complexity parameter is closely related to the Shannon entropy of the sequence distribution, $H_{(\mathrm{Shannon})} = -\Sigma p_k \log p_k$, where p_k is the probability of finding the system in state k [49]. Since the product TS is fixed by analogy to the spin system, the choice of the function corresponding to temperature allows some freedom and requires physical intuition.

In order to search for an interpretation of the error threshold relation by analogy to the spin system, we shall be more specific and consider binary sequences replicating with uniform error rates. Individual sequences are identified with the rows of a two-dimensional spin lattice. A genealogy corresponds to an entire, two-dimensional array of spins. We assign *spin values* ($s = \pm 1$) to the digits (0, 1). A sequence of ν digits is identified with a string of spin values:

$$I_k \equiv (s_1^{(k)}, s_2^{(k)}, \ldots, s_\nu^{(lk)}). \tag{III.21}$$

The Hamming distance $d(i, k)$ and the elements of the value matrix W can be expressed in terms of these spin values:

$$d(i, k) = \frac{1}{2}\left(\nu - \sum_{j=1}^{\nu} s_j^{(i)} s_j^{(k)}\right) \tag{III.22}$$

and

$$\begin{aligned} W_{ik} &= A_k [q(1-q)]^{\nu/2} \left(\frac{1-q}{q}\right)^{-1/2\Sigma_j s_j^{(i)} s_j^{(k)}} \\ &= A_k [q(1-q)]^{\nu/2} \exp\left(-K \sum_{j=1}^{\nu} s_j^{(i)} s_j^{(k)}\right), \end{aligned} \tag{III.23}$$

where $K = \frac{1}{2} \log [(1-q)/q]$.

Now we consider a two-dimensional spin lattice of dimension $\nu \times N$. The vector $\sigma^{(i)} = (\sigma_1^{(i)}, \sigma_2^{(i)}, \ldots, \sigma_\nu^{(i)})$ describes the spins in the ith row of the lattice. The spin variables can adopt the two values $\sigma_j^{(i)} = \pm 1$. The nonvanishing terms of the spin Hamiltonian in the nearest-neighbor approximation and in absence of an external magnetic field are

$$H = J_x \sum_{i=1}^{N} \sum_{j=1}^{\nu-1} \sigma_j^{(i)} \sigma_{j+1}^{(i)} + J_y \sum_{j=1}^{\nu} \sum_{i=1}^{N-1} \sigma_j^{(i)} \sigma_j^{(i+1)}. \tag{III.24}$$

Here we denote the constant of the nearest-neighbor spin interaction in the horizontal direction, that is, within a row, by J_x and that in the vertical direction by J_y. The partition function of the spin lattice can be factorized within the Ising model by means of a $2^\nu \times 2^\nu$ transfer matrix. The elements of this transfer matrix are obtained as Boltzmann weights of the energies of interaction between individual spin distributions:

$$T_{\sigma\sigma'} = \exp(-\beta E_{\sigma\sigma}) \exp\left(-\beta J_y \sum_{j=1}^{\nu} \sigma_j \sigma'_j\right). \qquad \text{(III.25)}$$

Herein, two neighboring rows of the two-dimensional lattice are labeled by $\sigma = (\sigma_1, \sigma_2, \ldots, \sigma_\nu)$ and $\sigma' = (\sigma'_1, \sigma'_2, \ldots, \sigma'_\nu)$, respectively. The energy $E_{\sigma\sigma}$ describes the interaction of spins within a row. It need not be restricted to nearest-neighbor terms and may also include interactions with a magnetic field. Temperature enters into the expression for $T_{\sigma\sigma'}$ through $\beta = 1/kT$. Macroscopic properties and spin distributions of the spin lattices can be obtained from eigenvalues and eigenvectors of the transfer matrix.

In the two-dimensional Ising model the critical temperature T_c for the order–disorder transition of the infinite system ($N \to \infty$, $\nu \to \infty$) is given by

$$\sinh \frac{2J_x}{kT_c} \sinh \frac{2J_y}{kT_c} = 1.$$

A plot of this relation is shown in Figure 9. The transition temperature

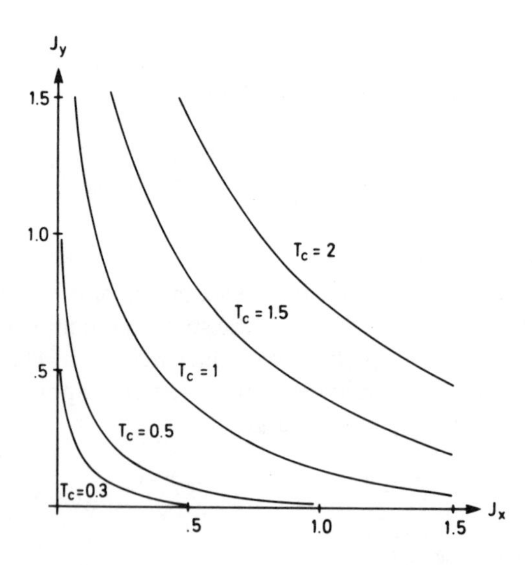

Figure 9. Critical temperture T_c of order–disorder transition in two-dimensional Ising lattice as function of coupling constants J_x and J_y. Note that transition temperature approaches zero when either of two coupling constants vanishes.

converges to $T_c = 0$ in case either of the two spin–spin coupling constants vanishes. In the one-dimensional Ising model we have only a "*degenerate*" phase transition at absolute zero.

Let us now turn to the analogy between the two matrices W and T on which the comparison between replication dynamics and spin lattice is based. The two last factors in Eqs. (III.23) and (III.25) show the same dependence on the spin variables and allow to identify K and βJ_y. This analogy provides a measure for "*temperature*" in replication dynamics:

$$\frac{1}{T} \propto \left| \log \frac{1-q}{q} \right| .$$

Replication degenerates to a random production of sequences in the limit $q \to \frac{1}{2}$ and corresponds to the limit $T \to \infty$, the case of maximum disorder. Direct and complementary replication (see also part C) are the analogs of ferro- and antiferromagnetic cases of the spin system. In the range $\frac{1}{2} < q \leqslant 1$ we have $K < 0$, which corresponds to the condition $\beta J < 0$ for ferromagnetic interaction. For complementary or plus–minus replication, $0 \leqslant q < \frac{1}{2}$, we have $K > 0$ and $\beta J > 0$, respectively, which is characteristic of an antiferromagnet.

A comparison of the first two factors on the right side of Eqn. (III.23) with the first factor in (III.25) is less straightforward. In the transfer matrix of the Ising model we have an "*in-row*" interaction energy of the type

$$E_{\sigma\sigma} = J_x \sum_{j=1}^{\nu-1} \sigma_j \sigma_{j+1},$$

which represents the first Boltzmann weight in the absence of a magnetic field. The energy $E_{\sigma\sigma}$ is an extensive quantity, and thus $|E_{\sigma\sigma}|$ increases linearly with ν. In addition this term shows a unique dependence on the sequence of digits, which is characteristic of spin–spin interactions. The analogy of the "*horizontal*" interaction factors of the spin lattice with replication kinetics requires

$$\exp(-\beta E_{\sigma\sigma}) = \exp\left(-\beta J_x \sum_{j=1}^{\nu-1} \sigma_j \sigma_{j+1} \right) \propto \exp(\log A_k)$$

apart from constant terms that can be annihilated by appropriate energy scaling; any constant term in the energy enters as a common factor into the partition function and hence has no influence on spin distributions or order–disorder transitions.

In the case of ultimate neutrality, all rate constants A_k are equal and analytical solutions can be derived, as is shown in Appendix 2. In terms of the

Ising model this case corresponds to the limit $J_x \to 0$, where the two-dimensional Ising lattice degenerates to the one-dimensional spin system. We are dealing with a degenerate phase transition:

$$\lim_{\nu\to\infty} T_c(\nu)=0 \quad \text{or} \quad \lim_{\nu\to\infty} q_c(\nu)=0,1.$$

There is no phase transition at finite temperatures in the asymptotic limit of infinite chains. This is no surprise since no deterministic selection occurs in the absence of differences in rate constants.

When strictly based on the Ising model the analogy puts a severe restriction on the choice of rate constants, and it is unlikely that any realistic system can meet it. We may, however, extend the comparison to more general spin systems. Many different spin Hamiltonians were proposed and analyzed. To give an example, we mention the Hopfield Hamiltonian [50], which was applied to biological problems, in particular to the study of neutral networks. A strategy proposed by Leuthäusser [46] starts from a given set of replication rate constants—from a "*value landscape*" of the replication system—and tries to find a Hamiltonian that is able to reproduce the rate constants as single row energies: $E_i = -(1/\beta)\ln A_i$. There are several ways to render the expression for the energy more flexible: the presence of an external magnetic field, further neighbor spin interactions, as well as many particle terms may be taken into account. It remains to be shown, however, that physically meaningful value landscapes indeed can be mapped onto more general spin Hamiltonians that will be susceptible to mathematical analysis.

Despite the fact that it may be very difficult to find spin coupling schemes that correspond to realistic value landscapes and can be directly analyzed, the analogy to spin lattices is of great heuristic value. It provides a straightforward explanation of the existence of well-defined error thresholds that sharpen with increasing chain length ν, just as cooperative transitions do in linear biopolymers.

IV. EXAMPLES OF FITNESS LANDSCAPES AND STATIONARY POPULATIONS

The quasi-species was introduced in Section II as a mutant distribution localized in a certain region of sequence space where it is centered around one or several degenerate master sequences. The transition from a nonlocalized to a localized distribution, or the transition among two localized distributions triggered by the appearance of an advantageous mutant, has been shown to be analogous to a phase transition in physical space. In Section III the threshold relations that govern such phase transitions in sequence space have been discussed in detail. In this section we shall see how population structures and phase transitions depend on the detailed properties of sequence space. Each point in sequence space characterizing a particular sequence is to be assigned a so-called fitness value that expresses the dynamic properties of the particular sequence, such as the replication and decomposition rate. These fitness values will show a more or less continuous short-range order building up to a value landscape that might be characterized by some fractal order, as do landscape topographies on earth. Evolution to optimal fitness may be regarded as a series of phase transitions in which the center of gravity of the localized mutant distribution moves through sequence space, approaching the highest reachable fitness peak. This motion depends not only on the fitness gradients among the 4^{ν} points in this space but also on the transition probabilities as expressed by the $4^{\nu} \times 4^{\nu}$ error or mutation matrix. Both dynamical parameters and the error matrix were used to construct the value matrix W that characterizes the systems of rate equations introduced in Section II.

In the following we discuss examples ranging from the simplest nontrivial cases, for which a full solution is possible, to rather realistic value matrices based on polynucleotide folding, where the analysis of Sections II and III must be applied. The latter is the subject of Section IV.3, while Sections IV.1 and IV.2 concentrate on examples for which an exact solution is numerically feasible.

In principle, stationary mutant distributions may be calculated from the value matrix W by standard diagonalization techniques. Accordingly, we have to compute the dominant eigenvector of a $4^{\nu} \times 4^{\nu}$ matrix (for the four bases occurring in present polynucleotides). For most cases of interest (e.g., $\nu > 20$) the problem is clearly intractable (but see the statistical analysis in Section III). This is particularly true for investigations of the error threshold, so in Section IV.1 we study a value matrix with high symmetry [41] for which longer sequences are tractable. In Section IV.2 we use general value matrices for short sequences to investigate the effect of neutral mutants and quasi-species with degenerate master sequences.

For the calculation of stationary mutant distributions we restrict attention to a uniform error rate per digit $(1-q)$ and assume equal degradation rate coefficients $D_1 = D_2 = \cdots = D_n = D$. Since the addition of a constant to all diagonal elements of a matrix just shifts the spectrum of eigenvalues and has no influence on the eigenvectors, we need only consider the case $D=0$ without loss of generality. Then the elements of the matrix W are determined by the replication rate coefficients A_k (as in Section III.2) and are of the form

$$W_{ik} = A_k q^{\nu - d(i,k)}(1-q)^{d(i,k)} = A_k q^{\nu}\left(\frac{1-q}{q}\right)^{d(i,k)}, \tag{IV.1}$$

where $d(i, k)$ is the Hamming distance introduced in Section II.1. As we mentioned in Section II.5, a value matrix whose elements are defined by Eqn. (IV.1) has exclusively real and positive eigenvalues. Finally, we choose two digits (0, 1) instead of the four bases occurring in natural polynucleotides in the following examples, as this simplifies both the representations and computations without detracting from the main features.

1. Error Threshold

One approach to calculating the stationary mutant distributions for longer sequences is to form classes of sequences within the quasi-species. These classes are defined by means of the Hamming distance between the master sequence and the sequence under consideration. Class 0 contains the master sequence exclusively, class 1 the ν different one-error mutants, class 2 all $\nu(\nu-1)/2$ two-error mutants, and so on. In general we have all $\binom{\nu}{k}$ k-error mutants in class k. In order to be able to reduce the 2^ν-dimensional eigenvalue problem to dimension $\nu+1$, we make the assumption that all formation rate constants are equal within a given class. We write A_0 for the master sequence in class 0, A_i for all one-error mutants in class 1, A_2 for all two-error mutants in class 2, and in general A_k for all k error mutants in class k.

New variables ξ_k are introduced for the relative concentrations of the sum of all k-error mutants:

$$\xi_k = \sum_j x_j, \quad I_j \in \text{Class } j, \quad \text{and hence} \quad \sum_k \xi_k = 1.$$

Finally, it remains to calculate the elements of the matrix Q' consisting of the mutation frequencies from one class of mutants into another. By straightforward combinatorics, we find

$$Q'_{ik} = q^\nu \sum_{j=0}^{m} \left(\frac{1-q}{q}\right)^{2j-|k-i|} \binom{\nu-k}{j+\frac{1}{2}(|k-i|-k+i)} \binom{k}{j+\frac{1}{2}(|k-i|+k-i)}$$

with $m = [\frac{1}{2}(\min\{k+i, 2\nu-(k+i)\} - |k-i|)]$. (IV.2)

Note that the matrix Q' is not symmetric, in contrast to the matrix of mutation frequencies between individual sequences [cf. Eqn. (IV.1)]. Multiplication by the corresponding rate constant A'_k yields the corresponding element of the matrix W'. This matrix has to be diagonalized in order to obtain the dominant eigenvector.

The fact that the matrix W' is related to the matrix W by lumping classes of mutants together suggests the existence of a method to relate W' to a symmetric matrix and directly exploit the positivity of the eigenvalue spectrum. In fact, the matrix Q' may be readily symmetrized by normalizing the classes by their number of elements $\binom{\nu}{k}$, and then the matrix W' is related to a symmetric matrix by the standard Jost procedure outlined by Rumschitzky [8]. This reduces the size of the computational problem.

The first special case to be discussed here considers oligonucleotide sequences of chain length $\nu = 5$. The stationary mutant distribution as a function of the mean single-digit accuracy, $\bar{\xi}_k(q)$ $(k = 1, \ldots, \nu)$, is plotted in Figure 10 for a particular choice of the A_ks. We distinguish three ranges of q with different properties of the stationary distribution:

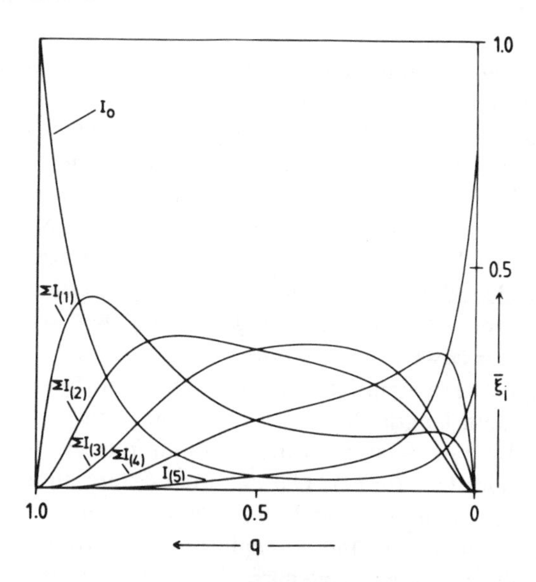

Figure 10. Quasi-species as function of single-digit accuracy of replication (q) for chain $\nu = 5$. We plot relative stationary concentration of master sequence ($\bar{\xi}_0$), sum of relative stationary concentrations of *all* one-error mutants ($\bar{\xi}_1$), of all two-error mutants ($\bar{\xi}_2$), etc. Note that we have only one five-error mutant $I_{(5)} = I_{31}$ in this particular example. We observe selection of master sequence at $q = 1$. Then relative concentration of master sequence decreases with decreasing q. At value $q = 0.5$ all sequences are present in equal concentrations. Hence, sums of concentrations of two- and three-error mutants are largest—they have statistical weight of 10—those of the one- and four-error mutants are half as large—they have statistical weight of 5—and finally master sequence I_0 and its complementary sequence, the five-error mutant I_{31}, are present in relative concentration of $\frac{1}{32}$ only. At $q = 0$ we have selection of "*master pair*", which consists of I_0 and I_{31} in our example. Thus we have direct replication with errors in range $1 > q > 0.5$ and complementary replication with errors in range $0 < q < 0.5$. Rate constants chosen as $A_0 = 10[t^{-1}]$ and $A_k = 1[t^{-1}]$ for all mutants $k \neq 0$. Here we denote arbitrary reciprocal time unit by $[t^{-1}]$. All degradation rate constants were put equal: $D_0 = D_1 = D_2 = \cdots = D_{31} = 0$.

1. At large values of the accuracy of replication ($q \simeq 1$) we observe a quasi-species characteristic for "direct" replication, $I_k \rightarrow 2I_k$ predominantly. The master sequence I_0 is most frequent, followed by some one-error mutants, two-error mutants, and so on.

2. In the middle of the plot shown in Figure 10, around the value $q=0.5$, incorporations of correct and incorrect digits occur with equal probabilities. Inheritance breaks down, and polynucleotide sequences are present according to their statistical weights: The sum of the relative concentrations of one- and four-error mutants, $\bar{\xi}_1$ and $\bar{\xi}_4$, is five times as large as that of the master sequence $\bar{\xi}_0$ or that of the five-error mutant $\bar{\xi}_5$. The statistical weights of two- and three-error mutants is 10. In order to point out the lack of sequence correlations between template and copy we called this process *random replication* [41].

3. At $q=0$ complementary digits are incorporated with complete accuracy. Thus, we notice "complementary" replication, $I_k^+ \rightarrow I_k^+ + I_k^-$ and $I_k^- \rightarrow I_k^- + I_k^+$ (where I_k^+ and I_k^- denote the plus and minus strand of the sequence I_k). Sequences are selected in pairs of complementary strands. We have a "master pair" that replaces the master sequence of direct replication. In our particular example this is the pair (I_0, I_5). At low nonzero values of the mean single-digit accuracy ($q \simeq 0$) we observe complementary replication with mutation.

For oligonucleotides of very small chain lengths, such as the ones we considered in Figure 10, the three ranges are not well separated. Random replication in a strict sense occurs at $q=0.5$ only. Indeed, there is practically no correlation between template and copy in the entire flat range $0.4<q<0.7$.

Increasing chain length changes the general features of the $\bar{\xi}_1$, q plots rather drastically. For $\nu=10$ the range of random replication appears to be substantially wider (Figure 11). The $\bar{\xi}_1$, q curves are almost horizontal on both sides of the maximum irregularity condition at $q=0.5$. In addition, the transitions from direct to random replication and from random to complementary replication are rather sharp. We are now in a position to compare the minimum accuracy of replication that we derived in Section III by perturbation theory with the exact population dependence on q. From Eqs. (III.1) and (III.4) we find ($D_k=0$; $k=0, 1, \ldots, n$)

$$q_{\min}=(\sigma_0)^{-1/\nu}=\left(\frac{\bar{A}_{-0}}{A_0}\right)^{1/\nu} \quad \text{with} \quad \bar{A}_{-0}=\sum_{k=1}^{\nu} A_k\bar{\xi}_k \Big/ (1-\bar{\xi}_0), \tag{IV.3}$$

and we may evaluate this directly because the A_k were chosen equal for $k \neq 0$.

The value of $q_{\min}$ calculated by perturbation theory in lowest order falls right into the center of the transition zone that separates the organized quasi-species from the uniform distribution. It represents a good approximation to

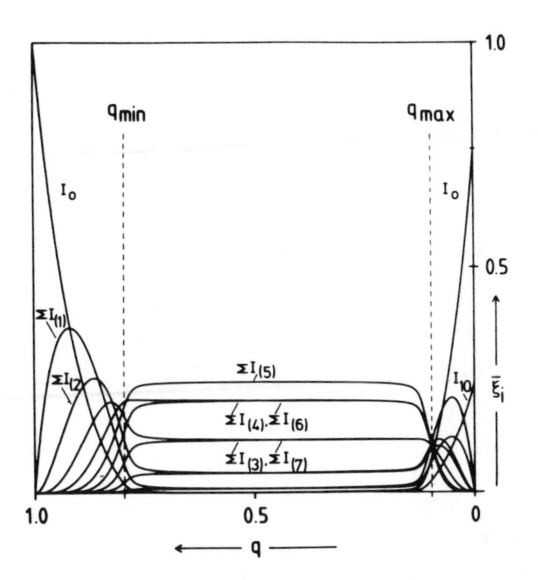

Figure 11. Quasi-species as function of single-digit accuracy of replication (q) for chain length $v = 10$. Computations were performed in complete analogy to those shown in Figure 10. Note that range of "*random replication*" has increased substantially compared to case $v = 5$. We observe fairly sharp transitions between direct and random replication at critical value $q = q_{min}$ and between random and complementary replication at $q = q_{max}$.

the exact error threshold. Indeed, the high-order perturbation result from Eqn. (III.10) agrees well with Eqn. (IV.3) in this special case.

On the other hand, high accuracy in *complementary* replication is characterized by low q values. Here, we have a maximum q value that sets a limit to the errors tolerated in complementary replication. From perturbation theory we derive

$$1 - q_{max} = (\sigma_{0,\bar{0}})^{-1/v} = \left(\frac{\bar{A}_{-0,\bar{0}}}{\sqrt{A_0 A_{\bar{0}}}} \right)^{1/v}$$

$$\text{with } \bar{A}_{-0,\bar{0}} = \sum_{k \neq 0} \sqrt{A_k^+ A_k^-} (\bar{x}_k + \bar{x}_{\bar{k}}) \Big/ \sum_{k \neq 0} (\bar{x}_k + \bar{x}_{\bar{k}}). \tag{IV.4}$$

Concentrations, rate constants, and other quantities referring to the complementary sequence are indicated by the minus symbol ($I_k^+ : \bar{x}_k, A_k^+, \ldots,$ and $I_k^- : \bar{x}_{\bar{k}}, A_k^-$). The value of q_{max} coincides with the center of the transition zone from random to complementary replication.

At still larger chain lengths (we used $v = 50$ in the calculations shown in Figure 12), the transitions become exceedingly sharp. The shape of the plots

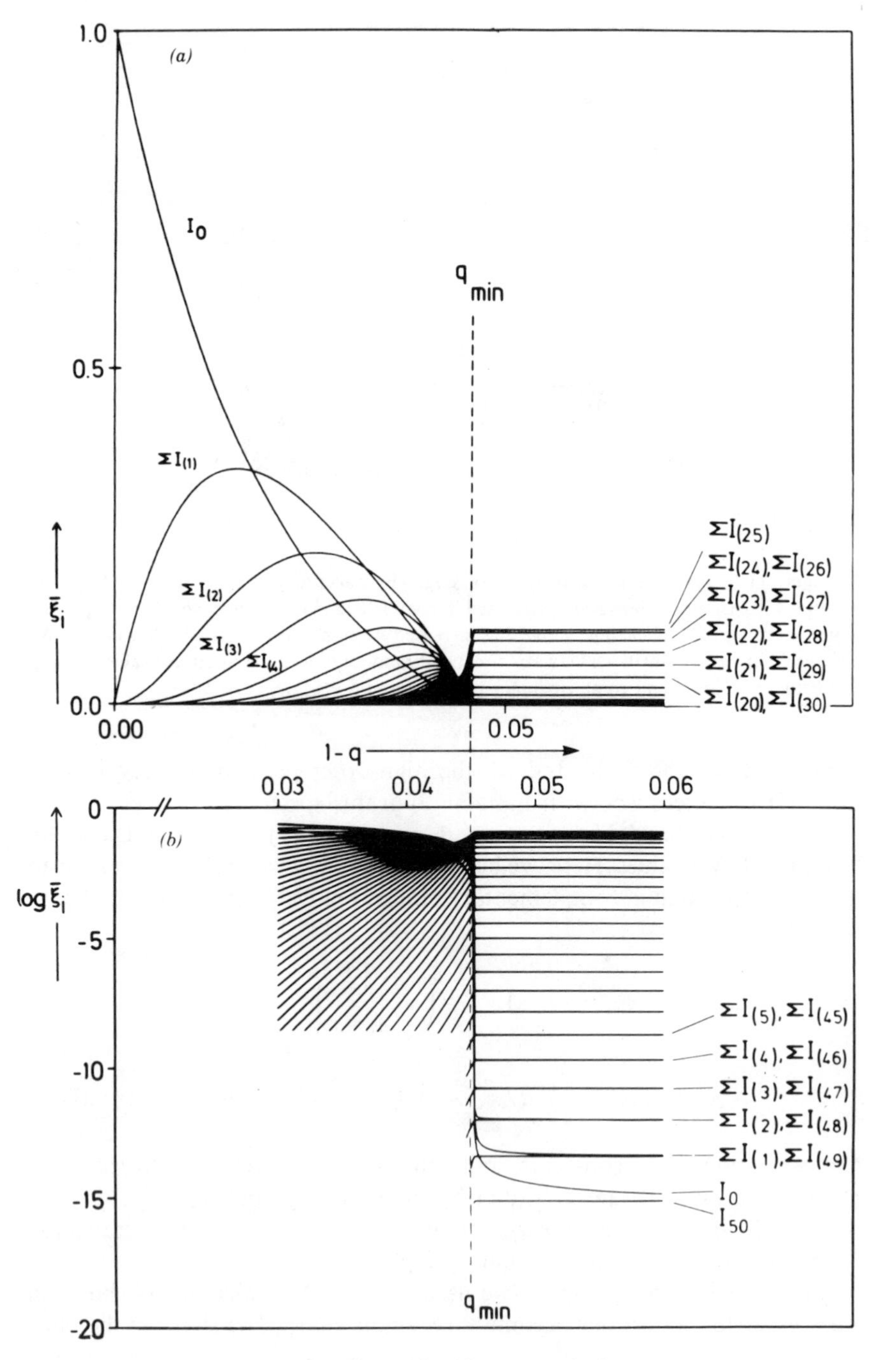

Figure 12. (*Caption overleaf*)

shown in the Figures 10–12 suggests that we are dealing here with an analog of a phase transition as discussed in detail in Section III.5.

2. Degenerate Quasi-species and Neutral Mutants

Mutants that have the same selective value as the master sequence are called (selectively) neutral. We distinguish two classes of neutral mutants:

1. neutral mutants that are "closely related" to the master sequence and
2. distant neutral mutants.

Closely related neutral sequences are present simultaneously in stationary populations. Distant sequences, on the other hand, occur together only occasionally in transient populations. Their mutant distributions are disjoint in the sense that they have no common mutant. Whether or not two selectively neutral sequences share mutants depends both on the Hamming distance and population size. In the first case, the mutant distribution is centered around a dominant, more or less connected group of polynucleotides (replacing the single master sequence considered so far). They are members of the same connected set of mutants and share a common stationary mutant distribution, which we call a *degenerate* quasi-species. However, for two or more neutral master sequences to be equally represented, their mutants have to give identical contributions too. This phenomenon is not considered in the classical study of neutral mutants, which neglects the mutant distribution. Symmetry is unlikely, so some degeneracies will depend on the error rate via a change in the mutant distribution. At different error rates, different neutral sequences are favored. In any case, the dominant eigenvector of the matrix W is representative for the actually observed distribution of sequences.

The second case shows very different behavior: The relative concentrations of the degenerate master sequences are subject to random drift, and the dominant eigenvector of W represents at best a time average of the mutant distribution. Then the dynamics can be modelled only by a stochastic process requiring careful choice of the appropriate mathematical technique and approximations in a hierarchy of equations (see refs. 48 and 51 and Section V.2). One difficulty here is that even very distant mutants contribute if sufficiently neutral. The results of Section III.2 indicate that there is (almost

---

Figure 12. Quasi-species as function of single-digit accuracy of replication (q) for chain length $\nu = 50$. Computations were performed in complete analogy to those shown in Figure 10. Transition is very sharp at this chain length already. We show linear plot (a) and, in order to demonstrate sharpness even more clearly, a logarithmic plot (b) of relative concentrations around critical point $q = q_{\min}$.

surely) a trade-off between distance and accurate neutrality that results in a preservation of the localization threshold behavior.

Here we restrict the discussion to examples of the first class (for further details see ref. 52) and begin with a simple example. Two sequences $I_0^{(1)}$ and $I_0^{(2)}$ have the same selective value W_0. This value is a maximum, $W_0 = \max [W_k; k=1, \ldots, n]$. We start the discussion of the mutant distribution around the two master sequences with the limit $q \rightarrow 1$. For $q=1$ we have a degenerate eigenvalue: $\lambda_1 = \lambda_2 = W_0$, and hence every combination of concentrations of the two master sequences with $\bar{x}_0^{(1)} + \bar{x}_0^{(2)} = 1$ represents an eigenvector of the matrix W. A small perturbation $(q=1) \rightarrow (q=1-\delta)$ destroys this degeneracy. From perturbation theory of degenerate states, which becomes exact in the limit considered, we derive that there exists a unique pair of stationary concentrations $(\tilde{x}_0^{(1)}, \tilde{x}_0^{(2)})$ that is the proper reference state. These relative stationary concentrations of the two master sequences are determined by the Hamming distance $d(1,2)$ and by the selective values of the neighboring mutants. It is useful to distinguish three subcases shown in Figure 13:

1. The two master sequences are nearest neighbors in sequence space ($d(1,2)=1$):

$$\lim_{q \rightarrow 1} \tilde{x}_0^{(1)}(q) = \lim_{q \rightarrow 1} \tilde{x}_0^{(2)}(q) = \tfrac{1}{2}.$$

   Both sequences are represented with equal frequency in the degenerate quasi-species.

2. The two master sequences have a Hamming distance $d(1,2)=2$. The normalized concentrations are then

$$\lim_{q \rightarrow 1} \tilde{x}_0^{(1)}(q) = \alpha, \quad \lim_{q \rightarrow 1} \tilde{x}_0^{(2)}(q) = 1 - \alpha \quad \text{with } 0 < \alpha < 1.$$

   The value of α is determined by the selective values of the mutants in the neighborhoods of $I_0^{(1)}$ and $I_0^{(2)}$.

3. The two master sequences have Hamming distances $d(1,2) \geqslant 3$. This leads to selection in the limit considered:

$$\lim_{q \rightarrow 1} \tilde{x}_0^{(1)}(q) = 1, \qquad \lim_{q \rightarrow 1} \tilde{x}_0^{(2)}(q) = 0$$

   or

$$\lim_{q \rightarrow 1} \tilde{x}_0^{(1)}(q) = 0, \qquad \lim_{q \rightarrow 1} \tilde{x}_0^{(2)}(q) = 1.$$

   The decision of which sequence is selected is made again by the neighboring mutants.

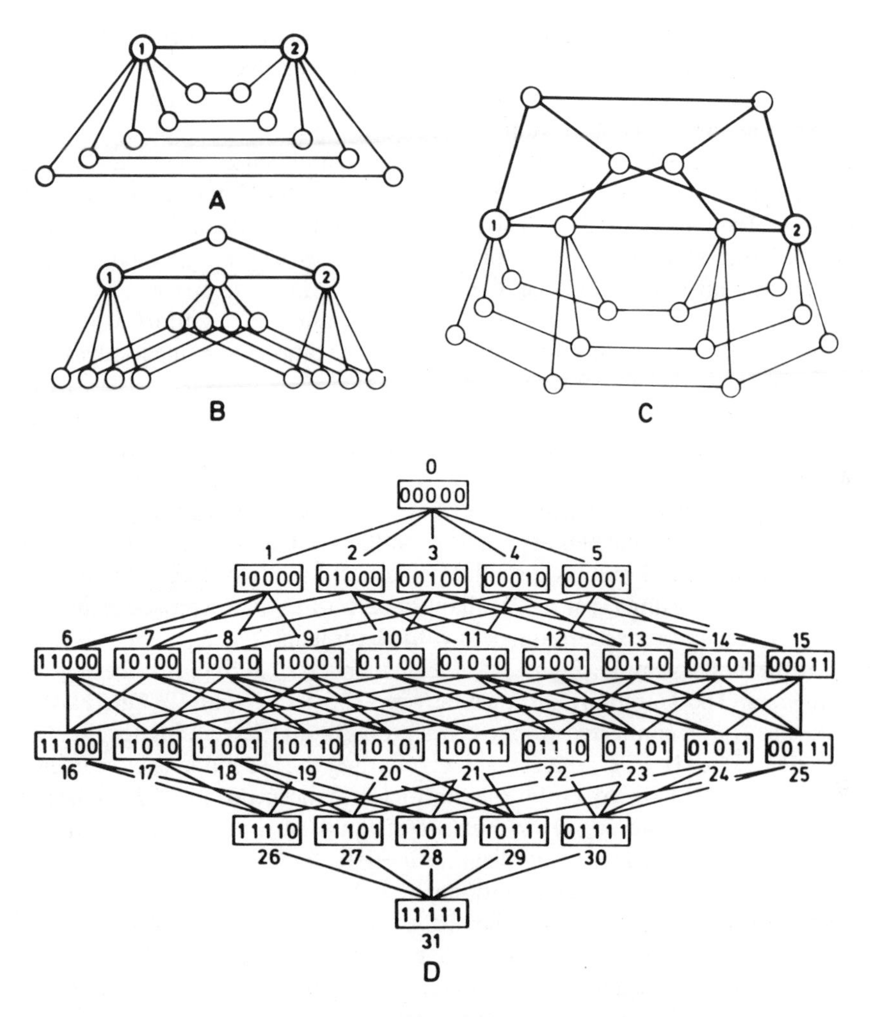

Figure 13. Neighborhood relations in sequence space. In A, B, and C we show neighborhoods of pair of sequences (I_1, I_2) with Hamming distances $d(1, 2)$ of 1, 2, or 3, respectively ($\nu = 6$). Part drawn in thick lines is general; connections in thin lines depend on chain length ν. To give example, for $\nu = 5$ one connection of equivalent set has to be eliminated. (D) Assignment of numbers to individual sequences of sequence space for $\nu = 5$ shown as used in Figures 14–30.

In the limit $q \rightarrow 0$ we observe an analogous situation. Here the degenerate master sequences are to be replaced by degenerate master pairs. A pair of complementary sequences always has a Hamming distance $d(I_k, I_{k^-}) = \nu$. As the Hamming distance between two pairs we use the smallest of the three distances computed for the four sequences. Then the selection behavior derived for degenerate sequences in the limit $q \to 1$ applies also for degenerate pairs in the limit $q \to 0$.

Around the point of maximum irregularity ($q=0.5$) the replication rates have no influence on the sequence distribution, and hence degeneracy of rate constants causes no change in the results. In the intermediate ranges, $0.5<q<1$ and $0<q<0.5$ we have to rely on computer calculations. There are two different scenarios: the "simple" case, where we observe gradual changes in the mutant distribution between error-free and error-prone replication, and the "complicated" case, which does not allow straightforward interpolation because drastic rearrangements of the stationary sequence distributions occur within narrow ranges of q values.

First we present examples of the simple case. In the domain of direct replication ($1>q>0.5$) the mutant distributions unfold smoothly around the degenerate sequences. The influence of the neighborhood of the two sequences grows with increasing error rates, that is, with decreasing q values. In the domain of complementary replication ($0<q<0.5$) we observe the analogous formation of mutant distributions around the two degenerate master pairs with increasing q values. We have chosen three characteristic examples of two degenerate master sequences with chain length $\nu=5$ (Figures 14–16).

In the first example (Figure 14) the sequences I_0 and I_6 are the master sequences. The Hamming distance between them is $d(1, 6)=2$. One master sequence (I_0) is connected to one-error mutants that have a higher average selective value ($\bar{W}=5q^{\nu}$) than those surrounding the second master sequence (I_6; $\bar{W}=2.6q^{\nu}$). Consequently, the stationary concentration of I_0 is larger than that of I_6. The one-error mutants of I_0 fall into two groups: mutants (I_1, I_2) that have a Hamming distance $d=1$ to both master sequences and hence are present at higher stationary concentrations than those of the second group (I_3, I_4, and I_5). They have Hamming distance $d=1$ to I_0 but Hamming distance $d=3$ to the other master sequence I_6. The mutant distribution thus nicely reflects the details of the structure of the matrix W. In the limit of ultimately precise complementary replication ($q=0$) we are also dealing with a case of degeneracy. We have two master pairs, (I_0, I_{31}) and (I_6, I_{25}) with identical mean rates of replication: $\sqrt{A_0 A_{31}}=\sqrt{A_6 A_{25}}=\sqrt{10}$. The two master pairs are separated by a Hamming distance $d(\{{}^{0;}_{31;}{}^{6}_{25}\})=2$. They have different neighborhoods and therefore provide an example of case 2 in the limit $q \to 0$.

The second example (Figure 15) considers two distant degenerate sequences I_0 and I_{30}, with $d(0, 30)=4$. Accordingly, we observe selection in the limit $q \to 1$. The sequence with more efficient one-error mutants (I_{30}) is selected. In the domain of complementary replication we are dealing with two

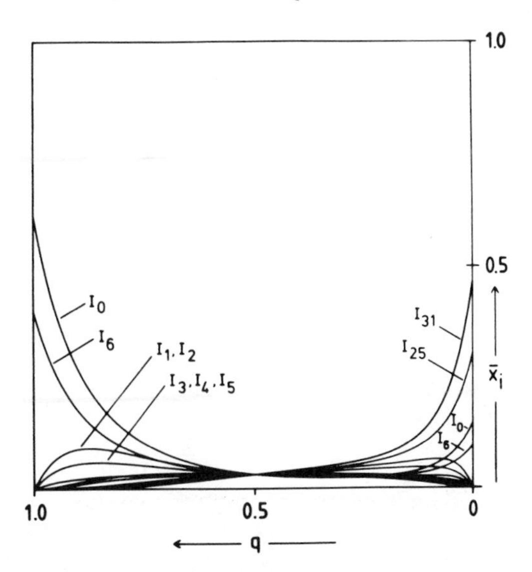

Figure 14. Degenerate quasi-species with $v = 5$. We plot relative concentrations ($\bar{x}_i$) as functions of single-digit accuracy of replication (q). Two degenerate master sequences I_0 and I_6 are of Hamming distance $d(1, 6) = 2$ in this example—for numbering of sequences see Figure 13. Rate constant: $A_0 = 10$, $A_1 = A_2 = \cdots = A_5 = 5$, $A_6 = 10$, $A_7 = \cdots = A_{31} = 1$. In limit $q \to 1$ we are dealing with two master sequences, whose concentrations approach certain ratio $0 < \alpha < 1$. Concentration of I_0 is higher since this sequence has more efficient neighbors. In limit $q \to 0$ we find two master pairs (I_0, I_{31}) and (I_6, I_{25}) that show essentially same qualitative behavior as two degenerate master sequences. They have Hamming distance $d = 2$ as well.

different selective values, and there the more efficient complementary pair (I_0, I_{31}) wins.

The third example (Figure 16) behaves in a very similar manner to the second in the range of direct replication. But the rate constants have been changed such that we now have two degenerate pairs of complementary sequences in the limit $q \to 0$, (I_0, I_{31}) and (I_1, I_{30}). These two pairs have a Hamming distance $d = 1$, and we expect equal concentrations of I_0 and I_{30} and of I_1 and I_{31}, respectively. It is interesting to note that these concentrations remain almost the same nearly for the whole domain of complementary replication.

Our last case (Figure 17) is an example of "complicated behavior" where the internal structure of the quasi-species is completely reorganized at some critical value of q. In the limit $q \to 1$ we are dealing with two nearly degenerate, distant sequences: $A_0 = 10$ and $A_{50} = 9.9$. Accordingly, we observe selection of the more efficient sequence I_0. The sequences in the neighborhood of I_{50},

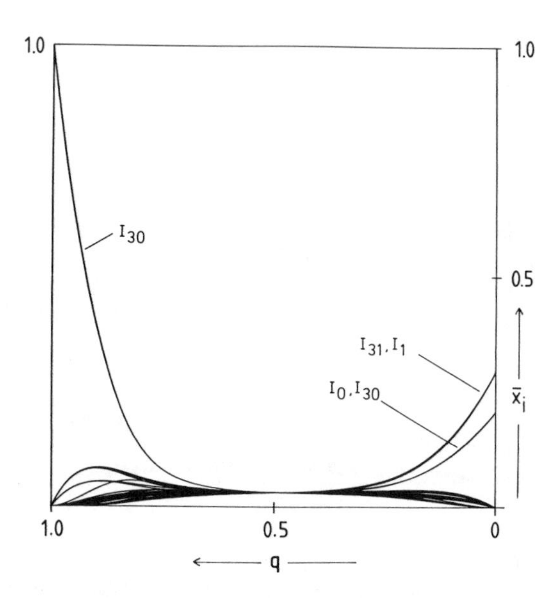

Figure 15. Degenerate quasi-species with $\nu = 5$. Plot is analogous to that shown in Figure 14. Rate constants A_k, in vectorial notation, are $\mathbf{A}$ = (10, 5, 5, 1, 1, 1, 2, 2, 2, 1, 1, 1, 1, 1, 1, 1, 5, 2, 2, 1, 2, 2, 2, 5, 2, 5, 5, 2, 2, 1, 10, 5). Two master sequences I_0 and I_{30} have Hamming distance $d(0, 30) = 4$. In limit $q \to 1$ sequence with more efficient neighborhood, I_{30} is selected. In limit $q \to 0$ we observe interesting phenomenon that both sequences are present in equal concentrations since two master pairs (I_0, I_{31}) and (I_{30}, I_1) are degenerate and have Hamming distance $d = 1$.

however, were chosen to be more efficient in replication than those in the neighborhood of I_0: $A_1 = 1$ and $A_{49} = 2$, respectively. Around the point $q \simeq 0.964$ the influence of the neighboring sequences becomes so strong that the less efficient sequence I_{50} outgrows I_0 and dominates the quasi-species. Hence for special choices of rate constants a situation where the most abundant sequence in the quasi-species is different from that with the highest selective value can occur. One may notice that this also means a new concept of neutrality that does not depend on simple degeneracy of the fitness values of two sequences but rather rates the complete fitness topography.

3. Conformation-dependent Value Functions and Fitness Landscapes

As outlined in previous sections, much depends on the properties of physically realistic sets of replication rate constants, despite the comparative insensitivity of error thresholds to details in their distributions. At present it is not feasible to measure or estimate real-valued landscapes empirically, not even in the most simple experimental systems like RNA replication in the Qβ

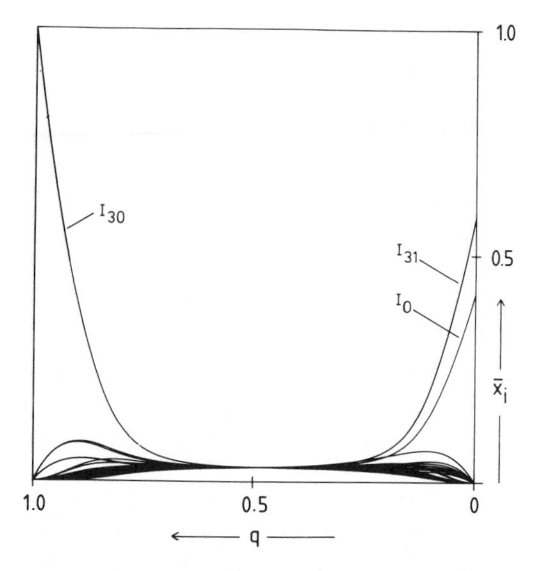

Figure 16. Degenerate quasi-species with $\nu = 5$. Plot is analogous to that shown in Figure 15. Rate constant of I_6 is reduced to $A_6 = 2$. Situation in limit $q \to 1$ is qualitatively same as that reported in Figure 15: Sequence with more efficient neighborhood, I_{30}, is selected. But in limit $q \to 0$ we are not dealing with degenerate master pairs any more: More efficient pair (I_0, I_{31}), which does not contain sequence I_{30}, wins.

system [10–12]. In this section we present some results of computer calculations on model fitness landscapes. In order to render the model tractable, simplifications are inevitable. Binary sequences I_k are considered instead of natural four-letter RNA sequences since this makes the algorithms much faster. We choose chain lengths of $\nu \approx 70$ and compute unknotted two-dimensional structures of folding patterns, $\mathscr{F}_k = \mathscr{F}(I_k)$, by means of a folding algorithm. Details are given in ref. 53. Any folding pattern of this kind is decomposed uniquely into ordered combinations of three structural elements:

1. double helical regions or stems containing "base pairs", which consist of complementary symbols on opposite strands, $0 \cdots 1$ or $1 \cdots 0$, respectively;
2. loops or reverse turns through which the string folds back on itself in order to form a stem; and
3. other unpaired regions of the sequence.

Folding, in essence, is based on a thermodynamic minimum free-energy criterion. The base pair is the most important stabilizing element of the

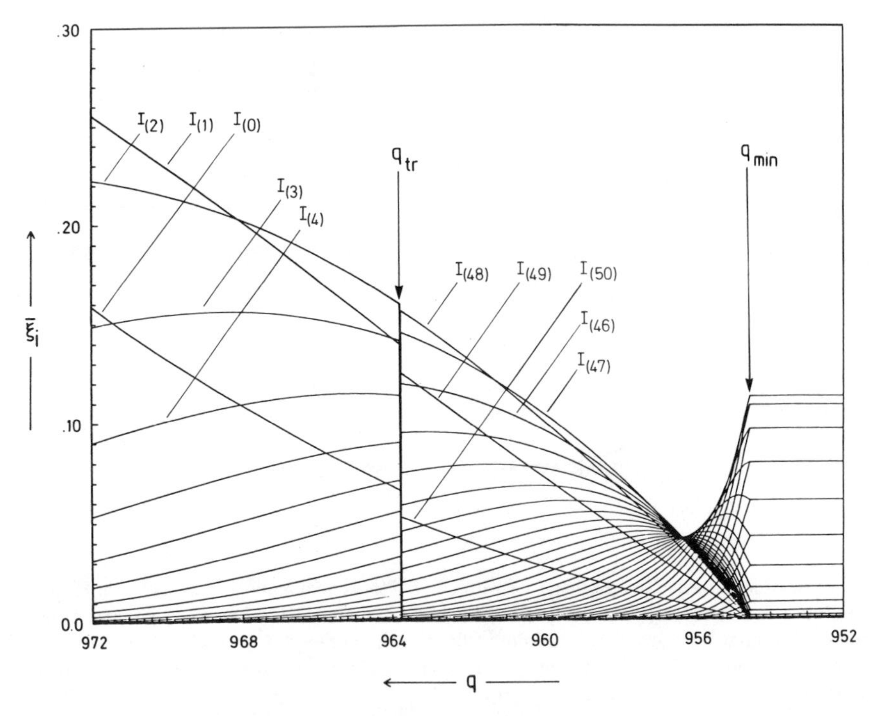

Figure 17. Example of almost degenerate quasi-species. For system of binary sequences with chain length $\nu = 50$ relative stationary concentrations of 51 mutant classes ($\bar{\xi}_i$) are shown in region $0.972 > q > 0.952$. Notice extremely sharp transition at $q_{tr} = 0.9638$ when new master sequence $I_{(50)}$ and its neighbors become dominant. At $q_{min} = 0.9546$ usual error threshold observed. Selective values used in this example are $A_{(0)} = 10$, $A_{(50)} = 9.9$, $A_{(49)} = 2$, and all other $A_{(k)} = 1$ for $k = 1, \ldots, 48$ (cf. figure 5).

pattern—more precisely there are two major contributions, base pairing and stacking of base pairs aligned in parallel planes—and hence the thermodynamic stability of the secondary structure is roughly proportional to the number of base pairs. In addition there are destabilizing contributions from loops due to their constrained configuration compared with other unpaired elements (element 3). The individual steps of the computer simulation are outlined in Table III.

The folding patterns or conformations, $\mathscr{F}_k$, are evaluated according to certain rules. Here we present three different examples of criteria for the rating of patterns:

1. similarity of folding patterns in comparison to a target structure,
2. thermodynamic stability of folding patterns, and

TABLE III
Computation of a Model Value Landscape

System	Realization	Process
Digit	$\{0, 1\}$	
	$\Downarrow$	*Ligation to String*
Binary Sequence String of $\nu \approx 70$ Digits	$\{2^{70} \approx 10^{21}$ Sequences$\}$	
Primary Structure	I_k : 011000110100...0	
	$\Downarrow$	*Folding through Base Pairing*
Secondary Structure	$\mathcal{F}_k(I_k)$: 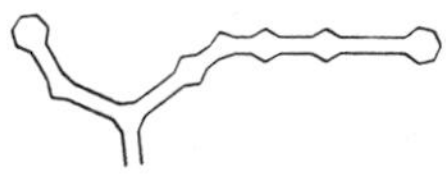	
	$\Downarrow$	*Evaluation*
Molecular Properties	A_k , D_k ; W_{kk}	

3. evaluation of replication and degradation rate constants according to model assumptions based on the folding patterns.

In the first case a measure of similarity between the conformation $\mathcal{F}_k = \mathcal{F}(I_k)$ and a predetermined "ideal" reference conformation $\mathcal{F}_0$ was employed to obtain selective values that depend on the folded conformations. In determining this function both the specific sequence *and* its folding play a role. We investigated several joint dependencies. At first it is important to obtain a workable similarity measure. This was achieved simply by encoding the two-dimensional folding pattern as a new sequence, the fold sequence f, which is a string built from symbols (0, 1, 2). Starting from the 5′ end, as usual, a 0 is assigned to a monomer if it is unbound, a 1 if it is part of the first strand of a double-helical region, and a 2 if it is part of the second strand. This suffices to determine the conformation because of the no-knots constraint. The map $I_k \rightarrow \mathcal{F}_k = \mathcal{F}(I_k) \rightarrow f(\mathcal{F}_k)$ is unique, and we define

$$f_k \doteq f(\mathcal{F}(I_k)). \qquad \text{(IV.5)}$$

Note that the inverse is not true: Several sequences I_j may lead to the same fold sequence f_k. A suitable measure of similarity of two folding patterns $\mathcal{F}_1$

and $\mathscr{F}_2$ is now just the Hamming distance between the two fold sequences $d(f_1, f_2)$, as defined in Section II.1. In fact, a more general distance function involving insertions and deletions [54] could be used, but the Hamming distance will suffice for the present discussion.

The most straightforward measure of the fitness of a conformation is to identify the dissimilarity from the reference conformation as an activation barrier for replication. Two such fitness assignments were applied: the first, indicated by superscript 1 in parentheses, is of a global nature, and it is based on the Hamming distance of fold sequences; the second, denoted by superscript 2 in parentheses, in addition evaluates similarity of structurally defined subsequences. Implicitly, we assume equal degradation rate constants in this model. Degradation then contributes just an additive constant to the selective values, and we may put $D_k = 0$ without losing generality. For the first assignment

$$A_k^{(1)} = A_0 \exp[-\alpha d(f_0, f_k)] \tag{IV.6}$$

is the rate constant of replication of the sequence I_k whose fold sequence is f_k. The reference fold sequence is f_0 here, and α is a real, positive constant determining the precision with which the reference fold sequence must be approximated for a given fitness to be achieved. The rate constant of replication of the reference sequence I_0 is A_0. This simple fitness function has a maximum of $A_k = A_0$ when $f_0 = f_k$, that is, when the sequence I_k has exactly the folding pattern specified by the fold sequence f_0. Because of the nontrivial relationship between sequence I_k and folding pattern or conformation (many sequences may yield the same conformation on folding), Eqn. (IV.6) sustains many local optima.

An additional dependence on segments of the base sequence defined relative to the conformation (such segments are exposed by certain loops, stems, branching points, etc.) may be introduced via the Hamming distance from reference segments in order to model more highly constrained fitness functions. Then the sequence plays a specific role relative to the conformation, and we obtain a second fitness function:

$$A_k^{(2)} = A_0 \exp\left\{-\alpha_1 d(f_0, f_k) - \alpha_2 \sum_{i=1}^{n} d(s_i(f_0), s_i(f_k))\right\}, \tag{IV.7}$$

where the segments of the sequence I_k are denoted by $s_i(f_k)$ and those of the reference fold sequence by $s_i(f_0)$; α_1 and α_2 are positive constants and A_0 is the replication rate constant of the reference sequence I_0. We also found it of interest of deal directly with the linearized forms of Eqs. (IV.6) and (IV.7), which radically changed the shape of the distribution of replication rate constants as well as the effective superiority of the wild type.

We compared the linearized form Eqn. (IV.6) with the long-range Ising-model-like fitness function (cf. the Hamiltonian in Section III.5)

$$A^{(s)} = \bar{w} + \sum_i h_i \sigma_i + \sum_{i<j}^{\nu} J_{ij} \sigma_i \sigma_j \tag{IV.8}$$

by generating for a larger number (10,000) of randomly chosen sequences the average values of the folding-based fitness function $A^{(1)}$ with

1. no sequence positions held fixed, $\bar{w} = \langle A^{(1)} \rangle$;
2. one sequence position k held fixed at σ_k,

$$h_k \sigma_k = \langle A^{(1)} | k \rangle - \bar{w};$$

and

3. two sequence positions k, l held fixed at σ_k, σ_l,

$$J_{kl} \sigma_k \sigma_l = \langle A^{(1)} | k \rangle - \bar{w} - h_k \sigma_k - h_l \sigma_l.$$

In this way the parameters $\bar{w}$, h_i, J_{ij} were deduced from the form of $A^{(1)}$. Using these values, the value of $A^{(s)}$ in Eqn. (IV.8) could be compared with $A^{(1)}$, and no significant correlation was found. The conclusion is that the folding fitness function cannot be decomposed into pairwise contributions from sequence positions. This was confirmed by the fact that the values of h_i, J_{ij} calculated using the different choices of σ_i, $\sigma_j = \pm 1$ gave uncorrelated results. Similar results pertain to $A^{(2)}$ in linearized form (and, of course, for the exponential versions).

In order to gain more insight into the nature of the fitness functions (IV.6) and (IV.7) as relevant for quasi-species structure, we employed a variety of representational tools:

1. The probability density for the varying fitness of sequences ($\sim 20{,}000$, chosen at random from the 2^{ν} possible with length $\nu = 72$) was calculated and is displayed in Figure 18 for the linearized form of Eqn. (IV.6) (with $\alpha = 1$). The density appears as a slightly skewed Gaussian about the mean distance from the goal fold. In general, other effects (including stochastic selection) require a truncation of the distribution above the generic or modal value when the fitness function has an exponential or other nonlinear (cooperative) form.

2. The correlation function reveals much more of the special nature of conformation-dependent fitness. However, in this case, just choosing a large

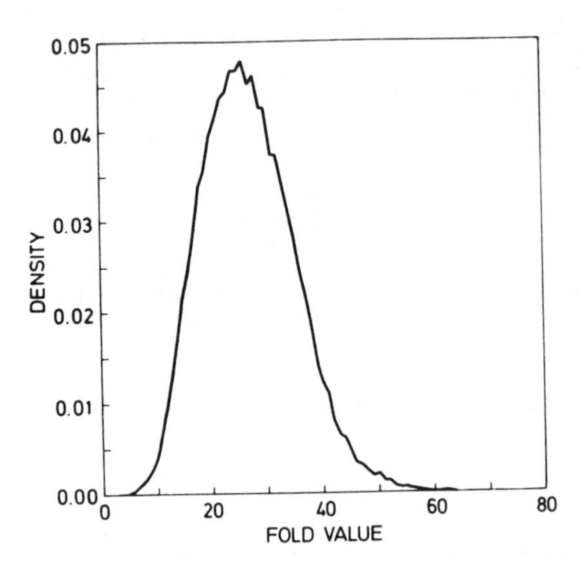

Figure 18. Probability density of folding fitness function. Curve is smoothed frequency plot showing probability density for integer-valued fitness function $A^{(1)}_{\text{linear}}(s) = \nu - d(f_0, f(s))$, for 21,610 sequences of length $\nu = 72$ (chosen at random). Here $d(f_0, f(s))$ measures distance between calculated folded form and reference fold f_0, which was taken to be particular nested pair of hairpin loops.

number of sequences at random can only give very well-separated sequences (Hamming distance) and does not provide information about shorter range correlations. We solved this sampling problem by examining sequences along ν random mutation pathways from a particular centre sequence. The resulting correlation functions when the center sequence was a local optimum for the two linearized value functions (using the same set of 72 mutant pathways of length 24) are shown in Figure 19. The important features are finite range and an approximately exponential form for both functions. Typical individual trajectories for the two fitness functions are shown in Figure 20.

3. The preceding results demonstrate that along a random mutant pathway the typical situation is a rapid decrease in the fitness level of sequences. However, the issue of the rare occurrence of long extended high-value ridges is left open. Indeed, one would like to know whether the high-value regions are isolated or whether they form a sparse but interconnected network through the mutant space? For the fitness function of type (IV.6), only those monomers directly involved in bonding (and a few additional ones near the boundaries of binding regions) were found to be important in determining fitness near an optimally folded sequence. Constraints on the remaining monomers are limited, and the picture of selection restricted to a roughly $\nu/2$

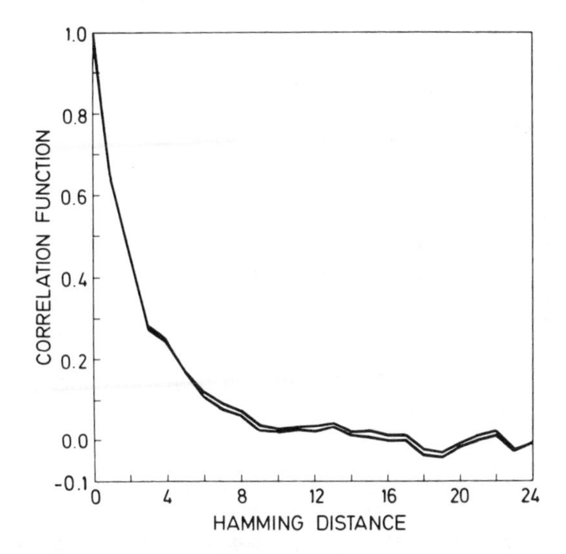

Figure 19. Correlation functions for fitness by folding. Correlation functions for linearized forms of folding fitness functions of type 1 [Eqn. (IV.6), upper curve] and type 2 [Eqn. (IV.7), lower curve] are displayed. Both are normalized by value at local optimum at distance 0. Seventy-two random paths radiating outward 24 shells and starting with distinct first mutants employed. Same paths and reference fold employed in both cases.

dimensional subspace of the ν-dimensional sequence space is qualitatively correct. For the fitness function of Eqn. (IV.7) the situation is different as the degeneracy has been removed. Here a plot of neutral mutants near a local optima reveals long ridges extending into the surrounding sequence space (Figure 21). This confirms, for the present conformational-dependent fitness function, the features that distinguish evolution in the quasi-species as guided in the choice of mutants and not purely random (cf. Section II.5).

The second model evaluation of folding patterns is based on the thermodynamic stability of polynucleotide conformations. Here, not only does this stability decide which of the many possible structures is formed (as in the preceding) but it also determines the fitness. The more stable is the optimal folding pattern $\mathscr{F}_k$ for a particular sequence, the higher is its selective value:

$$A_k^{(3)} = A_0 \exp(-\alpha\Delta(\Delta G)) \approx A_0(1 - \gamma\Delta(\Delta G)) \qquad \text{(IV.9)}$$

with $\Delta(\Delta G) = \Delta G(\mathscr{F}_k) - \Delta G_0$ and α and γ being some real and positive constants. The absolute minimum free energy of the folding patterns for all sequences of the given length, denoted by ΔG_0, is chosen as reference here. In

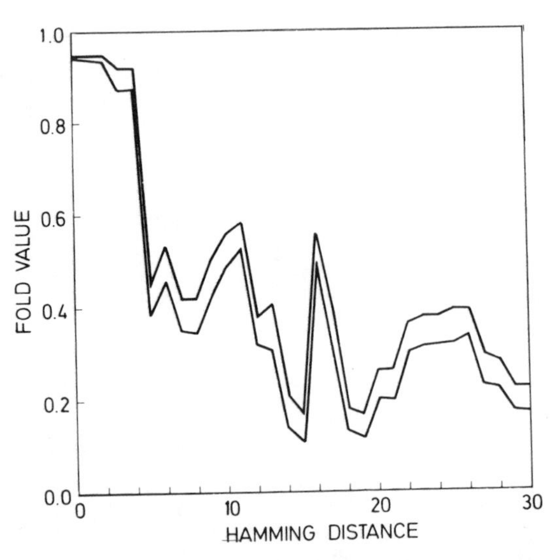

Figure 20. Values along radiating trajectory. Folding values for sequences lying on particular mutant path randomly radiating from locally optimal sequence. Linearized fitness functions of type 1 (upper curve) and type 2 (lower curve) employed as in Figure 19. Different paths yield very different jagged profiles.

contrast to the previous example, there is no unique reference conformation $\mathscr{F}_0$ because several folding patterns may have the lowest possible free energy. Since we are interested in qualitative features, we only use the linear approximation as indicated in the preceding. The basis for approximate free-energy estimates was worked out during the past two decades (see, e.g., ref. 55), and we are now in a position to assign energy increments to the basic substructures. In essence, these increments are based on the number of base pairs and their orientations. Here we use the energy weights for all (G, C) sequences as tabulated in ref. 56 and assign G to the digit 1 and C to 0, respectively. Again we assume that all degradation rate constants are equal, and we put $D_k = 0$ without losing generality. This fitness assignment to folding patterns, as for the previous one, is manifestly incomplete with respect to biophysical insight into replication kinetics since higher stability or smaller dissimilarity to a reference pattern do not alone entail faster replication. From the viewpoint of the mapping $I_k \rightarrow \mathscr{F}_k \rightarrow (A_k, D_k)$, however, we are dealing with a problem of the same "*universality class*" [57] as in case of the physically more appealing assignment of the third model. Moreover, we have the advantage of knowing a priori what the fittest folding patterns look like: Obviously these are the 32 (inverse) palindromes with complete parallel stacks and variations

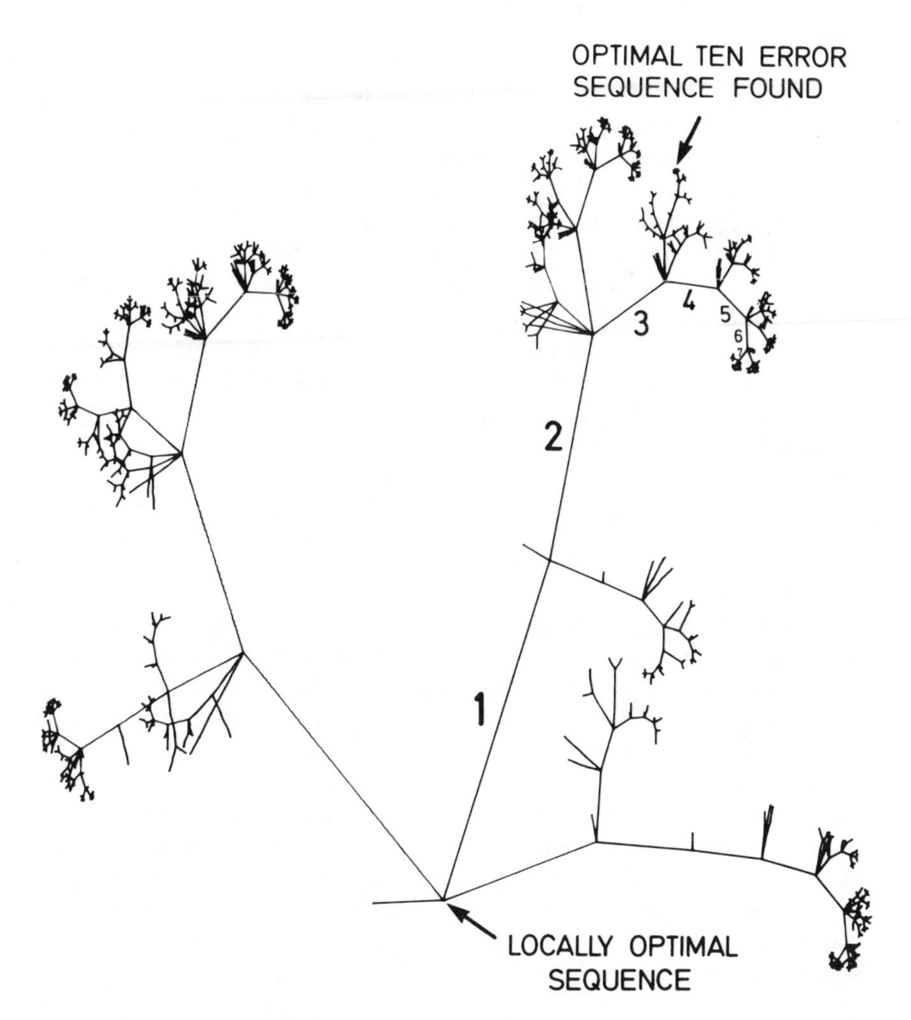

Figure 21. Mutant space high-value contour near local optimum. Diagram is multiply branched tree with different macromolecular sequences at vertices. Each line joins neighboring sequences whose values are within 0.5 of locally optimum sequence at lower center for linearized fitness function of type 2 [Eqn. (IV.7)] and reference fold that is cruciform, like tRNA, for sequence of length 72. Over 1300 branches shown extending up to 10 mutant shells away from central optimum. Better sequence (labeled optimum) was found in tenth mutant shell. Nonrandom sampling of mutant sequences demonstrated typical of population sampling in quasi-species model. Note small number of ridges that penetrate deeply into surrounding mutant space. (Additional connected paths due to hypercube topology of mutant space not shown.)

only in the four digits of the hairpin turn that form the most stable, "perfect" hairpin.

Since the third model is closely related to the second, we present its underlying assumptions first and discuss the results obtained for both value landscapes thereafter. In contrast to the two previous model landscapes we calculate both replication and degradation rate constants independently. Since up to now nobody knows how to compute replication rate constants from folding patterns of single-stranded polynucleotides, our model is essentially of heuristic value. It just tries to meet some biophysical constraints of replication and centers around the fact that virus-specific RNA replication operates on single strands [10] and that unzippering of double helical regions is cooperative [58]. We assume that every stacking region slows down the overall replication process in an additive fashion and that the term each stack contributes is a sigmoidal function that is reminiscent of the Monod–Wyman–Changeux model:

$$A_k^{(4)} = A_0 - A_\infty \sum_j \frac{\nu_j^{(k)}(1 + \nu_j^{(k)})^3}{(1 + \nu_j^{(k)})^4 + \mathscr{L}}. \qquad \text{(IV.10)}$$

The sum is to be taken over all stacking regions of the folding pattern $\mathscr{F}_k$; the $\nu_j^{(k)}$ values refer to the individual stacks; in particular, $\nu_j^{(k)}$ denotes the number of base pairs in the jth stack of the folding pattern $\mathscr{F}_k$. The parameter $\mathscr{L}$ is some large positive constant that determines the detailed shape of the sigmoidal curve. The parameters A_0 and A_∞ represent rate constants: A_0 is the replication rate constant of the linear unstacked chain that serves as the reference here. The difference $A_0 - A_\infty$ refers to an infinitely long, completely stacked conformation. Should a folding pattern yield a negative rate constant ($A_k < 0$), we would declare the sequence I_k as lethal and put $A_k = 0$. The two parameters A_0 and A_∞ may be used to control the fraction of lethal mutants.

The rate constants of the hydrolytic degradation are also made up of additive contributions—here from unpaired regions. Heavier penalties are assigned to free ends or joins than to loops. No cooperativity is assumed:

$$D_k^{(4)} = D_0 + D_l \sum_j \frac{l_j^{(k)}}{l_m} \exp\left[(l_j^{(k)} - l_m)/l_m\right] + D_u \sum_i \frac{\nu_i^{(k)}}{\nu}. \qquad \text{(IV.11)}$$

The first sum is taken over all loops; the second over all other external elements: tails, joins, and so on. Three rate constants D_0, D_l, and D_u are introduced to determine stability against hydrolysis. The number of unpaired bases in the jth loop of the folding pattern $\mathscr{F}_k$ is denoted by $l_j^{(k)}$, the length of the ith unpaired element in the same pattern by $\nu_j^{(k)}$; l_m is a weighting factor and ν the length of the sequence.

The selective values or fitness factors are readily obtained from the equation $W_{kk} = A_k q^\nu - D_k$. This simply introduces a weighting of replication against stability by means of the single-digit replication accuracy. For high accuracy it is better to replicate faster, whereas longevity or high stability against hydrolysis pays more at high error rates. The third model is an example of an assignment in which high fitness is based on a compromise between two contradictory trends: Long double-helical segments stabilize against hydrolysis, but they also reduce the rate of replication and vice versa. Such contradictory trends seem to be the rule in real biological systems.

In order to characterize the distributions of selective values in the second and the third model, we explored the value landscape by a Monte Carlo search. We created three random samples of 38,000 different sequences each (one repeat with 76,000 sequences gave essentially the same results) with predetermined ratios of probabilities for (0/1) digits, $\rho_1 = 0.2857$, $\rho_2 = 0.5$, and $\rho_3 = 0.7143$, which led to mutant distributions centered at the 20-, 35-, and 50-error mutants of the *all-zero* sequence I_0. Three different parts of the value landscapes determined by Eqs. (IV.9)–(IV.11) were explored in that way. The results are shown in Figure 22.

The value lanscapes created by the second and third model differ significantly. The selective values of the simple stability assignment (IV.9)—here computed for $q = 1$—show roughly a "noisy" Gaussian distribution. The mean selective values for ρ_1 and ρ_3 are about the same: $\bar{W} \approx 500\ [t^{-1}]$; selective values are rate constants by definition and are therefore given in reciprocal time units. The distribution around the 35-error mutant of I_0, obtained with ρ_2, is significantly superior: $\bar{W} = 750\ [t^{-1}]$. The interpretation of this finding is straightforward: The 35-error mutants have as many digits 0 as digits 1 and, provided the sequence admits it, form more base pairs than do the 20- or 50-error mutants. Thermodynamic stability in essence counts the numbers of base pairs, and therefore the 35-error mutants of I_0 are the most stable on average.

The selective values for the third model are shown for $q = 1$ in Figure 22 and hence represent excess productions, $E_k = A_k - D_k$. The distributions around the 20- and 50-error mutants of I_0 show interesting bimodal shapes. Further analysis shows that this bimodality is a result of subtle relations between the A_k and D_k according to Eqs. (IV.10) and (IV.11), and we shall not pursue this issue further here. It is useful to split the value landscape of the third model into different contributions resulting from replication and degradation rate constants (Figure 23). The distributions of the degradation rate constants have much in common with the thermodynamic stabilities used in the second model: They also resemble noisy Gaussian distributions. The distributions of replication rate constants, however, differ rather drastically from the others. They look much more bizarre. Due to the sigmoidal

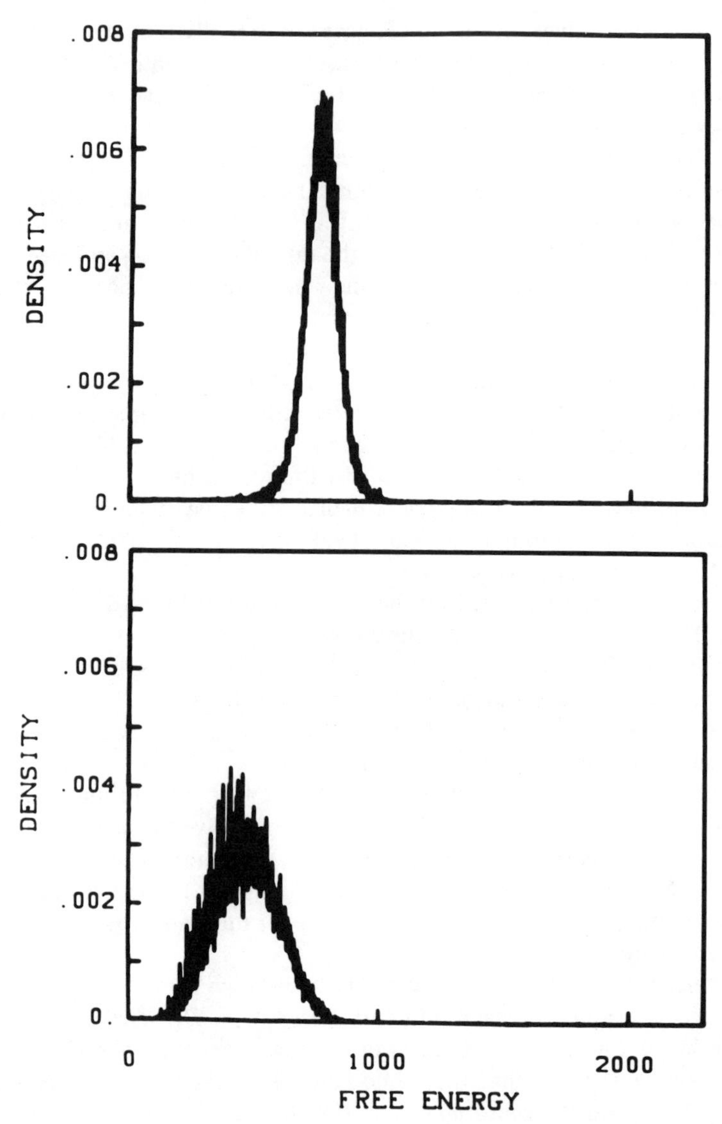

Figure 22. *(Continued overleaf)*

contribution of the lengths of double-helical regions, replication rates are extremely sensitive to minor details of the folding patterns, and this leads to enormous scatter in the distribution. Long stems are a disadvantage for replication, and hence, 20- or 50-error mutants of the reference sequence I_0 replicate faster on the average than do the 35-error mutants. The fraction of nonviable sequences in the sample centered around Hamming distance 35

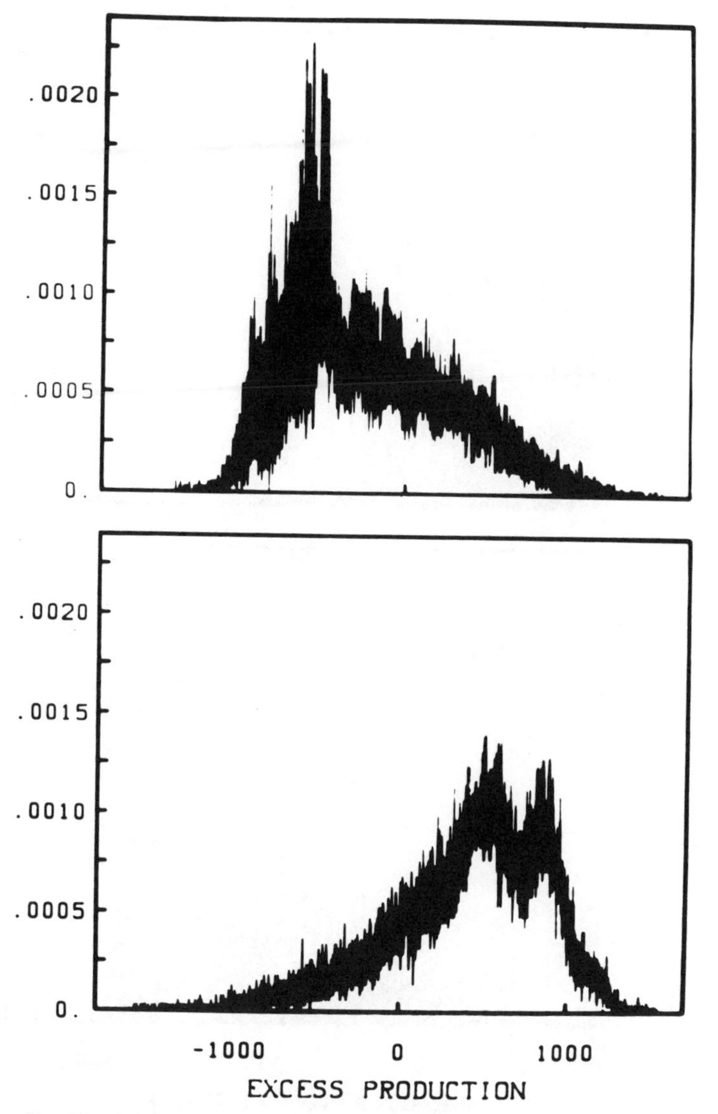

Figure 22. Distribution of selective values in sequence space. Here 38,000 different sequences of length $\nu = 70$ generated by introducing digits 1 at random with probability $\rho_1 = 0.2857$, $\rho_2 = 0.5$, or $\rho_3 = 0.7143$ into all-zero sequence I_0. This produces Gaussian shape samples centered around 20-, 35-, and 50-error mutants of all-zero sequence. Distributions of free energy $\Delta G(\mathscr{F}_k)$ and excess productions $E_k = A(\mathscr{F}_k) - D(\mathscr{F}_k)$ shown for regions located at mean Hamming distances 35 (upper plots) and 20 (lower plots) from I_0. Apart from scaling factor free-energy distribution constitutes fitness landscape of second model evaluation. Distribution of excess productions is representative for model 3 at high accuracy of replication. Densities in neighborhood of 50-error mutants (not shown) are essentially same as those around 20-error mutants.

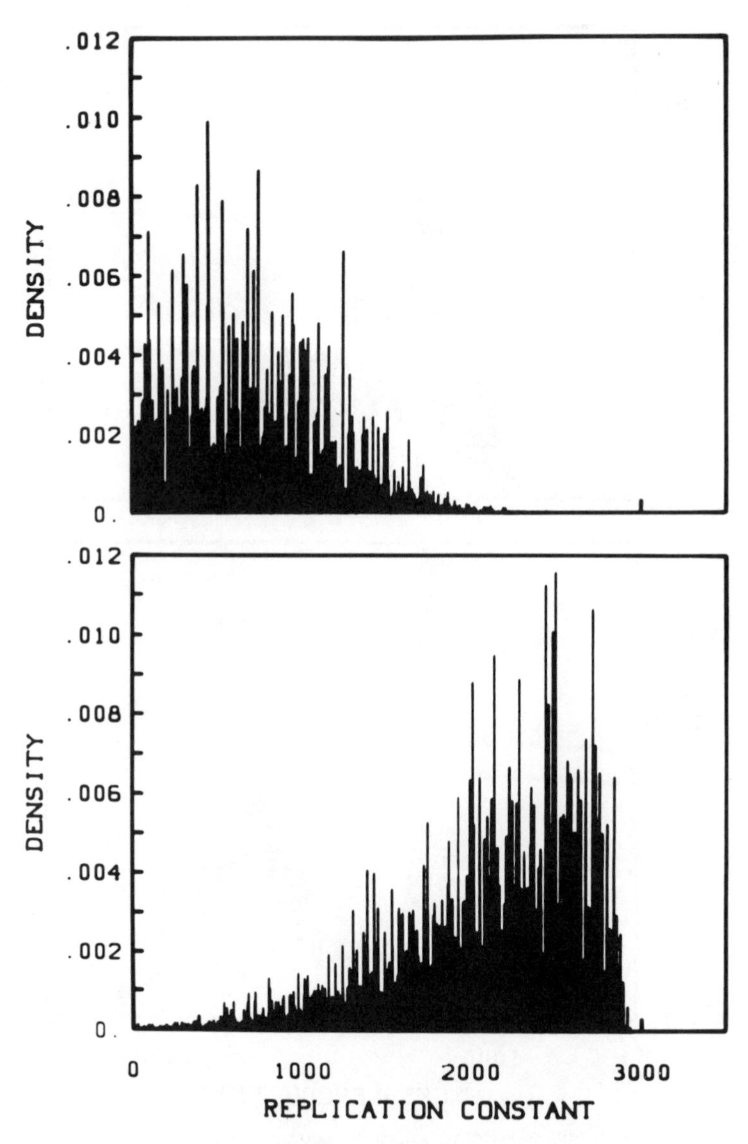

Figure 23. *(Continued overleaf)*

from I_0 is therefore higher than in the other two regions. It amounts to 0.18. As we mentioned in the preceding, the fraction of lethals is a consequence of the choice of A_0 and A_∞ in Eqn. (IV.10) and hence arbitrary. But, we believe this reflects the well-known, naturally arising situation where parts of the sequence space are particularly rich in lethal mutants.

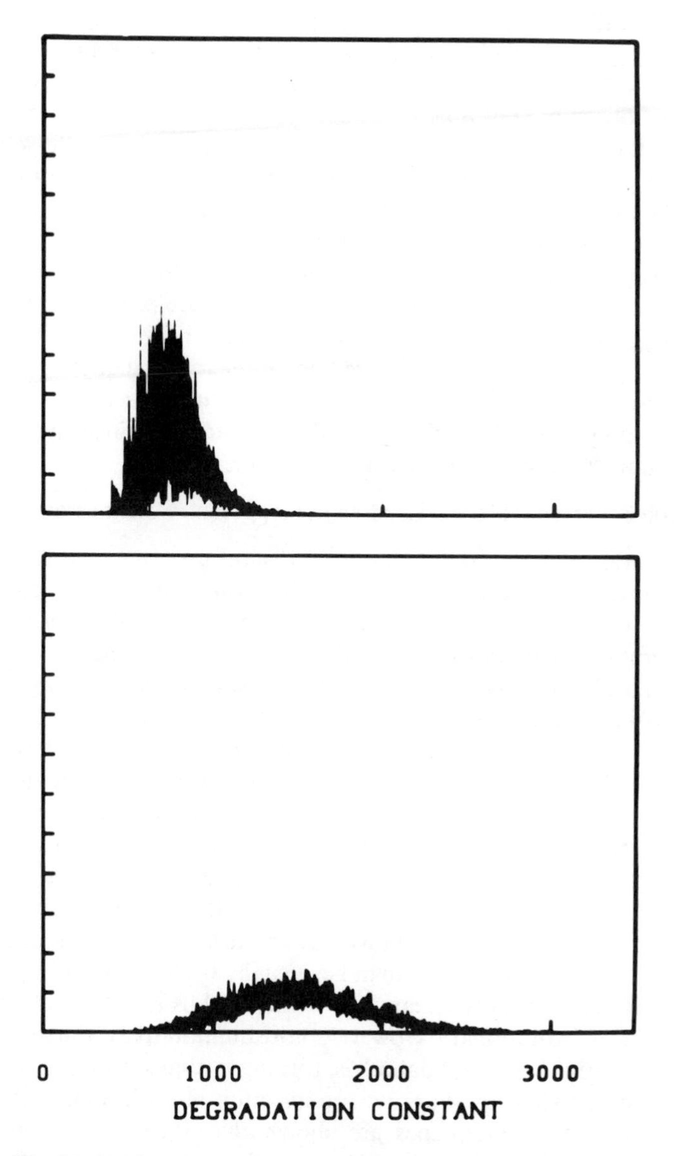

Figure 23. Partitioning of excess productions $E(\mathscr{F}_k)$ into replication (A_k) and degradation (D_k) rate constants. Upper and lower plots show distributions sampled from neighborhoods of 35- and 20-error mutants, respectively, as explained in the caption to Figure 22. Fraction of lethal variants in replication landscape A (this is the fraction of sequences with $A_k = 0$); it amounts to 0.18 in this particular case; has been cut off in order to show details of distribution at positive rate constants.

The overall scans of value landscapes presented in Figures 22 and 23 reveal only general trends persisting in certain regions. Nothing is said about the relative location of high-valued configurations and their connectivity. Investigations aiming at a better understanding of the topological details of the value landscapes of the second and third model, similar to those shown in Figure 21, are presently under way. A simple hint on this topology can be easily obtained, however, by exploring the local shape of the fitness surface. For this goal we calculated selective values of all sequences I_k surrounding some reference sequence I_j with Hamming distance $d(k, j) = 1$. The local scan through sequence space reflects the bizarre structure of realistic value landscapes (Figure 24). Nearby sequences may have very small, almost the same or higher selective values compared to a reference sequence of average properties. From molecular genetics we know that this is also the case in nature. A point mutation might be lethal or harmless since the translation into phenotype may depend crucially on single digits.

4. Asymmetry of Fitness Landscapes: Apparent Guidance of Evolution

The appearance of a mutant moderately distant from the wild type depends on the population of precursor mutants, that is, mutants that are closer to the wild type. It will be seen that this introduces an element of determinism into the generation of mutants despite the fact that the elementary process of mutation is of an entirely stochastic nature. "New" mutants always appear at the periphery of the populated mutant spectrum, which will be shown to be highly asymmetric with respect to the wild type, in a manner depending on the structure of the value landscape. Along ridges, the populated mutant spectrum may reach to regions quite distant from the wild type. Moreover, these are mountainous regions, that is, regions where, due to the correlated structure of the landscape, advantageous mutants are most likely to appear. Hence, the generation of mutants is a process that is highly biased toward success and, presuming some knowledge about the value landscape, appears to be by far less unpredictable than has usually been assumed in population genetics. In the light of the examples discussed thus far, this statement does not sound too revolutionary. However, more quantitative estimates will show how effective this "value guidance" is, turning it almost into a new principle rather than just some quantitative modification of evolutionary theory.

In Figure 5 two landscapes are shown that start at a binary master sequence ($\nu = 50$) and are followed up to the 12-error copies. One of the landscapes (left side) is the low-value plateau that has been considered already in previous examples, while the other resembles a mountain saddle as typical for any fractal type of hill country. The population numbers of mutants, relative to the population number of the wild type, were calculated by means of second-order perturbation theory [Eqs. (II.17) and (II.18)] for

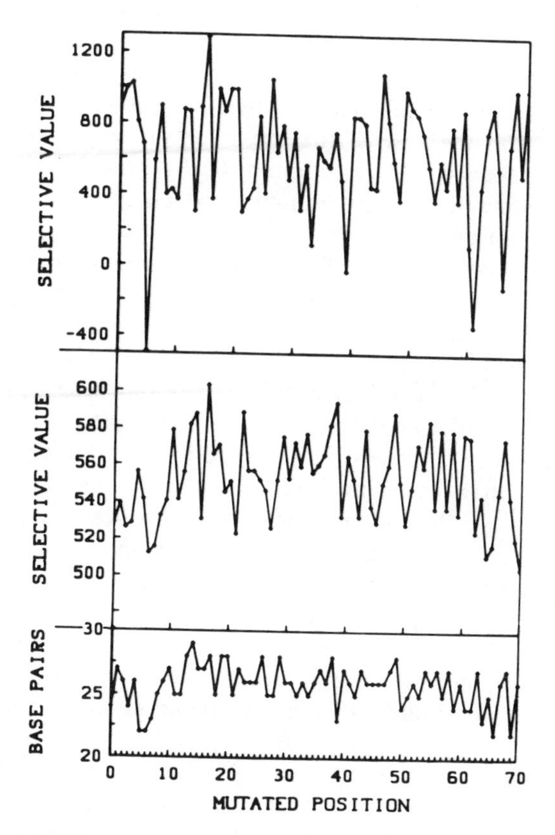

Figure 24. Local shape of value landscapes. Selective values as defined by Eqs. (IV.9)–(IV.11) of 70 nearest neighbors surrounding a given reference sequence I_R are shown. Upper curve refers to evaluation according to third model for $q = 1$ as in Figure 22. Middle curve represents free energies according to second model evaluation. Lower curve counts numbers of base pairs in different folding patterns $\mathscr{F}_k$. Free energy follows approximately number of base pairs, but excess production shows roughly opposite trend.

two cases: (i) The profile shown applies to only one individual chain of mutants up to the 12-error copy, all other mutants having a zero fitness value; and (ii) the profile applies to all precursors of the 12-error mutant, that is, all mutants directly between the wild type and a given 12-error mutant.

The results are very instructive. For the low-value plateau the probability of mutant appearance is essentially Poissonian for the single-mutant chain (i) and only slightly modified for the all-intermediate mutant plateau (ii), indicating that for typical population numbers of 10^{10}–10^{15} any given mutant that has a Hamming distance from the wild type larger than 4–5 is rarely populated. In contrast, the mountain saddle landscape shows probabilities that suggest in both cases a deterministic population for all mutants under consideration. In particular, the 12-error mutant here—supposing the extended landscape (case ii)—appears by 15–20 orders of magnitude more frequently than in the case of the low-value plateau, suggesting that the evolutionary process would be deterministically guided to the mountainous region surrounding the 12-error mutant, in which a more advantageous new mutant might appear.

In order to appreciate this effect adequately, we must be aware that a binary sequence of length 50 has $\binom{50}{12} \doteq 10^{11}$ twelve-error copies and hence correspondingly many (overlapping) precursor regions, among which all sorts of value profiles are to be expected. The consequence of value guidance is that large areas of sequence space will never be touched by the evolutionary process. Thus the trial-and-error process of mutation is concentrated on those regions in which the advantageous copy is most likely to appear. The large orders of magnitude involved offer a most efficient solution of the complexity problem that initially instigated those who have made quantitative estimates on mutant generation. What counts is not the number of possible mutants (which generally is hyperastronomically high) but rather the value topography of sequence space, that is, the presence or absence of sufficiently branched mountain ridges that pervade major parts of sequence space. They guarantee the teleonomic nature of evolution that was hard to imagine with a "blind filter" of selection acting upon the trial and error of mutation.

It is also obvious that the problem of neutral mutants appears in a new light. Neutral mutants are not restricted any more to the chance coincidence of degenerate wild types that differ in their relative fitness by less than the reciprocal population size. Mutants with a quite broad degree of variability of fitness are potential candidates when rated together with the fitness topography of their neighbors. Only in small populations of species of large genomes—that is, only where every mutant really is unique—would the stochastic trial-and-error model apply. Even in these cases, the long-term behavior would depend on the fitness topography of the kinship neighborhood of the wild type. The deterministic model, of course, represents a limiting case that has to be complemented by stochastic approaches (cf. Section V.5), just as the random-drift model of Kimura represents another limiting case.

V. CONCLUSION: REVIEW AND OUTLOOK

1. The Physical Basis of the Model

In this Chapter we have been dealing with the simplest model of a self-reproductive molecular system consistent with known [59–61] kinetic properties of RNA or DNA: the (molecular) quasi-species. In so far as they satisfy the prerequisites we have assumed, the model may well apply to more complex systems dealt with in molecular biology, including viruses [62, 63] or (vegetatively reproducing) cellular systems that we may collectively subsume as systems of individual "replicators." The model in its simple form does not apply to networks of interacting replicators, but it may be extended so as to include such properties. It is also not intended in this review to describe any realistic scenario of the early phase of the origin of life. These were "historical" rather than truly physical problems, requiring the knowledge of historical initial and boundary conditions. Although we may well guess about the general chemical nature of prebiotic earth, we entirely lack detailed knowledge about particular environmental conditions such as the presence of specific catalysts, be they of chemical (e.g., molecular complex) or of physical (e.g., interfacial) nature.

Hence we are dealing with a conditional model that becomes relevant wherever its prerequisites are fulfilled. Moreover, the model provides a framework for designing experiments dealing with (molecular) biological evolution that can be conducted under defined laboratory conditions.

Being one of the simplest molecular models that display Darwinian properties, it provides some features that may be taken as generally characteristic of Darwinian self-organization. We do not pretend that this simple model applies to all situations for which Darwinian reasoning has been used. This model, for instance, neglects any horizontal exchange of information such as regular and irregular crossing over, gene conversion, transposition, and so on, which are typical of higher organisms. Nevertheless, the rigorous treatment of the model reveals a much more subtle interpretation of what is usually referred to as Darwinian behavior, commonly described as an interplay of stochastic mutation and deterministic selection of statistical significant advantages (including the important modifications brought about by the theory of neutral selection). Evolutionary optimization based on neutral drift and natural selection as a hill-climbing process in the rugged terrain of a value space becomes understandable in a quantitative sense only by the subtleties the model involves. These effects have been tested and identified directly by experiments using molecular replicator systems. We are sure that these effects will carry through to more complex population systems of organisms and therefore must be taken into account for any quantitative description of Darwinian behavior based on natural selection and neutral drift, although they may be complemented by various other phenomena considered to be of importance in population biology, such as fluctuating environments, isolation, migration, and so on, in addition to the recombinative events typical of Mendelian populations.

The quasi-species model rests on essentially three prerequisites:

1. The major constituents of the system, for example, sequences of RNA or DNA, have to be inherently self-reproductive. This is expressed through positive diagonal terms (W_{ii}) in the kinetic equations.
2. Self-reproduction of the constituents is not entirely precise but is rather prone to mutation, being expressed by positive nondiagonal terms W_{ik}. (In other words; species i can come about not only through self-copying but also by miscopying of species k.)
3. The system has to stay far from equilibrium, or better; the turnover of the major constituents is not at equilibrium, where formation and decomposition would be microscopically reversible and detailed balance would hold. This requires a metabolism sustained by a supply of energy-rich monomeric building blocks of the polymeric constituents. (It should be stressed that at equilibrium the first two conditions are not sufficient to cause selective evolutionary self-organization.) As a consequence of microscopic reversibility, all eigenvalues will always turn out to be real and *negative*.

Apart from these three prerequisites, leading to the general form of kinetic equations as represented in Appendix 3, no specific interactions among the constituents—such as mutual enhancement or suppression, recombinative exchange, or other regulative couplings—have been assumed.

2. The Emerging Concepts of Molecular Evolution

The deterministic treatment of the model reveals three major regularities that may be best encompassed by the following terms:

Selection. Local stabilization of a quasi-species distribution in sequence space around one (or several degenerate) master species.

Evolution. Destabilization of the local quasi-species upon arrival of an advantageous mutant that establishes a new quasi-species.

Optimization. Tendency toward global stabilization guided by the population of nearly neutral mutants and their nonrandom distribution in sequence space.

To consider evolutionary optimization as a sequence of discrete selection steps requires some explanation. In fact, it is justified only on the basis of a hierarchical order of the rate terms, in which the off-diagonal terms usually are much smaller than the diagonal ones, decreasing in binomial or Poissonian progression with increasing Hamming distance between template and (erroneous) replica. As a consequence, the deterministic order around an advantageous master copy, that is, a local quasi-species, is established relatively quickly, while the arrival of a new mutant with selective advantage is a fairly slow process. Such a mutant originates from a precursor mutant at the periphery of the quasi-species distribution, that is, at a relatively large Hamming distance to the master copy, where population numbers relative to that of the master copy are rather small. Strictly speaking, we are dealing here with a stochastic rather than a deterministic event, and the deterministic model at this point needs a stochastic supplement (cf. what follows). However, deterministic population numbers near the periphery of the quasi-species distribution, even for a given error class, may differ by many orders of magnitude depending on the fitness (relative to the master copy) of that particular region in sequence space. A cumulative fitness distribution greatly enhances this effect, which provides a strong deterministic bias for the production of particular mutants at the periphery, preferentially those that are of high fitness or selectively advantageous. This effect provides such a strong guidance toward broad global fitness maxima (the examples given in Figure 5 showing how many orders of magnitude actually are involved) that the classical picture of an evolutionary process as a pure trial-and-error search for advantageous mutants (with its consequently low success probability) turns out to be obsolete. The target of selection is *not* the single advantageous copy, that is, an individual wild type. Instead, it is the total quasi-species distribution that is the target, and this is not of some simple narrow-banded Poissonian type but rather shows a fairly large nonsymmetrical dispersion with protrusions reaching far into the sequence space. These protrusions of the population distribution are found along the ridges of the fitness landscape bridging those areas with fitness values closely resembling that of the wild type.

A tremendous change in the deterministic concept of a Darwinian system is brought about by the quasi-species concept. Let us confront the current interpretation with the classical one.

1. The *wild type*, despite the fact that the whole system may be represented by a unique sequence, is not a single individual but rather a distribution having a defined consensus or master sequence. The individual (or group of individuals) that coincides with the consensus sequence may represent a peak in the population distribution. Its fraction of the total population, however, usually remains fairly small (in cases where it has been tested, even undetectably small).

2. The term *fittest*, correspondingly, is not related to any individual but rather to the complete quasi-species distribution acting as the target of selection. Mathematically, the fittest is characterized by a dominant eigenvector to which all populated states contribute. Only if one sequence clearly

dominates in fitness will its fitness value (diagonal coefficient W_{mm}) determine the maximum eigenvalue to a close approximation.

3. The *population of mutant states* in the quasi-species is strongly modulated by the fitness distribution. The effect is particularly strong for those mutants (i) whose fitness values resemble closely that of the wild type (m) due to the hyperbolic form $W_{ii}/(W_{mm} - W_{ii})$. Moreover, the population number of a given mutant is related not only to the wild-type fitness but also to the fitness values of its neighboring mutants, especially those situated in the subspace referring to all positions that differ from the master sequence. If, in such a domain, all fitness values (or a major portion thereof) are close to that of the master sequence, there will be a tremendous amplification of population numbers relative to domains of low fitness. Optimization routes are to this degree deterministically ordained.

4. *Neutral* or nearly neutral mutants appear in a new light. First, they appear to be of utmost importance in determining the population of states at the periphery of the mutant spectrum and hence in fixing the route of evolution. Second, being part of the quasi-species distribution, they are rated not only by their fitness relative to the master but also with respect to their own mutant environment. The uncertainty of classical theory, how closely the fitness of a mutant has to resemble that of the wild type in order to be called *neutral*, is now replaced by quantitative expressions. It turns out that mutants the fitness values of which deviate considerably more than by $1/n$ (n being the population number) from the wild-type fitness value still may play a very important role, while at the same time true degeneracies due to the rating of the mutant environment become quite rare. As a consequence, in macroscopic populations comprising something like 10^9–10^{12} individual replicators, we usually find only *one* defined consensus sequence, although many (nearly) neutral mutants tend to be populated.

5. *Stochastic theory* must supplement deterministic theory as in the classical wild-type model. This is a formidable task for the complete quasi-species model due to the nonlinear type of relations involved (cf. section 5.).

The results of *classical neutral theory* are valid only for systems of relatively low population numbers and large genomes. If the genome is large enough that even the 3ν one-error mutants cannot be populated because the population number n is smaller than 3ν, one may expect the results of so-called neutral theory to be representative. Otherwise, modifications due to the reproducible population of (nearly) neutral mutants, as indicated by the deterministic quasi-species model, pertain and finally destroy the basic assumption of the "blind" production of mutants at the periphery of the mutant spectrum.

6. In the limit of large populations and sufficiently small genomes (usually

fulfilled for single RNA or DNA molecules or viruses) we can now give a physical characterization of the terms *selection* and *evolutionary adaptation.*

Selection may be characterized physically as a condensation phenomenon (mathematically; localization of a distribution in sequence space). The analogy to ferromagnetism has been stressed, except that we are dealing with entirely different variables. The role of the Curie temperature is played by an error threshold. Far below the error threshold, the distribution contracts to populate the master species only. This state is determined by the fitness landscape's highest peak rather than cooperative forces, owing to the simple structure of the model. Surpassing the threshold means the "melting" of the quasi-species distribution due to accumulation of errors. Such an error catastrophy means a sharp loss of genetic information. Hence selection is a kind of phase transition in information space. The error threshold has been tested for various virus populations (Table IV) and found to be effective in nature (cf. what follows).

Evolution then may be viewed as a series of stabilizations and destabilizations of quasi-species distributions (i.e., a series of phase transitions) that, in a constant environment, are associated with an increase of fitness. This process is guided by the mutant distribution within each quasi-species. The advantageous mutant appears at the periphery of the populated mutant distribution usually in an area of relatively high fitness. The evolutionary route thus avoids the vast space with a low degree of fitness, and hence optimization proceeds along fairly limited pathways. How discrete the single steps are depends on the proportion of neutral and advantageous mutants

TABLE IV

Replication Error Rates (per Nucleotide) of RNA Viruses[a]

Coliphage $Q\beta$	3×10^{-4}
Foot-and-mouth disease	5×10^{-4}
Influenza	$7 \times 10^{-5} - 2.7 \times 10^{-4}$
Poliomyelitis	3×10^{-5}
Vesicular stomatitis	$3-5 \times 10^{-4}$
For Comparison (DNA Replication)	
Coliphage λ	2.4×10^{-8}
Coliphage T_4	1.7×10^{-8}
Escherichia coli	2×10^{-10}

[a] Values (literature cf. ref. 63) determined from reversion frequencies of mutants and hence represent *true* error rates (to be distinguished from population numbers of mutants, i.e., so-called mutant frequencies).

within a given quasi-species. If this degree is appreciable (relative to population size), selection and evolution may coalesce into one process approaching the global maximum of fitness. Again, the error threshold relation is of utmost importance. Evolutionary progress is expected to be greatest near the error threshold. Occasionally surpassing this limit may assist in the escape from metastable distributions to find a higher fitness peak, just as *simulated annealing* is used for efficient optimization.

Evolutionary optimization then is not just a blind stochastic trial-and-error search for a better adapted mutant but rather follows an inherent logic:

1. Selective advantage usually is to be expected in a distant mutant, that is, as the periphery of the distribution.
2. Mutants far distant from an established master sequence arise from those that are less distant, that is, from precursors "en route".
3. The probability of producing a distant mutant therefore depends critically on the population numbers of its precursors.
4. Population numbers of mutants are high relative to equidistant competitors if they are situated in a domain of high fitness.
5. A fractal clustering of the fitness distribution in domains is likely to occur.
6. Precursors of distant advantageous mutants therefore have a higher chance to be populated than precursors of deleterious mutants.
7. The high dimensionality of sequence space aids the connectivity of fitness domains of (nearly) neutral mutants.
8. Evolutionary optimization proceeds along defined pathways in sequence space. There are alternative routes, but their number is so highly restricted that one has the impression of a phenomenon of automatic guidance to higher fitness.

This logic, rather than the simplifying interpretation of Darwinian behavior as an interplay of chance (random) mutation and necessity (selective fixation of the advantageous mutant once produced), is the basis of evolutionary optimization or adaptation.

3. Experimental Evidence

Many of the classical observations of molecular genetics, especially those related to viruses, have to be reexamined in the light of the new theory. Virologists have accepted the concept of quasi-species (cf. articles by Domingo and Holland and others [64, 65]). What once had been interpreted as the wild type showing a defined individual sequence now has been identified to be a widely dispersed mutant spectrum with defined consensus sequence. One of the classical experiments was carried out by Weissmann and co-workers [66]. They cloned single particles of the Coli phage $Q\beta$, amplified the clones, and determined sequences or fingerprints. The result was that none of the different clones exactly agreed in their sequences but rather showed one or several differences. Weissmann estimated from his experiments that a true wild type as an individual sequence was present to an extent of less than 5%. Moreover, by evaluating the rate of revertant formation (occurring in artificially produced *site-specific* mutants) as well as the relative growth rates of the revertant (master copy) and the mutants, he could correlate sequence length with error and wild-type superiority, in agreement with the error threshold relation [Eqn. (III.5)]. The experiment is schematically represented in Figure 25. One may wonder why such an experiment works at all, that is, how a mutant clone can grow up before it is replaced by the revertant "wild type", which must appear soon in the mutant's quasi-species distribution. All clones would then show identical sequences, that is, the wild-type sequence. The explanation that mutant sequences can be identified in macroscopic amounts lies in the fact that what is being fished out in cloning single particles are nearly neutral mutants (being present in relatively high abundance). Since such a mutant (i) has a diagonal coefficient $W_{ii} \approx W_{mm}$ [i.e., being almost neutral with the master (m)], it grows

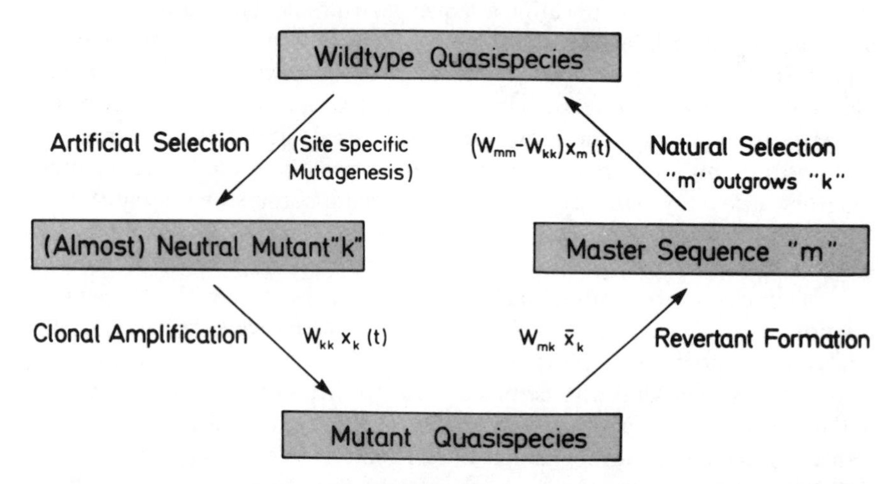

Figure 25. Scheme of Weissmann's experiment [66] for determining error rates of phage $Q\beta$. Success critically depends on finding nearly neutral mutant k, which grows up about as quickly as master copy m, i.e., $W_{kk} \approx W_{mm}$, while being replaced by revertant master copy with relatively small rate (i.e., $(W_{mm} - W_{kk}) \ll W_{kk}$). In natural population such nearly neutral mutants may be picked up frequently because they are abundantly populated. Weissmann chose an extracistronic site-specific mutant.

up in clone amplification as $e^{W_{ii}t} \approx e^{W_{mm}t}$, while the revertant replaces the mutant only according to $e^{(W_{mm} - W_{ii})t}$.

Error rates have also been determined for foot-and-mouth disease virus by Domingo [67], for influenza A virus by Palese [68], and for vesicular stomatitis virus by Holland [69] and their co-workers. All results show the correlation of error rate with sequence length. However, the factor $\ln \sigma$ may considerably influence the results if $\sigma \to 1$. In fact, some of those experiments—where single cloned mutants were amplified directly for the sake of sequence analysis—may produce misleading results since in these cases near-neutral mutants only are to be detected. (Mutants with W_{ii} considerably smaller than W_{mm} would appear with unduly large delay.) Influenza A seems to be a case where a large fraction of the infectious virus mutants are indeed (almost) neutral, possibly as a consequence of the fact that this virus contains eight separate RNA chains in its genome. On the other hand, poliovirus, which has been tested by the same method, did not show detectable (neutral) mutants. In these cases a quantitative study of the kinetics of revertant formation from site-directed mutants may offer a more direct access to the average error rate $(1 - \bar{q})$.

In carrying out such experiments on error rates, one has to specify carefully the conditions under which the results were obtained. In particular, one has to distinguish the (average) probability $(1 - \bar{q})$ of producing a substitution per site (in replicating a sequence) from the rate of appearance of a mutant in an established quasi-species distribution. Hence it makes a difference whether one observes one or a few replication rounds in an essentially nonstationary distribution or whether one counts frequencies of appearance at a steady state. Hot spots, that is, single sites showing extremely high rates of mutation, have been reported frequently in the literature, and their first discovery certainly may be considered a cornerstone in the rise of molecular virology [70, 71]. However, frequency of appearance may mean either a high rate of production (from wild type) or a high (almost neutral) rate of replication or both. Which interpretation is correct, that is, which fraction is to be associated with either of the two effects, remains to be analyzed for most cases reported.

In any case, working with complete virus particles requires some caution with respect to an evaluation of results. The observable amplification of many viruses appears in the form of single bursts defining infectious cycles. In each burst, however, many copies of virus particles are set free, of which only a minor fraction turns out to be infectious. Within a replication cycle inside the host cell the replication machinery is available to both viable and nonviable (or less efficiently) replicating virions, while after the burst only those particles will survive that have encoded the correct machinery for further infection. Moreover, one may start under conditions where single infections (i.e., one virus infecting a cell) are guaranteed. After several bursts, such conditions cannot be easily maintained. In the neighborhood of the site of a burst multiple infections will take place. It is important to carry out experiments in such a way that their interpretation either is not sensitive to the details of this mechanism or such that the exact mechanism, if known, is taken into account.

These difficulties can be avoided if replication experiments are carried out in vitro, that is, under cell-free conditions. The late Sol Spiegelman pioneered this kind of experiment [72–74] using the Coli phage $Q\beta$. The basis of these experiments was the observation that the isolated genome of phage $Q\beta$ can be replicated in a suitable reaction mixture, including the monomeric building blocks in an energy-rich form, that is, the four nucleoside triphosphates, the purified enzyme, and the $Q\beta$—replicase (which was isolated from infected cells), the mixture being adjusted to suitable pH and ionic strength conditions [72]. The first replication rounds produced infectious particles, that is, RNA chains that upon introduction to host cells were able to produce infectious phages. If, however, serial upgrowth and dilution was carried out over many generations, the acting selection pressure for efficient replication caused the sequences to delete those portions that are not essential for recognition by the replication enzyme and there grew out a particle only about one-tenth of the length of the intact phage genome but having a replication rate per site far in excess of the wild type [73]. One may call this process *degeneration* rather than *evolution* because the resulting particle entirely lost its capacity of infection. Nevertheless, with respect to replication by the $Q\beta$ enzyme, these particles were better adapted to their new environment, which did not require them to be infectious.

In another experiment using a small RNA variant of about 220 nucleotides (which, again, was not infectious but was efficiently recognized by the replication enzyme), Spiegelman et al. [74] demonstrated a true evolution process. They added ethidium bromide, an inhibitor of replication, to the reaction mixture and found the successive outgrowth of mutants with one, two, and three single-site substitutions, respectively (cf. Figure 26). The final product after about 40 generations was a three-error mutant that was about a factor of 2 more efficient in replication than the original wild type. When these experiments were carried out, the kinetics and mechanism of replication were not yet known in any detail. Such studies were recently done in collaboration with Biebricher and Gardiner [59–61]. The kinetic experiments were supplemented by computer simulations and by analytical treatment of simplified mechanisms. These studies led to a complete identification of all major reaction steps of enzyme RNA replication. The simplified scheme shown in Figure 27 indicates how complicated the true mechanism is and how important it is in evolutionary experiments to choose appropriate conditions. There are essentially three regimes of replication one has to

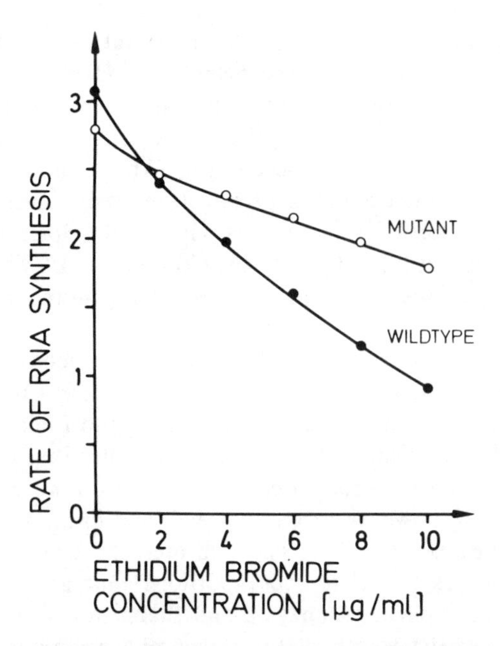

Figure 26. In serial dilution experiment, Spiegelman [73, 74] and co-workers obtained three successive mutants with increased adaptation to presence of ethidium bromide, a drug that interferes with replication. Experiment starts with population of 10^6 MDV strands (variant of Qβ–RNA comprising about 220 nucleotides that is well-adapted to Qβ–RNA–replicase). Population is amplified to about 10^{11} copies and subsequently diluted to initial concentration. Iteration of this procedure led to final product, three-error mutant that was obtained after about 40 iterations. As replication rate data show, mutant is slightly inferior to wild type in absence of ethidium bromide but twice as efficient as wild type at final concentration of ethidium bromide.

distinguish: (i) at low template-to-enzyme ratios where the reaction proceeds exponentially, rating the growth constant of plus and minus strands as a geometric mean; (ii) at template-to-enzyme ratios of 1 and moderately larger, where the enzyme gets saturated by template and the reaction proceeds essentially linearly with time according to some Michaelis–Menten scheme; and finally (iii) at large template-to-enzyme ratios where inhibitions (binding of template to replica site and template double-strand formation) cause the rate to reach a plateau (Figure 28). Selection experiments were carried out in the different regimes leading to differing results [75]. In the exponential growth range the most efficiently growing species (according to its overall rate) is selected while in the linear growth range a species grows out that wins the competition of binding to the enzyme (regardless of growth efficiency). It turns out that, with regard to the reaction mechanism, the original evolution

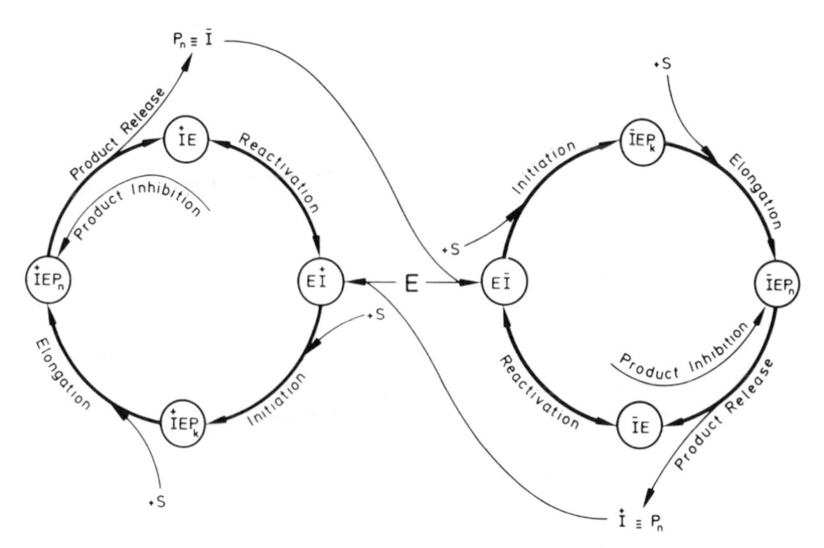

Figure 27. Detailed in vitro mechanism of RNA replication by Qβ–replicase [59]. RNA grows exponentially as long as template concentration is below enzyme concentration. Growth rate becomes constant and hence RNA concentration rises linearly when template concentration exceeds that of enzyme, while, finally, at large template excess, rate decreases down to zero due to enzyme inhibition and template double-strand formation. In these in vitro experiments, Qβ–replicase is present as environmental factor. In vivo the enzyme is formed during the first 20 min after infection of host cell followed by RNA replication during second half of infectious cycle. After about 40 min, about a thousand infectious phage particles per cell are released in burst. These thousand infectious particles usually are minor part of total burst size.

experiments were not carried out under uniquely defined (nor incidentally under optimal) conditions. It is important to optimize the population size with respect to mutant expectancy. Spiegelman's experiments were done under conditions far below the error threshold and therefore resembled quite inefficient adaptation.

The discovery of de novo synthesis [76–79] of RNA by Qβ replicase opened a new way of studying evolution processes in the laboratory under more favorable conditions. In the process of de novo synthesis many different templates are being formed by individual enzyme molecules and then compete for growth, leading finally to the outgrowth of a fittest sequence. These experiments can be done in the presence of conditions exerting some selection pressure. Ethidium bromide provides such a selection pressure [76], as in Spiegelman's experiments [74], and de novo synthesis in the presence of ethidium bromide does indeed reveal much more efficient products, as seen in Figure 29.

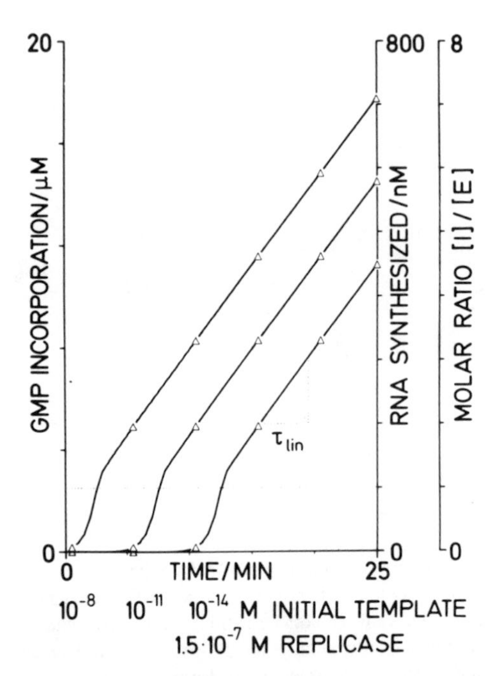

Figure 28. Rate profiles obtained in vitro experiments of RNA replication by Qβ–replicase [59]. Exponential growth range manifests itself by constant shift of growth curves (lag periods) for constant ratios of template numbers. Detectable growth range refers to template concentrations that are comparable to or higher than enzyme (replicase) concentration, while high template-to-enzyme ratios cause leveling of growth curve due to enzyme inhibition and template double-strand formation. Evolution experiments have been carried out with differing results in both exponential and linear growth range.

At present we are endeavoring to optimize these techniques so as to work close to the error threshold and use annealing procedures to speed up the evolutionary process. Automated machines for evolutionary adaptation of RNA sequences and of viruses have been constructed [80, 81].

In summary, experimental evidence available confirms the results of theory and opens some interesting aspects for future work on evolutionary optimization. In fact, very recently, Biebricher [82] was able to clone a quasi-species distribution of RNA and to demonstrate most directly the widely spread mutant spectrum that is characteristic of a molecular quasi-species.

4. Limitations of the Model

Major limitations have been mentioned already, as far as they are related to the particular mechanism of RNA replication or virus infection. The linear

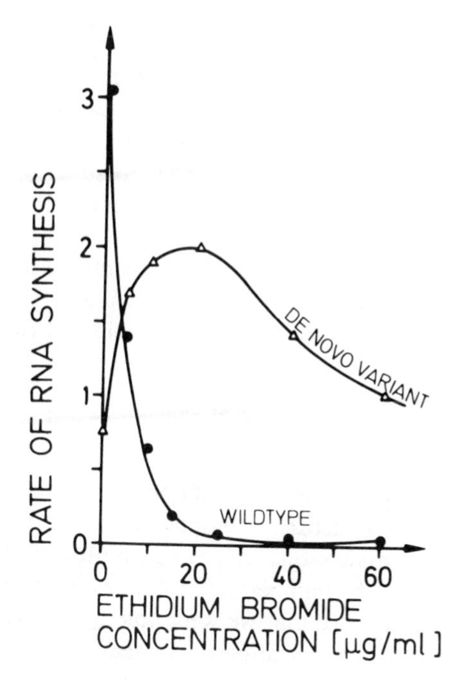

Figure 29. RNA sequences, obtained by de novo synthesis using Qβ–replicase, can be optimally adapted to presence of inhibitors [76]. Growth rate of such variant as function of ethidium bromide concentration indicates not only much higher degree of adaptation (as to be expected from broader quasi-species-like mutant spectrum obtained under de novo synthesis conditions, i.e., near error threshold) but also a drug addiction, i.e., higher synthesis rate at finite ethidium bromide concentration.

autocatalytic ansatz of replication (i.e., the term $W_{ii}x_i$) with uniform error rates ($W_{ik}x_k$) represents only the simplest possible model. Nevertheless, it does yield all the experimental features that have been observed in corresponding reaction systems. There is a wide range of conditions where the simple ansatz indeed proves to be a perfectly satisfactory approximation, as computer simulations of the more complex mechanisms have shown [73–75]. Moreover, knowing the particulars of any mechanism in more detail would allow one to adapt the equations correspondingly.

There are, however, more principal limitations, and it is important to stress that the quasi-species model is a particular model holding only where its prerequisites are fulfilled. The linear autocatalytic rate law is one of the prerequisities typical for the quasi-species nature. If rates become independent of the concentrations of growing substraces, coexistence may replace competition, or if, on the other hand, the autocatalytic rate law depends more strongly than linearly on the concentration of the growing entity (as is the case for hypercycles), selection may become of an entirely different, non-Darwinian nature than is normal for a quasi-species [4].

The typical outcome of quasi-species behavior is natural selection in the Darwinian sense. We believe that it should be possible to adapt the quasi-species model to any situation where natural selection in the Darwinian sense plays a major role. Hence the model may be generalized so as to include all kinds of horizontal gene transfer typical for recombination. The model as presented is essentially a deterministic model that holds only for a sufficiently large population size. This limitation and a possible way to overcome it will be the subject of the final section.

5. Replication as a Stochastic Process

Evolutionary phenomena by their very nature call for a stochastic description. The source of the undeniable, inherent importance of fluctuations can be located in the molecular structure of biopolymers. The numbers of potentially different polynucleotide sequences, as we pointed out in Section II, are always much larger than the numbers of molecules in realistic populations. So it came as rather a surprise that we were able to characterize stationary mutant distributions by the application of conventional reaction kinetics to polynucleotide replication. Error thresholds, derived by means of the quasi-species concept, brought us indeed to the limits of justification of the deterministic approach. Consider a population replicating with an error rate that exceeds the critical value. The model predicts an approximately uniform distribution of sequences at the stationary state. Such a state can never occur since we cannot have less than one molecule of every type. What happens at error rates above threshold is the breakdown of the stationary state. We are dealing with a changing ensemble of sequences that migrates through sequence space as a random walker. Around the error threshold—this is the range that appears to be most relevant for evolution—this walk is anything but trivial. The underlying value landscapes, as we pointed out in Section IV.3, are highly complicated, bizarre objects, and they determine where and how fast the population moves. The success of the quasi-species model is documented by the powerful localization concept [29] as outlined in Section III.2. It allowed us to extend the approach to such complicated cases where only statistical information on the distribution of selective values is available (Section III.3).

Other important questions, however, remain unanswerable within the deterministic theory. Among them are phenomena directly related to finite populations such as the dependence of error thresholds on population size or the mean lifetimes of mutants in populations. The latter quantity is of particular importance for highly fluctuating mutant distributions such as those occurring at high error rates or those encountered with distant, selectively neutral mutants. Then the distributions of polynucleotide sequences are essentially nonstationary and are characterized in terms of mean lifetimes rather than by relative concentrations. Another relevant problem concerns the probability of finite populations being caught in local optima of fitness landscapes. In principle, this question can be answered appropriately by means of first passage times that allow one to compute the average time a population will spend in a given region of sequence space before passing to a different region.

Several attempts to describe replication–mutation networks by stochastic techniques were made in the past. We cannot discuss them in detail here, but we shall briefly review some general ideas that are relevant for the quasi-species model. The approach that is related closest to our model has been mentioned already [51]: the evolutionary process is viewed as a sequence of stepwise increases in the populations' mean fitness. Fairly long, "quasi-stationary" phases are interrupted by short periods of active selection during which the mean fitness increases. The approach towards optimal adaptation to the environment is resolved in a manner that is hierarchical in time. Evolution taking place on the slow time scale represents optimization in the whole of the sequence space. It is broken up into short periods of time within which the quasi-species model applies only locally. During a single evolutionary step only a small part of sequence space is explored by the population. There, the actual distributions of sequences resemble local quasi-species confined to well-defined regions. Error thresholds can be defined locally as well.

In the case of selective neutrality—this means that all variants have the same selective values—evolution can be modeled successfully by diffusion models. This approach is based on the analysis of partial differential equations that describe free diffusion in a continuous model of the sequence space. The results obtained thereby and their consequences for molecular evolution were recently reviewed by Kimura [2]. Differences in selective values were found to be prohibitive, at least until now, for an exact solution of the diffusion approach. Needless to say, no exact results are available for value landscapes as complicated as those discussed in Section IV.3. Approximations are available for special cases only. In particular, the assumption of rare mutations has to be made almost in every case, and this contradicts the strategy basic to the quasi-species model.

Chemical reaction networks are frequently modeled by Markov processes and can be formulated as master equations. Commonly, it is straightforward to write down the master equation, but when it comes to derive solutions, hard-to-justify approximations are inevitable; see, for example, ref. 83. In essence, the same is true for polynucleotide replication described by a master equation. The master equation corresponding to the kinetic ansatz of Appendix 3 was first formulated by Ebeling and Feistel [84]. Several attempts were made to derive results from this master equation [85–89] or from a closely related Langevin equation [90], but none of these was successful in deriving a stochastic version of the error threshold.

A more restricted specification of the Markov process used in the modeling of replication and mutation yields more results for special cases. Birth-and-death processes were applied to error-free replication. We mention two examples [91, 92]. Probabilities of mutant extinction and first-passage times for the selectively neutral case can be derived immediately. Multitype branching processes allows mutations to be included [48, 93]. This approach corresponds to Eqn. (A3.1) without the flux term $\Phi(t)$. A stochastic error threshold was derived for this unconstrained replication–mutation system [48]. It is formally identical to the deterministic expressions derived in Section III.1, but the interpretation is different: the value matrix W is replaced by the mean matrix of the multitype branching process. At error rates smaller than the threshold the system sustains a master sequence and its mutant distribution with a certain probability that manifests itself in a finite probability of survival to infinite time. If the error rate exceeds the threshold value, the probability of survival for the master sequence and for all other sequences is zero. This implies that we are dealing with a changing ensemble of sequences just as in the deterministic model, where the quasi-species fails to localize under these conditions. The stochastic error threshold sharpens with increasing chain length ν—as it does in the deterministic case—and this implies here that the probability of survival to infinite time decreases sharply from almost 1 at error rates just below critical to almost zero just above the threshold. Attempts to incorporate the constraint $\Phi(t)$ into the multitype branching model by means of a combined branching and sampling technique are presently in progress.

A recent attempt at a direct stochastic theory by Weinberger [94] using the deterministic flow term as an external (precomputed) constraint should be mentioned here. The intractability of a large coupled system of second-order partial differential equations for the generating function is then reduced to a (nonlinearly coupled) system of ordinary differential equations. The price is the loss of proper population regulation and possible extinction.

At present the most that can be said analytically about the error threshold in a finite population is the necessary condition [50]

$$\frac{\sigma q^{\nu}}{\sigma q^{\nu} - 1} = \varepsilon \bar{x}_m, \qquad \varepsilon \ll 1, \tag{V.1}$$

which demands small fluctuations in the master sequence population. Master equations can be studied by computer simulation provided particle numbers are not too large. An algorithm particularly well suited for the investigation of stochastic chemical reactions networks was worked out by Gillespie [95]. It was used to compute error thresholds for finite populations on the simple value landscape we analyzed deterministically in Section IV.1. Chain lengths up to $\nu = 20$ and populations up to $N = 2000$ were simulated [96]. Some results are shown in Figure 30 for a superiority, $\sigma = 10$. In small populations the error threshold occurs at higher q values than in the infinite population, which is implicitly assumed in the deterministic approach. Fluctuations endanger the stability of stationary sequence distributions, and hence the smaller the population, the fewer errors it can tolerate. A comparison of Eqn. (V.1) with the numerical results with $\bar{x}_m$ determined by the deterministic second-order perturbation result,

$$\bar{x}_m = \frac{q^\nu - q^\nu_{\mathrm{det,cr}}}{1 - q^\nu_{\mathrm{det,cr}}} N, \tag{V.2}$$

is also shown in Figure 30 and yields good agreement for the conventional choice $\varepsilon = 0.1$. Further work along the lines suggested in the preceding will

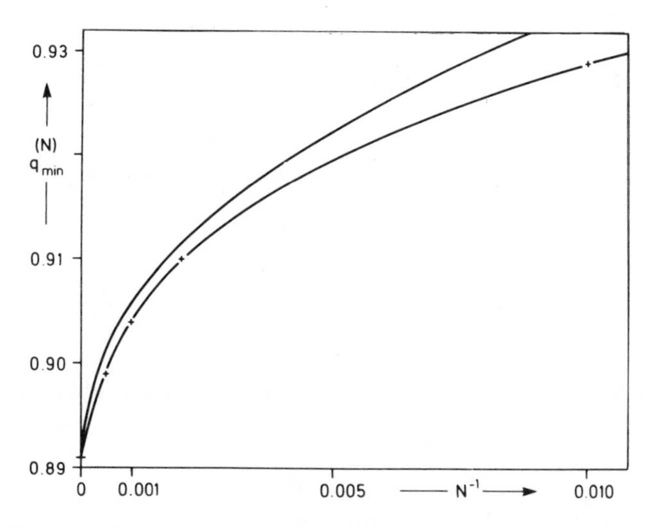

Figure 30. Error threshold as function of population size. Stochastic replication–mutation dynamics in ensemble of polynucleotide sequences with chain length $\nu = 20$ simulated by Gillespie's algorithm [95]. Critical single-digit accuracy of replication ($q_{\min}$) at which ordered quasi-species is converted into changing population of sequences with finite lifetimes is plotted as function of $1/N$, reciprocal population size (lower curve). For further details see ref. 96. Upper curve is theoretical prediction of Eqn. (V.1) based on ref. 51.

provide more information about this and the lifetime of metastable quasi-species. The above brief review of stochastic theory suffices to provide a suitable context for the embedding of the deterministic quasi-species model while substantiating its basic claims.

APPENDIX 1. MUTATION FREQUENCIES AND HAMMING DISTANCES

The Hamming distance $d(i, k)$ of two sequences I_i and I_k counts the number of digits in which they differ. The number of sequences in a mutant class, that is, the number of sequences with Hamming distance $d(0, k) = d$ from a reference sequence I_0, is

$$N_d = \binom{v}{d}(\kappa - 1)^d, \tag{A1.1}$$

where v is the number of symbols in the sequence class, that is, the length of the chain, and κ is the number of symbol classes, that is, the number of different digits. In binary sequences we have $\kappa = 2$, and in polynucleotides we have $\kappa = 4$. Summation over all possible Hamming distances $0 \leqslant d \leqslant 0$ yields

$$N = \sum_{d=0}^{v} \binom{v}{d}(\kappa - 1)^d = \kappa^d. \tag{A1.2}$$

The frequency at which the set of sequences with Hamming distance d is produced as error copies of the template I_0 is

$$Q_d = \binom{v}{d} q^{v-d}(1 - q)^d = \binom{v}{d} q^v (q^{-1} - 1)^d. \tag{A1.3}$$

The frequency of correct replication, Q_0, represents the quality factor of the replication process:

$$Q_0 = q^v. \tag{A1.4}$$

The frequency with which a given polynucleotide is produced as an error copy of the reference sequence I_0 is then obtained from

$$Q_{i0} = \frac{Q_{d(0,i)}}{N_{d(0,i)}} = Q_0 \left(\frac{q^{-1} - 1}{\kappa - 1} \right)^{d(0,i)}$$

or, in general, the frequency of producing I_i as an error copy of any sequence I_k is

$$Q_{ik} = \frac{Q_{d(k,i)}}{N_{d(k,i)}} = Q_0 \left(\frac{q^{-1} - 1}{\kappa - 1} \right)^{d(k,i)}. \tag{A1.5}$$

APPENDIX 2. RECURSIVE CONSTRUCTION OF THE MUTATION MATRIX Q

The mutation matrix for binary sequences of length $\nu + 1$, $Q(\nu + 1)$, is to be constructed from $Q(\nu)$ in such a way that the eigenvalues and eigenvectors are obtained recursively. We start with $\nu = 1$ and set

$$\varepsilon = (q^{-1} - 1) \quad \text{and} \quad B(\nu) = q^{-\nu} Q(\nu). \tag{A2.1}$$

Then we obtain for the two smallest B matrices (on the top and in the rightmost column of the matrix we indicate to which sequence the corresponding elements refer)

$$B(1) = \begin{matrix} & \begin{matrix} 0 & 1 \end{matrix} & \\ & \begin{pmatrix} 1 & \varepsilon \\ \varepsilon & 1 \end{pmatrix} & \begin{matrix} 0 \\ 1 \end{matrix} \end{matrix} \tag{A2.2}$$

$$B(2) = \begin{matrix} & \begin{matrix} 00 & 01 & 10 & 11 \end{matrix} & \\ & \begin{pmatrix} 1 & \varepsilon & \varepsilon & \varepsilon^2 \\ \varepsilon & 1 & \varepsilon^2 & \varepsilon \\ \varepsilon & \varepsilon^2 & 1 & \varepsilon \\ \varepsilon^2 & \varepsilon & \varepsilon & 1 \end{pmatrix} & \begin{matrix} 00 \\ 01. \\ 10 \\ 11 \end{matrix} \end{matrix} \tag{A2.3}$$

This recursion consists of duplication of sequence space as indicated in Figure 2. Note that the sequences in the rows and columns of the Q matrix are arranged in lexicographical order. The recursion can be generalized and yields

$$B(\nu + 1) = \begin{pmatrix} B(\nu) & \varepsilon B(\nu) \\ \varepsilon B(\nu) & B(\nu) \end{pmatrix} \tag{A2.4}$$

The eigenvalues $\lambda_k(\nu + 1)$ and the corresponding eigenvectors $\boldsymbol{\eta}_k(\nu + 1)$ are computed recursively too:

$$\lambda_k(\nu + 1) = \lambda_k(\nu)(1 + \varepsilon), \qquad \lambda_{k+2^\nu}(\nu + 1) = \lambda_k(\nu)(1 - \varepsilon) \tag{A2.5}$$

and

$$\boldsymbol{\eta}_k(\nu+1) = \begin{pmatrix} \boldsymbol{\eta}_k(\nu) \\ \boldsymbol{\eta}_k(\nu) \end{pmatrix}, \qquad \boldsymbol{\eta}_{k+2^\nu}(\nu+1) = \begin{pmatrix} \boldsymbol{\eta}_k(\nu) \\ -\boldsymbol{\eta}_k(\nu) \end{pmatrix}, \qquad k = 1, \ldots, 2^\nu. \tag{A2.6}$$

Hence, all eigenvalues are of the same general form:

$$\lambda_k(\nu) = \prod_{j=1}^{\nu} (1 + \gamma_j \varepsilon) \quad \text{with} \quad \gamma_j \in \{-1, +1\}. \tag{A2.7}$$

They are all real and positive and fulfill the inequality

$$\lambda_{\text{max}}(\nu) \geqslant \lambda_k(\nu) \geqslant \lambda_{\text{min}}(\nu)$$

with

$$\lambda_{\text{max}}(\nu) = (1 + \varepsilon)^\nu \quad \text{and} \quad \lambda_{\text{min}}(\nu) = (1 - \varepsilon)^\nu. \tag{A2.8}$$

Both the largest and the smallest eigenvalue are nondegenerate (the former result also following directly from the Frobenius theorem).

APPENDIX 3. THE RATE EQUATIONS

The differential equation for the time dependence of the population variable $n_i(t)$ is of the form (cf. Figure 2)

$$\dot{n}_i(t) = W_{ii} n_i(t) + \sum_{k \neq i} W_{ik} n_k(t) + \frac{n_i(t)}{\sum_k n_k(t)} \Phi(t). \tag{A3.1}$$

The number of variables is extremely large. If we admit all point mutations of sequences with length ν, we have $i = 1, 2, \ldots, \kappa^\nu$. The number of sequences per unit volume of a given type, I_i, are denoted by $n_i(t)$. The rate parameters are collected in the value matrix W. The diagonal elements W_{ii} are the rate parameters for the net or excess production of I_i through (correct) self-copying. They are composite quantities,

$$W_{ii} = A_i Q_{ii} - D_i \tag{A3.2}$$

with A_i and D_i being the rate constants of total (correct and erroneous) template- (I_i) induced replication and degradation, respectively. The frequency of correct copying, Q_{ii}, corresponds to the quality factor Q_0 as defined in Appendix 1. The off-diagonal elements W_{ik} are the rate parameters for mutational production of the sequence I_i through (imperfect) copying of the template I_k. They can be expressed in the form

$$W_{ik} = A_k Q_{ik}, \tag{A3.3}$$

where A_k is the total rate of replication on the template I_k and Q_{ik} is the frequency of the mutation $I_k \rightarrow I_i$ (Appendix 1). The (time-dependent) dilution flux is given by $\Phi(t)$. It represents the constraint for the particular model under consideration. The conservation law

$$\sum_k Q_{ik} = 1 \tag{A3.4}$$

allows one to define an average excess production $\bar{E}(t)$ that is of the simple form

$$\bar{E}(t) = \sum_k n_k(t)(A_k - D_k) \Big/ \sum_k n_k(t). \tag{A3.5}$$

The total production rate of the population can then be written as

$$\sum_k \dot{n}_k(k) = \bar{E}(t) \sum_k n_k(t) + \Phi(t) \tag{A3.6}$$

It will turn out to very useful to introduce relative fractions of population variables: $x_i(t) = n_i(t)/\Sigma_k n_k(t)$. Then, the equation for the relative growth rate of sequence I_i is

$$\dot{x}_i(t) = [W_{ii} - \bar{E}(t)]x_i + \sum_{k \neq i} W_{ik} x_k(t). \tag{A3.7}$$

APPENDIX 4. THE SOLUTION OF THE SYSTEM OF DIFFERENTIAL EQUATIONS (II.6)

Consider each variable $z_i(t)$ as the ith component of a column vector $\mathbf{z}$ and each rate coefficient W_{ik} as an element of the quadratic value matrix W. The system of rate Eqs. (II.6) then reads

$$\frac{d\mathbf{z}(t)}{dt} = W\mathbf{z}(t). \tag{A4.1}$$

Provided the matrix W is diagonalizable (which we shall assume throughout this Chapter; cf. also Appendix 2 and Section IV), the system of differential equations can be decoupled by means of the transformation

$$\mathbf{z}(t) = L\zeta(t) \quad \text{and} \quad \zeta(t) = L^{-1}\mathbf{z}(t), \tag{A4.2}$$

where $\zeta(t)$ represents again a column vector and L as well as its inverse L^{-1} are quadratic matrices. Combination of (A4.1) and (A4.2) yields

$$\frac{d\zeta(t)}{dt} = L^{-1}WL\zeta(t). \tag{A4.3}$$

The matrix $\Lambda = L^{-1}WL$ is diagonal as assumed in the preceding. Its elements are the eigenvalues of W: λ_k $(k = 0, 1, \ldots)$. The corresponding column vectors $\mathbf{l}_k = \{l_{1k}, l_{2k}, \ldots\}$ are stationary compositions (or "normal modes") for $\zeta(t)$, which grow in magnitude at the rates λ_k. They are found as (right-hand) eigenvectors of the matrix W. Provided the off-diagonal elements of the value matrix are strictly positive, the Frobenius theorem applies. Then the largest eigenvalue $\lambda_{\max} = \lambda_0$ is nondegenerate. Without loss of generality, we may order the eigenvalues

$$\lambda_0 > \lambda_1 \geqslant \lambda_2 \geqslant \lambda_3 \geqslant \lambda_4 \geqslant \cdots.$$

The eigenvector corresponding to λ_0 is called the dominant eigenvector $\mathbf{l}_0$. All its components are strictly positive: $l_{i0} > 0$ $(i = 1, 2, \ldots)$. The left-hand eigenvectors are given by the rows of the inverse matrix L^{-1}.

For each component of $\zeta(t)$ we obtain a solution curve of the general form

$$\zeta_k(t) = \zeta_k(0)\exp(\lambda_k t) \tag{A4.4}$$

wherein initial conditions are denoted by $\zeta_k(0) = \zeta_k(t = 0)$. The eigenvalues λ_k of the value matrix W are the solutions λ of the determinantal equation

$$\det(W - \lambda I) = 0, \tag{A4.5}$$

where det means determinant and I here denotes the unit matrix. Evaluation of Eqn. (A4.5) yields a polynomial in λ whose degree is given by the number of concentration coordinates. In order to determine the components of the right and left eigenvectors of the matrix W, we start from

$$WL = L\Lambda \tag{A4.6}$$

and

$$L^{-1}W = \Lambda L^{-1} \quad \text{or} \quad W'(L^{-1})' = (L^{-1})'\Lambda. \tag{A4.7}$$

The prime means here the transposed matrix. Explicitly in matrix elements, $L = \{l_{ij}\}$ and $L^{-1} = \{h_{ij}\}$, the last two equations read

$$\sum_j W_{ij} l_{jk} = l_{ik}\lambda_k, \tag{A4.8}$$

$$\sum_j W_{ji} h_{kj} = h_{ki}\lambda_k. \tag{A4.9}$$

The explicit solutions $z_i(t)$ of the linear differential Eqn. (A4.1) are obtained as superpositions of individual normal modes,

$$z_i(t) = \sum_k l_{ik} c_k \exp(\lambda_k t), \qquad i = 1, 2, \ldots . \tag{A4.10}$$

The constants c_k can be obtained from the initial conditions, $z_i(0) = x_i(0)$:

$$c_k = \sum_i h_{ki} x_i(0). \tag{A4.11}$$

[Note that the initial value of the function $f(t)$ in Eqn. (II.5) is $f(0) = 1$ by definition.] Explicit solutions for the relative populations variables $x_i(t)$ can be obtained from Eqs. (II.5) and (II.7):

$$x_i(t) = \frac{\sum_k l_{ik} c_k \exp(\lambda_k t)}{\sum_j \sum_k l_{jk} c_k \exp(\lambda_k t)}. \tag{A4.12}$$

Formally they can also be expressed as

$$x_i(t) = f(t) \sum_k l_{ik} c_k \exp(\lambda_k t) = \sum_k l_{ik} b_k y_k(t), \tag{A4.13}$$

where

$$y_k(t) = (c_k/b_k) f(t) \exp(\lambda_k t) \quad \text{or} \quad c_k/b_k = y_k(0) \tag{A4.14}$$

with

$$b_k = \left(\sum_i l_{ik}\right)^{-1} \quad \text{and} \quad \sum_k y_k(t) = 1. \tag{A4.15}$$

Note that the variables $y_k(t)$ are normalized, since

$$\sum_k \frac{c_k}{b_k} = \sum_k \sum_i h_{ki} x_i(0) \sum_j l_{jk} = \sum_k x_k(0) = 1 \tag{A4.16}$$

due to $LL^{-1} = I$.

APPENDIX 5. GRADIENT SYSTEMS AND SELECTION

Optimization processes are easily visualized in the case where the dynamics can be described by a "gradient system":

$$\dot{x}_k = \left(\frac{\partial V}{\partial x_k}\right)_{x_{j \neq k}}, \qquad k = 1, \ldots, n, \tag{A5.1}$$

or in vector notation, $\mathbf{x} = (x_1, \ldots, x_n)$:

$$\frac{d\mathbf{x}}{dt} = \operatorname{grad} V.$$

The potential is not an explicit function of time but depends on it implicitly: $V = V[\mathbf{x}(t)]$. The potential function V is nondecreasing along a trajectory $\mathbf{x}(t)$ since

$$\frac{dV}{dt} = \sum_k \frac{\partial V}{\partial x_k} \frac{dx_k}{dt} = \sum_k \left(\frac{\partial V}{\partial x_k}\right)^2 \geqslant 0. \tag{A5.2}$$

Gradient systems do not sustain oscillations in the variables $x_k(t)$: the Jacobian matrix $A = \{a_{ij}; i, j = 1, \ldots, n\}$where

$$a_{ij} = \frac{\partial}{\partial x_j} \frac{dx_i}{dt} = \left(\frac{\partial^2 V}{\partial x_j \, \partial x_i}\right)_{x_{k \neq i,j}} = \left(\frac{\partial^2 V}{\partial x_i \, \partial x_j}\right)_{x_{k \neq i,j}} = a_{ji} \tag{A5.3}$$

is symmetric and hence has no complex eigenvalues. Reaction–diffusion equations derived from gradient systems,

$$\frac{\partial x_k}{\partial t} = D_k \nabla^2 x_k + \left(\frac{\partial V}{\partial x_k}\right)_{x_{j \neq k}}, \qquad k = 1, \ldots, n,$$

do not form stable spatial patterns under no-flux boundary conditions [24].

First we consider error-free replication. Does it represent a gradient system? The differential equation for n competing sequences reads

$$\dot{x}_k = x_k(W_k - \bar{E}), \qquad k = 1, \ldots, n \quad \text{and} \quad W_k = A_k - D_k. \tag{A5.4}$$

Here $\bar{E}$ is a nondecreasing function of time provided the rate constants W_k are time independent:

$$\left(\frac{\partial \bar{E}}{\partial x_k}\right)_{x_{j \neq k}} = W_k - \bar{E},$$

and hence

$$\frac{d\bar{E}}{dt} = \sum x_k (W_k - \bar{E})^2 \geqslant 0. \tag{A5.5}$$

(This equation is always true since concentrations x_k are positive or zero and rate constants are real by definition.) The mean excess production $\bar{E}$, however, is not a potential function in the sense of Eqn. (A5.1) since the trajectories of Eqn. (A5.4) do not intersect with the constant level sets of $\bar{E}$ at right angles (Figure 4). Based on techniques originally introduced by Shahshahani [25], it is possible to transform Eqn. (A5.4) into a generalized gradient system [26, 27],

$$\frac{d\mathbf{x}}{dt} = \text{grad } V \tag{A5.6}$$

This generalized gradient is based on a Riemannian metric defined on the interior of the concentration space $\{x_k > 0; \Sigma_k x_k = 1, k = 1, \ldots, n\}$, which replaces the conventional Euclidean metric. We compare the definitions of the two inner products:

$$\text{Euclidean metric:} \qquad \langle \mathbf{x}, \mathbf{y} \rangle = \sum_{k=1}^{n} x_k y_k,$$

$$\text{Shahshahani metric:} \quad [\mathbf{x}, \mathbf{y}]_z = \sum_{k=1}^{n} \frac{1}{z_k} x_k y_k.$$

In the Shahshahani metric every component of the inner product is weighted by the corresponding coordinate of the point at which the inner product is formed. This weighting distorts the direction of the gradient of the potential

$$V(x_i, \ldots, x_n) = \bar{E} = \sum_{k=1}^{n} W_k x_k \tag{A5.7}$$

such that it coincides with the direction of the trajectories of Eqn. (A5.4). In Figure 4 we show a simple example of dimension $n = 3$.

Next we consider the equation with mutations in a coordinate system whose axes are spanned by the eigenvectors of the matrix W according to Eqn. (II.16). Then the selection equation is of the general form

$$\dot{y}_k = y_k(\lambda_k - \bar{E}), \qquad k = 1, \ldots, n, \tag{A5.4'}$$

wherein we denote the coefficient of the kth eigenvector ($\mathbf{l}_k$) by y_k and the corresponding eigenvalue by λ_k. Equation (A5.4'), although formally identical, differs from (A5.4) with respect to the domains of variables as well as the range of allowed parameter values. The relative concentrations $x_k(t)$ were nonnegative, whereas the y_k may be positive or negative depending on the initial conditions $y_k(0)$. The eigenvalues of the matrix W may be real or complex. Only the largest eigenvalue λ_0 is real and positive by the Frobenius theorem. Consequently, the mean excess production ($\Sigma x_k = \Sigma y_k = 1$)

$$\bar{E} = \sum_k x_k W_{kk} = \sum_k y_k \lambda_k \tag{A5.8}$$

is no longer a nondecreasing function of time since

$$\frac{d\bar{E}}{dt} = \sum_k y_k(\lambda_k - \bar{E})^2$$

may be either positive or negative. It is illustrative to split the concentration space around the asymptotically stable point ($y_0 = 1$, $y_k = 0$, $k = 1, \ldots, n - 1$) into orthants according to the signs of the variables (Figure 4). Since the differential Eqn. (A5.4') is invariant on the boundaries of the orthants,

$$y_k = 0 \rightarrow \dot{y}_k = 0 \rightarrow \ddot{y}_k = 0, \ldots$$

trajectories cannot cross these boundaries. The system thus will always stay inside the orthant to which it was assigned by the choice of initial conditions.There is one orthant, namely, the positive orthant ($y_k \geqslant 0$; $k = 0, 1, \ldots, n - 1$), in which $\bar{E}(t)$ is a nondecreasing function for real eigenvalues λ_k, and one orthant, the orthant ($y_0 > 1$, $y_k \leqslant 0$; $k = 1, \ldots, n - 1$), in which $\bar{E}(t)$ is nonincreasing for real eigenvalues λ_k. In the remaining orthants $\bar{E}$ may decrease, increase, or pass through an extremum.

Jones [19] derived a function

$$\bar{\varepsilon}(t) = \sum_k p_k \operatorname{Re} \lambda_k \tag{A5.9}$$

with $p_k(t) = |y_k(t)|/\Sigma_j|y_j(t)|$ and $\mathrm{Re}\,\lambda_k$ being the real part of a potentially complex pair of eigenvalues. It is easy to visualize that

$$\frac{d\bar{\varepsilon}}{dt} = \sum_k p_k(\mathrm{Re}\,\lambda_k - \bar{\varepsilon})^2 \geqslant 0 \tag{A5.10}$$

holds in full generality.

APPENDIX 6. BRILLOUIN–WIGNER PERTURBATION THEORY OF THE QUASI-SPECIES

The coordinates $z(t)$ of the differential Eqn. (II.6) can be expressed in the Laplace transform variable s, which is conjugate to the time t. With a single copy of sequence type I_k as initial conditions, they take the form

$$\tilde{z}_{ik}(s) = \frac{\delta_{ik}}{s - W_{ii}} + \sum_j \frac{1}{s - W_{ii}} W_{ij}\tilde{z}_{jk}(s). \tag{A6.1}$$

Here δ_{ik} denotes the Kronecker delta.

The perturbation series solution is then obtained by iteration as

$$\tilde{z}_{ik}(s) = \frac{\delta_{ik}}{s - W_{ii}} + \frac{1}{s - W_{ii}} {}^{c}W_{ik}\,\tilde{z}_{kk}(s), \tag{A6.2}$$

where

$${}^{c}W_{ik}(s) = \sum_j \frac{W_{ij}W_{jk}}{s - W_{jj}} + \sum_{j,l} \frac{W_{ij}W_{jl}W_{lk}}{(s - W_{jj})(s - W_{ll})} + \cdots. \tag{A6.3}$$

In particular,

$$\tilde{z}_{kk}(s) = [s - W_{kk} - {}^{c}W_{kk}(s)]^{-1}, \tag{A6.4}$$

and hence the perturbation solution is determined in closed form by the perturbation series of Eqn. (A6.3). The asymptotic behavior is exponential growth with the rate given by the dominant eigenvalue λ_0,

$$z_{ik}(t) \simeq a_{ik}\exp(\lambda_0 t) \quad \text{for large } t, \tag{A6.5}$$

where

$$a_{ik} = \lim_{s\to 0} s\tilde{z}_{ik}(s + \lambda_0) = \frac{1}{\lambda_0 - W_{ii}} {}^{c}W_{ik}(\lambda_0)\,a_{kk} \tag{A6.6}$$

and

$$a_{kk} = \lim_{s \to 0} \frac{s}{s + \lambda_0 - W_{kk} - {}^cW_{kk}(s + \lambda)}, \tag{A6.7}$$

which results from Eqs. (A6.2) and (A6.4). If $a_{kk} \neq 0$, then Eqn. (A6.7) yields directly the result of the Brillouin–Wigner perturbation theory for the dominant eigenvalue

$$\lambda_0 = W_{kk} + {}^cW_{kk}(\lambda_0), \tag{A6.8}$$

and the population variables x_k may be obtained in terms of Eqn. (A6.3) evaluated at $s = \lambda$:

$$(x_k)^{-1} = \sum_i \frac{a_{ik}}{a_{kk}} = 1 + \sum_{i \neq k} \frac{{}^cW_{ik}(\lambda)}{\lambda - W_{ii}}. \tag{A6.9}$$

APPENDIX 7. RENORMALIZATION OF THE PERTURBATION THEORY

Equation (A6.3) may be rewritten using the factorization of Eqn. (III.7) in the form

$${}^cW_{ik}(\lambda_0) = W_{kk}\left[\sum_j \frac{V_{ij}V_{jk}}{a_j} + \sum_{j,l} \frac{V_{il}V_{lj}V_{jk}}{a_l a_j} + \cdots\right], \tag{A7.1}$$

where

$$a_i = \frac{\lambda_0 - W_{ii}}{W_{ii}}, \tag{A7.2}$$

and V_{ik} is given by Eqn. (III.7) for $i \neq k$. Although the factors V_{ii} are trivially zero, the series may be renormalized to remove less immediate repetitions following the Watson [30] procedure exploited by Anderson [31]. The result is

$${}^cW_{ik}(\lambda_0) = W_{kk}\left[\sum_{j \neq k} \frac{V_{ij}V_{jk}}{e_j^k} + \sum_{j \neq k;}\sum_{l \neq j,k} \frac{V_{il}V_{lj}V_{jk}}{e_l^{kj} e_j^k} + \cdots\right], \tag{A7.3}$$

where

$$e_j^k = [\lambda_0 - W_{jj} - {}^cW_{jj}^k(\lambda_0)]/W_{jj},$$

$$e_l^{k,j} = [\lambda_0 - W_{ll} - {}^cW_{ll}^{k,j}(\lambda_0)]/W_{ll}, \ldots . \tag{A7.4}$$

Here the superscripts are suppressed in the summations involved in the corresponding perturbation expressions. Repeated indices have been removed (affecting first the fourth-order terms) at the expense of corrections to the denominators. The renormalized denominators may be simplified approximately to

$$e_i = \frac{\lambda_0}{W_{mm}} \frac{W_{mm} - W_{ii}}{W_{ii}} \tag{A7.5}$$

using a kind of mean-field approximation valid for high-dimensional mutant spaces [29]. The eigenvalue λ_0 is then to be obtained from a series of the self-consistent form:

$$\frac{\lambda_0}{W_{mm}} = 1 + \frac{W_{mm}}{\lambda_0}\left(\sum_{j \neq m} \frac{V_{mj}V_{jm}}{(W_{mm} - W_{jj})/W_{jj}} + \frac{W_{mm}}{\lambda_0} \times \sum_{j \neq m;\, l \neq j,m} \sum \frac{V_{ml}V_{lj}V_{jm}}{(W_{mm} - W_{ll})(W_{mm} - W_{jj})/W_{ll}W_{jj}} + \cdots\right), \tag{A7.6}$$

and the remaining expressions depend on λ_0.

APPENDIX 8. STATISTICAL CONVERGENCE OF PERTURBATION THEORY

Considering a typical high-order form in the renormalized perturbation theory expression for ${}^cW_{im}(\lambda)$, in particular for the sequence $\zeta = (j, k, \ldots, l)$ of $N - 1$ indices,

$$T^{(N)}_{mi\zeta} = \frac{V_{il} \cdots V_{kj}V_{jm}}{e_l \cdots e_k e_j} = V_{il} \cdots V_{kj}V_{jm}t^{(N)}_{\zeta}, \tag{A8.1}$$

direct averaging diverges. The logarithm of such a term, however, has mean and variance

$$\langle \ln t^{(N)}_{\zeta} \rangle = Nt \tag{A8.2}$$

and

$$\sigma^2\{\ln t^{(N)}_{\zeta}\} = N\sigma^2, \tag{A8.3}$$

where t and σ are asymptotically independent of N, provided there were no long-range correlations between the denominators e_i. Correlations as a result of repeated indices have been removed by the renormalization procedure outlined in Appendix 7. Equations (A8.2) and (A8.3) imply a sharpening probability distribution for $t_\zeta^{(N)}$:

$$\text{Prob}\{\exp[t(N - N^p)] < t_\zeta^{(N)} < \exp[t(N + N^p)]\} \to 1 \quad \text{as } N \to \infty, \tag{A8.4}$$

where p is a parameter satisfying $\frac{1}{2} < p < 1$. The preceding result summarizes the essential statistical argument employed by Economou and Cohen [33].

Since all the terms are positive rather then of random sign, Eqn. (A8.4) may be used to prove that the sum of all terms of Nth order has a value for larger N near

$$T_m^{(N)} = \sum_{i,\zeta} T_{mi\zeta}^{(N)} = e^{Nt} \sum_{i,l,\ldots,j} V_{il} \cdots V_{kj} V_{jm} = e^{Nt}(q^{-\nu} - 1)^N \tag{A8.5}$$

with probability 1. The last equality employs the closure summation (cf. Appendix 1):

$$\sum_{i \neq m} V_{im} = \sum_{d=1}^{\nu} N_d \left(\frac{q^{-1} - 1}{\kappa - 1} \right)^d = q^{-\nu} - 1. \tag{A8.6}$$

In the absence of systematic correlations between the replication rates of nearby mutants the average t may be obtained,

$$\langle t \rangle = -\langle \ln e_j \rangle = -\left\langle \ln \frac{W_{mm} - W_{jj}}{W_{jj}} \right\rangle - \langle \ln \lambda_0 \rangle + \ln W_{mm}, \tag{A8.7}$$

and the convergence of the perturbation series depends on the inequality

$$1 > (q^{-\nu} - 1) \frac{W_{mm}}{\langle \lambda_0 \rangle_{\ln}} \left\langle \frac{W_{jj}}{W_{mm} - W_{jj}} \right\rangle_{\ln}, \tag{A8.8}$$

where the logarithmic averages are defined by

$$\left\langle \frac{W_{jj}}{W_{mm} - W_{jj}} \right\rangle_{\ln} = \exp\left\{ \left\langle \ln \frac{W_{jj}}{W_{mm} - W_{jj}} \right\rangle \right\} \tag{A8.9}$$

and

$$\langle \lambda_0 \rangle_{\ln} = \exp\{\langle \ln \lambda_0 \rangle\}. \tag{A8.10}$$

APPENDIX 9. VARIABLES, MEAN RATE CONSTANTS, AND MEAN SELECTIVE VALUES FOR THE RELAXED ERROR THRESHOLD

The composite sequences (Figure 7) and its concentrations are defined by

$$I_{ij} = A_i - B_j\colon \quad [I_{ij}] = x_{ij}n; \quad n = \sum_{ij} n_{ij}; \quad \sum_{ij} x_{ij} = 1. \tag{A9.1}$$

This leads to the conventional selection equation

$$\dot{x}_{ij} = (W_{ij,ij} - \bar{E})x_{ij} + \sum_{ij \neq kl} W_{ij,kl}x_{kl}, \tag{A9.2}$$

with the selective value

$$W_{ij,kl} = A_{kl}Q_{ij,kl} - D_{ij}\delta_{ij,kl}. \tag{A9.3}$$

Averaging leads to

$$I_{i\cdot} \equiv A_i - B\colon \quad [I_{i\cdot}] = \left(\sum_{k=1}^{s} x_{ik}\right)n = x_{i0}\cdot n; \quad \sum_i x_{i0} = 1 \tag{A9.4}$$

and

$$I_{\cdot j} \equiv A - B_j\colon \quad [I_{\cdot j}] = \left(\sum_{k=1}^{r} x_{kj}\right)n = x_{0j}\cdot n; \quad \sum_j x_{0j} = 1 \tag{A9.5}$$

The same procedure for the rate constants yields

$$\bar{A}_{i0} = \frac{1}{x_{i0}} \sum_{k=1}^{s} A_{ik}x_{ik} \quad \text{and} \quad \bar{A}_{0j} = \frac{1}{x_{0j}} \sum_{k=1}^{r} A_{kj}x_{kj}, \tag{A9.6}$$

$$\bar{D}_{i0} = \frac{1}{x_{i0}} \sum_{k=1}^{s} D_{ik}x_{ik} \quad \text{and} \quad \bar{D}_{0j} = \frac{1}{x_{0j}} \sum_{k=1}^{r} D_{kj}x_{kj}, \tag{A9.7}$$

and

$$\bar{E} = \sum_{ij}(A_{ij} - D_{ij})x_{ij} = \sum_{i=1}^{r} (\bar{A}_{i0} - \bar{D}_{i0})x_{i0} = \sum_{j=1}^{s} (\bar{A}_{0j} - \bar{D}_{0j})x_{0j}. \tag{A9.8}$$

The only assumption made here concerns the mutation matrix: simultaneous mutations in both parts A and B are excluded,

$$Q_{ij,kl} = \begin{cases} Q_{ij,ij} = Q^A_{ii} Q^B_{jj} & \text{if } i = k \text{ and } j = l, \\ Q^A_{ik} & \text{if } j = l \text{ and } i \neq k, \\ Q^B_{jl} & \text{if } i = k \text{ and } j \neq l, \\ 0 & \text{if } i \neq k \text{ and } j \neq l, \end{cases} \tag{A9.9}$$

where

$$Q^A_{ii} = 1 - \sum_{k \neq i} Q^A_{ik} \quad \text{and} \quad Q^B_{jj} = 1 - \sum_{l \neq j} Q^B_{jl}. \tag{A9.10}$$

Finally we obtain the coupled selection equations

$$\dot{x}_{i0} = (\bar{A}_{i0} Q^A_{ii} - \bar{D}_{i0} - \bar{E}) x_{i0} + \sum_{k \neq i} Q^A_{ik} \bar{A}_{ik} x_{k0} \tag{A9.11}$$

and

$$\dot{x}_{0j} = (\bar{A}_{0j} Q^B_{jj} - \bar{D}_{0j} - \bar{E}) x_{0j} + \sum_{l \neq j} Q^B_{jl} \bar{A}_{0l} x_{0l} \tag{A9.12}$$

with $i, k = 1, 2, \ldots, r$ and $j, l = 1, 2, \ldots, s$.

Acknowledgment

This work was supported by the Max Planck Gesellschaft, Germany, by the Fonds zur Förderung der Wissenschaftlichen Forschung in Austria (Nr. 5286), and by the Stiftung Volkswagenwerk, Germany. We wish to thank Ruthild Winkler-Oswatitsch for her kind assistance in preparing the illustrations and for reading the manuscript and providing valuable suggestions.

References

1. E. P. Wigner, Proc. 11th Conf. Robert A. Welch Found. Houston (Texas) 1967.
2. M. Kimura, *The Neutral Theory of Molecular Evolution.* Cambridge University Press, Cambridge, U.K., 1983.
3. E. Domingo, D. Sabo, T. Taniguchi, and C. Weissmann, *Cell 13*:735–744 (1978).
4. M. Eigen and P. Schuster, *The Hypercycle—a Principle of Natural Selforganization.* Springer-Verlag, Berlin 1979. A combined reprint of *Naturwissenschaften 64*:541–565 (1977): *65*:7–41, 341–369 (1978).
5. M. Eigen, *Chem. Scrip. 26B*:13–26 (1986).
6. P. Schuster, *Chem. Scrip. 26B*:27–41 (1986).
7. M. Eigen and C. K. Biebricher, "Sequence Space and Quasispecies Distributions." In E.

Domingo, P. Ahlquist, and J. J. Holland (eds.), *RNA-Genetics, Vol. II: RNA Variability*. CRC Press, Baton Rouge, FL (in press).
8. D. S. Rumschitzki, *J. Math. Biol. 24*:667–680 (1987).
9. M. Eigen, *Naturwissenschaften 58*:465–526 (1971).
10. C. K. Biebricher, M. Eigen, and W. C. Gardiner, Jr., *Biochemistry 22*:2544–2559 (1983).
11. C. K. Biebricher, M. Eigen, and W. C. Gardiner, Jr., *Biochemistry 23*:3186–3194 (1984).
12. C. K. Biebricher, M. Eigen, and W. C. Gardiner, Jr., *Biochemistry 24*:6550–6560 (1985).
13. A. Kornberg, *DNA Replication*. Freeman, San Francisco, 1980.
14. I. R. Epstein, *J. Theor. Biol. 78*:271–298 (1979).
15. C. J. Thompson and J. L. McBride, *Math. Biosci. 21*:127–142 (1974).
16. B. L. Jones, R. H. Enns, and S. S. Rangnekar, *Bull. Math. Biol. 38*:15–18 (1975).
17. H. Haken and H. Sauermann, *Z. Physik 173*:261–275 (1963); *176*:47–62 (1963).
18. R. Feistel and W. Ebeling, *Stud. Biophys. (Berlin) 71*:139 (1976).
19. B. L. Jones, *J. Math. Biol. 6*:169–175 (1978).
20. R. A. Fisher, *The Genetical Theory of Natural Selection*. Oxford University Press, Oxford, 1930 [2nd rev. ed., Dover Publications, New York, 1958].
21. J. B. S. Haldane, *The Causes of Evolution*. Harper and Row, New York, 1932.
22. S. Wright, *Evolution and the Genetics of Populations*, Vols. I–IV. The University of Chicago Press, Chicago, 1968, 1969, 1977, and 1978.
23. B. B. Mandelbrot, *The Fractal Geometry of Nature*. Freeman, New York, 1983.
24. T. E. Creighton, *Proteins—Structures and Molecular Principles*. Freeman, New York, 1983.
25. R. G. Casten and C. J. Holland, *J. Diff. Eqs. 27*:266–273 (1978).
26. S. Shahshahani, *A New Mathematical Framework for the Study of Linkage and Selection*. Memoirs AMS, *211*, 1979.
27. K. Sigmund, "The Maximum Principle for Replicator Equations." In W. Ebeling and M. Peschel (eds.), *Lotka-Volterra-Approach to Cooperation and Competition in Dynamic Systems*. Akademie-Verlag, Berlin, 1985, pp. 63ff.
28. P. Schuster and K. Sigmund, *Ber. Bunsenges. Phys. Chem. 89*:668–682 (1985).
29. J. S. McCaskill, *J. Chem. Phys. 80*:5194 (1984).
30. K. M. Watson, *Phys. Rev. 105*:1388 (1957).
31. P. W. Anderson, *Phys. Rev. 109*:1492 (1958).
32. H. A. David, *Order Statistics*, Wiley, New York, 1970.
33. E. N. Economou and M. H. Cohen, *Phys. Rev. 135*:2931 (1972).
34. E. Batschelet, E. Domingo, and C. Weissmann, *Gene 1*:27 (1976).
35. E. Domingo, R. A. Flavell, and C. Weissman, *Gene 1*:3 (1976).
36. C. K. Biebricher, *Evolut. Bio. 16*:1 (1983).
37. R. K. Selander and B. R. Levin, *Science 210*:545 (1980).
38. D. E. Dykhuizen and D. L. Hartl, *Genetic 96*:801 (1980).
39. D. L. Hartl and D. E. Dykhuizen, *Proc. Natl. Acad. Sci. U.S.A. 78*:6344 (1981).
40. M. Eigen and P. Schuster (unpublished).
41. H. F. Schaefer III (ed.), *Modern Theoretical Chemistry, Vol. 3: Methods of Electronic Structure Theory*, Plenum Press, New York, 1977.
42. J. Swetina and P. Schuster, *Biophys. Chem. 16*:329 (1982).

43. R. Kindermann and J. L. Snell, *Contemporary Mathematics, Vol. 1: Markov Random Fields and Their Applications*, American Mathematical Society, Providence, R.I., 1980.

44. R. J. Baxter, *Exactly Solved Models in Statistical Mechanics*, Academic Press, London, 1982.

45. L. Demetrius, Polynucleotide Replication and Statistical Mechanics, preprint, 1987.

46. I. Leuthäusser, *J. Chem. Phys. 84*:1884 (1986).

47. I. Leuthäusser, Statistical Mechanics of Eigen's Evolution Model, *J. Stat. Phys. 48*:343 (1987).

48. L. Demetrius, P. Schuster, and K. Sigmund, *Bull. Math. Biol. 47*:239 (1985).

49. L. Demetrius, *J. Stat. Phys. 30*:709 (1983).

50. J. J. Hopfield, *Proc. Natl. Acad. Sci. U.S.A. 79*:2554 (1982).

51. J. S. McCaskill, *Biol. Cybern. 50*:63 (1984).

52. P. Schuster and J. Swetina, *Bull. Math. Biol. 50*:635 (1988).

53. W. Fontana and P. Schuster, *Biophys. Chem. 26*:123 (1987).

54. M. S. Waterman, T. F. Smith, and W. A. Beyer, *Adv. Math. 20*:367 (1976).

55. C. R. Cantor and P. R. Schimmel, *Biophysical Chemistry*, Vol. 3, Freeman, San Francisco, 1980, pp. 1183ff.

56. M. Zuker and D. Sankoff, *Bull. Math. Biol. 46*:591 (1984).

57. P. W. Anderson, *Proc. Natl. Acad. Sci. U.S.A. 80*:3386 (1983).

58. D. Pörschke, in I. Pecht and R. Rigler (eds.), *Chemical Relaxation in Molecular Biology*, Springer-Verlag, Heidelberg, 1977, pp. 191ff.

59. C. K. Biebricher, M. Eigen, and W. C. Gardiner, *Biochemistry 22*:2544–2559 (1983).

60. C. K. Biebricher, M. Eigen, and W. C. Gardiner, *Biochemistry 23*:3186 (1984).

61. C. K. Biebricher, M. Eigen, and W. C. Gardiner, *Biochemistry 24*:6550 (1985).

62. C. K. Biebricher, M. Eigen, in *RNA Genetics*, Vol. I, E. Domingo, J. J. Holland, and Ahlquist (eds.), CRC Press, Boca Raton, FL 1–21 (1988).

63. M. Eigen and C. K. Biebricher, in *RNA Genetics*, Vol. III, E. Domingo, J. J. Holland, and Ahlquist (eds.), CRC Press, Boca Raton, FL 211–245 (1988).

64. E. Domingo, E. Martina-Salas, F. Sobrino, J. C. de la Torre, A. Portela, J. Ortin, C. Lopez-Galindez, P. Perez-Brena, N. Villanueva, R. Najera, S. Van de Pol, S. Steinhauer, N. DePolo, and J. Holland, *Gene 40*:1 (1986).

65. J. Holland, K. Spindler, F. Horodyski, E. Grabau, S. Nichol, and S. Van de Pol, *Science 215*:1577 (1982).

66. E. Domingo, D. Sabo, T. Taniguchi and C. Weissmann, *Cell 13*:735 (1978). E. Domingo, A. Flavell, and C. Weissmann, *Gene 1*:3 (1976). E. Batschlelet, E. Domingo, and C. Weissmann, *Gene 1*:27 (1976).

67. E. Domingo, M. Davila, and J. Ortin, *Gene 11*:333 (1980).

68. J. D. Parvin, A. Moscona, W. T. Pan, J. Lieder, and P. Palese, *J. Virol. 59*:377 (1986).

69. K. R. Spindler, F. M. Horodyski, and J. J. Holland, *Virology 119*:96 (1982).

70. S. Benzer, *Proc. Natl. Acad. Sci. U.S.A. 47*:403 (1961).

71. S. Benzer, *Proc. Natl. Acad. Sci. U.S.A. 45*:1607 (1959).

72. D. R. Mills, R. L. Peterson, and S. Spiegelman, *Proc. Natl. Acad. Sci. U.S.A. 58*:217 (1967).

73. F. R. Kramer, D. R. Mills, P. E. Cole, T. Nishihara, and S. Spiegelman, *J. Mol. Biol. 89*:719 (1974).

74. D. R. Mills, F. R. Kramer, and S. Spiegelman, *Science 180*:916 (1973).

75. C. K. Biebricher, M. Eigen, and W. C. Gardiner, *Biochemistry* (submitted).
76. M. Sumper and R. Luce, *Proc. Natl. Acad. Sci., U.S.A. 72*:162 (1975).
77. C. K. Biebricher, M. Eigen, and R. Luce, *J. Mol. Biol. 148*:361 (1981).
78. C. K. Biebricher, M. Eigen, and R. Luce, *J. Mol. Biol. 148*:391 (1981).
79. C. K. Biebricher, M. Eigen, and R. Luce, *Nature 321*:89 (1986).
80. H. Otten, Dissertation, Braunschweig, 1988.
81. A. Schwienhorst, Diplom-Arbeit, Bielefeld, 1988.
82. C. K. Biebricher, *Cold Spring Harbor Symp., Quant. Biol. 52*: (in press).
83. C. W. Gardiner, *Handbook of Stochastic Methods*, Springer-Verlag, Berlin, 1983.
84. W. Ebeling and R. Feistel, *Ann. Physik 34*:81 (1977).
85. P. H. Richter, *Bull. Math. Biol. 37*:193 (1975).
86. B. L. Jones and H. K. Leung, *Bull. Math. Biol. 43*:665 (1981).
87. R. Heinrich and I. Sonntag, *J. Theor. Biol. 93*:325 (1981).
88. H. K. Leung, *Bull. Math. Biol. 46*:399 (1984).
89. H. K. Leung, *Bull. Math. Biol. 47*:231 (1985).
90. H. Inagaki, *Bull. Math. Biol. 44*:17 (1982).
91. A. F. Bartholomay, *Bull. Math. Biophys. 20*:97 (1958).
92. P. Schuster and K. Sigmund, *Bull. Math. Biol. 46*:11 (1984).
93. B. M. Pötscher, *Bull. Math. Biol. 47*:263 (1985).
94. E. D. Weinberger, Dissertation, New York University, 1987.
95. D. Gillespie, *J. Comp. Phys. 22*:403 (1976).
96. M. A. Nowak, Eine numerische Simulation der RNA-Replikation zur Berechnung der Stochastischen "Error-Threshold." Diploma Thesis, Universität Wien, 1987.

CHAIN CONFIGURATIONS AND DYNAMICS IN THE GAUSSIAN APPROXIMATION

GIUSEPPE ALLEGRA AND FABIO GANAZZOLI

Dipartimento di Chimica
Politecnico di Milano
Piazza Leonardo da Vinci, 32
I-20133 Milano (Italy)

CONTENTS

1. INTRODUCTION

It is known that the average configuration [1] of a long random-walk chain is well represented by Gaussian distribution laws for virtually all the interatomic distances [2, 3]. This is not the case, in general, when interactions among topologically distant chain atoms come into play, as in a polymer chain departing from the ideal state. Under these circumstances, in the limit of an infinite chain length a very powerful help to investigate chain equilibrium may be offered by the renormalization group (RG) approach [4, 5]. Among other results, several critical exponents were obtained, the most famous of which is probably the good-solvent exponent $2\nu = 1.176\ldots$ appearing in the expression $\langle r^2(N)\rangle \propto N^{2\nu}$ for the mean-square end-to-end distance of an N-segment chain ($N \rightarrow \infty$) [6, 7]. However, the pure RG approach has difficulties in describing the chain behavior in the intermediate range, where the asymptotic power laws do not apply yet. While it always regards the polymer chain as an ideal line with a definite persistence length and no internal friction, many current experiments are centered in the intermediate range; here portions of the chain under investigation are sufficiently short that the rigidity imparted by the statistical correlation among the skeletal rotations comes into play [2], together with the dynamic resistance to chain motion caused by the energy barriers to skeletal rotation (internal viscosity) [8]. Besides, the chain contraction in poor solvents is specifically influenced by medium-range repulsions depending on the thickness-to-persistence length ratio [9, 10], which defy the usual scaling arguments and consequently the current renormalization approaches. It is for these reasons that a flexible tool such as the Gaussian approximation coupled with the Fourier configurational decomposition along the chain coordinate and the self-consistent method for free-energy optimization may prove effective to incorporate all the preceding structural and dynamical features imparting each polymer its own "personality" in the intermediate range [10–15]. It should be clear that in the present context the "Gaussian approximation" implies that all the interatomic distances are expressed according to Gaussian laws, *no implication existing of the so-called Gaussian, or random-walk, chain.* In conclusion, the present approach is aimed to derive general laws of the configurational statistics and dynamics of polymer chains, accounting for their local stereochemical and conformational features.

When considering the long-range chain properties, we shall see arguments supporting the Gaussian approximation both in the unperturbed and in the poor-solvent case, respectively occurring at $T = \Theta$ and at $T < \Theta$, where Θ is the Θ temperature, analogous to the Boyle temperature for a real gas. In the good-solvent regime at $T > \Theta$ the long-range Gaussian approximation is incorrect, in principle; however, it was shown to yield [16] the Flory value [17] $2\nu = 1.2$ instead of the exact figure 1.176 for the previously quoted exponent, with a relatively small error. Moreover, in the intermediate good-solvent range we shall see that the Gaussian approximation is in good agreement with the most accurate results obtained through other approaches, which supports our confidence that it may be reasonably well extended to most cases of practical interest.

Chain dynamics will be addressed through the linear Langevin equation of motion [18], with the preaveraged hydrodynamic interaction [19, 20]. In our approach, the forces will always be considered as applied to the individual chain atoms, not to "segments" or "beads" [21, 22]. It is important to stress that the linear dependence of the intramolecular forces from the interatomic displacements is consistent with the Gaussian approximation concerning the intrachain distance probability distribution. The configurationally independent Fourier components obtained from optimization of the equilibrium free energy are very close to the eigenvectors of the dynamic equations. As usual, these should be regarded as an average description of chain motion; although as a rule the linear approximation does not lend itself to describe the propagation of specific conformational features such as kinks, jogs, and so on, we shall see this to be permitted to some average extent by our formulation of the internal viscosity.

In the following, chain equilibrium and dynamics will be separately considered, in this order. A fairly general presentation of the most relevant physical factors will be given. In particular, in the equilibrium section we shall generally limit our attention to the average mean square distance between atoms h and k, $\langle r^2(h, k)\rangle$, which gives the diffraction intensity, and to the mean square radius of gyration $\langle S^2\rangle$, whereas in the dynamic section the dynamic viscosity $\eta(\omega)$ and the dynamic structure factor $S(Q, t)$ will be our main object ($\mathbf{Q}$ is the scattering vector, t is time, ω is frequency). In general, we shall consider the dilute solution state, but polymer melt dynamics may also be tackled with suitable adaptation of the approach, as it will be shown. The chain will have an $-\!\!\![\mathrm{A}]\!\!\!-_N$ structure with N very large, but the case of the biatomic monomeric unit $-\!\!\![\mathrm{A}-\mathrm{B}]\!\!\!-_{N/2}$ will also be considered. The inevitable emphasis given to our work should not be taken as an implicit assessment of its relative importance. In this regard, although we shall generally refer to our previous papers concerning the self-consistent approach to configurational equilibrium, it should not be forgotten that, although unaware at the time, we were preceded by des Cloizeaux [23] and, in part, by Edwards and Singh [24].

We stress that the present approach is essentially self-contained. In addition to the self-consistent derivation of the average chain configuration at equilibrium, with the only exception of the internal viscosity, chain dynamics will be derived from the equilibrium results under classical assumptions (e.g., the preaveraged approximation, the linear Langevin equation). In particular, no a priori use will generally be made of scaling considerations.

To stress that the number of chain atoms N is always assumed to be very large, their index will be formally treated as a continuous variable so that, for example, $\mathbf{R}(h)$ will represent the position vector of the hth atom. Obviously enough, in the dynamic case the time variable will also be added.

A few among the results concerning the unperturbed and the good-solvent state (i.e., for $T \geqslant \Theta$) given in the dynamical section of this chapter were already reported by Doi and Edwards [25]. The results will be repeated here both for reasons of completeness and because they are derived from a different context.

2. CHAIN EQUILIBRIUM

2.1. The Unperturbed Chain

The essential feature of the ideal, or unperturbed, state resides in that two chain atoms do not interact if their separation along the chain sequence is sufficiently large. This will be expressed by saying that the sum of the binary cluster integral β and of a repulsive three-body contribution σ_3 is zero at the ideal temperature $T = \Theta$ [6]. We have [3], in $k_B T$ units,

$$e(\mathbf{r}) = 1 - \exp(-w(\mathbf{r})/k_B T), \qquad \beta = \int e(\mathbf{r})\, d^3\mathbf{r}, \tag{2.1.1}$$

$$\beta_{\text{eff}} = \int e(\mathbf{r})\, d^3\mathbf{r} + \sigma_3 = \beta + \sigma_3 = 0 \qquad (T = \Theta), \tag{2.1.2}$$

where $w(\mathbf{r})$ is the actual potential for two atoms at a distance $\mathbf{r}$ and $e(\mathbf{r})$ is the corresponding effective potential; σ_3 has a physical meaning analogous to β although originating from higher order contacts, mainly three-body interactions [26]. Deferring discussion of the three-body integral σ_3, suffices here to say that (i) it is positive (i.e., repulsive) and roughly independent of temperature in a reasonable vicinity of $T = \Theta$; (ii) its mathematical definition and value are quite different from their equivalent in the kinetic theory of gases since they are strictly dependent on the chain connectivity. Accordingly, in order to fulfil Eqn. (2.1.2), we must have $\beta < 0$ at $T = \Theta$. A suitable potential of the Lennard–Jones type may be a satisfactory representation of $w(\mathbf{r})$ (see Figure 1), and we shall assume it to be temperature independent for simplicity; it must be borne in mind that the attractive part is influenced by the solvent quality. At $T < \Theta$ the attractive, negative contribution will dominate and β_{eff} will also be negative; conversely, for $T > \Theta$ we shall have a positive value and hence a repulsive effect.

Referring to Figure 1, we shall assume at first that the distance $\bar{r}$ at which $w(\mathbf{r})$ attains its minimum is very small compared with the relevant average

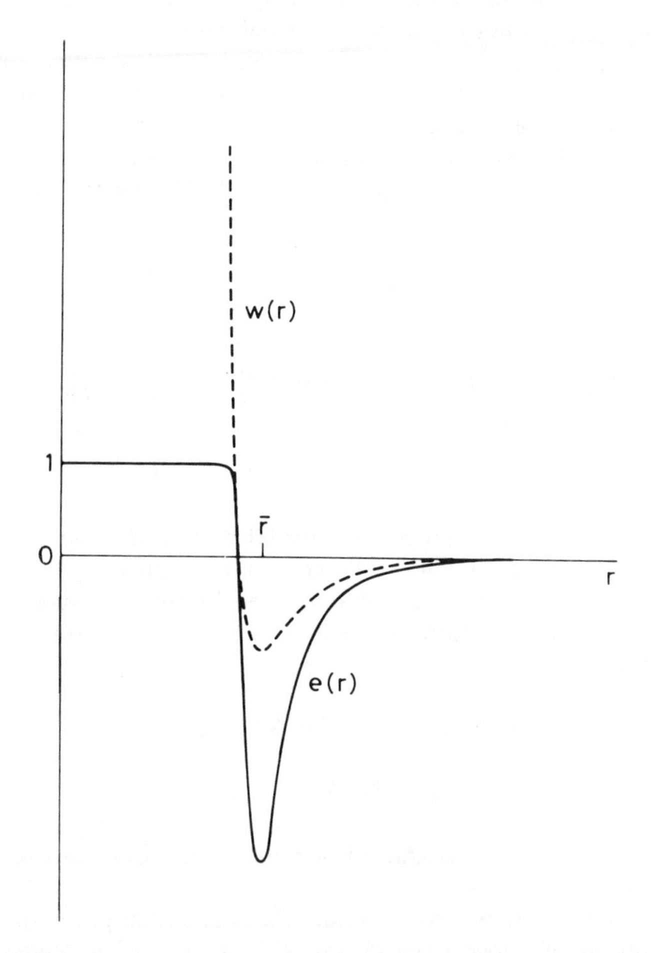

Figure 1. Schematic drawing of interatomic potential $w(r)$ (dashed line) in arbitrary units and of corresponding effective potential $e(r)$ (continuous line) as function of r. Both potentials are taken as spherically symmetrical and have minimum at distance $\bar{r}$. [see Eqn. (2.1.1)].

distances between chain atoms and that the potentials contributing to σ_3 will have nonvanishing values roughly in the same $\mathbf{r}$ range. As a consequence, on dealing with long interatomic distances the effective interaction potential may be expressed as $\beta_{\text{eff}}\delta(\mathbf{r})$ [see Eqn. (2.1.2)] and becomes identically zero at the Θ temperature. We are thus in the presence of the *unperturbed phantom chain* (henceforth *phantom chain* for brevity); except possibly for local elastic deformations and steric interactions between atoms separated by few bonds ($\sim$5–10), chain atoms are completely insensitive to one another. If $\mathbf{l}(h)$ is the vector associated with the hth chain bond, the average scalar product $\langle \mathbf{l}(h) \cdot \mathbf{l}(h+k) \rangle$ tends to vanish exponentially with large k. Conversely, if the interaction range is not negligible compared with the relevant interatomic distances so that the δ-function approximation is no longer acceptable, we have the real chain, and for $T = \Theta$ (i.e., $\beta_{\text{eff}} = 0$), the *unperturbed real chain* [27]. Now the average $\langle \mathbf{l}(h) \cdot \mathbf{l}(h+k) \rangle$ vanishes with large k as $k^{-\alpha}$, $\alpha > 1$.

Let us consider the general expressions of the mean-square end-to-end distance $\langle r^2(N) \rangle$ and radius of gyration $\langle S^2 \rangle$ of a chain with $N+1$ atoms (N bonds):

$$\langle r^2(N) \rangle = Nl^2 + 2 \sum_{h=1}^{N} \sum_{k=1}^{N-h} \langle \mathbf{l}(h) \cdot \mathbf{l}(h+k) \rangle, \tag{2.1.3}$$

$$\langle S^2 \rangle = (N+1)^{-2} \sum_{h=1}^{N} \sum_{k=1}^{N-h+1} \langle r^2(h, h+k) \rangle, \tag{2.1.4}$$

$$\mathbf{r}(h, h+k) = \mathbf{l}(h+1) + \mathbf{l}(h+2) + \cdots + \mathbf{l}(h+k), \tag{2.1.5}$$

where $\mathbf{r}(h, h+k)$ is the vector between the hth and the $(h+k)$th chain atoms. We see that, provided $\langle \mathbf{l}(h) \cdot \mathbf{l}(h+k) \rangle_0$ goes to zero more quickly than k^{-1} (as it happens in the unperturbed state denoted by the zero subscript), both $\langle r^2(N) \rangle_0$ and $\langle S^2 \rangle_0$ are proportional to N as $N \to \infty$ [28]. More precisely, in this limit we may write

$$\langle r^2(N) \rangle_0 = NC_\infty l^2 = Nl^2(2l_{\text{pers}}/l - 1), \tag{2.1.6}$$

$$\langle S^2 \rangle_0 = \tfrac{1}{6} \langle r^2(N) \rangle_0, \tag{2.1.7}$$

where C_∞ and l_{pers} are the characteristic ratio and the persistence length [29], respectively.

In the next section we shall consider the equilibrium properties of some typical models of unperturbed chains with an increasing degree of complexity. They are (i) the bead-and-spring phantom chain; (ii) the phantom chain with nearest-neighbor correlation; and (iii) the unperturbed real chain with a nonnegligible covolume and associated medium-range repulsions (screened interactions). In addition to the linear chain, the ring will also be considered.

2.1.1. *The Bead-and-Spring Phantom Chain*

The statistical independence among the N spring vectors $\mathbf{l}(1), \mathbf{l}(2), \cdots, \mathbf{l}(N)$ with $\langle l^2(h)\rangle = l^2$, is expressed by

$$\langle \mathbf{l}(h)\cdot\mathbf{l}(j)\rangle_0 = l^2\Delta(h-j), \tag{2.1.8}$$

where Δ is the Kronecker delta. The Boltzmann statistical weight of the general chain state is the product of N independent factors:

$$\begin{aligned}\mathscr{W}[\mathbf{l}(1), \mathbf{l}(2), \ldots, \mathbf{l}(N)] &= \prod_{h=1}^{N}\left\{\left(\frac{3}{2\pi l^2}\right)^{3/2}\exp\left[-\frac{3l^2(h)}{2l^2}\right]\lambda\right\}\\ &= \left(\frac{3}{2\pi l^2}\right)^{3N/2}\exp\left[-\sum_{h=1}^{N}\frac{3l^2(h)}{2l^2}\right]\lambda^N,\end{aligned} \tag{2.1.9}$$

where λ is the configurational partition function of each spring [3]. From the joint probability distribution (2.1.9), application of the Wang–Uhlenbeck method [30] allows us to derive any Gaussian probability for interatomic distances. In particular, imposing the constraint that the terminal atoms coincide in space, we get the mean-square distance $\langle r_R^2(k)\rangle_0$ between two atoms separated by k bonds in a ring chain. Compared with the corresponding mean-square distance $\langle r^2(k)\rangle_0$ for a linear chain, it reads

$$\langle r_R^2(k)\rangle_0 = k(1-k/N)l^2, \qquad \langle r^2(k)\rangle_0 = kl^2, \tag{2.1.10}$$

whereas the respective values of the radius of gyration are [from Eqn. (2.1.4)]

$$\langle S_R^2\rangle_0 = \tfrac{1}{12}Nl^2, \qquad \langle S^2\rangle_0 = \tfrac{1}{6}Nl^2. \tag{2.1.11}$$

From the preceding, by comparison with (2.1.6) and (2.1.7), we have

$$C_\infty = 1, \qquad l_{\text{pers}} = l. \tag{2.1.12}$$

Since $l^2(h) = l_x^2(h) + l_y^2(h) + l_z^2(h)$, where

$$l_x(h) = x(h) - x(h-1), \tag{2.1.13}$$

and so on, we have the result, generally valid within the Gaussian approximation,

$$\mathscr{W} = \mathscr{W}_x \cdot \mathscr{W}_y \cdot \mathscr{W}_z, \tag{2.1.14}$$

each component of $\mathscr{W}$ comprising the respective components of $\mathbf{l}(h)$. In view of the preceding, the intramolecular force acting on the hth bead has components

$$\begin{aligned} f_x(h) &= k_B T \frac{\partial \ln \mathscr{W}}{\partial x(h)} \\ &= \frac{3k_B T}{l^2}[x(h+1) - 2x(h) + x(h-1)]. \end{aligned} \tag{2.1.15}$$

It should be pointed out that these results also hold for the phantom freely jointed chain with fixed-length segments provided we neglect higher even-order terms in the exponent of (2.1.9) or, equivalently, higher odd-order terms in (2.1.15).

2.1.2. *The Phantom Chain with Nearest-Neighbor Interactions*

This is the model frequently adopted to describe the configurational statistics of polymer chains in the Θ state [2]. Intramolecular interactions are usually taken into account until the number of bonds separating the two interacting atoms is no larger than 5–10, which is the range responsible for the strongest effects. As a consequence, the statistical probability of a given rotational state around any chain bond is influenced by the rotations around one or two, rarely three, adjacent bonds on either side. Statistical correlation matrices are a convenient representation whether the correlation is considered between discrete sets of rotations around neighboring bonds (rotational isomeric state, or ris, scheme [2, 31]) or between the Fourier coefficients of the statistical weight in the continuum representation of skeletal rotations (all skeletal rotations, or asr, scheme [32–37]). The average scalar product between kth neighboring bonds turns out to be

$$\frac{\langle \mathbf{l}(h) \cdot \mathbf{l}(h+k) \rangle_0}{l^2} = \psi(k) = \sum_{n=1}^{v} \alpha_n \mu_n^k \tag{2.1.16}$$

where v is three times the order of the overall correlation matrix and the moduli of μ_n are all smaller than unity. If the largest of them is μ_1, say, we have

$$\psi(k) \propto |\mu_1|^k \cos(k\theta_1) \quad \text{as } k \to \infty, \qquad 0 \leqslant |\mu_1| < 1, \tag{2.1.16'}$$

where θ_1 is a suitable angle that reduces to zero if μ_1 is real (θ_1 is different from zero for polymers tending to have a helical conformation [22]). In view of the correlation between bonds, the chain statistical weight $\mathscr{W}\{\mathbf{l}(h)\}$ cannot be factorized into separate bond contributions, even in the Gaussian approximation. However, a Gaussian representation can still be achieved provided we consider suitable linear combinations of the bond vectors.

The Fourier Normal Modes. Let us suppose as an example that we select a set of vectors $\tilde{\mathbf{l}}_1, \tilde{\mathbf{l}}_2, \ldots, \tilde{\mathbf{l}}_N$ given by

$$\tilde{\mathbf{l}}_r = \sum_{h=1}^{N} \mathbf{l}(h) S_{hr}, \tag{2.1.17}$$

such that

$$\langle \tilde{\mathbf{l}}_r \cdot \tilde{\mathbf{l}}_s \rangle_0 = 0, \qquad r \neq s, \tag{2.1.18}$$

S_{hr} being a suitable set of constants with a broad distribution of their moduli. Then (i) in view of the central-limit theorem each of the $\tilde{\mathbf{l}}_r$ is Gaussian distributed and (ii) their joint distribution is simply the product of their Gaussian distributions with no cross factors if we neglect quartic and higher order interaction terms among the $\tilde{\mathbf{l}}_r$ [38]. Accordingly, the matrix $\mathbf{S}$ containing the S_{hr} elements diagonalizes the matrix $\mathbf{M}$ defined as follows:

$$M_{hj} = \langle \mathbf{l}(h) \cdot \mathbf{l}(j) \rangle_0. \tag{2.1.19}$$

To find the matrix $\mathbf{S}$, we must define suitable boundary conditions. It is expedient to consider a hypothetical chain with $2N$ bonds, instead of N, supposing it to be a portion of an infinite superchain repeating identically after each stretch of $2N$ bonds. Assuming $\langle \mathbf{l}(h) \cdot \mathbf{l}(h+N) \rangle_0 \ll l^2$, we shall focus our attention on two particular similarity operations by which the matrix $\mathbf{M}$ belonging to this hypothetical chain may be diagonalized. Letting

$$\mathbf{\Lambda}_1 = \mathbf{S}_1^{-1} \mathbf{M} \mathbf{S}_1, \qquad \mathbf{\Lambda}_2 = \mathbf{S}_2^{-1} \mathbf{M} \mathbf{S}_2, \tag{2.1.20}$$

where both $\mathbf{\Lambda}_1$ and $\mathbf{\Lambda}_2$ are diagonal, $\mathbf{S}_1$ and $\mathbf{S}_2$ will be ($\varphi = \pi/N$)

$$\mathbf{S}_1 = N^{-1/2} \times \begin{bmatrix} 1/\sqrt{2} & 1 & 1 & & 1 & +1 & 0 & 0 & \cdots & 0 \\ 1/\sqrt{2} & \cos\varphi & \cos 2\varphi & \cdots & \cos(N-1)\varphi & -1 & \sin\varphi & \sin 2\varphi & \cdots & \sin(N-1)\varphi \\ 1/\sqrt{2} & \cos 2\varphi & \cos 4\varphi & \cdots & \cos 2(N-1)\varphi & +1 & \sin 2\varphi & \sin 4\varphi & \cdots & \sin 2(N-1)\varphi \\ \vdots & \vdots & \vdots & & \ddots & \vdots & \vdots & \vdots & & \vdots \\ 1/\sqrt{2} & \cos\varphi & \cos 2\varphi & \cdots & & (-1)^{2N-1} & -\sin\varphi & -\sin 2\varphi & \cdots & \end{bmatrix}, \tag{2.1.21}$$

$$\mathbf{S}_2 = (2N)^{-1/2} \begin{bmatrix} 1 & 1 & 1 & \cdots & 1 \\ 1 & e^{i\varphi} & e^{i2\varphi} & \cdots & e^{i(2N-1)\varphi} \\ 1 & e^{i2\varphi} & e^{i4\varphi} & \cdots & e^{i2(2N-1)\varphi} \\ 1 & \ddots & & & \\ 1 & & & & \\ \vdots & & & & \end{bmatrix}. \tag{2.1.22}$$

Note that $\mathbf{S}_1$ and $\mathbf{S}_2$ are both unitary [i.e., $(\mathbf{S}_1^*)^T = \mathbf{S}_1^{-1}$]. In (2.1.20) the diagonal matrices $\mathbf{\Lambda}_1$ and $\mathbf{\Lambda}_2$ only differ in the sequential order of the eigenvalues. It should be pointed out that diagonalization of $\mathbf{M}$ by different transformations is made possible by the existence of pairwise degenerate eigenvalues, which is due to the *cyclic and symmetric structure* of $\mathbf{M}$. In turn, except for negligible end effects, $\mathbf{M}$ is indeed cyclic only in the assumption $\langle \mathbf{l}(h)\cdot\mathbf{l}(h+N)\rangle_0 \ll l^2$. The vectors $\tilde{\mathbf{l}}_r$ satisfying Eqn. (2.1.18) are contained in the array $\tilde{\mathbf{L}} = [\tilde{\mathbf{l}}_1, \tilde{\mathbf{l}}_2, \ldots, \tilde{\mathbf{l}}_{2N}]$ deriving from

$$\tilde{\mathbf{L}} = \mathbf{L}\cdot\mathbf{S}, \qquad \mathbf{L} = [\mathbf{l}(1), \mathbf{l}(2), \ldots, \mathbf{l}(2N)], \tag{2.1.23}$$

where $\mathbf{S}$ may be either $\mathbf{S}_1$ or $\mathbf{S}_2$.

We shall consider two special sets of arrays $\mathbf{L}$. The first set is antisymmetric and may be represented as

$$\begin{aligned}\mathbf{L}_1 = [&\mathbf{0}, \mathbf{l}(2), \mathbf{l}(3), \ldots, \mathbf{l}(N), \mathbf{0}, \\ &-\mathbf{l}(N), -\mathbf{l}(N-1), \ldots, -\mathbf{l}(3), -\mathbf{l}(2)].\end{aligned} \tag{2.1.24}$$

For reasons to be discussed later, it will be referred to as the *open-chain* set. The second set is characterized by a translational repetition between its two halves:

$$\mathbf{L}_2 = [\mathbf{l}(1), \mathbf{l}(2), \ldots, \mathbf{l}(N), \mathbf{l}(1), \mathbf{l}(2), \ldots, \mathbf{l}(N)], \tag{2.1.25}$$

and it will be defined as the *periodic-chain* set. We shall transform $\mathbf{L}_1$ and $\mathbf{L}_2$ through $\mathbf{S}_1$ and $\mathbf{S}_2$, respectively, obtaining

$$\tilde{\mathbf{L}}_1 = \mathbf{L}_1\cdot\mathbf{S}_1, \qquad \tilde{\mathbf{L}}_2 = \mathbf{L}_2\cdot\mathbf{S}_2, \tag{2.1.26}$$

the reason being that in doing so *both* $\tilde{\mathbf{L}}_1$ *and* $\tilde{\mathbf{L}}_2$ *only have N nonzero values.* Except for a constant factor, the elements of $\tilde{\mathbf{L}}_1$ and $\tilde{\mathbf{L}}_2$ will be denoted as $\tilde{\mathbf{l}}(q)$, where q is the variable conjugated to the index h of the general bond and the suffixes 1 and 2 will be dropped in $\tilde{\mathbf{l}}(q)$ for simplicity. In the following, the orthogonal transformations corresponding to Eqs. (2.1.26) will be given explicitly together with the similar ones involving the atomic position vectors $\mathbf{R}(h)$ of either set [connection between $\mathbf{R}(h)$ and $\mathbf{l}(h)$ is given by $\mathbf{R}(h)-\mathbf{R}(h-1)=\mathbf{l}(h)$]:

1. *The open-chain transform* (*i.e., set* 1):

$$\tilde{\mathbf{l}}(q)=\sqrt{2}\sum_{h=1}^{N}\mathbf{l}(h)\sin[q(h-1)], \qquad \mathbf{l}(h)=\frac{\sqrt{2}}{N}\sum_{\{q\}}\tilde{\mathbf{l}}(q)\sin[q(h-1)],$$
$$\tilde{\mathbf{R}}(q)=\sqrt{2}\sum_{h=1}^{N}\mathbf{R}(h)\cos[q(h-\tfrac{1}{2})], \qquad \tilde{\mathbf{R}}(0)=\sum_{h=1}^{N}\mathbf{R}(h),$$
$$\mathbf{R}(h)=\frac{\sqrt{2}}{N}\sum_{\{q\}}\tilde{\mathbf{R}}(q)\cos[q(h-\tfrac{1}{2})]+\frac{\tilde{\mathbf{R}}(0)}{N}, \tag{2.1.27}$$
$$\tilde{\mathbf{l}}(q)=-2\tilde{\mathbf{R}}(q)\sin(\tfrac{1}{2}q), \qquad \{q\}=\frac{\pi}{N},\frac{2\pi}{N},\frac{3\pi}{N},\ldots,\frac{(N-1)\pi}{N},$$
$$\Lambda(q)=N^{-1}\langle\tilde{\mathbf{l}}^2(q)\rangle_0=C(q)l^2.$$

2. *The periodic-chain transform* (*i.e., set* 2):

$$\tilde{\mathbf{l}}(q)=\sum_{h=1}^{N}\mathbf{l}(h)e^{iqh}, \qquad \mathbf{l}(h)=N^{-1}\sum_{\{q\}}\tilde{\mathbf{l}}(q)e^{-iqh},$$
$$\tilde{\mathbf{R}}(q)=\sum_{h=1}^{N}\mathbf{R}(h)e^{iqh}, \qquad \mathbf{R}(h)=N^{-1}\sum_{\{q\}}\tilde{\mathbf{R}}(q)e^{-iqh},$$
$$\tilde{\mathbf{l}}(q)=\tilde{\mathbf{R}}(q)(1-e^{iq}), \qquad \{q\}=0,\pm\frac{2\pi}{N},\pm\frac{4\pi}{N},\ldots, \tag{2.1.28}$$
$$\Lambda(q)=N^{-1}\langle\tilde{\mathbf{l}}(q)\cdot\tilde{\mathbf{l}}^*(q)\rangle_0=C(q)l^2.$$

The asterisk indicates the complex conjugate. The ring chain is the same as set 2 with the constraint $\sum_{k=1}^{N}\mathbf{l}(k)=\tilde{\mathbf{l}}(0)=0$. Note that formally for both sets we have

$$\langle|\tilde{\mathbf{R}}(q)|^2\rangle_0=\frac{\langle|\tilde{\mathbf{l}}(q)|^2\rangle_0}{4\sin^2(q/2)}=\frac{Nl^2}{\mu(q)}, \tag{2.1.29}$$

with

$$\mu(q)=\frac{4\sin^2(q/2)}{C(q)}. \tag{2.1.29'}$$

The Generalized Characteristic Ratio $C(q)$. For a very long chain ($N \to \infty$) the function $C(q)$ is the same for both transforms, and being a generalization of the classic characteristic ratio C_∞ in Fourier chain coordinate space, it is conveniently denoted as the generalized characteristic ratio [11, 21]. It is evaluated from

$$C(q) = 1 + 2\{\langle \mathbf{l}(h)\cdot\mathbf{l}(h+1)\rangle_0 \cos q + \langle \mathbf{l}(h)\cdot\mathbf{l}(h+2)\rangle_0 \cos 2q + \cdots\}/l^2, \quad [C(0) = C_\infty], \tag{2.1.30}$$

where all the configurational averages are also averaged over h. The general vector product average may be expressed as

$$\langle \mathbf{l}(h)\cdot\mathbf{l}(h+k)\rangle_0 = l^2[100](\mathbf{a}_R \otimes \mathbf{E}_3)\cdot\boldsymbol{\tau}^k\cdot(\mathbf{a}_c \otimes \mathbf{E}_3)\begin{bmatrix}1\\0\\0\end{bmatrix}, \tag{2.1.31}$$

where $\mathbf{E}_n$ is the unit matrix of order n and $\otimes$ is the symbol of the direct product, whereas

$$\boldsymbol{\tau} = (\mathbf{U} \otimes \mathbf{E}_3)\cdot\mathbf{T}_d/\lambda. \tag{2.1.31$'$}$$

In the ris scheme $\mathbf{U}$ is the matrix of the statistical weights for different pairs of neighboring rotational states [2, 39], $\mathbf{T}_d$ is the block diagonal matrix with the 3×3 rotation matrices, λ is the largest eigenvalue of $\mathbf{U}$ and $\mathbf{a}_R$, $\mathbf{a}_C$ are its row and column eigenvectors. In the asr scheme, $\mathbf{U}$ is built up with the Fourier coefficients of the Boltzmann statistical weight pertaining to a pair of neighboring rotations, $\mathbf{T}_d$ is modified accordingly, and $\mathbf{a}_R$, $\mathbf{a}_C$, and λ retain the same meaning as before [32, 33]. Replacing Eqn. (2.1.31) in (2.1.30) and performing the sum, we get ($N \to \infty$)

$$C(q) = [100](\mathbf{a}_R \otimes \mathbf{E}_3)\cdot(\mathbf{E} - \boldsymbol{\tau}^2)\cdot(\mathbf{E} + \boldsymbol{\tau}^2 - 2\boldsymbol{\tau}\cos q)^{-1}\cdot(\mathbf{a}_c \otimes \mathbf{E}_3)\begin{bmatrix}1\\0\\0\end{bmatrix}, \tag{2.1.32}$$

In the case of an $\left[\!\!-\mathrm{A}-\mathrm{B}-\!\!\right]_N$ polymer, the preceding formalism needs modification because odd and even bonds must be distinguished. Whereas the matrix $\mathbf{M}$ [see Eqn. (2.1.19)] is still periodic, its periodicity is now of order 2, that is, the kth row is obtained from the $(k-2)$th one after a cyclic shift. The eigenvalues of $\mathbf{M}$ belong to two sets, each containing N elements, where N is the number of monomeric units and $2N$ is the number of chain bonds:

$$\lambda(q) = \mathscr{A}(q) + |\mathscr{B}(q)|, \qquad \xi(q) = \mathscr{A}(q) - |\mathscr{B}(q)|,$$
$$\{q\} = 0, \pm\frac{2\pi}{N}, \pm\frac{4\pi}{N}, \ldots, \tag{2.1.33}$$

$$\mathscr{A}(q) = l^{-2} \sum_{k=1}^{N} \langle \mathbf{l}(1) \cdot \mathbf{l}(1+2k) \rangle_0 \exp(-iqk),$$

$$\mathscr{B}(q) = l^{-2} \sum_{k=1}^{N} \langle \mathbf{l}(2) \cdot \mathbf{l}(1+2k) \rangle_0 \exp(-iqk). \tag{2.1.34}$$

$\mathscr{A}(q)$ and $\mathscr{B}(q)$ may be also obtained in a closed form similar to (2.1.32) [14]. Given the sets of $\lambda(q)$ and $\xi(q)$, all the mean-square distances between chain atoms may be obtained as well as the corresponding intramolecular forces. Considering that k monomeric units amount to $2k$ bonds, $\lambda(q)$ takes the role of the function $C(q/2)$ in the long-distance limit or, in dynamics, in the long-time limit, corresponding to small values of q, whereas the contribution of $\xi(q)$ is minor and tends to vanish in these limits. It turns out that putting $\xi(q) \equiv 0$ amounts to considering an "average" chain where the configurational fluctuations due to the alternating A and B chain atoms are smoothed out. Accordingly, for an $\text{+A}-\text{B+}_N$ chain we shall put

$$C(q/2) \simeq \lambda(q), \tag{2.1.35}$$

and the mathematical formalism will be unchanged from that of an +A+_{2N} chain, at least if we consider distances or times of observation long enough that the internal structure of the monomeric unit may be ignored. Figure 2 shows the function $C(q)$ for three typical polymers.

A general remark is in order concerning the adoption of $C(q)$ in the framework of the present Fourier representation. The characteristic ratio is usually evaluated on a nonperiodic chain, in contrast with the Fourier representation. In the case of short chains this may entail minor errors deriving from end effects, which vanish if N is large enough that $\langle \mathbf{l}(1) \cdot \mathbf{l}(N) \rangle_0 \ll l^2$ [22].

The Mean-Square Distances. The mean-square distance between atoms h and j is given in a general form valid for any model by

$$\langle r^2(h,j) \rangle_0 = \frac{l^2}{N} \sum_{\{q\}} \mu(q)^{-1} |Q(q,h) - Q(q,j)|^2, \tag{2.1.36}$$

where the term with $q = 0$ is absent in the ring and in the open chain and must

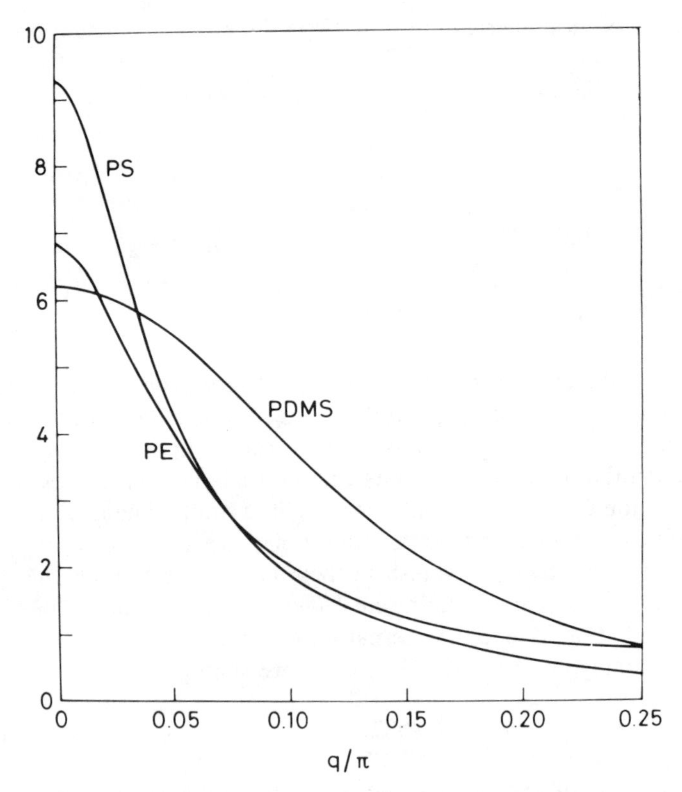

Figure 2. Generalized characteristic ratio $C(q)$ plotted vs. q/π for three typical polymers: polyethylene (PE) at 400 K [11, 21], atactic polystyrene (PS) at 300 K [14], and poly(dimethylsiloxane) (PDMS) at 350 K [15].

be included as the limit for $q \to 0$ for the periodic chain. In the preceding, $\mu(q)$ is defined in Eqn. (2.1.29′) and $Q(q, h)$ is the coefficient of the Fourier transform giving $\tilde{\mathbf{R}}(q)$, that is, either $\sqrt{2}\cos[q(h-\frac{1}{2})]$ for the open chain or e^{iqh} for the periodic and the ring chain.

The periodic chain offers the advantage that it is inherently uniform even in the perturbed state. As an example, the average distance between atoms separated by a fixed number k of skeletal bonds [$k = |j-h|$ in Eqn. (2.1.36)] is the same whether the two atoms are placed towards a chain end or in the middle:

$$\langle r^2(k)\rangle_0 = \frac{l^2}{N}\sum_{\{q\}} C(q)\frac{\sin^2(qk/2)}{\sin^2(q/2)}, \qquad q = \pm\frac{2\pi n_q}{N}. \tag{2.1.37}$$

In the case of the ring, it is easy to see that $\langle r^2(k)\rangle = \langle r^2(N-k)\rangle$ in view of the absence of the term with $q = 0$ and considering the permitted values of q; see Eqs. (2.1.28). Conversely, in the open chain the mean-square distance between atoms h and $j = h + k$ formally depends on both indices:

$$\langle r^2(h, h+k)\rangle_0 = \frac{2l^2}{N} \sum_{\{q\}} C(q) \frac{\sin^2(qk/2)}{\sin^2(q/2)} \sin^2\left[q\left(h + \frac{k-1}{2}\right)\right],$$

$$q = \frac{\pi n_q}{N}. \tag{2.1.37$'$}$$

The open-chain transform has the following advantages: (i) Unlike the periodic chain, as it will be shown, it may describe more accurately the perturbed (expanded or contracted) state because it does not *imply* the same average distance between all atom pairs separated by the same number of bonds and (ii) it is more suitable to describe the eigenfunctions of the dynamical modes of motion because the boundary constraint $\mathbf{l}(1) \equiv \mathbf{l}(N+1) \equiv 0$ is just the requirement needed if no intramolecular force is applied at, or transmitted by, the ends, as it happens in an actual open chain. However, when we consider "internal" modes, defined as $qN \gg 1$, or averages over all the modes, such as the dynamic structure factor (see in the following), the periodic transform may also be used, for the periodic constraint has a negligible effect on the modes consisting of many intrachain sinusoidal waves. Note that the mean-square distances between far-separated atoms [i.e., $k = O(N)$] depend heavily on the small-q modes. In the large separation limit we recover the simplicity of the bead-and-spring model, because then $C(q) \simeq C(0)$, and we get for both the open and the periodic chain

$$\langle r^2(h, h+k)\rangle_0 \sim \langle r^2(k)\rangle_0 \sim kC(O)l^2 \quad \text{as } k \to \infty. \tag{2.1.38}$$

Compare Eqn. (2.1.6); $C_\infty = C(0)$.

The mean-square radius of gyration may also be obtained from definitions (2.1.4) and (2.1.5) and transforms (2.1.27) and (2.1.28). We get for the two transforms

$$\langle S^2\rangle_0 = \begin{cases} \dfrac{l^2}{4N} \displaystyle\sum_{\{q\neq 0\}} \dfrac{C(q)}{\sin^2(q/2)} = \dfrac{l^2}{N} \displaystyle\sum_{\{q\neq 0\}} \mu(q)^{-1}, & q = \dfrac{\pi n_q}{N} \quad \text{(open chain)}, \\ \dfrac{l^2}{N} \displaystyle\sum_{\{q\neq 0\}} \mu(q)^{-1} + \tfrac{1}{12} N C(0) l^2, & q = \dfrac{2\pi n_q}{N} \quad \text{(periodic chain)}, \end{cases} \tag{2.1.39}$$

where the term with $q = 0$ is given separately in the second expression. As usual, the result for the ring is the same as for the periodic chain except that the zero-q term is missing. In the large-k and large-N limit, from the preceding we obtain the results (2.1.10) and (2.1.11) given for the bead-and-spring chain, except for an extra factor $C(0)$. The mean-square distance $\langle r_{Nh}^2 \rangle_0$ of the general hth atom from the chain center of mass may also be obtained. While it is given by $\langle S_R^2 \rangle_0$ for all the atoms in the ring, in the noncyclic chain we get (periodic chain, from Eqn. 16 of ref. 21)

$$\langle r_{Nh}^2 \rangle_0 = \frac{l^2}{N(N+1)^2} \sum_{\{q \neq 0\}} \mu(q)^{-1} [(N+1)^2 + 1 - 2(N+1)\cos q(h-1)] + \frac{l^2}{N} C(0) \frac{(2h-N-2)^2}{4}, \tag{2.1.40}$$

which in the large-N limit reduces to

$$\langle r_{Nh}^2 \rangle_0 \sim \tfrac{1}{3} N C_\infty l^2 \left[1 - 3\frac{h}{N}\left(1 - \frac{h}{N}\right)\right], \tag{2.1.41}$$

a result already reported [40].

The open-chain transform should be used when dealing with the overall size of the chain and, in dynamics, when separate consideration of the first, most collective modes of motion is required. In the following, the more expedient periodic-chain transform will generally be adopted unless specified otherwise.

2.1.3. *The Phantom Chain Configurational Potential*

Having found the independent linear combinations of bond vectors [see Eqs. (2.1.17), (2.1.27), and (2.1.28)], through the central-limit theorem it is easy to construct the Gaussian joint distribution and the associated quadratic potential. Adopting for simplicity the periodic-chain transform, we have

$$W\{\tilde{\mathbf{l}}(q)\} = \prod_{\{q\}} \left[\frac{3}{2\pi N C(q) l^2} \right]^{3/2} \exp\left(-\sum_{\{q\}} \frac{3\tilde{\mathbf{l}}(q)\cdot\tilde{\mathbf{l}}(-q)}{2NC(q)l^2} \right), \tag{2.1.42}$$

$$E\{\tilde{\mathbf{l}}(q)\} = -k_B T \ln W = \frac{3k_B T}{2Nl^2} \sum_{\{q\}} \frac{\tilde{\mathbf{l}}(q)\cdot\tilde{\mathbf{l}}(-q)}{C(q)} + \text{const.} \tag{2.1.43}$$

The intramolecular force component acting on the general chain atom with index h is

$$f_x(h) = \frac{-\partial E}{\partial x(h)} = -\frac{3k_B T}{Nl^2} \sum_{\{q\}} \mu(q) \tilde{x}(q) e^{-iqh}, \tag{2.1.44}$$

where $\tilde{x}(q)$ is the x projection of $\tilde{\mathbf{R}}(q)$, $(q \neq 0)$,

$$\tilde{x}(q) = \sum_{h=1}^{N} x(h)e^{iqh} = \frac{\tilde{l}_x(q)}{1 - e^{iq}}, \tag{2.1.45}$$

and $x(h)$ is the x component of $\mathbf{R}(h)$. Substitution of $\tilde{x}(q)$ in (2.1.44) with its Fourier transform containing the $x(k)$ shows that each atom is subject to elastic forces deriving from all its neighbors and that the force constants decrease with topological separation [21]. Only in the bead-and-spring limit we have $C(q) \equiv \text{const}\ (= 1)$, and the force constants are all zero except for first-neighboring atoms. The preceding potential may also be expressed as a sum over the chain bonds. Defining the following vector, associated with the general bond,

$$\mathbf{c}(h) = (Nl)^{-1} \sum_{\{q\}} \frac{\tilde{\mathbf{l}}(q)e^{-iqh}}{C(q)^{1/2}}, \tag{2.1.46}$$

we see that, apart from an additive constant, the potential E may be written as [41]

$$E = \tfrac{3}{2}k_B T \sum_{h=1}^{N} \mathbf{c}(h)\cdot\mathbf{c}^*(h) \simeq \tfrac{3}{2}k_B T \int_0^N \mathbf{c}(h)\cdot\mathbf{c}^*(h)\, dh, \tag{2.1.47}$$

the last expression being in the form of a Wiener integral [42]. The analogy with current formulations becomes more apparent after the following considerations. The generalized characteristic ratio $C(q)$ may be reasonably well approximated as

$$C(q) = \frac{a}{1 - b\cos q}, \qquad a, b > 0, \tag{2.1.48}$$

where $(1 - b) \ll 1$. Indicating with $q_{1/2}$ the Fourier coordinate which makes $C(q_{1/2}) = \frac{1}{2}C(0)$, we may write approximately

$$C(q) \simeq \begin{cases} \dfrac{a}{1-b} & \text{for } 0 \leqslant |q| \leqslant q_{1/2} = \left(\dfrac{2}{b} - 2\right)^{1/2}, \\[2ex] \dfrac{2a}{bq^2} & \text{for } q_{1/2} \leqslant |q| \leqslant \pi. \end{cases} \tag{2.1.49}$$

In an actual investigation one of the two q ranges is usually more important than the other depending on the distance of resolution D required for the problem at hand. Given D, we may define a corresponding Fourier coordinate q_D as follows: D^2 is the mean-square distance between two atoms separated by just one Fourier wave with a coordinate q_D, that is, by $2\pi/q_D$ chain bonds. Putting $k = 2\pi/q_D$ in Eqn. (2.1.37), we have the connection between D and q_D:

$$D^2 = \frac{l^2}{N}\sum_{\{q\}} C(q)\frac{\sin^2(\pi q/q_D)}{\sin^2(q/2)} \simeq \frac{l^2}{N}\sum_{|q|<q_D} C(q)\frac{\sin^2(\pi q/q_D)}{\sin^2(q/2)}, \qquad (2.1.50)$$

as all the Fourier components beyond q_D may be approximately neglected. If $D_{1/2}$ is the distance corresponding to $q_D = q_{1/2}$, from Eqn. (2.1.49) we may adopt either the upper or the lower expression for $C(q)$ depending on whether D is $\gg D_{1/2}$ or $\ll D_{1/2}$, respectively ($D_{1/2}$ is about 20–30 Å for ordinary hydrocarbon polymers [41]). After truncating at q_D the sum (2.1.46), from (2.1.47) and (2.1.49) we have, in either case,

$$E_D \simeq \begin{cases} \frac{3}{2}k_BT\dfrac{1-b}{al}\displaystyle\int_0^{Nl}\left(\frac{\partial \mathbf{R}}{\partial s}\right)^2 ds & (D \gg D_{1/2}), \\ \frac{3}{2}k_BT\dfrac{bl}{2a}\displaystyle\int_0^{Nl}\left(\frac{\partial^2 \mathbf{R}}{\partial s^2}\right)^2 ds & (D \ll D_{1/2}), \end{cases} \qquad (2.1.51)$$

where s is the curvilinear coordinate along the chain, and the equalities $s(h) = hl$, $\partial\mathbf{R}/\partial s = \mathbf{l}(h)/l = (Nl)^{-1}\sum_{\{q\}}\tilde{\mathbf{l}}(q)e^{-iqh}$ are used. The preceding equations are the Wiener integral representations of an elastic, ideally flexible chain and of a stiff chain, respectively. The latter model is analogous to the Porod–Kratky wormlike chain [43], which in turn is exactly reproduced at all distances of observation by Eqn. (2.1.48), assuming

$$l \to 0, \qquad 2a = 2\sqrt{2(1-b)} = \tau^2 \to 0, \qquad \tau^2/l = 4\lambda \quad \text{(const)};$$

$$C(q) = \frac{\tau^2}{(q^2+\tau^4/4)} = \frac{2k_p}{1+(q/k_p)^2}$$

$$\langle \mathbf{l}(h)\cdot\mathbf{l}(h+k)\rangle_0 = \frac{l^2}{2\pi}\int_{-\pi}^{\pi} C(q)\cos(qk)\,dq$$

$$= l^2 e^{-k\tau^2/2} = l^2 e^{-k/k_p}. \qquad (2.1.52)$$

In this case the crossover distance $D_{1/2}$ is $1.6/\lambda$, whereas $1/(2\lambda)$ is the persistence length and $k_p = 2/\tau^2$ is the number of bonds within one persistence length. We see that even the wormlike chain becomes ideally flexible if D is much larger than the persistence length. Note that the bond correlation behavior is qualitatively similar to the phantom chain's [see Eqn. (2.1.16′); $\theta_1 = 0$]

An important special case for which $C(q) = a/(1 - b\cos q)$ is exact is represented by the freely rotating chain with a fixed bond angle θ. Here $a = \sin^2\theta/(1 + \cos^2\theta)$ and $b = -2\cos\theta/(1 + \cos^2\theta)$ [21].

2.1.4. *The "Tube" as a Fuzzy Image of the Chain*

If the Fourier sum over q that gives $\mathbf{l}(h)$ [see Eqn. (2.1.28)] is truncated at $\pm q_D$, we obtain an unfocused, or fuzzy, chain image. Denoting with $\bar{\mathbf{l}}(h)$ the truncated bond vector,

$$\bar{\mathbf{l}}(h) = N^{-1} \sum_{|q| < q_D} \tilde{\mathbf{l}}(q) e^{-iqh}, \tag{2.1.53}$$

the line defined by the following vector $\bar{\mathbf{R}}(k)$ represents the average axis of the chain image as a function of $k (1 \leqslant k \leqslant N)$:

$$\bar{\mathbf{R}}(k) = \sum_{h=1}^{k} \bar{\mathbf{l}}(h). \tag{2.1.54}$$

The fuzzy image may be a simple representation of the tube concept that is useful in melt dynamics [44], the line $\bar{\mathbf{R}}(k)$ being reminiscent of the primitive tube path introduced by Doi and Edwards on intuitive grounds [45]. The smaller is $|q_D|$, the larger is the degree of fuzziness, whereas the tube becomes thicker and shorter; this situation corresponds to a decreasing concentration of entanglements with surrounding chains, which constrain the chain to "reptate" within the tube [46]. The average tube length L may be obtained after projecting the general bond vector $\mathbf{l}(h)$ along $\bar{\mathbf{l}}(h)$ and summing over all the bonds. Defining $\boldsymbol{\gamma}(h)$ as the unit vector parallel to $\bar{\mathbf{l}}(h)$, we have ($q_D \ll q_{1/2}$)

$$\boldsymbol{\gamma}(h) = (Nl)^{-1} \left(\frac{\pi}{q_D C(0)} \right)^{1/2} \sum_{|q| < q_D} \tilde{\mathbf{l}}(q) e^{-iqh}, \tag{2.1.55}$$

and L is given by

$$L = \sum_{h=1}^{N} \langle \mathbf{l}(h) \cdot \boldsymbol{\gamma}(h) \rangle_0 = Nl \left(\frac{q_D C(0)}{\pi} \right)^{1/2}. \tag{2.1.56}$$

We may also obtain the mean-square tube radius σ^2 by requiring that the mean-square end-to-end distance of the chain is given by a sum of independent contributions along the primitive path and between the chain ends and the terminal points of the tube. The result is

$$\sigma^2 = \frac{2}{3}\frac{C(0)l^2}{\pi q_D} \tag{2.1.57}$$

and Eqs. (2.1.56) and (2.1.57) provide two significant parameters of the tube related to a single quantity q_D.

2.1.5. *The Unperturbed Real Chain: The Self-Consistent Approach*

So far we assumed [see Eqs. (2.1.1) and (2.1.2)] that the overall binary potential may be expressed as a δ function, which reduces to the nul function in the unperturbed state (phantom chain). Remembering that in our picture the effective interactions are always referred to the backbone atoms for convenience, it is clear that in actual polymers the chain thickness [roughly equal to the distance $\bar{r}$ giving the minimum value of $e(\mathbf{r})$, see Fig. 1] may be nonnegligible in comparison with the average interatomic distances of interest. Under these circumstances, although at $T = \Theta$ the effective binary potential β_{eff} vanishes, a repulsive shorter range potential still exists between any two atoms [27]. In fact, let $W_k(\mathbf{r})$ be the Gaussian distance distribution between two atoms separated by k bonds and let us assume for sake of simplicity that the potential contributing to σ_3 be still represented by a δ function, meaning that any three-body repulsion is significantly large only if three atoms are exactly overlapping in space. Then we may write $\varepsilon(\mathbf{r}) = e(\mathbf{r}) + \sigma_3\delta(\mathbf{r})$ for the effective interatomic potential. The average interaction free energy is, in k_BT units,

$$\begin{aligned} a_2(k) &= \int \varepsilon(\mathbf{r})\, W_k(\mathbf{r})\, d^3\mathbf{r} \\ &= \left[\frac{3}{2\pi\langle r^2(k)\rangle}\right]^{3/2} \int \varepsilon(\mathbf{r}) \exp\left[-\frac{3r^2}{2\langle r^2(k)\rangle}\right] d^3\mathbf{r}. \end{aligned} \tag{2.1.58}$$

The root-mean-square distance $\langle r^2(k)\rangle^{1/2}$ will be assumed as much larger than the distances producing nonnegligible values of $\varepsilon(\mathbf{r})$, so that the exponential of $W_k(\mathbf{r})$ may be expanded to the r^2 term. The function $\varepsilon(\mathbf{r})$ is positive (i.e., repulsive) and negative (attractive) at smaller and larger distances, respectively, and we shall separately consider the integrals $\int \varepsilon(\mathbf{r})\, d^3\mathbf{r}$ and $\int r^2 \varepsilon(\mathbf{r})\, d^3\mathbf{r}$ over these two distance ranges. We get

$$\int \varepsilon(\mathbf{r})\, d^3\mathbf{r} = \int e(\mathbf{r})\, d^3\mathbf{r} + \sigma_3 = \beta + \sigma_3 = \beta_{\text{eff}} = v_c - v_a, \tag{2.1.59}$$

$$\int r^2 \varepsilon(\mathbf{r})\, d^3\mathbf{r} = \int r^2 e(\mathbf{r})\, d^3\mathbf{r} = v_c\langle r_c^2\rangle - v_a\langle r_a^2\rangle, \tag{2.1.60}$$

where v_c and $-v_a (v_c, v_a > 0)$ are the contributions to the integral of $\varepsilon(\mathbf{r})$ due to the repulsive and to the attractive range, respectively, whereas $\langle r_c^2 \rangle$ and $\langle r_a^2 \rangle$ are their second-moments ($\langle r_a^2 \rangle > \langle r_c^2 \rangle$). From the preceding, Eqn. (2.1.58) gives

$$a_2(k) = \left[\frac{3}{2\pi\langle r^2(k)\rangle}\right]^{3/2} \left\{\beta_{\text{eff}} + \frac{3}{2\langle r^2(k)\rangle} v_c[\langle r_a^2\rangle - \langle r_c^2\rangle] - \frac{3\beta_{\text{eff}}\langle r_a^2\rangle}{2\langle r^2(k)\rangle}\right\}$$

$$\simeq \left(\frac{3}{2\pi}\right)^{3/2} \frac{\beta_{\text{eff}}}{\langle r^2(k)\rangle^{3/2}} + \frac{3}{2}\left(\frac{3}{2\pi}\right)^{3/2} v_c \frac{\langle \Delta^2 r\rangle}{\langle r^2(k)\rangle^{5/2}}, \tag{2.1.61}$$

$$\langle \Delta^2 r\rangle = \langle r_a^2\rangle - \langle r_c^2\rangle, \tag{2.1.61'}$$

the approximate equality arising from $\langle r_a^2 \rangle \ll \langle r^2(k) \rangle$. The quantities v_c and $\langle \Delta^2 r \rangle$ may be respectively interpreted as the covolume per chain atom and the mean-square chain thickness. The preceding result shows that, even at the Θ temperature where $\beta_{\text{eff}} = 0$, the repulsion due to the chain thickness does not disappear. The residual potential is due to the mismatch between the repulsive domain and the surrounding, screening attraction domain, which justifies the term *screened interaction* [27]. It is worth pointing out that the dependence on $\langle r^2(k) \rangle^{-5/2}$ makes it a medium-range potential in that it does not alter the linear relationship between the mean-square radius of gyration $\langle S^2 \rangle_0$ and N, unlike the term depending on $\beta_{\text{eff}} \langle r^2(k) \rangle^{-3/2}$, which is long ranged. It might be shown that the limiting exponent of $\langle r^2(k) \rangle$ below which the medium-range character is maintained is -2.

A remark on the Gaussian approximation is in order at this point. It is well known that for topologically adjacent chain atoms of real chains the distribution of the interatomic distances deviates from the Gaussian distribution; in particular, their contact probability is much smaller than predicted for the Gaussian distribution yielding the correct mean-square value. This was shown by Flory and co-workers for the cases of polyethylene (PE) and poly(dimethylsiloxane) (PDMS) [47, 48]; as an example, for PE the contact probability at ordinary temperatures for $k = 10$ is virtually nil, and it is still below 50% of the Gaussian value for $k = 20$. We may account approximately for the small contact probability at short topological separation by retaining the Gaussian distribution for all the atom pairs, but introducing a cutoff $\bar{k}$ such that two atoms separated by $k < \bar{k}$ bonds are assumed to be incapable of contacting one another. The actual value of $\bar{k}$ is taken as equal to 40 for PE, as an average derived from the Yoon–Flory analysis, and may be adjusted for other polymers roughly in proportion to their characteristic ratio C_∞. It should be stressed that the preceding interactions do not encompass the short-range conformational energy contributions between neighboring units ($\lesssim$ 5–10 interatomic bonds), which are embodied in the function $C(q)$ characterizing the phantom chain; consequently, $\bar{k}$ is larger than the usual statistical segment length.

In order to derive the average chain dimensions, we need a suitable expression of the chain free energy wherefrom the relevant quantities will be obtained through a self-consistent procedure [10, 13, 23]. Since it can be extended to the perturbed case, we shall drop the zero suffix in the remainder of the section. The reduced excess free energy turns out to be

$$\mathscr{A} = (A - A_{\text{ph}})/k_B T = \mathscr{A}_{\text{el}} + \mathscr{A}_{\text{intra}}, \tag{2.1.62}$$

where A and A_{ph} are the actual and the phantom chain free energy and $\mathscr{A}_{\text{el}}$ and $\mathscr{A}_{\text{intra}}$ are the elastic and the intramolecular interaction components, respectively. The general expression of $\mathscr{A}_{\text{el}}$ is [10]

$$\mathscr{A}_{\text{el}} = \frac{3}{2}\sum_{\{q\}}[\tilde{\alpha}^2(q) - 1 - \ln \tilde{\alpha}^2(q)], \tag{2.1.63}$$

where $\tilde{\alpha}^2(q)$ is the strain ratio of the mean-square Fourier mode:

$$\tilde{\alpha}^2(q) = \frac{\langle|\tilde{\mathbf{l}}(q)|^2\rangle}{\langle|\tilde{\mathbf{l}}(q)|^2\rangle_{\text{ph}}} = \frac{\langle|\tilde{\mathbf{l}}(q)|^2\rangle}{NC(q)l^2}, \tag{2.1.64}$$

the subscript ph referring again to the phantom chain. The interaction free energy $\mathscr{A}_{\text{intra}}$ is, in the present case,

$$\mathscr{A}_{\text{intra}} = \sum_{k=\bar{k}}^{N}(N-k)a_2(k), \tag{2.1.65}$$

$a_2(k)$ being defined in (2.1.61), whereas the mean-square distance $\langle r^2(k)\rangle$ is still given by (2.1.37) except for replacing $C(q)$ with the effective value $\tilde{\alpha}^2(q)C(q)$:

$$\langle r^2(k)\rangle = \frac{l^2}{N}\sum_{\{q\}}\tilde{\alpha}^2(q)C(q)\frac{\sin^2(qk/2)}{\sin^2(q/2)}. \tag{2.1.66}$$

For a given chain, the strain ratios $\tilde{\alpha}^2(q)$ fully determine the mean-square dimensions of the chain. In the following, they will be derived from free-energy self-consistent optimization.

2.1.6. *Screened Interactions and Chain Expansion*

Let us put now $\beta_{\text{eff}} = 0$ in Eqn. (2.1.61) and therefore reinstate the zero subscript for the unperturbed state. After minimizing Eqn. (2.1.62), the general resulting equation is, taking $C(q) \simeq C(0)$ for simplicity [27],

$$\tilde{\alpha}_0^2(q) = \left\{1 - \tfrac{5}{3}\bar{K}[C(0)l^2]^{7/2}\left(\sin\frac{q}{2}\right)^{-2}\int_{\bar{k}}^{N} dk\left(1-\frac{k}{N}\right) \times \sin^2\left(\frac{qk}{2}\right)\langle r^2(k)\rangle^{-7/2}\right\}^{-1},$$

$$\bar{K} = \pi\left(\frac{3}{2\pi C(0)l^2}\right)^{5/2} v_c\langle\Delta^2 r\rangle, \tag{2.1.67}$$

whence we obtain, to within terms of the first order in $\bar{K}$ and for $k \gg 1$,

$$\tilde{\alpha}_0^2(q) = \tilde{\alpha}_0^2(0) - \frac{10}{3}\bar{K}Gq^{1/2};$$

$$\alpha_0^2(k) = \langle r^2(k)\rangle_0/\langle r^2(k)\rangle_{\text{ph}} = \tilde{\alpha}_0^2(0) - \frac{40}{3\sqrt{\pi}}\bar{K}Gk^{-1/2},$$

$$\tilde{\alpha}_0^2(0) = 1 + \frac{10}{3}\frac{\bar{K}}{(\bar{k})^{1/2}}, \tag{2.1.68}$$

$$G = 2^{-3/2}\int_0^\infty \frac{w^2 - \sin^2 w}{w^{7/2}}\,dw.$$

Equations (2.1.68) show that the long-range expansion limit $\alpha_0^2(k \to \infty)$ is finite so that the mean-square size of a large chain is still proportional to N. This confirms that we are indeed in the Θ state, although the chain is somewhat expanded with respect to the phantom state. Considering that the characteristic ratio C_∞ is given by $\langle r^2(N)\rangle_0/Nl^2$ [see (2.1.6)] and that $\mathbf{r}(N) = \tilde{\mathbf{l}}(q=0)$ [see (2.1.28)], we have

$$\alpha_0^2(k \to \infty) = \tilde{\alpha}_0^2(q=0) = \tilde{\alpha}_0^2(0),$$

$$C_\infty = \tilde{\alpha}_0^2(0)C(0). \tag{2.1.69}$$

According to Eqs. (2.1.68), the function $\tilde{\alpha}_0^2(q)$ displays a small, sharp peak at $q = 0$, otherwise being nearly stationary. Calculations performed on isotactic polypropylene, atactic polystyrene, and polyethylene with realistic parameters [27] lead to $\tilde{\alpha}_0^2(0)$ of 1.30, 1.20, and 1.05, respectively. This square expansion ratio increases with the thickness, or degree of bulkiness, of the chain and decreases with the length of the statistical segment if the two quantities are taken as proportional to $v_c\langle\Delta^2 r\rangle$ and to $C(0)l$, respectively.

In this connection, we mention Monte Carlo calculations performed by Mazur and McCrackin on self-avoiding walks on two different lattices [49]. Using an attractive potential to balance the self-avoiding repulsions, these authors were able to study the Θ state, among other things, finding that $C_\infty = \langle r^2(N)\rangle_0/Nl^2$ is larger than expected for a random walk with no step reversal ($\sim$ 5–10%). In the latter case the value of $\langle r^2(N)\rangle/Nl^2$ simply depends on the number of equiprobable choices available at each step, and therefore may be exactly evaluated through a simple formula. We believe this discrepancy is due to the effect of the screened interactions.

2.1.7. *Bond Correlation*

In analogy with Eqn. (2.1.3), in the periodic approximation the mean-square distance $\langle r^2(k)\rangle$ may be written, for large k and for any h, as

$$\begin{aligned}\langle r^2(k)\rangle &= kl^2 + 2\sum_{j=1}^{k}(k-j)\langle \mathbf{l}(h)\cdot\mathbf{l}(h+j)\rangle \\ &\simeq kl^2 + 2\int_0^k (k-j)\langle \mathbf{l}(h)\cdot\mathbf{l}(h+j)\rangle\, dj. \end{aligned} \qquad (2.1.70)$$

Differentiating twice with respect to k, we have

$$\langle \mathbf{l}(h)\cdot\mathbf{l}(h+k)\rangle = \frac{1}{2}\frac{d^2\langle r^2(k)\rangle}{dk^2}, \qquad (2.1.71)$$

a relationship that will be useful in the following. Let us consider Eqn. (2.1.66); although it applies to the periodic-chain transform, it reduces to the result for the ring chain once the term with $q = 0$ is subtracted. Then from Eqn. (2.1.71), remembering (2.1.69), we get the following result, valid for all the unperturbed models;

$$\langle \mathbf{l}(h)\cdot\mathbf{l}(h+k)\rangle_{0,\,\mathrm{ring}} = \langle \mathbf{l}(h)\cdot\mathbf{l}(h+k)\rangle_{0,\,\mathrm{periodic}} - C_\infty l^2/N. \qquad (2.1.72)$$

In turn, except for k very close to N, $\langle \mathbf{l}(h)\cdot\mathbf{l}(h+k)\rangle_{0,\,\mathrm{periodic}}$ is virtually identical to the result of the open chain, so that we drop the suffix henceforth. We have the following for the different unperturbed, nonring models ($k \gg 1$):

(a) bead-and-spring:

$$\langle \mathbf{l}(h)\cdot\mathbf{l}(h+k)\rangle_0 = 0 \quad \text{[see Eqn. (2.1.8)]};$$

(b) phantom (flexible or wormlike):

$$\langle \mathbf{l}(h)\cdot\mathbf{l}(h+k)\rangle_0 \propto \mu_1^k = e^{-k/k_p},$$

$$k_p = \frac{1}{\ln(1/\mu_1)}, \qquad 0 < \mu_1 < 1 \quad \text{[see Eqs. (2.1.16) and (2.1.52)];}$$

$$\tag{2.1.73}$$

(c) real:

$$\langle \mathbf{l}(h)\cdot\mathbf{l}(h+k)\rangle_0 \propto k^{-3/2} \quad \text{[see Eqs. (2.1.71) and (2.1.68)].}$$

As expected, in all cases the bond correlation decreases with k more quickly than k^{-1}, a result that ensures $\langle r^2(N)\rangle_0 \propto \langle S^2\rangle_0 \propto N$ [see Eqs. (2.1.6) and (2.1.7)].

2.2. The Perturbed Chain: Collapse and Expansion

At $T \neq \Theta$, β_{eff} differs from zero; see Eqn. (2.1.2). In this case we see from (2.1.61) that the two-body potential $a_2(k)$ contains a nonzero term proportional to $\beta_{\text{eff}}\langle r^2(k)\rangle^{-3/2}$, in addition to the one proportional to $\langle r^2(k)\rangle^{-5/2}$ produced by the screened interactions. As a result we have a long-range chain expansion at $T > \Theta$ (i.e., $\beta_{\text{eff}} > 0$) or contraction at $T < \Theta$ (i.e., $\beta_{\text{eff}} < 0$) having the character of a true transition in the thermodynamic limit [50]. Especially in the contraction case, three-body repulsive terms must be explicitly considered; the corresponding potential will be denoted as $a_3(k, k_1)$, where k and k_1 (both positive) are the bond separations between atom pairs 1–2 and 2–3, respectively, and the 1–3 separation is $k + k_1$. Since the physical origin of the three-body repulsion is not always agreed upon, it should be made clear that in this context it represents the extra contribution of two two-body interactions when involving a common atom. We shall then evaluate the intramolecular contact free energy $\mathscr{A}_{\text{intra}}$ as the sum of the two- and three-body terms. At any temperature we shall assume as an approximation the Fourier modes to be still orthogonal, which is strictly true of the unperturbed state only; therefore, the whole formalism developed to this point retains its validity. The two-body sum [see Eqs. (2.1.61) and (2.1.65)] will be performed with the coefficient β instead of β_{eff} because at this stage the σ_3 contribution [see Eqn. (2.1.2)] will remain incorporated into the overall three-body sum. We get

$$\begin{aligned}\mathscr{A}_{\text{intra}} = {} & \left(\frac{3}{2\pi}\right)^{3/2} \beta \sum_{k=\bar{k}}^{N} (N-k)\langle r^2(k)\rangle^{-3/2} \\ & + \frac{3}{2}\left(\frac{3}{2\pi}\right)^{3/2} v_c \langle \Delta^2 r\rangle \sum_{k=\bar{k}}^{N} (N-k)\langle r^2(k)\rangle^{-5/2} \\ & + \sum_{k=\bar{k}}^{N-\bar{k}} \sum_{k_1=\bar{k}}^{N-k} (N-k-k_1)\, a_3(k, k_1),\end{aligned} \tag{2.2.1}$$

where

$$a_3(k, k_1) = \bar{y}\left(\frac{3}{2\pi}\right)^3 [f(\langle r^2(k)\rangle, \langle r^2(k_1)\rangle, \langle r^2(k+k_1)\rangle)]^{-3/2}, \quad (2.2.2)$$

$$f(x, y, z) = \tfrac{1}{2}(xy + xz + yz) - \tfrac{1}{4}(x^2 + y^2 + z^2), \quad (2.2.3)$$

and $\bar{y}$ is a suitable positive coefficient proportional to v_c^2. The mean-square distances $\langle r^2(k)\rangle$ depend on the strain ratios $\tilde{\alpha}^2(q)$ as shown in Eqn. (2.1.66); $a_3(k, k_1)$ is proportional to the contact probability among three atoms separated by k, k_1, and $k+k_1$ bonds, as it may be checked by applying the Wang–Uhlenbeck method to the probability of three-body contact in our multivariate Gaussian distribution [51]. Optimization of the free energy $\mathscr{A} = \mathscr{A}_{\text{intra}} + \mathscr{A}_{\text{el}}$ [see Eqs. (2.2.1) and (2.1.63)] with respect to $\tilde{\alpha}^2(q)$ may now be carried out. The resulting equations show that the chain experiences a decreasing deformation with decreasing topological separation, so that for small values of $|T-\Theta|$ it is effectively unperturbed at the scale of a few interatomic bonds. We shall assume that significant expansion or contraction from the unperturbed state develops after an interatomic separation pN equal to $\sim$ 100 skeletal bonds at least, so that within this range the chain exactly behaves as in the unperturbed Θ state. It should be noted that for $N \to \infty$ this assumption is not in contrast with an arbitrarily large *overall* degree of expansion or contraction. From $\partial\mathscr{A}/\partial\tilde{\alpha}^2(q) = 0$, we get the following result for collective Fourier modes with $q \ll 1$:

$$\alpha^2(q) = \frac{\tilde{\alpha}^2(q)}{\tilde{\alpha}_0^2(0)}$$
$$= \left\{1 + \frac{4}{q^2}\int_{\bar{k}}^{N} dk\left(1 - \frac{k}{N}\right)\sin^2\left(\frac{qk}{2}\right)\varphi(k, T)J(k)^{-5/2} - \frac{20K}{3q^2}\int_{pN}^{N} dk\left(1 - \frac{k}{N}\right)\sin^2\left(\frac{qk}{2}\right)J(k)^{-7/2}\right\}^{-1}, \quad (2.2.4)$$

where

$$J(k) = \frac{\langle r^2(k)\rangle}{C_\infty l^2} = N^{-1}\sum_{\{q\}}\alpha^2(q)\sin^2\left(\frac{qk}{2}\right)\bigg/\left(\frac{q}{2}\right)^2, \quad (2.2.5)$$

$$C_\infty = \tilde{\alpha}_0^2(0)C(0), \quad (2.2.6)$$

$$\varphi(k, T) = -\beta_{\text{eff}}\left(\frac{3}{2\pi C_\infty l^2}\right)^{3/2} - K_1\int_{pN}^{N-k} dk_1\left(1 - \frac{k_1}{N-k}\right)\frac{J(k)^{5/2}[J(k_1) + J(k+k_1) - J(k)]}{f[J(k), J(k_1), J(k+k_1)]^{5/2}}, \quad (2.2.7)$$

$$\beta_{\text{eff}} = \beta + \frac{4\bar{y}}{\sqrt{\bar{k}}}\left(\frac{3}{2\pi C_{\infty} l^2}\right)^{3/2} = \beta + \sigma_3, \tag{2.2.8}$$

$$K = \bar{K}/\tilde{\alpha}_0^5(0), \qquad K_1 = \bar{y}\left[\frac{3}{2\pi C_{\infty} l^2}\right]^3. \tag{2.2.9}$$

In the preceding, $\alpha^2(q)$ (without the tilde) is the mean-square strain ratio of the q mode with respect to the unperturbed real chain with screened interactions (i.e., at $T = \Theta$), not with respect to the phantom chain; accordingly, $\alpha_0^2(q) \equiv 1$. In agreement with our assumption of a small solvent strength, β_{eff} will be proportional to $T - \Theta$, and we shall write

$$\beta_{\text{eff}}\left(\frac{3}{2\pi C_{\infty} l^2}\right)^{3/2} = \tau B, \qquad \tau = \frac{T - \Theta}{T}(|\tau| \ll 1), \tag{2.2.10}$$

where B is a suitable constant. The explicit expression of σ_3 is given in Eqn. (2.2.8). Although formally independent of pN, σ_3 is approximately given by the sum of those three-body contributions for which two atoms are separated by no more than pN bonds and the third one is further separated by k bonds from the second atom. As a consequence, the k dependence of these terms is similar to that of the two-body interactions [10]. Analogously, a substantial share of the higher order interactions is effectively embodied in the two- and three-body interactions [26]. It should also be noted that the function of interatomic distance $J(k)$ is so defined that $J(k) \simeq k$ for all k at the Θ temperature and for $k < pN$ at any temperature ($k \gg \bar{k}$).

It should be commented that two distinct readjustments of the reference Θ state are implicit in Eqn. (2.2.4), both relating to the change of the lower limit of integration from $\bar{k}$ to pN. When applied to the integral of the screened interactions [last integral in Eqn. (2.2.4)], this change reflects the adoption of the unperturbed *real* chain instead of the phantom chain. When applied to the three-body integral [see Eqn. (2.2.7)], it implies a small shift of the Θ temperature from the phantom chain value

$$\Delta\Theta = \Theta - \Theta_{\text{phantom}} = -\Theta \cdot 4K_1/B\sqrt{\bar{k}} \tag{2.2.11}$$

to compensate the residual three-body repulsions involving at least two neighboring atoms.

2.2.1. *The Chain Collapse*

Together with Eqn. (2.2.5), (2.2.4) represents a system of equations for the allowed q values or, alternatively, a functional equation if q is regarded as a continuous variable. After choosing the dimensionless parameters K, K_1, B, and N, we obtained the numerical solution through an iterative procedure for different values of the relative undercooling $\tau = (T - \Theta)/T\,(\tau < 0)$. As it will be discussed further, upon decreasing τ there may be either a first-order or a second-order transition depending upon the values of the parameters. The results may be well approximated by the following equation in the whole temperature range between the unperturbed coil and the collapsed globule [10, 52, 53]:

$$\alpha^2(q) = \frac{(Jq)^2}{1+(Jq)^2}, \tag{2.2.12}$$

where $J < N$ under collapse, whereas $J \to \infty$ in the unperturbed state. J is roughly equal to the number of chain bonds below which the chain is unperturbed, and we shall put henceforth

$$J = pN. \tag{2.2.12$'$}$$

If $J > N$, we are in the precollapse region; here the effect of the screened interactions and the three-body repulsions vanishes. This may be understood if we remember that the integrals respectively multiplying K and K_1 in Eqs. (2.2.4) and (2.2.7) reduce to zero because their lower limit ($= J$) exceeds the overall chain length. In the strong-compression limit $J \ll N$, substitution of Eqn. (2.2.12) into (2.2.5) gives

$$\langle r^2(k)\rangle = JC_\infty l^2[1 - \exp(-k/J)]. \tag{2.2.13}$$

From Eqn. (2.1.71) we get, from the preceding,

$$\langle \mathbf{l}(h)\cdot\mathbf{l}(h+k)\rangle = -\frac{1}{2}\frac{C_\infty l^2}{J}\exp(-k/J). \tag{2.2.14}$$

We see that the correlation law is of the same mathematical form as for the phantom chain [see Eqs. (2.1.73)], although with the unusual feature of the negative sign. Accordingly, in spite of the long-range attractive and repulsive forces, long-range correlation may be considered as absent, which is consistent with our Gaussian approximation approach.

Equations (2.2.12)–(2.2.14) define the so-called *random Gaussian globule* resulting from coil collapse; in words, all the long-range mean-square distances with $k \gg J$ are equal, whereas those between relatively close atoms

($k \ll J$) do not differ from the unperturbed value $kC_\infty l^2$. In the same limit of high compression the mean-square radius of gyration is

$$\langle S^2 \rangle = \tfrac{1}{2} J C_\infty l^2, \tag{2.2.15}$$

$$\alpha_S^2 = \langle S^2 \rangle / \langle S^2 \rangle_0 = 3J/N, \tag{2.2.15'}$$

and according to the preceding, we may say that the average distance between two atoms is unperturbed if it is smaller than the radius of gyration; otherwise it tends to a uniform value. Moreover, in this limit we have $\alpha_S^2 \propto \tau^{-2/3} N^{-1/3}$, which implies that, at a fixed τ, the globule's density tends to a constant value for $N \to \infty$.

Equation (2.2.12) may be directly obtained from minimizing the elastic free energy $\mathscr{A}_{\text{el}}$ under the constraint that the mean-square radius of gyration $\langle S^2 \rangle$ has a fixed value [see Eqs. (2.1.63) and (2.1.39), $C(q) \to \alpha^2(q) C(q)$] [10]. The physical meaning of this result is that under chain compression the free energy due to the interatomic contacts is basically a function of $\langle S^2 \rangle$ only, no matter what are the individual values of the $\alpha^2(q)$. As a consequence, all the mean-square distances $\langle r^2(k) \rangle$ may be expressed under a general form [53]. Defining

$$\zeta = k/N, \tag{2.2.16}$$

we have, remembering that $p = J/N$ from Eqn. (2.2.12),

$$\langle r^2(k) \rangle = \psi(\zeta, p) \tag{2.2.17}$$

with no formal dependence from the two- and three-body parameters. This implies that the integrals in Eqs. (2.2.4) and (2.2.7) may be carried out once p is known, which opens the way to a quasi-analytical approach. In fact, the value of p [$= f(\alpha_S^2)$] that minimizes the free energy may be derived from suitable universal plots and the knowledge of the two- and three-body parameters. The resulting equation is

$$\Delta_{\text{el}} - \tau B \sqrt{N} \Delta_2 + \frac{K}{\sqrt{N}} \Delta_{2S} + K_1 \Delta_3 = 0, \tag{2.2.18}$$

where [see also Eqn. (2.1.63)]

$$\Delta_{\text{el}} = \frac{\partial \mathscr{A}_{\text{el}}}{\partial \alpha_S^2}, \tag{2.2.19}$$

and Δ_2, Δ_{2S}, and Δ_3 are universal functions of α_S^2, proportional to the derivatives with respect to α_S^2 of the intramolecular contact free-energy contributions, whereas N is the number of chain atoms and B, K, and K_1 are defined in Eqs. (2.2.9) and (2.2.10). Figure 3 shows the plots of Δ_{el}, Δ_2, Δ_{2S}, and Δ_3 versus α_S. Knowing the parameters, we may derive $\tau = (T - \Theta)/T$ for any α_S from the preceding equation, which enables us to construct the plot of α_S versus τ. It should be mentioned that if one uses the periodic chain, the resulting mean-square end-to-end distance $\langle r^2(N) \rangle$ is much smaller than the plateau value $JC_\infty l^2$ [see Eqn. (2.2.13), $N \gg J$], whereas it is equal to $2JC_\infty l^2$

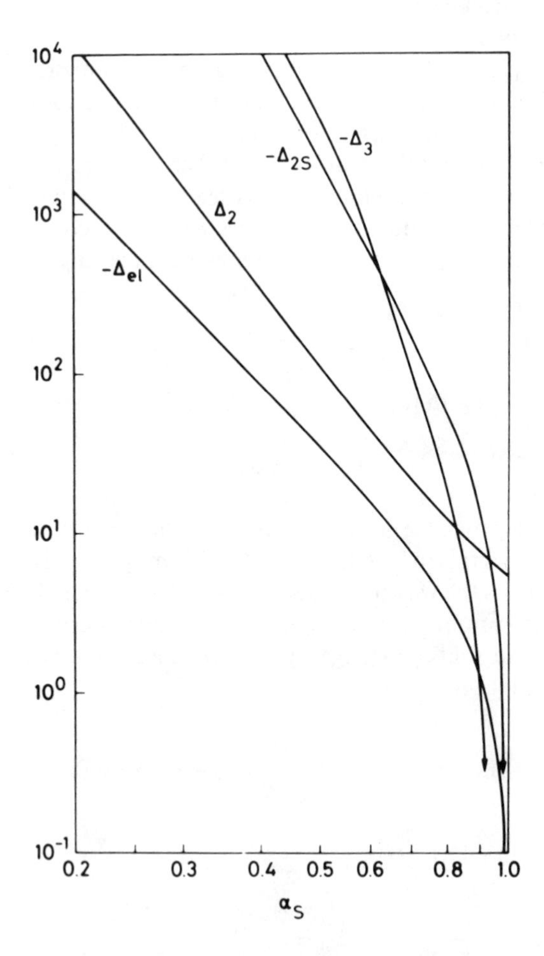

Figure 3. Universal plot of contributions to free-energy derivative [see Eqn. (2.2.18)] as function of α_s. For larger contractions, the four contributions are simple powers of α_s and may be directly extrapolated from plot or obtained analytically from ref. 53.

using the open-chain transform [54, 55]. The former result is to be considered as an artifact deriving from the periodic constraint, which may affect end-chain properties.

A comment on the dependence of $\mathscr{A}_{\text{el}}$ from α_S is in order. For small values of α_S we obtain [53]

$$\mathscr{A}_{\text{el}} \sim \frac{9}{4\alpha_S^2} \quad \text{as} \quad \alpha_S \to 0. \tag{2.2.20}$$

This result may be compared with that deriving from the *affine* mean-field approach, originally suggested by Flory to treat good-solvent chain expansion, that is [17],

$$\mathscr{A}_{\text{el}} = \tfrac{3}{2}[\alpha_S^2 - 1 - \ln \alpha_S^2] \sim \tfrac{3}{2} \ln \frac{1}{e\alpha_S^2} \quad \text{as } \alpha_S \to 0, \tag{2.2.21}$$

which indicates that the affine assumption underestimates the elastic free energy at high compression. Comparison between the two approaches may be carried further, and the equations expressing free-energy minimization are, in the limit $\alpha_S \to 0$,

(i) random Gaussian globule [54]:

$$-\alpha_S = \sqrt{3}\,\tau B N^{1/2} + 5\sqrt{3}\,K/(\alpha_S^2 N^{1/2}) + \frac{16}{\sqrt{3}} K_1/\alpha_S^3; \tag{2.2.22}$$

(ii) affine model [9, 56]:

$$\alpha_S^5 - \alpha_S^3 = \tau B N^{1/2} + y/\alpha_S^3, \tag{2.2.23}$$

(y is analogous, though not equal, to $\bar{y}$). While the left sides of both equations represent the elastic effect, the terms on the right side of (2.2.22) represent the attractive two-body, the repulsive, screened two-body, and the repulsive three-body interactions, respectively; in the other equation, the affine model has no screened interactions. The existence of at least one parameter for interatomic repulsions, in addition to the normalized undercooling $\tau B\sqrt{N}$, suggests that there is no universal curve describing chain collapse for all polymers. In qualitative analogy to what was shown by de Gennes for the affine model [9], we have either a first-order or a second-order transition if a suitable linear combination of $K/\sqrt{N}$ and of K_1 is below or above a critical value, respectively. Otherwise said, the first-order abrupt discontinuity is only possible if the local "bulkiness" of the chain is sufficiently small compared with Kuhn's statistical segment length $C_\infty l$. As an example, Figure 4 shows the calculated plot of α_S versus T for a set of parameters $K/\sqrt{N}$ and K_1 approximately fitting the experimental data obtained by Tanaka and co-workers from atactic polystyrene [57]*. Our results indicate a first-order discontinuity, in contrast with observation. However, if we account for the molecular weight distribution of the investigated sample ($M_w/M_n = 1.3$), the calculated curve is smoothened and the agreement is improved [10]. Unlike the affine approach, the α_S branch in the pretransition range between $\tau = 0$ and the critical undercooling is universal, being unaffected by the parameters of the two- and three-body repulsions. In fact, in this range we have $p > 1$ (i.e., $J > N$) so that the integrals accompanying K and K_1 in Eqs. (2.2.4) and (2.2.7) vanish, and so does the effect of these interactions.

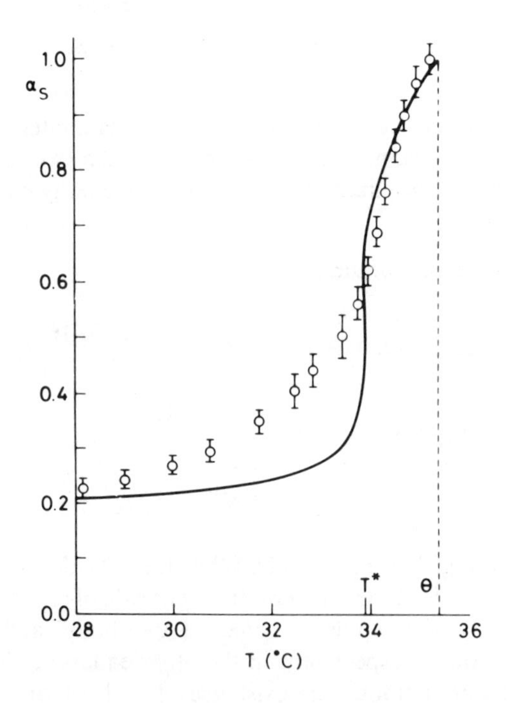

Figure 4. Experimental results of α_s as obtained by Tanaka and co-workers for atactic polystyrene in cyclohexane [57] and calculated plot from Eqn. (2.2.18) and Figure 3. [Parameters: $\Theta = 35.4\,^\circ\mathrm{C}$, $N = 9 \times 10^5$, $K = 0.37$, and $K_1 = 1.9 \times 10^{-3}$. Transition temperature $T^* = 33.9\,^\circ\mathrm{C}$ was fitted by choosing $B = 0.082$.]

* (Note added in proof) Recent experimental results by Chu and co-workers [see, e.g., *Macromolecules, 20*:1965 (1987); *21*:273, 1178 (1988)] suggest inaccuracy in Tanaka's results. This may imply that the numerical parameters employed by us should be modified.

A question may be whether the three-body or the two-body screened interactions are more effective in resisting chain collapse. An analysis based on Eqn. (2.2.22) shows that the screened interactions predominate for lower molecular weights whereas the three-body repulsions prevail for longer chains [52]. The number of chain atoms for which the two components have equal weight in the collapsed state just after transition is

$$\bar{N} = \tfrac{25}{64}\sqrt{5}\, 3^{7/4} K^2 K_1^{-3/2}. \tag{2.2.24}$$

With the present parameters (see caption of Figure 4) this chain length corresponds to a molecular weight of about 5×10^5 in the case of atactic polystyrene [10, 58].

Chain connectivity gives the first-order collapse transition an unusual character; namely, it is quasi-athermal in the thermodynamic limit, as first shown by Lifshitz, Grosberg, and Khokhlov through a different approach [59]. Indicating with ΔU^* the internal energy change at the transition for the whole chain, the energy change *per chain atom* $\Delta U^*/N$ scales as $N^{-1/2}$. As a consequence, the Boltzmann statistical fluctuation of the normalized under-cooling $[\tau B\sqrt{N}]^*$ at the collapse does not vanish in the thermodynamic limit. Denoting with σ the root-mean-square fluctuation, with α_S^* the average strain ratio after collapse and with C a constant of order 1, we get [10]

$$\sigma[\tau B\sqrt{N}]^*/[\tau B\sqrt{N}]^* = C(\alpha_S^*)^3, \tag{2.2.25}$$

also a result first obtained by Lifshitz et al. [59].

It should be stressed that in the preceding we have assumed throughout a polymer dilution large enough that the resulting process is always mono-molecular; that is, aggregation effects are excluded.

2.2.2. *The Chain Expansion*

We have seen that, as $T \to \Theta$ from below, the parameter p appearing in Eqs. (2.2.4) and (2.2.7) becomes larger than unity, thereby making the effect of both the screened and the three-body interactions vanishing. This conclusion may be considered as valid for $T > \Theta$ as well, so that Eqn. (2.2.4) reduces to ($q \ll 1$, $k \gg 1$)

$$\alpha^2(q) = \left\{1 - \frac{4}{q^2}\tau B \int_{\bar{k}}^{N} dk\left(1 - \frac{k}{N}\right)\sin^2\left(\frac{qk}{2}\right)J(k)^{-5/2}\right\}^{-1}, \tag{2.2.26}$$

$$\tau = \frac{T - \Theta}{T} > 0,$$

where $J(k)$ is defined in (2.2.5) and B in (2.2.10). We shall start from the assumption that $\alpha^2(q)$ obeys the following power law:

$$\alpha^2(q) = C_1 |q|^{-\gamma}, \tag{2.2.27}$$

with $\gamma > 0$. Accordingly, we obtain $J(k) \propto k^{1+\gamma}$. Let us write Eqn. (2.2.26) as $\alpha^2(q) = \{1 - U(q)\}^{-1}$; $U(q)$ must tend to unity from below for $q \to 0$. In this limit, assuming $qk \ll 1$, we have $\sin^2(qk/2)/q^2 \to k^2/4$, so that, performing the integral supposing at first $\gamma \neq \frac{1}{5}$,

$$\lim_{q \to 0} U(q) \propto \tau B \left\{ N^{1/2 - 5/2\gamma} \left[\frac{1}{\frac{1}{2} - \frac{5}{2}\gamma} - \frac{1}{\frac{3}{2} - \frac{5}{2}\gamma} \right] - \frac{(\bar{k})^{1/2 - 5/2\gamma}}{\frac{1}{2} - \frac{5}{2}\gamma} + \frac{(\bar{k})^{3/2 - 5/2\gamma}}{N[\frac{3}{2} - \frac{5}{2}\gamma]} \right\}. \tag{2.2.28}$$

It is apparent that γ cannot be less than $\frac{1}{5}$ because $U(q)$ would go to infinity as a positive power of N. On the other hand, if $\gamma > \frac{1}{5}$, $U(q)$ reduces to

$$\lim_{\substack{q \to 0 \\ N \to \infty}} U(q) \propto \frac{\tau B (\bar{k})^{1/2 - 5/2\gamma}}{\frac{5}{2}\gamma - \frac{1}{2}}, \tag{2.2.28$'$}$$

implying that the long-range chain expansion is dominated by the interaction between topologically adjacent atoms. In fact, if we write the integral in (2.2.26) as $\int_{\bar{k}}^{N} = \int_{\bar{k}}^{N/C} + \int_{N/C}^{N}$, where $1 < C \ll N$, we see that the last contribution is vanishingly small, with the result that the long-range interactions are substantially negligible, contrary to expectation. These considerations lead to the conclusion $\gamma = \frac{1}{5}$, corresponding to $\langle r^2(k) \rangle \propto J(k) \propto k^{6/5}$, that is, to the famous Flory $\frac{6}{5}$ exponent [17]. However, this choice still leads to a logarithmic divergence since $U(q) \propto \ln(N/\bar{k})$ for $q \to 0$; it may be eliminated by making recourse to factors depending on $\ln(1/q)$ on the right-hand side of Eqn. (2.2.27) [13, 16]. Based on a consistent conjecture that rests on continuity considerations, the final result is [16]

$$\begin{aligned} \alpha^2(q) &= \left[\frac{\tau B}{\sqrt{q}} H \prod_{n=1}^{\infty} L_n\left(\ln \frac{1}{q\bar{k}} \right) \right]^{2/5}, \\ \langle r^2(k) \rangle &= J(k) C_\infty l^2 \\ &= k C_\infty l^2 \left[\tau B \sqrt{k} I \prod_{n=1}^{\infty} L_n\left(\ln \frac{k}{\bar{k}} \right) \right]^{2/5}, \end{aligned} \tag{2.2.29}$$

where

$$L_1(x) = 1 + a_1 x, \qquad L_n(x) = 1 + a_n \ln L_{n-1}(x),$$
$$a_n < 1, \qquad \lim_{n\to\infty} a_n = 1, \qquad I, H \approx 1. \tag{2.2.30}$$

Writing $\langle r^2(k)\rangle \propto k^{2\nu(k)}$, the asymptotic value of the exponent $2\nu(k)$ for $k \to \infty$ is $\frac{6}{5}$, in agreement with Flory's result. Eqs. (2.2.29) are consistent both with numerical solutions of Eqn. (2.2.26) and, in a qualitative sense, with several Monte Carlo results from self-avoiding chains on three-dimensional lattices [60–63].

The preceding conclusions apply to very long chain portions within chains of infinite length. Unfortunately, no numerical results valid for this range appear to have been obtained with other methods. Conversely, several accurate calculations have been carried out in the last few years concerning whole chains, both open and ringlike, in the large-molecular-weight limit [64–67]. This prompted us to undertake an investigation of $\alpha_R^2 = \langle r^2(N)\rangle/\langle r^2(N)\rangle_0$, $\alpha_S^2 = \langle S^2\rangle/\langle S^2\rangle_0$ and $\alpha^2(q)$, where $\langle r^2(N)\rangle$ is the mean-square end-to-end distance and $\langle S^2\rangle$ is the mean-square radius of gyration. In the linear case, we used both the open- and the periodic-chain transform [68]. In order to avoid approximations in estimating the term with the integral in Eqn. (2.2.26), we used the exact double-sum term

$$U(q) = \frac{2\tau B}{N\sin^2(q/2)} \sum_{h=1}^{N-1}\sum_{k=1}^{N-h} \sin^2\left(\frac{qk}{2}\right) \times \sin^2\left[q\left(h + \frac{k-1}{2}\right)\right] J(h, h+k)^{-5/2}, \tag{2.2.31}$$

where

$$J(h, h+k) = \langle r^2(h, h+k)\rangle / C_\infty l^2 = 2N^{-1}\sum_{\{q\}} \frac{\alpha^2(q)\sin^2(qk/2)\sin^2\{q[h+(k-1)/2]\}}{\sin^2(q/2)}, \tag{2.2.32}$$

whereas in the periodic chain the factor $\sin^2\{q[h+(k-1)/2]\}$ is substituted by $\frac{1}{2}$ [see the analogous Eqs. (2.1.37) and (2.1.37′)]. After obtaining numerical solutions of Eqn. (2.2.26) with increasing N values and $\bar{k} = 1$, some of the universal results (i.e., $N \to \infty$) are shown in Figure 5 together with those obtained by other authors with different approaches [17, 64–67, 69]. Both in the linear and in the ring chain our results fit the exact perturbative expansion valid at low values of the solvent strength parameter $z = \tau B\sqrt{N}$ [3]. With the linear chain, around $z = 1$, where α_R^2 and α_S^2 are larger than 2, we have a

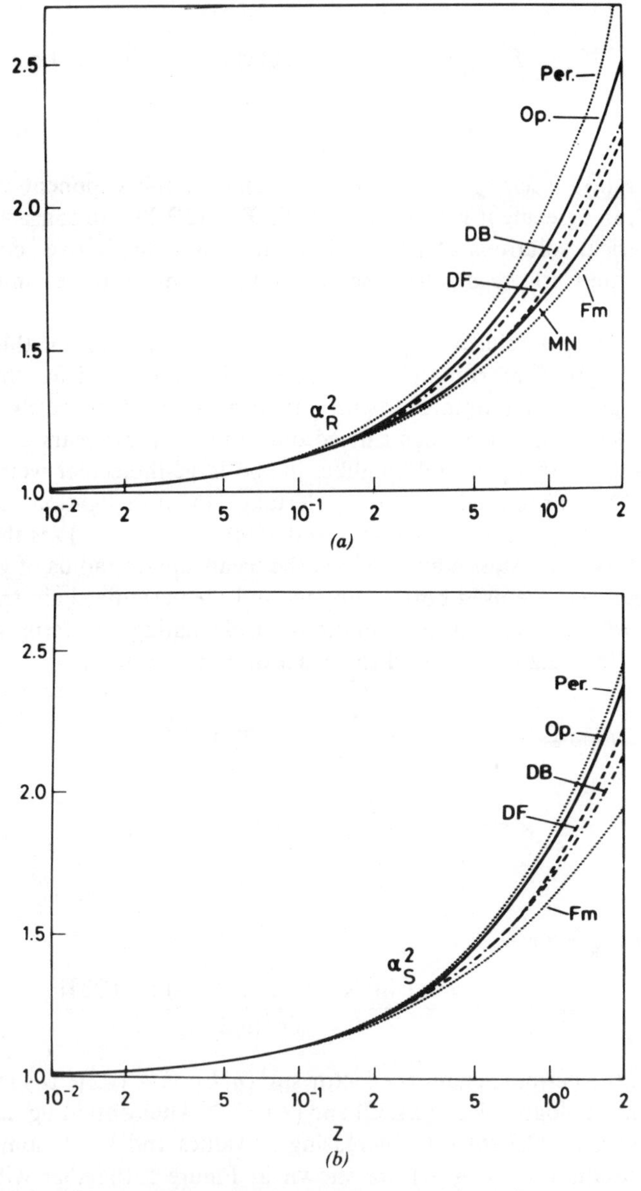

Figure 5. (*a*) Expansion factor of mean-square end-to-end distance α_R^2 and (*b*) radius of gyration α_s^2 as function of $z = \tau B\sqrt{N}$ for open (Op.) and periodic (Per.) chain. Other results shown here are Domb and Barrett (DB) [67], Douglas and Freed (DF) [66], Muthukumar and Nickel (MN) [64], and Flory modified (Fm) [17, 69]. (From ref. 68, by permission of the publishers, Butterworth & Co. (Publishers) Ltd. ©.)

relatively small discrepancy from the other results, among which those of Muthukumar and Nickel [64] and of des Cloizeaux et al. [65] for α_R^2 are perhaps the most accurate terms of reference. As expected, for α_R^2 the agreement is better with the open-chain than with the periodic-chain transform, which suffers from artificially induced end effects, whereas no remarkable difference in the results of α_S^2 is seen between the two transforms. Quite obviously, for large z our figures exceed those predicted from the theories that are consistent with the renormalization group exponent 1.176 instead of 1.2 as implicit in our approach. The reasonable accuracy of our results for α_R^2 and α_S^2 up to values around 2 implies that the whole pattern of the mean-square interatomic distances is correctly predicted within that range, including the mean-square strain ratios $\alpha^2(q)$ of the Fourier modes, which in turn are instrumental in determining the relaxation times (See following section 3.1).

A result emerging from our calculations is that chain portions distant from the ends are more extended than the end portions due to the repulsions involving outer atoms; for the same reason, chain portions are more extended than whole chains of the same length [70].

Although the exponent $2\nu = 1.2$ was obtained by us for chain portions within an infinite chain, we have reasons to believe it is to be extended to whole chains. The reason of the failure of the Gaussian approximation to obtain $2\nu = 1.176\ldots$, as established on field-theoretic, renormalization group grounds [7], is to be attributed to long-range correlation between chain bonds, which makes the Gaussian assumption incorrect. In fact, taking $\langle r^2(k)\rangle \propto k^{6/5}$, from Eqn. (2.1.71) we get

$$\langle \mathbf{l}(h)\cdot\mathbf{l}(h+k)\rangle \propto k^{-4/5}. \tag{2.2.33}$$

The result decreases with k *less* quickly than k^{-1}, contrary to the unperturbed and the collapsed chain [see Eqs. (2.1.73) and (2.2.14), respectively]; as a consequence, the sum of the long-range bond correlation contributions diverges and the mean-square chain size increases more quickly than N^1. In particular, the central-limit theorem cannot be applied and the Gaussian distribution for $W(k, \mathbf{r})$ is incorrect. Actually, it has been determined that for $k \gg 1$ this function obeys the generalized Domb–Gillis–Wilmers form [71, 72]

$$W(k, r) \propto \langle r^2(k)\rangle^{-3/2}\rho^l \exp(-C\rho^t), \qquad \rho = r/\langle r^2(k)\rangle^{1/2}, \tag{2.2.34}$$

where l is comprised between zero and 1, whereas t is larger than 2. It is implicit in the preceding expression that the interatomic contact probability is proportional to $\langle r^2(k)\rangle^{-(3+l)/2}$, whereas we *impose* the exponent $-\frac{3}{2}$ within the Gaussian approximation. Actually, as it may be shown from cluster expansion for each atom pair, we replace an average (i.e., mean-field) probability distribution of a Gaussian form to a linear combination of Gaussian distributions, each of which applies to a different set of fixed intramolecular contacts. A self-consistent-field approach would amount to solve the difficult problem of the self-consistent excluded volume interatomic potential in the multibody field of all skeletal atoms [5]. Without such a solution we believe the Flory exponent to be the best possible result of any self-consistent approach [see, e.g., Edwards [73], Gillis and Freed [74]].

2.3. Different Interaction Ranges: Scaling Laws

The conformations accessible to a polymer chain at the scale of a few consecutive chain bonds are mainly dictated by the rotational potentials and by the interaction energies between nonbonded atoms located in the neighborhood of the bonds undergoing rotation (short-range interactions). The interatomic distances determined by two or three, rarely more, neighboring rotation angles are used to calculate the nonbonding interaction energies, and these are summed up together with the rotational potentials to get the conformational energy maps. The procedure allows for interactions between atoms separated by no more than $\sim$5–10 bonds. The probability of each conformation is proportional to its Boltzmann statistical weight regardless of the interactions with more distant atoms that are supposed to contribute a sort of uniform average. At least for polymer chains where strong polar interactions are absent, such as the hydrocarbon polymers, the most important feature of a single interaction lies in its steep repulsive component, preventing any two atoms from getting too close. As an example, in the Lennard–Jones formulation this component changes as r^{-12}, r being the interatomic distance; the resulting energy increases so quickly with decreasing r that for some purposes it may be replaced with a hard-core potential. A change of temperature of a few degrees in the vicinity of the Θ temperature does not usually affect the local conformational energies and the corresponding probabilities; in other words, even though a long polyethylene chain in a dilute solution may change from expanded to collapsed, if regarded as a whole, the proportion of trans to gauche states may change very little, if at all. As a result of this conformational analysis, the characteristic ratio $C(0)$ or the persistence length may be evaluated [see Eqs. (2.1.6) and (2.1.30)], which summarize the local rigidity of the chain and, in case we are at the Θ temperature, may also give the average unperturbed dimensions, for example, $\langle S^2 \rangle_{\text{ph}} = \frac{1}{6}NC(0)l^2$. The subscript ph refers to the phantom chain to stress that we are neglecting any interaction beyond the shortest range.

A physically accessible conformation will not easily lead to ring closure with less than $\sim$10–20 bonds if followed by other accessible conformations. This problem was quantitatively investigated by Flory and co-workers for specific polymers, and these Authors also showed that in this range the chain is strongly departing from Gaussian statistics [47, 48]. Not until the number of intervening bonds is at least 30–50 is the Gaussian law a good representation of reality. We have seen in a previous section that this figure may also characterize the lower limit for the onset of the screened interactions, that is, interatomic repulsions arising from the chain thickness being not negligible in comparison with the average distance $\langle r^2 \rangle^{1/2}$ between the two atoms. In the Gaussian approximation, the corresponding energy changes as $\langle r^2 \rangle^{-5/2}$, thus vanishing quickly enough with large r that the mean-square radius of gyration depends linearly on the number N of chain atoms provided we are at the Θ temperature. The characteristic ratio C_∞ may be expressed as $\tilde{\alpha}_0^2(0)C(0)$, where $\tilde{\alpha}_0^2(0)$ may range between 1.05 and 1.30 for different, flexible hydrocarbon polymers according to our calculations. As an effect of the screened interactions, the ratio $\langle r^2(k) \rangle_0/kl^2$, where $\langle r^2(k) \rangle_0$ is the mean-square distance between atoms separated by k bonds, still increases when k is of the order of a few hundred, unlike the case of the phantom chain, devoid of intrinsic volume and thickness. The screened interactions may be regarded as *medium-range interactions.*

If $T \neq \Theta$, the overall interaction energy between two atoms is not zero if averaged through the whole space; in this case we have $\beta_{\text{eff}} \neq 0$ [see Eqn. (2.1.2)]. As a consequence, if the average distance between two atoms is large compared with the actual distances over which the interaction takes place, the effective interaction energy is proportional to $\langle r^2 \rangle^{-3/2}$ (*long-range interactions*). The simple proportionality between $\langle S^2 \rangle$ and N is lost. In the expansion regime at $T > \Theta$, if we are interested in long-range properties we may neglect the screened interactions, their effect being embodied in the characteristic ratio C_∞, and two parameters, such as $NC_\infty l^2$ and $\tau B\sqrt{N}$ ($\tau B \propto \beta_{\text{eff}}$), are sufficient to describe the average configurational properties in the infinite chain limit. If $T < \Theta$, the chain tends to collapse to a high-density state. Here the repulsive medium-range screened interactions come into play together with another type of medium-range interactions originating from simultaneous appearance of two two-body contacts involving one same atom. These are the so-called *three-body interactions*, which may be neglected for $T \geqslant \Theta$. Although in the thermodynamic limit $N \to \infty$ these interactions provide the dominant contribution to the collapse resistance, in many cases of practical interest the screened interactions may be more important. Accordingly, in the collapse regime we need two more parameters, one for the screened and one for the three-body interactions, which reduce to one in the thermodynamic limit. If the joint effect of these repulsive interactions is sufficiently small, we have a first-order, discontinuous drop of $\langle S^2 \rangle$ upon decreasing the temperature below $T = \Theta$; otherwise the transition is of the second order.

Terms depending on $\ln N$ have been proposed to account for the total number of three-body interactions in the vicinity of $T = \Theta$ [75, 76]. With our approach based on the lower cutoff $\bar{k}$, we have

Number of three-body interactions

$$\propto \sum_{k=\bar{k}}^{N} \sum_{k_1=\bar{k}}^{N-k} (N-k-k_1)\langle r^2(k)\rangle_0^{-3/2}\langle r^2(k_1)\rangle_0^{-3/2}$$

$$\propto \int_{\bar{k}}^{N} \frac{dk}{k^{3/2}} \int_{\bar{k}}^{N-k} \frac{dk_1(N-k-k_1)}{k_1^{3/2}} \simeq 4\left\{\frac{N}{\bar{k}} - 4\left(\frac{N}{\bar{k}}\right)^{1/2}\right\}, \tag{2.3.1}$$

whereas the corresponding number of two-body interactions is proportional to $(\bar{k})^{-1/2}[N^{1/2}-(\bar{k})^{1/2}]^2$. The two results are noticeably analogous; it should be remarked that the vast majority of the three-body interactions is due to contacts where at least two atoms are close along the chain. We have seen that these three-body interactions produce a lowering of the Θ temperature from the value expected for the phantom chain. This is due to the need to compensate the corresponding repulsion with some two-body attraction. We may qualitatively find the temperature shift by writing the proportionality

$$-\Delta\tau B \times (\text{number of two-body interactions})$$
$$\propto K_1 \times (\text{number of three-body interactions}), \tag{2.3.2}$$

where $\Delta\tau = (\Theta - \Theta_{\text{phantom}})/\Theta$. Except for numerical factors, Eqn. (2.2.11) follows if we take $N \gg \bar{k}$.

It is useful to summarize general results on polymer chains with appropriate scaling laws. Here, again, a distinction must be drawn between expansion and collapse. In the former regime a unique relationship for $\alpha^2(k) = \langle r^2(k)\rangle/\langle r^2(k)\rangle_0$ may be given, valid both for whole chains ($k = N$) and for internal portions of extremely long chains ($k \ll N$); it reads ($\langle r^2(k)\rangle_0 = kC_\infty l^2$)

$$\alpha^2(k) \propto k^{1/5}, \tag{2.3.3}$$

as it may be seen from Eqn. (2.2.29), neglecting the weaker logarithmic factors. Conversely, in the case of collapse we see from Eqn. (2.2.13) that $\alpha^2(k) \propto J/k$ ($k \gg pN$), and since at a given temperature $J \propto N^{2/3}$ [see the discussion after Eqn. (2.2.15′)], we get

$$\alpha^2(k) \propto N^{2/3}k^{-1}, \tag{2.3.4}$$

which reduces to $\alpha^2(N) \propto N^{-1/3}$ for $k = N$. We see that two different power laws apply, whether the base is N for the whole chain (exponent $-\frac{1}{3}$) or it is k

for a chain portion (exponent -1). This implies a word of caution in making use of scaling criteria in the collapse case.

3. CHAIN DYNAMICS

3.1. The Langevin Equation of Motion and the Spectrum of the Relaxation Times

We shall deal here with the thermal fluctuations of a polymer chain, in dilute solution unless otherwise stated, around the equilibrium configurations. At first, we shall neglect the internal viscosity, which may be relevant at high frequencies; it will be dealt with in Sections 3.3.2 and 3.3.3. The basic equations are quite general as they apply to both the open-chain and the periodic-chain transform. However, most results will be obtained through the much simpler periodic transform. Extension to the more accurate open-chain transform is conceptually straightforward although computationally tedious; it may be required when dealing with the most collective modes, and a few significant results thus obtained will be reported. Our first aim will be to solve the Langevin equation of motion; this will give us the spectrum of the relaxation times, which in turn contains all the dynamical information. As particularly significant aspects, we shall concentrate on the complex dynamic viscosity (which includes the zero-frequency intrinsic viscosity as a particular case) and on the interatomic correlation function $\langle [\mathbf{R}(h,t)-\mathbf{R}(j,0)]^2\rangle$, where $\mathbf{R}(h,t)$ is the vector position of atom h at time t. This will enable us to obtain the dynamic structure factor experimentally derivable from quasi-elastic scattering.

In the preaveraged approximation for the Oseen hydrodynamic tensor [19, 20], the linear Langevin equation may be written as

$$-\mathbf{f}(h,t)+\zeta[\dot{\mathbf{R}}(h,t)-\mathbf{v}^0(h,t)]-\frac{\zeta}{6\pi\eta_s}\sum_{j(\neq h)=1}^{N}\langle r^{-1}(h,j)\rangle \mathbf{f}(j,t)=\mathbf{\Gamma}(h,t), \tag{3.1.1}$$

where $-\mathbf{f}(h,t)$ is the force exerted by the hth chain atom upon the solvent and is equal and opposite to the intramolecular force at time t; $\mathbf{v}^0(h,t)$ is the solvent velocity in the absence of the chain at the position of the hth atom $\mathbf{R}(h,t)$; ζ and η_s, respectively, are the friction coefficient per chain atom and the solvent viscosity; $r(h,j)$ is the distance between atoms h and j, and $\mathbf{\Gamma}(h,t)$ is the Brownian stochastic force.

The linear form of the dynamic equation is consistent with the Gaussian distribution adopted throughout this chapter, which in turn produces a quadratic intramolecular potential of the form given in Eqn. (2.1.43) for the Θ state. However, in the more general case of a polymer chain not in the Θ state, we must adopt a more general quadratic potential. It was proven that the general Boltzmann statistical weight is [10]

$$\mathscr{W}\{\tilde{\mathbf{l}}(q)\} = \exp(-\mathscr{A})\prod_{\{q\}}\left\{\frac{3}{2\pi\tilde{\alpha}^2(q)NC(q)l^2}\right\}^{3/2} \times \exp\left[-\sum_{\{q\}}\frac{3|\tilde{\mathbf{l}}(q)|^2}{2\tilde{\alpha}^2(q)NC(q)l^2}\right]. \tag{3.1.2}$$

Considering that the reduced free energy $\mathscr{A}$ $(= A/k_BT)$ is stationary with respect to all the strain parameters $\tilde{\alpha}^2(q)$, the configurational, or elastic, force is given by [see also Eqn. (2.1.44)]

$$f_x(h, t) = k_BT\frac{\partial \ln \mathscr{W}\{\tilde{\mathbf{l}}(q, t)\}}{\partial x(h, t)} = -\frac{3k_BT}{Nl^2}\sum_{\{q\}}\frac{\mu(q)}{\tilde{\alpha}^2(q)}\tilde{x}(q, t)Q^*(q, h), \tag{3.1.3}$$

where

$$\begin{aligned} Q(q, h) &= \sqrt{2}\cos[q(h-\tfrac{1}{2})] \quad \text{(open chain);} \\ Q(q, h) &= \exp(iqh) \quad \text{(periodic chain).} \end{aligned} \tag{3.1.4}$$

Projecting (3.1.1) along x, multiplying by $Q(q, h)$, and summing over h, we get the transformed equation

$$\frac{3k_BT}{l^2}\frac{\mu(q)}{\tilde{\alpha}^2(q)}\tilde{x}(q, t) + \zeta(q)[\dot{\tilde{x}}(q, t) - \tilde{v}_x^0(q, t)] = \tilde{X}(q, t), \tag{3.1.5}$$

where [see Eqn. (2.1.29′)] [77, 78]

$$\mu(q) = 4\sin^2(q/2)/C(q), \tag{3.1.5′}$$

$$\zeta(q) = \zeta/\nu(q), \tag{3.1.5″}$$

with

$$\nu(q) = 1 + \frac{\zeta}{6\pi\eta_s}\sqrt{\frac{6}{\pi}}\frac{1}{N}\sum_{h\neq j=1}^{N}\sum Q(q, h)Q^*(q, j)\langle r^2(h, j)\rangle^{-1/2}. \tag{3.1.6}$$

In the preceding $\langle r^{-1}(h, j)\rangle$ is replaced by $\sqrt{6/\pi}\,\langle r^2(h, j)\rangle^{-1/2}$ in the Gaussian approximation. Also

$$\tilde{X}(q, t) = \tilde{\Gamma}_x(q, t)/\nu(q); \tag{3.1.7}$$

and we may define

$$R_{\text{eff}} = \zeta/6\pi\eta_s, \tag{3.1.7'}$$

where the parameter R_{eff} is a solvent-independent constant that may be interpreted as the effective radius per skeletal atom satisfying the Stokes-type law $\zeta = 6\pi\eta_s R_{\text{eff}}$. The quadratic average of $\tilde{X}(q, t)$ is specified by the fluctuation–dissipation theorem [79]

$$\langle \tilde{X}(q, t)\tilde{X}(q', 0)\rangle = 2Nk_B T\zeta(q)\Delta(q + q')\delta(t), \tag{3.1.8}$$

where Δ and δ are the Kronecker delta and the Dirac delta function, respectively.

Expression (3.1.6) is exact for ring chains, whereas it is only approximate for open and periodic chains, as off-diagonal elements depending on pairs of different q values are neglected according to the Zimm–Hearst approach [20, 80]. For the ring and the periodic chain, where $Q(q, h) = e^{iqh}$, Eqn. (3.1.6) reduces to

$$\nu(q) = 1 + \frac{\zeta}{3\pi\eta_S}\sqrt{\frac{6}{\pi}}\sum_{k=1}^{N-1}\left(1 - \frac{k}{N}\right)\langle r^2(k)\rangle^{-1/2}\cos(qk). \tag{3.1.9}$$

This equation implies perfect dynamical equivalence of all the chain atoms; this is physically true only for the ring, whereas it is a model assumption in the periodic case. However, it should be noted that, apart from the first few collective modes, the periodic chain gives a good description of the open-chain dynamics and may be safely retained when investigating local chain motions, as suggested by Akcasu, Benmouna, and Han [81] and shown by us [82, 83]. The general solution of Eqn. (3.1.5) may be cast in the form

$$\tilde{x}(q, t) = \tilde{x}(q, 0)\exp[-t/\tau(q)] + \zeta(q)^{-1}\int_0^t dt'[\tilde{X}(q, t') + \tilde{v}_x^0(q, t')\zeta(q)]\exp[-(t - t')/\tau(q)], \tag{3.1.10}$$

where the general relaxation time is given by

$$\tau(q) = \zeta(q)\bigg/\left[\frac{3k_B T}{l^2}\frac{\mu(q)}{\tilde{\alpha}^2(q)}\right] = \tfrac{1}{3}t_0\frac{\tilde{\alpha}^2(q)}{\mu(q)\nu(q)}, \tag{3.1.11}$$

$$t_0 = \zeta l^2/k_B T, \tag{3.1.11'}$$

and Eqn. (3.1.10) enables us to evaluate all the quadratic averages of interest *via* the fluctuation–dissipation theorem (3.1.8).

3.1.1. *The Dynamic Viscosity and the Dynamic Structure Factor*

The *dynamic viscosity* $\eta(\omega)$ is conveniently defined under alternating shear–stress conditions. Assuming the fluid velocity along x and the velocity gradient along z, the solvent velocity at the position of the hth atom in the absence of the chain is

$$v_x^0(h, t) = \gamma e^{i\omega t} z(h, t), \tag{3.1.12}$$

ω being the applied frequency. The general equation for the dynamic viscosity is [84]

$$\begin{aligned}\eta(\omega) &= \eta_s + Nc(\gamma e^{i\omega t})^{-1} \langle f_x(h, t) z(h, t)\rangle_h \\ &= \eta_s + c(N\gamma e^{i\omega t})^{-1} \sum_{\{q\}} \langle \tilde{f}_x(q, t) \tilde{z}(-q, t)\rangle,\end{aligned} \tag{3.1.13}$$

where c is the concentration as the number of chains per unit volume. Given the dynamic viscosity, the complex modulus $G(\omega)$ is defined as

$$G(\omega) = G'(\omega) + iG''(\omega) = i\omega\eta(\omega). \tag{3.1.14}$$

Using (3.1.3), (3.1.10), and (3.1.11), Eqn. (3.1.13) reduces to

$$\eta(\omega) = \eta_s + ck_BT \sum_{\{q\}} \frac{\tau(q)/2}{1 + i\omega\tau(q)/2} \tag{3.1.15}$$

For $\omega = 0$ we get the *intrinsic viscosity*

$$\begin{aligned}[\eta] &= \lim_{\rho \to 0} \frac{\eta(0) - \eta_s}{\rho\eta_s} \\ &= \frac{N_{Av} k_B T}{2M_0 N \eta_s} \sum_{\{q\}} \tau(q),\end{aligned} \tag{3.1.15'}$$

where ρ is the polymer concentration expressed in grams per volume, N_{Av} is Avogadro's constant, and M_0 is the molar mass per chain atom.

The *dynamic structure factor* is defined as [85]

$$\begin{aligned}S_{\text{coh}}(Q, t) &= N^{-2} \sum_{h,j=1}^{N}\sum \langle \exp\{-i\mathbf{Q}\cdot[\mathbf{R}(h, t) - \mathbf{R}(j, 0)]\}\rangle \\ &\simeq N^{-2} \sum_{h,j=1}^{N}\sum \exp\left[-\frac{Q^2}{6} B(h, j, t)\right],\end{aligned} \tag{3.1.16}$$

$$S_{\text{inc}}(Q, t) = \langle \exp\{-i\mathbf{Q}\cdot[\mathbf{R}(0, t) - \mathbf{R}(0, 0)]\}\rangle$$
$$\simeq \exp\left[-\frac{Q^2}{6} B(0, 0, t)\right], \tag{3.1.16'}$$

where S_{coh} and S_{inc} apply to the coherent and incoherent case, respectively, whereas $\mathbf{Q}$ is the scattering vector ($Q = 4\pi \sin(\theta/2)/\lambda$), and the dynamic correlation function $B(h, j, t)$ is

$$B(h, j, t) = \langle [\mathbf{R}(h, t) - \mathbf{R}(j, 0)]^2 \rangle$$
$$= \frac{l^2}{N} \sum_{\{q\}} \frac{\tilde{\alpha}^2(q)}{\mu(q)} \{|Q(q, j)|^2 + |Q(q, h)|^2$$
$$- |Q(q, j)Q^*(q, h) + Q^*(q, j)Q(q, h)| \exp[-t/\tau(q)]\}. \tag{3.1.17}$$

Note that the approximate equalities in Eqs. (3.1.16) and (3.1.16′) are due to the Gaussian approximation. Using the periodic-chain transform $Q(q, h) = e^{iqh}$, we have

$$B(h, j, t) \equiv B(k, t) = \frac{2l^2}{N} \sum_{\{q\}} \frac{\tilde{\alpha}^2(q)}{\mu(q)} \{1 - e^{-t/\tau(q)} \cos(qk)\}, \tag{3.1.18}$$

where $k = |h - j|$. The contribution with $q = 0$ to the sum (3.1.18) may be obtained as the limit for $q \to 0$, and we get

$$[B(k, t)]_{q=0} = \begin{cases} 6Dt + k^2\tilde{\alpha}^2(0)C(0)l^2/N & \text{(periodic chain)}, \\ 6Dt & \text{(ring chain)}, \end{cases} \tag{3.1.19}$$

where

$$D = k_B T / N\zeta(0) \tag{3.1.19'}$$

is the diffusion coefficient. It may be shown that using the open-chain transform the term with $q = 0$ in Eqn. (3.1.17) also gives $6Dt$. It is convenient to keep the diffusion term separate from the intramolecular contribution $\bar{B}(k, t)$:

$$B(k, t) = 6Dt + \bar{B}(k, t), \tag{3.1.20}$$

where $\bar{B}(k, t)$ is given by (3.1.18), the term $q = 0$ in the sum being replaced by $k^2\tilde{\alpha}^2(0)C(0)/2$ in the periodic chain and by zero in the ring. For both periodic and ring chains we have the simple result

$$\lim_{t \to \infty} \bar{B}(k, t) = 2S^2, \tag{3.1.21}$$

so that after a long time all the atoms are at the same average distance from their initial position, in a Cartesian framework with the origin placed at the chain's center of mass.

As a general remark, the most important terms contributing to both the coherent and the incoherent dynamic structure factor [see Eqs. (3.1.16)–(3.1.18)] are characterized by $Q^2 B(k, t) \approx 1$. Since $B(k, t)$ is monotonously increasing with t, long times of observation are best probed at small values of Q, and vice versa. In particular, for $t \to \infty$ comparison of Eqs. (3.1.20) and (3.1.21) leads to the important result $B(k, t) \to B(0, t)$ for any k.

Before proceeding to carry out calculations for specific models, it will be useful to examine a few scaling laws in the next section.

3.1.2. *Dynamic Scaling Laws*

The scope of the present section is to give a few scaling laws relating different equilibrium and dynamical quantities for purposes of understanding and rationalization. In general these scaling laws will not be used a priori to establish new results in what follows.

We shall focus our attention on the chain models already investigated in the equilibrium section in the time range $\tau_{max} \gg t \gg \tau_{min}$, where $\tau_{max(min)}$ is the largest (smallest) relaxation time. Under these conditions the periodic-chain transform may be used even for the open chain, a continuous description of the chain may be applied, and the sums may be changed to integrals. In the particular case of the wormlike chain τ_{min} vanishes because the chain bonds reduce to a zero length, whereas τ_{max} may be identified with $\tau(q^*)$, where $q^* = 2\pi/k_p$ [see Eqs. (2.1.52)]. In fact, for $q < q^*$ the wormlike model loses its specificity, being no different from the bead-and-spring model.

Let us first assume that in the range under consideration the following power law applies:

$$\tilde{\alpha}^2(q)C(q) \propto q^{-\gamma} \qquad (\gamma \geqslant 0). \tag{3.1.22}$$

Obviously enough, in the unperturbed state and $q \ll 1$, $\tilde{\alpha}^2(q) \equiv 1$, $C(q) \equiv C_\infty$, and $\gamma = 0$, whereas in the expanded state $\gamma = \frac{1}{5}$ [ignoring logarithmic factors, see Eqn. (2.2.29)]. In the wormlike chain $C(q) \propto q^{-2}$ [see Eqn. (2.1.52) with $q \gg q^* \propto 1/k_p$], and we have $\gamma = 2$. Note that the collapsed state with $\gamma = -2$ [see Eqn. (2.2.12) with $qJ \ll 1$] cannot be encompassed by this simple analysis. To evaluate the relaxation times, we must distinguish between the Rouse and the Zimm limits, specified by $v(q) \equiv 1$ and $v(q) \gg 1$, respectively.

In the *Rouse limit*, from Eqs. (3.1.11) and (3.1.5′), we have ($q \ll 1$)

$$v(q) \equiv 1,$$

$$\tau(q) \propto \frac{\tilde{\alpha}^2(q)C(q)}{v(q)q^2} \propto q^{-2-\gamma}. \tag{3.1.23}$$

Substituting this result into Eqn. (3.1.15) gives

$$\eta(\omega) \propto \omega^{-(1+\gamma)/(2+\gamma)}, \qquad G(\omega) = i\omega\eta(\omega) \propto \omega^{1/(2+\gamma)}. \tag{3.1.24}$$

Concerning the dynamic structure factor, we shall confine our attention to the incoherent case, where the self-correlation function $B(0, t)$ only is required; since we have $q \ll 1$, it may be shown that the results are essentially valid for the coherent case as well (long-time limit) [86]. From Eqn. (3.1.18) we get

$$B(0, t) \propto t^{(1+\gamma)/(2+\gamma)}, \tag{3.1.25}$$

and from Eqn. (3.1.16′)

$$S_{\text{inc}}(Q, t) = \exp\left[-\frac{Q^2}{6}B(0, t)\right]. \tag{3.1.26}$$

Therefore, the half-peak time $t_{1/2}$ defined as $S_{\text{inc}}\ (Q, t_{1/2}) = \frac{1}{2}S_{\text{inc}}(Q, 0)$ is linked to Q by the power law

$$t_{1/2}Q^{2(2+\gamma)/(1+\gamma)} = \text{const.} \tag{3.1.27}$$

In the *Zimm limit* we must evaluate $v(q)$ ($\gg 1$) from Eqn. (3.1.9), which requires prior calculation of $\langle r^2(k)\rangle$. From Eqn. (2.1.66), changing the sum to an integral we have

$$\langle r^2(k)\rangle \propto \begin{cases} k^{1+\gamma} & \text{if } \gamma < 1, \\ k^2 & \text{if } \gamma = 2 \text{ (wormlike chain).} \end{cases} \tag{3.1.28}$$

Accordingly, apart from logarithmic factors we have from (3.1.9)

$$v(q) \propto \begin{cases} q^{-(1-\gamma)/2} & \text{if } \gamma < 1, \\ \text{const} & \text{if } \gamma = 2, \end{cases} \tag{3.1.29}$$

and from (3.1.11) the relaxation times are

$$\tau(q) \propto \begin{cases} q^{-(3/2)(1+\gamma)} & \text{if } \gamma < 1, \\ q^{-4} & \text{if } \gamma = 2. \end{cases} \tag{3.1.30}$$

Corresponding power laws for $\eta(\omega)$ and $G(\omega)$ are

$$\eta(\omega) \propto \begin{cases} \omega^{-(1+3\gamma)/3(1+\gamma)} & \text{if } \gamma < 1, \\ \omega^{-3/4} & \text{if } \gamma = 2 \quad \text{ref. [87]}, \end{cases}$$
$$G(\omega) \propto \begin{cases} \omega^{2/3(1+\gamma)} & \text{if } \gamma < 1, \\ \omega^{1/4} & \text{if } \gamma = 2 \quad \text{ref. [87]}, \end{cases} \tag{3.1.31}$$

whereas $B(0, t)$ is

$$B(0, t) \propto \begin{cases} t^{2/3} & \text{if } \gamma < 1, \\ t^{3/4} & \text{if } \gamma = 2. \end{cases} \tag{3.1.32}$$

From the last result, the half-peak time $t_{1/2}$ for $S_{\text{inc}}(Q, t)$ is given by

$$t_{1/2}Q^3 = \text{const} \quad (\text{if } \gamma < 1), \qquad t_{1/2}Q^{8/3} = \text{const} \quad (\text{if } \gamma = 2). \tag{3.1.33}$$

Few considerations are in order. First, in the Zimm limit with $\gamma < 1$, the independence from γ of the exponents appearing in $B(0, \text{t}) \propto t^{\alpha}$ and $t_{1/2}Q^{\mathscr{B}} = \text{const}$ is noteworthy [81]. Second, α and $\mathscr{B}$ are always correlated by $\mathscr{B} = 2/\alpha$. Third, the exponent θ of $\tau(q) \propto q^{-\theta}$ is always related to that of $G(\omega) \propto \omega^{\zeta}$ by $\zeta = 1/\theta$. Fourth and last, it is remarkable that the power laws of the wormlike chain appear to be unaffected by the hydrodynamic interaction because $\nu(q) \cong \text{const}$; see Eqn. (3.1.29).

Concerning the intrinsic viscosity, we may derive its molecular weight dependence. In the Rouse limit, $\nu(q) \equiv 1$ and from Eqs. (3.1.11) and (3.1.15′), we get

$$[\eta] = \frac{N_{\text{Av}} k_B T t_0}{6 M_0 \eta_s} \frac{1}{N} \sum_{\{q\}} \frac{\tilde{\alpha}^2(q)}{\mu(q)} \tag{3.1.34}$$

and from Eqn. (2.1.39) and (3.1.11′) [open chain, $C(q) \rightarrow \tilde{\alpha}^2(q) C(q)$]

$$[\eta] = \frac{N_{\text{Av}} k_B T t_0}{6 M_0 \eta_s} \frac{\langle S^2 \rangle}{l^2} = \frac{N_{\text{Av}}}{M_0} \frac{\zeta}{6\eta_s} \langle S^2 \rangle. \tag{3.1.35}$$

Therefore, in the free-draining, Rouse limit, the molecular weight dependence of $[\eta]$ is the same as that of $\langle S^2 \rangle$.

In the Zimm limit, we shall extrapolate to the first modes the power laws given in the preceding. For flexible chains we may use Eqn. (3.1.29) and

(3.1.30) ($\gamma < 1$), wherefrom we obtain from Eqn. (3.1.15′), remembering that $q = (\pi/N)n_q$,

$$[\eta] \propto N^{1/2 + 3/2\gamma}. \tag{3.1.36}$$

In this case, Eqn. (3.1.29) ($\gamma < 1$) shows that $v(q)$ is a slowly decreasing function of q, which reduces the relative importance of the first modes compared to the internal ones. Therefore, the previously mentioned extrapolation is of little consequence as far as the general behavior is concerned.

3.2. The Bead-and-Spring Chain Model

3.2.1. *The Unperturbed Chain: The Rouse and Zimm Limits*

The equilibrium features of the unperturbed bead-and-spring chain are summarized by the relationships

$$C(q) \equiv C_\infty = 1; \qquad \tilde{\alpha}^2(q) \equiv \alpha^2(q) \equiv 1, \tag{3.2.1}$$

whence $\gamma = 0$ in Eqn. (3.1.22). As done in the previous section, the Rouse and the Zimm limits will be considered separately. The correlation function $B(k, t)$ is always evaluated within the periodic- chain transform as an approximation.

The Rouse limit is encountered when the chain is sufficiently short that $v(q) \simeq 1$ in Eqs. (3.1.6), so that we have $\zeta(q) \simeq \zeta$. In this case we get for the open chain from (3.1.11)

$$\tau(q) = \frac{\zeta l^2}{3k_B T q^2} = \frac{N^2 t_0}{3\pi^2 n_q^2} = \frac{\tau_{\max}}{n_q^2} \qquad \left(q = \frac{n_q \pi}{N}, t_0 = \frac{\zeta l^2}{k_B T}\right), \tag{3.2.2}$$

and for $\omega \ll \tau_{\min}^{-1}$ the dynamic viscosity may be obtained in the following closed form from Eqn. (3.1.15):

$$\eta(\omega) = \eta_s + \frac{\rho N_{\mathrm{Av}}}{M_0 N} \frac{k_B T}{\omega} \left\{ \frac{\pi b}{1+i} \frac{1 + i \tan(\pi b) \tanh(\pi b)}{\tanh(\pi b) + i \tan(\pi b)} + \frac{i}{2} \right\}, \tag{3.2.3}$$

where

$$b = \frac{N}{2\pi} \sqrt{\frac{\omega t_0}{3}} = \frac{1}{2} \sqrt{\omega \tau_{\max}}. \tag{3.2.3′}$$

In turn, for $\omega \gg \tau_{\max}^{-1}$ we may go to the limit of large ω, getting [86]

$$\eta(\omega) = \eta_s + \frac{\pi}{4}\frac{\rho N_{\mathrm{Av}}}{M_0 N} k_B T(1-i)\frac{1}{\omega}\sqrt{\omega\tau_{\max}}, \tag{3.2.4}$$

and therefore, from $G = G' + iG'' = i\omega\eta$,

$$G' = G'' \propto (\omega\tau_{\max})^{1/2}. \tag{3.2.4'}$$

The intrinsic viscosity $[\eta]$ is obtainable in the limit $\omega \to 0$ [86]:

$$[\eta] = \frac{1}{36}\frac{N_{\mathrm{Av}}}{M_0}\frac{k_B T}{\eta_s} N t_0. \tag{3.2.5}$$

Concerning the interatomic correlation function $B(k, t)$ we may transform Eqn. (3.1.18) into an integral from zero to infinity, provided $\tau_{\min} \ll t \ll \tau_{\max}$. This gives [88, 89]

$$B(k, t) = kl^2\left\{\operatorname{erf} w(k, t) + \frac{(1/\sqrt{\pi})e^{-w^2(k,t)}}{w(k, t)}\right\},$$

$$B(0, t) = l^2\left(\frac{12t}{\pi t_0}\right)^{1/2}, \qquad \tau_{\min} \ll t \ll \tau_{\max}, \tag{3.2.6}$$

where

$$w(k, t) = \frac{k}{(12t/t_0)^{1/2}}. \tag{3.2.6'}$$

In the preceding range the diffusion term gives a negligible contribution to $B(k, t)$. Conversely, for long times it becomes the major term, and we have from (3.1.19)–(3.1.21)

$$B(k, t) \simeq B(0, t) \simeq 6Dt = 6\frac{k_B T}{N\zeta}t = \frac{6l^2}{N}\frac{t}{t_0} \qquad (t \gg \tau_{\max}). \tag{3.2.7}$$

(It is noteworthy that in the Rouse limit the diffusion constant is independent of the coil's size.) From the preceding, in the long-time limit, the characteristic time $t_{1/2}$ of the incoherent dynamic structure factor $\exp[-\frac{1}{6}Q^2 B(0, t)]$ is linked to Q by the power law

$$t_{1/2}Q^2 = \text{const} \qquad (t \gg \tau_{\max}), \tag{3.2.8}$$

whereas at shorter times the power law given in Eqn. (3.1.27), $\gamma = 0$, is recovered [88]:

$$t_{1/2}Q^4 = \text{const} \qquad (\tau_{\min} \ll t \ll \tau_{\max}). \tag{3.2.9}$$

Just as in the Rouse limit the dominant term contributing to $v(q)$ in Eqn. (3.1.9) is unity, in the Zimm limit the sum over k is much larger than unity. Going to the integral for $q \ll 1$ and substituting $\langle r^2(k) \rangle$ with its unperturbed value kl^2, we have [82]

$$\begin{aligned} v(q) &\simeq \frac{\zeta}{3\pi\eta_s l}\sqrt{\frac{6}{\pi}}\int_0^N dk\left(1 - \frac{k}{N}\right)\frac{\cos(qk)}{\sqrt{k}} \\ &= \frac{2\zeta}{3\pi\eta_s l}\left(\frac{3}{q}\right)^{1/2}\left[C(qN) + \frac{S(qN)}{2qN}\right], \end{aligned} \tag{3.2.10}$$

where $C(x)$ and $S(x)$ are the Fresnel integrals defined as

$$\left.\begin{matrix} C(x) \\ S(x) \end{matrix}\right\} = (2\pi)^{-1/2}\int_0^x dz \left\{\begin{matrix} \cos z \\ \sin z \end{matrix}\right\} z^{-1/2}. \tag{3.2.10$'$}$$

For $qN = 2\pi n_q \gg 1$ this reduces to

$$v(q) = \frac{\zeta}{3\pi\eta_s l}\left(\frac{3}{q}\right)^{1/2}. \tag{3.2.11}$$

Accordingly, Eqn. (3.1.11) gives

$$\tau(q) = \frac{\eta_s l^3}{k_B T}(3\pi)^{-1/2}\left(\frac{N}{n_q}\right)^{3/2} = \tau_{\max}/n_q^{3/2}. \tag{3.2.12}$$

At a sufficiently large frequency the dynamic viscosity may be evaluated from Eqn. (3.1.15) as an integral over n_q getting [86]

$$\begin{aligned} \eta(\omega) &\simeq \eta_s + \frac{\rho N_{\text{Av}}}{M_0 N} k_B T \frac{1}{2}\int_0^\infty \tau(q)/[1 + i\omega\tau(q)/2]\, dn_q \\ &= \eta_s + \frac{\rho N_{\text{Av}}}{M_0 N}\frac{k_B T}{\omega}(\omega\tau_{\max})^{2/3}\frac{2^{4/3}\pi}{i^{1/3}3^{3/2}} \qquad (\tau_{\max}^{-1} \ll \omega \ll \tau_{\min}^{-1}) \end{aligned} \tag{3.2.13}$$

and

$$G' \propto G'' \propto (\omega\tau_{\max})^{2/3}, \tag{3.2.13$'$}$$

in agreement with Eqs. (3.1.31) with $\gamma = 0$. For $\omega = 0$, we get the intrinsic viscosity [86]

$$[\eta] \simeq \frac{1}{2}\frac{N_{Av}}{M_0 N}\frac{k_B T}{\eta_s}\tau_{max}[1 + 2^{-3/2} + 3^{-3/2} + \cdots]$$

$$= 0.425\frac{N_{Av}}{M_0}\sqrt{N}\, l^3. \tag{3.2.14}$$

[A more accurate evaluation of the first modes with the open-chain transform [83] leads to a coefficient 0.467 instead of 0.425.] The correlation function $B(k, t)$ cannot be easily evaluated in closed form for the Zimm limit. However, it may be seen from Eqs. (3.1.18) that for sufficiently short times, $B(k, t)/N$ is a universal function of the two variables k/N and t/τ_{max}. In particular, the time dependence of the self-correlation function $B(0, t)$ is again a simple power law if t is comprised between the two extremal relaxation times. We get, from (3.1.18) and (3.2.12) [90],

$$B(0, t) = \frac{Nl^2}{3\pi^2}\left(\frac{t}{\tau_{max}}\right)^{2/3}\int_0^\infty \frac{1 - e^{-x}}{x^{5/3}}\,dx \propto t^{2/3}, \qquad \tau_{min} \ll t \ll \tau_{max}. \tag{3.2.15}$$

in agreement with (3.1.32) with $\gamma < 1$; it should be noted that this result is independent of N in view of (3.2.12). Accordingly, we have for the characteristic time $t_{1/2}$ of the incoherent dynamic structure factor

$$t_{1/2}Q^3 = \text{const}, \tag{3.2.15$'$}$$

in agreement with (3.1.33). For $t \gg \tau_{max}$ the diffusive contribution prevails again, as in the Rouse limit, and we have

$$B(0, t) = 6Dt, \qquad t_{1/2}Q^2 = \text{const}, \qquad t \gg \tau_{max}, \tag{3.2.16}$$

$$D = \frac{k_B T}{N\zeta(0)} = \frac{k_B T}{6\pi\eta_s R_H};$$

$$R_H = \left[\frac{2}{N}\sum_{k=1}^{N-1}\left(1 - \frac{k}{N}\right)\langle r^{-1}(k)\rangle_0\right]^{-1} = \frac{3}{8}\left(\frac{\pi N}{6}\right)^{1/2} l, \tag{3.2.17}$$

where R_H is the hydrodynamic radius, which turns out to be smaller than the root-mean-square radius of gyration $S = \langle S^2\rangle_0^{1/2} = 6^{-1/2}\sqrt{N}\, l$ [see Eqn. (2.1.11)]. More precisely, their ratio is

$$\frac{R_H}{S} = \frac{3}{8}\sqrt{\pi} = 0.665. \tag{3.2.18}$$

It should be pointed out that in the Zimm, or *impermeable coil*, limit the specific value of R_{eff} is uninfluential since the friction coefficient $\zeta(q) = \zeta/\nu(q)$ is independent of R_{eff}, see Eqs. (3.1.7′) and (3.1.9), $\nu(q) \gg 1$. This is not true in the *partial-draining* case, in which $\nu(q)$ is of order unity although larger than the Rouse *free-draining* limit $\nu(q) = 1$, and its full expression must be considered.

3.2.2. *The Bead-and-Spring Model in Bad and Good Solvents*

When the bead-and-spring chain is not in the ideal state, the intramolecular force is given in Eqn. (3.1.3). As it may be seen, in general, the force is not simply transmitted by first-neighboring atoms, but it has a long-range character. The relaxation times are given by Eqn. (3.1.11); after they are known, the dynamic viscosity $\eta(\omega)$ and the atomic correlation function $B(k, t)$ are obtained from Eqs. (3.1.15) and (3.1.18) (for the periodic chain), and the complex modulus and dynamic structure factors are easily constructed.

We shall examine separately the chain dynamics in the collapsed and in the expanded state utilizing our previous knowledge of the equilibrium strain function $\alpha^2(q)$ as given by Eqs. (2.2.12) and (2.2.29). Considering the expression (3.1.11) for $\tau(q)$, let us point out that, in addition to appearing explicitly, $\alpha^2(q)$ indirectly affects $\nu(q)$ as it depends on the mean-square distances $\langle r^2(k)\rangle$, which in turn are a function of $\alpha^2(q)$ [see Eqs. (3.1.9) and (2.1.66)].

Dynamics of the Collapsed Chain. As done with the unperturbed chain, we shall treat the Rouse and the Zimm limits in order.

If $\nu(q) \equiv 1$ (Rouse limit), we have $\zeta(q) \equiv \zeta$, and the general relaxation time is, from Eqs. (3.1.11), combined with (2.2.12):

$$\tau(q) = \frac{J^2}{3(1 + J^2 q^2)}\, t_0, \quad t_0 = \frac{\zeta l^2}{k_B T}, \quad J = pN(\ll N), \quad q = \frac{n_q \pi}{N}, \tag{3.2.19}$$

and it should be noted that, for $Jq \ll 1$, we have $\tau(q) = \text{const}$, whereas for $Jq \gg 1$ we have the Rouse result, see Eqn. (3.2.2).

If $\omega \ll \tau_{min}^{-1}$, the sum giving $\eta(\omega)$ in Eqn. (3.1.15) may be carried out exactly, and we get

$$\eta(\omega) = \eta_s + \frac{\rho N_{A_v}}{M_0 N} k_B T \frac{N t_0}{12(J^{-2} + i\omega t_0/6)^{1/2}} \left\{ \coth\left(N\sqrt{J^{-2} + i\omega t_0/6}\right) - \frac{1}{N(J^{-2} + i\omega t_0/6)^{1/2}} \right\}. \tag{3.2.20}$$

For $\omega = 0$ this result reduces to

$$[\eta] = \frac{1}{12}\frac{N_{\rm Av}}{M_0}\frac{k_B T}{\eta_s} J t_0, \tag{3.2.21}$$

whereas for frequencies much larger than $\tau_{\max}^{-1}$ we obtain again Eqn. (3.2.4), in keeping with the notion that the localized configurations of the chain are unperturbed even under collapse. As ω approaches $\tau_{\max}^{-1}$, the collapsed chain shows the dynamic effect of the several collective modes with q comprised between 0 and $1/J$ ($\sim N/J$ in number), having approximately the same relaxation time $\tau(q) \simeq \tau_{\max} = J^2 t_0/3$ [see Eqs. (3.2.19)]. The interatomic correlation function $B(k,t)$ may be exactly evaluated from integration of Eqn. (3.1.18) provided the time is comprised between appropriate limits. The result is [54]

$$B(k,t) = Jl^2 \left\{ 1 - \tfrac{1}{2} e^{-k/J} \left[1 - \operatorname{erf}\left(\sqrt{\frac{3t}{t_0 J^2}} - k\sqrt{\frac{t_0}{12t}} \right) \right] - \tfrac{1}{2} e^{k/J} \left[1 - \operatorname{erf}\left(\sqrt{\frac{3t}{t_0 J^2}} + k\sqrt{\frac{t_0}{12t}} \right) \right] \right\}$$

$$\left(\frac{t_0}{3\pi^2} = \tau_{\min} \ll t \ll \tau_{\max} = J^2 \frac{t_0}{3} \right), \tag{3.2.22}$$

whereas for $t \gg \tau_{\max}$ we have the purely diffusive result reported in (3.2.7), independently of collapse. It is noteworthy that, before the diffusion sets in, $B(k,t)$ becomes virtually constant and equal to Jl^2 over a time interval that is longer the larger is N/J, as a consequence of the many degenerate modes. Correspondingly, the dynamic structure factor will also decrease to a stationary nonzero value. A result identical to (3.2.22) was obtained by Ronca [89] for a sufficiently small chain portion in a polymer melt within the so-called plateau region, where these portions are effectively confined inside a definite region of space.

Going now to the Zimm limit, $\nu(q) \gg 1$, from Eqs. (2.1.66) [with $C(q) \equiv 1$], (2.2.12), and (3.1.9), we get for the periodic chain

$$\nu(q) = \begin{cases} qJ \ll 1: & 2\ln 4 \sqrt{\dfrac{6J}{\pi}} \dfrac{\zeta}{6\pi\eta_s l} \\[2ex] qJ \gg 1: & 2\left(\dfrac{3}{q}\right)^{1/2} \dfrac{\zeta}{6\pi\eta_s l}. \end{cases} \tag{3.2.23}$$

Therefore, the relaxation times are given by

$$\tau(q) \simeq \begin{cases} qJ \ll 1: & \dfrac{\pi^{3/2} l^3 \eta_s}{\sqrt{6} \ln 4\, k_B T} J^{3/2}, \\[2ex] qJ \gg 1: & \dfrac{\pi l^3 \eta_s}{\sqrt{3}\, k_B T} q^{-3/2}. \end{cases} \tag{3.2.24}$$

As was the case with the free-draining model, the relaxation spectrum of the first, collective modes is flat, whereas it is unchanged with respect to the unperturbed state for more localized modes (see Figure 6). With the open chain, analytical difficulties prevent us from obtaining a closed-form solution. Numerical calculations show that the same results hold for the open chain

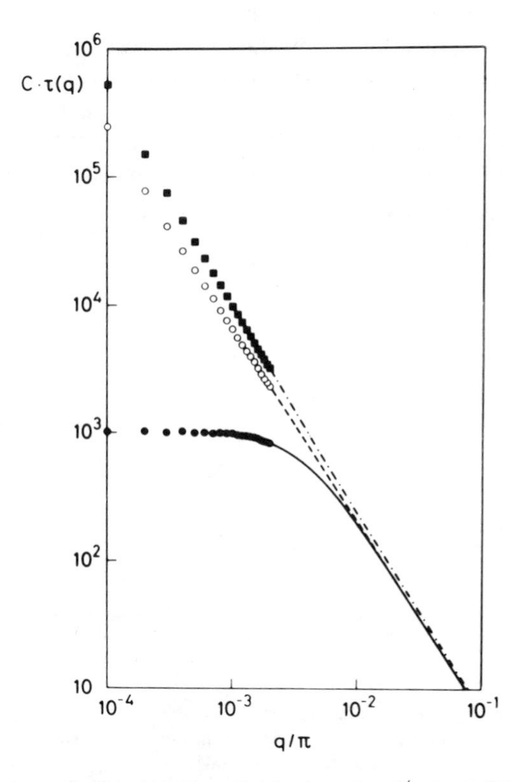

Figure 6. Spectrum of relaxation times $C\tau(q)$, where $C = \sqrt{12 k_B T / \pi^{3/2} \eta_s l^3}$ as function of q/π for bead-and-spring chain with $N = 10^4$ in Zimm limit. Empty circles and dashed line refer to unperturbed state; solid circles and continuous line to collapsed chain with $\alpha_s^2 = 0.03$; solid squares and dash-and-dot line to expanded chain with $\alpha_s^2 = 1.8$.

[55] at least if J/N is on the order of 10^{-2}, corresponding to $\alpha_S^2 \approx 0.03$. As a result, when ω^{-1} is in the vicinity of the multidegenerate $\tau_{\max}$, the results suggest the presence of a single relaxation time. The intrinsic viscosity turns out to be a function of $J^{1/2}$ only. The numerical result for the open chain was found to be [55]

$$[\eta]/[\eta]_0 = \alpha_\eta^3 = 1.52(J/N)^{1/2} = 0.88\,\alpha_S, \qquad (3.2.25)$$

showing that for strongly collapsed coils the intrinsic viscosity is proportional to the root-mean-square radius of gyration [see Eqs. (2.2.15) and (3.2.14)] unlike the good solvent state, see following.

The correlation function $B(k,t)$ cannot be evaluated in closed form in the present case. However, for t comprised between $\tau_{\min}$ and $\tau_{\max}$ numerical calculations [54] show that $B(0,t) \propto t^{2/3}$, whereas, for $t \gg \tau(q)$, $B(0,t) \propto t$ due to globule diffusion. Consequently, the $\mathscr{B}$ exponent of the relationship $t_{1/2}Q^{\mathscr{B}} = \text{const}$ has values 3 and 2 in the two respective regimes, as with the unperturbed chain [see Eqs. (3.2.15′) and (3.2.16)]. We do not obtain any plateau of $B(0,t)$ in this case, unlike the Rouse limit. Figure 7 shows the coherent dynamic structure factor $S(Q,t)$ as a function of t for two different Q values both for the unperturbed and for the collapsed chain. The two Q's correspond to observation distances $\simeq \sqrt{6}/Q$ [ref. 15, note 6] just below and

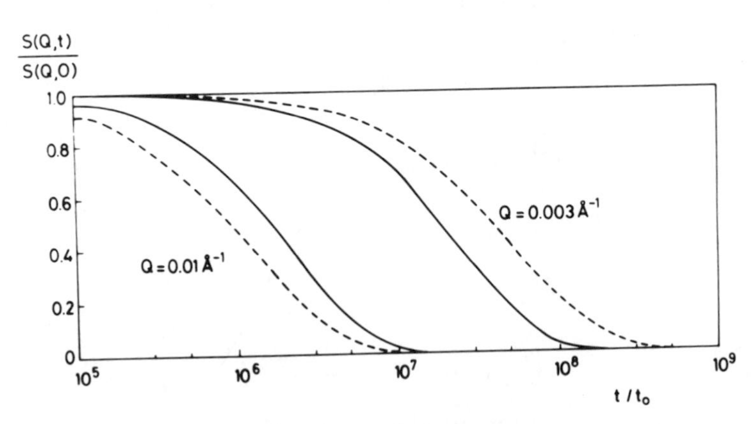

Figure 7. Coherent dynamic structure factor $S(Q,t)/S(Q,0)$ vs. t/t_0 ($t_0 = \zeta l^2/k_B T$) for atactic polystyrene in collapsed state (continuous lines) and in unperturbed state (dashed lines). The Q values correspond to observation distances just below ($Q = 0.01$ Å^{-1}) and above ($Q = 0.003$ Å^{-1}) radius of gyration of collapsed chain ($Q\langle S^2\rangle^{1/2}$ is 2.5 and 0.75, respectively). [Parameters: $N = 5 \times 10^5$ chain atoms, $J = 5 \times 10^3$, $\langle S^2\rangle_0^{1/2} = 1450$ Å, $\alpha_s = 0.17$, $R_{\text{eff}} = 0.38$ Å.] (Reprinted with permission from ref. 54, Copyright 1987, Società Chimica Italiana.)

above, respectively, the radius of gyration of the collapsed chain. In the former case the collapse slows down the chain dynamics due to the confining effect exerted by the globule walls; in the latter the collapse accelerates the dynamics because diffusion is attained quicker. In fact, before the diffusion sets in, the values of $B(0,t)$ are smaller than in the unperturbed case, although the $t^{2/3}$ dependence is essentially unchanged.

Dynamics of the Swollen Chain. Figure 8 shows the plots of $\alpha^2(q)$ for the open and the ring chain at different degrees of interaction strength $\tilde{z} = \tau B\sqrt{2\pi/q}$. The first collective modes are separately considered, whereas the following ones quickly merge on a universal curve. The results are consistent with those reported in Figure 5 and may be considered as accurate up to α^2 around 2.5. With the data reported in Figure 8 and the use of Eqs. (3.1.6) or (3.1.9) and (3.1.11), we obtain the relaxation times spectrum over this intermediate region (see Figure 6 for the Zimm limit, e.g.). It should be stressed that the present approach is suitable to address polymer configurations not reaching the asymptotic range of expansion as well as hydrodynamic regimes intermediate between the Rouse and Zimm limits, depending on the actual parameters. In the absence of analytical, closed-form expressions the results may be given numerically. The dynamic viscosity and

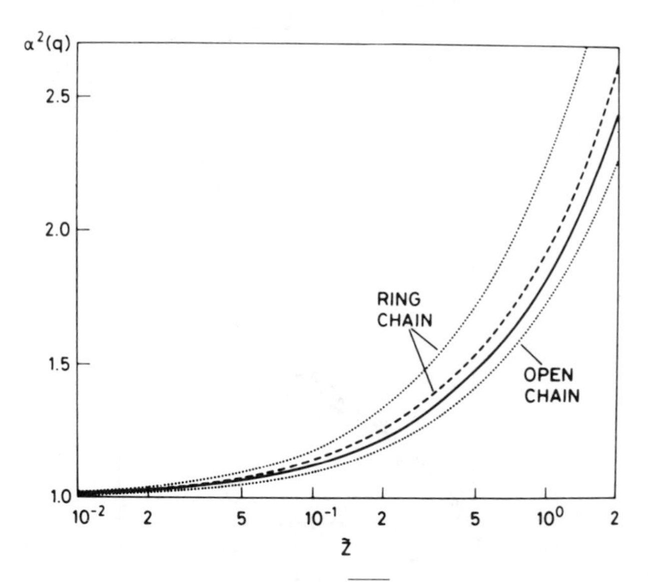

Figure 8. Expansion ratio $\alpha^2(q)$ vs. $\tilde{z} = \tau B\sqrt{2\pi/q}$. First modes are separately shown ($n_q = qN/2\pi = 1, 2$ for ring chain, $n_q = qN/\pi = 1$ for open chain), whereas following ones are well approximated by continuous line also valid for periodic chain. (From ref. 68, by permission of the publishers, Butterworth & Co. (Publishers) Ltd. ©.)

the dynamic structure factor are derivable from (3.1.15), (3.1.16)–(3.1.16′). As an example, Figure 9 shows the expansion ratio of the intrinsic viscosity for the infinite chain in the Zimm limit, as a function of $z = \tau B\sqrt{N}$, in comparison with experimental results [83].

We shall now consider the asymptotic region of large expansion; as a rough approximation, we take the logarithmic contribution to Eqs. (2.2.29) and (2.2.30) as unity [i.e., $L_n(x) \equiv 1$] and $I = H = C_\infty = 1$, so that

$$\alpha^2(q) \simeq [\tau B q^{-1/2}]^{2/5},$$
$$\langle r^2(k) \rangle \simeq (\tau B)^{2/5} k^{6/5} l^2 \qquad \left(\tau = \frac{T - \Theta}{T}\right). \tag{3.2.26}$$

From Eqs. (3.1.11) and (3.2.26), assuming $v(q) \equiv 1$, $\zeta(q) \equiv \zeta$ (Rouse limit), we have

$$\tau(q) = \tfrac{1}{3}(\tau B)^{2/5} t_0 \left(\frac{N}{\pi n_q}\right)^{11/5} = \tau_{\max} n_q^{-11/5}$$
$$\left(q = \frac{n_q \pi}{N}, \; n_q = 1, 2, 3, \ldots\right). \tag{3.2.27}$$

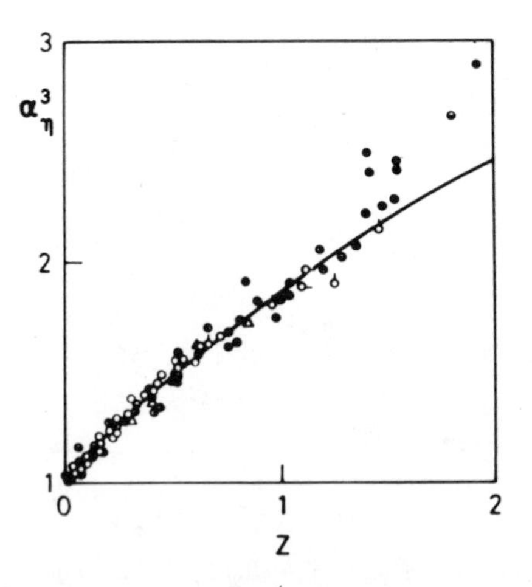

Figure 9. Plot of $\alpha_\eta^3 = [\eta]/[\eta]_0$ vs. $z = \tau B\sqrt{N}$ [from Figure 8 and Eqs. (3.1.15′), (3.1.11′), and (3.1.6)] for swollen chain in Zimm limit [83]. Experimental points from ref. 3, Section 41. (From ref. 83, by permission of the publishers, Butterworth & Co. (Publishers) Ltd. ©.)

The dynamic viscosity (3.1.15) in the two regimes $\omega = 0$ and $\tau_{\max}^{-1} \ll \omega \ll \tau_{\min}^{-1}$, respectively, is

$$[\eta] = \frac{1}{2}\frac{N_{\mathrm{Av}}}{M_0 N}\frac{k_B T}{\eta_s}\tau_{\max}[1 + 2^{-11/5} + 3^{-11/5} + \cdots]$$

$$= 0.020\,\frac{k_B T}{\eta_s}\frac{N_{\mathrm{Av}}}{M_0}(\tau B)^{2/5}\,t_0\,N^{6/5}, \qquad (3.2.28)$$

$$\eta(\omega) = \eta_s + \frac{1}{2}\frac{\rho N_{\mathrm{Av}}}{M_0 N}k_B T \int_0^\infty \tau_{\max} n_q^{-11/5} \Big/ [1 + i\omega\tau_{\max} n_q^{-11/5}/2]\, dn_q$$

$$= \eta_s + \frac{5}{11}\frac{\rho N_{\mathrm{Av}}}{M_0 N}\frac{k_B T}{\omega}\left(\frac{\omega\tau_{\max}}{2}\right)^{5/11}\left\{\frac{\pi}{2\sin(3\pi/11)} - i\frac{\pi}{2\cos(3\pi/11)}\right\}$$

$$= \eta_s + \frac{\rho N_{\mathrm{Av}}}{M_0 N}\frac{k_B T}{\omega}(\omega\tau_{\max})^{5/11}\{0.689 - i0.796\}$$

$$(\tau_{\max}^{-1} \ll \omega \ll \tau_{\min}^{-1}). \qquad (3.2.29)$$

Concerning the interatomic correlation function $B(k,t)$, we shall limit ourselves to the case with $k = 0$. From Eqs. (3.1.18), (3.2.26), and (3.2.27), we get

$$B(0,t) \propto \int_0^\infty \left\{1 - \exp\left[-\frac{3}{t_0}\left(\frac{\pi n_q}{N}\right)^{11/5}\frac{t}{(\tau B)^{2/5}}\right]\right\} n_q^{-11/5}\, dn_q$$

$$\propto t^{6/11}\int_0^\infty \frac{1 - \exp(-z)}{z^{17/11}}\, dz \propto t^{6/11}, \qquad (3.2.30)$$

whence the half-peak time $t_{1/2}$ is linked with Q by the equation $t_{1/2}Q^{\mathscr{B}} = \text{const}$, with $\mathscr{B} = \frac{11}{3}$ [91] [see Eqs. (3.1.27), $\gamma = \frac{1}{5}$].

Assuming $v(q) \gg 1$ (Zimm limit), from $\alpha^2(q) \propto q^{-1/5}$, we see from Eqn. (3.1.29) that

$$v(q) \propto q^{-2/5}, \qquad (3.2.31)$$

and from Eqn. (3.1.11)

$$\tau(q) \propto q^{-9/5}, \qquad (3.2.32)$$

equivalent to $\tau(q) = \tau_{\max} n_q^{-9/5}$, $n_q = 1, 2, 3 \ldots$. From the preceding and Eqs.

(3.1.15) and (3.1.15′) we have, performing the calculations in analogy with (3.2.28) and (3.2.29),

$$[\eta] \simeq \frac{N_{Av}}{M_0}(\tau B)^{3/5} l^3 N^{4/5} \frac{\pi^{-3/10} 6^{-1/2} 2^{-9/5}}{\varphi}(1 + 2^{-9/5} + 3^{-9/5} + \cdots)$$

$$= 0.087 \frac{N_{Av}}{M_0}(\tau B)^{3/5} l^3 N^{4/5}, \tag{3.2.33}$$

where

$$\varphi = \int_0^\infty \frac{\cos\omega}{\omega^{3/5}}\, d\omega = \Gamma(\tfrac{2}{5}) \cos\frac{\pi}{5} = 1.795;$$

$$\eta(\omega) = \eta_s + \frac{\rho N_A}{M_0 N} \frac{k_B T}{\omega}(\omega\tau_{max})^{5/9}(0.939 - i0.776) \quad \text{(ref. 86)}, \tag{3.2.34}$$

$$B(0, t) \propto \int_0^\infty \left\{1 - \exp\left[-\frac{t}{\tau_{max}} n_q^{9/5}\right]\right\} n_q^{-11/5}\, dn_q$$

$$\propto t^{2/3} \int_0^\infty \frac{1 - \exp(-z)}{z^{5/3}} dz \propto t^{2/3} \quad (\tau_{min} \ll t \ll \tau_{max}). \tag{3.2.35}$$

The last result is in agreement with the general scaling law of Eqn. (3.1.32), $\gamma = \frac{1}{5}$. The same conclusion was arrived at for the collapsed chain limited to sufficiently short times.

3.3. The Real Chains in Good Solvents: Intermediate Range

Heretofore we focused our attention on the dynamics of polymer chains under the following assumptions:

(i) Each macromolecule is a series of atomic "beads" connected by harmonic "springs."
(ii) The chain expansion in a good solvent reaches the asymptotic regime; that is, the power laws given in Eqs. (3.2.26) apply.
(iii) The hydrodynamic interaction is either absent or very strong; that is, either $\nu(q) = 1$ or $\nu(q) \gg 1$, respectively, in Eqs. (3.1.6) and (3.1.9).

In actual circumstances, these assumptions cannot be verified if the modes are sufficiently localized, as it happens whenever fast motions at a local scale are explored. Then we must recur to more realistic models that, although difficult to solve analytically in closed form, may be treated numerically with little problem. Typically, macromolecules do not lend themselves to simple dynamic modeling in the so-called *intermediate range*, which may be roughly comprised within the observation distances 5–50 Å. Above this range, the collective properties tend to prevail and the chains approach some kind of universal behavior. Below this range, the Gaussian approximation is grossly invalid, and analysis of single rotational jumps would probably be necessary, with serious theoretical and computational difficulties. Within the above range, the Gaussian approximation is acceptable [21], whereas the chains may display their "personal" behavior; namely, they are neither as flexible as a bead-and-spring chain nor as rigid as the wormlike model, neither devoid of solvent expansion nor in the asymptotic regime, and so on. Moreover, a specific dissipative effect related with the energy barriers to skeletal rotations may show up (*internal viscosity*). In conclusion, the dynamical behavior is dictated by the stereochemical structure in a subtler way than predicted by the classical theory based on two parameters (i.e., $NC_\infty l^2$ and $\tau B\sqrt{N}$ with the present symbols) plus the friction coefficient ζ. In terms of the scattering domain to be probed by quasi-elastic scattering, the intermediate range is roughly comprised between $Q = 0.05$ and $Q = 0.5$ Å^{-1}.

The equilibrium and dynamic factors that are responsible for the departure of a real polymer chain from the ideal flexibility of the bead-and-spring model will be addressed in the following sections. Then the specific cases of atactic polystyrene (PS) $\left[CH_2\text{–}CH(C_6H_5) \right]_N$ and of poly(dimethylsiloxane) (PDMS) $\left[Si(CH_3)_2\text{–}O \right]_N$, respectively regarded as examples of relatively rigid and flexible chains, will be discussed, and the importance of the preceding factors will be evaluated together with that of good-solvent expansion and of hydrodynamic interaction [14, 15].

3.3.1. *The Equilibrium Chain Rigidity as Embodied in the Generalized Characteristic Ratio*

Let us first consider the phantom chain. The sharper is $C(q)$ (see Figure 2), the larger is the number of neighboring atoms that contribute to the intramolecular force on any given atom. In particular, from Eqn. (2.1.44) we may prove that each atom exerts an elastic force on its kth neighbor with an elastic constant $12k_BT/Nl^2 \Sigma_{\{q\}} \cos(qk) \sin^2(q/2)/C(q)$, which decreases slowly with increasing k if $C(q)$ is sharply peaked. As a parallel effect, the proportionality between $\tau(q)$ and $C(q)$ shows that the relaxation times of the localized, larger-q modes decrease with increasing q much more strongly than with the bead-and-spring chain, which has $C(q) \equiv \text{const} = 1$. We may summarize the limits of highest and lowest chain flexibility in terms of $C(q)$ as given by, respectively,

$$
\begin{aligned}
C(q) &\equiv 1 && \text{(bead-and-spring chain),} \\
C(q) &= \lim_{\tau \to 0} [\tau^2/(q^2 + \tau^4/4)] && \text{(wormlike chain).}
\end{aligned}
\tag{3.3.1}
$$

It should be pointed out that, although the wormlike chain is originally defined assuming $l \to 0$, we are now retaining for l the meaning of the chain bond length; accordingly, the persistence length $2l/\tau^2$ is very large in a real chain having a wormlike behavior [92]. Equations (3.3.1) suggest again that a chain is stiffer the narrower is the origin peak of $C(q)$. Roughly speaking, the average vector product $\langle \mathbf{l}(h) \cdot \mathbf{l}(h+k) \rangle_0$ becomes small compared to unity when k is $\pi/q_{1/2}$ or more, $q_{1/2}$ being the half-peak half-width of $C(q)$ [see Eqs. (2.1.49)]; otherwise said, the longer is the range of bond correlation, the stiffer is the chain, and the sharper is $C(q)$. The usual matrix methods to account for bond rotation correlation indicate that $C(q)$ is a sum of contributions of the type $a/(1 - b\cos q)$. The term with the largest $|b|$ (< 1), usually real and positive, is the most important to determine the origin peak, whereas the other terms contribute predominantly to the background [21]. Accordingly, a reasonable representation of $C(q)$ for a few actual polymers turns out to be ($q < 1$)

$$C(q) \simeq \frac{a}{1 - b\cos q} + c \tag{3.3.2}$$

where a, b, and c are real, $a, b > 0$, and $|c| < 0.5$ (see Figure 2). Approximately incorporating the screened-interaction effect at the Θ state into C_∞, we may write $C(0) \simeq C_\infty$; from the equality $\int_{-\pi}^{+\pi} C(q)\, dq = 2\pi$ [see Eqn. (2.1.30)] and from (3.3.2), we have

$$a = 2v\sqrt{u}/(u+v), \qquad b = (u-v)/(u+v),$$

with

$$u = (C_\infty - c)^2, \qquad v = (1-c)^{1/2}. \tag{3.3.3}$$

These equations amount to two independent relationships, therefore enabling us to determine a and b if c and C_∞ are known. In view of its relative smallness, no serious error is entailed by putting $c = 0$, so that from C_∞ we have a, b and consequently the entire function $C(q)$. Our calculations on polyethylene [21], isotactic polypropylene [27], and atactic polystyrene [14] indicate $c = 0.43 \pm 0.07$ at ordinary temperatures; therefore it may be concluded that $c = 0.4$ is probably a good guess for poly-α-olefins, which leaves a single parameter to be determined, namely, the experimental value of C_∞. If $C(0) \simeq C_\infty$ is obtained from conformational analysis and matrix correlation methods, then $C(q)$ may be more accurately evaluated from Eqn. (2.1.32), with minor changes from the algorithm producing C_∞.

Since a sharp origin peak of $C(q)$ is usually associated with a large value of C_∞ (in fact, $C_\infty \simeq C(0) \propto q_{1/2}^{-1}$, if $C_\infty \gg 1$ and c is small), one might argue that

C_∞ itself is a good measure of the equilibrium rigidity. This is not accurate if chain bond angles larger than tetrahedral are present; as an example, poly(dimethylsiloxane) $\text{-}[\text{Si}(\text{CH}_3)_2\text{-O}]_N\text{-}$ has the relatively large value $C_\infty \cong 6.2$ at ordinary temperatures in spite of a large conformational flexibility as indicated by its large $q_{1/2}$ (21° vs. 8° for polystyrene). It was suggested that a more realistic figure of the chain conformational rigidity in the unperturbed state is $\sigma^2 = \langle r^2(N)\rangle_0/\langle L_{\text{fr}}^2\rangle$ instead of $C_\infty = \langle r^2(N)\rangle_0/Nl^2 (N \to \infty)$, L_{fr} being the end-to-end distance of the freely rotating chain with the same bond angles as the real chain [15]. For an $\text{-}[\text{A-}B]_{N/2}\text{-}$ chain with bond angles θ_1 and θ_2 we have [93]

$$\langle L_{\text{fr}}^2\rangle = Nl^2 (1 - \cos\theta_1)(1 - \cos\theta_2)/(1 - \cos\theta_1 \cos\theta_2). \qquad (3.3.4)$$

Comparing PDMS ($C_\infty = 6.2$, $\text{Si-}\widehat{\text{O}}\text{-Si} = \theta_1 = 143°$, $\text{O-}\widehat{\text{Si}}\text{-O} = \theta_2 = 110°$) with a hydrocarbon polymer with the same C_∞ and $\theta_1 \simeq \theta_2 = 111°$, we get σ^2 figures of 1.9 and 2.9, respectively. If the polyhydrocarbon has $C'_\infty = 9.4$, as is the case with atactic polystyrene, then σ^2 rises at 4.4, indicating a much larger equilibrium rigidity [15]. We see that $q_{1/2}^{-1}$ is roughly proportional to σ^2.

In the perturbed chain the characteristic ratio $C(q)$ is to be effectively multiplied by $\tilde{\alpha}^2(q)$, so that in the good-solvent case we have longer relaxation times [see Eqn. (3.1.11)] than in the unperturbed state, unlike the bad-solvent case. It must be remembered, though, that under collapse the Fourier modes that are significantly contracted are relatively few, being comprised between $n_q = 1$ and $n_q \simeq N/\pi J$ [see Eqn. (2.2.12), $q = n_q\pi/N$; see also Figure 6].

3.3.2. *The Dynamic Chain Rigidity: Internal Viscosity*

Whenever a chain undergoes a fast local change from a contracted to an extended conformation, an internal tension force arises that consists of a component with a constant time average and of a second component that decays quickly. As the decaying process goes on, the statistical distribution of the rotational states around chain bonds (usually gauche and trans in a hydrocarbon macromolecule) does change, approaching the equilibrium one. Departure from conformational equilibrium is intimately associated with the decaying force, henceforth denominated as the *internal viscosity force* [94, 95]. Conversely, the constant-average component is the equilibrium tension; as an example, if two chain atoms separated by k bonds are kept fixed at a distance r, the tension is $3k_B\text{T}r/\langle r^2(k)\rangle$.

If we stretch abruptly the chain ends from R to $R' > R$ through the application of an external force, supposing both R and R' to be much less than the largest chain extension, the pull will not be immediately transmitted to all chain segments, as it is apparent because the coiled configuration of the chain makes it a very soft object. As a result of an interplay between the equilibrium and the internal viscosity force, the chain tension will first be largest at the ends and will then proceed toward the chain center; since tension and distribution of rotational states are two aspects of a single phenomenon, the tension will travel at the velocity that is permitted by the rotational rearrangements [12, 94, 95]. The force and its time evolution may be obtained through nonequilibrium thermodynamical considerations, assuming a linear dependence between force and strain and their time derivatives [94]. Then, as usual in Brownian dynamics, the relationships may be extended to the spontaneous fluctuations of the chain, although being derived from assuming an external force. A Langevin equation is obtained in a form analogous to Eqn. (3.1.1) except for $\mathbf{f}(h,t)$ being changed into $\mathbf{f}(h,t) + \mathbf{\Phi}(h,t)$ in the left-hand side, where $\mathbf{\Phi}(h,t)$ accounts for the internal viscosity force. Projected along x, we have [see ref. 95, Eqs. (10), (15), and (18), $\Phi_0^*(t) \Rightarrow \Phi_x(h,t)$]

$$\dot{\Phi}_x(h,t) + \frac{\Phi_x(h,t)}{\tau_0} = \pm\frac{1}{2}V\frac{d}{dt}\left[\frac{\partial E}{\partial l_x(h,t)} + \frac{\partial E}{\partial l_x(h+1,t)}\right], \qquad (3.3.5)$$

where the sign ambiguity is related to the a priori unknown direction of propagation of the tension along the chain, V is a pure number around unity, τ_0 is the effective relaxation time for rotations around skeletal bonds, the configurational potential E is $-k_B T \ln \mathscr{W}$, and in turn the statistical weight $\mathscr{W}$ is given by Eqn. (3.1.2). Fourier transforming through $\Sigma\, e^{iqk}$, we obtain the following Langevin equation in the absence of a solvent flow gradient [12, 95]:

$$\frac{3k_B T}{l^2}\frac{\mu(q)}{\tilde{\alpha}^2(q)}\tilde{x}(q,t) \pm iV\frac{3k_B T}{C(q)l^2}\frac{\sin q}{\tilde{\alpha}^2(q)}\int_{-\infty}^{t} dt'\, e^{-(t-t')/\tau_0}\dot{\tilde{x}}(q,t') + \zeta(q)\dot{\tilde{x}}(q,t) = \tilde{X}(q,t), \qquad (3.3.6)$$

the term with the memory integral being the internal viscosity contribution $\tilde{\Phi}_x(q,t)$. The fluctuation–dissipation theorem dictating the statistical correlation among the Brownian forces reads $(t>0)$ [12]

$$\langle \tilde{X}(q,t)\tilde{X}(q',0)\rangle = Nk_B T\left\{2\zeta(q)\delta(t) \pm iV\frac{3k_B T}{C(q)l^2}\frac{\sin q}{\tilde{\alpha}^2(q)}e^{-t/\tau_0}\right\}\Delta(q+q'). \qquad (3.3.6')$$

The integral can be eliminated from Eqn. (3.3.6) after time differentiating both sides and substituting back. Then Fourier antitransformation through $N^{-1}\Sigma_{\{q\}}e^{-iqh}$ produces a differential equation in h space (remember h is the index of the general chain bond) that lends itself to an interesting interpretation. Assuming for simplicity $T = \Theta$, $\zeta(q) \equiv \zeta$ as in the free-draining model, and $q \ll 1$ so that h may be taken as a continuous variable, we get [95]

$$K\frac{\partial^2 x}{\partial h^2} - \zeta\frac{\partial x}{\partial t} \pm KV\tau_0\frac{\partial^2 x}{\partial h\,\partial t} + K\tau_0\frac{\partial^3 x}{\partial h^2\,\partial t} - \zeta\tau_0\frac{\partial^2 x}{\partial t^2} = X(h, t), \tag{3.3.7}$$

where $K = 3k_B T/(C_\infty l^2)$, $0 \leqslant h \leqslant N$. The first two terms on the left are the elastic force and the solvent friction, respectively. The third term is the most important contribution given by internal viscosity, resisting the propagation of the rotational rearrangement along the chain; the double sign refers to propagations toward either increasing or decreasing h. (A sign ambiguity is inherently associated with the first derivative with respect to the chain coordinate.) It may be shown that the Cerf–Peterlin theoretical results [96, 97] derive from making the third term imaginary with a definite sign. The fourth term may be viewed as the sum of two viscous forces along the two chain bonds adjoining the hth atom, each force having the form $K\tau_0[\dot{x}(k) - \dot{x}(h)]$, where k is either $h + 1$ or $h - 1$. This term is closely analogous to expressions proposed by Van Beek and Hermans [98], Bazua and Williams [99], and MacInnes [100] to account for internal viscosity. The last term on the left side has the same mathematical form as the inertial term for a vibrating string, although it physically derives from coupling between friction and internal viscosity. In conclusion, within the present theory the internal viscosity has a threefold effect: (i) It resists the propagation of conformational changes along the chain; (ii) it gives rise to a tensile force proportional to the strain rate; and (iii) it produces an inertialike effect.

The arbitrary sign in the last equations may be eliminated at the expense of ending up with a higher order differential equation [101]. Let us consider Eqn. (3.3.6) as an example, neglecting for a while the Brownian force $\tilde{X}(q, t)$ that will be reinstated later. If we time differentiate and multiply by τ_0, then add the result to the starting equation to get rid of the memory integral, we may write:

$$\pm\mathscr{F}\tilde{x}_\pm(q, t) = \mathscr{G}\tilde{x}_\pm(q, t), \tag{3.3.8}$$

where both $\mathscr{F}$ and $\mathscr{G}$ are linear differential operators and $\tilde{x}_+$ and $\tilde{x}_-$ are the solutions corresponding, respectively, to the plus and minus signs. It is easy to show that both of them satisfy $\mathscr{F}^2\tilde{x} = \mathscr{G}^2\tilde{x}$, or, considering again the Brownian force,

$$(\mathscr{F}^2 - \mathscr{G}^2)\tilde{x}_\pm(q, t) = \tilde{X}_\pm(q, t) \tag{3.3.9}$$

where $\tilde{X}_\pm(q, t)$ will obey a suitable form of the fluctuation–dissipation theorem, differing from (3.3.6′).

For current equilibrium applications we shall concentrate on the stationary-type function

$$\tilde{x}(q, t) = \frac{1}{\sqrt{2}}[\tilde{x}_+(q, t) + \tilde{x}_-(q, t)], \tag{3.3.10}$$

where the statistically independent components $\tilde{x}_+$ and $\tilde{x}_-$ are supposed to be identical except for the change of factors $e^{i\omega t}$ into $e^{-i\omega t}$ and vice versa. For the same reason as discussed before, $\tilde{x}(q, t)$ also obeys Eqn. (3.3.9), where $\mathscr{F}$ and $\mathscr{G}$ may be obtained from (3.3.6). It is possible to show that $\tilde{x}(q, t)$ is given by

$$\begin{aligned}\tilde{x}(q, t) = {} & A(q)\cos[\omega(q)t + \varphi(q)]\exp[-t/\tau(q)] \\ & + A'(q)\cos[\omega'(q)t + \varphi'(q)]\exp[-t/\tau'(q)],\end{aligned} \tag{3.3.11}$$

where all the quantities are real. Considering that $x(h, t) \propto \Sigma_{\{q\}}\tilde{x}(q, t)$ $Q^*(q, h)$, where $Q^*(q, h)$ is either e^{-iqh} or $\sqrt{2}\cos[q(h - \frac{1}{2})]$ and adopting the exponential form for the cosines, we see that the chain motion may be decomposed into damped oscillations traveling with a velocity $|\omega(q)$ [or $\omega'(q)]/q|$ along either chain direction. It turns out that $\tau(q) \geqslant \tau_0$, whereas $\tau'(q) \leqslant \tau_0$ [12]. For motions slow enough that $\theta \gg \tau_0$, where θ is the shortest time of observation relevant for the problem at hand, we may neglect the term accompanying $A'(q)$ in Eqn. (3.3.11). The traveling velocity $|\omega(q)/q|$ is never larger than $(V\tau_0)^{-1}$ bonds/s, which may be considered as the *natural* propagation velocity of a chain perturbation since it is reached whenever the solvent friction tends to zero [12, 94, 95]. For $|q| \ll 1$ we have [54]

$$\tau(q) \simeq \frac{\tilde{\alpha}^2(q)C(q)t_0}{3v(q)q^2}[1 + V^2\tau_0^2U^2(q)], \tag{3.3.12}$$

$$\omega(q) = V\tau_0 U(q)/\tau(q), \tag{3.3.13}$$

$$U(q) = 3qv(q)/[\tilde{\alpha}^2(q)C(q)t_0], \tag{3.3.14}$$

t_0 being given by (3.1.11′) and $v(q)$ by (3.1.6) or (3.1.9). The velocity of propagation is

$$v(q) = |\omega(q)/q| = V\tau_0 U(q)/|q|\tau(q), \qquad (3.3.15)$$

and it tends to its upper limit $(V\tau_0)^{-1}$ for $V\tau_0 U(q) \gg 1$ $(V \approx 1)$. Under this circumstance, it is interesting that the velocity is independent of q, so that any perturbation with an arbitrary shape may move coherently. With the hypothetical exception of strongly collapsed chains [54], for $q \to 0$ we have the important result $U(q) \to 0$, so that $\tau(q)$ *becomes independent of the internal viscosity*. Of course, in the limit $\tau_0 = 0$ we have $\omega(q) \equiv 0$ and $\tau(q)$ reduces to the form given in Eqn. (3.1.11) in the absence of internal viscosity. From Eqs. (3.3.12)–(3.3.15) we draw the conclusion that the internal viscosity effect is of the first order on $\omega(q)$ and $v(q)$, while being of the second order on $\tau(q)$.

The dynamic viscosity $\eta(\omega)$ may be obtained from Eqn. (3.1.13). Sending the reader to ref. 12 for the full result, we get for relatively small frequencies such that $\omega\tau_0 \ll 1$:

$$\eta(\omega) = \eta_s + \frac{\rho N_{\text{Av}}}{M_0 N} k_B T \sum_{\{q\}} \left[\frac{\tilde{\alpha}^2(q) C(q) t_0}{6 v(q) q^2} \frac{1 + i\omega V^2 \tau_0^2 U(q)/q}{1 + i\omega\tau(q)/2} \right], \qquad (3.3.16)$$

where in the periodic case the sum must be performed discarding the term $q = 0$. For $\omega = 0$ this result reduces to the usual form $[\eta] = (N_{\text{Av}} k_B T / M_0 N \eta_s) \Sigma_{\{q\}} \tau(q)/2$, but $\tau(q)$ is now given by Eqn. (3.1.11) and does not contain the internal viscosity contribution; accordingly, we conclude that the *intrinsic* viscosity $[\eta]$ is also independent of the *internal* viscosity. Conversely, for large ω the sum in Eqn. (3.3.16) tends to a real quantity, indicating that the polymer may become very stiff under high frequencies (provided $\omega < \tau_0^{-1}$), in agreement with several experimental results [8]. As an example, Figure 10 shows the full results of our calculations of the real and imaginary components of the complex modulus $G(\omega) = i\omega\eta(\omega)$ for atactic polystyrene, compared with experimental results obtained by Massa, Osaki, Schrag, and Ferry [102, 103].

By comparison of the dynamic viscosity expressions without and with internal viscosity [see Eqs. (3.1.15) and (3.3.16), respectively], we see that in the former case the sum tends to zero for large ω, unlike in the latter; we conclude that in the presence of internal viscosity the dynamic viscosity deviates from what is commonly regarded as a general law. The reason lies in the fact that with internal viscosity the intramolecular tension contains a contribution depending on $\dot{x}(h, t)$, unlike the other models where it depends on the elastic force only, that is, on $x(h, t)$.

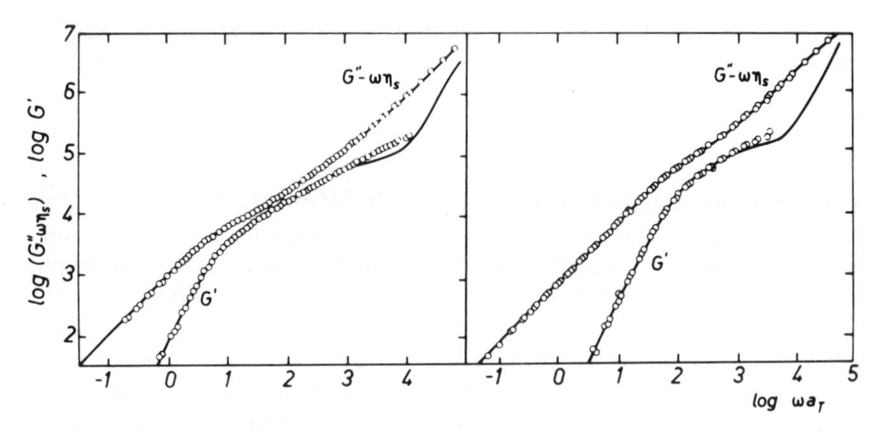

Figure 10. Real and imaginary part of complex modulus, G' and $G'' - \omega\eta_s$, vs. ωa_T for atactic polystyrene in dilute solution; a_T is shift factor. Experimental points are from refs. 102 and 103; best-fit continuous lines [from Eqn. (34) of ref. 12] superimposed on experimental points after rigid, parallel shift. [Model assumptions and parameters: unperturbed periodic chain, $N = 8000$ (left) and $N = 1300$ (right), $\tau_0/t_0 = 47$, $R_{eff} = 0.125$ Å.] (Reprinted with permission from ref. 12, Copyright 1981, American Chemical Society.)

The interatomic correlation function $B(k, t) = \langle [\mathbf{R}(k, t) - \mathbf{R}(0, 0)]^2 \rangle$ has a relatively simple form if the shortest time interval of actual interest $\theta \gg \tau_0$. Then we may neglect the term accompanying $A'(q)$ in Eqn. (3.3.11) and get [14]

$$B(k, t) = 3\langle [x(k, t) - x(0, 0)]^2 \rangle = \frac{2l^2}{N} \sum_{\{q\}} \tilde{\alpha}^2(q) C(q)$$

$$\times \{1 - \exp[-t/\tau(q)] \cos[\omega(q)t] \cos(qk)\}/q^2. \qquad (3.3.17)$$

By comparison with Eqn. (3.1.18), we see that the internal viscosity not only modifies $\tau(q)$, but also produces the new term $\cos[\omega(q)t]$ originating from the oscillatory relaxation mechanism. Figure 11 shows results of $B(k, t)$ for PS corresponding to the data of Figure 10 for different values of k (the full expression may be found in ref. 12); results in the absence of internal viscosity are also reported. The slowing of the intramolecular motion produced by internal viscosity for $t \approx \tau_0$ is evident.

It has often been remarked from experimental data that the internal viscosity effect appears to be approximately proportional to the solvent viscosity [102, 103]. This suggests that the molecular mechanism of the rotational rearrangements may be considered as an example of Kramer's

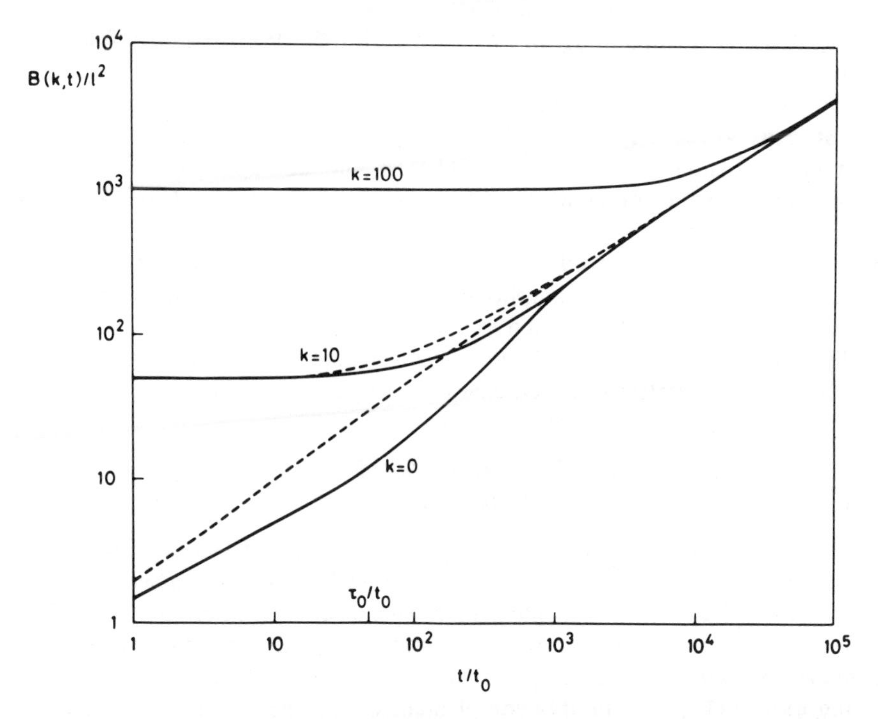

Figure 11. Dynamic correlation function $B(k, t)$ [from Eqn. (41) of ref. 12] as function of t/t_0 for three values of k as indicated on curves in presence of internal viscosity. Result with $\tau_0 = 0$ (i.e., with no internal viscosity) are given by dashed lines. [Model assumptions and parameters: periodic chain same as in Figure 10, $N \gg 100$.]

kinetic theory in the diffusive limit [104], which predicts

$$\tau_0 \propto \eta_s \exp(\Delta/k_B T), \tag{3.3.18}$$

where Δ is the effective energy barrier. Careful studies of the most probable such mechanisms operating in a long chain were performed by Helfand and co-workers [105–107], one important result being that Δ turns out to be on the order of the barrier height for a single skeletal rotation, unlike the prediction of the so-called crankshaft model, for example, which suggests twice as much. This result was experimentally confirmed by Morawetz with spectroscopic measurements on suitably labelled polymers [108, 109]. Concerning other predictions of Helfand's theory, it may be pointed out that the chain relaxation mechanism following a single skeletal rotation tends to a classical time exponential [107] in the long-time limit, in basic agreement with Eqn. (3.3.5) if $\Phi_x(k, t)$ is considered as linearly related to the atomic coordinates even in the case of single-rotation steps. It should be pointed out that the Kramers-activated temperature dependence of τ_0 gives rise to a new temperature effect in addition to the usual dependence of the friction coefficient *via* solvent viscosity. As a consequence, departures from the Williams–Landel–Ferry equation [110] should be expected whenever we enter the frequency range where polymer stiffness is significantly increased by internal viscosity.

As a concluding remark on this subject, we point out that the present approach rests upon the simplifying assumption that the chain rotational rearrangements may be reduced to a single, "average" mechanism displaying the characteristic time τ_0. Reality is likely to be more complex, and the existence of a spectrum of such characteristic times should be accounted for in a more sophisticated theoretical approach.

3.3.3. *Examples of a Rigid and a Flexible Polymer: Polystyrene and Poly(dimethylsiloxane)*

Both these polymers have an $-[A-B]_N-$ structure [14, 15], so their characteristic ratio $C(q)$ was obtained through the procedure outlined in Section 2.1.2; see Eqs. (2.1.33)–(2.1.35) in particular. Since we were interested in stereoirregular (i.e., atactic) polystyrene for comparison with experimental data, the matrix procedure based on parameters proposed by Yoon, Sundararajan, and Flory [111] was suitably complemented with the pseudostereochemical equilibrium algorithm, which allows units of opposite configuration to be formally interconvertible with fixed relative amounts [36]. In the temperature range 30–70°C the results may be fairly well expressed by the following analytical forms:

$$\begin{aligned} \text{PS}(-[\text{CH}_2\text{–CH}(\text{C}_6\text{H}_5)]_N-)\text{:}\quad & C(q) = \frac{0.0899}{1 - 0.990\cos q} + 0.365 \\ & [C(0) = 9.36,\ q_{1/2} = 8°], \\ \text{PDMS}(-[\text{O–Si}(\text{CH}_3)_2]_N-)\text{:}\quad & C(q) = \frac{0.4312}{1 - 0.932\cos q} - 0.152 \\ & [C(0) = 6.19,\ q_{1/2} = 21°]. \end{aligned} \tag{3.3.19}$$

(The amount of meso and racemic dyads in PS was 0.43 and 0.57, respectively. The conformational parameters for PDMS were taken from ref. 112.) The strong difference in the half-peak widths $q_{1/2}$ is a clear indication of a higher conformational rigidity for PS, which is also suggested by the characteristic ratios relative to the freely rotating chain $\sigma^2 = \langle r^2(N)\rangle_0/\langle L_{\text{fr}}^2\rangle$ [see Eqn. (3.3.4)], the value of which is 4.4 and 1.9 for PS and PDMS, respectively.

The excluded-volume parameter β_{eff} for PS was derived from viscosimetric data in good solvent collected by Nyström and Roots for different molecular masses [113]. The internal viscosity characteristic time τ_0 was obtained from best-fit data [12] of mechanical–dynamical results due to Massa, Schrag, Ferry, and Osaki [102, 103] (see Figure 10). In analogy with what was previously found by other authors, notably Kirkwood and Riseman on the same polymer (PS) with different solvents [19], fitting the experimental data seems to require two different values of $R_{\text{eff}} = \zeta/6\pi\eta_s$, one to obtain $v(q)$ through Eqn. (3.1.9) and a second one to evaluate $t_0 = \zeta l^2/k_B T$; the latter figure is about three times larger. Neglecting possible concentration effects, this finding suggests that either the preaveraged approximation for the hydrodynamic interaction or Stokes' law or both are not fully adequate to describe the motion of polymer segments through solvent molecules of a comparable size. The Stokes' law problem was theoretically investigated by Brey and Gómez Ordóñez [114], among others; they found that the friction coefficient ζ of a solute molecule with a radius R is less than expected from $\zeta = 6\pi\eta_s R$ if the solvent molecules have a radius $R' \approx R$. However, it is noteworthy that a single parameter $R_{\text{eff}} = 0.38$ Å gives values of t_0 that fit the theoretical results [from Eqs. (3.1.16) and (3.3.17)] with $S_{\text{coh}}(Q, t)$ neutron-scattering data over about one Q decade ($0.03\ \text{A}^{-1} < Q < 0.3\ \text{Å}^{-1}$) and two time decades.

Half-peak times of $S_{\text{coh}}(Q, t)$ for PS in different solvents and at different temperatures are plotted versus Q in Figure 12; both $\zeta(= 6\pi\eta_s R_{\text{eff}})$ and τ_0 [see Eqn. (3.3.18)] are suitably adjusted for the different conditions. Assuming now as a first approximation $S_{\text{coh}}(Q, t) \propto S_{\text{inc}}(Q, t) = \exp[-\frac{1}{6}Q^2 B(0, t)]$ and taking $B(0, t) \propto t^{\alpha}$, we have

$$\mathscr{B} = \frac{2}{\alpha} = \frac{-\partial \ln t_{1/2}}{\partial \ln Q}, \qquad t_{1/2} Q^{\mathscr{B}} = \text{const.} \tag{3.3.20}$$

This result gives the exponent $\mathscr{B}$ as the slope of the curves reported in Figure 12. It is remarkable that $\mathscr{B}$ is everywhere smaller than 3, thus being outside the range 3–3.7, characteristic of the swollen bead-and-spring chain in the Zimm and Rouse limits, respectively [see Eqs. (3.1.33) and (3.1.27), $\gamma = \frac{1}{5}$]. Only for the wormlike chain may $\mathscr{B}$ become as small as $\frac{8}{3} = 2.67$ [see Eqn. (3.1.33)], that is, well above the value of 2 observed for PS in the higher Q range. In principle, such a small value might be explained by the hypothesis that we are observing the purely diffusive motion of the individual scatterers; however, this is in contrast with the average observation length $d = \sqrt{6}/Q$ being much larger than the C–C bond length ($d > 8$ Å for $Q_{\min} = 0.3\ \text{Å}^{-1}$, $l_{\text{C–C}} = 1.54$ Å). Actually, internal viscosity offers the only possible explanation for this low $\mathscr{B}$ value; we stress the conclusion that both the equilibrium

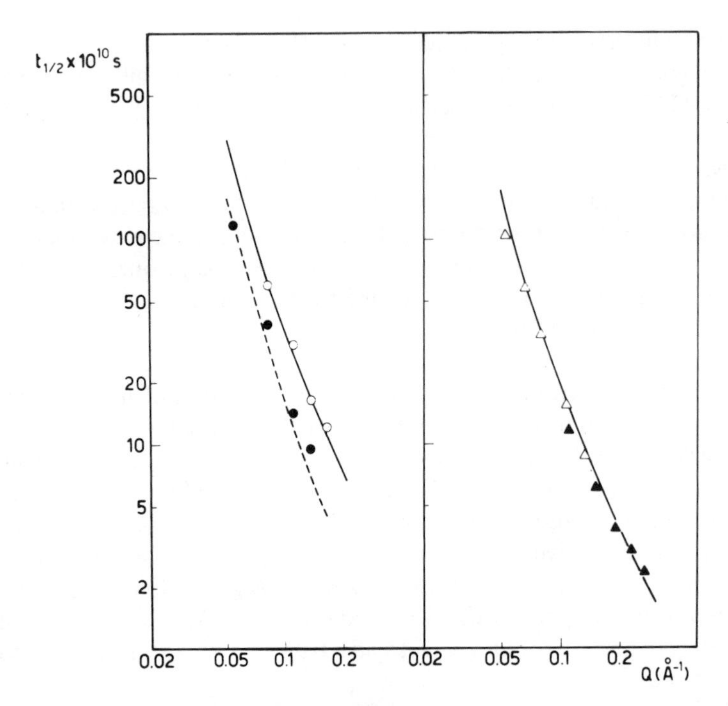

Figure 12. Half-peak time $t_{1/2}$ of coherent dynamic structure factor for atactic polystyrene plotted vs. Q. Left, C_6D_6 solution; right, CS_2 solution; continuous lines, calculated results at $T = 30\,°C$; dashed line, at $T = 70\,°C$. Experimental points from ref. 14. [Model assumptions and parameters: same as in Figures 10 and 11, $\tau B = 0.012$.] (Reprinted with permission from ref. 14, Copyright 1984, American Chemical Society.)

and the dynamic chain rigidity [i.e., $C(q)$ and τ_0, respectively] lead to a decrease of $\mathscr{B}$ below the values expected for the bead-and-spring model.

The preceding conclusions may be suitably checked upon comparison with PDMS. We send the interested reader to ref. 15 for the choice of the parameters. Unlike the case of PS, a molten polymer sample was also considered, in which case the hydrodynamic interaction was assumed to vanish [i.e., $v(q) \equiv 1$] because of the hydrodynamic screening exerted by the polymer chains. In view of the apparently low energy barriers to the rotation around Si–O chain bonds, we assumed the internal viscosity to be absent, that is, $\tau_0 = 0$. Incidentally, we remark the difference from the case of polystyrene where, in addition to the intrinsic rotation barrier around C–C bonds adjoining tetrahedral-coordinated atoms (~ 3 kcal/mol), the side phenyl rings contribute significantly to the rotational hindrance. In Figure 13 the characteristic times $t_{1/2}$ [$t_{3/4}$ for the melts [115]] are plotted versus Q.

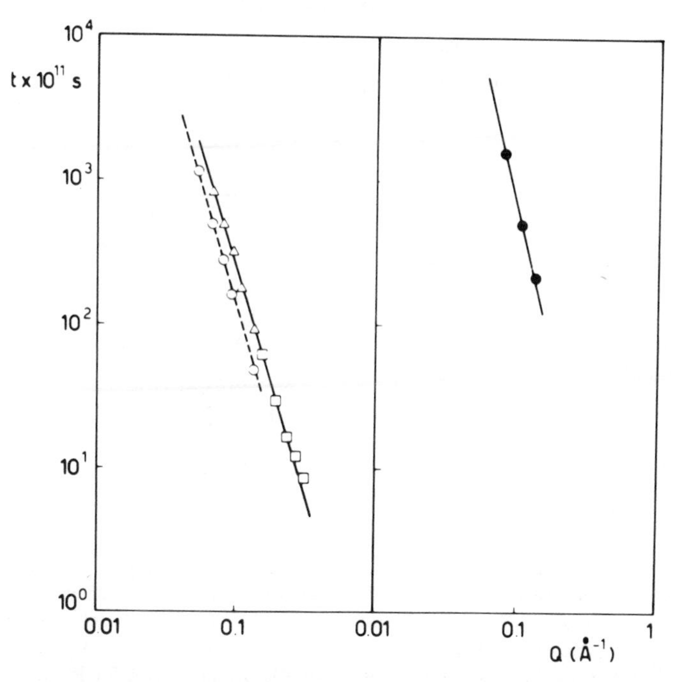

Figure 13. Left: half-peak time $t_{1/2}$ of coherent dynamic structure factor plotted vs. Q for poly(dimethylsiloxane) in C_6D_6. Continuous line, calculated results at $T = 30\,°C$; dashed line, at $T = 70\,°C$. Experimental points are from ref. 115. Right: characteristic time $t_{3/4}$ vs. Q for molten sample at $T = 100\,°C$. Experimental points are from ref. 115. [Model assumptions and parameters: periodic chain, $\tau_0 = 0$, $R_{eff} = 0.30$ Å, $\tau B = 0.018$ for solutions, $R_{eff} = 0$Å, $\tau B = 0$ for melt.] (Reprinted with permission from ref. 15, Copyright 1985, American Chemical Society.)

The exponent $\mathscr{B}$ is around 3 for the solutions and around 4 for the melts; comparison with Figure 12 shows that these exponents are larger than for PS at corresponding ranges of Q, in agreement with the higher conformational and dynamic rigidity of the latter polymer. Figure 14 shows a comparison between the plots of the characteristic time $t_{3/4}$ defined as $S_{coh}(Q, t_{3/4}) = \frac{3}{4}S_{coh}(Q, 0)$, versus a suitably normalized Q coordinate for melts of PS, PDMS, and PE (polyethylene). The last polymer is reported for comparison and is to be considered as intermediate between the other two; in fact, $C(q)$ is closer to that of PS, whereas τ_0 is put equal to zero, like PDMS and unlike PS. The slopes of the curves give $\mathscr{B} - 2$; the joint effect of the conformational and dynamic rigidity makes $\mathscr{B}(\mathrm{PS}) < \mathscr{B}(\mathrm{PE}) < \mathscr{B}$ (PDMS) in the intermediate range, represented in Figure 14 by the values of Q close to $Q\,l_n = 1$.

An important issue is what is the relative weight of the different factors in determining $\mathscr{B}$: (i) solvent expansion; (ii) hydrodynamic interaction; (iii)

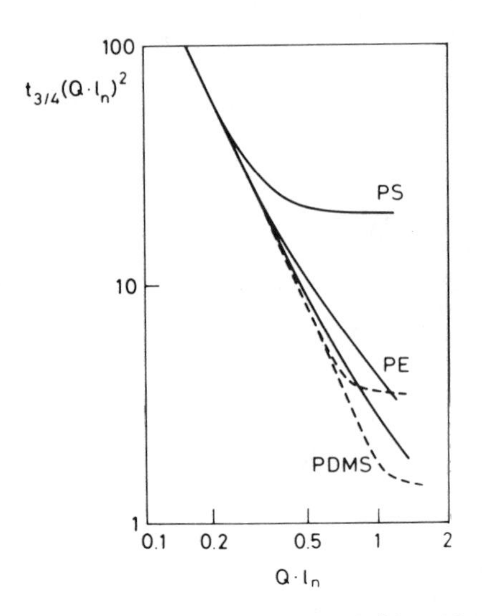

Figure 14. Characteristic times $t_{3/4}$ plotted as $t_{3/4}\,(Q\,l_n)^2$ in arbitrary units as function of normalized scattering vector $Q\,l_n$ for coherent scattering from melts. Dashed lines, calculated Rouse Results. Curves shifted so that they merge at low $Q\,l_n$. [Model assumptions and parameters: suppression of hydrodynamic interaction and of good solvent expansion, $l_n = [\langle L_{fr}^2 \rangle / N]^{1/2}$, see Eqn. (3.3.4), l_n is 2.99 Å for PDMS and 2.24 Å for PS and PE; Rouse model, $C(q) \equiv$ const, statistical segment length 11.5 Å for PDMS and 12.8 Å for PE.] (Reprinted with permission from ref. 15, Copyright 1985, American Chemical Society.)

deviation of $C(q)$ from a constant value, that is, equilibrium deviation from the bead-and-spring model; and (iv) internal viscosity, that is, dynamic deviation from the bead-and-spring model; note that the last two factors are nothing but the conformational and the dynamic rigidity, respectively. Table 1 shows the effect upon $\mathscr{B}$ due to the suppression of a single factor at a time for the case of PS [14]. The modest importance of the excluded-volume expansion and of $C(q)$ in this intermediate range was checked in the PDMS case and may be considered as a general result. [$C(q)$ would become significant for PS toward $Q = 0.5$ Å^{-1}.] Hydrodynamic interaction is especially important toward the lower Q end, whereas τ_0 gives an average decrease of $\mathscr{B}$ of about 0.5 units, the decrease being more pronounced at higher Q.

It must be noted that differences of the exponent between a flexible and a rigid chain are also reflected in the *shape of a single line* $S(Q, t)$ versus t at constant Q [here $S(Q,t) = S_{coh}(Q,t) \approx S_{inc}(Q,t)$]. From $S(Q,t) = \exp[-\frac{1}{6}Q^2 B(0,t)]$ and $B(0,t) \propto t^{2/\mathscr{B}}$, we get

$$\ln S(Q,t) = -Q^2 C t^{2/\mathscr{B}}. \tag{3.3.21}$$

TABLE 1
Values of Calculated Exponent $\mathscr{B}$ in $t_{1/2}Q^{\mathscr{B}}$ = const for coherent Scattering from Polystyrene Solutions at $T = 30\,°C$ and Variations ($\Delta\mathscr{B}$) Induced By Suppression of Single Factors

	$\mathscr{B}$	$\Delta\mathscr{B}$			
Q Range ($Å^{-1}$)	(All Factors Included)	No Good-Solvent Expansion	No Hydrodynamic Interaction	No Variation of C(q) [$C(q)$ = const]	No Internal Viscosity
$0.03 < Q < 0.1$	2.9	~0	+0.8	~0	+0.3
$0.1 < Q < 0.2$	2.3	~0	+0.5	+0.1	+0.6
$0.2 < Q < 0.3$	2.0	~0	+0.2	+0.2	+0.7

The value of $\mathscr{B}$ is smaller for a rigid polymer than for a flexible polymer; accordingly, it produces a smaller upward concavity of the plots of $\ln S(Q, t)$ versus $t(\mathscr{B} \geqslant 2)$. This is apparent from Figure 15, where two experimental lineshapes relating to PS and PDMS are reported for the same Q. The PS line is nearly straight, in agreement with the value of $\mathscr{B}$ being close to 2.

Equations (3.3.12)–(3.3.14) show that in the presence of internal viscosity the relaxation times are not merely proportional to t_0 [i.e., to the ratio $\zeta/T \propto \eta_s/T$, see Eqs. (3.1.7') and (3.1.11')] but depend on τ_0/t_0 as well. Otherwise said, the logarithmic spectrum of the relaxation times is not merely shifted by a term $\ln(\eta_s/T)$ after a temperature change. From Eqn. (3.3.18) we expect that a temperature increase makes $\tau_0/t_0 \propto T\exp(\Delta/k_B T)$ to decrease (for $\Delta > k_B T$), thereby leading to an *increase* of the exponent $\mathscr{B}$. This agrees with neutron-scattering results by Allen, Higgins, Maconnachie, and Ghosh on melts of PDMS, poly(propylene oxide) (PPO), poly(tetrahydrofuran) (PTHF), and polyisobutylene (PIB) [116]; see Table 2. Within the same Q range from 0.1 to 1 $Å^{-1}$, for all the polymers with the exception of PDMS, an increase with temperature of the average $\mathscr{B}$ value is observed, which is largest for PIB. While it is reasonable to expect the rotational energy barrier for PDMS to be very small, it is consistent with stereochemical expectation that it should be highest with the strongly hindered PIB. The observed increase of $\mathscr{B}$ for PIB is roughly of the same amount as we calculate for PS (Δ = 3.7 kcal/mol) for the same temperatures.

3.4. Dynamics of Melts: Chain Reptation

In a melt or in a concentrated solution each polymer chain is basically unperturbed and shielded from hydrodynamic interaction by the surrounding chains. Accordingly, if the medium embedding each chain could be assimilated to a structureless continuum, polymer dynamics should be well described by the so-called Rouse model. Actually, this is true only if the

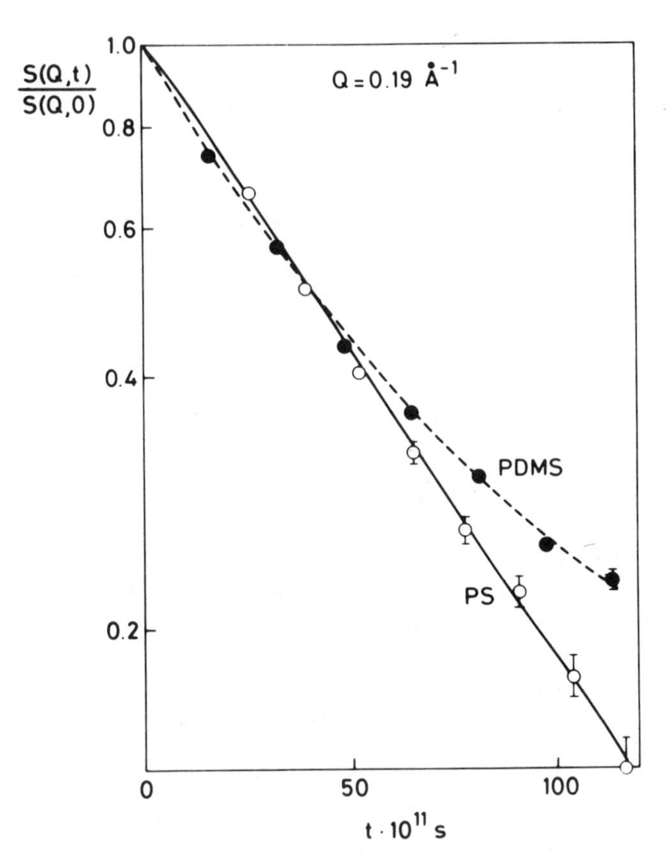

Figure 15. Dynamic coherent-scattering lineshapes for PS and PDMS at $Q = 0.19$ Å^{-1} in different solvents. Lines are guides for eye. Time scale on abscissa refers to PS in CS_2 at $T = 30$ °C, whereas that of PDMS in C_6D_6 at the same temperature scaled so that the two half-peak times coincide.

molecular length does not exceed a limit, say N_c, beyond which the chain self-diffusion and the longest relaxation time change as N^{-2} and $(N)^{\sim 3}$, respectively, instead of N^{-1} and N^2. These results may be rationalized by the "reptation" model, assuming that each chain may only slide along the axis of an embedding "tube" the diameter of which is determined by the concentration of the surrounding entanglements. More specifically, we assume that all the (average) distances entering the dynamic definitions be evaluated along the tube axis, that is, the only free coordinate permitted to the chain as a whole, according to the formalism developed in Section 2.1.4. Let us write the relaxation time $\tau(q)$ of a collective mode from Eqn. (3.1.11), considering that

TABLE 2
Experimental Exponent $\mathscr{B}$ for Different Polymer Melts at Several Temperatures[a]

Polymer	T(°C)	$\mathscr{B}$
PDMS	20	4
	150	4
PPO	67	3.0
	150	3.4
PTHF	73	3.1
	150	3.5
PEO	68	3.2
PIB	73	2.2
	150	2.9

[a] From ref. 116. The Q range is centered at 0.5 Å^{-1}. Abbreviations: PDMS, poly(dimethylsiloxane); PPO, poly(propylene oxide); PTHF, poly(tetrahydrofuran); PEO, poly(ethylene oxide) PIB, polyisobutylene.

$\mu(q) \simeq q^2/C_\infty$ and $v(q) \equiv 1$ in the present case; we have

$$\tau(q) = \frac{t_0 \alpha^2(q) C_\infty}{3q^2}. \tag{3.4.1}$$

According to the preceding, if the motion is restricted to sliding along the tube, $\alpha^2_{\text{tube}}(q)$ must be evaluated using the Fourier modes constructed with the bond projections $\lambda(h) = \mathbf{l}(h) \cdot \boldsymbol{\gamma}(h)$ (see Section 2.1.4) along the tube axis. In analogy with the open-chain transform and using the approximation $\sin[q(h-1)] \simeq \sin(qh)$ because $q \ll 1$, we have

$$\begin{aligned}\tilde{\lambda}(q) &= \sqrt{2} \sum_{h=1}^{N} \lambda(h) \sin(qh) \\ &= \sqrt{2} \sum_{h=1}^{N} [\mathbf{l}(h) \cdot \boldsymbol{\gamma}(h)] \sin(qh),\end{aligned} \tag{3.4.2}$$

whence

$$\alpha^2_{\text{tube}}(q) = \langle \tilde{\lambda}^2(q) \rangle / [\tfrac{1}{3} N C_\infty l^2], \tag{3.4.3}$$

the denominator being the chain mean-square end-to-end distance projected along any Cartesian coordinate. Carrying out the calculations using Eqs. (2.1.27) and (2.1.55), where e^{-iqh} is substituted by $\sqrt{2}\sin(qh)$ in the open-chain transform, remembering $q = n_q\pi/N$, we get

$$\alpha^2_{\text{tube}}(q) \begin{array}{ll} \simeq 24Nq_D/(n_q^2\pi^3) & \text{for } n_q = 1, 3, 5, \ldots, \\ = O(1) & \text{for } n_q = 2, 4, 6, \ldots, \end{array} \tag{3.4.4}$$

where the Fourier cutoff q_D is directly and inversely proportional to the tube mean-square length and thickness, respectively. Accordingly, we have for the relaxation times from (3.4.1)

$$\tau(q) \propto \begin{cases} N^3 q_D/n_q^4 & \text{for } n_q = 1, 3, 5, \ldots, \\ N^2/n_q^2 & \text{for } n_q = 2, 4, 6, \ldots. \end{cases} \tag{3.4.5}$$

Let us define as N_e the number of bonds in a chain segment comprised between two consecutive "entanglement" points, namely, two points where we assume the chain is forced to pass through because of the constraints exerted by the surrounding chains. Since the resolution distance D of our tube image [see Eqn. (2.1.50)] is conceptually close to the distance between entanglement points, we may write as a definition

$$q_D = \pi/N_e, \tag{3.4.6}$$

D being the distance between atoms separated by $2\pi/q_D = 2N_e$ chain bonds. With this substitution, Eqs. (3.4.5) agree with Eqn. (15) of ref. 117 in the limit of large N.

Following Kavassalis and Noolandi [117], let us now consider another, equivalent parameter $\bar{N}$ instead of N_e, $\bar{N}$ being the number of nontail segments belonging to surrounding chains that effectively constrain a chain segment comprised between consecutive entanglement points. It seems reasonable to expect that $\bar{N}$ is independent of the molecular length N; as a consequence, if N is small, the entanglement length N_e must increase because the surrounding segments are more likely to belong to chain tails, in which case they are ineffective as a constraint. As a result, given N and $\bar{N}$, N_e must obey the equation

$$\bar{N} + 1 = \tfrac{1}{6}\pi\phi N_e^{1/2}(1 - N_e/N), \tag{3.4.7}$$

where ϕ is the polymer volume fraction. The equation shows that there is a critical chain length, say N_c, below which no solution exists for N_e. We have

$$N_c = 3^5(\bar{N} + 1)^2/(\pi\phi)^2. \tag{3.4.8}$$

If N is just larger than N_c, the entanglement length N_e turns out to be larger than the asymptotic value for $N \to \infty$, and the dependence between N_e and N may be expressed by $N_e \propto N^{-\sigma(N)}$, where σ (> 0) decreases with N. This modifies the asymptotic dependence $\tau(q) \propto N^3$ suggested by Eqs. (3.4.5). Since $q_D \propto N_e^{-1} \propto N^{+\sigma(N)}$, we have

$$\tau(q = \pi n_q/N) \propto N^{3+\sigma(N)}/n_q^4 \quad (n_q \text{ odd}). \tag{3.4.9}$$

In conclusion, the exponent of N is > 3, in agreement with experimental results [8, 25], and it is expected to approach that value in the thermodynamic limit. Conversely, for $N < N_c$ the chain behaves according to the Rouse predictions in the unperturbed case (see Section 3.2.1).

4. CONCLUDING REMARKS

A consideration to be stressed after the present analysis is that the classical two-parameter model of chain configurations is only valid at temperatures equal to or above the Θ point and at large-scale distances; in the present notation the two parameters may be conveniently chosen as $NC_\infty l^2$ and $\tau B\sqrt{N}$, $\tau = (T - \Theta)/T$. If distances between atoms separated by less than about 100 skeletal bonds are investigated, the local chain rigidity, controlled by short-range forces and embodied within this theory in the generalized characteristic ratio $C(q)$, should be taken into account (see Section 2.1.2). This is especially true if the chain is very rigid, approaching the wormlike behavior (Section 2.1.3). At this scale of distances the chain is little influenced by a temperature change of a few degrees around the Θ point. In the respective cases of expansion ($T > \Theta$) and of collapse ($T < \Theta$), for $N \to \infty$, we have $\langle S^2 \rangle \propto N^{6/5}$, apart from logarithmic factors, and $\langle S^2 \rangle \propto N^{2/3}$, S being the chain radius of gyration, whereas $\langle S^2 \rangle \propto N$ at $T = \Theta$. Both expansion and collapse are produced by two-body, solvent-mediated long-range forces, but in the latter case shorter-range repulsive forces resist the collapse; these repulsions arise from three-body interactions and from two-body screened interactions due to the chain thickness. Their overall strength determines whether the collapse transition is of the first order or of the second order in the high-dilution limit. In the expansion case the whole set of the mean-square distances between atoms separated by k bonds (k large) change according to the power dependence $k^{6/5}$; this is not so with the collapse, in which case they are either uniform if k is larger than a limit dictated by the globule's size or they are proportional to k as in the unperturbed state if k is smaller than this limit. This strong dissymmetry between expansion and collapse is illustrated in Figure 6 from the dynamical viewpoint. During expansion all the relaxation times are increased according to the same power law. Conversely, in the collapse case all the relaxation times belonging to short-range, large-q Fourier modes are unchanged from the unperturbed values, whereas they merge into a single value if the mode wavelength encompasses chain portions long enough to reach opposite walls of the globule.

The chain equilibrium properties are obtained from a self-consistent approach through free-energy minimization. Once the equilibrium properties are given, the chain dynamics may be derived from the additional knowledge of the friction coefficient, thus making the theory fully self-contained. Only in case the rotational barriers are sufficiently large, such as in sterically hindered polymers like polystyrene and polyisobutylene, the theory needs further specification of the internal viscosity characteristic time τ_0 (see Section 3.3.2). In the intermediate range of observation distances roughly comprised between 5 and 50 Å the equilibrium rigidity, corresponding to a sharply peaked $C(q)$, adds to the dynamic rigidity, corresponding to large τ_0, in producing the overall chain stiffness. Considering the power law $t_{1/2}Q^{\mathscr{B}} = \text{const}$, where $\mathbf{Q}$ is the scattering vector and $t_{1/2}$ is the half-peak time of the dynamic structure factor, the larger is the chain stiffness, the smaller is the value of $\mathscr{B}$, which may be as small as 2.

As a last remark, we remind that the Gaussian approximation is strictly consistent with the linear character of the dynamic equations, from a logical viewpoint. In turn, the Gaussian approach is essentially rigorous both at $T = \Theta$ and at $T < \Theta$ in the long-distance range, whereas it is more or less inexact at $T > \Theta$ over long distances or at any temperature for an interatomic separation smaller than ~ 50 bonds. Concerning the Gaussian approximation in dynamics, in analogy with the equilibrium case we may roughly accept it whenever the observation time t is long enough that the investigated root-mean-square distances $\langle [\mathbf{R}(h,t) - \mathbf{R}(j,0)]^2 \rangle^{1/2}$ are larger than 5–10 Å at least.

Acknowledgments

We gratefully thank Dr. Robert Ullman for stimulating us to prepare this chapter and Dr. Monica A. Fontelos for helping us with several calculations. We also thank Larry Marks of the Xerox Research Centre, Mississauga (Toronto, Canada) for very valuable discussions and suggestions. We acknowledge the financial support received from Progetto Finalizzato Chimica Fine e Secondaria, Consiglio Nazionale delle Ricerche and Ministero Pubblica Istruzione (40%), Italy.

References and Notes

1. According to Flory (ref. 2), we use the terms *configuration* and *conformation* to mean the overall state of the macromolecule and the local shape as defined by few rotation angles, respectively. In this sense, configuration should not be confused with the *stereochemical configuration* of asymmetric atoms, e.g.

2. P. J. Flory, *Statistical Mechanics of Chain Molecules*, Interscience, New York, 1969.
3. H. Yamakawa, *Modern Theory of Polymer Solutions*, Harper and Row, New York, 1971.
4. J. des Cloizeaux and G. Jannink, *Les Polymères en Solution: leur Modélisation et leur Structure*, Les Editions de Physique, Paris, 1987.
5. K. F. Freed, *Renormalization Group Theory of Macromolecules*, Wiley-Interscience, New York, 1987.
6. P. G. de Gennes, *Scaling Concepts in Polymer Physics*, Cornell University Press, Ithaca, 1979.
7. J. C. Le Guillou and J. Zinn-Justin, *Phys. Rev. Lett.*, *39*:95 (1977).
8. J. D. Ferry, *Viscoelastic Properties of Polymers*, Wiley, New York, 1970.
9. P. G. de Gennes, *J. Phys. (Paris) Lett.*, *36*:L55 (1975).
10. G. Allegra and F. Ganazzoli, *J. Chem. Phys.*, *83*:397 (1985).
11. G. Allegra, *J. Chem. Phys.*, *68*:3600 (1978).
12. G. Allegra and F. Ganazzoli, *Macromolecules*, *14*:1110 (1981).
13. G. Allegra and F. Ganazzoli, *J. Chem. Phys.*, *76*:6354 (1982).
14. G. Allegra, J. S. Higgins, F. Ganazzoli, E. Lucchelli, and S. Brückner, *Macromolecules*, *17*:1253 (1984).
15. F. Ganazzoli, G. Allegra, J. S. Higgins, J. Roots, S. Brückner, and E. Lucchelli, *Macromolecules*, *18*:435 (1985).
16. G. Allegra and F. Ganazzoli, *J. Chem. Phys.*, *87*:1817 (1987).
17. P. J. Flory, *J. Chem. Phys.*, *17*:303 (1949).
18. M. Bixon and R. Zwanzig, *J. Chem. Phys.*, *68*:1896 (1978).
19. J. G. Kirkwood and J. Riseman, *J. Chem. Phys.*, *16*:565 (1948).
20. B. H. Zimm, *J. Chem. Phys.*, *24*:269 (1956).
21. G. Allegra and F. Ganazzoli, *J. Chem. Phys.*, *74*:1310 (1981).
22. A. Perico, F. Ganazzoli, and G. Allegra, *J. Chem. Phys.*, *87*:3677 (1987).
23. J. des Cloizeaux, *J. Phys. (Paris)*, *31*:715 (1970).
24. S. F. Edwards and P. Singh, *J. Chem. Soc. Faraday Trans. 2*, *75*:1001 (1979).
25. M. Doi and S. F. Edwards, *The Theory of Polymer Dynamics*, Oxford University Press, Oxford, 1986.
26. A. R. Khokhlov, *J. Phys. (Paris)* *38*:845 (1977).
27. G. Allegra, *Macromolecules*, *16*:555 (1983).
28. Actually, it is possible for $\langle \mathbf{l}(h) \cdot \mathbf{l}(h+k) \rangle$ to decrease exponentially with k from the negative side at $T < \Theta$, so that $\langle r^2(N) \rangle$ and $\langle S^2 \rangle$ become of order N^α, $\alpha < 1$, as in the case of the collapse [see section (2.2.1)].
29. Chap. IV of ref. 2, this list.
30. M. C. Wang and G. E. Uhlenbeck, *Rev. Mod. Phys.*, *17*:323 (1945).
31. P. J. Flory, *Macromolecules*, *7*:381 (1974).
32. G. Allegra and A. Immirzi, *Makromol. Chem.*, *124*:70 (1969).
33. S. Brückner, *Macromolecules*, *14*:449 (1981).
34. G. Allegra, M. Calligaris, and L. Randaccio, *Macromolecules*, *6*:390 (1973).
35. G. Allegra, M. Calligaris, L. Randaccio, and G. Moraglio, *Macromolecules*, *6*:397 (1973).
36. G. Allegra and S. Brückner, *Macromolecules*, *10*:106 (1977).

37. S. Brückner, L. Malpezzi Giunchi, and G. Allegra, *Macromolecules, 13*:946 (1980).
38. S. A. Adelman and K. F. Freed, *J. Chem. Phys., 67*:1380 (1977).
39. S. Lifson, *J. Chem. Phys., 30*:964 (1959).
40. Sect. 7 of ref. 3, this list.
41. G. Allegra, *J. Chem. Phys., 79*:6382 (1983).
42. R. A. Harris and J. E. Hearst, *J. Chem. Phys., 44*:2595 (1966).
43. O. Kratky and G. Porod, *Rec. Trav. Chim. Pays Bas, 68*:1106 (1949).
44. J. Noolandi, G. W. Slater, and G. Allegra, *Makromol. Chem. Rapid Commun, 8*:51 (1987).
45. M. Doi and S. F. Edwards, *J. Chem. Soc., Faraday Trans. 2, 74*:1789 (1978).
46. P. G. de Gennes, *J. Chem. Phys., 55*:572 (1971).
47. D. Y. Yoon and P. J. Flory, *J. Chem. Phys., 61*:5366 (1974).
48. P. J. Flory and V. W. C. Chang, *Macromolecules, 9*:33 (1976).
49. J. Mazur and F. L. McCrackin, *J. Chem. Phys., 49*:648 (1968).
50. Chap. 10 of ref. 6, this list.
51. The positivity requirement of $f(x, y, z)$, see Eqn. (2.2.3), leads to $(x^{1/2} + y^{1/2}) \geqslant z^{1/2}$, $|(x^{1/2} - y^{1/2})| \leqslant z^{1/2}$. In other words, $\langle r^2(k)\rangle^{1/2}$, $\langle r^2(k_1)\rangle^{1/2}$ and $\langle r^2(k + k_1)\rangle^{1/2}$ must obey the same inequalities as the edges of a triangle, in order for the chain connectivity to be maintained.
52. G. Allegra and F. Ganazzoli, *Macromolecules, 16*:1311 (1983).
53. G. Allegra and F. Ganazzoli, *Gazz. Chim. It., 117*:99 (1987).
54. G. Allegra and F. Ganazzoli, *Gazz. Chim. It., 117*:599 (1987).
55. F. Ganazzoli and M. A. Fontelos, *Polymer Comm. 29*:269 (1988).
56. O. B. Ptitsyn, A. K. Kron, and Y. Y. Eizner, *J. Polym. Sci., C, 16*:3509 (1968).
57. S. T. Sun, I. Nishio, G. Swislow, and T. Tanaka, *J. Chem. Phys., 73*:5971 (1980).
58. The numerical coefficient in Eqn. (2.2.24) is different from that reported in ref. 10 because of a more rigorous evaluation of the three-body interactions (see also ref. 53).
59. I. M. Lifshitz, A. Yu. Grosberg, and A. R. Khokhlov, *Rev. Mod. Phys., 50*:683 (1978).
60. C. Domb, in *Advances in Chemical Physics*, Vol. 15, Interscience, New York, 1969, pp. 229–259.
61. F. T. Wall, L. A. Hiller, Jr., and W. F. Atchison, *J. Chem. Phys., 23*:913 (1955).
62. F. T. Wall and J. J. Erpenbeck, *J. Chem. Phys., 30*:634 (1959).
63. F. T. Wall, L. A. Hiller, Jr., and W. F. Atchison, *J. Chem. Phys., 26*:1742 (1957).
64. M. Muthukumar and B. G. Nickel, *J. Chem. Phys., 86*:460 (1987).
65. J. des Cloizeaux, R. Conte, and G. Jannink, *J. Phys. Lett., 46*:L595 (1985).
66. J. F. Douglas and K. F. Freed, *Macromolecules, 17*:2344 (1984).
67. C. Domb and A. J. Barrett, *Polymer, 17*:179 (1976).
68. F. Ganazzoli and G. Allegra, *Polymer, 29*:651 (1988).
69. W. H. Stockmayer, *J. Polym. Sci., 15*:595 (1955).
70. F. Ganazzoli, *Makromol. Chem., 187*:697 (1986).
71. C. Domb, J. Gillis, and G. Wilmers, *Proc. Phys. Soc., 85*:625 (1965); *86*:426 (1965).
72. M. E. Fisher, *J. Chem. Phys., 44*:616 (1966).
73. S. F. Edwards, *Proc. Phys. Soc., 85*:613 (1965).

74. H. P. Gillis and K. F. Freed, *J. Chem. Phys.*, *63*:852 (1975).
75. Chap. 4 of ref. 6, this list.
76. B. Duplantier, *J. Phys.* (*Paris*), *43*:991 (1982).
77. G. Ronca, private communication.
78. S. F. Edwards and K. F. Freed, *J. Chem. Phys.*, *61*:1189 (1974).
79. H. Mori, *Progr. Theor. Phys.* (*Kyoto*), *33*:423 (1965).
80. J. E. Hearst, *J. Chem. Phys.*, *37*:2547 (1962).
81. A. Z. Akcasu, M. Benmouna, and C. C. Han, *Polymer*, *21*:866 (1980).
82. G. Allegra, F. Ganazzoli, and R. Ullman, *J. Chem. Phys.*, *84*:2350 (1986).
83. F. Ganazzoli and M. A. Fontelos, *Polymer*, *29*:1648 (1988).
84. Chap. 6 of ref. 3, this list.
85. B. J. Berne and R. Pecora, *Dynamic Light Scattering with Applications to Chemistry, Biology and Physics*, Wiley-Interscience, New York, 1976.
86. Sect. 4.5 of ref. 25, this list.
87. J. E. Hearst, R. A. Harris, and E. Beals, *J. Chem. Phys.*, *45*:3106 (1966).
88. P. G. de Gennes, *Physics*, *3*:37 (1967).
89. G. Ronca, *J. Chem. Phys.*, *79*:1031 (1983).
90. E. Dubois-Violette and P. G. de Gennes, *Physics*, *3*:181 (1967).
91. M. Benmouna and A. Z. Akcasu, *Macromolecules*, *11*:1187 (1978).
92. G. Allegra, S. Brückner, M. Schmidt, and G. Wegner, *Macromolecules*, *19*:399 (1986).
93. Sect. 7 of Chap. I in ref. 2, this list.
94. G. Allegra, *J. Chem. Phys.*, *61*:4910 (1974).
95. G. Allegra, *J. Chem. Phys.*, *84*:5881 (1986).
96. R. Cerf, *J. Polym. Sci.*, *23*:125 (1957); *J. Phys.* (*Paris*), *38*:357 (1977).
97. A. Peterlin, *J. Polym. Sci.*, *Part A-2*, *5*:179 (1967); *Colloid Polym. Sci.* *260*:278 (1982).
98. L. K. H. Van Beek and J. J. Hermans, *J. Polym. Sci.*, *23*:211 (1957).
99. E. R. Bazua and M. C. Williams, *J. Chem. Phys.*, *59*:2858 (1973).
100. D. A. Mac Innes, *J. Polym. Sci.*, *Polym. Phys. Ed.*, *15*:465 (1977).
101. G. Allegra and F. Ganazzoli, *Macromolecules*, *16*:1392 (1983).
102. D. J. Massa, J. L. Schrag, and J. D. Ferry, *Macromolecules*, *4*:210 (1971).
103. K. Osaki and J. L. Schrag, *J. Polym. Sci.*, *2*:541 (1971).
104. H. A. Kramers, *Physica*, *7*:284 (1940).
105. E. Helfand, *J. Chem. Phys.*, *54*:4651 (1971).
106. J. Skolnick and E. Helfand, *J. Chem. Phys.*, *72*:5489 (1980).
107. C. K. Hall and E. Helfand, *J. Chem. Phys.*, *77*:3275 (1982).
108. D. T.-L. Chen and H. Morawetz, *Macromolecules*, *9*:463 (1976).
109. T.-P. Liao and H. Morawetz, *Macromolecules*, *13*:1228 (1980).
110. Chap. 11 of ref. 8, this list.
111. D. Y. Yoon, P. R. Sundararajan, and P. J. Flory, *Macromolecules*, *8*:776 (1975).
112. P. J. Flory, V. Crescenzi, and J. E. Mark, *J. Am. Chem. Soc.*, *86*:146 (1964).
113. B. Nyström and J. Roots, *Prog. Polym. Sci.*, *8*:333 (1982).
114. J. J. Brey and J. Gómez Ordóñez, *J. Chem. Phys.*, *76*:3260 (1982).

115. D. Richter, A. Baumgärtner, K. Binder, B. Ewen, and J. B. Hayter, *Phys. Rev. Lett.*, *47*:109 (1981).
116. G. Allen, J. S. Higgins, A. Maconnachie, and R. E. Ghosh, *J. Chem. Soc., Faraday Trans. 2*, *78*:2117 (1982).
117. T. A. Kavassalis and J. Noolandi, *Phys. Rev. Lett.*, *59*:2674 (1987).

QUANTUM CRYOCHEMICAL REACTIVITY OF SOLIDS

V. I. GOL'DANSKII, V. A. BENDERSKII and L. I. TRAKHTENBERG

Academy of Sciences of the USSR Institute of Chemical Physics Moscow, USSR

CONTENTS

1. INTRODUCTION

Arrhenius law (1889) describing the dependence of a chemical reaction rate constant on temperature T is one of the most fundamental laws of chemical kinetics. The law is based on the notion that reacting particles overcome a certain potential barrier with height E_a, called the *activation energy*, under the condition that the energy distribution of the particles remains in Boltzmann equilibrium relative to the environment temperature T. When these conditions are satisfied, the Arrhenius law states that the rate constant K is proportional to $\exp[-E_a/K_B T]$, where K_B is the Boltzmann constant. It follows that, for $E_a > 0$, K tends to zero as $T \to 0$.

Quantum mechanics has introduced four very important new notions into chemical kinetics:

1. The existence of an activation barrier is determined by restructuring of the valence bonds in a chemical transformation, that is, the value of E_a is related to the electron-vibrational structure of the molecules and the

chemical reaction should be regarded as a nonradiative transition between initial and final states.

2. The existence of vibrational and rotational degrees of freedom with energies typically exceeding the heat energy leads to the substitution of the discrete energy distribution for the continuous distribution and also to the presence of zero-point energy which does not disappear as $T \to 0$.
3. The transition from the initial state to the final one is related to electron-vibrational relaxation processes in the crossing region; this depends on the separation of electron and nuclear motions according to the Born–Oppenheimer principal.
4. The activation barrier can be overcome not only by thermally activated "corpuscular" transitions but also by tunneling "wave" penetration across the barrier.

A broad range of phenomena caused by such tunneling and observed during the last several decades in chemical physics, particularly in the kinetics of chemical conversions, was described and analyzed recently by two authors of this survey (V. I. G. and L. I. T.) and Fleurov [1].

The concept of tunneling in chemical kinetics was initiated by Hund [2] in discussing the problem of delocalization of electron between two potential wells. It was stated for the first time in this work that for a "double-well" potential with typical vibration frequency ω in the wells and barrier height V_m, the tunneling transfer probability decreases exponentially with growing $2V_m/\hbar\omega$, and therefore, for typical values of V_m of 0.2–2 eV and $\hbar\omega = 0.1$ eV, it can change in the range of almost 8 orders of magnitude.

Later, several papers appeared [3, 4] dedicated directly to the analysis of the possible role of tunneling transitions in chemical reactions. For the following decades the tunneling concept in chemical transformations was mainly developed by Bell [5–7], the first to treat the problem for H-atom penetration across one-dimensional potential barriers of various shapes, particularly the parabolic and Eckart barriers. In his well-known books [6, 7] Bell gave a detailed analysis of the possible contribution of tunneling of H atoms and ions in transfer reactions in liquid and gas phases and explained on this basis the existing deviations from the Arrhenius law (decrease of apparent activation energy) and the growth of the kinetic isotope effect (KIE) (K decrease under hydrogen substitution for deuterium) with decreasing temperature. Nevertheless, the noticeable role of tunneling was for a long time considered a privilege of the lightest particle—hydrogen—and was regarded rather as an exotic and badly identified phenomenon since chemical reactions at low temperatures were not systematically studied.

Only in 1959 was it stated by Gol'danskii [8] that there is a certain "tunneling" temperature T_t below which the role of tunneling transitions becomes predominant and the reaction rate constant becomes independent of T, reaching the low-temperature quantum limit. The paper [8] discussed the transport along reaction coordinate x of a particle with energy E and reduced mass m across the one-dimensional potential barrier $V(x)$. The transfer rate under the quasi-classical treatment of this model is determined by the Gamov factor:

$$w = \omega \exp\left[-\frac{2}{\hbar}\int_a^b |p|\,dx\right], \tag{1}$$

where

$$p(x, E) = [2m\{V(x) - E\}]^{1/2}, \tag{2}$$

a and b are the turning points, and ω is the vibration frequency in the initial state corresponding to the frequency of reaching a turning point. Relations (1) and (2) are true when the de Broglie wavelength $\lambda = \hbar/\sqrt{2mE}$ is small compared to the barrier width $d = b - a$. Besides, to secure sufficient smallness of the rates of both tunneling and thermally activated transitions, compared to ω, it is necessary that the barrier height V_m exceed $K_B T$ manyfold:

$$1 \gg \lambda/d \gg K_B T/(V_m - E). \tag{3}$$

Tunneling temperature T_t was introduced by Gol'danskii as a characteristic of the switch from tunneling transition to the thermally activated one. At $T = T_t$ the rate constant is two times that for tunneling transition at $T = 0$:

$$T_t \simeq \frac{\hbar}{\pi K_B d}\sqrt{\frac{2(V_m - E)}{m}}. \tag{4}$$

It was shown [8] that in the case of initial-state continuous spectra this temperature for parabolic barrier is the same as for a discrete spectrum corresponding to the intersection of two Morse curves with similar parameters and shifted equilibrium positions.

The reaction rate constant is determined with statistic averaging of transfer rate $w(E)$ across the initial-state energy spectrum. When this spectrum is continuous, the averaging comes down to integration [6, 7]:

$$K(T) = \frac{1}{K_B T}\int_0^\infty \exp(-E/K_B T)\,w(E)\,dE. \tag{5}$$

Since at $E \geqslant V_m$ the probability of transition becomes equal to ω, the integral (5) is divided into two summands; the second is proportional to exp $(-V_m/K_B T)$ and determines the constant value of the activation energy in the Arrhenius region where all transitions are the "over"-barrier ones. With falling temperature the contribution of transitions corresponding to the region $w(E) \ll \omega$ increases, and apparent activation energy E^*,

$$E^* = -K_B T^2 \, d \ln K/dT, \tag{6}$$

turns out to be lower than V_m. At $E^* < V_m$ the main contribution in $K(T)$ is made by the energy region near E^*, which is related to the barrier parameters by the relationship [9]

$$\frac{\hbar}{2K_B T} = \mathscr{T}(E^*) = \int_a^b \left(\frac{2m}{V(x) - E^*} \right)^{1/2} dx, \tag{7}$$

where $\mathscr{T}(E^*)$ is the typical time of particle motion between the turning points in the potential well formed by the turned-over barrier $V(x)$. Formally, Eqn. (7) describes the classical motion of a particle with pure imaginary time $i\mathscr{T}(E)$ across the barrier $V(x)$. It follows from Eqn. (7) that a change in E^* occurs in the T_t region: At $T > T_t$, $E^* \simeq V_m$, and at $T < T_t$, $E^* \to 0$, so that the reaction rate constant reaches the low-temperature limit equal to [10]

$$K(0) = \omega \exp[-(V_m/E_d)^{1/2}], \tag{8}$$

where typical energy E_d is equal to

$$E_d = \xi \hbar^2/(md^2) \tag{9}$$

and ξ is a coefficient that depends on the barrier shape. For instance, $\xi = \frac{1}{8}$ and $\xi = 2/\pi^2$ for rectangular and parabolic barriers, respectively. Typical relationships for $K(T)$ and $E^*(T)$ are shown in Figure. 1.

The main significance of the works [8] was in revealing the existence, irrespective of the barrier shape, of the finite low-temperature limit of the rate constant $K(0)$. Even for Eckart barrier $V(x) = V_m/ch^2(2x/d)$, having an infinite width at $E = 0$, the tunneling probability remains finite due to the existence of zero-point vibrations.

Since issues of *Doklady AN SSSR* were not regularly translated into English in 1959, the papers [8] were not considered in Western scientific literature. Only after a number of experimental papers were published confirming the existence of the low-temperature limit of the rate constants of

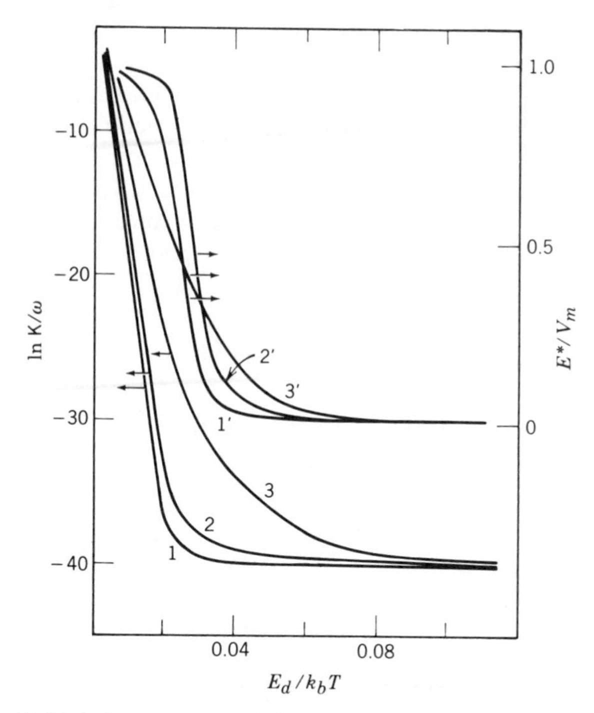

Figure 1. $K(T)$ and $E^*(T)$ relationships (curves 1–3 and 1′–3′, respectively) for one-dimensional penetration of barrier of different shapes; $K(0)/\omega$ value assumed equal: (1, 1′) rectangular, (2, 2′) parabolic, and (3, 3′) diabatic barriers: $V(x) = V_m(1 - 2|x|/b)$

various reactions there appeared several interesting additions to the one-dimensional tunneling model verifying the basic deductions in ref. 8. LeRoy, Murai, and Williams [11] emphasized the role of zero-point vibrations (first accounted for in ref. 8) in the emergence of the low-temperature limit. Buhks and Jortner [12] introduced the relations equivalent to Eqs. (4) and (8) based on the nonradiative multiphonon transitions theory and showed that the general character of these relations stems from the equivalence of the Gamov tunneling factor and the asymptotic behavior of Franck–Condon factors characterizing the vibrational wavefunction overlap of the initial and final states under large shifts of equilibrium position, that is, within the strong bond.

To each of the two electron states between which the nonradiative transition occurs there corresponds its Hamiltonian function describing nuclei motion:

$$H_{1(2)} = T + U_{1(2)}(R) + \varepsilon_{1(2)},$$

where T is the operator of nuclei kinetic energy; $U_x + E_x$ $(x = 1, 2)$ are the electron terms under fixed coordinates of R nuclei. The potential energy surfaces $U_{1(2)}(R)$ correspond to electron wave functions $\phi^e_{1(2)}(r, R)$ and can be defined with account for complete electron interaction (adiabatic terms) or with neglect of its part (diabatic terms). In each case there exists a perturbation V causing the transition. In the chemical reactions considered in what follows such perturbation is the neglected part of electron interaction V_e determined by the electron wavefunction overlap localized at different sites in initial and final states:

$$V = \langle \psi^e_1 | V_e | \psi^e_2 \rangle. \tag{10}$$

In first order of the perturbation theory the transition probability per unit time is described by the well-known golden rule

$$W_{1,2} = \frac{2\pi}{\hbar} \sum_{i,f} |\langle \psi_{1i}(R) | V | \psi_{2f}(R) \rangle|^2 e^{-E_i/K_B T} \, \delta(E_f - E_i), \tag{11}$$

where $\psi_{1(2)}(R)$ are the vibration wavefunctions of states 1 and 2. Relation (11) expresses the Gibbs distribution averaging of the Franck–Condon factors of initial states with the energies E_i multiplied by the density of the final states with the energies $E_f = E_i$.

The temperature relationship of the nonradiative transition probability, comprised of the Arrhenius region going over—with decreasing T—into the low-temperature plateau [similar to the relationship of $K(T)$ shown in Figure 1], stems from the theory developed by Kubo and Toyazawa [13] for two intersected terms of the same intrinsic frequencies at the interaction parameter V independent of T. The diagram of the terms is shown in Figure 2. In the case of a strong bond $(E_s/\hbar\omega \gg 1)$, the transition probability in the Arrhenius region $(K_B T \gg \hbar\omega/2)$ is equal to [13–15]

$$W_{1,2} = \frac{V^2}{\hbar} \left(\frac{\pi}{K_B T E_s} \right)^{1/2} \exp\left[-\frac{(E_s - \Delta)^2}{4 E_s K_B T} \right]. \tag{12}$$

Relation (12) describes a thermally activated transition with activation energy $E_a = (E_s - \Delta)^2 / 4E_s$ equal to the potential at the point of intersection. The preexponential factor in Eqn. (12) is equal to the vibration frequency $\omega/2\pi$ multiplied by the transmission coefficient P representing the Landau–Zener factor [16] for the transition between the terms in the region of their

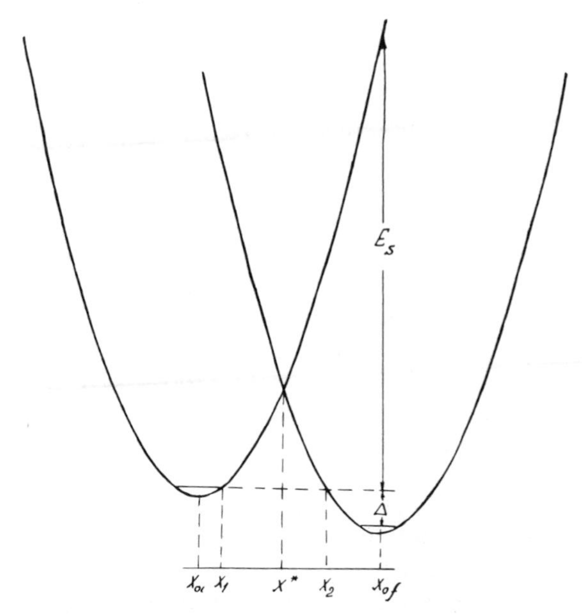

Figure 2. Harmonic terms with equal frequencies within the strong bond $E_s/\hbar\omega \gg 1$: X_{0i} and X_{0f}, coordinates of equilibrium positions in initial and final states; x_1, x_2, turning points; x^*, intersection point of terms; E_s, reorganization energy, $E_s = \frac{1}{2}m\omega^2(X_{0f} - X_{0i})^2 = \frac{1}{2}\hbar\omega [(X_{0f} - X_{0i})/\delta_0]^2$; $\delta_0^2 = \hbar/m\omega$, squared amplitude of zero-point vibrations.

intersection under Maxwellian distribution by rate:

$$P = 4\pi \frac{V^2}{\hbar\bar{v}|F_2 - F_1|}, \tag{13}$$

where $\bar{v} = (8K_BT/\pi m)^{1/2}$ is the mean thermal rate; F_1 and F_2 are the slopes of the terms at the intersection point. Relations (12) and (13) have the form typical of the transition state theory [17, 18].

At low temperatures ($K_BT < \hbar\omega/2$) with the same strong bond approximation ($E_s/K_BT \gg 1$), the following relation holds:

$$W_{12}(0) = \frac{2\pi V^2}{\hbar^2\omega} \exp\left(-\frac{E_s - \Delta}{\hbar\omega} + \frac{\Delta}{\hbar\omega} \ln\frac{E_s}{\Delta}\right), \tag{14}$$

corresponding to the tunneling quantum transition from the zeroeth vibrational sublevel of the initial state to the resonant sublevel of the final state. Relation (14) can be directly obtained by integration of Eqn. (1) in the interval from x_1 to x_2. The transmission coefficient in Eqn. (14) also corresponds to

Landau–Zener factor at particle rate $\bar{v} = (2E_a/m)^{1/2}$ when its energy is equal to the barrier height. The dependence of W_{12} on T and Δ for parabolic terms of different frequencies is discussed elsewhere [19].

Earlier, in chemical transformations related to valence bond restructuring, the close analogy between the nonradiative relaxation processes and chemical reactions was revealed in the electron transfer reactions. The necessity to account for nuclear shifts in these reactions, emphasized for the first time by Libby [20], was then analyzed in detail by Marcus, Levich, Dogonadze, Ovchinnikov, and Ovchinnikova [14, 15, 21–25] in work dedicated to redox reactions in polar media and electrode reactions. We should note the conceptual proximity between the model of the dynamic polarization of continuous medium with electron transfer and the process of small-radius polaron formation (autolocalization) discussed for the first time by Landau and Pekar [26]. The separation of electron and nuclear motions enables one to present the transition probability as a product of the squared-matrix element of respective electron transfer V_e and the Franck–Condon factor. Since the typical energies of electron interactions normally are much higher than $K_B T$, the V_e does not depend on T, and the temperature dependence of the reaction rate constant is defined in accordance with relation (10) by vibration spectrum statistical averaging, involving both high-frequency intramolecular and low-frequency intermolecular vibrations.

In reactions of outer-sphere electron transfer between the ions in polar liquids, the main role is played by the low-frequency polarization vibrations with typical energies much smaller than $K_B T$. In the case of classical nuclear motions the medium reorganization energy is the sum of reorganization energies of the set of normal classical oscillators ($\hbar\omega_k < K_B T$) formed by shifts similar to the one shown in Figure 2:

$$E_s = \frac{1}{2}\sum_K m_K \omega_K^2 x_K^2. \tag{15}$$

As was first shown by Marcus [21], the transition probability is described by relation (12). In case of the strong bond, an essential peculiarity of the nonradiative transitions is their statistical cooperative character determined by participation of many vibrational degrees of freedom in the transition. Due to this the summary strong bond with the medium is provided by the large number of relatively weak interactions with its individual modes; for each interaction the equilibrium position shift is comparable to the zero-point vibration amplitude δ_K:

$$S_K = \frac{1}{2\hbar} m_K \omega_K x_K^2 = \frac{1}{2}\left(\frac{x_K}{\delta_K}\right)^2 \lesssim 1, \qquad \delta_K^2 = \hbar/m_K\omega_K. \tag{16}$$

Condition (16) enables one to consider such transitions in harmonic approximation. It is shown elsewhere [25] that the description of the polar medium by the set of classical harmonic oscillators is equivalent to the supposition of linear polarizability, that is, the applicability of Maxwell equations. Arrhenius temperature dependence $K(T)$ is typical of the overwhelming majority of the outer-spheric reactions of electron transfer in polar liquids [27, 28].

Violation of the Arrhenius law was observed for the first time in the reaction of electrochemical hydrogen evolution, where the temperature dependence of the rate constant of H-ion discharge with formation of the hydrogen atom adsorbed at the metal,

$$H^+ + e^- \rightarrow H_{ads}$$

(e^- is the electron passing from the metal), is described by [29]

$$K(T) = K_0 \exp\left[-\frac{E_a}{K_B(T + T_0)}\right]. \tag{17}$$

In refs. 30–32 this dependence was explained by participation of the proton quantum shift—along with reorganization of classical medium—in the reaction. The dependence of W_{12} on Δ was thoroughly analyzed elsewhere [14, 27, 32] in connection with the theory of electrode reactions for which the dependence of the rate constant on the energy difference of initial and final states (equal to the overvoltage) can be measured directly. As follows from relations (12) and (14), the probability of the nonradiative transition is maximal when the final-state potential curve crosses the initial-state term close to its minimum. Such reactions were called "activationless". The transition probability is described by

$$W_{12} \simeq W_0 \exp\left(-\frac{(\Delta - E_s)^2}{\sigma(T)}\right), \tag{18}$$

where W_0 weakly depends on T and

$$\sigma(T) = \sum_K S_K(\hbar\omega_K)^2 \tanh^{-1}\frac{\hbar\omega_K}{2K_B T} \simeq a + bK_B T. \tag{19}$$

The linear dependence $\sigma(T)$ is determined by summation of both quantum and classical degrees of freedom numbered by the indexes $K \leqslant K_0$ and $K > K_0$, respectively:

$$a = 2\sum_{K > K_0} S_K(\hbar\omega_K)^2, \qquad b = 4\sum_{K \leqslant K_0} S_K\hbar\omega_K. \tag{20}$$

Relations (18) and (19) are in good agreement with the experimental dependence (17). Such consideration of the activationless reactions is given elsewhere [31, 33]. An example of the tunneling electrode reaction found in ref. 34 is the electrochemical desorption of hydrogen:

$$H_{ads} + H^+ + e^- \rightarrow H_2,$$

with the rate constant of $10^{-14} - 10^{-11}$ cm^3/s which does not depend on temperature and, in spite of high values approaching the diffusion limit, changes severalfold under light-hydrogen substitution by deuterium. By estimations [32–33], the value of T_0 characterizing the relative contribution of quantum and classical degrees of freedom to the reorganization energy is 10^3 K in this reaction. In H-ion discharge reactions $T_0 \sim 100$ K. It is emphasized in ref. 30 that the interaction parameter V is a function of the shifts along the coordinates of the medium.

At longer distances of electron transfer, d_e, compared to typical length of electron density localization, the electron wave function is described by an asymptotic expression,

$$\phi_e(z) \sim \exp(-\alpha z), \tag{21}$$

where $\alpha = \sqrt{2m_e I}/\hbar$, I is the ionization energy of this state, and m_e is the electron mass. It follows from ref. 21 that at $\alpha d_e > 1$ the matrix element of the electron transfer has the form of the Gamov factor:

$$V_e \sim \exp\left\{-\frac{\xi_e}{\hbar} d_e (m_e I)^{1/2}\right\}, \tag{22}$$

where ξ_e, as in relation (9), depends on the barrier shape. Proceeding from relations (9) and (22), the electron transfer over long distances, accompanied with nuclei quantum shifts at low temperatures, was termed electron–nuclear tunneling. The first example of this tunneling type was by DeVault and Chance [35, 36] published in 1966; in the range 4–120 K they found a low-temperature plateau of the electron transfer rate under cytochrome-*c* oxidation by chlorophyll in the process of bacterial photosynthesis. The value of $K(0)$ was 300 s^{-1}. This work became the basis for the rapid development of notions about electron tunneling in biological processes, biopolymers, and membranes (see refs. 37, 38).

Another type of tunneling redox process is related to the long-distance electron transfer from donor to acceptor in the solid γ-irradiated polar media. This type of reaction, in which the rate constant low-temperature plateau was also observed, plays an important role in radiation chemistry. Research into these processes were started with the ESR observation of K constancy under stabilized electrons trapping by O^- ions in alkali water–alcohol glasses at 4–120 K [39, 40]. A detailed review of this research branch is given by Zamaraev and co-workers [41, 42].

References 39–42 are also interesting due to the fact that, typical for solid-state reactions, the "polychronicity" (nonexponentiality) of the kinetics in tunneling reactions arises due to a distribution of the barrier parameters I and d_e in the non-crystalline solid media. According to refs. 39 and 40, the concentration of the reagents decreases linearly with logarithm of time. The analysis of kinetic curves enables one to determine the parameters of the spatial and energy distribution of the reagents stabilized in traps of different depths. After the discovery by Chance and DeVault there appeared a number of theoretical models [43–46] of electron–nuclear tunneling in the redox transformations of biomolecules, and in particular, the anharmonicity and change in vibration frequencies under transition were accounted for.

The consistent theory of the tunneling transfer of electrons was developed by Ivanov and Kozhushner [47]. The following relation was obtained for the transition probability:

$$W_e = \omega_e \exp\left\{-\frac{2d_e}{\hbar}\sqrt{2m_e(I-\varepsilon)}\right\}\exp\left\{-\frac{(E_{s1}+E_{s2}-\Delta)^2}{\sigma_1+\sigma_2}\right\}$$
$$\times \exp\left\{\frac{m_e d_e^2}{2\hbar^2(I-\varepsilon)}\,\frac{\sigma_1\sigma_2}{\sigma_1+\sigma_2}\right\}, \tag{23}$$

where σ_1 and σ_2 are determined using relation (19) and are related to the vibrational subsystems of donor and acceptor having the reorganization energies under electron transfer equal to E_{s1} and E_{s2}. The first exponent in Eqn. (23) is the Gamov factor for the electron with effective energy ε:

$$\varepsilon = \frac{E_{s1}\sigma_2 + (\Delta - E_{s2})\sigma_1}{\sigma_1+\sigma_2}; \tag{24}$$

the second one, as in relation (18), is the quantum analog of the Marcus formula. The latter cofactor accounts for the dependence of ε on the transfer distance d_e. Electron transfer is accompanied by the vibrational transitions in both sites; the energy ε of the electron tunneling depends on the ratio of the electron–phonon bond in the donor and acceptor. If the bond in the donor is stronger than in the acceptor, ε is close to the energy level of the acceptor, and the energy is evolved on the vibration modes of the donor. When the transition probability is described, as usual, by the relation

$$W_e = \omega_e \exp\left\{-\frac{2d_e}{a}\right\}, \tag{25}$$

the relationship of ε to the vibrational subsystem reorganization parameters, determined by relation (24), leads to the dependence of the indicator a on the same parameters.

Thus, the discovery of the low-temperature limit of the electron transfer rate in the redox processes became the first convincing proof of the existence of electron–nuclear tunneling, and further research confirmed the universal character of this phenomenon.

However, such electron transfer can hardly be called a *chemical reaction* in the full sense of this term; rather it is closer to the class of tunneling phenomena in the physics of solids and electronics. A chemical reaction, by definition, must include the reconstruction of molecules, the spatial rearrangement of atoms, and changes of lengths and angles of valence bonds; the definition simply does not apply to small shifts over distances of the order of the zero-point vibration amplitude.

This review pays primary attention to chemical reactions related to heavy-particle transfer, particularly atoms and molecular fragments.

Section 2 contains a brief review of experimental data showing that temperature dependences of $K(T)$, comprised of the Arrhenius region going into the low-temperature plateau, are typical of a considerably large number of chemical reactions in which the transfer of not only hydrogen atoms but also the much heavier particles takes place. The data show vividly that the chemical bond restructuring in the solid-state chemical reactions cannot be regarded as the transfer of relatively light molecular fragments within a fixed array of heavy reagents, secured in the lattice of a solid which plays only the role of a thermostat (i.e., provides storage of energy and momentum in the transformation process). Development of the notion about the active role of the solid matrix is based on the rejection of an assumption that the static potential barrier is independent of the dynamic properties of the medium surrounding the reagents.

The importance of the barrier in solid-state reactions is considered to be due to the securing of reagents in lattice sites in which, due to long distances between the lattice sites, the interaction energy is low and the barrier should be close in magnitude to the energy of the chemical bond. Evidently, the probability of tunneling penetration of heavy particles through such high and wide barriers should be vanishingly small. At the same time, in the gas phase, where the reagents are not fixed, the same reactions in many cases turn out to be activationless. Therefore, for theoretical description of solid-state reactions it was first necessary to establish the mechanism of barrier lowering, ensuring—just as in the case of gas phase reactions—the horizontal shift of the terms in Figure 2.

The first step in that direction was accounting for fluctuations in the atomic configuration surrounding the chemical bond under rearrangement. In the low-temperature plateau region these fluctuations result from the zero-point intermolecular vibrations and have a quantum character. In the Arrhenius region they become thermal. The typical temperature T_t is related to the Debye frequency of the solid. The probability amplitude of such fluctuations, naturally, does not depend on the tunneling particle mass, and therefore the fluctuation lowering of the barrier turns out to be more considerable for small transition probability and, in particular, for large mass of the tunneling particle.

The models of "fluctuation preparation" of the potential barrier of solid-state chemical reactions were suggested by Klochikhin, Pshezhetsky, and Trakhtenberg [48], Ovchinnikova [49], and Benderskii, Gol'danskii, and Ovchinnikov [50]. These models made possible the quantitative description of low-temperature reactions with H-atom transfer [48, 49] and also the qualitative explanation of the dependence of solid-state reaction rate constants on matrix properties [48, 50]. It should be noted that the role of phonons in the process of fluctuation preparation of the barrier proves radically different than in the reorganization of the medium: The existence of intermolecular vibrations creates additional reaction paths with much lower potential barriers than in the case of fixed reagents. While the medium reorganization characterizes, as seen from relation (15), the change of phonon subsystem resulting from the valence bond rearrangement (i.e., under the action of the intramolecular subsystem), the fluctuation preparation of the barrier describes the reverse action of the phonon subsystem on the intramolecular one [i.e., it represents the mechanism of so-called phonon-assisted tunneling (PAT) most important for the chemical reactivity of solids].

Transfer of heavier particles with abnormally high rate constants, compared to that expected according to the Arrhenius law, was observed in the structurally nonequilibrious glasslike matrices as well as in chain reactions in the reagent crystals. To explain these data, it was assumed [50, 51] that in the initial state the reagents are at shorter intermolecular distances than those in equilibrium; this creates a lowering of the potential barrier. Formation of these states occurs under the action of local mechanical stresses typical of disordered solids. These stresses may result from both external mechanical effects and a previous transformation which causes local changes in density and temperature. Notions about the accelerating mechanical effects led to the development of the Semenov idea [52] about the existence of energetic chains in solid-state chemical reactions. According to [52], the slow relaxation of the energy evolved as a result of a chemical reaction can, at low temperatures, ensure barrier lowering for consequent transformation. Notions about the formation of an atomic configuration—favorable for heavy-particle transfer—at the expense of the fraction of the chemical reaction thermal effect causing local slow-relaxing deformations of the solid lattice, are closely connected with the model of plastic deformation and low-temperature fracture of solids. Similar to other notions, they are at present in the initial stage of development.

The current status of the models of fluctuational and deformational preparation of the chemical reaction barrier is discussed in the Section 3. Section 4 is dedicated to the quantitative description of H-atom transfer reactions. Section 5 describes heavy-particle transfer models for solids, conceptually linked with developing notions about the mechanism of low-temperature solid-state chemical reactions. Section 6 is dedicated to the macrokinetic peculiarities of solid-state reactions in the region of the rate constant low-temperature plateau, in particular to the emergence of non-thermal critical effects determined by the development of energetic chains.

In essence, the goal of this review is to describe the evolution of notions about tunneling, from the Gamov one-dimension model potential barrier penetration to multidimensional quantum motion with account for an active role of the medium. Such evolution goes on in various branches of modern chemical physics and the physics of solids. Therefore, the analysis of the common features of developing theories seems timely and important.

2. LOW-TEMPERATURE LIMIT OF CHEMICAL REACTION RATES

The low-temperature limit of chemical reaction rates was discovered as a result of a research cycle carriered out in 1970–1973 [53–58] and described in refs. 57 and 58 and illustrated by the spontaneous chain growth in the radiation-induced polymerization of solid formaldehyde (see also additional data and reviews [59–64]). The experiments were carried out at 4.2–140 K; the polymerization chain length and growth rate were determined with high-sensitivity calorimeters of adiabatic and diathermic types with a lag of 0.3–1 s [53, 54]. It was found that the radiation yield of polymerization G, equaling the number of transformed molecules per 100 eV of absorbed energy of penetrating radiation, was 10^7, 10^5, and 10^3 at 140, 77, and 4.2 K, respectively. Since $G = g\nu$, where g is the initiation radiation yield equaling ~ 1 in the absence of chain growth and ν is the chain length, the observed values of G are of the same order of magnitude as the kinetic length of the polymerization chain. For the first time the high G values vividly showed the existence of spontaneous chemical transformations even at the temperature of liquid helium. Later the existence of such transformations was confirmed by the registration of formaldehyde postpolymerization at 5 K upon γ-irradiation cutoff [59]. The kinetic measurements [54–58] enabled us to establish the polymer chain growth time τ and the rate constant $K = \nu/\tau$ corresponding to the addition of one unit to the chain. The temperature relationship of K (Figure 3) comprises the Arrhenius region at $T \geqslant 20$ K with apparent activation energy increasing up to $E_m \simeq 0.1$ eV in the 140-K region and a low-temperature plateau at $T \sim 12$ K; that is, it corresponds to the expected relationship for the preceding models of molecular tunneling. The value of $K(0)$ is $10^2\ s^{-1}$, while the extrapolation by the Arrhenius law with the measured E value would have led to values (per Kelvin) tremendously exceeding, at $T < 10$ K, the age of the universe (10^{30} and 10^{100} years at 10 and 4.2 K, respectively). The formaldehyde polymerization is in all aspects a typical example of a solid-state chemical reaction. As a result of transformation, the triangular $H_2C{=}O$ molecules arranged in parallel planes in the crystal at van der Waals distances of 3.3–3.5 Å are rearranged into the long chain of tetrahedral chemical bonds C–O–C–O of 1.4 Å length. Polymerization is accompanied with a considerable ($\sim 40\%$) density increase.

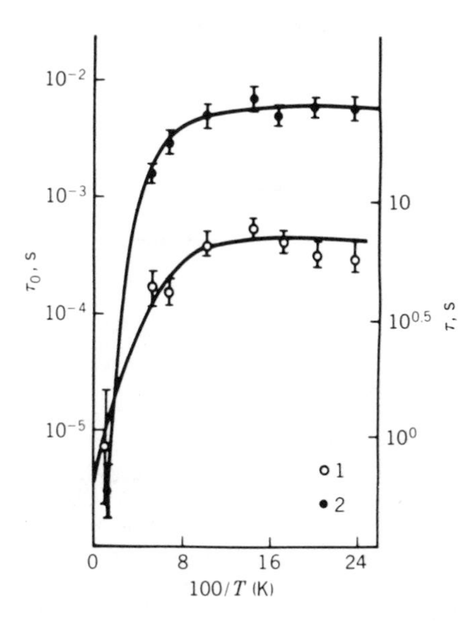

Figure 3. Temperature dependence (in Arrhenius coordinates) of polymer chain growth time τ (curve 1) and mean time of one unit attachment to this chain τ_0 (curve 2) under radiation-induced polymerization of formaldehyde.

Later experimental researches have shown that the temperature relationship of $K(T)$ comprised of the Arrhenius region and low-temperature plateau is typical of various solid-state chemical reactions. The set of kinetic data is given in Figure 4.

Fraunfelder and co-workers [65–68] have found the same relationship in the reduction kinetics of CO and O_2 ligand coordination bonds with the complex-forming Fe ions of the heme group myoglobin upon photodissociation of these bonds induced by laser photolysis of $\sim 10^{-6}$ s duration. The formation of Fe–CO and Fe–O_2 bonds after photolysis was registered with a spectrophotometer by restoration of the mother compound absorption spectrum in the wide range of times (10^{-6}–10^3 s) and temperatures (2–300 K). Before photolysis the six-coordinate Fe^{2+} ion is in the heme plane. The coordination bond break causes not only the ligand shift but also

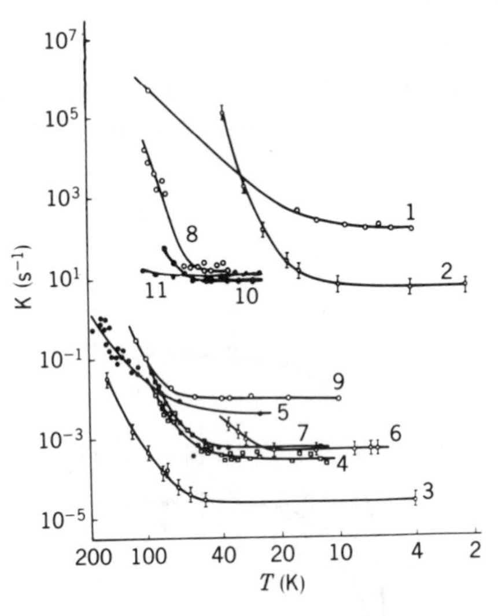

Figure 4. Examples of low-temperature limit of rate constant of solid-state chamical reactions obtained in different laboratories of the USSR, United States, Canada, and Japan: (1) formaldehyde polymerization chain growth (USSR, 1973 [56]); (2) reduction of coordination Fe–CO bond in heme group of mioglobin broken by laser pulse (United States, 1975 [65]); (3) H-atom transfer between neighboring radical pairs in γ-irradiated dimethylglyoxime crystal (Japan, 1977, [72], (4, 5) H-atom abstraction by methyl radicals from neighboring molecules of glassy methanol matrix (4) and ethanol matrix (5) (Canada, United States, 1977 [11, 78]); (6) H-atom transfer under sterically hampered isomerization of aryl radicals (United States, 1978 [73]); (7) C–C bond formation in cyclopentadienyl biradicals (United States, 1979 [111]); (8) chain hydrobromination of ethylene (USSR, 1978 [119]); (9) chain chlorination of ethylene (USSR, 1986 [122]); (10) organic radical chlorination by molecular chlorine (USSR, 1980 [124,125]); (11) photochemical transfer of H atoms in doped monocrystals of fluorene (B. Prass, Y. P. Colpa, and D. Stehlik, *J. Chem. Phys.*, in press.).

rearrangement of the whole coordination sphere at which the Fe^{2+} ion leaves the heme plane. The reverse exothermal ligand attachment is related to the potential barrier penetration separating the configuration formed after photolysis from the initial one. It was found that the low- temperature plateau emerges below 10 K, and $K(0)$ is $\sim 10\,s^{-1}$. The estimation of tunneling parameters for such a plateau gave the values of the height ($E_a \approx 0.04$ eV) and the width ($d \approx 0.5$ Å) of the potential barrier [67].

Subsequently the circle of methods applied to the studies of MbCO dissociation and rebinding was broadened by the EXAFS (Extended X-Ray absorption fine structure) technique. At first the EXAFS measurements detected at 4 K of only a very small change, Δd, of the distance between the Fe atom and the CO ligand in the photoinduced dissociation: $\Delta d = 0.05 \pm 0.03$ Å [69]. However, later [70], much larger displacements of CO were recorded, and as Britton Chance wrote to one of us (V. I. G.) on April 9, 1986; "We have been fortunate enough to have detected an additional state in myoglobin at 40 K, where the ligand apparently accumulates in the crevice in the protein about 0.7 Å farther from the iron than in our first determination of 0.05 Å. Both forms are valid; they occur at different temperatures, 4 K and 40 K respectively." Discussion on the details of Fe–CO rebinding in heme-containing proteins is still not completed; among its main points is the question of whether one should treat the system formed after the MbCO photodissociation as a continuum of states, which leads to the nonexponential, or so-called polychronic (polychromatic) kinetics of rebinding, or as only two well-defined states. Depending on the chosen approach, one should give different emphasis to the motion along the Fe–CO reaction coordinate ("pure" molecular tunneling) and to the reorganization of other bonds in heme (e.g., between the central Fe atoms and the four pyrrol-nitrogen atoms), which lower the reaction potential barrier by other (than along the reaction coordinate) modes of motion, as low-frequency intermolecular motions [48–50] or intraglobular rearrangements.

A quite definite reference of molecular shifts takes place under dimethylglyoxime (DMGO) crystal radiolysis. Lebedev and Yakimchenko [71] have found using electron spin resonance (ESR) that two types of radical pairs are formed in γ-irradiated crystals. The change in ESR spectrum observed after radiolysis is determined by the transformation of radical pairs related with H-atom transfer along the straight line connecting the neighboring pairs O-atoms. Since the structures of crystal and radical pairs are known, one can determine the O–H bond length in the initial and final states ($\geqslant 2.5$ and 1.0 Å, respectively). The H-atom shift length exceeds 1.5 Å (Figure 5). Yakimchenko and Lebeded [71] note a considerable deviation of the transfer rate from the Arrhenius law as well as KIE growth with a temperature decrease from 200 to 77 K. Toriyama and colleagues [72], with the same measure-

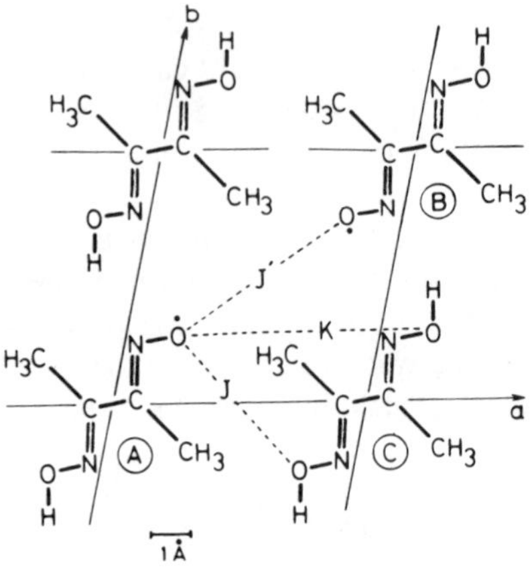

Figure 5. Spatial arrangement of reactants under H-atom transfer in γ-irradiated dimethylglyoxime crystal. Radical pair J transfers into radical pair K as result of H-atom transfer from molecule C to radical B [72].

ments performed down to 4.2 K, found the low-temperature plateau below 50 K corresponding to $K(0) = 10^{-5}\ s^{-1}$. Above 150 K the activation energy is ~ 0.43 eV; these authors compared the experimental relationship of $K(T)$ with the one-dimensional tunneling model. Although with a chosen height of the asymmetric Eckart barrier equal to $V_m = 0.65$ eV and a width of $d = 0.44$ Å, relations (5)–(8) at $\omega = 8 \times 10^{12}\ s^{-1}$ describe the $K(T)$ change at $T \geqslant 50$ K well enough, the deviation at 4.2 K exceeds an order of magnitude.

The H-atom intramolecular transfer leading to the isomerization of aryl radicals, particularly in the reaction

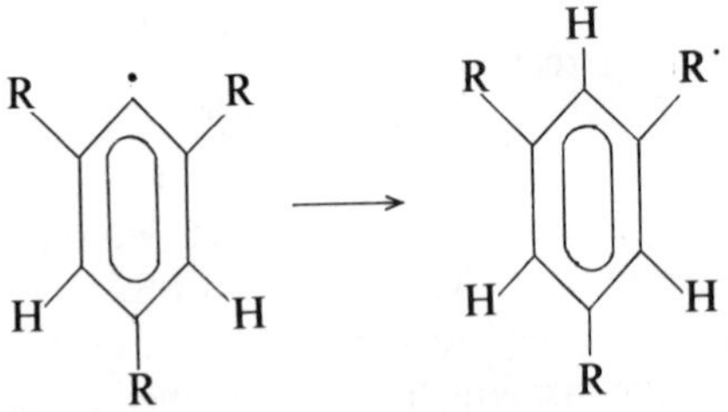

was studied by Brunton et al. [73, 74] using ESR in the temperature range 28–300 K. At the same temperatures the reaction rates are the same in the solid and liquid matrices. The isotope H–D effect is $\sim 10^2$ and $\sim 10^4$ at 230 and 120 K, respectively. The reaction rate constant ceases to depend on temperature below 100 K; the $K(T)$ value is $\sim 7 \times 10^{-4}\ s^{-1}$. An attempt to coordinate the experimental and computed $K(T)$ relationships applying relations (1) and (5) for one-dimensional potential barriers of different shape at a preset H-atom shift failed.

Abstraction of H atoms by the methyl radical from glasslike matrices of aliphatic alcohols was thoroughly studied using ESR [11, 75–82]. The kinetics of CH_3 radical transformation into the matrix radical (CH_2OH and CH_3CHOH for methanol and ethanol, respectively) is nonexponential. Therefore, in refs. 75–78 and 11, the H-atom transfer rate constant was determined by the initial section of kinetic curves. The low-temperature plateau emerges below 40 K, where $K(0)$ is two to three orders of magnitude higher than in dimethylglyoxime crystals (at values of $E_a = 0.4$ eV) in the region 100–150 K where the Arrhenius law is satisfied. It is noted in refs. 78 that the one-dimensional tunneling model does not allow one to describe the observed relationship of $K(T)$ for reasonable parameters of the potential barrier.

The reaction

$$CH_3 + CH_3OH \rightarrow CH_4 + CH_2OH$$

is characterized by the isotope H–D effect of $\sim 10^3$ at 77 K; this clearly indicates the quantum character of the hydrogen transfer. At the same time, CH_3 radical decay in the crystalline matrix occurs approximately 10-fold faster than in the glasslike matrix [76]. The reaction

$$CH_3 + C_2H_5OH \rightarrow CH_4 + CH_3CHOH$$

has the same rate constant as the previous one [11]. With substitution of H atoms by D atoms in the methylene group of ethanol, the CH_3 radical decay rate at 77 K decreases almost 40-fold. Also, H-atom abstraction occurs from the methyl group with formation of CH_2CD_2OH radicals [83].

As shown elsewhere [79–82], the kinetic curves in the glasslike matrices consist of the initial section with the preceding parameters and the further section of more complete transformation described by the relation

$$R(t) = R_0 \exp[-Kt^{\xi}], \tag{26}$$

where $R(t)$ is the CH_3 radical concentration. Parameter ξ [26] decreases with temperature, tending to 1 above the matrix glass transition temperature where the kinetic curves become exponential. Estimation of the matrix molecules available for the CH_3 radical [82] from the comparison of kinetic curves in matrices of different isotopic composition has shown that H-containing molecules of the second coordinate sphere can be involved in the reaction if the first sphere is formed of only deutero-substituted molecules. According to refs [80–82], in the process of H-atom tunneling transfer, rearrangement of a radical-surrounding matrix should occur, ensuring the formation of a configuration necessary for transfer, in which the radical is close to the required CH bond. Naturally, such rearrangement is determined by the matrix properties, and in disordered glasslike media a wide variety of radical-surrounding configurations is possible; this is what results in the nonexponentiality of kinetic curves.

Neiss and Willard [84] note that structural relaxation of the nonequilibrium glasslike matrix of 3-methylpentane caused by temperature annealing leads to a lowered decay rate of CH_3 radicals.

Similar reactions of H-atom abstraction by a CH_3 radical were observed in γ-irradiated crystals of acetonitrile [75]. In the region 69–112 K the apparent activation energy increases from 0.03 to 0.1 eV and is much lower than in the gas phase ($\sim$ 0.4 eV). Under consecutive substitution of one, two, and all three H atoms by deuterium in the methyl group of acetonitrile, the CH_3 decay rate decreases by factors of 2, 3, and $\geqslant 10^3$, respectively. Hydrogen atom tunneling transfer in this reaction results in a large-isotope H–D effect found by Sprague [85] to exceed 2.8×10^4 at 77 K, while the maximum possible isotope effect (IE) determined by the difference in the zero-point vibration energies of CH and CD bonds does not exceed 10^3. Below 113 K an acetonitrile crystal exists in two modifications. At 77 K H-atom abstraction in the low-temperature stable modification is 10-fold slower than in the metastable high-temperature one [86].

An indication of the tunneling mechanism of transfer was also obtained in the reaction of H-atom abstraction from methyl isocyanide molecules by methyl radical, where a decrease in the apparent activation energy from 0.2 to 0.06 eV in the range 125–77 K and the absence of CH_3 decay under complete deuteration of the matrix were observed [11].

In γ-irradiated crystals of sodium acetate trihydrate the reaction rate of

$$CH_3 + CH_3COO^- \rightarrow CH_4 + CH_2COO^-$$

decreases fourfold under substitution of crystal water molecules, H_2O, by D_2O; this indicates the effect of the mass of the surrounding molecules on the tunneling transfer rate of the H atom not linked to them by valence bonds [87].

A common feature of the rate constant–temperature relationship of the reactions between CH_3 radicals and organic compounds is a rather wide temperature range above T_t in which the $K(T)$ is described by

$$\ln K \sim T + \text{const} \tag{27}$$

obtained theoretically elsewhere [48, 88, 89]. Recently, the linear temperature relaionship of $\log K$ was also established experimentally in reactions of H-atom transfer from fluorene to acridine molecules [90] and proton (in $CsHSO_4$ [91]).

The low apparent activation energies indicating the transition from an Arrhenius relationship to a low-temperature plateau were found by Dubinskaya and Butyagin [92, 93] in solid-state reactions of H atoms with organic polymers, *n*-hexadecane and malonic acid ($\sim$ 0.09, 0.05, and 0.07 eV, respectively, in the range 100–150 K). In the latter case they observed the activation energy growth to 0.26 eV in the range 200–250 K. In these reactions, H_2 and the matrix radical are formed; therefore, H-atom transfer kinetics is reliably indicated by the intensity change of well-defined lines of H atoms and free radicals in ESR spectra. Under polymethylmetacrylate deuteration the reaction rate decreases approximately 300-fold. In the reactions of atomic hydrogen with isopropanol and isopropanol-d_7 studied by Dainton and coworkers [94], the IE of H_2 and HD formation increases with temperature, decreasing from 300 to $\sim$150 K, where H_2 formation is 10-fold faster than that of HD. However, at 77 K there is no IE. It is assumed in ref. 94 that at low temperatures the reaction rate is limited by the H-atom diffusion rate, leading to the formation of the configuration required for their transfer. With rising temperature, the limiting stage turns out to be a transfer process itself characterized, unlike diffusion, by a considerable IE.

The IE growth with a temperature rise from 4.2 to 50 K was observed by Iwasaki and co-workers in the reactions of H- and D-atom abstraction from saturated hydrocarbon molecules in crystalline xenon matrices [95]. The temperature rise causes a growing selectivity of various CH-bond breaks under H-atom abstraction. The ratio of the accumulation rates of radicals formed under a break of primary, secondary, and tertiary CH bonds in isobutane is 1:1:1 at 4.2 K and 1:6:14 at 30 K [96–98]. These data also indicate the existence of the pretunneling molecular shifts of the reagents within the nearest coordination spheres. It is assumed in ref. 82 that these shifts are quasi-reversible, and the kinetic pattern of H-atom transfer reactions has the form

$$RH + R_1 \underset{k_{-1}}{\overset{k_1}{\rightleftharpoons}} [RH \ldots R_1] \overset{k}{\rightarrow} R + HR_1$$

where $[RH \ldots R_1]$ means a reagent configuration suitable for tunneling reaction.

Similar reactions between the stabilized H atoms and aliphatic hydrocarbons were observed by Wang and Willard [99] in glasslike matrices at 5–29 K. Activated transfer of H atoms occurs above 18 K; in the range 10–18 K the reaction rate is temperature independent, and it increases with further temperature lowering (such an increase is absent in D-atom reactions).

In reactions between carbenes and organic matrices studied by Platz and colleagues [100, 101],

$$\overset{R}{\underset{R}{>}}C: + HR_1 \longrightarrow \overset{R}{\underset{R}{>}}\dot{C}H + \dot{R}_1,$$

low apparent activation energies (~ 0.04 eV in the region 77–150 K) and a high isotope H–D effect (up to 10^3) were observed. At close values of the rate constant of H-atom transfer, IE changes over a wide range for different matrices (10–20 in toluene and methylcyclohexane to 500 in isopropanol).

In addition to the preceding reactions of H-atom transfer between nearest neighbors in a solid lattice, detailed studies were done on the diffusion-controlled reactions of H and D atoms in mixed molecular crystals of H isotopes in which the atoms were formed by H-plasm discharge condensation [104] and γ-irradiation or photolysis [105–109]. After formation of crystals, there is a decrease of D concentration with time and a growth of H concentration; the remaining amount of atoms in the sample (Figure 6) signifies that there is no recombination of atoms. The complementary change in concentrations is determined by the reaction

$$HD + D \rightarrow H + D_2;$$

the exothermicity of the reaction is ensured by the difference of zero-point vibration energies of HD and D_2, which equals 0.087 eV. This reaction rate

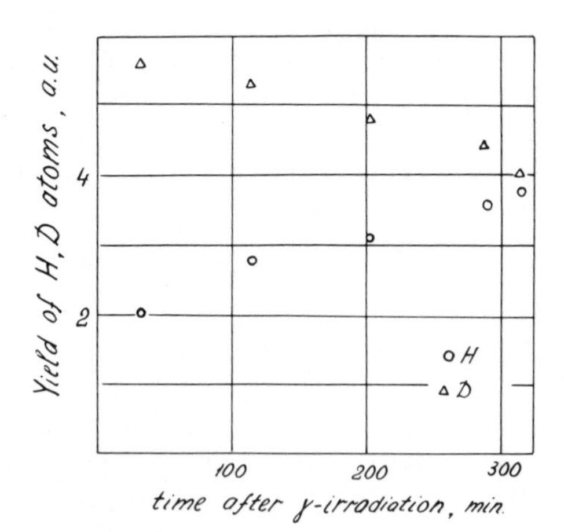

Figure 6. A change with time of H- and D-atom concentrations (in arbitrary units) after cutoff of γ-irradiation of mixed crystal D_2–HD (15.6 mol %) at 4.2 K determined by the reaction $HD + D \rightarrow H + D_2$.

constant is 4.2×10^{-27} cm^3/s at 1.9 and 4.2 K [109]. Collision of the reaction partners occurs due to D-atom quantum diffusion in the D_2 lattice to HD impurity molecules. The diffusion coefficient is 2×10^{-20} cm^2/s at 1.9–4.2 K and corresponds to the probability of D-atom transition between the neighboring traps ($W = 6D/a^2$, where $a = 3.6$ Å is the distance between the neighboring sites of D localization in the D_2 lattice) equaling $\sim 10^{-4}$ s^{-1}.

From the discussion of chemical reactions related to H-atom transfer let us now switch to reactions in which the low-temperature transfer of heavier atoms and molecular fragments occurs. As Lukas and Pimentel have shown [110], in a solid mix of NO, O_3, and N_2 with molar ratio 1:30:25, even at 10–20 K a reaction starts:

$$NO + O_3 \rightarrow NO_2 + O_2.$$

Its apparent activation energy is only 4.3×10^{-3} eV, while in the gas phase, depending on the electron state of the formed nitrogen dioxide, the activation energy is 0.10 or 0.18 eV. The reaction rate constant in the preceding temperature range is $\sim 10^{-4}$ s^{-1}.

After the formaldehyde polymerization described in the beginning of this section, the next, example of a low-temperature kinetic plateau in reactions related to heavy molecular fragment transfer was observed by Buchwalter and Closs [111] in the photoisomerization of a 1,3-cyclopentadiyl (CPDY) biradical isomerized into a bicyclo-(2,10)-pentane (BCP):

The final product was identified with chromatography. The isomerization rate was indicated by a decrease in the time of the biradical ESR signal. Due to the kinetic curve nonexponentiality the value of K was determined by the initial section of these curves. At temperatures above 20 K, the Arrhenius relationship of $K(T)$ was observed with an activation energy of $\sim$0.1 eV and a preexponential time of $\sim 10^8$ s^{-1}. The $K(0)$ value at $T \leqslant 20\,K$ is $\sim 10^{-3}$ s^{-1}. Though the $K(T)$ relationship is almost a duplicate of that mentioned for H-atom abstraction by the CH_3 radical from the surrounding matrix molecules (cf. curves 4 and 8 in Figure 5), the isotope H–D effect in reaction (17) is small, proving the absence of a considerable contribution of individual H-atom quantum shifts to this reaction. It was assumed elsewhere [111] that biradical isomerization is related to CH_3 group tunneling shift across a parabolic barrier 0.64 Å wide and 0.1 eV high.

The probability of the tunneling intramolecular rearrangement of 2-norboryl carbocation was considered in connection with the symmetry of its

NMR spectrum remaining up to 4.2 K and signifying the existence of fast (compared to Zeeman proton frequency, $\sim 10^9\ s^{-1}$) intramolecular transitions between the two asymmetric structures [112, 113]:

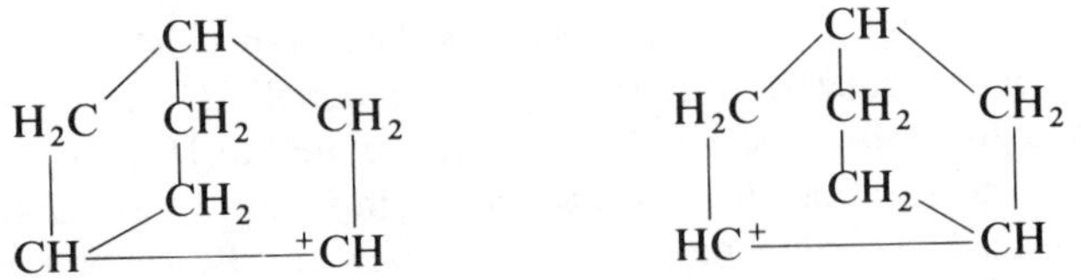

```
          CH                                  CH
        /  |  \                             /  |  \
   H2C    CH2   CH2                    H2C    CH2   CH2
    |      |      |                     |      |      |
    |     CH2     |                     |     CH2     |
    |   /         |                     |         \   |
   CH ---------- +CH                   HC+ ---------- CH
```

Fong [114] and Brickmann [115] suggested that interconversion between these states of the ion occurs by tunneling. This assumption was based on the fact that classical penetration of the energy barrier would be relatively fast only if its height is lower than 10^{-2} eV. In fact, the energy barrier in this system, as is shown with ESCA, is noticeably higher. The same results were also obtained by Carpenter [116] in theoretical studies of the intermolecular quantum and classical transitions in antiaromatic annulenes. The calculations [116] were confirmed in theoretical works by Dewar and Merz [117, 118], who showed that at the barrier height agreeing with experimental data ($\geqslant 0.1$ eV) CH_2 group tunneling transitions between the two bilaterally symmetric structures should prevail over the classical ones.

Studies of chain reactions of hydrobromination [119–121] and halidization [122, 123] of ethylene and chlorination of various saturated hydrocarbons and their derivatives [124–128] confirmed the existence of the low-temperature limit of the rate constant. The pattern of chain growth under hydrobromination of ethylene after γ-irradiation-induced HBr decomposition is as follows:

$$Br^0 + C_2H_4 \rightarrow C_2H_4Br^0$$

$$C_2H_4Br + HBr \rightarrow C_2H_5Br + Br^0$$

Both chain growth reactions are exothermal, their enthalpies being -0.35 and -0.45 eV, respectively. The activation energy in the region of 90 K is 0.17 eV; the plateau is observed below 50 K, where $K(0) = 20\ s^{-1}$. The absence of $K(T)$ changes under HBr substitution by DBr shows that H-atom transfer does not limit the rate of reactions.

The long-chain grown ($\nu \gtrsim 10^2$) in the mixed 1:1 crystals of ethylene with chlorine and bromine was studied elsewhere [122, 123]. The peculiarity of these crystals is the alternating quasi-one-dimensional packing of ethylene and halogen molecules determined by donor–acceptor interaction. The axis of halogen molecules is orthogonal to the parallel planes of ethylene molecules [129, 130]. At temperatures above 90 K a spontaneous molecular attachment of Cl_2 to C_2H_4 is observed related to the molecular rearrangement in the

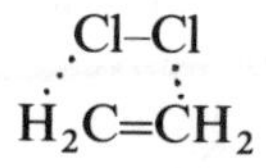

$$\begin{array}{c} Cl{-}Cl \\ \vdots \quad \vdots \\ H_2C{=}CH_2 \end{array}$$

complex [131]. The product of molecular attachment is gauche-1,2 dichloroethane, which at 115 K isomerizes into the transform [132]. The attachment rate constant is $\sim 10^{-5}\,s^{-1}$ at 90 K and obeys the Arrhenius law with activation energy 0.24 eV. According to Titov et al. [122], at lower temperatures the chain growth goes on only after photolysis of the halogen. The IR spectra changes in the process of halidization consist in the disappearance of the stretching vibration bands of ethylene and halogen and the appearance of the C–Cl (or C–Br) vibration bands and bridge vibration bands of *trans*-dichloroethane (or the transform $C_2H_4Br_2$). The stereospecificity of the reaction is determined by the fact that the radical forming under halogen atom attachment to ethylene can transform right into the $C_2H_4Cl_2$ or $C_2H_4Br_2$ trans forms, abstracting one more halogen atom from the neighboring molecule:

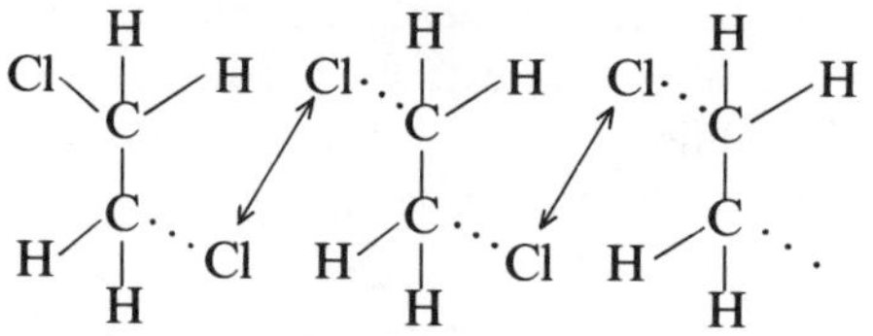

```
     H           H           H
Cl\  |  /H  Cl·. |  /H  Cl·. |  /H
    C            C           C
    |            |           |
    C·.          C·.         C·. .
H/  |  ·Cl  H/   |  ·Cl  H/  |
    H            H           H
```

In fact, the mixed crystal of the reagents is a very convenient prefabrication for chain solid-state transformation, ensuring the spatial transfer of free valence, which is not related to translational molecule shifts comparable to the lattice transition. As shown elsewhere in [123], the chain length at low temperatures is determined by the structural perfection of the crystal and seems to be limited by the existence of defects which break the reagent packing suitable for the reaction. Changing the conditions of crystal formation, it is possible to change the chain length from 50 to 250. Changes of ν are accompanied by a proportional change in the chain growth time, τ, proving the legitimacy of the relation $\tau = \nu/K$, which holds true under the condition that chain initiation is a faster process than chain growth. The $K(T)$ reaches the low-temperature plateau at temperatures below 45 K. Growth of $K(T)$ in the region 50–90 K corresponds to the apparent activation energy 0.035–0.040 eV. The $K(0)$ values in reactions of C_2H_4 with Cl_2 and Br_2 are 12 and $8\,s^{-1}$; that is, $K(0)$ does not depend on the halogen mass.

A series of works [124–128] studied the effect of the structural non-equilibrium of saturated hydrocarbon glasslike mixes with chlorine on the rate of chain photochlorination. By the beginning of this research much evidence has been obtained on the active role of the noncrystalline medium in the solid-state transformations, a part of which, referring to H-atom transfer reactions, is mentioned in the preceding. Peculiarities of these media, compared to the crystalline ones, are their excess free energy and the configurational entropy accumulated in the glass transition temperature region T_g, which slowly relaxes at $T < T_g$ (see, e.g., ref. 133). Evidently, chemical transformations in such a slowly relaxing medium should be accompanied by changes in its properties. A question arises whether these changes affect the rate of the chemical reactions causing them. The following experimental data, though obtained at higher temperatures, gave ground for an affirmative answer to this question:

1. The reaction rate depends considerably on sample preparation conditions, in particular the cooling rate near T_g, that also affect the excess energy of glasses [134].
2. The reaction rate sharply increases (up to several orders of magnitude) in the phase transition region [135, 136].
3. The kinetics of even the elementary solid-state reactions is normally nonexponential; this indicates the existence of a set of configurations characterized by different values of rate constants. The existence of this set of configurations distinguishes the glasslike state from the crystalline one.
4. The low-temperature reaction rate constants are most sensitive to outer mechanical stresses causing, in particular, the plastic deformation and fracture [137].
5. The autocatalytic character of many solid-state reactions indicates the medium change in the process of transformation [136].

A series of works [124–128] applied a combination of the chlorine consumption spectrophotometry and organic radical concerntration change evaluation by ESR spectra in the process of photolysis and after it.

The photochlorination of saturated compounds includes the cellular reaction (I) of particles formed under photolysis with their closest neighbors and the chain growth (II) initiated by the outgoing chlorine atoms and radicals:

$$[Cl_2 \ldots RH] \overset{h\omega}{\rightleftarrows} [Cl_2^* \ldots RH] \begin{array}{l} \overset{I}{\nearrow} RCl + HCl \\ \underset{II}{\searrow} R^0 + Cl^0 + HCl \end{array}$$

$$R^0 + Cl_2 \rightarrow RCl + Cl^0, \qquad Cl^0 + RH \rightarrow R^0 + HCl$$

The ultimate quantum yield of chlorine decomposition is

$$Y = Y_0 + \nu Y_r$$

where Y_0 and Y_r are the quantum yields of cellular reaction and radicals, respectively.

The glasslike mixes of methane and chlorine with molar ratio 5:1 prepared by joint reagent condensation (studied in ref. 122) are structurally non-equilibrium. Their structural relaxation lifetimes 10^4–10^2 s at 22–28 K is experimentally detected by the light diffraction change determined by the coalescence of submicroheterogeneities typical of glasses with internal stresses. The temperature relationship of methane photochlorination quantum yield changes sharply in the process of glass relaxation (Figure 7). For the initial, mostly nonequilibrium glasses the value of Y in the low-temperature plateau region is ~ 0.2 and decreases with further relaxation by more than 10^2 times. The apparent activation energy in the Arrhenius region increases from 0.1 to 0.28 eV.

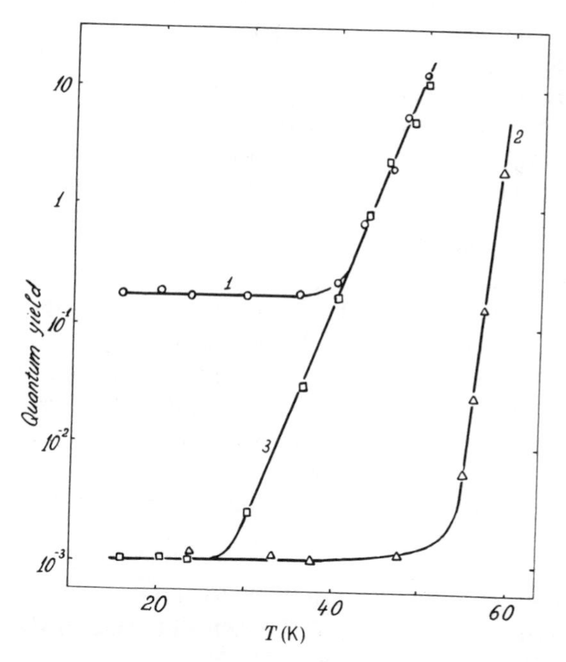

Figure 7. Temperature dependence of quantum yield of chlorine consumption in reaction of methane photochlorination in structurally nonequilibrium glassy mixes of reactants: (1) structurally nonequilibrium mix of reactants prepared by rapid spraying at 22 K; (2) same mix after thermal annealing at 30–40 K; (3) same mix after ultrasonic treatment at 22 K.

The $K(T)$ relationships for photochlorination of butyl chloride [124], methylcyclohexane [125], and other hydrocarbons [126] are composed of the low-temperature plateau at $T \lesssim 50$ K and the Arrhenius region with $E_a \simeq 0.2$ eV. The $K(0)$ value changes as a result of the thermal annealing of glasses from 10^{-1} to $10^{-3}\,\mathrm{s}^{-1}$.

Thus, the data of refs. 119–128 confirmed the existence of the low-temperature limit of the rate constant of solid-state chemical reactions found in refs. 57 and 58 and demonstrated an even stronger dependence of reaction rate on the properties of the medium than in the earlier discussed reactions with H-atom transfer.

The authors of ref. 51, where nonthermal critical effects typical of the low-temperature limit region were found, assumed that acceleration of reactions was determined by the energy barrier lowering due to the excess free energy of structurally nonequilibrium medium G and suggested a phenomenological relation:

$$V_m = V_m^0 - \alpha G, \tag{28}$$

in which V_m^0 relates to the equilibrium medium referring to which G is countes and α is a coefficient that has the significance of the Brensted coefficient relating the change of activation energy to the free energy of the reaction. The typical values of the excess energy of organic glasses are 30–50 J/cm^3 [139], and their change in the process of structural relaxation goes up by 10–20% [140], so the values of G for glasses turn out to be of the same order of magnitude as the activation energies of their low-temperature chemical reactions. The possibility of the autocatalytic acceleration of solid-state chemical reactions due to G accumulation as a result of local deformations during previous transformation leads to the formation of the energetic chains predicted elsewhere [52]. As follows from the set of data shown in Figure 4, the values of the rate constants in the low-temperature $K(0)$ plateau region for the studied chain reactions are several orders of magnitude higher than in the reactions with H-atom transfer, in spite of the considerably greater mass of the transferred particles. On this basis it was assumed in ref. 123 that these reactions, as well as explosive low-temperature reactions [51], run by the energetic chain mechanism. The macrokinetic peculiarities of such reactions are discussed in Section 6.

3. DEVELOPMENT OF THE NOTIONS ABOUT THE MECHANISM OF LOW-TEMPERATURE SOLID-STATE REACTIONS

The data in Figure. 4 show that the transition from an Arrhenius $K(T)$ relationship to the low-temperature $K(0)$ plateau predicted by various quantum models is actually a common feature of solid-state chemical reactions.

The common character of the phenomenon stimulated the introduction of quantum notions into the theory of chemical reactions, notions that up to the second half of the 1970s were based exclusively on the one-dimensional tunneling model. Though in the frameworks of this model one may attempt to explain a strong dependence of reaction rates on the medium properties by the existence of the barrier whose parameter spectrum is determined by the distribution of reagent configurations, the insufficiency of such treatment is shown by quantitative comparison with experiment even in the case of reactions with H-atom transfer.

Under one-dimensional tunneling across the static potential barrier, its height V_m is related to the transition temperature T_t from an Arrhenius relationship to the plateau by relation (4) at $E = 0$, and at the same time it determines the $K(0)$ value:

$$K(0) = \omega \exp\left(-\frac{V_m}{K_B T_t}\right). \tag{29}$$

Relation (29) holds true for the parabolic barrier. For approximation of the $K(T)$ relationship in the Arrhenius region it is required that V_m should at least exceed the maximum value of the apparent activation energy E_m. Therefore, applying relation (5), assuming $V_m = E_m$, we can find the maximum barrier width d_m; its correspondence to the van der Waals reagent radii sum is the most evident test of the model's fitness, since these are the equilibrium distances at which the reagents should be in the initial state in the solid phase. Besides, the V_m value should tally with the chemical bond break energy taking into account the reagent interaction in the initial state. Thus, actually, the $K(T)$ relationship in the one-dimensional tunneling model is determined by only one parameter, the equilibrium distance between the reagents, R_e.

For H-atom transfer in a dimethylglyoxime crystal the experimental value $T_t = 50$ K corresponds to the parabolic barrier height 0.18 eV, while $E_m = 0.4$ eV. Assuming $V_m = E_m$, the value of T_t becomes almost 20-fold greater than the one observed, and it is impossible to correctly describe the $K(T)$ relationship in the Arrhenius region. The Eckart barrier ($d = 0.44$ Å wide and $V_m = 0.65$ eV high), providing better agreement with experiment, also does not tally with crystallography data.

In the H-atom abstraction by the methyl radical in various matrices, experimental $K(T)$ relationships can be satisfactorily described by relation (5) for Gauss and Eckart barriers at V_m values from 0.48 to 0.65 eV and d_m values from 1.7 to 2.3 Å [11]. The d_m values correspond to the equilibrium intermolecular distances. However, at such distances the interaction between the reagents is weak, and there are no reasons for such sharp V_m decrease compared with the C–H bond energy in the matrix molecules. Calculated with the semiempirical LEPS method (see, e.g., ref. 18), the potential energy surfaces for reactions $RH + CH_3 \rightarrow R + CH_4$ correspond to V_m values of 2.5–3.0 eV, that is, several times higher than the experimental value [88, 89, 141]. This discrepancy is not accidental but a direct consequence of the main assumption of the barrier statistical model, ignoring the dependence of V_m on the distance between reagents. As an illustration, Figure 8 shows a two-dimensional potential energy surface (PES) for the reaction between CH_3 and CH_3CN [1]. The distances r_{AH} and r_{HB} between the tunneling atoms of hydrogen and carbon in the acetonitrile and radical molecules, respectively, are indicated on the coordinate axes. The reaction coordinate—the path of the fastest descent from reagent valley to product valley—is shown by the solid line. Along this coordinate incorporating the inflection point $R^{\neq}$ occurs the motion in gas phase reactions at high temperatures. The one-dimensional tunneling model assumes that the $R^{\neq}$ point is not reached and the transition runs under fixed arrangement of the molecule and radical when the value of $r_{AH} + r_{HB}$ is constant, that is, along the dashed line connecting points 2 and 3. High V_m values are determined by the choice of reaction path. The sectional view of the potential energy surface at $r_{AH} + r_{HB} = 4$ Å is also shown in Figure 8.

It is even more difficult to apply the one-dimensional tunneling model to reactions with heavy-particle transfer, where the barrier heights, judging by the apparent activation energy values, differ only very slightly from those of the H-transfer reactions. In spite of the difference in the masses, these reactions are characterized by almost the same temperature T_t and even higher values of $K(0)$ than H-transfer reactions. If V_m values are fixed, the comparable $K(0)$ values may be obtained if, in accordance with relation (2), the tunneling distance d of a particle with mass $m \gg m_H$ is shorter than that of the H-atom (d_H) by a factor $(m/m_H)^{1/2}$. Therefore, at $d_H \lesssim 1.5$ Å for CH_2 fragments, formaldehyde radicals, CO, NO, and halogen atoms involved in the studied reactions, d values should not exceed 0.2–0.4 Å; this does not in any case correspond to the initial equilibrium arrangement of the reactants in the solid lattice.

A natural development of the tunneling notions, eliminating the preceding drawback, are the multidimensional models accounting, along with the tunneling particle transition, for reactant convergence as a whole resulting from intermolecular vibrations which ensure the fluctuation preparation of the barrier [48–50]. The ideas of multidimensional tunneling emerged in the gas phase reaction research. Jonston and Rapp [142] seem to be the first to note, in the example of the reactions of H-atom abstraction by the methyl radical, that the isolated reaction coordinate was not separated if the de Broglie wavelength for the transferred particle exceeded the typical length of the region of fastest descent from reactant valley to product valley. Due to

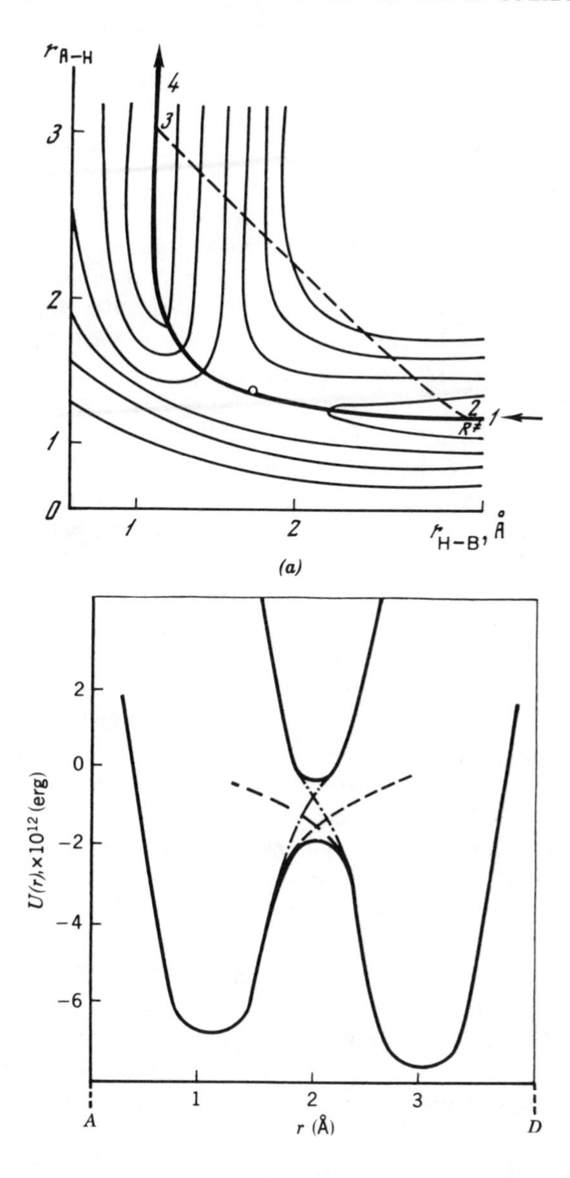

Figure 8. (*a*) Two-dimensional potential surface of $CH_3 + CH_3CN$. Dashed line shows H-atom tunneling at the fixed distance between C atoms of reactants of 4.0 Å. Solid (1–4) and dashed lines show the reaction coordinate and H-atom tunneling direction (from initial state 2 to final state 3). (*b*) Change of potential energy along this line. Dashed lines are Morse potentials of nonexcited CH chemical bonds in CH_3 and CH_3CN. The dash-dot lines show potential energy of C . . . H–C chemical bonds with account for a diagonal correction for term interaction. Solid lines show computation by LEPS method.

tunneling, the reaction path (trajectory leading to transition) does not pass across the saddle point $R^{\neq}$, and the tunneling paths "cutting the angle" become preferable, as shown in Figure 8.

In the derivation of relation (7) use was made of the fact that in the framework of the quasi-classical approximation the tunneling transition probability is an exponentially decreasing function of the imaginary part of the classical action $S(E)$ along the mechanical trajectory. George and Miller [9] generalized this treatment for the case of multidimensional PESs. It was noted [9] that, unlike the case of one-dimensional tunneling, in the "multi-dimentional" case the barrier becomes dynamical. A detailed review of the method of the quasi-classical S matrix was made by Miller [143]. It was shown [9, 144, 145], with the example of the $H + H_2$ exchange reaction, that at the low energy of reactants (i.e., below 1000 K) the set of paths making the basic contribution into the transition probability is located in the classically forbidden region, and with lowering temperature the tunneling effective distance increases. For instance, at 240 K the distance between H atoms in the symmetric state is 1.09 Å instead of 0.95 Å in the saddle point.

For the low-temperature solid-state reactions the multidimensional models suggested elsewhere [48–50] allowed coordination of a greater barrier height corresponding to the initial arrangement of reactants at equilibrium intermolecular distances, with the experimental $K(0)$ and T_t values; this was, as already mentioned, impossible in the one-dimensional models. Besides, the relationship between the reaction rate and the vibration spectrum of a solid which underlies these models led to a natural explanation of the role of the medium in low-temperature reactions. As an example illustrating the role of intermolecular vibrations, let us consider (following refs. 10 and 146) a collinear model of the exothermal reaction $AB + C \rightarrow A + BC$. Let us assume that the final-state spectrum is close to the continuous one and its slope is so large that a change of coordinate from the barrier top to the final-state minimum is small compared to a complete B-atom shift, equaling b (Figure 9). In this case the transition probability does not depend on the final-state parameters. Though this simplification is not of a principal character, it enables one to get an expression for the reaction rate constant in the analytical form, writing the transition conditions in the form

$$q_1 - q_2 = b \tag{30}$$

where q_1 and q_2 are the shifts along the intra- and intermolecular coordinates A–B and B–C, respectively. In a harmonic approximation, the potential energy surfaces in the reactant valley are a family of concentric ellipses. Condition (30) in Eyring coordinates, diagonalizing the kinetic energy operator [17], determines the straight line separating the reactant valley from the

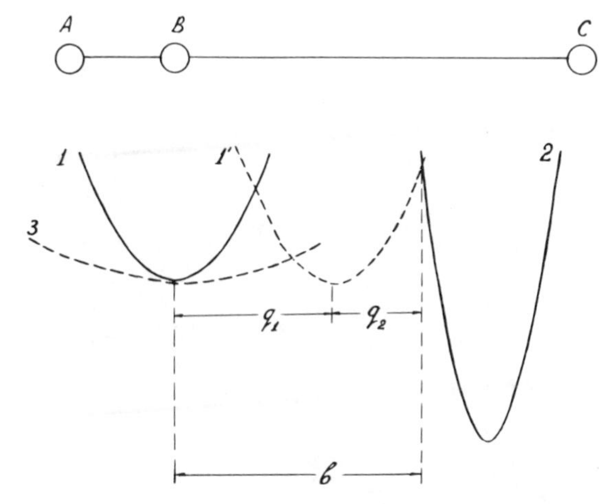

Figure 9. Model of two harmonic oscillators for analysis of AB + C → A + BC reaction: (1) term of initial AB state; (2) term of final BC state; (3) term of molecular vibrations.

product valley (Figure 10). The existence of two motions changes the picture of transition. Under constancy of the intra- or intermolecular coordinate the transition should have gone along lines OA_1 and OA_2, respectively, parallel to the coordinate axes. Due to a smaller value of the force constant of the intermolecular vibration compared to the intramolecular one, the barrier in point A_1 is lower than in point A_2. The barrier height corresponds to the saddle point A^* and is determined by a simultaneous shift along both coordinates:

$$V_m = \frac{K_{12}K_{23}}{K_{12}+K_{23}}\frac{b^2}{2}. \qquad (31)$$

However, the transition is not equivalent to one-dimensional tunneling along the potential surface section containing the saddle point since under two-dimensional tunneling the law of energy conservation is not satisfied along any one-dimensional path, and the path corresponding to the minimum action is not one-dimensional (Figure 10). The transition probability at $T = 0$ can be presented in the form

$$K(0) \sim \exp\left\{-\frac{b}{\hbar}\sqrt{2\mu V_m}\,F(\eta)\right\}, \qquad (32)$$

where $\mu^{-1} = m_A^{-1} + m_H^{-1}$ (μ is reduced mass) and $\eta = K_{13}/K_{12}$. The $F(\eta)$ function is shown in Figure 11 and demonstrates the difference between the

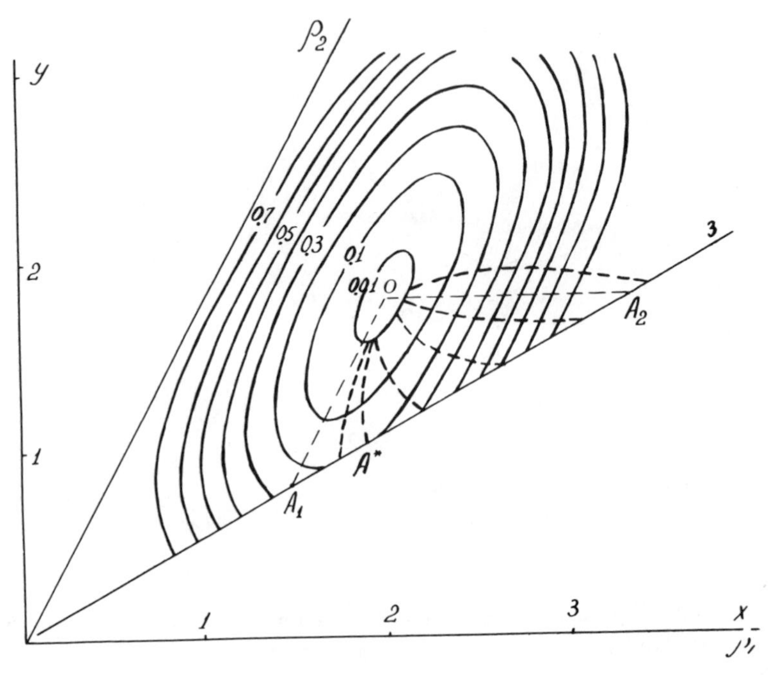

Figure 10. Equipotential curves for system of two coupled oscillators regarded as simplified collinear model of AB + C → A + BC reaction shown in Figure 9:

$$U = \frac{K_{12}}{2}(\rho_1 - R_1)^2 + \frac{K_{23}}{2}(\rho_2 - R_2)^2, \qquad \rho_1 = R_1 + q_1, \qquad \rho_2 = R_2 + q_2.$$

Condition of reaction (30) corresponds to curve 3, separating the reactant and product valleys (latter not shown). Values of energies to which equipotential curves correspond are given in following units: $U_0 = \frac{1}{2}K_{12}b^2$, $\eta = K_{23}/K_{21} = 0.2$, $R_1/b = 1$, $R_2/b = 2$, and $m_1 = m_2 = m_3$. The axes of the ellipses ρ_1 and ρ_2 are determined by relation $y = 0$ and $y = x\sqrt{3}$; x and y are Eyring coordinates [17]. Saddle point A^* corresponds to barrier minimum. Points A_1 and A_2 are intersection point of straight line 3 with ellipses axes. Dashed lines show transition paths.

two-dimensional model and the one-dimensional one for which $F(\eta) = 1$. Taking into account intermolecular vibrations, the $F(\eta)$ function decreases; thus, to the experimental $K(0)$ values there correspond reasonable V_m values, exceeding severalfold those calculated in the one-dimensional model. In the case $\eta \ll 1$ corresponding to experiment, relation (32) may be presented in the form

$$K(0) \sim \exp\left\{-\frac{b^2}{\delta_1^2 + (x/2 - 1)^2\delta_0^2}\right\}, \tag{33}$$

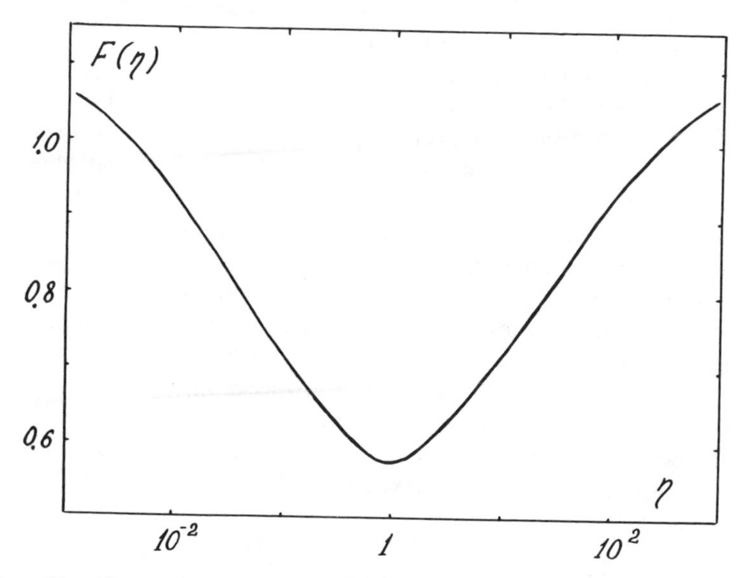

Figure 11. Change of exponent of Gamov factor for collinear harmonic model of two-dimensional tunneling in relation (32). the horizontal axis is the ratio of force constants of inter- and intramolecular vibrations: $\eta = K_{23}/K_{12}$, $m_1 = m_2 = m_3$.

where $\delta_0^2 = \hbar/\sqrt{\mu_1 K_{12}}$, $\delta_1^2 = \hbar/\sqrt{MK_{23}}$ are the amplitudes of the zero-point intra- and intermolecular vibrations, $M = (m_A + m_B)m_C/(m_A + m_B + m_C)$, $x = 2\mu_1/m_B$, and $\mu_1^{-1} = m_B^{-1} + m_C^{-1}$. Relation (33) shows that at $\delta_1 \gtrsim \delta_0$ the saddle point on the reaction coordinate is reached due to preferential shift along the intermolecular coordinate and not as a result of $\bar{A}$–B chemical bond stretching in the initial state at a fixed position of A and C. Generalization of relation (33) for a motion of the reactants located in neighboring sites of the solid lattice [10, 50] relates the transition probability to the mean-square amplitude of the reactant relative shift resulting from a totality of lattice vibrations of various frequencies equal, at $T = 0$, to

$$\delta_1^2(0) = \langle x^2(0) \rangle = \frac{\hbar}{M} \int_0^{\omega_m} \rho(\omega) \frac{d\omega}{\omega}, \tag{34}$$

where $\rho(\omega)$ is the frequency spectrum density of intermolecular vibrations (phonons) taking into account local modes emerging upon introduction of A into the BC lattice [10]; ω_m is the maximum frequency of the phonon spectrum. The simplified model considered shows effective decrease of the potential barrier in Figure 8 due to zero-point intermolecular vibrations. The reactant shift it creates leads to a shift in the reaction path in Figure 8 from the dashed curve corresponding to the constancy of the distance between reactants (islands) A and C to the solid curve corresponding to the gas phase reaction coordinate. The smallness of the barrier along this curve, typical of the discussed low-temperature exothermal reactions, explains the low values of activation energy. The analogy between the paths of the gas phase and solid state reactions, taking into account intermolecular vibrations, is however far from complete. In the latter case the motion along the reaction coordinate, in spite of the barrier smallness, is not free. Since the shifts along this coordinate are large compared to the amplitude of intermolecular vibrations, the $K(0)$ turns out to be 10–12 orders of magnitude lower than in the same reactions in the gas phase. The role of intermolecular vibrations turns out to be so significant because the two exponentially small probabilities—reactant convergence and tunneling—are comparable.

At temperatures comparable to $\hbar\omega_m/K_B$, growth of the intermolecular vibration amplitude is observed, described by the relation

$$\delta_1^2(T) = \langle x^2(T) \rangle = \frac{\hbar}{M} \int_0^{\omega_m} \rho(\omega) \coth\left(\frac{\hbar\omega_m}{2K_B T}\right) \frac{d\omega}{\omega} \tag{35}$$

and leading to the increased reaction rate constant. As noted elsewhere [50], the mean-square amplitude of the reactant vibrations $\langle x^2(T) \rangle$ is directly related to the Debye–Waller factor of the matrix. From relations (33) and (35) it follows directly that the $K(T)$ transition from the low-temperature plateau to the Arrhenius relationship occurs at $T_t \simeq \hbar\omega_m/K_B$ and satisfies the additional condition $2K_B T/M\omega_m^2 \gg \delta_0^2$, signifying that the amplitude of the classical intermolecular vibrations with frequency ω_m exceeds the amplitude of intramolecular vibrations. So, the transition temperature T_t is not related to the intramolecular vibration frequency along the reaction coordinate but is determined by the phonon spectrum of the matrix surrounding the reactants [48, 50, 102]. If lattice deformation by reactant molecules is not large, the ω_m value is close to the Debye frequency, and T_t corresponds to the Debye temperature θ_D. The θ_D independence of the tunneling particle mass and close values of θ_D for most organic matrices ($\theta_D \sim 100$ K) are the reasons for the observed approximate coincidence of T_t(see Figure 4) for various solid-state reactions.

Participation of low-frequency intermolecular vibrations in the fluctuation preparation of the barrier changes principally the mechanism of temperature dependence $K(T)$ compared to the one-dimensional tunneling model. According to relation (5), to the Arrhenius relationship there corresponds the predominance of thermally activated "over-the-barrier" transitions over the tunneling ones, which is determined in the harmonic terms model (Figure 1) by the thermal occupation of the highest vibrational sublevels of the initial state. Transitions from these sublevels become predominant at $T \gg T_t = \hbar\omega_i / K_B$, where ω_i is the intramolecular vibration frequency (see, e.g., ref. 147). High values of intramolecular frequencies and their dependence on the tunneling particle mass lead to T_t values contradicting experimental data. It is shown in the literature [30, 31, 148, 149] that at $T < \hbar\omega_i / K_B$ the contribution of the tunneling transitions from the highest vibrational sublevels of the initial state is insignificant. Increasing probability of transition from these sublevels does not compensate the low probability of their occupation in the studied temperature range.

The potential energy of two coupled oscillators (Figure 9) is a squared function of the shift along the intermolecular coordinate R:

$$U(R, q) = \tfrac{1}{2} K_0 q^2 + \tfrac{1}{2}\gamma q (R - R_0) + \tfrac{1}{2}\gamma_1 (R - R_0)^2, \tag{36}$$

where $R = R_0$ corresponds to the equilibrium position. At $K_{12} \gg K_{13}$, $K_0 \simeq K_{12}$, and $\gamma \simeq \gamma_1 \simeq K_{23}$. Substituting Eqn. (36) into relations (1) and (2), we can present the transition probability in the form

$$\omega(R) = \omega \exp[-J(R)] \tag{37}$$

where the function $J(R)$, being an imaginary part of the action, expands in a power series of $R - R_0$:

$$J(R) = J(R_0) + J'(R_0)(R - R_0) + \tfrac{1}{2} J''(R_0)(R - R_0)^2. \tag{38}$$

Expansion (38) (suggested in ref. 48) is applicable for arbitrary-shape terms, and in the case of parabolic terms it is precise. Under classical vibrations of reactants the probability of convergence of potential wells corresponding to the initial and final states of the tunneling particle equals

$$P(R) = \pi^{-1/2} \delta_1^{-1/2} \exp[-(R - R_0)^2 / \delta_1^2], \qquad \delta_1^2 = 2K_B T / \gamma. \tag{39}$$

Averaging Eqn. (39) by reactants shifts [50, 141], we obtain the relation

$$K(T) = \omega \left(1 + \frac{J''(R_0)}{\gamma}\right)^{-1/2} \exp\left\{-J(R_0) + \frac{[J'(R_0)]^2 K_B T}{2(\gamma + J''(R_0) K_B T)}\right\}, \tag{40}$$

which is a generalization of relation (33) for the case of arbitrary PESs. To the normally applied approximation of a linear bond (see, e.g., ref. 150) in expansion (38) there corresponds $J''(R_0) = 0$ when interaction of the intra- and intermolecular vibrations does not lead to changes in the frequencies and spectrum of the latter but only causes a shift of equilibrium positions. At $J''(R_0) = 0$ the linear relationship (27), of $\ln K$ and temperature in agreement with experiment, follows from Eqn. (40). The $K(T)$ relationship becomes Arrheniusian when the following conditions are satisfied:

$$\frac{J'(R_0)}{J''(R_0)} \gg \delta = \left(\frac{2K_B T}{\gamma}\right)^{1/2}, \qquad J''(R_0) > \frac{\gamma}{2K_B T}, \tag{41}$$

corresponding to the amplitudes of the thermal intermolecular vibrations exceeding the amplitude of the quantum intramolecular ones. The activation energy in the Arrhenius region is equal to $(J'/J'')^2\,(\gamma/2K_B)$.

Relations (33) and (40) show that the kinetic isotope H–D effect in H-transfer reactions is determined by the ratio of $\delta_0^2 \sim M^{1/4}$ and δ_1^2, which leads to its decrease with rising temperature and increasing dependence on matrix properties.

We should dwell on a comparison of the model suggested in refs. 48–50 with the models of medium reorganization [14, 24, 28], discussed in Section 1, in which the effect of fluctuation preparation of the barrier was not revealed. The main difference between the models, noted first by Flynn and Stoneham [151] in research on H diffusion in metals, lies in the fact that the models of medium reorganization [14, 24, 28] postulate constancy of the matrix element interaction. This corresponds to the Condon approximation (see also [33]), according to which the transition along the intramolecular coordinate occurs under fixed shifts along intermolecular coordinates. In this case, the fixed shifts create only the reorganization energy E_s but do not change the barrier parameters. As seen from Figure 10 the paths leading to fluctuation preparation of the barrier do not correspond to the Condon approximation. Evidently, not all normal lattice vibrations can cause barrier lowering; therefore, some of the vibrations, determined by the reactants geometry and surroundings, contribute only to reorganization energy (for example, the transverse vibrations of the reactants in the collinear model). As shown elsewhere [151], for impurity atoms positioned in the octahedral sites of the face-centered lattice, a fluctuation preparation of the barrier is possible, while it is impossible for tetrahedral sites in the body-centered lattice.

The main point of the preceding discussion is an assumption about the adiabaticity of intramolecular motion with reference to the intermolecular one, that is, division of the system into two subsystems: the fast one involving electrons and intramolecular vibrations and the slow one incorporating the intermolecular vibrations. This division of the motions was called a double adiabatic approximation (DAA) and was applied earlier in the theory of proton transfer in the reorganizing medium (see, e.g., refs. 10, 14, 28, and 31–33). The wavefunctions in DAA are presented as the products of the wavefunctions of the fast and slow subsystems:

$$\psi_{m,v}(\bar{q}, \bar{R}) = \phi_m(\bar{q}, \bar{R})\,\varphi_{m,v}(\bar{R}). \tag{42}$$

The condition of applicability of DAA is the smallness of the matrix elements of the nonadiabacity operator $\hat{T}_N$ (the kinetic energy of a slow subsystem) compared to the difference in energies of slow subsystem levels:

$$|\langle \psi_{m,v}| T_N(R) |\psi_{m',v}\rangle| \ll |E^0_{m,v} - E^0_{m',v}|. \tag{43}$$

An estimation corresponding to Eqn. (43) can be easily obtained for the preceding collinear model of two coupled oscillators [88]. The variables diagonalizing the potential energy of the system are $q + \frac{1}{2}\eta(R - R_0)$ and $R - R_0$, where $\eta = K_{23}/K_{12}$ is the ratio of force constants of inter- and intramolecular vibrations. Therefore, for the derivative of the intramolecular wavefunction $\phi_m(q, R)$ along the intermolecular coordinate a relation is true:

$$\frac{\partial \phi_m}{\partial R} = \frac{\eta}{2}\frac{\partial \phi_m}{\partial q}.$$

Applying a normal estimation of matrix elements,

$$\left\langle \phi_m(q)\frac{\partial \phi_{m'}(q)}{\partial q}\right\rangle \sim \delta_0^{-1}, \qquad \left\langle \varphi_{m,v}(R)\frac{\partial \varphi_{m,v}}{\partial R}\right\rangle \sim \delta_1^{-1},$$

$$\left\langle \varphi_{m,v}\frac{\partial^2 \varphi_{m,v}}{\partial R^2}\right\rangle \sim \delta_1^{-2},$$

where $|E_{m,v} - E_{m',v}| \geqslant \hbar\omega_0$, we obtain

$$\frac{\omega_1}{2\omega_0} + \eta\sqrt{\frac{\mu}{M}\frac{\omega_1}{\omega_0}} \ll 1. \tag{44}$$

Relation (44) shows that the DAA is true when the frequencies and force constants of intramolecular vibrations are considerably higher than those of intermolecular ones.

At present, the model of solid-state chemical reactions suggested in the literature [48–50] has won certain recognition. McKinnon and Hurd [152] and Siebrand and co-workers [153] compare the mechanism of rate constant temperature dependence by the occupation of the highest vibrational sublevels of the tunneling particle to that of fluctuation preparation of the barrier; the latter Siebrand et al. is preferred. [153] particularly emphasize the experimental proof of the linear dependence of the rate constant logarithm on the temperature predicted by this model. The importance of an account for intermolecular vibrations in the problem of heavy-particle tunneling [48–50] is also noted elsewhere [103].

4. ROLE OF INTERMOLECULAR VIBRATIONS IN SOLID-STATE REACTION KINETICS

The evaluation made in the preceding section shows the possibility of a natural explanation of the regulation of low-temperature solid-state reactions in a model accounting for barrier parameter oscillations resulting from intermolecular vibrations. A consistent analysis of such a model is required . A mathematical body used for this purpose is, conceptually, close to the common theory of nonradiative transitions, but unlike the latter it enables us to exceed the limits of the one-dimensional Franck–Condon approximation which is inapplicable in treatment of heavy-particle transfer.

We shall regard a chemical reaction as a nonradiative transition from the term of the initial bound state $\varepsilon_i(\bar{\Gamma})$, where $\bar{\Gamma}$ is the aggregate of nuclear coordinates, to the term of the final state, $\varepsilon_f(\bar{r})$, which can be either bound or dissociate. The terms of initial and final states are obtained from solution of Schrödinger equations with the electron–nuclear system Hamiltonian function in adiabatic approximation and without taking into account interaction between the terms. Accounting for the interaction the crossing terms ε_i and ε_f transform into the noncrossing adiabatic terms ε_l and ε_u; this is shown schematically in Figure 8*b*. Depending on the value of the Landau–Zener factor [13], the transitions are divided into adiabatic ($P \gg 1$) and nonadiabatic ($P \ll 1$) ones. At $P \gg 1$ the ε_i and ε_f term splitting is high, and transition occurs by the ε_1 term. When $P \ll 1$, nonadiabatic transition from the ε_i to the ε_f term occurs. After these preliminary remarks we obtain, following the work by Ovchinnikova [49], a general expression for $J(R)$ in relation (37) not based, unlike previously discussed, on any specific model of transition. It should be noted, however, that the results in refs. 49 and 50 are true only when the final intramolecular states belong to the continuous spectrum (see ref. 161). Let us consider H-atom transfer between heavy reactants A and C, assuming PESs of the AHC collinear model as known. Under the classical motion of A and C in a potential V_a with energy E the H-atom transfer probability equals

$$\omega(R) = \Omega(R)\exp[-2\tilde{S}(R, \varepsilon(R))], \tag{45}$$

where $\tilde{S}$, the imaginary part of the action, is calculated along the cross section at a constant R:

$$\varepsilon(R) = V_a(R) + \tfrac{1}{2}\hbar\omega_0 \tag{46}$$

where ω_0 is the HC-bond vibration frequency. The motion of reactants remains classical up to the turning point R_t, where $V_a(R_t) = E$. The reaction however can also run at $R < R_t$, that is, in the classically inaccessible region, with a probability

$$\omega(R)|\psi_E(R)|^2 \sim \omega(R)\exp\left\{-2\int_{-R_t}^{R} P(R')\,dR'\right\},$$
$$P(R) = \sqrt{2M(V_a(R) - E)}. \tag{47}$$

It follows from Eqn. (47) that the reaction path is determined by the minimum of function

$$\phi(R) = \tilde{S}(R, \varepsilon(R)) + \frac{1}{\hbar}\int_{R_t}^{R} \sqrt{2M(V_0(R) - E)}\,dR' \tag{48}$$

that is, by the ratio of the probabilities of H-atom tunneling motion and A and C reactant convergence (classical or quantum). At $R = R^*$, where $\partial\phi/\partial R = 0$,

$$K(T) \sim \exp(-\phi(R^*) - E^*/K_B T), \tag{49}$$

apparent activation energy is determined by this equation, which is a generalization of Eqn. (7) for the case of multidimensional motion (see also refs. 9, 143).

The rate constant of adiabatic as well as nonadiabatic transitions may be presented by the same relation:

$$K = \frac{2\pi}{\hbar} A v_i \sum_f |\langle f|\hat{H}|i\rangle|^2 \delta(E_f - E_i), \tag{50}$$

where $\langle f|$ and $|i\rangle$ are the wavefunctions of the initial and final states of the system and $\hat{H}$ is its full Hamiltonian function. Due to a difference in the terms of adiabatic and nonadiabatic transitions, the matrix elements, in this case $\langle f|\hat{H}|i\rangle$, are different. The $A v_i$ signifies statistical averaging by initial states.

In DAA the intramolecular wavefunctions depend on intermolecular coordinates $\bar{R}$, as on a parameter; and in the matrix elements involved in Eqn. (50) we can separate a matrix element at the electron–vibration intramolecular wavefunctions $V(\bar{R})$, and consequently, for both adiabatic and nonadiabatic intramolecular transitions the following equation holds:

$$\langle f|\hat{H}|i\rangle = \langle v_f|V(R)|v_i\rangle, \tag{51}$$

where $|v\rangle$ and v are the wavefunctions and quantum numbers of the slow subsystem. If the matrix element $V(R)$ depends only weakly on R, it can be taken out of the integral (51), and the rate constant is then determined by the Franck–Condon factors of the intermolecular subsystem. This approximation holds, in particular, under electron transfer, whose localization radius is much greater than the intermolecular vibration amplitude. However, in heavy-particles transition, where, as mentioned in Section 3, the effect of intermolecular vibrations is significant because the amplitudes of the intra- and intermolecular vibrations are of the same order [see relation (33)], the $V(R)$ is an exponential function of intermolecular coordinates, and the account for this dependence is the basis of the model.

In the theory of nonradiative transitions, the following methods were applied for the sums in Eqn. (50): the method of generating polynominals [154], operational calculus [155], and the matrix density method [13]. The latter seems to be the most justifiable in the case of deviations from Condon approximation, which are important in this problem. It should be emphasized that the problem is not confined to the account of some members of the fast subsystem matrix element expansion in the slow subsystem coordinates [156]. The $V(R)$ function is too sharp for such expansion to be satisfactory, though the expansion of the exponent index (38) remains tolerable.

The result of summations in Eqn. (50) is largely determined by the spectrum of the intramolecular subsystem final state. The spectrum can be either discrete or continuous. For instance, in the case of H-atom abstraction from the molecule by the methyl radical the value of the thermal effect of the reaction is such that the energy evolved may be sufficient only to excite the fourth vibration level of the C–H bond of the CH_4 molecule. In such small molecules the quasi-continuum region lies much higher [157]. For the symmetric reaction of radical pair transformation in dimethylglyoxime the thermal effect is nought, and the discreteness of the final intramolecular spectrum, in this case, is evident. If, however, as a result of the reaction, highly excited multiatom molecules are formed or dissociation of the excited molecules occurs, the intramolecular subsystem final state spectrum is continuous.

To clarify the question of the chemical reaction heat distribution in the vibrational degrees of freedom of the product, let us compare the matrix elements of the transition from the fundamental initial state to various final vibrational states, assuming for the sake of definiteness that the transition is nonadiabatic. Applying the known expressions for the Franck–Condon factors of harmonic oscillators, we obtain

$$\langle f|V|i\rangle \sim \exp\left[-\frac{\xi_0^2}{2}+\frac{1}{2}\sum_j \xi_j^2\right]\frac{\xi_0^{2n_0}}{n_0!}\prod_j \frac{\xi_j^{2v_{fi}}}{v_{fj}!}, \qquad (52)$$

where ξ_0 and ξ_j are the shifts along the intra- and jth intermolecular coordinates in units of mean-squared amplitudes of the respective vibrations; n_0 and v_{fj} are their quantum numbers. It follows from Eqn. (52) that excitation of intramolecular vibrations with the highest possible value is the most probable of intramolecular vibrations with the highest possible value $n_{0m} = [\Delta/\hbar\omega]$ (the brackets indicate that this is the integer part of a number) is the most probable since with decreasing n_0 the low-frequency intermolecular vibrations, according to the law of energy conservation, should be excited for which $\xi_j \ll \xi_0$. Due to this the heat of the chemical reaction is almost completely transferred to the valence bond formed, and only the energy defect is compensated by the excitation of low-frequency vibration modes, which is why in Eqn. (50) only summations by intermolecular subsystem states (v_i, v_f) remain. Here, it is convenient to rewrite the argument of the δ function, introducing the energy of the intermolecular subsystem, $E_{i,f}$, and the energy defect, ΔE, of the intramolecular subsystem.

$$E_f - E_i = E_{v_f} - E_{v_i} - \Delta E. \tag{53}$$

Appplying the integral representation of the δ function and taking into account that the intramolecular subsystem energy defect is independent of intermolecular coordinates, following ref. 13, we obtain

$$\begin{aligned} K &= (1/\hbar) \int dt\, e^{-i\Delta E t}\, \mathrm{Sp}\{V(R)\rho^i_{\beta+it}(R,R')\,V(R') \\ &\quad \times \rho^f_{-it}(R,R')\}/\mathrm{Sp}\{\rho^i_\beta(R,R')\}, \\ \beta &= 1/K_\beta T. \end{aligned} \tag{54}$$

Thus, knowing the expressions for density matrices of the intermolecular subsystem in the initial $[\rho^i_{\beta+it}(R,R')]$ and final $[\rho^f_{-it}(R,R')]$ states as well as the form of the matrix elements, $V(R)$, calculation by Eqn. (54) presents no principal difficulties.

Normally, in solid-state problems a harmonic approximation is used enabling one to introduce normal coordinates for the intermolecular subsystem. The density matrix in these coordinates equals the product of the density matrices of independent oscillators, the expressions for which are well-known [158].

In Eqn. (54) the density matrices of the initial and final states of the intermolecular system are present. It should be taken into consideration that the equilibrium coordinates, natural frequencies, and directions of the principal axes may not coincide for the initial and final states; consequently, the normal coordinates in these states may also be different. At the same time the relative density matrices must be written in the same coordinates; thus, the derivation of an expression for the transition rate constant was made [88] in an arbitrary system of coordinates not necessarily coinciding with the normal ones. For this purpose (as in ref. 13), the coordinates Q were derived and normalized so that the Hamiltonian function of the intermolecular subsystem in the harmonic approximation has the form

$$H^{i,f} = \frac{1}{2}\left[(Q - Q_0^{i,f})(\Omega_{i,f})^2 (Q - Q_0^{i,f}) + \frac{\partial^2}{\partial Q^2} \right], \tag{55}$$

where $Q_0^{i,f}$ and $\Omega_{i,f}$ are the equilibrium values of coordinates and frequency tensors in the initial and final states.

Since the matrix elements $V(Q)$ depend exponentially on the intermolecular coordinates, they can be conveniently represented in the form

$$V(Q) = V(Q_0) \exp[-\tfrac{1}{2}J(Q)]. \tag{56}$$

Further, the function $J(Q)$ is expanded in a shift series from the equilibrium state, retaining two terms:

$$\begin{aligned} J(Q) &= J'(Q_0)(Q - Q_0) + (Q - Q_0)[\tfrac{1}{2}J''(Q_0)](Q - Q_0), \\ Q_0 &= \tfrac{1}{2}(Q_0^i + Q_0^f), \end{aligned} \tag{57}$$

where $J'(Q_0)$ is the vector and $J''(Q_0)$ is the tensor. Thus, the integrand in Eqn. (54) is an exponential function having a quadratic form of coordinates as the exponent; simple transformations lead, as shown elsewhere [88], to the following expression for the transition rate constant:

$$\begin{aligned} K &= \frac{1}{\hbar}|V(Q_0)|^2 \int_{-\infty}^{\infty} dt' \{\det \phi\}^{-1/2} \exp\{-i\Delta E t' + \tfrac{1}{4}[(T_i - T_f)\bar{\Delta} + J'] \\ &\quad \times [T_i + T_f + \tfrac{1}{2}J'')]^{-1}[(T_i - T_f)\bar{\Delta} + J'] - \tfrac{1}{4}\bar{\Delta}(T_i + T_f)\bar{\Delta}\}, \\ T_i &= \Omega_i \tanh[\tfrac{1}{2}(\beta + it')\Omega_i], \qquad C_i = \Omega_i \coth[\tfrac{1}{2}(\beta + it')\Omega_i], \\ T_f &= \Omega_f \tanh[\tfrac{1}{2}it'\Omega_f], \qquad C_f = \Omega_f \coth[\tfrac{1}{2}it'\Omega_f], \\ \Phi &= [2\cosh(\tfrac{1}{2}\beta\Omega_i)]^{-2}(\Omega_i)^{-1} \sinh[(\beta + it')\Omega_i] \\ &\quad \times (T_i + T_f + \tfrac{1}{2}J'')(C_i + C_f + \tfrac{1}{2}J'')(\Omega_f)^{-1} \cosh(-it'\Omega_f). \end{aligned} \tag{58}$$

From ref. 58 it follows that the rate constant value and its temperature dependence are considerably affected by three factors: potential barrier parameter oscillations (their role is determined by the J' and J'' values), reorganization of medium resulting from a transition that can be described by the displacement vector of equilibrium coordinates ($\bar{\Delta}$), and a fast subsystem energy defect (ΔE).

The expression for the rate constant is much simpler when the intermolecular vibration frequencies and the system of normal coordinates do not change after transition. Even the failure of normal vibrations in the initial and final states to coincide is neglected since the frequency effect manifests itself in a higher order of adiabatic approximation than in the change of equilibrium coordinates [159].

The $T_{i,f}$, $C_{i,f}$, and $\Omega_{i,f}$ tensors in normal coordinates are diagonal, and expression (58) is converted to

$$K = \frac{1}{\hbar} |V(R_0)|^2 \int_{-\infty}^{\infty} dt' \{\det \phi\}^{-1/2} \exp\left\{ \frac{\Delta E}{2K_B T} - i\Delta E t' + \frac{1}{4} \sum_{\mu,\mu'} A_\mu (B^{-1})_{\mu\mu'} A_{\mu'} - \frac{1}{2} \sum_\mu (\Delta q_\mu)^2 \times \frac{\sinh(\beta\hbar\omega_\mu/2)}{\cosh(\beta\hbar\omega_\mu/2) + \cos(\hbar\omega_\mu t')} \right\},$$

$$A_\mu = \frac{\partial J}{\partial q_\mu} + 2i\Delta q_\mu \frac{\sin(\hbar\omega_\mu t')}{\cosh(\beta\hbar\omega_\mu/2) + \cos(\hbar\omega_\mu t')}, \tag{59}$$

$$B_{\mu\mu'} = \frac{1}{2} \frac{\partial^2 J}{\partial q_\mu \partial q_{\mu'}} + \frac{2\delta_{\mu\mu'} \sinh(\beta\hbar\omega_\mu/2)}{\cosh(\beta\hbar\omega_\mu/2 + \cos(\hbar\omega_\mu t')},$$

$$\Delta q_\mu = q_\mu^{0f} - q_\mu^{0i},$$

where q_μ and $q_\mu^{0i,f}$ are the phonon coordinates and their equilibrium values in the initial and final states corresponding to the frequency.

Equation (59) enables us to analyze the rate constant in the high- and low-temperature regions and for different models of a solid [88, 160, 161].

At low temperatures ($T \ll \hbar\omega_D/2K_B$, $T \ll \gamma/J'' K_B$, where ω_D is the Debye frequency), the expression for the rate constant has a rather simple form [88]:

$$K \sim \exp\left\{ \frac{\Delta E - |\Delta E|}{2K_B T} + \frac{1}{2} \sum_\mu \left[\frac{(J'_\mu)^2}{4} - (\Delta q_\mu)^2 \right] \coth \frac{\beta\hbar\omega_\mu}{2} \right\},$$

$$J'_\mu = \frac{\partial J}{\partial q_{\mu'}} \tag{60}$$

The low-temperature limit of the rate constant is obtained from Eqn. (60) when $\coth(\beta\hbar\omega_\mu/2)$ is replaced by unity. It is clear that the low-temperature limit exists only for nonendothermal reactions. This condition is satisfied by the first term in the exponent, which at $\Delta E > 0$ turns into nought and at $\Delta E < 0$ (endothermal process) leads to an activation dependence of the rate constant on temperature. The next term is responsible for the effective decrease of the potential barrier due to lattice vibrations. The last term corresponds to tunneling rearrangement of the medium accompanying particle transfer.

With rising temperature $[\sinh(\frac{1}{2}\beta\hbar\omega_D) \ll (\hbar\omega_D/E)\{(\hbar J'^2/4M\omega_D) + \Sigma_\mu(\Delta q_\mu)^2\}$, but $T \ll \gamma/J'' K_B]$, the expression for rate constant (60) transforms into

$$K \propto \exp\left\{\frac{\beta\Delta E}{2} + \frac{1}{8}\sum_\mu (J')^2 \coth\frac{\beta\hbar\omega_\mu}{4} - \frac{1}{2}\sum_\mu (\Delta q_\mu)^2 \tanh\frac{\beta\hbar\omega_\mu}{4}\right\}. \quad (61)$$

Proceeding from Eqs. (60) and (61), it is possible to obtain the dependence of the tunneling rate constant on temperature; it is only necessary to specify the model of a solid. If the reactants differ only slightly from the medium molecules in mass and interaction constant, as, for instance, in the case of radical pair transformation in dimethylglyoxime, then the medium may be regarded as an ideal crystal lattice. Besides, to simplify treatment by Eqs. (60) and (61), we can assume $\Delta q_\mu = 0$, considering the reorganization of the medium to be small.

As is known, at low temperatures a Debye model of a solid is accurate; applying it, we obtain from Eqn. (60), at $\Delta q_\mu = 0$ for the three-dimension lattice,

$$K \propto \exp[\tfrac{1}{2}\beta(\Delta E - |\Delta E|) + \alpha' T^4], \qquad \alpha' = \frac{\pi^6}{60}\frac{J'^2 K_B^4}{\hbar^3 \omega_D^5}. \quad (62)$$

It is clearly seen that at low temperatures the tunneling rate constant depends rather weakly on temperature; only at $T \simeq \hbar\omega_D/4K_B$ [see Eqn. (61)] there occurs a "defreezing" of temperature dependence and the rate constant starts to grow rapidly with temperature. Expanding the function $\coth(\beta\hbar\omega_\mu/4)$ in Eqn. (56) at $\Delta q_\mu = 0$, we get

$$K \propto \exp[\tfrac{1}{2}\beta\Delta E + \alpha'' T], \qquad \alpha'' = \frac{K_B}{2\hbar}\sum_\mu \frac{|J'_\mu|^2}{\omega_\mu}. \quad (63)$$

Furthermore, assuming $T > \Delta E/2K_B$ the temperature dependence of the rate constant would be almost completely determined by the linear term of the exponent. This result tallies with estimate (40) to within the preexponential factor.

The preceding results were obtained when the quadratic term of expansion (57) was neglected. At the same time, the treatment of the transfer rate at temperatures comparable to the Debye temperature and transition to the activation dependence are possible only when taking into account the quadratic term of the expansion [50, 88]. However, it seems unreal to obtain a clear-cut form of the rate constant temperature dependence for the general case taking into account $J''_{\mu\mu'}$ due to the difficulties in calculating the inverse matrix $B^{-1}_{\mu\mu'}$ [see Eqn. (58)]. At the same time, in the studied temperature range the Einstein approximation is valid for a solid lattice description; in this approximation the molecular displacements from equilibrium positions are used as the normal coordinates. The matrices A_μ and $B^{-1}_{\mu\mu'}$ in this case acquire a simple form, and as shown in ref. 161, we have the following relation for the transition rate constant:

$$\begin{aligned} K &= \frac{2\pi}{\hbar}|V(R_0)|^2\left[\left(1+\frac{\bar{J}''}{2}\coth\frac{\beta\hbar\Omega}{4}\right)\left(1+\frac{\bar{J}''}{2}\tanh\frac{\beta\hbar\Omega}{4}\right)^{-1/2}\right] \\ &\times \exp\left[\frac{\beta\Delta E}{2}+\varphi_1(1-\varphi_2)-\frac{E_s}{\hbar\Omega}\coth\frac{\beta\hbar\Omega}{2}\right] \\ &\times \frac{I_{\Delta E}}{\hbar\Omega}\left(\varphi_1\varphi_2+\frac{E_s/\hbar\Omega}{\sinh(\beta\hbar\Omega/2)}\right)\rho_f, \\ E_s &= \frac{\hbar\Omega}{2}\sum_\mu(\Delta q_\mu)^2, \qquad \bar{J}' = J'\left(\frac{\hbar}{M\Omega}\right)^{1/2}, \qquad \bar{J}'' = J''\frac{\hbar}{M\Omega}, \\ \varphi_1 &= \frac{\bar{J}'^2}{4}\left[\tanh\frac{\beta\hbar\Omega}{4}+\frac{\bar{J}''}{2}\right]^{-1}, \qquad \rho_f = (\hbar\Omega)^{-1}, \\ \varphi_2 &= \frac{\sinh(\beta\hbar\Omega/4)/\cosh^3(\beta\hbar\Omega/4)}{\tanh(\beta\hbar\Omega/4)+\bar{J}''/2}. \end{aligned} \tag{64}$$

Here Ω is the Einstein frequency, E_s is the energy of the medium reorganization, $I_{\Delta E/\hbar\Omega}$ is the Bessel function of imaginary argument, and ρ_f is the final-state density.

At higher temperatures, when the Bessel function argument becomes large, it is possible to apply its asymptotic expansion. Furthermore, for temperatures at which intermolecular vibrations are classical ($T > \hbar\Omega/4K_B$), the rate constant takes the form

$$K = \frac{2\pi}{\hbar} |V(R_0)|^2 \left[4\pi E_s K_B T \left(1 + \frac{J'' K_B T}{\gamma} \right) \right]^{-1/2} \exp\left\{ -\frac{(\Delta E - E_s)^2}{4K_B T E_s} + \frac{J'^2 K_B T}{2(\gamma + J'' K_B T)} \right\}; \qquad \varphi_1 \varphi_2 \ll \frac{E_s}{\hbar\Omega \sinh(\beta\hbar\Omega/2)}. \tag{65}$$

It is easily seen that contributions to the temperature dependence have been separated from rearrangement of the medium and the reactants. The barrier, stipulated by medium organization, is determined by the first term of the Eqn. (60) and the second term accounts for the medium effect on the barrier penetrated by the particle. If $\Delta E = E_s$, then only this second term determines the temperature dependence of the rate constant. At $T < \gamma / J'' K_B$, the second term in Eqn. (65) gives a linear dependence of the exponent on temperature typical for the tunneling of a particle [88]. In case of higher temperatures, $T > \gamma / J'' K_B$, the temperature dependence of the rate constant is totally of the activation type, and the activation energy equals the sum of two terms, the first determined by the medium rearrangement and fast subsystem energy defect [the first term of Eqn. (65)] and the second (as shown in ref. 161) determined by the energy spent on convergence of potential wells to complete disappearance of the potential barrier. In this case the rate constant equals the product of the transition rate constant in the absence of the barrier and the probability of such a fluctuation.

Thus, expression (59) enables us to describe the solid-state reaction rate constant dependence on the parameters of the potential barrier and medium properties in a wide temperature range, from liquid helium temperatures when the reaction runs by a tunneling mechanism to high temperatures (naturally, not exceeding the melting point) when the transition is of the activation type.

The cited formulas enable one to describe the kinetic isotope effect (KIE), which, as was mentioned in the preceding section, depends on the temperature and medium properties. Since the KIE is independent of the reorganization of the medium, the effect of the latter is completely confined to the potential barrier parameter oscillations under intermolecular vibrations. An increase in the amplitude of thermal vibrations with temperature leads to a decreased KIE [162].

Let us switch to a discussion of particle transfer when the final states of the intramolecular subsystem belong to the continuous spectrum [161]. This is different from the case of discrete final states, where the transition from the initial formula describing the rate constant in DAA to expression (54) is executed taking into account only one initial and one final—closest by energy—state of the intramolecular (fast) subsystem.

Summation over the initial intramolecular states in Eqn. (50) can be omitted in this case too; as for the final states, their energy is determined by the relation

$$E_{n_f} = E_{n_i} + (E_{v_i} - E_{v_f}), \tag{66}$$

where $E_{n_{i,f}}$, $n_{i,f}$ are the energies and quantum numbers of the initial and final states of the intramolecular subsystem, numbered by the index pairs (n_i, v_i) and (n_f, v_f).

Going, in Eqn. (50), from summation to integration by E_{n_f} (in this case the wavefunctions of the final state of the intramolecular subsystem are normalized by the energy), we come to

$$K = \frac{2\pi}{\hbar z_i} \sum_{v_i, v_f} \exp[-\beta E_{v_i}] \langle v_i | V(E_{n_f}, R) | v_f \rangle \langle v_f | V(E_{n_f}, R') | v_i \rangle,$$
$$z_i = \sum_{v_i} \exp[-\beta E_{v_i}]. \tag{67}$$

In the case of weak dependence of the matrix element $V(E_{n_f}, R)$ on the energy transferred to the intermolecular subsystem, it is possible to show [161] that the rate constant is expressed in terms of the diagonal elements of the density matrix of the intermolecular subsystem initial state [50]:

$$K = \frac{2\pi}{\hbar \, \mathrm{Sp}\{\rho^i_\beta(R, R')\}} \mathrm{Sp}\{|V(E_{n_i}, R)|^2 \rho^i_\beta(R, R')\}. \tag{68}$$

The matrix element $V(E_{n_f}, R)$ depends exponentially on phonon coordinates [see Eqs. (56) and (57)] and energy transferred to the phonon subsystem $(E_{v_f} - E_{v_i})$. It is convenient, therefore, to use the representation

$$V(E_{n_f}, R) = V(E_{n_i}, R_0) \exp[-\tfrac{1}{2} J(E_{n_f}, R)], \tag{69}$$

where the function $J(E_{n_f}, R)$, as earlier, is expanded in a series, retaining only three terms:

$$J(E_{n_f}, R) = -J'_E(E_{v_i} - E_{v_f}) + J'_\mu(q_\mu - q^0_\mu) + \tfrac{1}{2}(q_\mu - q^0_\mu) J''_{\mu\mu'}(q_{\mu'} - q^0_{\mu'}). \tag{70}$$

Substituting expressions (69) and (70) into Eqn. (67), we can write the transition rate constant in terms of the density matrices of phonon subsystem initial and final states, as in (54). Following the respective manipulations of ref. 161, we come to the final expression for the rate constant:

$$K = \frac{2\pi}{\hbar}\{\det\tilde{\phi}\}^{-1/2}\exp\left\{\frac{1}{4}\sum_{\mu\mu'}\tilde{A}_\mu\tilde{B}^{-1}_{\mu\mu'}\tilde{A}_{\mu'} - \frac{1}{4}\sum_\mu(\Delta q_\mu)^2 \times \sinh\left(\frac{\beta\hbar\omega_\mu}{2}\right)\cosh\left[\frac{(\beta - J'_E)\hbar\omega_\mu}{2}\right]\cosh\left[\frac{J'_E\hbar\,\omega_\mu}{2}\right]\right\},$$

$$\begin{aligned}
\tilde{A}_\mu &= J'_\mu + \Delta q_\mu \sinh[(\tfrac{1}{2}\beta - J'_E)\hbar\omega_\mu]\cosh[\tfrac{1}{2}(\beta - J'_E)\hbar\omega_\mu]\cosh(\tfrac{1}{2}J'_E\hbar\omega_\mu),\\
\tilde{B}_{\mu\mu'} &= \tfrac{1}{2}J''_{\mu\mu'} + \delta_{\mu\mu'}\sinh(\tfrac{1}{2}\beta\hbar\omega_\mu)/\cosh(\tfrac{1}{2}J'_E\hbar\omega_\mu)\cosh[\tfrac{1}{2}(\beta - J'_E)\hbar\omega_\mu],\\
\tilde{\phi} &= \{\delta_{\mu\mu'} + \tfrac{1}{2}J''_{\mu\mu'}\cosh[\tfrac{1}{2}(\beta - J'_E)\hbar\omega_\mu]\cosh(\tfrac{1}{2}J'_E\hbar\omega_\mu)/\sinh(\tfrac{1}{2}\beta\hbar\omega_\mu)\}\\
&\quad\times\{\delta_{\mu\mu'} + \tfrac{1}{2}J''_{\mu\mu'}\sinh[\tfrac{1}{2}(\beta - J'_E)\hbar\omega_\mu]\sinh(\tfrac{1}{2}J'_E\hbar\omega_\mu)/\sinh(\tfrac{1}{2}\beta\hbar\omega_\mu)\}.
\end{aligned} \tag{71}$$

At low temperatures, when the quadratic term in expansion (70) can be neglected, the exponent in expansion (71) is simplified ($\tilde{\phi} = 1$), and the expression for the rate constant takes the form

$$\begin{aligned}
K = \frac{2\pi}{\hbar}|V(E_{n_i}, R_0)|^2\exp\Bigg\{&\frac{1}{4}\sum_\mu(J'_\mu)^2\\
&\times\frac{\cosh(J'_E\hbar\omega_\mu/2)\cosh[(\beta - J'_E)\hbar\omega_\mu/2]}{\sinh(\beta\hbar\omega_\mu/2)} - \sum_\mu(\Delta q_\mu)^2\\
&\times\frac{\sinh(J'_E\hbar\omega_\mu/2)\sinh[(\beta - J'_E)\hbar\omega_\mu/2]}{\sinh(\beta\hbar\omega_\mu/2)}.
\end{aligned} \tag{72}$$

Equation (72) enables us to consider the low-temperature limit of the chemical reaction, which can be obtained setting $T = 0$. The first term of the exponent ensures an increase in the rate constant resulting from intermolecular vibrations; the second term describes the reorganization of the medium. It is clearly seen that the contribution of the latter depends on medium properties as well as on potential barrier parameters.

When reorganization of the medium is small, the rate constant in the Debye approximation is described (as shown in ref. 161) by expression (62) at $\Delta E = 0$, but with doubled α', the contribution of zero-point vibrations is also doubled.

At higher temperatures, the rate constant is described by expression (63), and the defreezing of temperature dependence occurs, however, at $T \simeq \hbar\omega_D/2K_B$, not at $T = \hbar\omega_D/4K_B$, as in the case of reactions in which products with discrete states are formed.

At temperatures where the Einstein model of a solid is valid, the matrices $\tilde{A}_{\tilde{\mu}}$ and $\tilde{B}_{\mu\mu'}$ appearing in Eqn. (71) are considerably simplified, and for the rate constant we have the following expression:

$$K = \frac{2\pi}{\hbar} |V(E_{n_i}, R_0)|^2 \{\det \tilde{\phi}\}^{-1/2}$$

$$\times \exp\left\{\frac{1}{2} \frac{\bar{J}'^2}{[\sinh(\beta\hbar\Omega/2)]/\cosh(J'_E\hbar\Omega/2)\cosh[(\beta - J'_E)\hbar\Omega/2]}\right.$$

$$\left. - \frac{(2E_s/\hbar\Omega)\sinh(J'_E\hbar\Omega/2)\sinh[(\beta - J'_E)\hbar\Omega/2]}{\sinh(\beta\hbar\Omega/2)}\right\}, \tag{73}$$

enabling us to analyze the high-temperature limit of the rate constant of particle transfer as well as the KIE, which in this case depends on the medium reorganization energy.

As we have seen, the expressions for the rate constant obtained for different models describing the lattice vibrations of a solid are considerably different. At the same time in a real situation the reaction rate is affected by different vibration types. In low-temperature solid-state chemical reactions one of the reactants, as a rule, differs significantly from the molecules of the medium in mass and in the value of interaction with the medium. Consequently, an active particle involved in reaction behaves as a point defect (in terms of its effect on the spectrum and vibration dynamics of a crystal lattice). Such a situation occurs, for instance, in irradiated molecular crystals where radicals (defects) are formed due to irradiation. Since a defect is one of the reactants and thus lattice regularity breakdown is within the reaction zone, the defect of a solid should be accounted for even in cases where the total number of radiation (or other) defects is small.

As is known, point defects affect the spectrum and vibration dynamics of the solid lattice [163, 164]. The frequencies inside the allowed bands drift slightly in the dense part of the phonon spectrum, and one or several frequencies that prior to defect introduction lay near the edges of the allowed band may transfer into the forbidden band. These frequencies correspond to peculiar local vibrations when the amplitudes of atomic shifts decrease with increasing distance to the defect; that is, only a small number of the neighboring atoms of the lattice is involved in vibrations with the defect.

The analysis of the solutions of equations for the motion of a crystal lattice with defects showed [10, 89] that there occurs a kind of displacement from the reaction zone of the vibrations corresponding to the dense part of the phonon spectrum, and the main contribution to the rate constant is made by local vibration.

The preceding results were obtained under an assumption of harmonicity of intermolecular vibrations, which is valid only at sufficiently low temperatures. Application of this approximation made possible the analytical solution of the problem. In the case of an arbitrary potential for the intermolecular subsystem the rate constant is found numerically [49].

If the anharmonicity is small, it can be accounted for in the framework of the body of mathematics used in harmonic approximation. Such analysis was made elsewhere [88], and it was shown that the formulas for the rate constant remain unchanged, but the parameters involved—equilibrium coordinates and phonon frequencies—turn out to be temperature-dependent effective values. The applicability criteria for harmonic approximation were also obtained.

Passing over to the computation of the rate constants of specific reactions, we again emphasize that the $J(R)$ expansion from (37) in a power series of R is not necessary. It only enables one to obtain analyzable relations through application of different models of a solid. In the general case the problem of the calculation of low-temperature chemical reaction rate constants requires consecutive solution of two problems: search of convenient PESs and averaging of the imaginary part of the action along the optimal path from relation (49).

The H-transfer reactions are discussed in the literature [49, 88, 89, 160–162]. Since the overlap of the intermolecular wavefunctions of A and C skeletons is small, the skeletons in these works were regarded as massive structureless particles, reciprocally vibrating with frequency Ω. The dependence of the matrix element transition on intermolecular coordinates was regarded as determined by only the overlap of vibration wavefunctions of the breaking (AH) and forming (HC) bonds. The reorganization energy did not count. Thus, the calculation contains two characteristic parameters, the equilibrium distance between skeletons, R_0, and their vibration frequency Ω. The adiabatic term of the intramolecular subsystem, $\varepsilon_i(\Gamma)$, was constructed by the LEPS method when the potential energy of a three-body collinear system AHC is written in the form of the empirical London relation (see, e.g., ref. 18):

$$\varepsilon_i = \frac{1}{1+P}\{Q_1 + Q_2 + Q_3 - 2^{-1/2}[(I_1 - I_2)^2 + (I_1 - I_3)^2 + (I_2 - I_3)^2]^{1/2}\}, \tag{74}$$

where Q_1, Q_2, Q_3 are the Coulomb integrals and I_1, I_2, I_3 the exchange integrals of AH, CH, and AC bonds, respectively. Their dependence on nuclear spacing is described by the Morse potential with experimental values of dissociation energies and frequencies of characteristic stretching vibrations of respective bonds. The fitting parameter P, a squared overlap integral, is chosen based on gas phase reaction computations [165]. The terms for nonadiabatic transitions were taken in the form of Morse potentials of AH and HC bonds with the minima at the distance R_0. An example of computed PESs is shown in Figure 8. The exponent in relation (37), which equals

$$J(R) = \frac{2}{\hbar}\int_a^b dr\sqrt{2m_H(\varepsilon_i(z, R) - E)}, \tag{75}$$

was found by numerical integration. The $J(R)$ dependences are shown in Figure 12, and they show, in particular, that a change in R of 0.1 Å leads to a

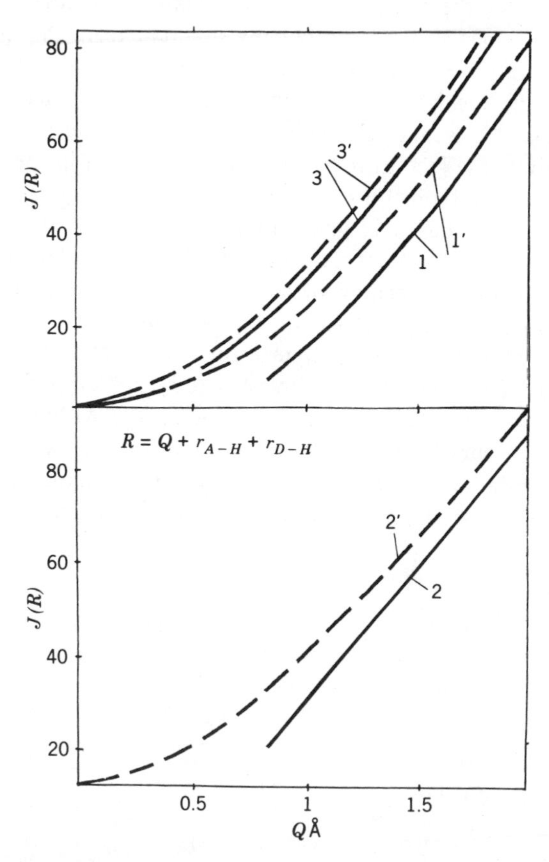

Figure 12. The $J(R)$ function dependence in Eqn. (37) on A-to-C distance in H-atom abstraction by methyl radicals from matrices of methanol and ethanol (1, 1′) acetonitrile and methylisocyanide (2, 2′), and radical pair transformation in dimethylglyoxime (3, 3′). Solid lines (1–3), computation results for adiabatic transitions; dashed lines, (1′–3′), nonadiabatic transitions.

change in the transition probability of almost two orders of magnitude. Values of R_0 and Ω at which the computed $K(T)$ dependences tally with experimental ones are shown in Table 1. Conformity between computation and experiment is shown in Figure 13.

Crystallography data for dimethylglyoxime crystals (Figure 5) correspond to O-to-O distances of 3.1–3.2 Å, in agreement with the found R_0 value. For other matrices there are no structural data. The data in Table 1 correspond to van der Waals radii of C and H. Higher $K(0)$ values in these matrices compared to DMG at greater A-to-C distances are determined by smaller barrier height.

It was assumed for the reaction in DMG crystals that Ω corresponds to the Debye frequency since it is not practically different from the matrix molecules. In reactions of H-atom abstraction by methyl radicals the Ω was identified with the local vibration frequency. This difference between the models can explain a weaker dependence of $K(T)$ at the initial section of growth in the first case [see Eqn. (62)] compared to that in the second. The found values of Ω are typical for organic solids.

Description of the paths is given in ref. 49. Figure 14 illustrates the vibrational character of reactant motion under H transfer. The barrier height is 0.65 eV. Due to intermolecular vibrations the optimum distance between reactants turns out to be close to the saddle point ($R - R^{\neq} = 0.2$–0.3 Å) instead of 0.9 Å at equilibrium configuration.

Using found PES values it is possible to study the temperature relationship of the kinetic isotope H–D effect (relation k_H/k_D). At low temperatures, when transition occurs by the tunneling mechanism and with the distance between the minima of AH and HC terms equalling 1.5 Å, the Franck–Condon factor for the H atom is 10^{10}–10^{12}-fold greater than for the

TABLE 1

Values of A-to-C Distances and Typical Frequencies of Intermolecular Vibrations in Adiabatic Reactions,[a] AH + C → A + HC

A	C	R_0 (Å)	Ω ($\times 10^{13}$ s^{-1})
DMG[b]	DMG	3.17 (3.19)	2.9 (3.1)
CH_3	CH_3OH	3.66 (3.71)	3.6 (3.7)
CH_3	CH_3CH_2OH	3.64 (3.71)	3.5 (3.7)
CH_3	CH_3C(I)[c]	3.82 (3.93)	3.3 (3.5)
CH_3	CH_3C(II)[c]	3.89 (4.01)	3.1 (3.2)
CH_3	CH_3C	3.86 (3.97)	2.9 (3.4)

[a] Values for nonadiabatic transitions are in parentheses.

[b] Dimethylglyoxime.

[c] Two different crystallographic modifications of acetonitrile.

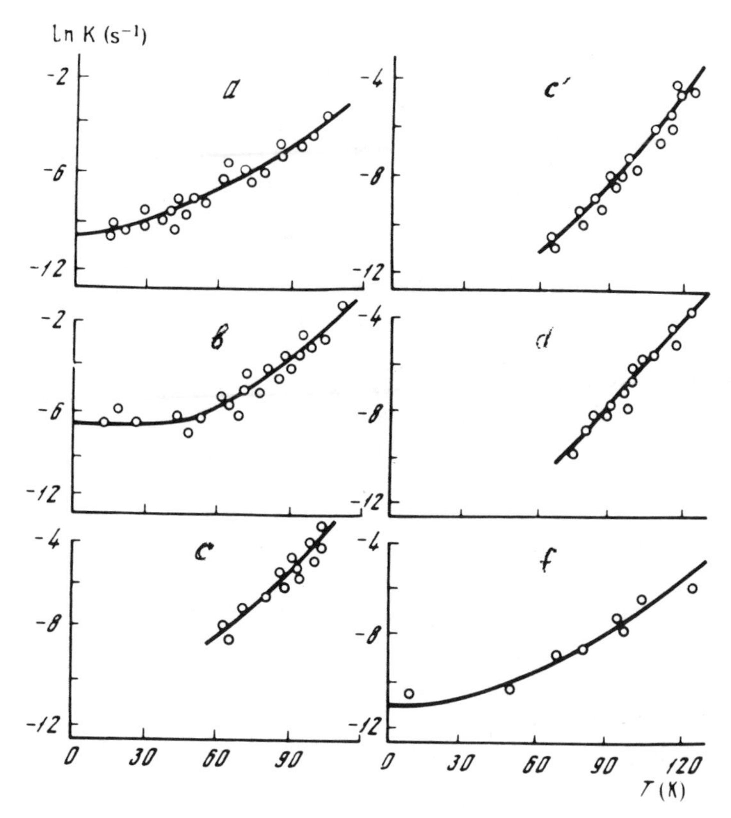

Figure 13. Comparison of computed and experimental temperature relationships of H-atom transfer rate constants in reactions of methyl radicals with methanol (*a*), ethanol (*b*), acetonitrile (*c*, *c*′), methylisocyanide (*d*), and of radical pair transformation in dimethylglyoxime (*f*). Solid lines, computation; circles, experiment.

D atom. This value would have determined the KIE under one-dimensional tunneling. The potential barrier oscillations independent of the tunneling particle mass lead to a 10^5–10^6-fold decrease in the isotope effect. Growth of the intermolecular vibration amplitude with temperature causes a decrease of k_H/k_D, whose dependence on T, computed for the reaction between the methyl radical and CH_3CN, is shown in Figure 15. At 77 K, the computed value, 5×10^5, is close to the experimental one ($\geqslant 3.10^4$).

Of course, a comparison of the computation to experiment in which R_0 and Ω play, in fact, the role of fitting parameters, ensuring a conformity of computed $K(T)$ and k_H/k_D curves to experimental values, can only define the correctness of the qualitative description of solid-state reactions. For further development it is necessary to have, on the one hand, the data on the

Figure 14. The solid lines demonstrate H-atom motion trajectories in reaction of H abstraction by CH_3 radical from CH_3OH molecule with energies of relative motion equal 4.6 and 2.4 kcal/mol respectively. Vibrational motion in reactant valley stipulated by intermolecular vibrations (from ref. 49). The dashed lines refer to fixed distances between CH_3 and CH_3OH.

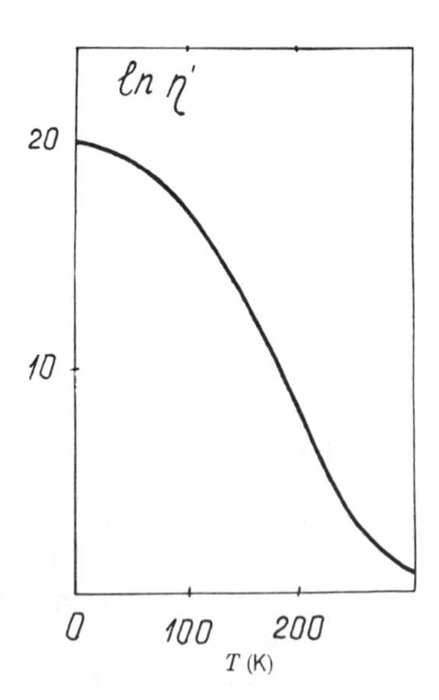

Figure 15. Computed temperature relationship of kinetic isotope H-D effect in reactions of methyl radicals with acetonitrile; $\eta' = k_H/k_D$.

configuration and molecular dynamics of reactants and, on the other, a study of the intermolecular vibration which exceed the frameworks of DAA, since among them there are rather low-frequency ones close to phonon spectrum, and the reaction barrier is determined not only by localization of reactants in the lattice sites, but also by intramolecular reorganization. For instance, in the reaction of methyl radicals the change of configuration under transition from the plane CH_3 radical to the tetrahedral CH_4 molecule creates a barrier of ~ 0.5 eV [165].

Naturally, the preceding peculiarities of solid-state reactions acquire even greater significance in the reactions of particles heavier than H atoms due to the fact that intermolecular vibration amplitudes, as a rule, become greater than those of intramolecular vibrations. Reactant convergence, compared to the mean R_0 values, may be determined by dynamic effects, related to greater amplitudes of intermolecular vibrations, δ_1^2, as well as by static effects, related to the reactant location at shorter distances resulting from local distortions of the lattice. As was first noted elsewhere [50], a sharp growth of δ_1^2 in the region of structural phase transitions explains the observed reaction rate peaks close to the transition temperature T_c. Computation of the $K(T)$ behavior in the T_c region in the framework of considered notions is given in ref. 167. The dynamic model of structural transitions [168–170] assumes that due to the order parameter fluctuations in the T_c region, a soft mode is split from the phonon spectrum; this mode frequency may be presented in the form

$$\omega_K^2 = \alpha(T - T_c) + \beta \sum_{x, y, z} (k_i - k_{0i})^2, \tag{76}$$

where α is the Landau parameter [171], β has the order of a squared sound speed, and k_{i0} are the components of the wave vector in the Brillouin zone point where the soft mode is formed. A mean-square displacement of the neighboring lattice sites is expressed by the formula

$$\delta_1^2(T) = \delta_{10}^2(T) + \delta_c^2(T), \tag{77}$$

in which δ_{10} is determined by the regular part of the phonon spectrum remaining in the transition region, and δ_c^2 is related to the soft mode. Considering $\omega_K \ll K_B T_0/\hbar$, we find an expression for δ_c^2 under reactant arrangement along the z axis:

$$\delta_c^2(T) = \frac{v_0 K_B T_c}{\pi^2 M \beta^{3/2}} \int_0^{q_m} [1 + \varphi(q)] \frac{q^2\, dq}{\omega_0^2 + q^2}, \tag{78}$$

where

$$\varphi(q) = -\cos k_z a \frac{\sin(a\beta^{-1/2}q)}{a\beta^{-1/2}q},$$

$$q_m^2 = \beta \sum_{x,y,z} \left(\frac{\pi}{a} - k_{id}\right)^2, \qquad \omega_0^2 = \alpha(T - T_c),$$

k_{id} are values of the wave vector at the zone boundary, a is the lattice spacing, and V_0 is the elementary cell volume. Since in transition only a small part of the phonon spectrum is disturbed close to $\bar{k}_0$, $a\beta^{-1/2}q_m \ll 1$ and relation (78) can be simplified to

$$\delta_c^2 \simeq \frac{v_0 K_B T_c}{\pi^2 M \beta^{3/2}} (1 - \cos k_{0z} a)\left(q_m - \frac{\pi}{a}\omega_0\right). \tag{79}$$

From (79) it follows that δ_c^2 depends on the point of k space where the soft-mode splitting-out occurs. In ferrodistortion transition, when $k_{0z} = 0$, that is, in the center of the Brillouin zone, $\delta_c^2 = 0$, since the neighboring site displacement for long-wave phonons is of a small order $(aq_m)^2$. The highest δ_c^2 value emerges in an antiferrodistortion transition, where the soft-mode condensation occurs at the zone boundary ($k_{0z} \simeq \pi/a$). Such transition occurs, in particular, in a deuteromethane crystal at 27 K [172, 173]. Using the transition parameters defined in these papers, $K(T)$ increase in the T_c region at formation of the soft mode with various k_{0z} was computed in ref. 166 with relation (33). The results are shown in Figure 16. The $K(T)$ peak width is determined by the soft-mode decay.

There are no $K(T)$ computations for specific reactions with heavy-particle transfer at present. In ref. 174 it is only noted that experimental values of $K(0)$ ($\sim 10^{-3}\ s^{-1}$) for the reaction $R + Cl_2 \rightarrow RCl + Cl$ correspond to the reasonable values of barrier height (0.2 eV) and intermolecular vibration frequencies ($2 \times 10^{13}\ s^{-1}$) at $R_0 = 3$ Å. The higher $K(0)$ values, compared to H-transfer reactions, are explained by the decreased barrier height due to higher exothermicity of the reaction. Another example of the heavy-particle quantum transition is discussed in Section 5.

In the preceding discussion the character of intermolecular vibrations was not specified; therefore, the results are equally applicable for translational as well as for orientational vibrations. The role of the latter becomes determinate when barrier penetration is coupled with reactant rotation. As shown in ref. 175, the difference in the spectrum of orientational vibrations causes the difference in $K(T)$ relationships. According to ref. 176, the dispersion relation takes the form

$$\omega(x) = \Omega_1[1 - q\cos(xR_0)]^{1/2}, \qquad |q| \leqslant 1, \tag{80}$$

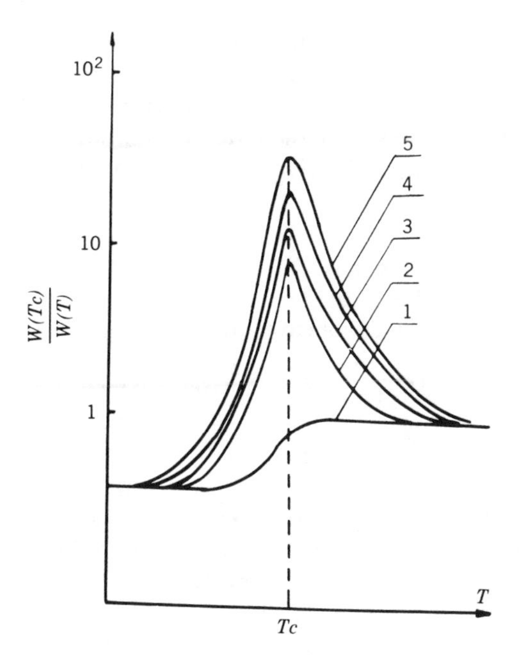

Figure 16. Illustration of change of solid-state reaction rate constant in phase transition. Soft-mode splitting out occurs at point K_0, equalling 0, $\pi/4a$, $\pi/2a$, $3\pi/4a$ and π/a for curves 1–5, respectively. $K(T)$ jump at $K_0 = 0$ determined by difference in amplitude in two phases.

that is, the frequencies of orientational vibrations changing the distances between reactants lie in the range between $\omega_{\min} = \Omega_1\sqrt{1-|q|}$ and $\omega_{\max} = \Omega_1\sqrt{1+|q|}$, and $\omega_{\max}$ is smaller than the Debye frequency of translational phonons. That is why the values $T_t \sim \hbar\omega_{\min}/k_B$ turn out to be smaller. The $K(T)$ peculiarities are manifested in the region $T < \hbar\omega_{\min}/K_B$, where the exponent in relation (37) takes the form $A + B\exp(-\hbar\omega_{\min}/K_BT)$, as in the case of local translation vibrations. The mean-square displacements of reactants under orientational vibrations are ω_D/Ω_1-fold greater than under translational vibrations; according to Eqn. (33), this increases the transition probability.

The reason for the nonexponential kinetics of solid-state chemical reactions lies in the existence of the rate constant distribution determined by the set of different configurations of the reactants, incommensurate to the solid lattice (see, e.g., ref. 177). In the case of low-temperature reactions the existence of the set of configurations can be phenomenologically accounted for by introduction of the equilibrium distance distribution R_0. As shown in the literature [138, 178, 179], introduction of this distribution into the discussed model of low-temperature reactions enables us to quantitatively describe the transformation kinetics satisfying the empiric relation (26) and to explain the change in the ξ' degree by the change of this distribution's width with temperature. If we take the distribution function $f(R)$ in the form of a rectangle with the width ΔR, then the $K(R)$ relationship specified by relations (37) and (38) leads to the following expression for the change of reactant concentration over time:

$$C(t)/C(0) = (J'\Delta R)^{-1}\{E_i(k_{\min}t) - E_i(k_{\max}t)\} \tag{81}$$

where $E_i(x)$ is the integral exponential function, $k_{\min}$ and $k_{\max}$ correspond to $R = \bar{R} \mp \Delta R/2$. In the experimentally realized region of the $C(t)/C(0)$ change from 0.5 to 0.03 relation (81) is well approximated by Eqn. (26), the value of ξ' decreasing with growing $J'\Delta R$, but at $\Delta R = 0$ the $\xi' = 1$ and kinetic curves are exponential. Figure 17 shows how relation (81) describes the experimental $C(t)$ relationships for methyl radical reactions with methanol at different temperatures. The temperature relationship of ΔR corresponding to the experimental function of $\xi'(T)$ is shown in Figure 18. The distribution width growth with lowering temperature seems to be related to the peculiarities of the structure of amorphous solids and the distribution of reactants in them. In the theory developed by Cohen and Grest [180, 181] based on percolation theory (see, e.g., ref. 182), an amorphous solid is regarded as a combination of

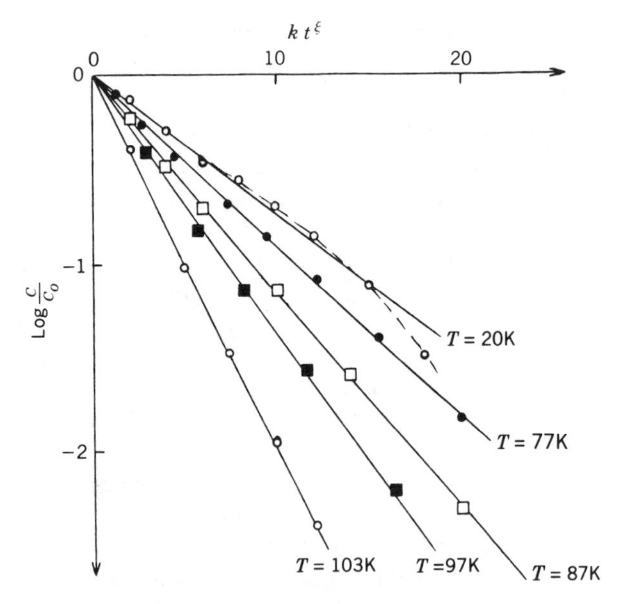

Figure 17. Description of kinetic curves by relation (80). Coordinates chosen to give straight lines for relationships $C(t)/C(0) = \exp[-kt^{\xi'}]$.

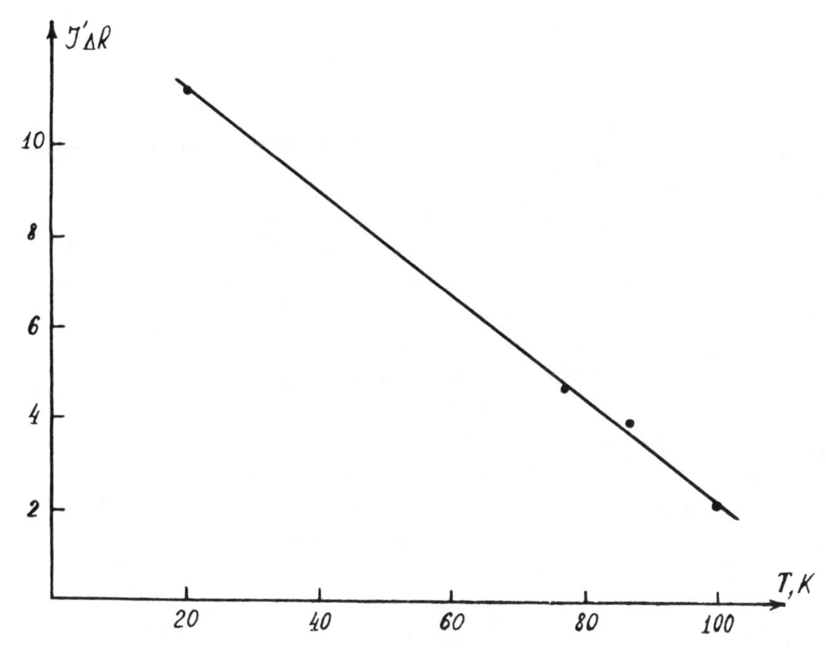

Figure 18. Temperature relationship of width of distribution of distances between reactants in H abstraction from methanol molecules by CH_3 radicals.

crystallike and liquidlike formations characterized by different times of structural relaxation and their different temperature dependences. Near T_g the crystallike regions form a continuous frame (infinite cluster) ensuring the absence of macroscopic flow. At the same time, the liquidlike regions trapped in this frame retain a relatively high molecular mobility. Though with lowering temperature the fraction of liquidlike regions decreases, the relaxation times in them remain finite even considerably below T_g. A particular observation in favor of this model is a sharp deviation of the temperature dependence of rotational relaxation times from the Arrhenius law in the T_g region and their weak temperature dependence below T_g [183, 184].

Due to large anisotropic widening of ESR lines in the randomly oriented crystallike regions the main contribution to the observed signal is made by the radicals located in the liquidlike regions. The increase of the structural relaxation times with lowering temperature creates a set of nonequilibrium configurations typical of the amorphous state, that is, an increase of ΔR. It cannot be ruled out that the weak temperature dependence of the low-temperature relaxation rate in glasses might be explained in the models accounting for, as those previously discussed, the quantum character of intermolecular vibrations.

5. LOW-TEMPERATURE NONCHEMICAL TRANSFER OF HEAVY PARTICLES IN SOLIDS

The tunneling states of impurity sites in crystals emerge if an atom or molecule has several equilibrium positions in the lattice. When the frequencies of tunneling between these states exceeds 10^9–10^{10} s^{-1}, splittings corresponding to them are observed in the high-resolution optical spectra and ESR spectra. A review of the results is found in ref. 185. For instance, an Li^+ ion substituting for a K^+ ion in the KCl lattice has eight equilibrium positions laid out in the vertices of a cube with a 1.4-Å edge and separated by ~ 0.02 eV potential barrier. The tunneling splitting for ^7Li and ^6Li is 0.77 and 1.03 cm^{-1}, respectively. A peculiarity of the impurity tunneling states is extremely high sensitivity of splitting to lattice deformations: Splitting increases by an order of magnitude at compressive stress of ~ 10^7 N/m^2.

Results of studies of the impurity states in crystals made possible an assumption that the peculiarities of the glasslike state are determined by structural defects formation having at least two configurations with similar energies. First, these notions were applied to an explanation of the thermodynamic properties of glasses at temperatures close to T_g, where transitions between configurations are thermally activated [186]. Such a study of thermal, electrical, and acoustic parameters of glasses of various composition in the region 0.1–10 K revealed several regularities distinguishing the amorphous state of a substance from the crystal one. Since results of these studies are discussed in a number of reviews and monographs (see, e.g., refs. 187, 188), here we only note some of the found abnormalities: a linear temperature dependence of heat capacity considerably exceeding the Debye heat capacity for which $C \sim T^3$; a lower thermal conductivity proportional to T^2; and a strong damping of ultrasound and electromagnetic waves. The low-temperature properties of the glasses were explained in the framework of the two-level system hypothesis suggested by Phillips [190] and Anderson and co-workers [189]. According to this hypothesis, the glasses contain localized states having tunneling frequencies of 10^8–10^{10} s^{-1}. At the present stage of glassy state research the microscopic structure of these hypothetical states remains unknown in most cases, and in fact, they are deduced phenomenologically. This in no way means that there is any ground to doubt the reality of tunneling states. On the contrary, the possibility of a general explanation of the totality of glass properties strongly implies the necessity of the notions about the so-called phonon-assisted tunneling (PAT) of heavy particles.

The model of two-level systems was suggested for amorphous silicon dioxide (fused quartz) where the increase of the distance R between Si atoms compared to that of the crystal was found [191]. Due to the extention of bonds, the O atom in the chain Si–O–Si has several equilibrium positions.

The design of the structure of the cluster composed of two SiO_3 tetrahedrons linked by the O atom was done elsewhere [192, 193]. A double-well potential (DWP) of the central O atom is shown in Figure 19. When R is close to equilibrium for crystal, the barrier would not emerge ($E = 0$), and the O atom has one equilibrium position. Tunneling states are formed at $R \geqslant 4.2$ Å. According to ref. 193, the DWP energy levels are described by an approximate relation:

$$\varepsilon_n = \frac{2\hbar}{a}\sqrt{\frac{2V_m}{m}}\,(n+\tfrac{1}{2}). \tag{82}$$

The barrier penetration for $n = 0$ equals

$$J(R) = \frac{8a\sqrt{2mE_m}}{3\hbar}\sqrt{1+P_0}\,[E(l) - K(l)], \tag{83}$$

where $P_0 = \sqrt{\varepsilon_0/V_m}$, $l = \sqrt{(1-P_0)/(1+P_0)}$, and $E(z)$ and $K(z)$ are the complete elliptic integrals of the first and second kind. The DWP parameters computed in ref. 193 are shown in Table 2. At $R > 4.2$ Å the barrier height increases sharply and its transparency lowers. The computation in ref. 193 confirms a more general consideration of the DWP formation process suggested by Klinger and co-workers [194–196]. Let us discuss their results

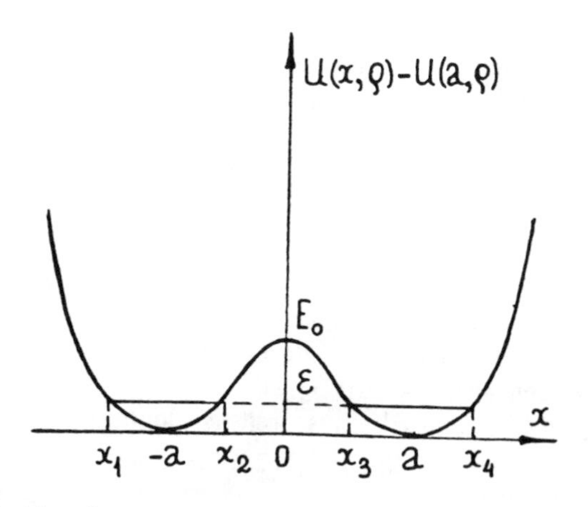

Figure 19. Double-well potential of central oxygen atom in O_3Si–O–SiO_3 cluster imitating structure of fused quartz.

TABLE 2
DWP Parameters of Oxygen Atom in Amorphous SiO_2 Lattice

R(Å)	a(Å)	V_m(eV)	ε_0(eV)	J
4.0	0.00	0.000	$> E_0$	0
4.2	0.18	0.013	$> E_0$	16.7
4.4	0.42	0.26	0.028	43.9
4.6	0.57	0.73	0.034	107
4.8	0.70	1.29	0.037	181

using a particular example of convergence of two equal counter Morse potentials, with the distance between their minima equaling R:

$$U(x, R) = U_0(x + \tfrac{1}{2}R) + U_0(-x + \tfrac{1}{2}R), \qquad U(x) = D(1 - e^{-\alpha x})^2. \quad (84)$$

The two minima $U(x, R)$ emerge under the condition

$$R > R_c = \frac{1}{2\alpha} \ln 2 \quad (85)$$

and the barrier between them is

$$V_m = \left(\frac{\xi}{1+\xi}\right)^2 D, \qquad \xi = \tfrac{1}{2}(e^{\alpha R/2} - 1) \quad (86)$$

The natural frequency of vibrations in DWP minima is related to the vibration frequency in an individual well, ω_0, by

$$\omega^2 = \omega_0^2 \frac{\xi(2+\xi)}{(1+\xi)^2}. \quad (87)$$

A condition for two-level system emergence, $V_m \geqslant \frac{1}{2}\hbar\omega$, corresponds to ξ values exceeding critical $\xi_c = (\hbar\omega_0/2D)^{2/3}$. From (86) and (87) it follows that at $\xi \simeq \xi_c$ two-level systems emerge with a very weak bond characterized by low natural frequencies $\omega_c = \omega_0(\hbar\omega_0/2D)^{2/3}$; great amplitudes of zero-point vibrations, $\delta_c^2 = \delta_0^2(2D/\hbar\omega_0)^{1/3}$; as well as high deformation potential, $B = R(d/dR)(\hbar\omega) = \hbar\omega_0(2D/\hbar\omega_0)^{1/3}(\ln 2/\sqrt{2})$. Such DWPs are called *soft*. The existence of soft DWPs, in which tunneling frequencies equaling $\omega_0 e^{-J}$ lie in the mentioned interval, explains the properties of glasses below 10 K. At higher temperatures the role of soft DWPs is insignificant, and the most important are hard DWPs with low tunneling frequencies. Unlike soft DWPs, whose existence in glasses as well as in crystals can be detected in high-resolution spectra and thermal capacity abnormalities, the hard DWPs manifest themselves, as in the case of low-temperature chemical reactions, only in the kinetic parameters of the glasses.

The two-level systems are related to the "frozen" interval in the glass transition point fluctuations of the spatial arrangement of atoms that takes place in the liquid phase. The probability of such fluctuations P decreases exponentially with their scale increasing; thus,

$$P \sim \exp\left(-k\frac{a^2}{2K_B T_g}\right), \tag{88}$$

where $\frac{1}{2}ka^2$ is the lattice deformation energy under particle displacement from the equilibrium position at a distance a and k is the corresponding force constant. As shown in the preceding, J is rapidly growing with a, and from (88) it follows that the probability of two-level system formation with high J values (i.e., low tunneling frequencies) is exponentially small. An assumption about the existence of the two groups of two-level systems in glasses was stated elsewhere [197, 198]. To explain the non monotonic temperature dependence of changes in the rate and coefficient of ultrasound absorption in glasses, it was assumed [199, 200] that there is a slit in the two-level system spectrum separating soft and hard DWPs.

The relaxational sound absorption in glasses is stipulated by the modulation of the levels of two-level systems in the sound wave field and is determined by their relaxation time τ, which depends on the difference in energies, ΔE. Large absorption in soft DWPs is stipulated by the high values of deformation potential; this was noted for the first time in refs. 199 and 200. At a fixed sound frequency ω_s, the absorption maximum is observed at temperature T_m, at which $\omega_s\tau(\Delta E)|_{\Delta E = K_B T_m} = 1$. With growing T_m the relaxation time of soft DWPs rapidly decreases, and their contribution to absorption becomes small. The observed absorption growth is stipulated by the interaction between the acoustic vibrations and hard DWPs. It is noted [199, 200] that under the activation-type relaxation mechanism of the hard DWPs is impossible to explain the existence of the plateau in the temperature dependence of thermal conductivity in the region of 10–50 K (see, e.g., ref. 188). Besides, the activation-type relaxation mechanism of hard DWPs leads to too low values of barrier height and does not explain the low values of τ^{-1} typical for metallic glasses as compared to the vibration frequencies [201].

As in the description of low-temperature chemical reactions, these difficulties can be overcome by accounting for the barrier fluctuations in the

DWP due to lattice vibrations. In both cases DAA is applied, which is valid if the barrier penetration time $\tau = ma/\hbar J'$ is shorter than the typical time of change of its shape (i.e., vibration period Ω^{-1}). An appropriate condition may be represented in the form

$$\delta_1 J' \gg \frac{m}{M}\frac{a}{\delta_1}. \tag{89}$$

Since for heavy-particle transfer $m/M \sim 1$, condition (89) is satisfied only under strong connection of the two-level systems with specific modes ensuring high J' values. It is these modes that make the basic contribution to the transition probability. In the opposite case of high-frequency vibrations, when $\tau\Omega \gg 1$, the J value is averaged by vibration period and not the amplitude of transition probability (proportional to e^{-J}), as at $\tau\Omega \ll 1$. Therefore, for the high-frequency modes the fluctuation preparation of the barrier is insignificant. The two-level system itself "chooses" an appropriate mode for which $\delta_1 J'$ is the highest, and the DAA for this mode turns out valid.

As an illustration, let us consider O-atom tunneling in a SiO_2 cluster at $V_m = 0.5$ eV, $a = 0.5$ Å, and $J = 50$ corresponding to the data of Table 2 at $R = 4.3$ Å, setting $\hbar\Omega/K_B = 50$ K. In this case $\delta_1 J' \simeq 2.3$ and $a/\delta_1 \simeq 2.2$; that is, condition (89) is well satisfied. The transition probability is described by relations (33) and (40), from which it follows that due to the zero-point vibrations of the lattice at $T \leqslant \hbar\Omega/2K_B$ the effective value of the exponent of barrier permeability, that is,

$$J(R) - \frac{\delta_1^2 (J')^2}{2[\delta_1^2 J'' + 2\tanh(\hbar\Omega/2K_B T)}, \tag{90}$$

decreases to 12; that is, the transition probability increases approximately 10^{16}-fold at the expense of potential barrier width fluctuations. These estimations are naturally applicable both to kinetic processes in glasses and low-temperature chemical reactions and, in spite of their rather approximate character, show extremely high efficiency of the mechanism discussed. The relaxation rate of two-level systems and the related coefficient of sound absorption are described by Eqn. (40). Their temperature dependence consists of the Arrhenius section passing over into the low-temperature plateau.

The history of the development of the theory of low-temperature plasticity of solids resembles very much the development of tunneling notions in cryochemistry. This resemblance is not casual; it is related to the similarity of the elementary act pictures; this was noted by Eyring, who successfully applied the theory of absolute rates to a description of fracture kinetics [202]. Plastic deformation at constant stress (creep) is stipulated by dislocation slip in the potential pattern formed by the crystal discreteness (Paierls pattern) and point defects (intersections with Franck–Read net) (see, e.g., refs. 203, 204). Interaction between the defects and the dislocations forms potential barriers pinning the dislocations, and the barrier height decreases linearly with growing applied stress σ. Detachment of a dislocation from the defect is a thermally activated process whose probability is described by

$$w = \omega_0 \exp[U/K_B T], \qquad U = U_0 - v(\sigma - \sigma_i), \tag{91}$$

where σ_i is the internal stress depending, under stationary creeping, on deformation, and v is the activation volume. Relation (91) describes well the plastic deformation of various solids at normal temperatures (lower than half of the melting point), and stresses much lower than critical $\sigma_c \sim U_0/v$. Detection of an abnormally high and temperature-independent metallic creeping below 10–30 K at $\sigma \ll \sigma_c$ [205] gave Mott [206] the grounds to assume that dislocation detachment from pinning sites at low temperatures is stipulated by tunneling penetration of the barrier. Mott obtained the estimation of penetration probability regarding the element of the slipping dislocation line, which interacts with the defect, as a localized mass penetrating a one-dimensional barrier. Though this simple model of one-dimensional tunneling leads to finite values of creeping at $T = 0$, the values of tunneling temperature T_t, at which the tunneling mechanism becomes predominant, turn out to be lower than 0.1–1 K. Further studies showed that preservation of final plasticity at $T \lesssim$ 10–30 K is typical of quite a number of crystals, both metallic and dielectric (see refs. 207, 208). The quantitative discrepancy between the Mott model and experimental data inspired creation of more realistic models. Gilman [209, 210] gave a more detailed account for the dependence of transition probability on the stress. The potential of a local site with size a across a one-axis outer stress, directed along the x axis, has the form

$$V(x, \sigma) = V(x, 0) - a\sigma x. \tag{92}$$

The stress not only causes a decrease in the barrier height and displacement of the equilibrium position in the initial state, but also forms the linear term of the final state (Figure 20). The probability of resonance tunneling in the state of this term is described by the expression

$$\begin{aligned} \omega(\sigma) &\sim \exp\left\{-\frac{4}{3}\frac{U_0}{\hbar}\frac{\sqrt{2MU_m}}{a\sigma}\right\}, \\ U_m &= U_0\left(1 - \frac{v\sigma}{U_0}\left[1 - \ln\frac{v\sigma}{U_0}\right]\right), \end{aligned} \tag{93}$$

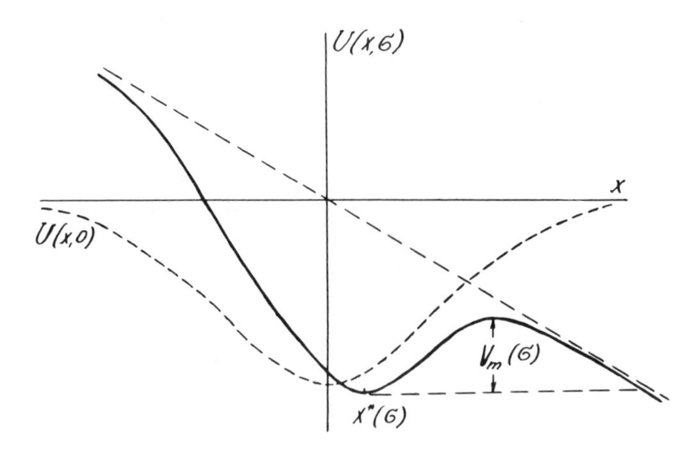

Figure 20. Potential of local center pinning dislocation under one-axis deformation. Barrier parameters depend on outer stress σ. Model potential $V(x, 0)$ in relation (92) chosen in form $V(x, 0) = -V_0(1 + \alpha|x|)e^{-\alpha|x|}$.

where U_0 is the barrier height at $\sigma = 0$. The logarithmic part of the correction to U_m corresponds to the model potential chosen in ref. 210 and characterizes growth of the barrier transparency due to its width decrease with growing σ. From comparison of $\omega(\sigma)$ with the thermoactivated transition probability follows an estimation of T_t:

$$K_B T_t \simeq \frac{3}{4} \frac{\hbar}{\sqrt{2MU_m}} a\sigma, \tag{94}$$

showing that the low-temperature plateau widens with growing σ. Gilman [209] cited an overestimation of T_t having assumed that $\frac{3}{4}(\hbar/\sqrt{2MU_0}) \simeq v/a$. From Eqn. (94) it follows that at typical stresses of 10^7–10^8 N/m^2 and $U_0 \geqslant 0.1$ eV the expected T_t values, as in the Mott model, turn out to be noticeably lower than the experimental values. The one-dimensional tunneling models, as in the case of low-temperature chemical reactions, do not explain the high values of T_t and creeping.

The effect of dislocation line oscillations on the probability of its separation from the defect was accounted for by Natsik [211, 212], who has considered the following model: A dislocation line moves in the slipping plane under the action of the one-axis stress. The dislocation is characterized by the linear mass density ρ and tension coefficient C. Pinning in A and B sites is considered to be rigid, and the dislocation detachment from the defect located in the zero point is considered (Figure 21). At $\sigma \ll \sigma_c$, l_1, $l_2 \ll d$ (d is

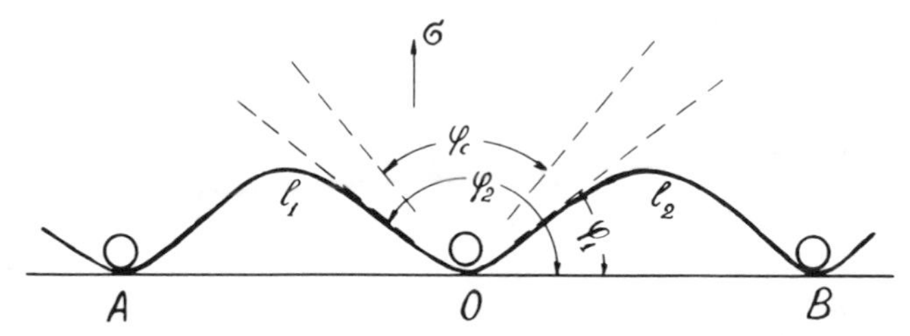

Figure 21. Model of dislocation string detachment from point defect. Pinning in A and B points supposed rigid; l_1, l_2 are lengths of vibrating dislocation segments. Arrow shows direction of dislocation slipping and outer stress; φ_c is critical detachment angle.

the typical size of the defect potential), the dislocation has an equilibrium configuration. The dislocation element d interacts directly with the defect. The attached segments l_1 and l_2 acquire string configuration with the ends fixed in the defect region. This configuration is characterized by the angle φ_0 independent of σ:

$$\varphi_0 = \varphi_1 - \varphi_2 = \pi - \frac{b(l_1 + l_2)}{2C}\sigma, \tag{95}$$

where b is the value of the Bürgers dislocation vector. It is detached at the critical value $\varphi_c = \pi - U_0/\pi bc$, that is, the critical stress $\sigma_c = 2U_0C/b(l_1 + l_2)$. At $\sigma \ll \sigma_c$, when $\varphi \ll \varphi_c$, dislocation detachment is determined by the fluctuation motion of the segments and pinned element. Motion of the segments relative to the equilibrium configuration is described by the totality of the dislocation phonons with the finite frequency Ω_m. The pinned element has mass $M = \rho d$ and natural vibration frequency $\omega_0 \sim [(1/M)(\partial^2 U/\partial y^2)]^{1/2} \gg \Omega_m$. Thus, the system is divided into the fast (pinned element) and slow (segment) subsystems, and we discuss the effect of the slow subsystem, composed of a set of harmonic oscillators, on the fast one—a particle with mass M and vibration frequency near equilibrium ω_0 in the field of potential $U(x, \varphi)$ parametrically dependent on the slow subsystem coordinates. This problem is similar to the one discussed in preceding sections.

A parameter

$$p = \frac{16\pi}{3}\frac{\Omega_m d}{\omega_0 \lambda}\left(1 - \frac{\sigma}{\sigma_c}\right)^{-3/4} \tag{96}$$

where $\lambda = u(\Omega_m/2)$ and u is the speed of sound, determine whether the predominate mechanism is tunneling of the particle, ($p \ll 1$), or transition of the particle under the action of segments. Parameter p is related to the squared ratio of the vibration amplitudes of segments and particles $(\delta_1/\delta_0)^2$, as in relation (37). At $p \ll 1$

$$\omega(\sigma) \sim \exp\left\{-\frac{\pi v_0}{\hbar\omega_0}\left(1-\frac{\sigma}{\sigma_c}\right)^{5/4}\right\}, \tag{97}$$

in good agreement with Mott and Gilman formulas taking into account the difference in barrier shape. At $p \gg 1$ the critical value φ_c creates vibrations of segments that at $T < \theta = \hbar\Omega_m/K_B$ are quantum vibrations:

$$\omega(\sigma) \sim \exp\left\{-\frac{\lambda_c\varphi_c^2}{4\hbar\Omega_m}\frac{(1-\sigma/\sigma_c)^2}{1+T^2/\theta^2}\right\}. \tag{98}$$

The temperature dependence of the probability of transition turns out to be the same as underbarrier oscillations in the Debye model of a solid, discussed in Section 4. It is noted in ref. 213 that in a one-dimensional lattice the exponent in relation (62) contains not T^4 but T^2; that is, it corresponds to Eqn. (98). At $p \gg 1$ and $T > \theta$ the $\omega(\sigma)$ is described by the activation relation. Since θ is of the same order as the Debye temperature, T agrees with experiment.

The quantum motion of a dislocation in the Paierls potential profile was studied by Pokrovsky and Petukhov [214, 215]. A dislocation string penetrates the barrier separating the pattern valleys due to fluctuations related to its lateral vibrations. As a result of transition, a double bend of the string (Figure 21) is formed when a dislocation segment passes over to the neighboring valley. The length of the segment must exceed the critical value. The role of quantum effects increases with the amplitude of zero-point vibrations. It is noted in ref. 215 that the quantum motion of dislocations is typical, primarily, of simple molecular crystals.

Tunneling in potential (92) was discussed in ref. 210 as the reason for deviations of the fracture rate temperature dependence from that expected in thermal fluctuation theory [216]. To the low-temperature fracture stipulated by the stressed lattice vibrations there corresponds a transition, in the θ_D region, from the Arrhenius relationship to the low-temperature plateau [217]. Experimental data are explained in terms of these notions [218].

Concluding this section we note that the relation between the quantum motion of heavy particles in solids and solid-state chemical reactions is not confined to the similarity of the notions developed above. If the limiting stage of transformation is the structural relaxation of the medium, the quantum character of the discussed processes is the cause of the emergence of the rate constant low-temperature plateau.

6. MACROKINETIC PECULIARITIES OF SOLID-STATE CHEMICAL REACTIONS IN THE REGION OF LOW-TEMPERATURE LIMIT OF RATE CONSTANT

As was noted, the low-temperature chemical reactions run, normally, in the thermodynamically nonequilibrium glassy mixes of reactants, where the excess—compared to crystals—free energy is comparable to the activation energy of typical radical solid-state reactions. This excess energy of glass is stipulated by the static set of configurations formed at the expense of the spatial fluctuations of equilibrium positions fixed in the glass transition point. The percolation of chemical reactions in such microheterogeneous media not only is determined by the rate constant averaging by these configurations (see Section 4), but also can—due to low rates of structural relaxation below T_g—lead to an increase of local nonequilibrium in the process of transformation. With lowering temperature, the role of the latter should increase due to both increased relaxation time and deviation of the reaction rate constant from the Arrhenius law. In the low-temperature limit region, where the temperature ceases to affect the reaction rate, the structural defects caused by chemical transformation become the determining factor. Reaction product formation causing the local deformation of the lattice can bring about a decrease in the reaction barrier in the same degree as that resulting from the detachment of dislocations from defects under the action of outer stresses, discussed in the preceding section. Acceleration of solid-state reactions should first exist in chain reactions when successive reaction acts are spatially related by the formation of active particles acting as the chain material carriers directly in the region of lattice deformation in the preceding reaction acts.

The low-temperature acceleration of solid-state reactions was observed elsewhere [51] in the study of methylcyclohexane photochlorination proceeding the nonbranched chain growth mechanism discussed in the Section 2. The reactant nonequilibrium mixes prepared by rapid cooling were inflamed upon accumulation of critical concentrations of R_c radicals as a result of chlorine photolysis. In the range 10–40 K the value of R_c increases with temperature, and above 60 K no inflammation occurs in spite of the remaining rate of radical accumulation. Inflammation can also be induced by cooling the reactant mix after photolysis at higher temperature, where the mix remains stable. As in the preceding experiments, the temperature of inflammation T_0 increases with radical concentration. This phenomenon is called the *upper temperature limit of inflammation* (UTLI), in contrast to the lower temperature limit characteristic of thermal inflammation. Further experiments [219–224] (see also reviews 225–227) helped find another series of examples of low-temperature inflammation and showed that critical parameters of UTLI do not depend on the sample shape and size, which determine the conditions of heat removal, but depend on the structural nonequilibrium of the mix. With further thermal annealing in the T_g region, lowering the nonequilibrium of glass, the T_0 and R_c values increase. Under deep annealing the critical phenomena disappear, though the conditions of radical accumulation remain unchanged. It is shown in ref. 221 that the UTLI is characterized by the absence of preexplosion warming up, which is the main indication of thermal inflammation. The temperature range where the non thermal critical phenomena emerge corresponds to the low-temperature plateau of the rate constants, while in the Arrhenius region of $K(T)$ the inflammation of the same mixes becomes thermal; that is, two inflammation regions are observed [226] with different critical temperature dependences on radical concentration (Figure 22).

Depending on the heat removal conditions the explosive reaction occurs in two regimes. In massive samples inflammation is accompanied by the mix

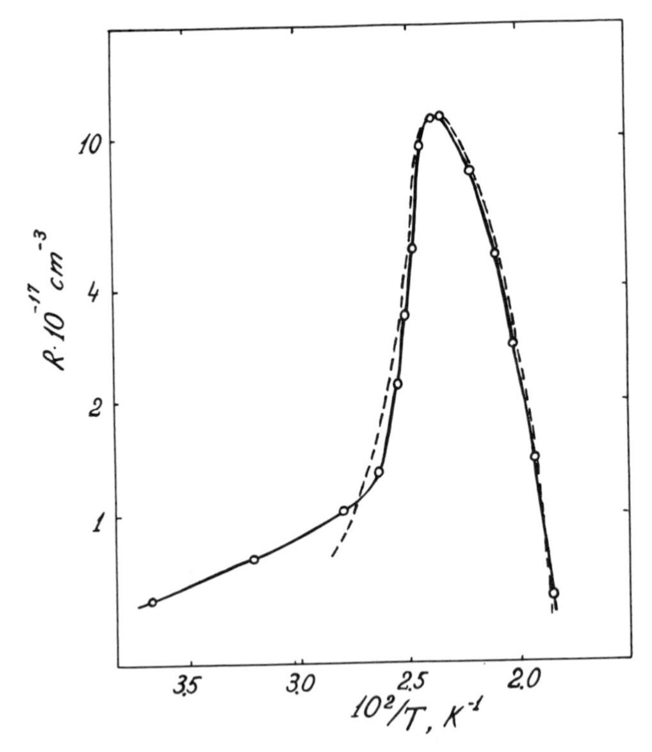

Figure 22. Dependence of inflammation temperature of equimolar mixes of ethylene with chlorine on radical concentration, showing existence of two regions corresponding to upper and lower inflammation limits. Dashed line shows computed relationship at $\xi = 0.7$, $q_1/q = 4 \times 10^{-2}$, $K_g = 3 \times 10^{-4}\ s^{-1}$, $K = 3 \times 10^{-2}\ s^{-1}$ at 37 K, $V_m^0 = 0.09$ eV.

heating up by 150–200 K (above T_g) under conditions close to adiabatic. The main transformation, whose degree η reaches 0.5–1.0, proceeds in the liquid phase, and its characteristics do not differ from those of thermal explosion. In thin samples, due to rapid heat removal, low-temperature inflammation leads only to an insignificant transformation ($\eta < 0.05$). Heating of the mix turns out insufficient for transition into the Arrhenius region, and the reaction proceeds in the solid phase below T_g, not transforming into a thermal explosion, as in the case of massive samples. A typical time of solid-state transformation is 10^{-2}–10^{-3} s, with corresponding rate constants of explosive reactions equaling 10^4–10^5 s^{-1} at 40 K.

It was found [219] that inflamation in the $K(T)$ plateau can be induced by an outer pulse of nondestructive compression with 3×10^6–3×10^7 N/m^2 amplitude. The initiation of reaction occurs simultaneously with pulse application. In a sample with radical concentration of $\sim 10^{17}$ cm^{-3} at the initial temperature of 40 K in 10^{-2} s, a 0.02 transformation degree is reached, corresponding to a chain length of $\sim 10^2$. Its growth rate constant 10^4 s^{-1}, which is five orders of magnitude higher than in the mix of the same composition after thermal annealing, when critical phenomena disappear. After the initial pulse a slower reaction is observed for 3–5 s, determined by the stresses emerging in the very process of transformation. Its rate constant is $\sim 10^1$ s^{-1}, and $\eta \simeq 0.03$. The low degree of transformation is not related to active particle exhaustion since it is possible to initiate the reaction by the repeated impulse of compression without additional photolysis. With higher radical concentration the reaction passes over to the regime of thermal explosion with a respective increase of η. This transition occurs in a narrow interval of the change in R, when the initial adiabatic heating is sufficient for transition to the Arrhenius region.

Since low degrees of transformation under nonthermal inflammation conditions do not rule out the possibility of an in-site reaction course with high local heating, a special study of the spatial distribution of reaction products was carried out [219]. Estimation of sizes of the reaction zones is based on measurements of the angular distribution of the Rayleigh light scattering intensity prior to and after the fast stage of transformation initiated by the loading impulse. The change of light scattering is determined by the difference in the refractive indices of reactants and products. The upper estimate of the product particle size was found not to exceed 0.14 μm. From the balance condition of heat release and removal it follows that local heating of zones of that size at the mentioned reaction rate constant values cannot exceed 1 K. Thus, the possible spatial heterogeneity of the reaction cannot be the cause of its acceleration.

The nonthermal character of the low-temperature critical phenomena is also manifested in the recently revealed autowave propagation of cryochemical conversions in mixtures of frozen reactants [228–231]. Authors of refs. 228–231 have shown that a peculiarity of the reaction front propagation is the absence of thermal heating (see also refs. 232, 233). Its formation and propagation is determined by emergence, in the course of chemical transformation, of critical mechanical stresses sufficient for an adiabatically fast growth of main-line cracks, the zones of which are reaction sites. Acceleration of solid-state reactions in the damaged zones and its dependence on crack growth rates and the local temperature of neighboring plastic zones was treated in ref. 222, where, in particular, it was shown that UTLI is not related to the formation and growth of the submicrocracks, microcracks, and main-line cracks. We do not discuss here the phenomenon of autowave propagation of cryochemical solid-state reactions in more detail since that topic is treated in another survey written for the special synergetics issue of *Advances in Chemical Physics* prepared by V. I. Krinskii.

From the preceding considerations it follows that for the explanation of macrokinetic peculiarities of solid-state chemical reactions in the low-temperature $K(\mathrm{T})$ plateau region, it is necessary to introduce, in addition to active particle concentration and temperature, a parameter characterizing the local, structural nonequilibrium state of the medium capable of changing in the course of the reaction. In ref. 51, the excess free energy of glass, G, was chosen as such a parameter. It is well-known [139, 140, 188] that G can be varied in a rather wide range by changing the conditions of glass preparation or with mechanical action, both outer and internal. The phenomenological model of nonthermal critical phenomena suggested in ref. 51 was developed in ref. 236. The model is based on two assumptions. The chemical reaction barrier is regarded as a linear function of G [cf. relation (28)], so the $K(T)$ relationship may be presented in the form

$$K(T) = K_0 \exp\left[-\frac{V_m^0 - \alpha G}{K_B(T + T_k)} \right], \tag{99}$$

where the characteristic temperature T_k determines the transition from the Arrhenius relationship to the low-temperature plateau. The second assumption of this model states that a part, q_1, of the thermal effect of reaction q is spent on the increase in G in the course of transformation. The system of macrokinetic equations describing the homogeneous inflammation has the form

$$\frac{dT}{dt} = \frac{q}{C_v \rho} K_0 R \exp\left\{ \frac{V_m^0 - \alpha G}{K_B(T + T_k)} \right\} - K_t(T - T_c), \tag{100}$$

$$\frac{dG}{dt} = q_1 K_0 R \exp\left\{ \frac{V_m^0 - \alpha G}{K_B(T + T_k)} \right\} - K_g G, \tag{101}$$

where the radical concentration R is considered constant in the process of inflammation, in agreement with experimental data; T_0 is the thermostat temperature; k_t and k_g are the rates of heat removal and relaxation of G. The latter is described by the relation

$$k_g = k_g^0 \exp\left(-\frac{U}{K_B(T+T_r)} \right), \tag{102}$$

accounting for the decrease in the apparent activation energy of the relaxation processes with temperature, noted in Section 4. Since it is impossible to explain inflammation under cooling and formation of UTLI in the framework of the notions about thermal inflammation under any temperature relationships of $K(T)$ and K_t, we can—without loss of generality—regard the latter as temperature independent. Equation (100) at constant G and $T_k = 0$ is a common Semenov equation for thermal inflammation [235] (cf. also ref. 236), and Eqn. (101) phenomenologically describes the change of the reactive medium in the course of the reaction. Introducing dimensionless variables,

$$\theta = U\frac{T-T_0}{(T_r+T_0)^2}, \qquad g = \frac{\alpha G}{T_k+T_0}, \qquad \tau = K_t \tag{103}$$

we transform Eqs. (101) and (102):

$$\begin{aligned} \frac{d\theta}{d\tau} &= \gamma \exp[\xi\theta + g - 1] - \theta, \\ \chi\frac{dq}{d\tau} &= \gamma \exp[\xi\theta + g - 1] - \beta e^{\theta} g. \end{aligned} \tag{104}$$

The dimensionless parameters are

$$\begin{aligned} \beta &= \frac{q}{\alpha q_1}\frac{U(T_k+T_0)}{C_v\rho(T_r+T_0)^2}\frac{k_g^0}{k_t}\exp\left[-\frac{U}{T_r+T_0} \right], \\ \gamma &= \frac{U}{T_r+T_0}\frac{q}{C_v\rho(T_r+T_0)^2}\frac{k_0}{k_r}R\exp\left[-\frac{V_m^0}{T_k+T_0}+1 \right], \\ \xi &= \frac{V_m^0}{U}\left(\frac{T_r-T_0}{T_k+T_0} \right)^2, \\ \chi &= \frac{q}{\alpha q_1}\frac{U(T_k+T_0)}{(T_r+T_0)^2}. \end{aligned} \tag{105}$$

The problem is reduced to finding the phase trajectories of the equation system (104) at the (g, θ)-plane at different γ values (dimensionless reaction rate) and values of β (relationship of the rates of relaxation g and heat removal at $T = T_0$). Dependence of the solution on x and ξ in the physically justified ranges of their variation ($\eta \gg 1$ at $q \gg q_1$; $\xi < 1$) turns out to be relatively weak. The authors of ref. 234 applied the well-known method of analysis of specific trajectories changing at the bifurcational values of parameters [237]. In the general case, the system of equations (104) has four singular points. The inflammation condition has the form

$$\gamma = \gamma_c = \varphi(\theta^*) = \theta^* \exp\left(1 - \xi\theta^* - \frac{\theta^*}{\beta} e^{-\theta^*}\right), \qquad g^* = \frac{\theta^*}{\beta} e^{-\theta^*} \tag{106}$$

and corresponds to the tangency of the $\varphi(\theta)$ function plot by the horizontals, which is similar to the thermal inflammation condition. Unlike the latter, the $\varphi(\theta)$ function in the general case has two maxima; therefore, two inflammation regimes are possible. It is clearly seen from Figure 23 that the division of

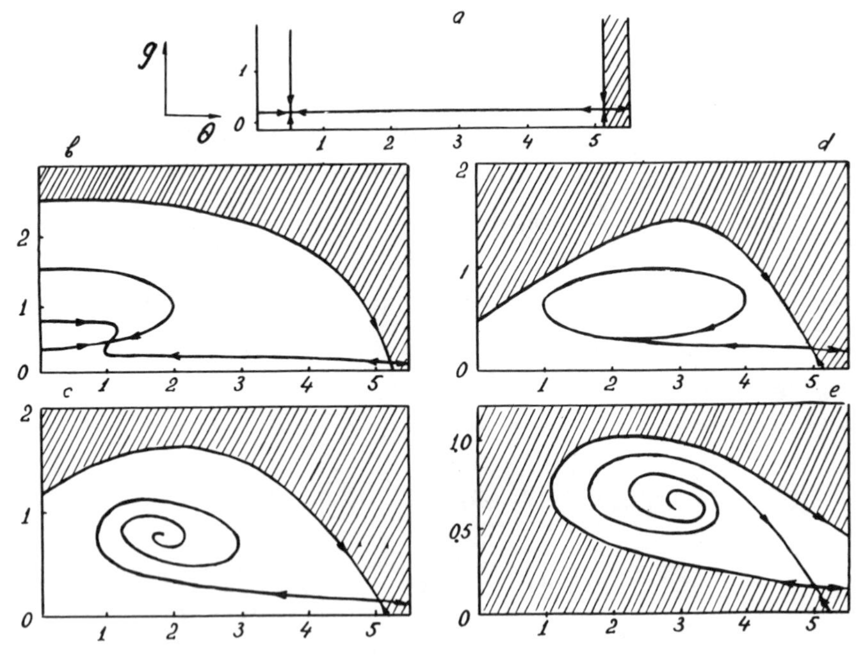

Figure 23. Singular trajectories of system of equations (104) with existence of two singulars at different values of β: (a), 10; (b), 1; (c), 0.35; (d), 0.30; (e), 0.27. $\gamma = 1$, $\xi = 0.5$, $\eta = 10$. Singularities are site-saddle (a, b), stable focus-saddle (c), unstable focus with stable limiting cycle-saddle (d), and unstable focus-saddle (e).

the (g, θ)-plane by the singular trajectories represented in the case of two singulars of system (104) by the saddle separatrices tend to a stable site or to a focus as well as to infinity. At $\beta > \eta$, $\gamma < 1/\xi$, inflammation becomes thermal. The singulars of the system are the stable site and the saddle (Figure. 23*a*), corresponding to the stable and unstable points in the Semenov diagram; they are positioned close to the θ axis. The saddle separatrices are almost parallel to the coordinate axes. Inflammation (transition to the region of unlimited trajectories) is determined by θ growth and is independent of g. Decreasing β increases the role of G. At

$$1/\chi > \beta > \beta_m = p^2 e^{-\beta}, \qquad p = \frac{1-(1-\xi)^{1/2}}{\xi}, \tag{107}$$

the slope of a separatrix dividing the stable and unstable regions becomes negative (Figure 23*b*). At sufficiently high g values the unlimited trajectories emerge in the absence of heating up (at $\theta = 0$). The lower is β, the smaller are the g values at which inflammation occurs. When a stable point turns from the site into a focus, the slope of the separatrix at low θ becomes noble (Figure 23*c*, *d*), and a low-temperature region emerges. With further β lowering the singular point lying in the θ region between the $\varphi(\theta)$ function maxima becomes an unstable focus surrounded by the stable limiting cycle growing in size (Figure 23*e*). The autovibrational states emerging under formation of the limiting cycle have the corresponding oscillations of the reaction rate: Temperature rise causes a simultaneous increase of the rates of reaction and relaxation, G, and the latter, in its turn, diminishes the reaction rate. Two transitions are possible under existence of four singulars. In the low-temperature region under transition from the stable site to an unstable focus or limiting cycle surrounding the unstable focus, the system remains stable, but the reaction rate increases stepwise. Transition occurs at g values limited both from above and below, the latter meaning that heating diminishes G, so that the system returns into the state with a low reaction rate. It explains low degrees of transformation under UTLI. At higher g values the system is not "jammed" in a valley between the $\varphi(\theta)$ function maxima, and low-temperature inflammation passes into thermal explosion.

The bifurcational values of γ and β determining the confluence line of sites and focuses are the solutions of the equations

$$\beta = \theta^* \frac{1-\theta^*}{1-\xi\theta^*} e^{-\theta^*}, \qquad \gamma = \theta^* \exp\left[-\theta^* \frac{1-\xi\theta^*}{1-\theta^*}\right]. \tag{108}$$

At $\theta^* \ll 1$, $\beta < \beta_m$ (i.e., in the case of four singulars), γ and β are related by

$$\gamma \simeq \beta + (3-\beta)\beta^2. \tag{109}$$

From Eqs. (105) and (109) it follows that the critical concentration of radicals corresponding to inflammation under the thermostat temperature T_0 equals

$$R_c \sim \exp\left[-(1-\xi_1)\frac{U}{T_k+T_0}\right], \qquad \xi_1 = \xi\frac{T_k+T_0}{T_r+T_0}. \tag{110}$$

According to Eqn. (110), the UTLI emerges when $\xi_1 < 1$, that is,

$$\frac{\partial \ln K_g}{\partial T} > \frac{\partial \ln K}{\partial T} \tag{111}$$

Relation (111) signifies that the UTLI is possible if the rate constant of structural relaxation changes with temperature faster than the reaction rate constant. This evidently occurs in the region of low-temperature limit. Under UTLI, the reaction rate increases proportionally to $-(1/\beta)\ln\beta$ only at the expense of a G increase under isothermal conditions ($T^* = T_0$), that is, as a result of the energetic chain development described by Eqn. (101). Temperature rise under UTLI turns from the accelerating to the decelerating factor of chemical reaction. In another limiting case, $\beta \gg \xi \simeq 1/\gamma$ inflammation becomes thermal ($G^* = 0$), and the relation between R_c and T_0 corresponds to the lower temperature limit of inflammation:

$$R_c \sim \exp\left[\frac{V_m^0}{T_k+T_0}\right] \tag{112}$$

The autooscillations of the reaction rate predicted by the discussed phenomenological model were found elsewhere [220]. They occur under radical concentrations corresponding to the transition from the limited low-temperature transformation to the thermal explosion. After loading impulse the sample heats up by 10–20 K; then the reaction stops, the sample cools, which brings about another rate increase, and so on. It is possible to observe four to seven decaying impulses of reaction with a period exceeding severalfold the time of the thermal relaxation of the sample (Figure 24). Autooscillations occur in a solid phase and demonstrate the main peculiarity of the low-temperature critical phenomena: cooling accelerates the reaction, while a temperature rise (at low η) inhibits it. These macrokinetic peculiarities, seeming at first glance paradoxical, are a direct consequence of the reaction rate constant temperature independence and sharp dependence on molecular distances discussed in the preceding sections.

The phenomenological model explains the totality of the data on the thermal critical phenomena. A linear relationship between V_m and G underlying the model was confirmed by Ovchinnikov and Shamovsky [238], who

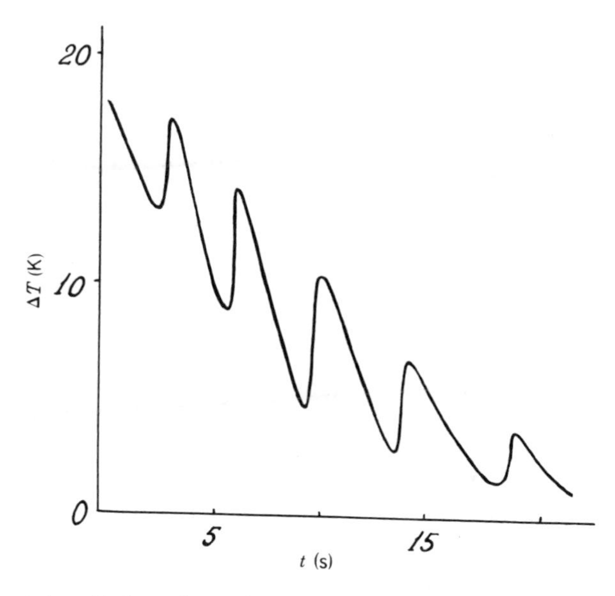

Figure 24. Autooscillations of rate of methylcyclohexane chlorination in its equimolar mix with molecular chlorine. Initial temperature is 45 K. Radical concentration created with pre-photolysis is 4×10^{17} cm^{-3}.

analyzed using mathematical modelling the changes of the excess free energy and the mean height of the barriers separating the local minima of the potential energy under disordering of a two-dimensional lattice. The relaxation process for glass was described by Ovchinnikov and Onischuk [239, 240] in a model of the dynamic molecular field. It was shown that part of the thermal effect is really accumulated in the form of local excess energy of glass near the chemical transformation site.

Further development of the notions about the nonthermal critical phenomena in the region of the low-temperature limit of the rate constant should be based on quantum-chemical computations of local deformation of the solid lattice and its effect on the consequent reaction act rate.

7. CONCLUSION

Cryochemical research in the past 15 years has established the existence of the low-temperature limit of rate constants of various solid-state chemical reactions with transfer of atoms and molecular fragments of different masses over distances comparable with intermolecular ones, from H-atoms transfer under intramolecular rearrangement to organic radicals and halogen atom transfer in chain reactions. Though the values of rate constants in the region of this limit, $K(0)$, equalling 10^{-5}–10^{2} s^{-1}, are smaller than for common chemical reactions proceeding in the kinetic regime, the very possibility of their observation at superlow temperatures absolutely contradicts the Arrhenius law and the underlying notions about the thermoactivated character of chemical transformations. A strong dependence of low-temperature solid-state reactions on the structural and dynamic parameters of a solid and their variations under mechanical impacts [manifested e.g., by emergence of non-thermal critical phenomena in the $K(0)$ limit region] violates the traditional idea that the medium around a reactant acts as a thermostat, ensuring only a quick setting of thermal equilibrium. The medium rearrangement becomes the factor determining the reaction rate, and the limit itself emerges when the medium motion turns from the classical into the quantum one. Another feature of low-temperature reactions emphasizing their quantum nature is the strong kinetic isotope H–D effect, exceeding by several orders of magnitude the one at higher temperatures. The totality of data on the specific non-Arrhenius cryochemical reactivity of the solids finds its natural explanation in the framework of the new notions, presented in this review, based on the quantum dynamics of solids.

Thus, the observed features of low-temperature reactions and the need of quantum notions for their explanation are the main features of the chemistry of the deep cold.

The problems of quantum cryochemical reactivity are closely related to a wide range of modern problems of solid-state physics; this was emphasized in our review. In both cases we speak about the evolution from simple one-dimensional tunneling models to multidimensional ones, accounting for the quantum character of solid-state dynamics. We would like to particularly emphasize the conceptual closeness with the problems of autolocalization and heavy-particle transfer in disordered solids and the role of lattice vibrations in these processes. At the same time, this chapter has left untouched the works dedicated to the coherent processes of heavy-particle zone transfer significant, particularly, for the treatment of quantum crystal dynamics [241, 242], quantum diffusion [243, 244], and phase transitions in quantum crystals [245]. Very important questions of the effect of underbarrier scattering on transfer probability [246] and their relation with the phenomenological models of the Cramers type [247] were also left out. Though the tunneling frequencies typical of chemical transformations are several orders of magnitude lower than in the purely physical processes of heavy-particle transfer, the similarity of the phenomenon in both cases can be related not only to the fluctuational restructuring of the barrier, but also to participation of tunneling states in the stage of formation of the reactant configuration required for chemical reaction. A possible role of tunneling states in the dynamics of proteins and other biopolymers [248, 249], adjacent to the scope of the discussed problems, was not included in the present review either.

The quantum cryochemistry is rapidly developing, from the viewpoint of both the scope of the studied objects and theoretical notions. That is why this chapter should not be regarded as summarizing any specific stage of research. In fact, only the first step is made, showing the specificity of the novel field of chemical kinetics. At present, continuously arising questions outnumber the answers, and it would be more proper to speak about the prospects of development, not about its accomplishments. That is why we point out, not claiming completeness, some of the tasks of future research. The most acute among them is a development of quantum-chemical models of specific reactions, lagging, at present, behind development of a dynamic theory of gas phase reactions. These models should be based on data on the spatial structure of reactants and should account for the vibrational spectrum of both the reactants and the surrounding lattice. A more consistent analysis is required of inter- and intramolecular motions, both translational and orientational, and their role in the formation and fluctuations of the potential barrier of the reaction. Implementation of this program requires wide cryochemical application of the modern studies of crystalline and glassy solids. For instance, the high-resolution optical spectroscopy, particularly in combination with the matrix isolation of reactant pairs, may enable one to detect and investigate the relations among the existence of high-frequency tunneling states, the phonon spectrum of the matrix, and the chemical transformation rate. The same aims can be pursued by the study of low-temperature anomalies of specific heats, (damping of electromagnetic waves and ultrasound) and of acoustoemission.

As any review dedicated to a rapidly developing field, our review is subjective and bears traces of the scientific interests and preferences of its authors. This, inter alia, has affected the choice of references from adjacent fields, where we have intended to cite only the most important original papers.

Acknowledgment

Concluding, the authors consider it a pleasant duty to express gratitude to Yu. M. Kagan, A. A. Ovchinnikov, and M. A. Kozhushner for fruitful discussions and I. P. Kim for assistence in finalizing the manuscript.

References

1. V. I. Goldanskii, L. I. Trakhtenberg, and V. N. Fleurov (1986). *Tunneling Phenomena in Chemical Physics.* Nauka, Moskow, (in Russian); Gordon and Breach, New York, (1987) (in English); *Sov. Sci. Rev. B.*, *9*:59–124 (1987).

2. F. Hund, (1927) *Z. Phys.*, *43*:805.
3. S. Z. Roginsky and L. V. Rosenkevitsch (1930) *Z. Phys. Chem.* (*B*) *10*:47.
4. E. Z. Wigner (1932). *Phys. Chem.* (*B*), *19*:1903.
5. R. P. Bell (1933). *Proc. Roy. Soc., London* (*A*), *139*:466; (1935) *148*:241; (1937) *158*:128.
6. R. P. Bell (1973). *The Proton in Chemistry*, 2nd ed. Cornell University Press, Ithaca, N.Y.
7. R. P. Bell (1980). *The Tunnel Effect in Chemistry*. Chapman and Hall, London.
8. V. I. Goldanskii (1959). *Dokl. Acad. Nauk SSSR*, *124*:1261; (1959) *127*:1037.
9. W. H. Miller and T. F. George (1972). *J. Chem. Phys.*, *56*:5668.
10. V. A. Benderskii, Yu. I. Dakhnovskii, A. A. Ovchinnikov, and P. G. Filippov (1982). *Chem. Phys.*, *67*:301.
11. R. I. LeRoy, H. Murai, and F. Williams (1980). *J. Am. Chem. Soc.*, *102*:2325.
12. E. Buhks and J. Jortner (1980). *J. Phys. Chem.*, *84*:3370.
13. R. Kubo and Y. Toyazawa (1955). *Progr. Theoret. Phys.*, *13*:160.
14. R. R. Dogonadze and A. M. Kuznetsov (1975). *Progr. Surf. Sci.*, *6*:1.
15. A. A. Ovchinnikov and M. Ya. Ovchinnikova (1982). *Adv. Quant. Chem.*, *16*:161.
16. L. D. Landau and E. M. Lifshits (1963). *Quant. Mechanic.* GIFML. Moscow, p. 385.
17. S. Glasstone, K. I. Laidler, and H. Eyring (1941). *The Theory of Rate Process*. McGraw-Hill, New York, ch. 4.
18. H. Eyring, S. H. Lin, and S. M. Lin (1980). *Basic Chemical Kinetics. Wiley-Interscience*, New York, ch. 4.
19. M. Ya. Ovchinnikova (1968). *Theor. Exper. Chim.*, *4*:575.
20. W. F. Libby (1952). *J. Phys. Chem.*, *56*:833.
21. R. A. Marcus (1956). *J. Chem. Phys.*, *24*:966; (1956) *24*:979; (1957) *26*:867; (1963) *38*:1858; (1963) *39*:1734; (1965) *43*:679.
22. R. A. Marcus (1964). *Ann. Rev. Phys. Chem.*, *15*:155.
23. V. G. Levich and R. R. Dogonadze (1959). *Dokl. Akad. Nauk SSSR*, *124*:123.
24. V. G. Levich (1965). *Adv. Electrochem. Eng.*, *4*:249.
25. A. A. Ovchinnikov and M. Ya. Ovchinnikova (1969). *Zh. Eksp. Teor. Fiz.*, *56*:1278.
26. L. D. Landau and S. I. Pekar (1948). *Zh. Eksp. Teor. Fiz.*, *18*:419.
27. J. O'M. Bockris and S. U. M. Khan (1979). *Quantum Electrochemistry*. Plenum Press, New York, ch. 6.
28. J. Ulstrup (1979). *Charge Transfer Process in Condenced Media*. Springer-Verlag, Berlin.
29. B. Conway and M. Solomon (1964). *J. Chem. Phys.*, *41*:3169.
30. V. A. Benderskii and A. A. Ovchinnikov (1978). *Dokl. Akad. Nauk SSSR*, *243*:393.
31. A. A. Ovchinnikov and V. A. Benderskii (1979). *J. Electroanal. Chem.*, *100*:563.
32. V. A. Benderskii and A. A. Ovchinnikov (1980). *Phys. Chem. Contemporary Problems*. vol. 1, *Chemistry*, Moscow., p. 159.
33. V. A. Benderskii, Yu. I. Dakhnovskii, and A. A. Ovchinnikov (1983). *J. Electroanal. Chem.*, *148*:161.
34. A. A. Ovchinnikov, V. A. Benderskii, and A. G. Krivenko (1978). *J. Electroanal. Chem.*, *91*:321.
35. D. DeVault and B. Chance (1966). *Biophys. J.*, *6*:825.
36. B. Chance, D. DeVault, and V. Legallais (1967). In: *Fast Reaction and Primary Processes in Chemical Kinetics* (Ed.) S. Classon, Almqvist and Wiksell, Stockholm, p. 437.

37. B. Chance (Ed.) (1979). *Tunneling in Biological Systems.* Academic Press, New York, p. 4.
38. D. De Vault (1980). *Quart. Rev. Chem. Soc. London, 13*:387.
39. K. I. Zamaraev, R. F. Khairutdinov, A. I. Mikhailov, and V. I. Goldanskii (1971). *Dokl. Akad. Nauk SSSR, 199*:640.
40. J. R. Miller (1972). *J. Chem. Phys. 56*:5173.
41. K. I. Zamaraev and R. F. Khairutdinov (1980). *Sov. Sci. Rev. 2*:357.
42. K. I. Zamaraev, R. F. Khairutdinov, and V. P. Zhdanov (1985). *Tunneling Electron in Chemistry.* Novosibirsk, Nauka.
43. L. N. Grigorov and D. S. Chernavskii (1972). *Biofiz., 17*:195.
44. L. A. Blumenfeld and D. S. Chernavskii (1973). *J. Theor. Biol., 39*:1.
45. J. J. Hopfield (1977). *Biophys. J., 18*:311.
46. J. Jortner (1976). *J. Chem. Phys., 64*:4860.
47. G. K. Ivanov and M. A. Kozhushner (1982). *Khim. Fiz., 1*:1039.
48. V. L. Klochikhin, S. Ya. Pshezhetskii, and L. I. Trakhtenberg (1978). *Dokl. Akad. Nauk SSSR, 239*:879.
49. M. Ya. Ovchinnikova (1979). *Chem. Phys., 36*:85.
50. V. A. Benderskii, V. I. Goldanskii, and A. A. Ovchinnikov (1980). *Chem. Phys. Lett., 73*:492.
51. V. A. Benderskii, A. A. Ovchinnikov, P. G. Filippov, and E. Ya. Misochko (1980). *Lett. J. Eksp. Teor. Phys., 32*:429.
52. N. N. Semenov (1962). *Pure Appl. Chem., 5*:353.
53. D. P. Kiryukhin, I. M. Barkalov, V. I. Goldanskii, and A. M. Kaplan (1970). *Vysokomol. Soedinen., 12*:491.
54. D. P. Kiryukhin, A. M. Kaplan, I. M. Barkalov, and V. I. Goldanskii (1971). *Dokl. Akad. Nauk SSSR, 199*:857.
55. D. P. Kiryukhin, A. M. Kaplan, I. M. Barkalov, and V. I. Goldanskii (1972). *Vysokomol. Soedinen., A14*:2115..
56. D. P. Kiryukhin, A. M. Kaplan, I. M. Barkalov, and V. I. Goldanskii (1972). *Dokl. Akad. Nauk SSSR, 206*:147.
57. V. I. Goldanskii, M. D. Frank-Kamenetskii, and I. M. Barkalov (1973). *Dokl. Akad. Nauk SSSR, 211*:133.
58. V. I. Goldanskii, M. D. Frank-Kamenetskii, and I. M. Barkalov (1973). *Science, 182*:1344.
59. D. P. Kiryukhin, I. M. Barkalov, and V. I. Goldanskii (1974). *Vysokomol. Soedinen., 16*:565.
60. V. I. Goldanskii (1975). *Usp. Khim., 44*:2121.
61. V. I. Goldanskii (1976). *Ann. Rev. Phys. Chem., 27*:85.
62. V. I. Goldanskii (1979). In: *Tunneling in Biological Systems.* B. Chance et al. (Eds.) New York, Academic Press.
63. V. I. Goldanskii (1978–79). *Chem. Scripta, 13*:1.
64. V. I. Goldanskii (1979). *Nature, 279*:109.
65. R. H. Austin, K. W. Beeson, L. Eisenstein, H. Frauenfelder, and I. Gunsalus (1975). *Biochem., 14*:5355.
66. N. Alberding, R. H. Austin, S. S. Chan, L. Eisenstein, H. Frauenfelder, I. C. Gunsalus, and T. M. Nordlung (1976). *J. Chem. Phys., 65*:4701.
67. N. Alberding, R. H. Austin, K. W. Beeson, S. S. Chan, L. Eisenstein, H. Frauenfelder, and T. M. Nordlung (1976). *Science, 192*:1002.

68. J. O. Alben, D. Beece, S. F. Bowne, L. Eisenstein, H. Frauenfelder, D. Good, M. C. Marden, P. P. Moh, L. Reinisch, A. H. Reynolds, and T. Yuek (1980). *Phys. Rev. Lett.*, *44*:1157.
69. B. Chance, R. Fischetti, and L. Powers (1983). *Biochem.*, *22*:3820.
70. P. Debrunner and H. Frauenfelder (1982). *Experience in Biochemical Perception.* Academic Press, New York, p. 327.
71. O. E. Yakimchenko and Ya. S. Lebedev (1971). *Intern. J. Radiat. Phys. Chem.*, *3*:17.
72. K. Toriyama, K. Nunome, and M. Iwasaki (1977). *J. Am. Chem. Soc.*, *99*:5823.
73. G. Brunton, D. Griller, L. D. C. Barclay, and K. V. Ingold (1976). *J. Am. Chem. Soc.*, *98*:6803.
74. G. Brunton, I. A. Gray, D. Griller, L. R. C. Barclay, and K. V. Ingold (1978). *J. Am. Chem. Soc.*, *100*:4197.
75. E. D. Sprague and F. Williams (1971). *J. Am. Chem. Soc.*, *93*:787.
76. A. Campion and F. Williams (1972). *J. Am. Chem. Soc.*, *94*:7633.
77. A. R. I. Le Roy, E. D. Sprague, and F. Williams (1972). *J. Phys. Chem.*, *76*:546.
78. P. L. Hudson, M. Shiotani, and F. Williams (1977). *Chem. Phys. Lett.*, *48*:193.
79. B. V. Bolshakov and V. A. Tolkatchev (1976). *Chem. Phys. Lett.*, *40*:486.
80. B. V. Bolshakov, A. B. Doctorov, V. A. Tolkatchev, and A. I. Burstein (1979). *Chem. Phys. Lett.*, *64*:113.
81. B. V. Bolshakov, A. A. Stepakov, and V. A. Tolkatchev (1980). *Int. J. Chem. Kinet.*, *12*:271.
82. V. M. Zaskulnikov, V. L. Vyazovkin, B. V. Bolshakov, and V. A. Tolkatchev (1981). *Intern. J. Chem. Kinet.* *13*:707.
83. H. S. Judeikis and S. Siegel (1965). *J. Chem. Phys.*, *43*:3625.
84. M. A. Neiss and I. E. Willard (1975). *J. Phys. Chem.*, *79*:783.
85. E. D. Sprague (1977). *J. Phys. Chem.*, *81*:516.
86. E. D. Sprague, K. Takeda, I. T. Wang, and F. Williams (1974). *Canad. J. Chem.*, *52*:2840.
87. M. P. Fuks, B. V. Bolshakov, and V. A. Tolkatchev (1975). *React. Kinet. Intern. Lett.*, *3*:349.
88. L. I. Trakhtenberg, V. L. Klochikhin, and S. Ya. Pshezhetskii (1981). *Chem. Phys.*, *59*:191; (1982) *69*:121.
89. L. I. Trakhtenberg (1982). *Khim. Fis.*, *1*:53.
90. I. P. Colpa, B. Prass, and D. Stehlik (1984). *Chem. Phys. Lett.*, *107*:469.
91. N. G. Khainovskii, Yu. T. Pavlyukhin, R. F. Khairutdinov, and V. V. Boldirev (1985). *Dokl. Akad. Nauk SSSR*, *283*:172.
92. A. M. Dubinskaya and P. Yu. Butyagin (1973). *Dokl. Akad. Nauk SSSR*, *211*:141.
93. A. M. Dubinskaya (1978). *Usp. Khim.*, *47*:1169.
94. F. S. Dainton, B. I. Holt, A. Philipson, and M. I. Pilling (1976). *J. Chem. Soc. Farad. Trans. I*, *72*:257.
95. K. Toriyama, K. Nunome, and M. Iwasaki (1980). *J. Phys. Chem.*, *84*:2374.
96. D. Bhattacharya and I. E. Willard (1981). *J. Phys. Chem.*, *85*:154.
97. M. Iwasaki, K. Toriyama, and H. Muto (1981). *Intern. J. Radiat. Phys. Chem.*, *17*:304.
98. M. Iwasaki, K. Toriyama, and H. Muto (1981). *J. Phys. Chem.*, *85*:1326.
99. H. Y. Wang and I. E. Willard (1979). *J. Phys. Chem.*, *83*:2585.
100. V. P. Senthilnathan and M. S. Platz (1980). *J. Am. Chem. Soc.*, *102*:7637.
101. M. S. Platz, V. P. Senthilnathan, B. B. Wright, and C. W. McCardy (1982). *J. Am. Chem. Soc.*, *104*:6494.

102. V. L. Klochikhin, S. Ya. Pshezhetskii, and L. I. Trakhtenberg (1980). *Zh. Fis. Khim.*, *54*:1324.

103. P. H. Gribb (1984). *Chem. Phys.*, *88*:47.

104. E. B. Gordon, A. A. Pelmenev, O. F. Pugatchev, and V. V. Hmelenko (1983). *Pis'ma Zh. Eksp. Teor. Fiz.*, *37*:237.

105. A. Ya. Katunin, I. I. Lukashevich, S. I. Orozmanatov, V. V. Sklyarevskii, V. V. Filippov, and V. A. Shevtsov (1980). *Pis'ma Zh. Eksp. Teor. Fiz.*, *34*:375.

106. A. F. Iskovskikh, A. Ya. Katunin, I. I. Lukashevich, and V. A. Shevtsov (1986). *Zh. Eksp. Teor. Fiz.*, *91*:1832.

107. H. Tsuruta, T. Miyazaki, T. Fueki, and N. Azuma (1983). *J. Phys. Chem.*, *87*:5422.

108. T. Miyazaki and K. P. Lee (1986). *J. Phys. Chem.*, *90*:400.

109. K. P. Lee, T. Miyazaki, K. Fueki, and K. Cotoh (1987). *J. Phys. Chem.*, *91*:180.

110. D. Lukas and G. G. Pimentel (1979). *J. Phys. Chem.*, *83*:2311.

111. S. L. Buchwalter and G. H. Closs (1978). *J. Am. Chem. Soc.*, *101*:4688.

112. C. S. Yannoni, V. Macho, and P. C. Myhre (1982). *J. Am. Chem. Soc.*, *104*:7380.

113. P. C. Myhre, K. L. MeLaren, and C. S. Yannoni (1985). *J. Am. Chem. Soc.*, *122*:59.

114. F. K. Fong (1974). *J. Am. Chem. Soc.*, *96*:7638.

115. I. Brickmann (1981). *Ber. Bunsengeselsch. Phys. Chem.*, *85*:106.

116. B. K. Carpenter (1983). *J. Am. Chem. Soc.*, *105*:1700.

117. M. I. S. Dewar and K. M. Merz (1985). *J. Phys. Chem.*, *89*:4739.

118. M. I. S. Dewar and K. M. Merz (1986). *J. Am. Chem. Soc.*, *108*:5634.

119. D. P. Kiryukhin, I. M. Barkalov, and V. I. Goldanskii (1978). *Dokl. Akad. Nauk SSSR*, *238*:388.

120. D. P. Kiryukhin, I. M. Barkalov, and V. I. Goldanskii (1979). *J. Chim. Phys.*, *76*:1013.

121. I. M. Barkalov, V. I. Goldanskii, D. P. Kiryukhin, and A. M. Zanin (1980). *Chem. Phys. Lett.*, *73*:273.

122. V. A. Titov, P. G. Filippov, E. Ya. Misochko, R. D. Usmanov, and V. A. Benderskii (1986). *Dokl. Akad. Nauk SSSR*, *290*:1414.

123. V. A. Titov, E. Ya. Misochko, P. G. Filippov, V. A. Benderskii, and V. A. Filippov, (1988). *Khim. Fiz.*, *7*:327.

124. E. Ya. Misochko, P. G. Filippov, and V. A. Benderskii (1980). *Dokl. Akad. Nauk SSSR*, *253*:163.

125. P. G. Filippov, E. Ya. Misochko, A. V. Kulikov, V. A. Benderskii, and A. A. Ovchinnikov (1981). *Dokl. Akad. Nauk SSSR*, *256*:1173.

126. V. A. Benderskii, E. Ya. Misochko, A. A. Ovchinnikov, and P. G. Filippov (1983). *J. Phys. Chem.*, *57*:1079.

127. V. A. Benderskii, Yu. A. Galchenko, E. Ya. Misochko, P. G. Filippov, and Yu. L. Moskvin. (1983). *Kinetika i kataliz.*, *24*:19.

128. V. A. Benderskii, V. A. Titov, and P. G. Filippov (1984). *Dokl. Akad. Nauk SSSR*, *278*:1157.

129. O. Hassel and C. Romming (1962). *Quart. Rev.*, *14*:1.

130. L. Fredin and B. Nelander (1973). *J. Molec. Struct.*, *16*:205.

131. I. M. Kimelfeld, E. V. Lumer, and A. P. Shwedchikov (1973). *Chem. Phys. Lett.* *21*:429.

132. G. B. Sergeev, V. V. Smirnov, and M. I. Shilina (1982). *Kinetika i kataliz*, *23*:322.

133. I. H. Gibbs (1960). *Modern Aspects of Vitreous State.* Batterworth London, ch. 7.

134. G. Adams and I. H. Gibbs (1965). *J. Chem. Phys., 43*:139.
135. D. A. Kritskaya, A. N. Ponomarev, and V. L. Talroze (1968). *Khim. Vysok. Energ. 2*:61.
136. V. A. Kargin, V. A. Kabanov, and I. M. Papisov (1964). *J. Polym. Sci., C4*:1967.
137. V. N. Kapustyan, A. A. Zharov, and N. S. Enikolopyan (1968). *Dokl. Akad. Nauk SSSR, 179*:627.
138. V. L. Klochikhin and L. I. Trakhtenberg (1984). *Zh. Fiz. Khim., 58*: 2877
139. S. S. Chang and A. B. Bestul (1978). *J. Chem. Phys., 56*:503.
140. P. P. Kobeko (1952). *Amorfnie Veschestva.* Nauka, Moscow, chs. 7 and 8.
141. V. L. Klochikhin and L. I. Trakhtenberg (1981). *Dokl. Akad. Nauk SSSR, 259*:399.
142. H. S. Jonston and D. Rapp (1961). *J. Am. Chem. Soc., 83*:1.
143. W. H. Miller (1975). *Adv. Chem. Phys., 30*:77.
144. R. A. Marcus and M. E. Coltrin (1977). *J. Chem. Phys., 67*:2609.
145. R. I. Altkorn and G. G. Schatz (1980). *J. Chem. Phys., 75*:3337.
146. V. A. Benderskii, Yu. I. Dakhnovskii, and A. A. Ovchinnikov (1981). *Dokl. Akad. Nauk SSSR, 261*:653.
147. J. Jortner and I. Ulstrup (1979). *Chem. Phys. Lett., 63*:236.
148. V. L. Klochikhin, S. Ya. Pshezhetskii, and L. I. Trakhtenberg (1979). *Vysokomol. Soedinen., 27A*:2792.
149. V. L. Klochikhin, L. I. Livshin, S. Ya. Pshezhetskii, and L. I. Trakhtenberg (1980). *Khim. Vysok. Energ., 14*:329.
150. T. Holstein (1959). *Ann. Phys. N.Y., 8*:325.
151. C. P. Flynn and A. M. Stoneham (1970). *Phys. Rev. B., 1*:3966.
152. W. R. McKinnon and C. M. Hurd (1983). *J. Chem. Phys., 87*:1283.
153. W. Siebrand, I. A. Wildman, and M. Z. Zgierski (1983). *Chem. Phys. Lett., 98*:108.
154. S. I. Pekar (1953). *Usp. Fiz. Nauk., 50*:197.
155. R. Feynmann (1957). *Phys. Rev., 84*:108.
156. N. Kh. Bikbaev, A. I. Ivanov, G. S. Lomakin, and O. A. Ponomarev (1978). *Zh. Eksp. Teor. Fiz., 74*:2054.
157. V. N. Bagratashvili, V. S. Letokhov, A. A. Makarov, and E. A. Ryabov (1981). *Itogi nauki i tekhniki. Fiz. atom. mol.*, vol. 2, VINITI, Moscow.
158. R. P. Feynman (1972). *Statistical Mechanics.* Benjamin, New York.
159. Yu. E. Perlin (1963). *Usp. Fiz. Nauk., 80*:553.
160. L. I. Trakhtenberg and V. L. Klochikhin (1983). *Electrokhim., 19*:1212.
161. L. I. Trakhtenberg (1983). *Khim. Fiz., 2*:1683.
162. V. L. Klochikin and L. I. Trakhtenberg (1983). *Khim. Fiz., 2*:810.
163. A. A. Maradudin, E. A. Montrell, and G. H. Weiss (1965). *Theory of Lattice Dynamics in the Harmonic Approximation.* Academic Press, New York.
164. A. M. Stoneham (1975). *Theory of Defects in Solids.* Clarendon Press, Oxford.
165. V. N. Kondratev, E. E. Nikitin, A. I. Reznikov, and S. Ya. Umanskii (1976). *Termicheskie bimolek. reak. in gase.* Nauka, Moscow.
166. S. P. Walch and T. H. Dunning (1980). *J. Chem. Phys., 72*:3221.
167. Yu. I. Dakhnovskii, A. A. Ovchinnikov, and V. A. Benderskii (1982). *Chem. Phys., 66*:93.
168. V. G. Vaks, V. M. Galitskii, and A. I. Larkin (1966). *Zh. Eksp. Teor. Fiz. 51*:1522.

169. V. G. Vaks (1973). *Introduction in Mikroskopical Theory of Segnetoelectrik.* Nauka, Moscow.

170. A. Z. Patashinskii and V. L. Pokrovskii (1982). *Fluctuation Theory of Fase Transitions.* Nauka, Moscow.

171. L. D. Landau and E. M. Lifshits (1964). *Statistic Physics.* Nauka, Moscow. p. 138.

172. W. Press (1972). *J. Chem. Phys.*, *56*:2597.

173. W. Press, A. Huller, H. Siller, W. Stirling, and R. Currant (1974). *Phys. Rev. Lett.*, *32*:1354.

174. P. G. Filippov, E. Ya. Misochko, A. V. Kulikov, and V. A. Benderskii (1981). *Dokl. Akad. Nauk. SSSR*, *256*:1173.

175. L. I. Trakhtenberg and N. Ya. Shteinshneider (1986). *Zh. Fiz. Khim.*, *60*:1409.

176. A. I. Anselm and N. N. Porfireva (1949). *J. Exp. Teor. Phys.*, *19*:438.

177. Ya. S. Lebedev and O. E. Yakimchenko (1978). *Usp. Chem.*, *47*:1018.

178. A. I. Burshtein, V. A. Klochikhin, and L. I. Trakhtenberg (1984). *Khim. Fiz.*, *3*:155.

179. V. L. Klochikhin and L. I. Trakhtenberg (1988). *Khim. Fiz.*, (in press).

180. M. H. Cohen and G. S. Grest (1980). *Phys. Rev. Lett.*, *45*:1271.

181. M. H. Cohen and G. S. Grest (1979). *Phys. Rev. B.*, *20*:107; (1980) *21*:411.

182. B. I. Shklovskii and A. A. Efros (1979). *Electronnie Svoistva Legirovannikh Poluprovodnikov.* Nauka, Moscow.

183. J. I. Spielberg and E. Gelerinter (1984). *Phys. Rev. B.*, *30*:2319.

184. V. A. Benderskii and N. P. Piven (1977). In: *Phys. Chem. Contemporary Problems*, vol. 6. Chemistry, Moscow, p. 137.

185. V. Narayanamurthi and R. O. Pohl (1970). *Rev. Mod. Phys.*, *42*:201.

186. O. L. Anderson and H. E. Bomnel (1955). *J. Am. Ceram. Soc.*, *38*:125.

187. W. A. Phillips (1978). *J. Non-Crystal. Solids*, *31*:267.

188. W. A. Phillips (Ed.) (1981). *Amorphous Solids.* Springer-Verlag, Berlin.

189. P. W. Anderson, B. I. Halperin, and C. M. Varma (1972). *Philos. Magaz.*, *25*:1.

190. W. A. Phillips (1972). *J. Low-Temp. Phys.*, *7*:381.

191. R. E. Strakna (1961). *Phys. Rev. Lett.*, *123*:2020.

192. M. R. Vakcevich (1972). *J. Non-Crystal. Solids*, *11*:25.

193. V. N. Fleurov, I. D. Mikheikin, and L. I. Trakhtenberg (1985). *Solid State Commun.*, *55*:537.

194. M. I. Klinger and V. G. Karpov (1980). *Pis'ma Zh. Thekhnich. Fiz.*, *6*:1473.

195. V. G. Karpov, M. I. Klinger, and F. N. Ignatiev (1983). *J. Non-Crystal. Solids*, *55*:307.

196. M. I. Klinger (1985). *Usp. Fiz. Nauk.*, *146*:105.

197. J. L. Black and B. I. Halperin (1977). *Phys. Rev. B.*, *16*:2879.

198. J. L. Black (1978). *Phys. Rev. B.*, *17*:2740.

199. V. N. Fleurov and L. I. Trakhtenberg (1982). *Solid State Commun.*, *44*:187.

200. V. N. Fleurov and L. I. Trakhtenberg (1986). *J. Phys. C.*, *19*:5529.

201. S. Tyagi and A. E. Lord (1986). *J. Non-Crystal. Solids*, *19*:5529.

202. A. S. Krayss and H. Eyring (1975). *Deformation Kinetics.* Wiley-Interscience, New York.

203. J. Friedel (1964). *Dislocations.* Pergamon Press, Oxford.

204. F. Garafalo (1963). *Fundamentals of Creep and Creep-Rapture in Metals.* MacMillan, New York.

205. I. M. Glen (1956). *Philos. Mag.*, *1*:400.

206. N. Mott (1956). *Philos. Mag., 1*:568.

207. V. I. Startsev, V. P. Soldatov, and A. I. Osetskii (1978). *Fiz. Nizk. Temp., 4*:1347.

208. V. I. Startsev (1983). *Dislocations in Solids*, F. R. N. Nabarro (Ed.), North-Holland, Amsterdam. ch. 28.

209. I. I. Gilman (1968). *J. Appl. Phys., 39*:6086.

210. I. I. Gilman and H. C. Tong (1971). *J. Appl. Phys., 42*:3479.

211. V. D. Natsik, A. I. Osetskii, V. P. Soldatov, and V. I. Startsev (1972). *Phys. Stat. Solidi* (*b*), *54*:99.

212. V. D. Natsik (1979). *Fiz. Nizk. Temp., 5*:400.

213. L. I. Trakhtenberg and V. N. Fleurov (1982). *Zh. Eksp. Teor. Fiz., 83*:1908.

214. B. V. Petukhov and V. L. Pokrovskii (1972). *Pis'ma Zh. Eksp. Teor. Fiz., 15*:63.

215. B. V. Petukhov and V. L. Pokrovskii (1972). *Zh. Eksp. Teor. Fiz., 63*:634.

216. V. P. Regel, A. I. Slutsker, and E. E. Tomashevskii (1974). *Kineticheskaya Teoriya Prochnosty Tverdykh Tel.* Nauka, Moscow.

217. R. L. Salganic (1970). *Int. J. Fracture Mechan., 6*:1.

218. A. I. Slutsker, *Kh. Aidarov* (1983) *Fiz. Tverd. Tela., 25*:477.

219. V. A. Benderskii, V. M. Beskrovnii, E. Ya. Misochko, and P. G. Filippov (1982). *Dokl. Akad. Nauk SSSR, 265*:1161; (1983) *273*:364.

220. V. A. Benderskii, E. Ya. Misochko, A. A. Ovchinnikov, and P. G. Filippov (1982). *Dokl. Akad. Nauk SSSR, 266*:144.

221. V. A. Benderskii, A. V. Kulikov, E. Ya. Misochko, A. A. Ovchinnikov, and N. P. Piven (1983). *Kinet. Kataliz., 24*:27.

222. V. A. Benderskii, V. M. Beskrovnii, E. Ya. Misochko, and P. G. Filippov (1984). *Khim. Fiz. 3*:1172.

223. V. A. Benderskii, E. Ya. Misochko, and P. G. Filippov (1985). *Khim. Fiz. 4*:409; (1986) *5*:955.

224. V. A. Beskrovnii, E. Ya. Misochko, V. A. Titov, and P. G. Filippov (1985). *Zh. Fiz. Khim., 59*:1755.

225. V. A. Benderskii, E. Ya. Misochko, A. A. Ovchinnikov, and P. G. Filippov (1983). *Zh. Fiz. Khim., 57*:1079.

226. V. A. Benderskii and P. G. Filippov (1984). *Khim. Fiz., 3*:147.

227. V. A. Benderskii, A. A. Ovchinnikov, and P. G. Filippov (1988), *Reactivity of Solids 4*:228.

228. A. M. Zanin, D. P. Kiryukhin, I. M. Barkalov, and V. I. Goldanskii (1981). *Pis'ma Zh. Eksp. Teor. Fiz., 33*:336 (Sov. Phys. J. ETP Lett. 1981, *33*:320); (1981) *Dokl. Akad. Nauk SSSR, 260*:1171; (1981) *260*:1397; (1982) *Vysokomol. Soyedin.* (*B*), *24*:243.

229. A. M. Zanin, D. P. Kiryukhin, V. V. Barelko, I. M. Barkalov, and V. I. Goldanskii (1981). *Dokl. Akad. Nauk SSSR, 261*:1367; (1982), *264*:99; (1983), *268*:1146; (1986), *288*:406; *Khim. Fiz.* (1982) *1*:265.

230. I. M. Barkalov, V. I. Goldanskii, D. P. Kiryukhin, and A. M. Zanin (1983). *Int. Rev. Phys. Chem., 3*:247–262.

231. I. M. Barkalov and V. I. Goldanskii (1987). *Usp. Fiz. Nauk., 151*:174.

232. G. B. Sergeev, V. V. Zagorskii, and A. M. Kosolapov (1982). *Khim. Fiz., 1*:1719; (1985), *4*:246.

233. V. A. Kabanov, G. B. Sergeev, G. M. Lykovkin, and V. Yu. Baranovskii (1982). *Dokl. Akad. Nauk SSSR, 266*:1410.

234. V. A. Benderskii, E. Ya. Misochko, A. A. Ovchinnikov, and P. G. Filippov (1982). *Khim. Fiz.*, *1*:685.

235. N. N. Semenov (1940). *Usp. Fiz. Nauk*, *23*:251, 443.

236. D. A. Frank-Kamenetskii (1947). *Diffuziya i Teploperedatcha v khimicheskoy Kinetike.* Akad. Nauk. SSSR, Moscow.

237. A. A. Andronov, E. A. Leontovitch, I. I. Gordon, and A. G. Mayer (1967). *Teoriya bifurkatsii dinamitcheskikh sistem na ploskosti.* Nauka, Moscow, ch. 8.

238. A. A. Ovchinnikov, and I. L. Shamovskii (1987). *Dokl. Akad. Nauk SSSR*, *293*:910.

239. A. A. Ovchinnikov and V. A. Onishchuk (1983). *Dokl. Akad. Nauk SSSR*, *269*:635.

240. A. A. Ovchinnikov and V. A. Onishchuk (1984). *Khim. Fiz.*, *4*:512.

241. A. F. Andreev and I. M. Lifshitz (1969). *Zh. Eksp. Teor. Fiz.*, *56*:2057.

242. A. F. Andreev (1976). *Usp. Fiz. Nauk.*, *118*:251.

243. Yu. Kagan and L. A. Maksimov (1973). *Zh. Eksp. Teor. Fiz.*, *65*:622.

244. Yu. Kagan and M. I. Klinger (1976). *Zh. Eksp. Teor. Fiz.*, *70*:255.

245. A. F. Andreev and A. Ya. Parshin (1978). *Zh. Eksp. Teor. Fiz.*, *75*:1510.

246. A. O. Caldeira and A. I. Leggett (1983). *Ann. Phys.*, *149*:374.

247. H. Kramers (1940). *Physica*, *7*:284.

248. F. Parak, E. N. Frolov, R. L. Mössbauer, and V. I. Goldanskii (1981). *J. Molec. Bio.*, *145*:825.

249. V. I. Goldanskii, Yu. F. Krupyanskii, and V. N. Fleurov (1983). *Dokl. Akad. Nauk SSSR*, *272*:978.

COLLISION-INDUCED LIGHT SCATTERING: A BIBLIOGRAPHY*

ALEKSANDRA BORYSOW[†] AND LOTHAR FROMMHOLD

Physics Department,
University of Texas at Austin,
Austin, Texas 78712-1081.

CONTENTS

1. Introduction

This bibliography briefly reviews over 700 studies in collision-induced light scattering (CILS) of gases, liquids, and solids. We consider work concerning the spontaneous Raman spectra of transitory, nonreactive complexes of atoms or simple molecules. Much of the literature deals with spectra that are "forbidden" in the separated molecular partners, but studies of collision-induced spectroscopic components associated with allowed spontaneous

---

* This work is dedicated to Dr. George Birnbaum on the occasion of his 70th birthday.

[†] Present address: Joint Institute for Laboratory Astrophysics, University of Colorado, Boulder, CO 80309-0440.

Raman transitions are of considerable interest as well. Furthermore, in the Appendix we consider briefly the literature concerning collision-induced electronic, reactive, and nonlinear light scattering processes.

Collision-induced light scattering is the name for the Raman spectroscopy that arises from the interaction-induced variations of the polarizability of a sample. For a description of the scope of CILS, we consider briefly two somewhat extreme cases, a tenuous gas and a liquid.

On the one hand, when the duration of a collision is much less than the time between collisions, that is, for gases in the low-density limit, binary collisional complexes of molecules are well-defined entities, and pair properties are more or less directly measurable. We are interested in nonreactive atoms or molecules in thermal (i.e., nonviolent) collisions, which largely preserve the character of the constituents throughout the encounter. In much of the ongoing research concerning van der Waals pairs, it is common practice to view the complex as a kind of transient *supermolecule* that for the duration of the interaction possesses various properties of its own. Specifically, the total polarizability of the complex differs slightly from the sum of the permanent polarizabilities of its parts, for various reasons. For example, in the presence of a polarized neighbor, local electric fields typically differ from the externally applied field of the light wave. Since local field strengths vary with the laser field and yet differ from these, one can describe the situation classically in terms of a temporary polarizability increment. Furthermore, at near range, exchange and dispersion forces cause a temporary rearrangement of electronic charge that affects the "permanent" polarizabilities of the collisional partners. With increasing separation, the induced polarizability rapidly falls off to negligibly small values.

For example, in the case of two interacting spherical atoms, an induced optical anisotropy of the supermolecule arises, given by

$$\beta(R_{12}) = \alpha_{\parallel}(R_{12}) - \alpha_{\perp}(R_{12}). \tag{1}$$

Here, $\alpha_{\parallel}$ and $\alpha_{\perp}$ are the polarizabilities of the diatomic complex parallel and perpendicular to the internuclear separation, R_{12}. A purely classical theory, which accounts for the electrostatic distortion of the local field by the proximity of a point dipole (the polarized collisional partner), suggests that $\beta(R_{12}) \approx 6\alpha_0^2/R_{12}^3$; with α_0 designating the permanent polarizability of an unperturbed atom. This expression is known to approximate the induced anisotropy of such diatoms fairly well. This anisotropy gives rise to the much studied pressure-induced depolarization of scattered light and to depolarized CILS spectra in general. The depolarization of light by dense systems of spherical atoms or molecules has been known as an experimental fact for a long time. It is, however, discordant with Smoluchowski's and Einstein's celebrated theories of light scattering formulated in the early years of this century that consider the effects of fluctuation of density and other thermodynamic variables.*

The concept of supermolecules allows a rather rigorous treatment of the aspects of CILS in the binary regime. The frequency dependence of the scattered light is described by the differential photon scattering cross section per unit angular frequency and per pair,

$$\left.\frac{\partial^2 \sigma}{\partial \Omega \, \partial \omega}\right|_{\hat{\mathbf{n}}_0 \hat{\mathbf{n}}_s} = k_0 k_s^3 g(\omega; T), \tag{2}$$

where $k_0 = 2\pi/\lambda_0$ and $k_s = 2\pi/\lambda_s$ are the magnitudes of incident and scattered wave vectors. The spectral function is given by the "golden rule",

$$g(\omega; T) = \sum_{i, f} P_i(T) |\langle i | \hat{\mathbf{n}}_0 \cdot \overleftrightarrow{\boldsymbol{\Delta\alpha}} \cdot \hat{\mathbf{n}}_s | f \rangle|^2 \, \delta(\omega_{if} - \omega). \tag{3}$$

Here, $|i\rangle$ and $|f\rangle$ are initial and final states of the supermolecular complex, $\overleftrightarrow{\boldsymbol{\Delta\alpha}}$ is the induced incremental polarizability (a second-rank tensor), $\hat{\mathbf{n}}_0$ and $\hat{\mathbf{n}}_s$ are unit vectors in the direction of the electric polarization of incident and scattered waves, which are often specified in the form of subscripts V and H, for vertical and horizontal, $\omega_{if} = (E_f - E_i)/\hbar$ is the energy difference of initial and final state in units of angular frequency, $P_i(T)$ is the population of the initial state (a function of temperature), $\delta(\omega)$ is Dirac's delta function, and the summation is over all initial and final states of the collisional complex.

On the other hand, in the condensed phases the concept of supermolecules is not useful because every atom or molecule interacts simultaneously with many neighbors. The many-body nature of the induction process, combined with the statistical mechanics of liquids and a complex local field problem have been serious difficulties for a quantitative description of CILS in dense matter. Furthermore, for an accurate modeling, irreducible (i.e., the pairwise nonadditive) contributions of the intermolecular interaction and induction mechanisms may be significant, which complicate the problem even more. Most treatments of CILS in the dense phases have been undertaken in the time domain, based on correlation functions of the type

$$\Phi(t) = \sum_{i \neq j} \sum_{k \neq l} \langle (\hat{\mathbf{n}}_0 \cdot \overleftrightarrow{\boldsymbol{\Delta\alpha}}(R_{ij}(0)) \cdot \hat{\mathbf{n}}_s)(\hat{\mathbf{n}}_0 \cdot \overleftrightarrow{\boldsymbol{\Delta\alpha}}(R_{kl}(t)) \cdot \hat{\mathbf{n}}_s) \rangle \tag{4}$$

---

*M. Smoluchowski, *Annalen der Physik, 25* (1980): 205; A. Einstein, *Annalen der Physik, 33* (1910): 1275.

that for practical applications are often more or less simplified. Here, $R_{ij} = R_{ij}(t)$ designates the separation of particles i and j as a function of time. The angular brackets indicate an ensemble average. Molecules k and l may be the same as or different from i and j, which thus gives rise to two-, three-, and four-body contributions in Eqn. (4). CILS spectra are obtained as the Fourier transforms of the correlation functions.

Several types of CILS spectra are known. We have already mentioned the depolarized translational spectra. Polarized translational spectra have also been observed. Alternatively, these translational spectra are called Rayleigh spectra because they appear at small frequency shifts (up to tens, perhaps hundreds of reciprocal centimeters); in the low-density limit, terms like *intracollisional spectra* have been used to distinguish these from the intercollisional components that arise from correlations that exist in successive collisions. Various rotational and rotovibrational spectra that are forbidden in noninteracting molecules have been observed. Furthermore, induced Raman components have been seen in the far wings (i.e., in the higher spectral moments) of certain allowed transitions and are widely thought to be quite common. Induction also gives rise to simultaneous transitions in two or more interacting molecules that occur at sums and differences of the rotovibrational frequencies of the molecules involved.

Collision-induced "lines," and the induced far wings of allowed lines, are usually very diffuse, typically of a width of $\Delta\omega \approx 1/\Delta t \approx 10^{12}$ Hz reflecting the short duration ($\Delta t \approx \rho/v_{\rm rms} \approx 10^{-12}$ s) of intermolecular interactions. Here, ρ designates a range of the induction mechanism on the order of the collision diameter, and $v_{\rm rms}$ is the root-mean-square velocity. In the low-density limit, induced intensities vary as the *square* of the gas density, which reflects the supermolecular origin. With increasing density, cubic and higher intensity components have also been identified. Induced spectra are related to various virial properties of real gases and may be considered virial properties themselves. Collision-induced light scattering is important when dealing with the Raman spectra of dense matter such as compressed gases, liquids, and solids, especially of bands that are Raman inactive in the isolated molecule.

We mention that, by contrast, the more familiar Raman spectra arising from the permanent polarizabilities of the individual (noninteracting) molecules of the complex are not considered a part of the supermolecular spectra, or of CILS. In ordinary Raman spectroscopy of rarefied gases the invariants of the permanent molecular polarizability tensor are conveniently considered to be not affected by intermolecular interactions, an approximation that is often justified because induced spectral components are usually much weaker than ordinary "allowed" Raman bands.

This bibliography collects the references to the literature that has evolved in the field of collision-induced light scattering in the roughly 25 years of its existence. We arrange the bibliography in two parts, spontaneous spectroscopy of gases (I) and of the condensed phases (II).

Part I is divided into sections dealing with induced polarizabilities, depolarization ratios, and rotovibrational and rototranslational spectra. Part II is similarly divided into general references related to light scattering in the dense states such as the local field problem and the general theory, and the spectra of solids and liquids. Molecular dynamics studies are usually aimed toward modeling the dense states and are, therefore, included in Part II. Some overlap between these various sections is, of course, unavoidable, but we have made an attempt at cross-referencing these; multiple listings are avoided. Relevant papers that review a certain aspect of collisional induction are pointed out in all sections where these exist. In the Appendix, electronic and nonlinear CILS is considered.

We note that in the infrared region of the spectrum, another collision-induced spectroscopy is well known that in the laboratory is usually observed in absorption. CILS is related to collision-induced absorption as ordinary Raman spectroscopy is related to the usual rotovibrational spectroscopy in the infrared. An extensive bibliography and a recent update of the literature concerning collision-induced absorption have been compiled previously.*

This bibliography evolved from a literature database in use at the University of Texas for over 10 years. In the spring of 1987, a rough draft of this bibliography was mailed to about 80 active workers in the field with a request to check and supplement where necessary the entries especially of their own contributions. About as many replies were received, many carefully correcting our draft. We express our appreciation and gratitude for the help received. Most review papers and many articles quoted were checked for references and a computer search was attempted for the broadest possible scope. We are also grateful to our friends in Florence who graciously gave permission to scan their personal literature datacard files. Most entries in this bibliography were actually looked up in the source journals and books and numerous small errors or inaccuracies of citations have been corrected. Any remaining errors and omissions the reader may notice are, however, the fault of the authors who request that these be brought to their attention so that corrections can be made in the computer-readable copy of this bibliography.

Database. A computer-readable form of this bibliography exists and is available on request. The format is as described by L. Lampert in LATEX—*A Document Preparation System*, Addison-Wesley, 1986. The file is about 3500

---

* N. H. Rich and A. R. W. McKellar, *Canadian J. Phys.*, *54* (1976): 486; J. L. Hunt and J. D. Poll, *Molec. Phys.*, *59* (1986): 163.

lines long and is in the ordinary 80-column ASCII punch file format. There is no charge if a BITNET address is specified when requesting a copy of the database; we ask that you not request computer-readable copies on floppy disks or tape unless BITNET is absolutely not accessible in your country.

PART I. INDUCED SCATTERING OF GASES*

In the low-density limit, collision-induced light scattering arises from binary complexes. In this case, a rigorous theory of induced polarizabilities, spectral moments, and spectral line shapes exists. In recent years, many measurements have been shown to be in agreement with the fundamental theory. With increasing gas density, ternary and higher spectral contributions (which may be associated with negative integrated intensities) are superposed with the binary spectra and modify these more or less. Part I of this bibliography is concerned with the spontaneous Raman process, primarily the binary induced spectra and the associated pair polarizabilities, but work at moderate densities showing the onset of ternary (and perhaps higher) contributions is also considered. By contrast, Part II deals with induced spectra of the condensed phases that are many-body spectra. Electronic and nonlinear CILS are considered in the Appendix.

1. Induced Polarizabilities**

In general, the polarizability is a tensor whose invariants, trace and anisotropy, give rise to polarized and "fully" depolarized light scattering, respectively. Collision-induced light scattering is caused by the excess polarizability of a collisional pair (or a larger complex of atoms or molecules) that arises from the intermolecular interactions. In Section I.1, we are concerned with the definition, measurement, and computation of interaction-induced polarizabilities and their invariants.

Induced polarizabilities differ from zero for a variety of reasons. A most important influence is the deviation of the local electric field from the externally applied laser field if another polarized atom or molecule is in close proximity. The classical dipole-induced dipole (DID) model considers the interacting pair as point dipoles [135, 136] and leads to separation-dependent invariants of the incremental polarizabilities [51, 62, 87, 115]. For the more highly polarizable particles, the DID anisotropy model is fairly accurate. Even for the less polarizable atoms such as helium and neon, at most separations of interest, the DID component of the anisotropy is known to be

* References 1–358.

** References 1–152.

more important than all other induced contributions combined. However, a quantum treatment is indispensable and has pointed out various corrections to the DID model, especially for the induced trace; the shape and intensity of the far wings of depolarized CILS spectra also indicate the presence of significant nonclassical contributions to the induced anisotropy. Deviations from the DID model arise at near range (where the electron shells of interacting molecules overlap); electron exchange lowers the polarizability of the pair [78, 82, 105]. At distant range, dispersion forces enhance the polarizabilities [21, 32, 75, 82]. If molecules are involved, collisional frame distortion [132, 134], multipole polarizabilities, and so on [1, 79, 240, 241] may also be important factors to be considered. Nonlinear polarization in the presence of strong electric fields from the permanent multipole moments of neighboring molecules is also significant [23, 25, 75, 95, 192]. The induced dipole is linear in the electric field, but the hyperpolarizability contributes because of the strong fields of surrounding molecules. Nonuniform fields from neighboring induced dipoles interacting with higher multipole polarizabilities also contribute [241]. From measurements as well as theory, much is known about the invariants of the incremental pair polarizability. In contrast, our knowledge of the irreducible components of the incremental polarizability of higher than binary complexes is much sparser [126, 165, 175, 423].

Previous Reviews. A general survey of the effects of molecular interactions on the optical properties of matter was recently given by Buckingham [435]. The work concerning ab initio and approximate computations of pair polarizabilities has recently been reviewed by Hunt [80]; a careful comparison of the data available from various measurements reveals a high degree of consistency with the fundamental theory. Hunt has also reviewed the utility of the DID model and its limitations [79]. The results of measurements of the polarizability invariants of rare-gas pairs have been reviewed by one of the authors [271]. Substantial discussions of induced polarizabilities can be found in a number of review articles on CILS and dielectric properties [11, 27, 143, 274, 343, 376].

Measurements. The *anisotropy* of the pair polarizability has been determined from three types of measurements: depolarized CILS spectra, pressure dependent depolarization ratios, and second virial Kerr coefficients.

Spectral moments [as in Eqn. (5)] of the depolarized CILS spectra can be determined from measurements. These are related to the anisotropy (and the interaction potential) by well-known sum formulas [326]. If a suitable analytical function with a few adjustable parameters is adopted, such as the DID model [near Eqn. (1)], supplemented by some exponential overlap term, empirical anisotropy models can be obtained that are consistent with the measured spectral moments [104, 292, 307]. Alternatively, if spectral lineshapes are calculated from models of the anisotropy and pair potential, the parameters of empirical anisotropy models can be adjusted until agreement of computed and measured profiles is obtained [59, 291]. The moment and lineshape methods are somewhat complementary and should perhaps be combined for a maximum of dependability of the inferred data [196]. Certain ambiguities that have occasionally been encountered when the method of moments is used are best dealt with by computing the associated lineshapes. Lineshapes will usually differ dramatically for different input, and only one of the several lineshapes thus obtained will agree with the measurement [59, 267]. Empirical anisotropy models obtained from depolarized CILS for the rare gases and a few molecular gases have been compiled [7–9, 114, 315, 316, 331, 348]. Numerous papers that discuss both depolarized CILS spectra and pair polarizabilities are listed in Section I.3 [213, 268, 269, 271, 315, 329, 331].

In recent years, backed by ample evidence, a consensus has evolved that the DID model of the anisotropy is a useful approximation for the induced pair anisotropy[41, 59, 187, 267]. For most of the simple gases, empirical corrections of the usually small deviations from the DID model are known. However, one must keep in mind the fact that these corrections are obtained with the assumption of a particular model of the interaction potential. While for many systems very accurate potential models exist, small differences remain between even the best models. Especially the uncertainties at near range (at separations around the collision diameter) influence the inferred data. Furthermore, because of the small magnitude of these corrections, relative uncertainties are high even if the potentials were known exactly. As a consequence, the corrections obtained by different authors often differ substantially. Some caution is necessary, especially if one wants to transfer these to conditions that are very different from what the authors have considered. We note that in a few instances, the differences among empirical corrections are greater than necessary; see the discussion concerning the lesser suited symmetrization procedures of measured lineshapes near Eqns. 6 and 7.

Empirical models of the induced anisotropy have also been obtained from measurements of the pressure- and field-induced optical birefringence (Kerr effect) [20]. While these are not spectroscopic procedures, we include such references here because of their significance for CILS [20, 24–26, 52, 53]. Other modeling attempts are based on measurements of depolarization ratios as a function of pressure [163, 178, 179]. In recent work satisfactory consistency of the anisotropies derived from second virial Kerr coefficients, pressure-induced depolarization ratios, and depolarized CILS has been reported [11, 80]. We note that the confusion that has existed in the early years of CILS studies is now understood to have been due to the previous lack of accurate potential models and the use of experimental data that were more or less affected by ternary contributions.

The induced *trace* has been modeled from measurements of polarized CILS spectra [131, 331]. Integrated intensities of induced trace spectra are typically much weaker than those of the induced anisotropy, and since they are superposed with the latter, they cannot be determined with the same high accuracy. Consequently, the induced trace models thus obtained are somewhat more uncertain than the corresponding models of the anisotropy. Nevertheless, some data for the rare-gas diatoms are available for comparison with other measurements and theory [271]; reasonable consistency is observed.

Empirical models for the induced trace have also been obtained from (nonspectroscopic) measurements of the second virial dielectric coefficient of the Clausius–Mosotti and Lorentz–Lorenz expansions [30]. Excellent surveys with numerous references to the historical as well as the modern dielectric research activities were given by Buckingham [27], Kielich [89], and Sutter [143] in 1972; see also a recent review with a somewhat more spectroscopic emphasis [11].

Computations. Efforts to compute pair polarizabilities from first principles, using perturbation techniques or modern quantum chemical methods, have been known for many years and are reviewed by Hunt [80] in the desirable detail. Amos's review of ab initio methods applied to the computation of molecular properties considers supermolecular properties, too [2].

The helium pair polarizability increment has been studied extensively [19, 29, 39, 41, 48, 56, 57, 63]. We mention in particular the most recent work by Dacre [48], which includes a careful review of previous results. Systems such as H–H in the $^3\Sigma_u$ and $^1\Sigma_g$ states may be considered as tractable examples representative of various types of real collisional pairs [6, 28, 54, 55, 66, 74, 85, 144, 145, 147]. Elaborate self-consistent field (SCF) calculations, supplemented by configuration interaction (CI) corrections are also known for the neon diatom [50]. For atom pairs with more electrons, attempts have been made to correct the SCF data in some empirical fashion for CI effects [47, 49]. Ab initio studies of molecular systems, such as H_2–H_2 and N_2–N_2 have been communicated [16, 18].

Approximate methods such as nonlocal polarizability density models or label-free exchange perturbation theory are useful near the minimum of the interaction potential and go beyond the classical DID model [77]. Several papers deal with "effective" polarizabilities in liquids [88, 95–98, 140, 152, 391]; ionic melts [14, 15, 58]; and solids [64, 66, 67, 137–139, 661–663]. The case of charge transfer in atomic collisions has also been considered [71].

General Theory. In addition to the work already quoted, a few papers considering various aspects of the general theory of polarizabilities are also included. We mention the efforts to consider the effects of the polarizability of relatively crude models (such as metallic spheres) [101, 103]. The tensor representations of molecular polarizabilities have been carefully discussed [84, 141, 362]. More general procedures applicable to a variety of systems (including big molecules and molecular complexes) that are often based on classical electrostatics have been proposed [4, 21, 118, 119, 127, 135, 136]. Various books and articles are concerned with the definition of the terms in common use [12, 13, 20, 91, 109]. Other articles deal with relevant general properties of matter [22, 113] and molecules in electric and magnetic fields [31, 239], triplet polarizabilities [10, 61, 102, 126, 208], and the calculations of polarizabilities of molecules in the liquid.

2. Depolarization Ratio*

Suppose a beam of light is incident along a horizontal axis with the electric polarization along the vertical axis. In a horizontal direction, the scattered intensities, I_{HV} and I_{VV}, are observed with horizontal and vertical polarization of the detector, respectively. The quotient $\eta = I_{HV}/I_{VV}$ is called the depolarization ratio. For optically anisotropic systems, it is greater than zero.

For some time, it was known [159, 160, 390, 405] that the depolarization of fluids composed of isotropic atoms or molecules is nonvanishing if the particle density is sufficiently high. The density-dependent depolarization of light by isotropic fluids is now understood to be caused by the anisotropy of the collision-induced polarizability increment of two or more interacting atoms or molecules [177, 178, 376, 390]. Depolarization ratios are often expressed in terms of a virial series whose nth term accounts for the n-body interactions, with $n=2, 3 \ldots$ [161, 164, 165, 167, 170, 175]. Fluids composed of optically anisotropic molecules show a variation of the depolarization ratio with density that is understood as arising from the interplay of the permanent and induced anisotropies; the anisotropy of the interaction potential also plays a role [161, 188]. The literature concerned with density-dependent depolarization of light is included in Section I.2. By contrast, CILS *spectroscopy* is considered in Section I.3.

The depolarization of gases composed of isotropic molecules at intermediate densities is reviewed by Thibeau and associates [182, 187]. For argon and methane in the low-density limit, a comparison is made of the data obtained from the measurements of the depolarization ratios, CILS line-

*References 153–191.

shapes, and the Kerr effect; theoretical predictions based on the DID model are also considered.

3. Induced Spectra of Gases*

CILS spectroscopy is concerned with the frequency, density, polarization, and temperature dependences of supermolecular light scattering. Pure and mixed gases are considered, that is, complexes of like and of dissimilar molecules are of interest. In Section I.3, we consider ordinary collision-induced Raman scattering; nonlinear and electronic collision-induced Raman processes are the subject of the Appendix; CILS of liquid and solid samples are considered in Part II, together with molecular dynamics simulations.

Reviews and Monographs. The volumes edited by van Kranendonk [354] and Birnbaum [228] contain reviews written by a selection of the most active researchers and reflect the knowledge of the time. Several monographs on Raman spectroscopy have more or less detailed sections devoted to CILS [160, 224]; see also the monographs mentioned in Part II. Articles of broad scope by Gelbart [376] and Tabisz [343] review depolarized Rayleigh and Raman scattering. Other articles are concerned with CILS spectral lineshapes [227, 231]. We mention furthermore Kielich's detailed article on intermolecular spectroscopy [288] with more than 250 references and Knaap and Lallemand's review on light scattering by gases [289]. The discussions of the 22th Faraday Symposium of the Chemical Society (London), held in December 1987, were devoted to "Interaction induced spectra in dense fluids and disordered solids" and have recently appeared in print [435].

Measurements. CILS spectra are described by a photon scattering cross section, σ [Eqn. (2)], the spectral function, $g(\omega; T)$ [Eqn. (3)], or the Fourier transformation of a correlation function, $\Phi(t)$ [Eqn. (4)]. For the most part, CILS spectra are diffuse and show little structure so that previously it was often considered sufficient to specify a few spectral moments,

$$M_n(T) = \int \omega^n g(\omega; T)\, d\omega, \tag{5}$$

with $n = 0, 1, 2 \ldots$ In this expression, the range of integration covers the full width of the spectrum and includes positive and negative frequency shifts ω. Subscripts such as $\hat{\mathbf{n}}_0$, $\hat{\mathbf{n}}_s$ [as in Eqn. (2)] or HV or VV (for vertical and horizontal) that specify the polarization of incident beam and detector are

*References 192–358.

attached to M_n, σ, $g(\omega)$, or $\Phi(t)$, or are specified in some other way. The determination of the integrated intensity ($n=0$) implies a calibration of the spectra in absolute intensity units, which for some time has become a standard procedure.

The Raman intensities of several rotational and rotovibrational lines of H_2 or N_2 are accurately known, and it is straightforward to express CILS intensities relative to these known intensities. This can be done by either adding a small amount of the calibration substance (H_2) to the gas under investigation ("internal standard") or by using separate gas cells for the two ("external standard").

The ratio of second and zeroth moment, M_2/M_0, defines some average squared width of the spectrum. Spectral moments are related to the induction operator and interaction potential by well-known sum formulas [200, 215, 235, 292, 318, 319, 326, 350, 351, 422] that permit a comparison of the measurements with the fundamental theory [196, 208, 209, 293, 296, 307, 316, 335, 357].

We note that in recent years, some CILS spectra have been measured with accuracy over a peak-to-wing intensity range exceeding $10^6:1$ [219, 331], and the information contained in the far wing is often not well described by the lower spectral moments ($n \leqslant 2$). Higher moments have, therefore, been determined [326], and lineshape analyses based on numerical computations of Eqn. (3) have been developed [202, 227, 231, 264, 266, 271, 276, 290, 316, 322, 337, 421] for more fully analyzing such measurements. Lineshape computations also permit the analysis of certain details of CILS profiles such as dimer structures, the intercollisional effects, logarithmic curvatures, and the shapes of the far wings where valuable information concerning certain quantum effects of the intermolecular interactions is available [232, 266, 271, 306, 328].

In gases near the low-density limit, at all frequencies where significant CILS intensities are observed, the spectral intensity is proportional to the *square* of the density, $g(\omega; T) \propto n^2$, in harmony with the bimolecular origin (see, however, Ref. 326). For an analysis of the measurements in terms of binary encounters, it is essential to demonstrate that the measured intensities are exactly proportional to the squared density at all frequencies. Under such circumstances, the normalized spectra, $g(\omega; T)/n^2$, that is, the *spectral shapes*, do not depend on density.

At higher densities the shapes of most CILS spectra vary with density as it had been noticed since the first experimental studies [307, 308]. This fact reveals the presence of three-body and possibly higher CILS components. The onset of discernible many-body spectral components is best dealt with in the form of a virial expansion of the spectral moments [208, 209, 326] that at least in principle permits the separation of the binary, ternary, and so on spectral components, but virial expansions of spectral lineshapes have also been reported. Most interesting attempts have been made at intermediate densities to separate the ternary from the binary (and sometimes, from the higher) contributions [208, 212, 221, 309, 357].

In experimental studies of CILS, a right-angle scattering geometry is often employed with a vertically polarized laser source. The incident and scattered beams define a horizontal plane. Because scattered signals are typically weak, the detector is usually unpolarized and thus monitors the sum of the scattered signals I_{VV} and I_{HV}. If the polarization of the incident beam is rotated into the horizontal scattering plane by means of a half-wave plate, an unpolarized detector records the sum of the signals I_{VH} and I_{HH} that are equal. In other words, even with an unpolarized detector, it is straightforward to separate experimentally the Raman spectra arising from the isotropic part of the polarizability tensor (the trace), $I_{Tr} \cong I_{VV} - \frac{4}{3} I_{HV}$, from those of the anisotropic part, $I_{An} \cong I_{HV}$. Corrections due to the finite aperture of the light-collecting lens may be necessary [331]. Collision-induced trace and anisotropy spectra have thus been experimentally separated [131, 270, 329, 330]. For the rare gases at the highest frequency shifts, the collision-induced trace spectra are usually more intense than the induced anisotropy spectra. At the lower frequency shifts, the anisotropy spectra typically dominate. In the molecular gases like hydrogen, the induced trace spectra are even harder to separate from the induced anisotropy spectra due to the feebleness of the trace and the presence of the allowed $S_0(0)$ line that makes the recordings of the high-frequency wings of the translational CILS spectra impossible.

Collision-induced light scattering spectra vary with temperature. These variations are of interest because at high temperatures the interactions are probed at smaller separations than at the lower temperatures. An accurate knowledge of the induction operators (i.e., induced trace and anisotropy) over a maximal range of separations is most desirable. Such temperature-controlled studies are just now emerging [197].

In mixtures of two gases A and B, "enhanced" CILS components appear that in the limit of binary interactions are proportional to the product of the densities n_A and n_B. These may be considered the Raman spectra of the transient supermolecule composed of dissimilar parts. Such spectra have been observed and analyzed [324, 334] and typically require accurate knowledge of the induced CILS spectra of the gases A and B that must be subtracted from the intensities measured with the mixtures.

It is interesting to consider the CILS spectra of isotopes. For example, the spectra of ^{3}He pairs differ from those of ^{4}He pairs. One reason is related to the fact that ^{3}He is a fermion but ^{4}He is a boson; these require different exchange symmetries of the pair wavefunctions. While at room temperature (where the spectra have been taken) much of the differences that arise from the different exchange symmetries have disappeared, small but discernible differences remain. These are related to the fact that the spectroscopic interactions tend to emphasize the near range, that is, the lower partial waves where the effects of exchange symmetry are more noticeable. Furthermore, the mass difference of the isotopes is roughly 30%, which has a noticeable effect on the spectral widths. It is therefore significant that the measured CILS spectra of both ^3He–^3He and ^4He–^4He pairs were shown to be accurately reproduced from theory when the same induction operators were used as input [330, 331].

Much of the early work in CILS was done with the rare gases. The reasons are conceptual simplicity as well as fairly well established data, such as the permanent polarizabilities and interaction potentials. The absence of allowed Raman spectra further simplifies the analysis; only the translational CILS spectra are observed in this case. However, since the earliest days of CILS studies, the optically isotropic gases such as CH_4 and SF_6 have also been of considerable interest. Besides translational CILS, these show collision-induced rotational and rotovibrational bands [240, 241, 258, 324, 338, 342, 347].

We mention that molecules like CH_4 are optically isotropic only if they are not rotating. If rotationally excited, centrifugal forces appear that distort the spherical symmetry and "rotation-induced" Raman lines arise* whose intensities are proportional to the density at any fixed temperature. These are usually superimposed with the collision-induced rotational components whose intensities vary with a higher than the first power of density.

In more recent years, CILS spectra of pairs of optically anisotropic molecules such as H_2, D_2, CO_2, and N_2 have been investigated. Besides the allowed rotational and rotovibrational Raman bands of such molecules, induced rotational and rotovibrational bands occur, and interesting interference of allowed and induced spectra are observed [33, 258, 324, 342, 347]. The situation is complicated by the fact that the allowed rotational lines are pressure broadened and extend substantially into the very broad induced wings. While the induced components are typically weak when compared with the peak of the allowed rotational intensity, significant CILS contributions have been seen in the second moment that are related to the fact that CILS spectra are much broader than ordinary Raman spectra and are almost certainly significant in the wings of allowed lines if the densities are high enough. Vibrational CILS has been reported by Le Duff [294, 297].

The analysis of molecular CILS spectra is more involved and often less certain than the analysis of the rare-gas spectra because of their complexity.

---

* A. Rosenberg, I. Ozier, and A. K. Kudian, *J. Chem. Phys.*, *57* (1972):568; A. Rosenberg and I. Ozier, *J. Molec. Spectrosc.*, *56* (1975):124; I. Ozier, *Phys. Rev. Lett.*, *27* (1971):1329.

The induced polarizability is also more complex to model especially for the big polyatomic molecules. Insufficient data concerning the interaction potentials are often a hindrance and make an accurate analysis less certain. To complicate matters even more, the anisotropy of the intermolecular interaction is often insufficiently known, or it is hard to account for even if it is known.

Because of these problems, it is important to study CILS of the simplest molecules thoroughly. An obvious example is the case of hydrogen molecules. At low temperatures ($\leqslant 200$ K), most *para*-hydrogen molecules are in the rotational ground state ($J=0$). Under such circumstances, pairs of *para*-H_2 molecules interact with an isotropic potential and, unless a rotational or rotovibrational excitation is involved, the permanent polarizability is purely isotropic. In other words, the translational spectra of *para*-hydrogen pairs at low temperatures are much like those of helium pairs because of the undefined rotational orientation of the $J=0$ state. At elevated temperatures (>200 K), or in the usual mixture of *para*- and *ortho*-hydrogen, the anisotropies of the permanent polarizability and the interaction potential may have a discernible influence. Observed differences between *para*- and normal hydrogen translational CILS spectra at low temperature are, therefore, of the greatest interest as they must arise from these anisotropies.

Le Duff has reported such a study of depolarized CILS at 123 K in *para*- and normal hydrogen [299]. At frequency shifts above 50 cm^{-1}, the spectra of *para*- and normal hydrogen are practically identical, which suggests a negligible role of the anisotropy of the potential at the higher frequencies. In normal hydrogen at low frequency, a roughly Lorentzian wing is observed that is absent in *para*-hydrogen. This component was assumed and has now been shown to represent the pressure-broadened wing of an allowed orientational transition of the hydrogen $J=1$ state with $\Delta M_J = \pm 2$ at zero frequency shift [320]. Recent work free from three-body contributions, and work studying the onset of three-body contributions, has recently been undertaken to further elaborate the point made [199]. The isotopic CILS spectra of deuterium pairs have also been investigated [237].

An interesting CILS pròcess involves simultaneous transitions in two (or more) interacting particles that appear at sums or differences of the rotovibrational transition frequencies of the molecules involved [243, 347].

A barely touched field of research is the use of Raman spectroscopy for the study of van der Waals dimers (and of higher bound complexes). In the low-density limit, to a large extent, CILS is the spectroscopy of the *unbound* molecular pairs in collisional interaction. However, it is well known that theoretically for most gases, even at room temperature, a small percentage of the observable intensities of the so-called collision-induced Raman spectra arises from bound–bound and bound–free transitions involving van der

Waals dimers [266, 267, 302, 328]. The theoretically predicted dimer Raman spectra [262, 263] have hitherto not been observed in the laboratory, presumably because of the substantial width (≈ 10 GHz) of the 5145- and 4880-Å argon laser lines commonly employed in such work and because the high gas densities used tend to broaden the rotational features and render them indiscernible [262]. Nevertheless, the envelopes of the rotational van der Waals dimer bands have been seen in several gases and mixtures [229, 230, 266, 267, 328]. Under more favorable jet expansion conditions, high-resolution argon dimer spectra have been obtained recently [280]. There seems to be little room for doubt that eventually van der Waals dimer spectra will be obtained with ordinary Raman or third-order nonlinear techniques [676].

Theory. If the invariants of the pair polarizability are known, along with a refined model of the intermolecular interaction potential, the lineshapes of binary spectra can be computed quite rigorously [227, 231, 271]. Lineshape computations based on exact or approximate classical trajectories are known [196, 264, 276, 316, 337]. Such computations generate spectral functions that are symmetric, $g(-\omega)=g(\omega)$. For massive pairs at high enough temperatures, such "classical" profiles are often sufficient at frequency shifts much smaller than the average thermal energy, $|\hbar\omega| \ll kT$, albeit special precaution is necessary when the system forms van der Waals dimers [302].

At greater frequency shifts, the asymmetry arising from detailed balance,

$$g(-\omega)=e^{-\hbar\omega/kT}\, g(\omega), \tag{6}$$

must be artificially accounted for even if "classical" systems are considered, that is massive pairs at temperatures where many partial waves are necessary for a description of the spectra (correspondence principle).

It has been pointed out that there are many ways to generate symmetric profiles from the observed asymmetric ones or from profiles that obey Eqn. (6) [217, 326]. The results of such symmetrization are by no means unique and do not necessarily agree with the classical profile. An illustration of this fact was given by Borysow et al. [232] who, for a given induction operator and interaction potential, compute both an exact quantum and a classical profile; the quantum profile was previously shown to be in close agreement with the measurement. Employing several well-known symmetrization procedures, these authors were able to show that the symmetric profiles thus obtained from the *same* quantum profile differ substantially at the higher frequency shifts. Only one, which was obtained with the help of the Egelstaff symmetrization procedure, is in close agreement with the classically computed profile. In other words, especially the higher moments ($n > 0$) one computes from a symmetrized measurement differ more or less from the classical moments unless the Egelstaff procedure is employed. An inconsistency of a symmetrized profile with the classical moment expressions is undesirable and may, for example, adversely affect the dependability of the small empirical overlap corrections to the DID anisotropy if the less well suited symmetrization procedures, combined with the classical sum formulas, are employed in the evaluations.

Quantum calculations of spectral profiles based on Eqn. (3) are known; isotropic interaction is commonly assumed. Such quantum calculations account for detailed balance in exact ways and consider not only the free-state to free-state transitions of binary complexes but also the bound–free and bound–bound transitions involving bound van der Waals pairs that are an inseparable part of the CILS spectra. These bound–free and bound–bound components can be quite significant, especially at the lower temperatures and for massive systems [266, 302, 328]; see also related work on dimer spectra [262, 263], and the review papers on CILS lineshapes [227, 231, 271]. Lineshape calculations are useful for the detailed comparison of the fundamental theory with spectroscopic measurements. For helium pairs, a close agreement between the fundamental theory and the recent measurements is now observed; similarly for the neon pair [45].

For the vibrational bands, the detailed balance condition, Eqn. (6), is valid as long as the variation of the interaction potential with the vibrational coordinates of the molecules involved can be ignored. For most systems of interest, however, the dependence of the interaction on the vibrational state cannot be ignored. In that case, the spectral profiles $g_{vv'}(\omega)$ depend on the vibrational quantum numbers v, v' of initial and final states of the molecules of the collisional complex. Spectral moments (for $n>0$), too, depend on the v, v'. In this case, the symmetry is given by

$$g_{v'v}(-\omega)=e^{-\hbar\omega/kT}\,g_{vv'}(\omega), \tag{7}$$

which may differ substantially from Eqn. (6) as Moraldi et al. have shown.*

As a historical aside, we mention that during the first half of the century the *idea* of van der Waals molecules and their possible spectroscopic discovery was firmly established in the leading spectroscopy laboratories. Accordingly, various attempts have been made in the infrared and with Raman techniques

* M. Moraldi, J. Borysow, and L. Frommhold, *Phys. Rev. A*, *36* (1987):4700, and M. Moraldi, A. Borysow, and L. Frommhold, *Phys. Rev. A*, 38 (1988) 1839. Note that these articles concern collision-induced absorption; analogous work related to CILS has not yet been communicated but very similar conclusions are to be expected for that case.

to record such bound–bound spectra of molecular pairs. Such work is difficult and only in relatively recent years the necessary modern tools (low noise photomultiplier tubes, lasers, holographic, gratings etc.) have become available for successful work of this kind—work that is still very scarce (i.e., difficult [280]) in spite of the vastly improved situation of the experimentalist. Consequently, we may assume that for some time, historical attempts of recording dimer spectra remained unsuccessful and thus unrecorded. In 1949, Harry Welsh and his associates discovered a new spectroscopy—collision-induced absorption—in an attempt to record oxygen dimer spectra in the infrared. While the bound–bound absorption spectra of $(O_2)_2$ were not recorded for another 25 years [259], the much stronger spectra arising from free–free transitions had thus finally been obtained and understood in terms of the basic processes involved. The discovery of CILS could similarly have taken place during the search for dimer bands using Raman spectroscopy, but the fact is that the first measurement was quite independent of any such motivation. Rather, it was carefully designed to demonstrate the collision-induced Raman effect in the rare gases [104, 307]. Early theoretical work by Levine fully appreciates the close connection of dimer and diatom spectra (i.e., of bound and of free pairs) [302]. We note that in the very early publications by Finkelnburg some of the essential ideas of the collision-induced spectroscopies have been expressed in lucid but qualitative terms [688, 689]; the close relationship of collision-induced and dimer spectra is there clearly exposed in very general terms even if the discussions seem to imply electronic transitions (which is not the case for much of the work considered here).

Lineshape calculations have a reputation of being involved, and the required computer codes are not widely available. It has, therefore, always been thought worthwhile to consider sum formulas (spectral moments) that are much more straightforward to compute, especially the classical expressions, which can be readily corrected to the order $\hbar^2$ for quantum effects. Expressions for the even spectral moments $n=0$, 2, 4, and 6 are known.

Classical moment expressions are usually sufficient for the massive systems at high enough temperatures, that is, under conditions where many partial waves are required to describe the collisional interactions. For long-range interaction (like DID), quantum corrections are generally less substantial than for short-range (overlap) induction. The second moments require more substantial quantum corrections than zeroth moments under comparable conditions.

Quantum sum formulas based on exact pair distribution functions (obtained in the isotropic potential approximation) are also known for $n=0$, 1, 2, and 3 [318, 319]; we mention also unpublished work by J. D. Poll. Levine has given detailed estimates based on classical moments, assessing the bound dimer contributions [302].

We note that it is often possible to approximate exact lineshapes fairly closely if suitable analytical model profiles are selected whose lowest two or three spectral moments are matched to those of the measurement [231]. Various suitable model profiles are known, but certain three-parameter models approximate the exact shapes so closely that lineshape calculations may be dispensible for some applications. Other analytical profiles are, however, less than convincing for the purpose [314].

Theoretical attempts to deal with complexes of more than two atoms (molecules) are scarce. One notable exception is the intercollisional process [303, 304, 306], which models the existing correlations of subsequent collisions. Intercollisional effects are well known in collision-induced absorption, but in CILS not much experimental evidence seems to exist. Three-body spectral moment expressions have been obtained under the assumptions of pairwise interactions [198, 200, 208, 209, 212, 218, 340, 422]; see also references in Part II. Multiple scattering will depolarize light and has been considered in several depolarization studies of simple fluids [273, 274, 290, 376].

PART II. CILS OF THE CONDENSED STATES*

In Part II we are concerned with CILS of liquids and solids. Computer simulation of molecular dynamics has emerged as a most powerful technique for the study of simple liquids and solids, and we include here such work as far as it is related to CILS because molecular dynamics studies are usually aimed at simulating the dense phases. This part is divided into three sections: the first one considers the general theory of light scattering in the dense states related to CILS, the internal field problem, and so on. Section 2 is concerned with the CILS spectra of liquids, mostly, of course, of ordinary liquids and solutions, but work concerning superfluids and ionic melts is also included. Section 3 deals with the CILS-related spectra of amorphous and crystalline solids that have been prominently featured in the recent 22nd Faraday Symposium [435].

The dense phases of interest here differ significantly from the (mostly) dilute gases considered in Part I. Every atom (molecule) interacts significantly with several neighboring molecules at any time. The difficulties arising from the many-body nature of the induction process in liquids have been a serious barrier to a quantitative theoretical description of the complete spectra. The treatment of light scattering in general, and of CILS in particular, resorts to correlation functions [as in Eqn. (3)] whose Fourier transforms describe the

* References 359–664.

spectra. Correlation functions are rigorously defined* in time-dependent statistical mechanics and provide an important link between theoretical and experimental work.

Topics not covered in this bibliography here are, for example, the hydrodynamic theory; fluctuations of density, anisotropy, concentration; fluid shear waves; and critical phenomena.

1. General Theory**

Historically, light scattering has been described in terms of a "molecular" or "electronic" theory† on the one hand, and a "phenomenological" (i.e., thermodynamic) theory‡ on the other. For our present interest (i.e., CILS) an understanding at the molecular level is usually attempted.

It has been known for a long time that the fluctuation theory of light scattering is unable to explain the depolarized component observed in dense fluids composed of isotropic atoms or molecules [160]. The pioneering work by Buckingham and Stephen [159] explains the depolarization of light in terms of the induced anisotropies. Mazur [113] gives an early review of the electromagnetic properties of the light scattering by many-body systems. Kielich [390, 391] develops the general principles of a statistical-molecular theory of light scattering; he writes down expressions for the second virial coefficient of "isotropic" and "anisotropic" light scattering,§ which he computes for spherical molecules of variable polarizability; following Buckingham, he also considers the effects of hyperpolarizability on the pressure-induced depolarization ratio. Theimer and Paul [405] conclude in a theoretical study that for *very* dense, monatomic fluids induced depolarization cannot be ignored. Especially with the experimental findings of depolarized scattering of tenuous isotropic fluids in the late sixties by Birnbaum and collaborators, and by Thibeau and associates, new theoretical and experimental studies of CILS and CILS-related processes were stimulated. Tanaka develops a molecular theory of light scattering that considers the time dependence of the density fluctuation for a theoretical description of the depolarized spectrum of

* See, for example, R. G. Gordon in *Adv. Magn. Reson.*, *3* (1968):1; B. J. Berne in *Physical Chemistry: An Advanced Treatise*; *VIII B*, Academic Press, New York, 1971, p. 539; D. A. Kivelson, in *Symposia of the Faraday Soc.*, *11* on "Newer Aspects of Molecular Relaxation Processes", Chem. Soc. London, 1977, p. 7–25; also Refs. 370, 374, and 443.

** References 359–410.

† Lord Rayleigh, *Phil. Mag.*, *47* (1899):375; M. Born, *Verh. deutsch. phys. Gesellsch.* (1917):243 and (1918):16; *Optik*, Springer, Berlin, 1933; M. Born and E. Wolf, *Principles of Optics*, Pergamon, New York, 1964; Refs. 62, 160, 375, 404, 409, and 410.

‡ Fluctuation theory by Smoluchowski and Einstein; see the first footnote of Part I.

§ The terms *isotropic* and *anisotropic* are meant to refer to the invariants of the polarizability tensor, not to scattering per se; we prefer generally terms such as *trace* and *anisotropy scattering*.

liquids [404]. Fleury and Boon review the recent theoretical, experimental and instrumentation advances in the field of inelastic light scattering from fluid systems [373]. That review contains a bibliography of more than 500 entries of which about 50 are more or less directly related to CILS. Gelbart reviews the existing theories and attempts to provide a unified molecular treatment of depolarized scattering of light by simple fluids (i.e., gases and liquids made up of atoms or isotropically polarizable molecules) [376]. He points out the equivalence of the multiple scattering theories [375] and Fixman's anisotropic local field considerations [372]. The *nonlinear*-induced polarization effects, which Gelbart also discusses, will interest us in the Appendix.

Current theories as well as the historical developments are reviewed in detail in a number of monographs [224, 399, 473], conference proceedings [228, 354, 407], reviews, many of which quote numerous references [288, 365, 369, 394, 401, 406], and other articles [360–363, 367, 368, 371, 375, 384, 385, 388, 391].

Serious criticism of the early theories of light scattering related to the local field problem has been voiced for some time [372, 408–410]. For a study of the *band shape* of liquids it is usually possible to ignore the effect of the internal field correction, which often does not vary significantly over the frequency range considered. For the interpretation of integrated intensities in terms of pair-correlations, on the other hand, a knowledge of the internal field is indispensable. The effective field sensed by a "test molecule" differs from the applied laser field by the vector sum of the induced fields surrounding the neighbor molecules. The problem is to find a self-consistent solution to Maxwell's equations in the presence of polarizable neighbors whose optical properties undergo thermal fluctuation. In principle, the problem can be solved by iteration if the molecular distribution functions are known. In practice, various approximations have been made and refined over the years. Quantitative theories are now available for polarized scattering by simple fluids, and for depolarized scattering by rigid, nonspherical molecules and chain molecules of constant polarizability [389]. However, to this day, the local field problem is not without controversy as a number of articles [372, 374, 381, 382, 384, 386, 396, 398, 443, 559, 560, 561] and recent reviews by Keyes and Ladanyi [389] and Ladanyi [396] demonstrate.

2. Liquids*

Traditionally, Raman studies have been concerned with the determination of molecular transition frequencies, but in more recent years an interest in vibrational band shapes and the striking broad low-frequency features of

* References 411–624.

liquids has emerged. These are related to the thermal molecular motions in liquids that have been modeled in more or less simplified ways but are at present not very well understood. The simplicity of the gaseous state and the symmetries of crystalline solids, which make the many-body problem more tractable, are absent in liquids. Consequently, the molecular dynamics of liquids is theoretically much less developed, and spectroscopic studies based on band shapes are thus of the greatest interest. Correlation functions provide the link between molecular dynamics and the observable spectra, which are the Fourier transforms of these. Some of the best spectroscopic studies are paired with computer simulations for the development of a better understanding on a molecular level.

Computer simulation of molecular dynamics is concerned with solving numerically the simultaneous equations of motion for a few hundred atoms or molecules that interact via specified potentials. One thus obtains the coordinates and velocities of the ensemble as a function of time that describe the structure and correlations of the sample. If a model of the induced polarizabilities is adopted, the spectral lineshapes can be obtained, often with certain quantum corrections [425, 426]. One primary concern is, of course, to account as accurately as possible for the pairwise interactions so that by carefully comparing the calculated with the measured band shapes, new information concerning the effects of irreducible contributions of intermolecular potential and cluster polarizabilities can be identified eventually. Pioneering work has pointed out significant effects of irreducible long-range forces of the Axilrod–Teller "triple-dipole" type [10]. Very recently, on the basis of combined computer simulation and experimental CILS studies, claims have been made that irreducible three-body contributions are observable, for example, in dense krypton [221].

Reviews and Monographs. Clarke has reviewed the band shapes and molecular dynamics in liquids; many references are cited [443]. Buckingham has given the general introduction of the 22nd Faraday Symposium on Interaction Induced Spectra in Dense Fluids and Disordered Solids [435]. Madden has reviewed a variety of highly relevant topics, such as vibrational spectra and the interference of allowed and induced spectral components [531, 533–535]. The progress in liquid-state physics and CILS has been reviewed [505, 597]. Several tractates on molecular dynamics and computer simulation studies have appeared [415, 443, 463, 510, 592, 603, 604, 616]. The induced anisotropy and light scattering in liquids and various studies of molecular motions in liquids are the subject matter of other reviews [472, 490, 491, 517, 522, 586, 587]. The separation of allowed and induced spectral components is the subject matter of great interest [443, 534, 535]; very interesting studies in the time domain have been reported [465]. Molecular spectroscopy of the dense phases is discussed in various books and reviews [431, 477, 579]. Litovitz and his associates have compared CILS and collision-induced absorption in liquids [451, 521, 525, 593, 613]. Strauss reviews the literature of light scattering for the study of molecular liquids [591]. Keyes describes his CILS studies of growing clusters [501]. Pecora considers induced scattering from macromolecular liquids [562]. Oxtoby reviews vibrational lineshapes of charge transfer complexes [558]. In most reviews, extensive citations of relevant work are made.

(Mostly) Experimental Studies. Many of the most enlightening studies of CILS in liquids combine theory, computation, and/or observation in various ways; purely experimental studies are relatively rare. Nevertheless, it seems natural to divide the wealth of material again into (mostly) experimental, computational, and theoretical studies. Under the circumstances, the boundaries are, of course, somewhat diffuse.

One class of very basic investigations is concerned with the rare-gas liquids and the very dense monatomic gases [424, 444, 446, 459, 461, 483, 514, 526, 527, 544, 567, 580, 583, 607, 608, 609, 624] and fluids composed of spherical molecules [423, 433, 434, 464, 488, 519, 566–569, 580, 583, 595, 596, 621]. Many aspects of the fundamental studies of condensed helium and superfluidity are closely related [437, 438, 439, 475, 476, 480, 482, 492, 503, 504, 553, 554, 589, 594, 599, 619]. Light scattering studies of gas and liquid helium [565], liquid and solid helium [585, 615], liquid and solid hydrogen [633], and liquid and solid argon, krypton, and xenon [604, 624, 630, 634] have been reported. Furthermore, CILS from melts and ionic liquids have been reported [436, 442, 460, 469, 470, 543, 546, 572].

Liquids composed of optically anisotropic molecules have been studied quite intensely. In most of these studies, the interplay of allowed and induced components is of some concern. Dense hydrogen [419], nitrogen [453, 454, 575], carbon disulfide [447–450, 484–487, 518, 529, 629], carbon dioxide [489, 537, 552, 610, 618], clorine [471, 601], hydrogen sulfide [542], bromine [520], other simple, linear molecules [432, 499, 606, 622], water and aqueous solutions [474, 538, 545, 547, 548], chloroform and similar molecules [516, 567–569, 576, 577, 595, 596, 617], and various organic liquids have been studied [452, 493, 502, 523, 524, 563, 564, 578, 581, 605, 611]. The spectra of small diatomics dissolved in argon have been reported [481]. Solutions of carbon tetrachloride have also been considered [551]. Hydrogen chloride has been studied [417]. The effects of pressure on the dynamic structure of liquids has been investigated [494–496].

Theory. Dense rare-gas fluids have been considered by Guillot, Bratos, and Birnbaum [479]. Mixed liquids are considered by Keyes [500]. Detailed theoretical studies of CS_2 have been reported [416, 507, 508, 518, 598].

(Mostly) Molecular Dynamics. Light scattering in the low-density limit has been compared with molecular dynamics results [420, 600]. Usually, of course, in such computer experiments, much higher densities are of interest. Quantum corrections of spectral moments and pair distribution functions computed in molecular dynamics studies have been obtained [425, 426]. Boundary problems have been considered [445]. Most thoroughly studied are, of course, the monatomic fluids (often with data input for argon) [411, 412, 414, 421, 428, 512, 514, 515, 570, 571]. Light scattering from linear nonpolar molecules such as N_2 [440, 441], O_2, CS_2, CO_2 [468, 509, 511, 529, 530, 536, 537, 552, 623], and F_2, Cl_2, Br_2, [584, 592, 622] has been reported.

3. Solids*

The Raman spectra of solids have a more or less prominent collision-induced component. Rare-gas solids held together by van der Waals interactions have well-studied CILS spectra [656, 657]. The face-centered, cubic lattice can be grown as single crystals. Werthamer and associates [661–663] have computed the light scattering properties of rare-gas crystals on the basis of the DID model. Helium as a quantum solid has received special attention [654–658] but other rare-gas solids have also been investigated [640]. Molecular dynamics computations have been reported for rare-gas solids [625, 630, 634].

Induced scattering from rock salt [641, 650] and CS_2 [629] has been studied. Disordered solids were reviewed by Signorelli et al. [648, 651]; studies of partially disordered molecular solids have been reported in ice [642, 646], α-AgI [628, 643, 644, 647], and other solids [627, 639, 645, 664]. Lennard–Jones glasses [649] and various plastics [632, 637] have also been studied.

Stoicheff and his collaborators have investigated Brillouin scattering by rare-gas crystals, which is not really an "induced" process. However, for a quantitative description, significant induced contributions to the polarizability must be taken into account, and it is for that reason that we list relevant work in Section II.1. For a review of these efforts and the results obtained see Ref. 139.

The simplest molecular solids are those of hydrogen and its isotopes, which are discussed in a recent monograph [659] and in various review articles [626, 633, 635, 638, 652, 653]. The spectra are related to the zero-point motion of the lattice, rotation, and translation. The phonon spectra observed in these quantum solids have been of special interest.

---

* References 625–664.

Depolarized Raman scattering from amorphous solids have recently been considered [649, 651], and a formalism for the interpretation of the Raman spectra of disordered solids has been proposed [648]. Raman scattering in other molecular crystals have also been reported [629, 637, 641, 646].

APPENDIX: ELECTRONIC AND NONLINEAR CILS*

In recent years, mostly in connection with the ever expanding laser studies, a number of new collision-induced light scattering processes were described in the literature. We cannot claim a complete survey of such activities, which sometimes are very closely related or are analogous to the main subject of this bibliography. But we have tried to give here references to studies that the community of CILS specialists may be interested to know.

There are, for example, the fields of redistribution of radiation [677–682, 690, 692, 719] for which several reviews have appeared [683, 694], electronic CILS [686, 693], and collision-induced fine structure transitions [670]. Laser-assisted collisions are the topic of various monographs [716, 717] and review articles [667, 710]. Light scattering from chemically reacting systems has been studied [671, 684, 720, 721]. Also mentioned were nonlinear processes such as second-harmonics generation [691, 696, 698, 702, 718], hyper-Raman scattering [668, 669, 672, 685, 711], collision-induced coherent Raman scattering processes [676, 707, 709, 715], multiphoton scattering [695, 700, 701, 708], and pressure-induced extra resonances (PIER4) in four-wave mixing [417, 665, 673–675, 714].

Acknowledgment

The support of the National Science Foundation, grant AST-8613085, and of the Robert A. Welch Foundation, grant F-988, is acknowledged.

Bibliography

I.1. *Pair Polarizabilities*

1. R. D. Amos. An accurate *ab initio* study of the multipole moments and polarizability of methane. *Molec. Phys., 38*:33–45 (1979).
2. R. D. Amos. Molecular property derivatives. *Adv. Chem. Phys., 67*:99–153 (1987).
3. J. Applequist, J. R. Carl, and K.-K. Fung. An atom dipole interaction model for molecular polarizabilities: Application to polyatomic molecules and determination of atom polarizabilities. *J. Am. Chem. Soc., 94*:2952–2960 (1972).
4. J. Applequist. An atom-dipole interaction model for molecular optical properties. *Accounts Chem. Res., 10*:79–85 (1977).

* References 665–721.

5. G. P. Arreghini, C. Guidotti, and U. T. Lamanna. Refractivity of helium gas: An estimate of the 2nd virial coefficient. *Chem. Phys.*, *16*:29–40 (1976).
6. M. D. Balmakov and A. V. Tulub. Nonadditive polarizability of a system of interacting atoms. *Optika i Spektr.* (*U.S.S.R.*), *32*:871–874 (1972); Engl. transl. in *Optics and Spectr.*, *32*:462–464 (1972).
7. F. Barocchi and M. Zoppi. Determination of collision induced polarizability in Ar, Kr, and Xe by means of collision induced scattering: Analysis and empirical pair potentials. *J. Chem. Phys.*, *65*:901–905 (1976).
8. F. Barocchi, M. Neri, and M. Zoppi. Determination of the pair polarizability of argon from the two-body CILS spectrum. *Chem. Phys. Lett.*, *59*:537–540 (1978).
9. F. Barocchi, M. Zoppi, U. Bafile, and R. Magli. The pair polarizability anisotropies of Kr and Xe from depolarized interaction induced light scattering spectra. *Chem. Phys. Lett.*, *95*:135–138 (1983).
10. B. A. Baron and W. M. Gelbart. The triplet polarizability and depolarization virial data for atomic gases. *Chem. Phys.*, *6*:140–143 (1974).
11. T. K. Bose. "A Comparative Study of the Dielectric, Refractive and Kerr Virial Coefficients." In G. Birnbaum (ed.), *Phenomena Induced by Intermolec. Interactions*, Plenum Press, New York, 1985, pp. 49–66.
12. C. J. F. Böttcher. *Theory of Electric Polarization*, Vol. 1, Elsevier, New York, 1973.
13. C. J. F. Böttcher and P. Bordewijk. *Theory of Electric Polarization*, Vol. 2, Elsevier, New York, 1978.
14. D. G. Bounds, J. H. R. Clarke, and A. Hinchcliffe. The polarizability of LiCl as a function of bond distance. *Chem. Phys. Lett.*, *45*:367–369 (1977).
15. D. G. Bounds and A. Hinchcliffe. The pair polarizability of Li^+–Li^+ and Cl^-–Cl^-. *Chem. Phys. Lett.*, *56*:303–306 (1978).
16. D. G. Bounds. The interaction polarizability of two hydrogen molecules. *Molec. Phys.*, *38*:2099–2106 (1979).
17. D. G. Bounds and A. Hinchliffe. The shapes of pair polarizability curves. *J. Chem. Phys.*, *72*:298–303 (1980).
18. D. G. Bounds, A. Hinchliffe, and C. J. Spicer. The interaction polarizability of two nitrogen molecules. *Molec. Phys.*, *42*:73–82 (1981).
19. L. W. Bruch, P. J. Fortune, and D. H. Berman. Dielectric properties of helium: Quantum statistical mechanics. *J. Chem. Phys.*, *61*:2626–2636 (1974).
20. A. D. Buckingham. Theoretical studies of the Kerr effect: II. The influence of pressure. *Proc. Phys. Soc.*, *A68*:910–919 (1955).
21. A. D. Buckingham. The polarizability of a pair of interacting atoms. *Trans. Faraday Soc.*, *52*:1035–1041 (1956).
22. A. D. Buckingham and J. A. Pople. Electromagnetic properties of compressed gases. *Discuss. Faraday Soc.*, *22*:17–21 (1956).
23. A. D. Buckingham. The molecular refraction of an imperfect gas. *Trans. Faraday Soc.*, *52*:747–753 (1956).
24. A. D. Buckingham and B. J. Orr. Electric birefringence in molecular hydrogen. *Proc. Roy. Soc.* (*London*), *A 305*:259–269 (1968).
25. A. D. Buckingham and D. A. Dunmur. Kerr effect in inert gases and sulfur hexafluoride. *Trans. Faraday Soc.*, *64*:1776–1783 (1968).
26. A. D. Buckingham and B. J. Orr. Kerr effect of methane and its four fluorinated derivatives. *Trans. Faraday Soc.*, *65*:673–681 (1969).

27. A. D. Buckingham. *Theory of Dielectric Polarization,* Vol. 2 of *Series One*, Butterworths, London, 1972, pp. 241–261.

28. A. D. Buckingham, P. H. Martin, and R. S. Watts. The polarizability of a pair of hydrogen atoms at long range. *Chem. Phys. Lett.*, *21*:186–190 (1973).

29. A. D. Buckingham and R. S. Watts. The polarizability of a pair of helium atoms. *Molec. Phys.*, *26*:7–15 (1973).

30. A. D. Buckingham and C. Graham. The density dependence of the refractivity of gases. *Proc. Roy. Soc. (London)*, *A 337*:275–291 (1974).

31. A. D. Buckingham. Gaseous molecules in electric and magnetic fields. *Ber. Bunsen Ges. Phys. Chem.*, *80*:183–187, (1976).

32. A. D. Buckingham and K. L. Clarke. Long-range effects of molecular interactions on the polarizability of atoms. *Chem. Phys. Lett.*, *57*:321–325 (1978).

33. A. D. Buckingham. "Basic Theory of Intermolecular Forces." In B. Pullman (ed.), *Intermolecular Interactions from Diatomics to Biopolymers,* Wiley, New York, 1978, Chapter 1.

34. A. D. Buckingham. The effects of collisions on molecular properties *Pure Appl. Chem.*, *52*:2253–2260 (1980).

35. A. D. Buckingham and K. L. Clarke-Hunt. The pair polarizability anisotropy of SF_6 in the point-atom-polarizability approximation. *Molec. Phys.*, *40*:643–648 (1980).

36. A. D. Buckingham. "Small Molecules in Electric and Optical Fields." In S. Krause, (ed.), *Molecular Electro-Optics: Electro-optic Properties of Macromolecules and Colloids in Solution,* Plenum Press, New York, 1981, pp. 61–73.

37. A. D. Buckingham, P. A. Galwas, and Liu Fan-Chen. Polarizabilities of interacting polar molecules. *J. Molec. Struct.*, *100*:3–12 (1983).

38. R. C. Burns, C. Graham, and A. R. M. Weller. Direct measurement and calculation of the second refractivity virial coefficients of gases. *Molec. Phys.*, *59*:41–64 (1986).

39. P. R. Certain and P. J. Fortune. Long-range polarizability of the helium diatom. *J. Chem. Phys.*, *55*:5818–5821 (1971).

40. V. Chandrasekharan and G. C. Tabisz. Resonance and dispersion energies and the point-dipole model of atomic pair polarizability. *Canad. J. Phys.*, *58*:1420–1423 (1980).

41. K. L. Clarke, P. A. Madden, and A. D. Buckingham. Collison induced polarizabilities of inert gas atoms. *Molec. Phys.*, *36*:301–316 (1978).

42. R. Coulon, G. Montixi, and R. Occelli. Détermination experimentale des coefficients du viriel de la réfractivité des gaz: étude de l'argon. *Canad. J. Phys.*, *59*:1555–1559 (1981).

43. P. D. Dacre. A calculation of the helium pair polarizability including correlation effects. *Molec. Phys.*, *36*:541–551 (1978).

44. P. D. Dacre. A calculation of the neon pair polarizability including correlation effects. *Canad. J. Phys.*, *59*:1439–1447 (1981).

45. P. D. Dacre and L. Frommhold. Spectroscopic examination of *ab initio* neon diatom polarizability invariants. *J. Chem. Phys.*, *75*:4159–4160 (1981).

46. P. D. Dacre and L. Frommhold. Rare gas diatom polarizabilities. *J. Chem. Phys.*, *76*:3447–3460 (1982).

47. P. D. Dacre. An SCF calculation of the pair polarizability of argon. *Molec. Phys.*, *45*:1–15 (1982).

48. P. D. Dacre. On the pair polarizability of helium. *Molec. Phys.*, *45*:17–32 (1982).

49. P. D. Dacre. Pair polarizabilities of the heavy inert gases: II. SCF calculations of the pair polarizabilities of krypton and xenon. *Molec. Phys.*, *47*:193–208 (1982).

50. P. D. Dacre. A calculation of the neon pair polarizability including correlation effects: II. Inclusion of the configuration interaction basis extension corrections. *Canad. J. Phys.*, *60*:963–967 (1982).

51. S. R. de Groot and C. A. ten Seldam. Influence of pressure on the polarizability of molecules. *Physica*, *13*:47–48 (1947).

52. D. A. Dunmur and N. E. Jessup. The influence of intermolecular interactions on the Kerr effect in gases: I. Statistical theory for spherical top molecules. *Molec. Phys.*, *37*:697–711 (1979).

53. D. A. Dunmur, D. C. Hunt, and N. E. Jessup. The influence of intermolecular interactions on the Ker effect in gases: II. Experimental results for spherical top molecules. *Molec. Phys.*, *37*:713–724 (1979).

54. D. B. DuPré and J. P. McTague. Pair polarizability of hydrogen atoms. Relation to collision induced light scattering and dielectric phenomena in rare gases. *J. Chem. Phys.*, *50*:2024–2028 (1969).

55. A. L. Ford and J. C. Browne. Direct-resolvent-operator computations on the hydrogen-molecule, dynamic polarizability, Rayleigh, Raman scattering. *Phys. Rev. A*, *7*:418–426 (1973).

56. P. J. Fortune and P. R. Certain. Dielectric properties of helium: The polarizability of diatomic helium. *J. Chem. Phys.*, *61*:2620–2625 (1974).

57. P. J. Fortune, P. R. Certain, and L. W. Bruch. Helium diatom polarizability and the dielectric properties of fluid helium. *Chem. Phys. Lett.*, *27*:233–236 (1974).

58. P. W. Fowler and P. A. Madden. Fluctuating dipoles and polarizabilities in ionic materials: Calculations on LiF. *Phys. Rev. B*, *31*:5443–5455 (1985).

59. L. Frommhold and M. H. Proffitt. About the anisotropy of the polarizability of a pair of argon atoms. *Molec. Phys.*, *35*:681–689 (1978).

60. L. Frommhold and M. H. Proffitt. Concerning the anisotropy of the helium diatom polarizability. *J. Chem. Phys.*, *70*:4803–4804 (1979).

61. H. W. Graben, G. S. Rushbrooke, and G. Stell. On the dielectric constant of non-polar Lennard-Jones fluids according to the Kirkwood-Yvon theory. *Molec. Phys.*, *30*:373–388 (1975).

62. H. S. Green. *The Molecular Theory of Fluids*, Interscience, New York, 1952.

63. R. A. Harris, D. F. Heller, and W. M. Gelbart. *A priori* calculation of collisional polarizabilities: He_2, Ne_2, Ar_2. *J. Chem. Phys.*, *61*:3854–3855, 1974.

64 J. Heinrichs. Contribution of dipole fluctuations in the quantum theory of electronic polarizability of crystals. *Phys. Lett.*, *18*:251–252 (1965).

65. J. Heinrichs. Influence of molecular interactions on the Clausius-Mosotti function for nonpolar gases. *Chem. Phys. Lett.*, *1*:467–472 (1967); Erratum, *1*:560 (1968).

66. J. Heinrichs. Effect of molecular interactions on the dielectric constant for multi-level systems. *Chem. Phys. Lett.*, *4*:151–154 (1969).

67. J. Heinrichs. Theory of the dielectric constant in the individual-ion model of ionic and rare-gas crystals. *Phys. Rev.*, *179*:823–836 (1969).

68. J. Heinrichs. Role of quantum dipole fluctuations in the theory of excitations and of the dielectric constant of crystals. *Phys. Rev.*, *188*:1419–1430 (1969).

69. D. F. Heller and W. M. Gelbart. Short range electronic distortion and the density dependent dielectric function of simple gases. *Chem. Phys. Lett.*, *27*:359–364 (1974).

70. D. F. Heller, R. A. Harris, and W. M. Gelbart. Density functional formulation of collisional

polarizabilities: Application to homonuclear noble gas diatoms. *J. Chem. Phys.*, *62*:1947–1953 (1975).

71. P. R. Hilton and D. W. Oxtoby. Charge transfer in atomic collisions: The pair polarizability anisotropy. *J. Chem. Phys.*, *72*:473–477 (1980).

72. P. R. Hilton and D. W. Oxtoby. Collisional pair polarizabilities calculated by exchange perturbation theory using electric field variant basis sets. *J. Chem. Phys.*, *74*:1824–1829 (1981).

73. J. S. Høye and D. Bedeaux. On the dielectric constant for a nonpolar fluid, a comparison. *Physica A*, *87*:288–301 (1977).

74. K. L. C. Hunt and A. D. Buckingham. The polarizability of H_2 in the triplet state. *J. Chem. Phys.*, *72*:2832–2840 (1980).

75. K. L. C. Hunt. Long-range dipoles, quadrupoles, and hyperpolarizabilities of interacting inert-gas atoms. *Chem. Phys. Lett.*, *70*:336–342 (1980).

76. K. L. C. Hunt, B. A. Zilles, and J. E. Bohr. Effects of *van der Waals* interactions on the polarizability of atoms, oscillators, and dipolar rotors at long range. *J. Chem. Phys.*, *75*:3079–3086 (1981).

77. K. L. C. Hunt. Nonlocal polarizability densities and *van der Waals* interactions. *J. Chem. Phys.*, 78:6149–6155 (1983).

78. K. L. C. Hunt. Nonlocal polarizability densities and the effects of short-range interactions on molecular dipoles, quadrupoles and polarizabilities. *J. Chem. Phys.*, *80*:393–407 (1984).

79. K. L. C. Hunt. "Classical Multipole Model: Comparison with *ab initio* and Experimental Results." In G. Birnbaum (ed.), *Phenomena Induced by Intermolecular Interactions*, Plenum Press, New York, 1985, pp. 1–28.

80. K. L. C. Hunt. "*Ab initio* and Approximate Calculations of Collision Induced, Polarizabilities." In G. Birnbaum (ed.), *Phenomena Induced by Intermolecular Interactions*, Plenum Press, New York, 1985, pp. 263–290.

81. K. L. C. Hunt and J. E. Bohr. Field-induced fluctuation correlations and the effects of van der Waals interactions on molecular polarizabilities. *J. Chem. Phys.*, *84*:6141–6150 (1986).

82. L. Jansen and P. Mazur. On the theory of molecular polarization in gases: I. The effect of molecular interactions on the polarizability of spherical nonpolar molecules. *Physica*, *21*:193–207 (1955).

83. M. Jaszuński. Perturbation theory of pair polarizabilities. *Chem. Phys. Lett.*, *135*:565–570 (1987).

84. M. Kaźmierczak and T. Bancewicz. On the spherical representation of the polarizability tensor of an ensemble of interacting molecules within the DID model: Comparison with previous results. *J. Chem. Phys.*, *80*:3504–3505 (1984).

85. S. P. Keating. Optics and intermolecular forces. *Molec. Phys.*, *58*:33–52 (1986).

86. S. Kielich. Multipolar theory of dielectric polarization in dense mixtures. *Molec. Phys.*, *9*:549–564 (1965).

87. S. Kielich. Théorie de l'anisotropie optique des mélanges de molécules différentes avec interactions. *J. Phys.* (*Paris*), *29*:619–630 (1968).

88. S. Kielich. Kerr effect induced in liquid argon and carbon tetrachloride by fluctuational-statistical processes. *Acta Phys. Polonica, A*, *41*:653–656 (1972).

89. S. Kielich. "General Molecular Theory and Electric Field Effects in Isotropic Dielectrics." In M. Davies (ed.), *A Specialist Periodical Report—Dielectric and Related Molecular Processes*, Vol. 1, Chemical Society, London, 1972, pp. 192–387.

90. A. L. Kielpinski and S. Murad. The effect of molecular polarizability on the properties of fluids: I. Dilute gases. *Molec. Phys.*, *61*:1563–1573 (1987).

91. J. G. Kirkwood et al. *Dielectrics, Intermolecular Forces, Optical Rotation.* Vol. 2 of *J. G. Kirkwood: Collected Works*, Gordon and Breach, New York, 1965.

92. S. Kirouac and T. K. Bose. Polarizability and dielectric properties of helium. *J. Chem. Phys.*, *64*:1580–1582 (1976).

93. J. W. Kress and J. J. Kozak. Determination of the pair polarizability tensor for the neon diatom. *J. Chem. Phys.*, *66*:4516–4519 (1977).

94. R. A. Kromhout and B. Linder. Effect of van der Waals interaction on Raman transitions: I. *Molec. Phys.*, *62*:689–699 (1987).

95. B. M. Ladanyi and T. Keyes. Theory of the static Kerr effect in dense fluids. *Molec. Phys.*, *34*:1643–1659 (1977).

96. B. M. Ladanyi and T. Keyes. Effect of internal fields on depolarized light scattering from *n*-alkane gases. *Molec. Phys.*, *37*:1809–1821 (1979).

97. B. M. Ladanyi and T. Keyes. The influence of intermolecular interactions on the Kerr constant of simple liquids. *Canad. J. Phys.*, *59*:1421–1429 (1981).

98. B. M. Ladanyi and T. Keyes. Effective polarizabilities of *n*-alkalenes: Intramolecular interactions in solution and in the gas phase. *J. Chem. Phys.*, *76*:2047–2055 (1982).

99. P. Lallemand, D. J. David, and B. Bigot. Calculs de polarisabilités atomiques et moléculaires I. Méthodologie et application à l'association Ar–Ar. *Molec. Phys.*, *27*:1029–1043 (1974).

100. M. Lallemand and D. Vidal. Variation of the polarizability of noble gases with density. *J. Chem. Phys.*, *66*:4776–4780 (1977).

101. Y. Le Duff, A. Chave, B. Dumon, H. Idrissi, and M. Thibeau. Two-body collision induced light scattering: Effect of the spatial extension of the polarizability for conducting spheres. *Phys. Lett. A*, *73*:307–308 (1979).

102. H. B. Levine and D. A. McQuarrie. Second and third ordinary and dielectric virial coefficients for nonpolar axial molecules. *J. Chem. Phys.*, *44*:3500–3505 (1966).

103. H. B. Levine and D. A. McQuarrie. Dielectric constant of simple gases. *J. Chem. Phys.*, *49*:4181–4187 (1968).

104. H. B. Levine and G. Birnbaum. Collision induced light scattering. *Phys. Rev. Lett.*, *20*:439–441 (1968).

105. H. B. Levine and G. Birnbaum. Determination of models for collision induced polarizability by the method of moments. *J. Chem. Phys.*, *55*:2914–2917 (1971).

106. D. E. Logan and P. A. Madden. On the dielectric constant of nonpolar fluids. *Molec. Phys.*, *46*:715–742 (1982).

107. D. E. Logan and P. A. Madden. On the second dielectric virial coefficients of methane and the inert gases. *Molec. Phys.*, *46*:1195–1211 (1982).

108. W. C. Mackrodt. The coupling of intermolecular interactions to a static electric field. *Molec. Phys.*, *27*:933–948 (1974).

109. P. Madden and D. Kivelson. "A Consistent Molecular Treatment of Dielectric Phenomena." In I. Prigogine and S. A. Rice (eds), *Advances in Chemical Physics*, Vol. 56, Wiley, New York, 1984, pp. 467–566.

110. P. Mazur and L. Jansen. On the theory of molecular polarization: II. Effects of molecular interactions on the Clausius-Mosotti function for systems of spherical nonpolar molecules. *Physica*, *21*:208–218 (1955).

111. P. Mazur and M. Mandel. On the theory of the refractive index of nonpolar gases I. *Physica*, *22*:289–298 (1956).

112. P. Mazur and M. Mandel. On the theory of the refractive index of nonpolar gases II. *Physica*, *22*:299–310 (1956).

113. P. Mazur. On statistical mechanics and electromagnetic properties of matter. *Adv. Chem. Phys.*, *1*:309–360 (1958).

114. N. Meinander, G. C. Tabisz, and M. Zoppi. Moment analysis in depolarized light scattering: Determination of a single-parameter empirical pair polarizability anisotropy for Ne, Ar, Kr, Xe and CH_4. *J. Chem. Phys.*, *84*:3005–3013 (1986).

115. A. Michels, J. de Boer, and A. Bijl. Remarks concerning molecular interactions and their influence on the polarizability. *Physica*, *4*:981–994 (1937).

116. M. Neumann. Collision induced light scattering by globular molecules: Applequist's atom dipole interaction model and its implementation in computer simulation. *Molec. Phys.*, *53*:187–202 (1984).

117. E. F. O'Brien, V. P. Gutschick, V. McKoy, and J. P. McTague. Polarizability of interacting atoms: relation to collision induced light scattering and dielectric models. *Phys. Rev. A*, *8*:690–696 (1973).

118. W. H. Orttung. Direct solution of the Poisson equation for biomolecules of arbitrary shape, polarizability density, and charge distribution. *Ann. N. Y. Acad. Sci.*, *303*:22–37 (1977).

119. W. H. Orttung. Extension of the Kirkwood-Westheimer model of substituent effects to general shapes, charges, and polarizabilities. Application to the substituted Biocyclo[2.2.2]octanes. *J. Am. Chem. Soc.*, *100*:4369–4375 (1978).

120. V. D. Ovsyannikov. Atomic susceptibilities for collision induced polarizability and the corrections to the dispersion forces in the field of a light wave. *Zh. Eksp. Teor. Fiz.* (*U.S.S.R.*), *82*:1749–1761 (1982) [*Sov. Phys. JETP* (*USA*), *55*:1010–1016 (1982)].

121. D. W. Oxtoby and W. M. Gelbart. Collisional polarizability anisotropies of the noble gases. *Molec. Phys.*, *29*:1569–1576 (1975).

122. D. W. Oxtoby and W. M. Gelbart. Collisional polarizabilities of the inert gases: second-order overlap, exchange and correlation effects. *Molec. Phys.*, *30*:535–547 (1975).

123. D. W. Oxtoby. The calculation of pair polarizabilities through continuum electrostatic theory. *J. Chem. Phys.*, *69*:1184–1189 (1978).

124. D. W. Oxtoby. Local polarization theory for field-induced molecular multipoles. *J. Chem. Phys.*, *72*:5171–5176 (1980).

125. E. W. Pearson, M. Waldman, and R. G. Gordon. Dipole moments and polarizabilities of rare gas atom pairs. *J. Chem. Phys.*, *80*:1543–1555 (1984).

126. J. J. Perez, J. H. R. Clarke, and A. Hinchliffe. Three-body contributions to the dipole polarizability of He_3 clusters. *Chem. Phys. Lett.*, *104*:583–586 (1984).

127. P. L. Prasad and L. A. Nafie. The atom dipole interaction model of Raman optical activity: Reformulation and comparison to the general two-group model. *J. Chem. Phys.*, *70*:5582–5588 (1979).

128. P. L. Prasad and D. F. Burrow. Raman optical activity, computation of circular intensity differentials by the atom-dipole interaction model. *J. Am. Chem. Soc.*, *101*:800–805 (1979).

129. P. L. Prasad and D. F. Burrow. Raman optical activity, computation of circular intensity differentials for bromochlorofluoromethane. *J. Am. Chem. Soc.*, *101*:806–812 (1979).

130. M. H. Proffitt and L. Frommhold. Concerning the anisotropy of the polarizability of pairs of methane molecules. *Chem. Phys.*, *36*:197–200 (1979).

131. M. H. Proffitt and L. Frommhold. New measurement of the trace of the helium diatom polarizability from collision induced polarized Raman spectra. *Phys. Rev. Lett.*, *42*:1473–1475 (1979).

132. D. Robert and L. Galatry. Influence of molecular non-rigidity on the infrared absorption and Raman scattering line shape in dense media. *J. Chem. Phys.*, *64*:2721–2734 (1976).

133. D. W. Schaefer, R. E. J. Sears, and J. S. Waugh. Second virial coefficients of the Kerr effect in methyl chloride, methyl fluoride and fluoroform. *J. Chem. Phys.*, *53*:2127–2128 (1970).

134. D. P. Shelton and G. C. Tabisz. Molecular frame distortion and molecular pair polarizability. *Chem. Phys. Lett.*, *69*:125–127 (1980).

135. L. Silberstein. Molecular refractivity and atomic interaction. *Phil. Mag.*, *33*:92–128 (1917).

136. L. Silberstein. Molecular refractivity and atomic interaction II. *Phil. Mag.*, *33*:521–533 (1917).

137. J. E. Sipe and J. van Kranendonk. Limitations of the concept of polarizability density as applied to atoms and molecules. *Molec. Phys.*, *35*:1579–1584 (1978).

138. J. E. Sipe. Macroscopic theory of dielectric solids. II. The theory of Brillouin scattering from rare gas crystals. *Canad. J. Phys.*, *56*:199–215 (1978).

139. J. E. Sipe. "Brillouin Scattering from Rare Gas Crystals: A Theory for the Pockels Coefficients". In J. van Kranendonk (ed.), *Intermolecular Spectroscopy and Dynamical Properties of Dense Systems—Proceedings of the International School of Physics "Enrico Fermi," Course LXXV*, North-Holland, Amsterdam, 1980, pp. 430–464.

140. G. Stell, G. N. Patey, and J. S. Høye. Dielectric constants of fluid models: Statistical mechanical theory and its quantitative implementation. *Adv. Chem. Phys.*, *48*:183–328 (1981).

141. A. J. Stone. Transformation between cartesian and spherical tensors. *Molec. Phys.*, *29*:1461–1471 (1975).

142. A. J. Stone. Distributed multipole analysis, or how to describe a molecular charge distribution. *Chem. Phys. Lett.*, *83*:233–239 (1981).

143. H. Sutter. "Dielectric Polarization in Gases." In M. Davies (ed.), *A Specialist Periodic Report—Dielectric and Related Molecular Processes*, Vol. 1, Chemical Society, London, 1972, pp. 65–99.

144. L. G. Suttorp and M. A. J. Michels. Retardation in the atomic pair polarizability. *Chem. Phys. Lett.*, *46*:391–394 (1977).

145. L. G. Suttorp and M. A. J. Michels. The retarded energy shift and pair polarizabilities of interacting atoms in an external field: Application of resummed field-theoretical perturbation theory. *Molec. Phys.*, *40*:1089–1105 (1980).

146. G. C. Tabisz and V. Chandrasekharen. The point-dipole model of atomic pair polarizability. *Canad. J. Phys.*, *58*:1420–1423 (1980).

147. A. V. Tulub, M. D. Balmakov, and S. A. Khallaf. Interaction energy of neutral atoms in an external electric field. *Doklady Akad. Nauk SSSR*, *196*:75–77 (1971) [*Sov. Phys. DOKLADY*, *16*:18–20 (1971)].

148. J. Vermesse, D. Levesque, and J. J. Weis. Influence of the potential on the dielectric constant of inert gases. *Chem. Phys. Lett.*, *80*:283–285 (1981).

149. D. Vidal and M. Lallemand. Evolution of the Clausius-Mosotti function of noble gases and nitrogen, at moderate and high density, near room temperature. *J. Chem. Phys.*, *64*:4293–4302, (1976).

150. J. F. Ward and C. K. Miller. Measurements of nonlinear optical polarizabilities for twelve small molecules. *Phys. Rev. A*, *19*:826–833 (1979).

151. S. Woźniak and S. Kielich. Effective optical anisotropy of polar molecules from Rayleigh light scattering studies. *J. Phys. (Paris)*, *36*:1305–1315 (1975).

152. S. Woźniak and S. Kielich. Influence of interaction between non-dipole solvent and dipole solute molecules on anisotropic light scattering. *Acta Phys. Polonica A*, *52*:863–873 (1977).

I.2. *Depolarization Ratio*

153. J. Berrué, A. Chave, B. Dumon, and M. Thibeau. Effets des chocs multiples sur la diffusion de la lumière dans l'argon, le méthane, l'azote et l'ethyléne gazeux. *J. Phys. (Paris)*, *37*:845–854 (1976).

154. J. Berrué, A. Chave, B. Dumon, and M. Thibeau. Dépolarisation de la lumière diffusée par des molécules anisotropes: dépendance avec la densité. *J. Phys. (Paris)*, *37*:L.165–L.168 (1976).

155. J. Berrué, A. Chave, B. Dumon, and M. Thibeau. High pressure cell to measure the depolarization ratio of the light scattered by gases. *Rev. Phys. Appl.*, *12*:1743–1746 (1977).

156. J. Berrué, A. Chave, B. Dumon, and M. Thibeau. Mesure du dépolarisation dans l'azote comprimé et évaluation théorique des différents processus qui y contribuent. *J. Phys. (Paris)*, *39*:815–823 (1978).

157. J. Berrué, A. Chave, B. Dumon, and M. Thibeau. A new apparatus for measuring the depolarization ratio of light scattered by gaseous argon, methane and tetrafluoromethane. *Opt. Communic.*, *31*:317–320 (1979).

158. J. Berrué, A. Chave, B. Dumon, and M. Thibeau. Density dependence of light scattering from nitrogen: Permanent and collisional effects and their interferences. *Canad. J. Phys.*, *59*:1510–1513 (1981).

159. A. D. Buckingham and M. J. Stephen. A theory of the depolarization of light scattered by a dense medium. *Trans. Faraday Soc.*, *53*:884–893 (1957).

160. J. Cabannes. *La Diffusion de la Lumière*. Presses Universitaires de France, Paris, 1929.

161. E. Dayan, D. A. Dunmur, and M. R. Manterfield. Depolarized light scattering from gases of anisotropic molecules at intermediate pressures. *J. Chem. Soc. Faraday Trans. 2*, *76*:309–315 (1980).

162. S. Dumartin, M. Thibeau, and B. Oksengorn. Dépolarisation du rayonnement Rayleigh diffusé par un milieu formé de molécules isotropes: essai de calcul de l'effet de symmétrie croissante de l'environnement moléculaire avec densité. *Compt. Rend. Acad. Sci. Paris, ser. B*, *271*:884–887 (1970).

163. D. A. Dunmur, M. R. Manterfield, and D. J. Robinson. Depolarized light scattering studies of the collision induced polarizability anisotropy of atoms and spherical top molecules. *Molec. Phys.*, *50*:573–583 (1983).

164. W. M. Gelbart. Many-body effects and the depolarization of light by simple gases. *J. Chem. Phys.*, *57*:699–701 (1972).

165. G. C. Gray and H. I. Ralph. Virial expansion for the depolarization ratio of Rayleigh scattering from monatomic gases. *Phys. Lett. A*, *33*:165–166 (1970).

166. R. H. Huiser, A. C. Michels, H. M. J. Boots, and N. J. Trappeniers. "The Depolarization Factor of Light Scattered from Xenon near Its Critical Point." In B. Vodar and Ph. Marteau (ed.,) *High Pressure Science and Technology*, Pergamon, New York, 1980, pp. 727–729.

167. R. A. Huijts, J. C. F. Michielsen, and J. van der Elsken. Anomalous density dependence in depolarized light scattering by simple fluids. *Phys. Lett. A*, *112*:230–233 (1985).

168. S. Kielich. Molecular light scattering in dense mixtures. *Acta Phys. Polonica*, *33*:63–79 (1968).

169. Y. Le Duff. Facteur de dépolarisation de l'hélium gazeux. *J. Phys. (Paris)*, *40*:L267–L269 (1979).

170. Y. Le Duff, A. Gharbi, and T. Othman. Depolarization ratios of gas mixtures. *Phys. Rev. A*, *28*:2714–2721 (1983).

171. B. Oksengorn. Dépolarisation du rayonnement Rayleigh diffusé par des gaz rares comprimés: Considérations sur l'anisotropie de la polarisabilité induite par collisions d'un couple d'atomes en interaction. *Compt. Rend. Acad. Sci. Paris, ser. B*, *279*:275–278 (1974).

172. B. Oksengorn. Dépolarisation induite par la pression du rayonnement Rayleigh diffusé par SF_6 dans la domaine des faibles densités. *Compt. Rend. Acad. Sci. Paris, ser. B*, *282*:223–226 (1976).

173. B. Oksengorn. Collision induced depolarization of the Rayleigh light scattering by argon at low density. *Chem. Phys. Lett.*, *102*:429–432 (1983).

174. F. J. Pinski and C. E. Campbell. Depolarization of light and pair-pair correlations of argon gas. *Chem. Phys. Lett.*, *56*:156–160 (1978).

175. H. I. Ralph and C. G. Gray. Calculation of the third depolarization virial coefficient for monatomic gases. *Molec. Phys.*, *27*:1683–1685 (1974).

176. R. L. Rowell, G. M. Aval, and J. J. Barrett. Rayleigh-Raman depolarization of laser light scattered by gases. *J. Chem. Phys.*, *54*:1960–1964 (1971)

177. M. Thibeau, B. Oksengorn, and B. Vodar. Étude expérimentale de la dépolarisation du rayonnement Rayleigh diffusé par l'argon comprimé jusqu' à 800 bars. *Compt. Rend. Acad. Sci. Paris, ser. B*, *263*:135–138 (1966).

178. M. Thibeau, B. Oksengorn, and B. Vodar. Étude théorique de la dépolarisation du rayonnement Rayleigh diffusé par un gaz rare comprimé. *Compt. Rend. Acad. Sci. Paris, ser. B*, *265*:722–725 (1967).

179. M. Thibeau, B. Oksengorn, and B. Vodar. Étude théorique et expérimentale de la dépolarisation du rayonnement Rayleigh diffusé par les gaz rares comprimés. *J. Phys. (Paris)*, *29*:287–296 (1968).

180. M. Thibeau and B. Oksengorn. Étude théorique de la largeur spectrale de la partie dépolarisée du rayonnement diffusé par un gaz rare comprimé. *Molec. Phys.*, *15*:579–586 (1968).

181. M. Thibeau, B. Oksengorn, and B. Vodar. Application d'un modéle de cage a l'étude de la dépolarisation du rayonnement Rayleigh diffusé par un gaz rare sous trés haute pression. *J. Phys. (Paris)*, *30*:47–52 (1969).

182. M. Thibeau, G. C. Tabisz, B. Oksengorn, and B. Vodar. Pressure induced depolarization of the Rayleigh light scattered from a gas of optically isotropic molecules. *J. Quant. Spectro. Rad. Trans.*, *10*:839–856 (1970).

183. M. Thibeau and B. Oksengorn. Some recent results about the depolarization of the Rayleigh light scattered by optically isotropic molecules. *J. Phys. (Paris)*, *33*:C1:247–C1:255 (1972).

184. M. Thibeau, A. Gharbi, Y. Le Duff, and V. Sergiescu. A comparative study of Rayleigh and Raman depolarized light scattering by a pure fluid of isotropic molecules. *J. Phys. (Paris)*, *38*:641–651 (1977).

185. M. Thibeau. "Rayleigh and Raman Depolarized Light Scattering by Isotropic Molecules." In B. Vodar and Ph. Marteau, (eds.), *High Pressure Science and Technology*, Pergamon, New York, 1980, pp. 724–726.

186. M. Thibeau, B. Dumon, A. Chave, and V. Sergiescu. A dynamical Onsager model approach of field fluctuations and light scattering in a gas of isotropic molecules. *Physica A*, *105*:219–228 (1981).

187. M. Thibeau, J. Berrué, A. Chave, B. Dumon, Y. Le Duff, A. Gharbi, and B. Oksengorn. "Depolarization Ratio of Light Scattered by a Gas of Isotropic Molecules." In G. Birnbaum (ed.), *Phenomena Induced by Intermolecular Interactions*, Plenum Press, New York, 1985, pp. 291–309.

188. N. J. Trappeniers, V. A. Kuz, and C. A. ten Seldam. Depolarization of light scattered by systems of anisotropic molecules. *Physica*, *57*:294–305 (1972).

189. A. Triki, B. Oksengorn, and B. Vodar. Étude expérimentale de la dépolarisation du rayonnement Rayleigh diffusé par l'hexafluorene de soufre gazeaux à densités modérés. *Compt. Rend. Acad. Sci. Paris, ser. B*, *274*:783–786 (1972).

190. A. Triki, B. Oksengorn, and B. Vodar. Effets de densité et de températur sur la dépolarisation du rayonnement Rayleigh diffusé par des gaz rares comprimés. *Compt. Rend. Acad. Sci. Paris, ser. B*, *277*:411–414 (1973).

191. R. C. Watson and R. L. Rowell. Depolarized light scattering from Ar, Kr, CH_4, SF_6, and $C(CH_3)_4$. *J. Chem., Phys.*, *61*:2666–2669 (1974).

I.3. *Spectra of Gases*

192. R. D. Amos, A. D. Buckingham, and J. H. Williams. Theoretical studies of the collision induced Raman spectrum of carbon dioxyde. *Molec. Phys.*, *39*:1519–1526 (1980).

193. S.-C. An, C. J. Montrose, and T. A. Litovitz. Low-frequency structure in the depolarized spectrum of argon. *J. Chem. Phys.*, *64*:3717–3719 (1976).

194. S.-C. An, L. Fishman, T. A. Litovitz, C. J. Montrose, and H. A. Posch. Depolarized light scattering from dense noble gases. *J. Chem. Phys.*, *70*:4626–4632 (1979).

195. O. I. Arkhangelskaya and N. G. Bakhshiev. Effect of intermolecular interactions on the vibrational spectra of molecules: IX: Method for determining the dipole moment and polarizability functions of diatomic molecules in solution. *Optika i Spektr. (U.S.S.R.)*, *31*:53–57 (1971) [*Opt. Spectroscopy (USA)*, *31*:26–28 (1971)].

196. U. Bafile, R. Magli, F. Barocchi, M. Zoppi, and L. Frommhold. Line shape and moment analysis in depolarized collision induced light scattering. *Molec. Phys.*, *49*:1149–1166 (1983).

197. U. Bafile, L. Ulivi, M. Zoppi, and F. Barocchi. Collision induced Raman scattering of hydrogen at various temperatures. *Chem. Phys. Lett.*, *117*:247–250 (1985).

198. U. Bafile, L. Ulivi, M. Zoppi, and F. Barocchi. The three-body correlation spectral moments in depolarized interaction induced light scattering of H_2 at 297 K. *Chem. Phys. Lett.*, *138*:559–564 (1987).

199. U. Bafile, L. Ulivi, M. Zoppi, and F. Barocchi. Depolarized light scattering spectrum of normal gaseous hydrogen at low density and 297 K. *Phys. Rev. A*, *37:4133* (1988).

200. U. Balucani and R. Vallauri. Collision induced light scattering at intermediate densities: I. The integrated intensity. *Molec. Phys.*, *38*:1099–1113 (1979).

201. U. Balucani and R. Vallauri. The role of relative radial motion in collision-induced light scattering. *J. Phys. Chem.*, *12*:L915–L919 (1979).

202. U. Balucani, V. Tognetti, and R. Vallauri. "Memory Function Approach to the Line Shape Problem in Collision Induced Light Scattering." In J. van Kranendonk (ed.), *Intermolecular Spectroscopy and Dynamical Properties of Dense Systems—Proceedings of the International School of Physics "Enrico Fermi," Course LXXV*, North-Holland, Amsterdam, 1980, pp. 275–293.

203. U. Balucani and R. Vallauri. Tow-body dynamics in fluids as probed by interaction-induced light scattering. *Canad. J. Phys.*, *59*:1504–1509 (1981).

204. T. Bancewicz and S. Kielich. Isotropic Raman scattering for non-totally symmetric vibrations of correlated molecules with intrinsic optical anisotropy. *J. Chem. Phys.*, *75*:107–109 (1981).

205. F. Barocchi and J. P. McTague. Collision induced light scattering in gaseous CH_4. *Phys. Lett. A*, *53*:488–490 (1975).

206. F. Barocchi, R. Vallauri, and M. Zoppi. Collision induced spectrum decay constant for Lennard-Jones interaction potential. *Phys. Lett. A*, *50*:451–452 (1975).

207. F. Barocchi, M. Zoppi, D. P. Shelton, and G. C. Tabisz. A comparison of the spectral features of the collision induced light scattering by the molecular gases CH_4 and CF_4 and argon. *Canad. J. Phys.*, *55*:1962–1969 (1977).

208. F. Barocchi, M. Neri, and M. Zoppi. Derivation of three-body collision induced light scattering spectral moments for argon, krypton, xenon. *Molec. Phys.*, *34*:1391–1405 (1977).

209. F. Barocchi, M. Neri, and M. Zoppi. Moments of the three-body collision induced light scattering spectrum. *J. Chem. Phys.*, *66*:3308–3310 (1977).

210. F. Barocchi, P. Mazzinghi, and M. Zoppi. Collision induced light scattering in gaseous helium. *Phys. Rev. Lett.*, *41*:1785–1788 (1978).

211. F. Barocchi and M. Zoppi. Collision induced light scattering spectra and pair polarizability of gaseous argon. *Phys. Lett. A*, *66*:99–102 (1978).

212. F. Barocchi and M. Zoppi. Collision induced light scattering: Three- and four-body spectra of gaseous argon. *Phys. Lett. A*, *69*:187–189 (1978).

213. F. Barocchi and M. Zoppi. "Experimental Determination of Two-Body Spectrum and Pair Polarizability of Argon." In J. van Kranendonk, (ed.), *Intermolecular Spectroscopy and Dynamical Properties of Dense Systems—Proceedings of the International School of Physics "Enrico Fermi," Course LXXV*, North-Holland, Amsterdam, 1980, pp. 237–262.

214. F. Barocchi and M. Zoppi. "Experimental Determination of Two-Body Collision Induced Light Scattering of Helium." In J. van Kranendonk (ed.), *Intermolecular Spectroscopy and Dynamical Properties of Dense Systems—Proceedings of the International School of Physics "Enrico Fermi," Course LXXV*, North-Holland, Amsterdam, 1980, pp. 263–274.

215. F. Barocchi, M. Moraldi, M. Zoppi, and J. D. Poll. Quantum mechanical approximation for collision induced light scattering: Spectral Moments. *Molec. Phys.*, *43*:1193–1197 (1981).

216. F. Barocchi, M. Zoppi, M. H. Proffitt, and L. Frommhold. Determination of the collision induced light scattering cross section of the argon diatom. *Canad. J. Phys.*, *59*:1418–1420 (1981).

217. F. Barocchi, M. Moraldi, and M. Zoppi. Comment on the detailed balance symmetrization procedure of classical spectra. *Molec. Phys.*, *45*:1285–1289 (1982).

218. F. Barocchi. Interaction induced light scattering as a probe of many-body correlations in fluids. *J. Phys. (Paris)*, *46*:C9.123–C9.128 (1985).

219. F. Barocchi and M. Zoppi. "Depolarized Interaction Induced Light Scattering Experiments in Argon, Krypton, Xenon." In G. Birnbaum (ed.), *Phenomena Induced by Intermolecular Interactions*, Plenum Press, New York, 1985, pp. 311–343.

220. F. Barocchi and M. Zoppi. Collision induced light scattering: three- and four-body spectra of argon. *Phys. Lett. A*, *69*:187–189 (1987).

221. F. Barocchi, M. Celli, and M. Zoppi. Interaction-induced translational Raman scattering in dense krypton gas: Evidence of irreducible many-body effects. *Phys. Rev. A*, *38:in press* (1988).

222. A. Barreau, J. Berrué, A. Chave, B. Dumon, and M. Thibeau. Spectrum of collision induced scattering: Role of multipolar polarizabilities. *Optics Communic.*, *55*:99–104 (1985).

223. M. Bérard and P. Lallemand. Collision induced light scattering in low density neon gas. *Molec. Phys.*, *34*:251–260 (1977).

224. B. J. Berne and R. Pecora. *Dynamic Light Scattering*, Wiley, New York, 1976.

225. J. Berrué, A. Chave, B. Dumon, and M. Thibeau. Lumière diffusée par une assemblée de sphères conductrices a basse densité. *J. Phys.* (*Paris*), *36*:1079–1088 (1975).

226. G. Birnbaum. "Study of Atomic and Molecular Interactions from Collision-Induced Spectra." In J. V. Sengers (ed.), *Thermophysical Properties of Fluids*, American Society of Mechanical Engineers, New York, 1981, pp. 8–17.

227. G. Birnbaum, B. Guillot, and S. Bratos. "Theory of Collision Induced Line Shapes—Absorption and Light Scattering at Low Densities." In I. Prigogine and S. A. Rice (ed.) *Advances in Chemical Physics*, Vol. 51, Wiley–Interscience, New York, 1982, pp. 49–112.

228. G. Birnbaum (ed.). *Phenomena Induced by Intermolec. Interactions.* Plenum Press, New York, 1985.

229. A. Borysow and T. Grycuk. Raman spectrum of Hg_2 *van der Waals* molecules. *Acta Phys. Polonica, A*, *60*:129–139 (1981).

230. A. Borysow and T. Grycuk. Collision induced Raman spectra of Hg-rare gas *van der Waals* complexes. *Physica C*, *114*:414–422 (1982).

231. J. Borysow and L. Frommhold. "The Infrared and Raman Line Shapes of Pairs of Interacting Molecules." In G. Birnbaum (ed.), *Phenomena Induced by Intermolecular Interactions*, Plenum Press, New York, 1985, pp. 67–93.

232. J. Borysow, M. Moraldi, and L. Frommhold. The collision induced spectroscopies: concerning the desymmetrization of classical line shape. *Molec. Phys.*, *56*:913–922 (1985).

233. S. Bratos, B. Guillot, and G. Birnbaum. "Theory of Collision Induced Light Scattering and Absorption in Dense Rare Gas Fluids." In G. Birnbaum (ed.), *Phenomena Induced by Intermolecular Interactions*, Plenum Press, New York, 1985, pp. 363–381.

234. G. Briganti, V. Mazzacurati, G. Ruocco, G. Signorelli, and M. Nardone. Collision induced light scattering in gaseous H_2S. *Molec. Phys.*, *49*:1179–1186 (1983).

235. G. Briganti, D. Rocca, and M. Nardone. Interaction induced light scattering: First and second spectral moments in the superposition approximation. *Molec. Phys.*, *59*:1259–1272 (1986).

236. C. Brodbeck and J.-P. Bouanich. "Simultaneous Transitions in Compressed Gas Mixtures." In G. Birnbaum (ed.), *Phenomena Induced by Intermolecular Interactions*, Plenum Press, New York, 1985, pp. 169–192.

237. M. S. Brown, M. H. Proffitt, and L. Frommhold. Collision induced Raman scattering by anisotropic molecules, H_2 and D_2. *Chem. Phys. Lett.*, *117*:243–246 (1985).

238. M. S. Brown and L. Frommhold. Dimer features in the translational depolarized Raman spectra of molecular hydrogen pairs. *Chem. Phys. Lett.*, *127*:197–199 (1986).

239. A. D. Buckingham and R. E. Raab. Electric field induced differential scattering of right and left circularly polarized light. *Proc. Roy. Soc.* (*London*), *A*, *345*:365–377 (1975).

240. A. D. Buckingham and G. C. Tabisz. Collision induced rotational Raman scattering. *Opt. Lett.*, *1*:220–222 (1977).

241. A. D. Buckingham and G. C. Tabisz. Collision-induced Raman scattering by tetrahedral and octahedral molecules. *Molec. Phys.*, *36*:583–596 (1978).

242. A. D. Buckingham and G. C. Tabisz. Collision induced rotational Raman scattering. *Opt. Lett.*, *1*:220–222 (1978).

243. M. O. Bulanin and A. P. Kouzov. Simultaneous transitions in light-scattering spectra induced by collisions between molecules. *Optika i Spektr.* (*U.S.S.R.*), *53*:266–269 (1982).

244. V. G. Cooper, A. D. May, E. H. Hara, and H. F. P. Knaap. Depolarized Rayleigh scattering in gases as a new probe of intermolecular forces. *Phys. Lett. A*, *27*:52–53 (1968).

245. A. De Lorenzi, A. De Santis, R. Frattini, and M. Sampoli. Induced contributions in the Rayleigh band of gaseous H_2S. *Phys. Rev. A*, *33*:3900–3912 (1986).

246. A. De Santis, M. Sampoli, G. Sartori, P. Morales, and G. Signorelli. Density evolution of Rayleigh and Raman depolarized scattering in fluid N_2. *Molec. Phys.*, *35*:1125–1140 (1978).

247. A. De Santis, E. Moretti, and M. Sampoli. Rayleigh spectra of N_2 fluid: Temperature and density behavior of the second moment. *Molec. Phys.*, *46*:1271–1282 (1982).

248. A. De Santis and M. Sampoli. Induced contributions in isotropic and anisotropic Rayleigh spectra of fluid H_2S. *Chem. Phys. Lett.*, *102*:425–428 (1983).

249. A. De Santis and M. Sampoli. Raman spectra of fluid N_2: Temperature and density behavior of the second moment. *Molec. Phys.*, *51*:97–117 (1984).

250. A. De Santis and M. Sampoli. A light scattering study of rotational bands in fluid CO_2. *Molec. Phys.*, *53*:717–729 (1984).

251. A. De Santis and M. Sampoli. Depolarized Rayleigh bands of fluid CO at room temperature: Induced effects on the second moment. *Phys. Lett. A*, *100*:25–27 (1984).

252. A. De Santis and M. Sampoli. Induced interactions on the depolarized Rayleigh bands of fluid CO_2. *Phys. Lett. A*, *101*:131–133 (1984).

253. A. De Santis and M. Sampoli. "Raman scattering from linear molecules." In G. Birnbaum (ed.), *Phenomena Induced by Intermolecular Interactions*, Plenum Press, New York, 1985, pp. 627–642.

254. A. De Santis, R. Frattini, and M. Sampoli. Rayleigh spectra of fluid N_2: Mean square torques and induced contributions to the fourth moment. *Chem. Phys. Lett.*, *114*:71–73 (1985).

255. A. De Santis, R. Frattini, and M. Sampoli. On the sixth moment of Rayleigh spectra of fluid N_2. *Chem. Phys. Lett.*, *114*:297–300 (1985).

256. A. De Santis, R. Frattini, and M. Sampoli. A comparative study of fluid carbon monoxide and fluid nitrogen by Rayleigh scattering. *Phys. Lett. A*, *109*:28–30 (1985).

257. A. De Santis, R. Frattini, M. Sampoli, and R. Vallauri. Raman scattering from CO_2: A computer simulation investigation of the effects of polarizability and potential models. *Europhys. Lett.*, *2*:17–22 (1986).

258. S. M. El-Sheikh, N. Meinander, and G. C. Tabisz. Induced rotational Raman scattering and the dipole-octopole polarizability of SF_6. *Chem. Phys. Lett.*, *118*:151–155 (1985).

259. G. E. Ewing. The spectroscopy of *van der Waals* molecules. *Canad. J. Phys.*, *54*:487–504 (1976).

260. P. A. Fleury, W. B. Daniels, and J. M. Worlock. Density and temperature dependence of intermolecular light scattering in simple fluids. *Phys. Rev. Lett.*, *27*:1493–1496 (1971).

261. P. A. Fleury. Dynamics of simple fluids by intermolecular light scattering. *J. Phys. (Paris)*, *33*:C1:264 (1972).

262. D. Frenkel and J. P. McTague. On the Raman spectrum of argon dimers. *J. Chem. Phys.*, *70*:2695–2699 (1979).

263. L. Frommhold. Interpretation of Raman spectra of *van der Waals* dimers in argon. *J. Chem. Phys.*, *61*:2996–3001 (1974).

264. L. Frommhold. About the collision induced Raman spectrum of gaseous argon. *J. Chem. Phys.*, *63*:1687–1688 (1975).

265. L. Frommhold and R. Bain. Comments concerning the Raman spectra of *van der Waals* dimers in argon. *J. Chem. Phys.*, *63*:1700–1702 (1975).

266. L. Frommhold, K. H. Hong, and M. H. Proffitt. Absolute cross sections for collision induced scattering of light by binary paris of argon atoms. *Molec. Phys.*, *35*:665–679 (1978).

267. L. Frommhold, K. H. Hong, and M. H. Proffitt. Absolute cross sections for collision induced depolarized scattering of light in krypton and xenon. *Molec. Phys.*, *35*:691–700 (1978).

268. L. Frommhold and M. H. Proffitt. Raman spectra and polarizability of the neon diatom. *Chem. Phys. Lett.*, *66*:210–212 (1979).

269. L. Frommhold and M. H. Proffitt. Collision induced Raman spectra of neon and the diatom polarizability. *Phys. Rev. A*, *21*:1249–1255 (1980).

270. L. Frommhold and M. H. Proffitt. The polarized Raman spectrum of the argon diatom. *J. Chem. Phys.*, *74*:1512–1513 (1981).

271. L. Frommhold. "Collision Induced Scattering of Light and the Diatom Polarizabilities." In I. Prigogine and S. Rice (ed.) *Advances in Chemical Physics*, Vol. *46*, Wiley, New York, 1981, pp. 1–72. See also an update in *Canad. J. Phys.*, *59*:1459 (1981).

272. W. M. Gelbart. On the translational Raman effect. *Molec. Phys.*, *26*:873–886 (1973).

273. W. M. Gelbart. Collision induced and multiple light scattering by simple fluids. *Phil Trans. Royal Soc.* (*London*) *A*, *293*:359–375 (1979).

274. W. M. Gelbart. "Intermolecular Spectroscopy." In J. van Kranendonk (ed.), *Intermolecular Spectroscopy and Dynamical Properties of Dense Systems—Proceedings of the International School of Physics "Enrico Fermi," Course LXXV*, North-Holland, Amsterdam, 1980, pp. 1–44.

275. J. I. Gersten, R. E. Slusher, and C. M. Surko. Relation of collision induced light scattering in rare gases to atomic colllision parameters. *Phys. Rev. Lett.*, *25*:1739–1742 (1970).

276. J. I. Gersten. Asymptotic line shape in collision induced light scattering. *Phys. Rev. A*, *4*:98–108 (1971).

277. A. Gharbi and Y. Le Duff. Collision induced scattering for molecules. *Physica A*, *87*:177–184 (1977).

278. A. Gharbi and Y. Le Duff. Line shape of the collision induced scattering in CF_4. *Physica A*, *90*:619–625 (1978).

279. A. Gharbi and Y. Le Duff. Many-body correlations for Raman and Rayleigh collision induced scattering. *Molec. Phys.*, *40*:545–552 (1980).

280. H. P. Godfried and I. F. Silvera. Raman spectroscopy of argon dimers. *Phys. Rev. Lett.*, *48*:1337–1340 (1982).

281. R. A. Harris. On the theory of light scattering from fluids and polymers. *Chem. Phys. Lett.*, *19*:49–52 (1973).

282. W. Holzer and Y. Le Duff. Collision induced light scattering observed at the frequency region of vibrational Raman bands. *Phys. Rev. Lett.*, *32*:205–208 (1974).

283. W. Holzer and R. Ouillon. Forbidden Raman bands of SF_6: Collision induced Raman scattering. *Chem. Phys. Lett.*, *24*:589–593 (1974).

284. R. A. Huijts, J. C. F. Michielsen, and J. van der Elsken. The density dependence of the depolarized light scattering at high frequency in rare gas fluids. *Chem. Phys. Lett.*, *92*:649–652 (1982).

285. D. J. G. Irwin and A. D. May. The complete spectrum of collision induced light scattering in compressed argon. *Canad. J. Phys.*, *50*:2174–2177 (1972).

286. S. Kielich, T. Bancewicz, and S. Woźniak. Spectral distribution of scattered light by fluid mixtures of correlated atoms and molecules. *Canad. J. Phys.*, *59*:1620–1626 (1981).

287. S. Kielich. Time-correlation functions for new cross-multipole field fluctuations in binary light scattering by unlike polar molecules. *J. Phys.* (*Paris*), *43*:L.389–L.394 (1982).

288. S. Kielich. Intermolecular light scattering. *Proc. Indian Acad. Sci. (Chem. Sci.)*, *94*:403–448 (1985).

289. H. F. P. Knaap and P. Lallemand. Light scattering by gases. *Ann. Rev. Phys. Chem.*, *26*:59–81 (1975).

290. P. Lallemand. Spectral distribution of double light scattering by gases. *Phys. Rev. Lett.*, *25*:1079–1081 (1970).

291. P. Lallemand. Spectre de la lumière doublement diffusée par un gaz. *J. Phys. (Paris)*, *32*:119–127 (1971).

292. P. Lallemand. Diffusion de la lumière—calcul des moments du spectre de la lumière diffusée par une paire de molécules. *Compt. Rend. Acad. Sci. Paris, ser. B*, *273*:89–92 (1971).

293. P. Lallemand. Diffusion dépolarisée de la lumière due aux collisions. *J. Phys. (Paris)*, *33*:C1:257–C1:263 (1972).

294. Y. Le Duff and W. Holzer. Collision induced light scattering observed in the frequency region of the vibrational Raman bands. *Phys. Rev. Lett.*, *32*:205–208 (1974).

295. Y. Le Duff and A. Gharbi. Collision induced scattering at a vibrational Raman frequency. *Phys. Rev. A*, *17*:1729–1734 (1978).

296. Y. Le Duff. Collision induced scattering in helium. *Phys. Rev. A*, *20*:48–53 (1979).

297. Y. Le Duff. Depolarized scattering from gaseous molecular hydrogen. *J. Phys. B*, *14*:55–61 (1981).

298. Y. Le Duff and V. Sergiescu. On a pair-potential-independent criterion for DID light scattering. *Physica A*, *127*:347–353 (1984).

299. Y. Le Duff and R. Ouillon. Depolarized scattering from para- and normal-hydrogen. *J. Chem. Phys.*, 82:1–4 (1985).

300. Y. Le Duff. "Light Scattering in Compressed Gases: A Workshop Report." In G. Birnbaum (ed.), *Phenomena Induced by Intermolecular Interactions*, Plenum Press, New York, 1985, pp. 359–361.

301. Y. Le Duff, R. Ouillon, and B. Silvi. Low temperature Raman spectrum of xenon diatoms. *Molec. Phys.*, *62*:1065–1077 (1987).

302. H. B. Levine. Spectroscopy of dimers. *J. Chem. Phys.*, *56*:2455–2473 (1972).

303. J. C. Lewis and J. van Kranendonk. Intercollisional interference effects in collision induced light scattering. *Phys. Rev. Lett.*, *24*:802–804 (1970).

304. J. C. Lewis and J. van Kranendonk. Theory of intercollisional interference effects: III. Collision induced light scattering. *Canad. J. Phys.*, *50*:2902–2913 (1972).

305. J. C. Lewis. "Intercollisional Interference Effects." In J. van Kranendonk (ed.), *Intermolecular Spectroscopy and Dynamical Properties of Dense Systems—Proceedings of the International School of Physics "Enrico Fermi," Course LXXV*, North-Holland, Amsterdam, 1980, pp. 91–110.

306. J. C. Lewis. "Intercollisional Interference—Theory and Experiment." In G. Birnbaum (ed.), *Phenomena Induced by Intermolecular Interactions*, Plenum Press, New York, 1985, pp. 215–257.

307. J. P. McTague and G. Birnbaum. Collision induced light scattering in gaseous Ar and Kr. *Phys. Rev. Lett.*, *21*:661–664 (1968).

308. J. P. McTague and G. Birnbaum. Collision induced light scattering in gases: I. Rare gases Ar, Kr and Xe. *Phys. Rev. A*, *3*:1376–1383 (1971).

309. J. P. McTague, W. D. Ellenson, and L. H. Hall. Spectral virial expansion for collision induced light scattering in gases. *J. Phys. (Paris)*, *33*:C1:241–Cl:246 (1972).

310. F. D. Medina and W. B. Daniels. Low frequency Raman spectrum of methane at high densities. *J. Chem. Phys.*, *66*:2228–2229 (1977).

311. F. D. Medina and W. B. Daniels. Collision induced light scattering from fluid methane at high densities. *Phys. Rev. A*, *17*:1474–1477 (1978).

312. F. D. Medina. Collision induced light scattering from fluid methane: Temperature dependence of the low-frequency decay constants. *Phys. Lett. A*, *81*:139–140 (1981).

313. F. D. Medina and J. M. Dugas. Density dependence of the light scattering spectrum of fluid nitrogen. *J. Chem. Phys.*, *75*:3252–3257 (1981).

314. N. Meinander and G. C. Tabisz. Information theory and line shape engineering approaches to spectral profiles of collision induced depolarized light scattering. *Chem. Phys. Lett.*, *110*:388–394 (1984).

315. N. Meinander, A. R. Penner, U. Bafile, F. Barocchi, M. Zoppi, D. P. Shelton, and G. C. Tabisz. The spectral profile of the collision induced translational light scattering by gaseous CH_4: Determination of the pair polarizability anisotropy. *Molec. Phys.*, *54*:493–503 (1985).

316. N. Meinander and G. C. Tabisz. Moment analysis and line shape calculations in depolarized induced light scattering: Modeling empirical pair polarizability anisotropy. *J. Quant. Spectros. Rad. Transfer*, *35*:39–52 (1986).

317. C. E. Morgan and L. Frommhold. Raman spectra of *van der Waals* dimers in argon. *Phys.Rev. Lett.*, *29*:1053–1055 (1972); see correction in *J. Chem. Phys.*, *61*:2996 (1974).

318. M. Moraldi. Quantum mechanical spectral moments in collision induced light scattering and absorption in rare gases at low densities. *Chem. Phys.*, *78*:243–250 (1983).

319. M. Moraldi, A. Borysow, and L. Frommhold. Quantum sum formulae for the collision induced spectroscopies: Molecular systems as H_2–H_2. *Chem. Phys.*, *86*:339–347 (1984).

320. M. Moraldi, A. Borysow, and L. Frommhold. Depolarized Rayleigh scattering in normal and para-hydrogen. *J. Chem. Phys.*, *88*:5344–5351 (1988).

321. M. Neumann and H. A. Posch. "Interaction Induced Light Scattering from Tetrahedral Molecules.", In G. Birnbaum (ed.), *Phenomena Induced by Intermolecular Interactions*, Plenum Press, New York, 1985, pp. 475–495.

322. V. D. Ovsyannikov. Asymptotic line shape for collision induced depolarized light scattering in atomic gases. *Phys. Lett. A*, *85*:275–277 (1981).

323. R. A. Pasmanter, R. Samson, and A. Ben-Reuven. Molecular theory of optical polarization and light scattering in dielectric fluids. II. Extensions and applications. *Phys. Rev. A*, *14*:1238–1250 (1976).

324. A. R. Penner, N. Meinander, and G. C. Tabisz. The spectral intensity of the collision induced rotational Raman scattering by gaseous CH_4 and CH_4-inert gas mixtures. *Molec. Phys.*, *54*:479–492 (1985).

325. M. Perrot and J. Lascombe. "Collision Induced Effects in Allowed Infrared and Raman Spectra of Molecular Fluids". In G. Birnbaum (ed.), *Phenomena Induced by Intermolecular Interactions*, Plenum Press, New York, 1985, pp. 613–625.

326. J. D. Poll. "Intermolecular Spectroscopy of Gases." In J. van Kranendonk (ed.), *Intermolecular Spectroscopy and the Dynamical Properties of Dense Systems*, North-Holland, Amsterdam, 1980, pp. 45–76.

327. H. A Posch. Collision induced light scattering from fluids composed of tetrahedral molecules: III. Neopentane vapor. *Molec. Phys.*, *46*:1213–1230 (1982).

328. A. T. Prengel and W. S. Gornall. Raman scattering from colliding molecules and van der Walls dimers in gaseous methane. *Phys. Rev. A*, *13*:253–262 (1976).

329. M. H. Proffitt and L. Frommhold. Collision induced polarized and depolarized Raman spectra of helium and the diatom polarizability. *J. Chem. Phys.*, *72*:1377–1384 (1980).

330. M. H. Proffitt, J. W. Keto, and L. Frommhold. Collision induced spectra of the helium isotopes. *Phys. Rev. Lett.*, *45*:1843–1846 (1980).

331. M. H. Proffitt, J. W. Keto, and L. Frommhold. Collision induced Raman spectra and diatom polarizabilities of the rare gases—an update. *Canad. J. Phys.*, *59*:1459–1474 (1981).

332. R. Samson and A. Ben-Reuven. Theory of collision induced forbidden Raman transitions in gases: Application to SF_6. *J. Chem. Phys.*, *65*:3586–3594 (1976).

333. M. Sampoli, A. De Santis, and M. Nardone. Rotational and collisional contributions in the Rayleigh and Raman spectra of fluid N_2. *Canad. J. Phys.*, *59*:1403–1407 (1981).

334. D. P. Shelton, M. S. Mathur, and G. C. Tabisz. Collision induced light scattering by compressed CF_4 and CF_4–He mixtures. *Phys. Rev. A*, *11*:834–840 (1975).

335. D. P. Shelton and G. C. Tabisz. Moment analysis of collision induced light scattering from compressed CF_4. *Phys. Rev. A*, *11*:1571–1575 (1975).

336. D. P. Shelton, M. S. Mathur, and G. C. Tabisz. "Collision Induced Light Scattering by Compressed CF_4 and SF_6." In M. Grossmann, G. Elkomoss, and J. Ringeisen (ed.) *Molecular Spectroscopy of Dense Phases—Proceedings of 12th European Congress on Molecular Spectroscopy*, Elsevier, Amsterdam, 1976, pp. 555–558.

337. D. P. Shelton and G. C. Tabisz. Binary collision induced light scattering by isotropic molecular gases: I. Details of the experiment and the argon spectrum. *Molec. Phys.*, *40*:285–297 (1980).

338. D. P. Shelton and G. C. Tabisz. Binary collision induced light scattering by isotropic molecular gases: II. Molecular spectra and induced rotational Raman scattering. *Molec. Phys.*, *40*:299–308 (1980).

339. D. P. Shelton and G. C. Tabisz. A comparison of the collision induced light scattering by argon and by isotropic molecular gases. *Canad. J. Phys.*, *59*:1430–1433 (1981).

340. D. P. Shelton, G. C. Tabisz, F. Barocchi, and M. Zoppi. The three-body correlation spectrum in collision induced light scattering by isotropic molecular gases. *Molec. Phys.*, *46*:21–31 (1982).

341. H. K. Shin. Effects of 3-particle collisions on collision induced light scattering. *Chem. Phys. Lett.*, *35*:218–221 (1975).

342. G. C. Tabisz and A. D. Buckingham. Remarks concerning CILS by tetrahedral molecular pairs. *Faraday Symposia Chem. Soc.*, *11*:170–172 (1977).

343. G. C. Tabisz. "Collision Induced Rayleigh and Raman Scattering." In R. F. Barrow, D. A. Long, and J. Sheridan (eds.) *Specialist Periodical Report—Molecular Spectroscopy VI*, Chemical Society, London, 1979, pp. 136–173.

347. G. C. Tabisz, N. Meinander, and A. R. Penner. "Interaction Induced Rotational Light Scattering in Molecular Gases." In G. Birnbaum (ed.), *Phenomena Induced by Intermolecular Interactions*, Plenum Press, New York, 1985, pp. 345–358.

349. R. C. H. Tam and A. D. May. The collision induced contribution to the depolarized Raman spectrum of compressed HCl, CO, N_2 and CO_2. *Canad. J. Phys.*, *61*:1571–1578 (1983).

350. R. Vallauri and U. Balucani. Collision induced light scattering at intermediate density I. Integrated intensities. *Molec. Phys.*, *38*:1099–1113 (1979).

351. R. Vallauri and U. Balucani. Collision induced light scattering at intermediate densities II. Second moment. *Molec. Phys.*, *38*:1115–1125 (1979).

352. A. van der Avoird, P. E. S. Wormer, F. Mulder, and R. M. Berns. *Ab initio* studies of the

interactions of *van der Waals* molecules. *Topics in Current Chem. — van der Waals Systems*, *93*:1–51 (1980).

353. H. Versmold. Depolarized Rayleigh scattering: Molecular reorientation of CO_2 in a wide density range. *Molec. Phys.*, *43*:383–395 (1981).

354. J. van Kranendonk (ed.). *Intermolecular Spectroscopy and Dynamical Properties of Dense Systems—Proceedings of the International School of Physics "Enrico Fermi," Course LXXV*, North-Holland, Amsterdam, 1980.

355. V. Volterra, J. A. Bucaro, and T. A. Litovitz. Two mechanisms for depolarized light scattering from gaseous argon. *Phys. Rev. Lett.*, *26*:55–57 (1971).

356. H. Vu. "Recent progress in infrared, Raman and Rayleigh spectroscopy of molecular interactions in dense fluids." In *High-Pressure Science and Technology—Proceedings of 7th International AIRAPT Conference Le Creusot (France), 1979, pt. II*, Pergamon, Oxford, 1980, pp. 730–738.

357. M. Zoppi, F. Barocchi, D. Varshneya, M. Neumann, and T. A. Litovitz. Density dependence of the collision induced light scattering spectral moments of argon. *Canad. J. Phys.*, *59*:1475–1480 (1981).

358. M. Zoppi, M. Moraldi, F. Barocchi, R. Magli, and U. Bafile. Two-body depolarized CILS spectra of krypton and xenon at 295 K. *Chem. Phys. Lett.*, *83*:294–297 (1981).

II.1. *Condensed Phases—General Theory*

359. B. J. Alder, J. C. Beers II, H. L. Strauss, and J. J. Weis. Dielectric constant of atomic fluids with variable polarizability. *Proc. Nat. Acad. Science (U.S.A.)*, *77*:3098–3102 (1980).

360. M. P. Allan, J. P. Chappell, and D. Kivelson. A three-variable theory of flow birefringence and its relation to VH depolarized light scattering. *J. Chem. Phys.*, *74*:5942–5945 (1981).

361. T. Bancewicz. Molecular-statistical theory of the influence of molecular fields in liquids on the spectral distribution and intensity of Rayleigh scattered light in the approach of Racah algebra. *Acta Phys. Polonica, A*, *56*:431–438 (1979).

362. T. Bancewicz. Rayleigh light scattering by liquids composed of interacting anisotropic molecules: Spherical tensor approach within the second-order approximation of the DID model. *Molec. Phys.*, *50*:173–191 (1983).

363. T. Bancewicz, S. Kielich, and W. A. Steele. Interaction induced Rayleigh light scattering from molecular fluids by projection operator technique. *Molec. Phys.*, *54*:637–649 (1985).

364. F. Barocchi, M. Moraldi, and M. Zoppi. "Almost classical" many-body systems: Quantum mechanical corrections to the moments of a general spectrum. *Phys. Rev. A*, *26*:2168–2177 (1982).

365. D. R. Bauer, J. I. Brauman, and R. Pecora. "Depolarized Light Scattering from Liquids." In B. S. Rabinovitch (ed.), *Annual Review of Physical Chemistry*, Vol. 27, Annual Reviews, Palo Alto, Calif., 1976, pp. 443–463.

366. S. Bratos and B. Guillot. Theoretical investigation and experimental detection of rattling motions in atomic and molecular fluids. *J. Molec. Struc.*, *84*:195–203 (1982).

367. G. Breuer. Contribution to the time independent theory of Rayleigh scattering in liquids: II Second order approximation in Kielich's molecular scattering theory. *Molec. Phys.*, *45*:349–371 (1982).

368. J. V. Champion, G. H. Meeten, and C. D. Whittle. Electrooptic Kerr effect in *N*-alkane liquids. *Trans. Faraday Soc.*, *66*:2671–2680 (1970).

369. P. J. Chappell, M. P. Allen, R. I. Hallem, and D. Kivelson. Theory of depolarized light scattering. *J. Chem. Phys.*, *74*:5929–5949 (1981).

370. M. W. Evans. "Correlation and Memory Function Analysis of Molecular Motion in Fluids." In M. Davies (ed.), *A Specialist Periodic Report—Dielectric and Related Molecular Processes*, Vol. 3, Chemical Society, London, 1975, pp. 1–44.

371. B. U. Felderhof. On the propagation and scattering of light in fluids. *Physica*, *76*:486–502 (1974).

372. M. Fixman. Molecular theory of light scattering. *J. Chem. Phys.*, *23*:2074–2079 (1955).

373. P. A. Fleury and J. P. Boon. "Laser Light Scattering in Fluid Systems." In I. Prigogine and S. A. Rice (ed.) *Advances in Chemical Physics*, Vol. 24, Wiley-Interscience, New York, 1973, pp. 1–93.

374. P. A. Fleury. "Light Scattering Determinations of Four Point Correlation Functions." In J. Woods Halley (ed.), *Proceedings of the NATO Advanced Study Institute on Correlation Functions and Quasiparticle Interactions*, Plenum, New York, 1978, pp. 325–361.

375. H. L. Frisch and J. McKenna. Double scattering of electromagnetic radiation by a fluid. *Phys. Rev.*, *139*:A:68–A:77 (1965).

376. W. M. Gelbart. "Depolarized Light Scattering by Simple Fluids." In I. Prigogine and S. A. Rice (eds.), *Advances in Chemical Physics*, Vol. 26, Wiley-Interscience, New York, 1974, pp. 1–106.

377. W. M. Gelbart. "Lecture Notes of NATO Advanced Study Institute." In J. Woods Halley (ed.), *Correlation Functions and Quasi Particle Interactions in Condensed Matter*, Plenum Press, New York, 1978, p. 389.

378. N. D. Gershon, E. Zamir, and A. Ben-Reuven. Rayleigh wing scattering from liquids of anisotropic molecules. *Ber. Bunsen Ges. Phys. Chem.*, *75*:316–319 (1971).

379. R. W. Hellwarth. Theory of molecular light scattering spectra using the linear dipole approximation. *J. Chem. Phys.*, *52*:2128–2138 (1970).

380. T. Keyes, D. Kivelson, and J. P. McTague. Theory of k-independent depolarized Rayleigh wing scattering in liquids composed of anisotropic molecules. *J. Chem. Phys.*, *55*:4096–4100 (1971).

381. T. Keyes and B. M. Ladanyi. The role of local fields and interparticle pair correlations in light scattering by dense fluids: II. Depolarized spectra for non-spherical molecules. *Molec. Phys.*, *33*:1099–1107 (1977).

382. T. Keyes and B. M. Ladanyi. The role of local fields and interparticle pair correlation in light scattering by dense fluids: IV. Removal of the point-polarizability approximation. *Molec. Phys.*, *33*:1271–1285 (1977).

383. T. Keyes and B. M. Ladanyi. Can dielectric theories and measurements be used to predict depolarized light scattering intensities? *Molec. Phys.*, *34*:765–771 (1977).

384. T. Keyes. Depolarized light scattering by simple liquids. *J. Chem. Phys.*, *70*:5438–5441 (1979).

385. T. Keyes, B. M. Ladanyi, and P. A. Madden. Is depolarized light scattered from simple liquids mainly double-scattered? *Chem. Phys. Lett.*, *64*:479–484 (1979).

386. T. Keyes and B. M. Ladanyi. The relation of the Kerr effect to depolarized Rayleigh scattering. *Molec. Phys.*, *37*:1643–1647 (1979).

387. T. Keyes. A particle-hole model for light scattering by simple fluids. *Chem. Phys. Lett.*, *70*:194–198 (1980).

388. T. Keyes and P. A. Madden. Exact results for light scattering by dense systems of isotropic

particles: How good is the DID model? *Canad. J. Phys.*, *59*:1560–1562 (1981).

389. T. Keyes and B. M. Ladanyi. "The Internal Field Problem in Depolarized Light Scattering." In I. Prigogine and S. A. Rice (eds.) *Advances in Chemical Physics*, Vol. 56, Wiley, New York, 1984, pp. 411–465.

390. S. Kielich. A molecular theory of light scattering in gases and liquids. *Acta Phys. Polonica*, *19*:149–178 (1960).

391. S. Kielich. Role of molecular interactions in anisotropic light scattering by liquids. *J. Chem. Phys.*, *46*:4090–4099 (1967).

392. S. Kielich. Many-molecular correlation-induced anisotropic light scattering and birefringence in simple fluids. *Optics Communic.*, *4*:135–138 (1971)

393. S. Kielich. Depolarization of light scattering by atomic and molecular solutions with strongly anisotropic translational-orientational fluctuations. *Chem. Phys. Lett.*, *10*:516–521 (1971); Erratum, *19*:609 (1973).

394. D. Kivelson and P. A. Madden. Light scattering studies of molecular liquids. *Ann. Rev. Phys. Chem.*, *31*:523–558 (1980).

395. B. M. Ladanyi and T. Keyes. New method for the calculation of light-scattering intensities: Application to depolarized scattering from simple fluids. *Molec. Phys.*, *31*:1685–1701 (1976).

396. B. M. Ladanyi. "Local Fields in Liquids." In G. Birnbaum (ed.), *Phenomena Induced by Intermolecular Interactions*, Plenum Press, New York, 1985, pp. 497–523.

397. P. A. Madden. In J. Lascombe and P. V. Hong (eds.) *Raman Spectroscopy—Linear and Nonlinear*, Wiley, New York, 1982, pp. xxx–xxx.

398. G. D. Patterson. Rayleigh scattering in a dense medium. *J. Chem. Phys.*, *63*:4032–4034 (1975).

399. W. G. Rothschild. *Dynamics of Molecular Liquids*. Wiley, New York, 1984.

401. R. Samson, R. A. Pasmanter, and A. Ben-Reuven. Molecular theory of optical polarization and light scattering in dielectric fluids: I. Formal theory. *Phys. Rev. A*, *14*:1224–1237 (1976).

402. V. Sergiescu. On the usefulness of the lattice-gas model in the theory of the DID Rayleigh and Raman depolarized scattering by a fluid of isotropic molecules. *Physica C*, *97*:292–298 (1979).

403. W. A. Steele. A theoretical approach to the calculation of the time correlation functions of several variables. *Molec. Phys.*, *61*:1031–1043 (1987).

404. M. Tanaka. Molecular theory of scattering of light from liquids. *Progr. Theor Phys.*, *40*:975–988 (1968).

405. O. Theimer and R. Paul. Anisotropic light scattering by inner-field fluctuations in a dense, monatomic gas. *J. Chem. Phys.*, *42*:2508–2517 (1965).

406. J. van Kranendonk and J. E. Sipe. Foundations of the macroscopic electromagnetic theory of dielectric media. *Progr. Optics*, *XV*:245–350 (1977).

407. B. Vodar and Ph. Marteau (eds.). *High Pressure Science and Technology*. Pergamon, New York, 1980. Proceedings of the VIIth International AIRAPT Conference, France 1979.

408. J. Yvon. *Compt. Rend. Acad. Sci. Paris*, *202*:35–37 (1936).

409. J. Yvon. "Recherches sur la théorie cinétique des fluides: II. La propagation et la diffusion de la lumière." In *Actualités Scientifiques et Industrielles*, Hermann and Cie, Paris, 1937. No. 543; see the discussion by W. F. Brown, Jr., *J. Chem. Phys.*, *18*:1193 (1950).

410. J. Yvon. "Recherches sur la théorie cinétique des fluides." In *Actualités Scientifiques et Industrielles*, Hermann and Cie, Paris, 1937. No. 652.

II.2. *Liquids*

411. B. J. Alder, H. L. Strauss, and J. J. Weis. Studies in molecular dynamics: XII. Band shape of the depolarized light scattered from atomic fluids. *J. Chem. Phys.*, *59*:1002–1012 (1973).

412. B. J. Alder, J. J. Weis, and H. L. Strauss. Depolarization of light in atomic fluids. *Phys. Rev. A*, *7*:281–284 (1973).

413. B. J. Alder, H. L. Strauss, and J. J. Weis. Dielectric properties of fluids composed of spheres of constant polarizability. *J. Chem. Phys.*, *62*:2328–2334 (1975).

414. B. J. Alder, J. C. Beers II, H. L. Strauss, and J. J. Weis. Depolarized scattering of atomic fluids with variable polarizability. *J. Chem. Phys.*, *70*:4091–4094 (1979).

415. B. J. Alder and E. L. Pollock. Simulation of polar and polarizable fluids. *Ann. Rev. Phys. Chem.*, *32*:311–329 (1981).

416. R. D. Amos, A. D. Buckingham, and J. H. Williams. Theoretical studies of the collision induced Raman spectrum of CS_2. *Molec. Phys.*, *39*:1519–1526 (1980).

417. C. Andreani, P. Morales, and D. Rocca. On the dynamical properties of liquid hydrogen chloride: A light scattering experiment. *Molec. Phys.*, *44*:445–457 (1981).

418. H. Aoki, Y. Kakudate, S. Usuba, M. Yoshida, K. Tanaka, and S. Fujiwara. High pressure Raman study of liquid and crystalline C_2H_2. *J. Chem. Phys.*, *88*:4565 (1988).

419. U. Bafile, L. Ulivi, M. Zoppi, and F. Barocchi. Depolarized interaction induced light scattering in dense phases. *J. Chem. Soc. Faraday Trans. 2*, *83*:1751–1758 (1987).

420. U. Balucani, V. Tognetti, and R. Vallauri. Tow-body collision induced light scattering: Comparison between a memory function approach and molecular dynamics experiments. *Phys. Lett. A*, *64*:387–389 (1978).

421. U. Balucani, V. Tognetti, and R. Vallauri. Line shape theory and molecular dynamics in collision-induced light scattering. *Phys. Rev. A*, *19*:177–186 (1979).

422. U. Balucani and R. Vallauri. Collision induced light scattering at intermediate densities: II. The second frequency moment. *Molec. Phys.*, *38*:1115–1125 (1979).

423. F. Barocchi and J. P. McTague. Binary collision induced light scattering in liquid CH_4. *Optics Communic.*, *12*:202–204 (1974).

424. F. Barocchi, G. Spinelli, and M. Zoppi. Empirical polarizability anisotropy and the DILS spectrum in dense, gaseous and liquid argon. *Chem. Phys. Lett.*, *90*:22–26 (1982).

425. F. Barocchi, M. Zoppi, and M. Neumann. First-order quantum corrections to depolarized interaction induced light scattering spectral moments: Molecular dynamics calculation. *Phys. Rev. A*, *27*:1587–1593 (1983).

426. F. Barocchi, M. Neumann, and M. Zoppi. Molecular dynamics calculation of the quantum corrections to the pair distribution function for a Lennard-Jones fluid. *Phys. Rev. A*, *29*:1331–1334 (1984).

427. M. R. Battaglia, T. I. Cox, and P. A. Madden. The orientational correlation parameter for liquid CS_2, C_6H_6 and C_6F_6. *Molec. Phys.*, *37*:1413–1427 (1979).

428. B. J. Berne, M. Bishop, and A. Rahman. Depolarized light scattering from monatomic fluids. *J. Chem. Phys.*, *58*:2696–2698 (1973).

429. B. J. Berne and R. Pecora. Laser light scattering from liquids. *Ann. Rev. Phys. Chem.*, *25*:233–253 (1974).

430. J. Berrué, A. Chave, B. Dumon, and M. Thibeau. Influence de la nature de l'anisotropie induite par collision sur la diffusion collisionelle de la lumière dans les fluides denses. *Compt. Rend. Acad. Sci. Paris, ser. B*, *287*:269–272 (1978).

431. G. Birnbaum. "Collision Induced Vibrational Spectroscopy in Liquids." In S. Bratos and R. M. Pick (ed.) *Vibrational Spectroscopy of Molecular Liquids and Solids*, Plenum, New York, 1980, pp. 147–165.

432. J. Bruining and J. H. R. Clarke. Molecular orientation correlations and reorientational motions in liquid carbon monoxide, nitrogen and oxygen at 77 K: A Raman and Rayleigh light scattering study. *Molec. Phys.*, *31*:1425–1446 (1976).

433. J. A. Bucaro and T. A. Litovitz. Rayleigh scattering: Collisional motions in liquids. *J. Chem. Phys.*, *54*:3846–3853 (1971).

434. J. A. Bucaro and T. A. Litovitz. Molecular motions in CCl_4: Light scattering and infrared absorption. *J. Chem. Phys.*, *55*:3585–3588 (1971).

435. A. D. Buckingham. General Introduction—Faraday Symposium 22 on Interaction Induced Spectra in Dense Fluids and Disordered Solids. *J. Chem. Soc. Faraday Trans. 2*, *83*(10):1743–1939 (1987), University of Cambridge (England), held on Dec. 10–11, 1986.

436. R. A. J. Bunten, R. L. McGreevy, E. W. J. Mitchell, C. Raptis, and P. J. Walker. Collective modes in molten alkaline-earth chlorides: I. Light scattering. *J. Phys. Chem.*, *17*:4705–4724 (1984).

437. C. E. Campbell and F. J. Pinski. Theory of fourth-order structure and Raman scattering in liquid ^{4}He. *J. Phys. (Paris)*, *39*:C6.233–C6.234 (1978).

438. C. E. Campbell and F. J. Pinski. The structure of the ground state and low excited states of quantum fluids. *Nuc. Phys. A*, *328*:210–239 (1979).

439. C. E. Campbell, J. W. Halley, and F. J. Pinski. Integrated polarized light scattering from liquid ^{4}He. *Phys. Rev. B*, *21*:1323–1325 (1980).

440. P. S. Y. Cheung and J. G. Powles. The properties of liquid nitrogen: IV. A computer simulation. *Molec. Phys.*, *30*:921–949 (1975).

441. P. S. Y. Cheung and J. G. Powles. The properties of liquid nitrogen: V. Computer simulation with quadrupole interaction. *Molec. Phys.*, *32*:1383–1405 (1976).

442. J. H. R. Clarke and L. V. Woodcock. Light scattering from ionic liquids. *J. Chem. Phys.*, *57*:1006–1007 (1972).

443. J. H. R. Clarke. "Band Shapes and Molecular Dynamics in Liquids." In R. J. H. Clark and R. E. Hester (eds.), *Advances in Infrared and Raman Spectroscopy*, Vol. 4, Heyden, London, 1978, pp. 109–193.

444. J. H. R. Clarke and J. Bruining. The absolute intensity of depolarized light scattering from liquid argon. *Chem. Phys. Lett.*, *80*:42–44 (1981).

445. J. H. R. Clarke and L. V. Woodcock. Boundary problems in the calculations of light scattering intensities from liquids using computer simulation. *Chem. Phys. Lett.*, *78*:121–124 (1981).

446. J. H. R. Clarke and J. Bruining. The absolute intensity of depolarized light scattering from liquid argon. *Chem. Phys. Lett.*, *80*:42–44 (1982).

447. T. I. Cox and P. A. Madden. Collision induced Raman spectra of liquid CS_2. *Chem. Phys. Lett.*, *41*:188–191 (1976).

448. T. I. Cox, M. R. Battaglia, and P. A. Madden. Properties of liquid CS_2 from the allowed light scattering spectra. *Molec. Phys.*, *38*:1539–1554 (1979).

449. T. I. Cox and P. A. Madden. A comparative study of the interaction induced spectra of liquid CS_2: I. Intensities. *Molec. Phys.*, *39*:1487–1506 (1980).

450. T. I. Cox and P. A. Madden. A comparative study of the interaction induced spectra of liquid CS_2: III. Solutions. *Molec. Phys.*, *43*:307–319 (1981).

451. H. Dardy, V. Volterra, and T. A. Litovitz. Molecular motions in liquids: Comparison of light scattering and infra-red absorption. *Faraday Symposia Chem. Soc.*, *6*:71–81 (1972).

452. H. D. Dardy, V. Volterra, and T. A. Litovitz. Rayleigh scattering: Orientational motion in highly anisotropic liquids. *J. Chem. Phys.*, *59*:4491–4500 (1973).

453. A. De Santis, M. Nardone, M. Sampoli, P. Morales, and G. Signorelli. Raman spectra of fluid nitrogen: Intermolecular torques and orientational correlation times. *Molec. Phys.*, *39*:913–921 (1980).

454. A. De Santis, M. Nardone, and M. Sampoli. Correlation functions from depolarized Raman and Rayleigh spectra of N_2 at high density and 150 K. *Molec. Phys.*, *41*:769–777 (1980).

455. A. De Santis and M. Sampoli. Induced isotropic scattering from liquid carbon dioxide. *Chem. Phys. Lett.*, *96*:114–118 (1983).

456. A. De Santis and M. Sampoli. Induced contributions in isotropic and anisotropic Rayleigh spectra of fluid H_2S. *Chem. Phys. Lett.*, *102*:425–428 (1983).

457. A. De Santis, M. Sampoli, and R. Vallauri. Raman bands of fluid N_2: A molecular dynamics and experimental study. *Molec. Phys.*, *53*:695–715 (1984).

458. A. De Santis, R. Frattini, M. Sampoli, V. Mazzacurati, M. Nardone, and M. A. Ricci. Raman spectra of water in the translational and librational regions. I. Study of the depolarization ratios. *Molec. Phys.*, *61*:1199–1212 (1987).

459. I. M. de Schepper and A. A. van Well. Structural slowing down and depolarized light spectra in dense noble gas fluids. *J. Chem. Soc. Faraday Trans. 2*, *83*:1759–1764 (1987).

460. M. Fairbanks, R. L. McGreevy, and E. W. J. Mitchell. Depolarization ratios for Raman scattering from molten alkali and alkali-earth halides. *J. Phys. Chem.*, *19*:L53–L58 (1986).

461. P. A. Fleury and J. P. McTague. Effects of molecular interactions on light scattering by simple fluids. *Optics Communic.*, *1*:164–166 (1969).

462. D. Frenkel and J. P. McTague. Molecular dynamics study of orientational and collision-induced light scattering in molecular fluids. *J. Chem. Phys.*, *72*:2801–2818 (1980).

463. D. Frenkel. "Intermolecular Spectroscopy and Computer Simulations." In J. van Kranendonk (ed.), *Intermolecular Spectroscopy and Dynamical Properties of Dense Systems—Proceedings of the International School of Physics "Enrico Fermi," Course LXXV*, North-Holland, Amsterdam, 1980, pp. 156–201.

464. H. S. Gabelnick and H. L. Strauss. Low-frequency motions in liquid carbon-tetrachloride; II. The Raman spectrum. *J. Chem. Phys.*, *49*:2334–2338 (1968).

465. G. M. Gale, C. Flytzanis, and M. L. Geirnaert. "Time Domain Separation of Collision Induced and Allowed Raman Spectra." In G. Birnbaum (ed.), *Phenomena Induced by Intermolecular Interactions*, Plenum Press, New York, 1985, pp. 739–748.

466. Y. Garrabos, R. Tufeu, and B. Le Neindre. Étude expérimentale de la dépolarisation du rayonnement Rayleigh diffusé au voisinage du point critique du xenon. *Compt. Rend. Acad. Sci. Paris, ser. B*, *282*:313–316 (1976).

467. Y. Garrabos, R. Tufeu, and B. Le Neindre. Depolarized light scattered near the gas-liquid critical point of Xe, SF_6, CO_2, C_2H_4, and C_2H_6. *J. Chem. Phys.*, *68*:495–503 (1978).

468. L. C. Geiger and B. Ladanyi. Higher-order interaction-induced effects on Rayleigh light scattering by molecular liquids, part IV. *J. Chem. Phys.*, *87*:191–202 (1987).

469. J. Giergiel, P. C. Eklund, and K. R. Subbaswamy. Raman scattering from potassium iodide near the melting point. *Solid State Commun.*, *40*:139–143 (1981).

470. J. Giergiel, K. Subbaswamy, and P. C. Eklund. Light scattering from molten alkali halides. *Phys. Rev. B*, *29*:3490–3499 (1984).

471. E. B. Gill and D. Steele. The Raman spectrum of liquid chlorine—a study of the molecular motion. *Molec. Phys.*, *34*:231–239 (1977).

472. W. S. Gornall, H. E. Howard-Lock, and B. P. Stoicheff. Induced anisotropy and light scattering in liquids. *Phys. Rev. A*, *1*:1288–1290 (1970).

473. C. G. Gray and K. E. Gubbins. *Theory of Molecular Fluids*. Clarendon Press, Oxford, 1984.

474. M. A. Gray, T. M. Loehr, and P. A. Pincus. Depolarized Rayleigh-wing scattering in water and aqueous solutions. *J. Chem. Phys.*, *59*:1121–1127 (1973).

475. T. J. Greytak and J. Yan. Light scattering from rotons in liquid helium. *Phys. Rev. Lett.*, *22*:987–990 (1969).

476. T. J. Greytak, R. Woerner, J. Yan, and R. Benjamin. Experimental evidence for a two-phonon bound state in superfluid helium. *Phys. Rev. Lett.*, *25*:1547–1550 (1970).

477. M. Grossman, S. G. Elkomoss, and J. Ringeisen (eds.). *Molecular Spectroscopy of Dense Phases*, Elsevier, Amsterdam, 1979.

478. F. Guillaume, J. Yarwood, and A. H. Price. Infrared, Raman and microwave studies of the molecular dynamics and interactions in liquid benzonitrile. *Molec. Phys.*, *62*:1307–1321 (1987).

479. B. Guillot, S. Bratos, and G. Birnbaum. Theoretical study of spectra of depolarized light scattered from dense rare-gas fluids. *Phys. Rev. A*, *22*:2230–2237 (1980).

480. J. W. Halley. "Theory of Optical Processes in Liquid Helium." In G. B. Wright (ed.), *Light Scattering in Solids*, Springer, Berlin, 1969, p. 175.

481. F. E. Hanson and J. P. McTague. Raman studies of the orientational motions of small diatomics dissolved in argon. *J. Chem. Phys.*, *72*:1733–1740 (1980).

482. R. Hastings and J. W. Halley. Quantitative study of the Zawadowski–Ruvalds–Solona model of the dynamics of liquid helium. *Phys. Rev. A*, *10*:2488–2500 (1974).

483. G. Hauchecorne, F. Kerhervé, and G. Mayer. Mesure des interactions entre ondes lumineuses dans diverses substances. *J. Phys. (Paris)*, *32*:47–62 (1971).

484. B. Hegemann and J. Jonas. Reorientational motion, collision induced scattering, and vibrational relaxation in liquid carbonyl sulfide. *J. Chem. Phys.*, *79*:4683–4693 (1983).

485. B. Hegemann, K. Baker, and J. Jonas. Temperature and density effects on the collision induced depolarized Rayleigh line shapes of liquid carbon disulfide. *J. Chem. Phys.*, *80*:570–571 (1984).

486. B. Hegemann and J. Jonas. Temperature study of Rayleigh and Raman line shapes in liquid carbonyl sulfide. *J. Phys. Chem.*, *88*:5851–5855 (1984).

487. B. Hegemann and J. Jonas. Separation of temperature and density effects on collision induced Rayleigh and Raman line shapes of liquid carbon disulfide. *J. Chem. Phys.*, *82*:2845–2855 (1985).

488. J. H. K. Ho and G. C. Tabisz. Collision induced light scattering in liquids and the binary collision model. *Canad. J. Phys.*, *51*:2025–2031 (1973).

489. W. Holzer and R. Ouillon. Collision induced Raman scattering in CO_2 at the frequency of the ν_2 inactive vibrations. *Molec. Phys.*, *36*:817–826 (1978).

490. H. E. Howard-Lock and R. S. Taylor. "Induced Anisotropy and Light Scattering in Liquids." In *Advances in Raman Spectroscopy*, Heyden and Son, London, 1972, p. x.

491. H. E. Howard-Lock and R. S. Taylor. Induced anisotropy and light scattering in liquids II. *Canad. J. Phys.*, *52*:2436–2444 (1974).

492. F. Iwamoto. Raman scattering in liquid helium. *Progr. Theor. Phys.*, *44*:1121–1134 (1970).

493. D. A. Jackson and B. Simic-Glavaski. Study of depolarized Rayleigh scattering in liquid benzene derivatives. *Molec. Phys.*, *18*:393–400 (1970).

494. J. Jonas. Pressure effects on the dynamic structure of liquids. *Acc. Chem. Res.*, *17*:74–80 (1984).

495. J. Jonas. "Pressure—An Essential Experimental Variable in Spectroscopic Studies of Liquids." In G. Birnbaum (ed.), *Phenomena Induced by Intermolecular Interactions*, Plenum Press, New York, 1985, pp. 525–547.

496. J. Jonas. Density effects on collision induced spectra in fluids. *J. Chem. Soc. Faraday Trans. 2*, *83*:1777–1789 (1987).

497. K. Jurkowska and R. Mierzecki. The influence of intermolecular interactions on the invariants of the Raman scattering tensors, II. The $CDCl_3$ tensors in solutions with acetone, dioxane, benzene and mesythylene. *Acta Phys. Polonica, A*, *63*:701–706 (1983).

498. K. Jurkowska and R. Mierzecki. The influence of intermolecular interactions on the invariants of the Raman scattering tensors, III. The $CDCl_3$ tensors in solution with pyridine, diethylamine and triethylamine. *Acta Phys. Polonica, A*, *67*:649–652 (1985).

499. R. A. J. Keijser, M. Jansen, V. G. Cooper, and H. F. P. Knaap. Depolarized Rayleigh scattering in carbon dioxide, carbon oxysulfide and carbon disulfide. *Physica*, *51*:593–600 (1971).

500. T. Keyes and B. M. Ladanyi. Light scattering from two-component systems: an analysis of the dilution experiment. *Molec. Phys.*, *38*:605–610 (1979).

501. T. Keyes, G. Seeley, P. Weakliem, and T. Ohtsuki. Collision induced light scattering from growing clusters: Depolarization by fractals. *J. Chem. Soc. Faraday Trans. 2*, *83*:1859–1866 (1987).

502. S. Kielich, J. R. Lalanne, and F. B. Martin. Apport de la diffusion Rayleigh dépolarisée et de l'effet Kerr optique a l'étude des corrélations radiales et d'orientation dans quelques liquides. *J. Phys. (Paris)*, *33*:C1:191–C1:205 (1972).

503. P. Kleban and R. Hastings. Excluded volume conditions and Raman scattering in He II. *Phys. Rev. B*, *11*:1878–1883 (1975).

504. D. H. Kobe and S. T. Cheng. Light scattering from bound roton pairs in He II. *J. Low-Temp. Phys.*, *19*:379–396 (1975).

505. F. Kohler, E. Wilhelm, and H. Posch. Recent advances in the physical chemistry of the liquid state. *Adv. Molec. Relaxation Processes (Netherlands)*, *8*:195–239 (1976).

506. B. M. Ladanyi and T. Keyes. The role of local fields and interparticle pair correlations in light scattering by dense fluids: I. Depolarized intensities due to orientational fluctuations. *Molec. Phys.*, *33*:1063–1097 (1977).

507. B. M. Ladanyi and T. Keyes. The intensity of light scattered by liquid CS_2. *J. Chem. Phys.*, *68*:3217–3221 (1978).

508. B. M. Ladanyi, T. Keyes, D. J. Tildesley, and W. B. Streelt. Structure and equilibrium optical properties of liquid CS_2. *Molec. Phys.*, *39*:645–659 (1980).

509. B. M. Ladanyi. Molecular dynamics study of Rayleigh light scattering from molecular fluids. *J. Chem. Phys.*, *78*:2189–2203 (1983).

510. B. M. Ladanyi and N. E. Levinger. Computer simulation of Raman scattering from molecular fluids. *J. Chem. Phys.*, *81*:2620–2633 (1984).

511. B. M. Ladanyi. Higher-order interaction-induced effects on depolarized light scattering from fluids of optically anisotropic molecules. *Chem. Phys. Lett.*, *121*:351–355 (1985).

512. B. M. Ladanyi, A. Barreau, A. Chave, B. Dumon, and M. Thibeau. Collision induced light

scattering by fluids of optically isotropic molecules: Comparison of results of two model studies. *Phys. Rev. A*, *34*:4120–4130 (1986).

513. B. M. Ladanyi, L. C. Geiger, T. W. Zerda, X. Song, and J. Jonas. Experimental and molecular dynamics study of the pressure dependence of Raman spectra of oxygen. *J. Chem. Phys.*, *89*:660–672 (1988).

514. A. J. C. Ladd, T. A. Litovitz, and C. J. Montrose. Molecular dynamics studies of depolarized light scattering from argon at various fluid densities. *J. Chem. Phys.*, *71*:4242–4248 (1979).

515. A. J. C. Ladd, T. A. Litovitz, J. H. R. Clarke, and L. V. Woodcock. Molecular dynamics simulations of depolarized Rayleigh scattering from liquid argon at the triple point. *J. Chem. Phys.*, *72*:1759–1763 (1980).

516. P. Lallemand. Comparison of depolarized Raman and Rayleigh lines scattered by chloroform. *Compt. Rend. Acad. Sci. Paris, ser. B*, *72*:429–432 (1971).

517. P. Lallemand. "Étude des mouvements moléculaires par diffusion de la lumière." In J. Lascombe (ed.), *Molecular Motion in Liquids*, D. Reidel, Dordrecht, 1974, pp. 517–534.

518. H. Langer and H. Versmold. Depolarized Rayleigh scattering: Orientational correlation functions of acetonitrile and carbon disulfide. *Ber. Bunsen Ges. Phys. Chem.*, *83*:510–517 (1979).

519. Y. Le Duff and A. Gharbi. Diffusion collisionnelle induite pour les bandes Rayleigh et ν_1 Raman du liquide isotopique $C^{35}Cl_4$. *Optics Communic.*, *30*:369–372 (1979).

520. M. D. Levenson and A. L. Schawlow. Depolarized light scattered from liquid bromine. *Optics Communic.*, *2*:192–195 (1970).

521. T. A. Litovitz, C. J. Montrose, R. A. Stuckart, and T. G. Copeland. Collision induced far infrared absorption and depolarized light scattering in liquids. *Molec. Phys.*, *34*:573–578 (1977).

522. T. A. Litovitz and C. J. Montrose. "Interaction Induced Light Scattering in Liquids: Molecular Motion and Structural Relaxation." In J. van Kranendonk (ed.), *Intermolecular Spectroscopy and Dynamical Properties of Dense Systems—Proceedings of the International School of Physics "Enrico Fermi," Course LXXV*, North-Holland, Amsterdam, 1980, pp. 307–324.

523. H. C. Lucas, D. A. Jackson, J. G. Powles, and B. Simic-Glavaski. Temperature variation of polarized and depolarized scattered light spectra from liquid benzene derivatives. *Molec. Phys.*, *18*:505–521 (1970).

524. H. C. Lucas and D. A. Jackson. The intensity and spectra of depolarized light scattered from benzene nitrobenzene mixtures. *Molec. Phys.*, *20*:801–810 (1971).

525. P. A. Lund, O. F. Nielsen, and E. Praestgaard. Comparison of depolarized Rayleigh-wing scattering and far-infrared absorption in molecular liquids. *Chem. Phys.*, *28*:167–173 (1978).

526. P. A. Madden. The line shape of the depolarized Rayleigh scattering from liquid argon. *Chem. Phys. Lett.*, *47*:174–178 (1977).

527. P. A. Madden. The depolarized Rayleigh scattering from fluids of spherical molecules. *Molec. Phys.*, *36*:365–388 (1978).

528. P. A. Madden. Light scattering studies of the dynamics and structure of liquids. *Phil. Trans. Royal Soc. (London), A*, *293*:419–428 (1979).

529. P. A. Madden and T. I. Cox. A comparative study of the interaction induced spectra of liquid CS_2: II. Line shapes. *Molec. Phys.*, *43*:287–305 (1981).

530. P. A. Madden and D. J. Tildesley. The interaction induced spectra of liquid CS_2: A computer simulation study. *Molec. Phys.*, *49*:193–219 (1983).

531. P. A. Madden. "Interaction Induced Phenomena in Molecular Liquids." In A. J. Barnes, W. J. Orville-Thomas, and J. Yarwood (eds.), *Molecular Liquids*, D. Reidel, Dordrecht, 1984.

532. P. A. Madden. "Interaction-induced Subpicosecond Phenomena in Liquids." In D. H. Auston and K. B. Eisenthal (eds.), *Ultrafast Phenomena IV*, Springer-Verlag, Berlin, 1984, pp. 244–257.

533. P. A. Madden. "Interaction Induced Vibrational Spectra in Liquids." In G. Birnbaum (ed.), *Phenomena Induced by Intermolecular Interactions*, Plenum Press, New York, 1985, pp. 399–413.

534. P. A. Madden. "Interference of Molecular and Interaction Induced Effects in Liquids." In G. Birnbaum (ed.), *Phenomena Induced by Intermolecular Interactions*, Plenum Press, New York, 1985, pp. 643–659.

535. P. A. Madden. "The Interference of Allowed and Induced Molecular Moments in Liquids: A Workshop Report." In G. Birnbaum (ed.), *Phenomena Induced by Intermolecular Interactions*, Plenum Press, New York, 1985, pp. 695–697.

536. P. A. Madden and D. J. Tildesley. Interaction induced contributions to Rayleigh and allowed Raman bands: A simulation study of CS_2. *Molec. Phys.*, *55*:969–998 (1985).

537. P. A. Madden and T. I. Cox. Interaction induced spectra of CO_2. *Molec. Phys.*, *56*:223–235 (1985).

538. P. A. Madden and R. W. Impey. On the infrared and Raman spectra of water in the region 5–250 cm^{-1}. *Chem. Phys. Lett.*, *123*:502–506 (1986).

539. J. R. Magana and J. S. Lannin. Role of density in Raman scattering of iodine. *Phys. Rev. B*, *37*:2475–2482 (1988).

540. G. D. Mahan. Light scattering from hard-sphere fluids. *Phys. Lett. A*, *44*:287–288 (1973).

541. M. S. Malmberg and E. R. Lippincott. Evidence of molecular interactions in Rayleigh light scattering. *J. Colloid Interface Sci.*, *27*:591–607 (1969).

542. V. Mazzacurati, M. A. Ricci, G. Ruocco, and M. Nardone. Isotropic induced scattering in liquid H_2S. *Molec. Phys.*, *50*:1083–1087 (1983).

543. R. L. McGreevy. Interaction induced spectra of molten alkali-metal and alkaline-earth halides. *J. Chem. Soc. Faraday Trans. 2*, *83*:1875–1889 (1987).

544. J. P. McTague, P. A. Fleury, and D. B. Du Pré. Intermolecular light scattering in liquids. *Phys. Rev.*, *188*:303–308 (1969).

545. R. Mierzecki. Influence of molecular interactions on the isotropy of molecules as studied by Raman scattering. *J. Molec. Struc.*, *47*:53–58 (1978).

546. E. W. J. Mitchell and C. Raptis. Raman scattering from molten alkali halides. *J. Phys. Chem.*, *16*:2973–2985 (1983).

547. C. J. Montrose, J. A. Bucaro, J. Marshall-Coakley, and T. A. Litovitz. Depolarized Rayleigh scattering and hydrogen bonding in liquid water. *J. Chem. Phys.*, *60*:5025–5029 (1974).

548. C. J. Montrose, T. G. Copeland, T. A. Litovitz, and R. A. Stuckart. Collision induced far infrared absorption and depolarized light scattering in liquid water. *Molec. Phys.*, *34*:573–578 (1977).

549. N. E. Moulton, G. H. Watson, Jr., W. B. Daniels, and D. M. Brown. Raman scattering from fluid hydrogen to 2500 amagats. *J. Chem. Phys.* *89*:in press (1988).

550. R. D. Mountain. Temperature dependence of depolarized scattered light near the critical point. *J. Phys. (Paris)*, *33*:Cl:265–Cl:268 (1972).

551. W. F. Murphy, M. V. Evans, and P. Bender. Measurement of the depolarization ratio for the

A_1 Raman line of carbon tetrachloride in binary mixtures. *J. Chem. Phys.*, *47*:1836–1839 (1967).

552. C. S. Murthy, R. Vallauri, H. Versmold, U. Zimmermann, and K. Singer. Depolarized Rayleigh scattering from CO_2: An experimental and molecular dynamics investigation. *Ber. Bunsen Ges. Phys. Chem.*, *89*:18–20 (1985).
553. S. Nakajima. Liquid helium I. *Butsuri (Japan)*, *26*:736–745 (1971); see also part II, *26*:897–905 (1971).
554. S. Nakajima. Elementary quantum theory of light in liquid helium. *Progr. Theor. Phys.*, *45*:353–364 (1971).
555. P. T. Nikolaenko and A. I. Frorvin. Oscillation of liquid molecules and the wing of the Rayleigh line. *Izv. Vuz. Fiz.*, *8*:107–111 (1967) [in Russian].
556. D. W. Oxtoby and W. M. Gelbart. Depolarized light scattering near the gas-liquid critical point. *J. Chem. Phys.*, *60*:3359–3367 (1974).
557. D. W. Oxtoby. Remarks on the presentation by Stuckart, Montrose and Litovitz. *Faraday Symposia Chem. Soc.*, *11*:173–174 (1977).
558. D. W. Oxtoby. "Vibrational Spectral Line Shapes of Charge Transfer Complexes." In G. Birnbaum (ed.), *Phenomena Induced by Intermolecular Interactions*, Plenum Press, New York, 1985, pp. 715–725.
559. R. A. Pasmanter, R. Samson, and A. Ben-Reuven. Light scattering by dense dielectric media. *Chem. Phys. Lett.*, *16*:470–472 (1972).
560. G. D. Patterson and P. J. Flory. Depolarized Rayleigh scattering and the mean-squared optical anisotropies of n-alkalenes in solution. *J. Chem. Soc. Faraday Trans. 2*, *68*:1098–1110 (1972).
561. G. D. Patterson and P. J. Flory. Optical anisotropies of polyethylene oligomers. *J. Chem. Soc. Faraday Trans. 2*, *68*:1111–1116 (1972).
562. R. Pecora. Laser light scattering from macromolecules. *Ann. Rev. Biophys. Bioeng.*, *1*:257 (1972).
563. M. Perrot, M. Bouachir, and J. Lascombe. Orientational and structural dynamics of liquid allene at various temperatures from depolarized light scattering. *Molec. Phys.*, *42*:551–564 (1981).
564. M. Perrot, M. Besnard, J. Lascombe, and M. Bouachir. Orientational and induced depolarized Rayleigh scattering of some anisotropic molecules in their liquid state. *Canad. J. Phys.*, *59*:1481–1486 (1981).
565. E. R. Pike and J. M. Vaugham. High frequency depolarized light scattering from liquid helium I and helium gas. *J. Phys. Chem.*, *4*:L362–L366 (1971).
566. H. A. Posch and T. A. Litovitz. Depolarized light scattering in liquid SF_6. *Molec. Phys.*, *32*:1559–1575 (1976).
567. H. A. Posch. Kollisionsinduzierte Lichtstreuung an einfachen atomaren und molekularen Flüssigkeiten. *Acta Phys. Austriaca, Suppl. XX* :157–166 (1979).
568. H. A. Posch. Collision induced light scattering from fluids composed of tetrahedral molecules I. *Molec. Phys.*, *37*:1059–1075 (1979).
569. H. A. Posch. Collision induced light scattering from fluids composed of tetrahedral molecules: II. Intensities. *Molec. Phys.*, *40*:1137–1152 (1980).
570. H. A. Posch, F. Vesely, and W. Steele. Atomic pair dynamics in a Lennard-Jones fluid: Comparison of theory with computer simulation. *Molec. Phys.*, *44*:241–264 (1981).

571. H. A. Posch, U. Balucani, and R. Vallauri. On the relative dynamics of pairs of atoms in simple liquids. *Physica A*, *123*:516–534 (1984).

572. C. Raptis, R. A. J. Bunten, and E. W. J. Mitchell. Raman scattering from molten alkali iodides. *J. Phys. Chem.*, *16*:5351–5362 (1983).

573. L. A. Reith and H. L. Swinney. Depolarized light scattering due to double scattering. *Phys. Rev. A*, *12*:1094–1105 (1975).

574. W. G. Rothschild. "Bandshapes and Dynamics in Liquids." In J. R. Durig (ed.), *Vibrational Spectra and Structure*, Elsevier, Amsterdam, 1986, Chapter 2.

575. P. E. Schoen, P. S. Y. Cheung, D. A. Jackson, and J. G. Powles. The properties of liquid nitrogen: III. The light scattering spectrum. *Molec. Phys.*, *29*:1197–1220 (1975).

576. J. Schroeder and J. Jonas. Density effects on depolarized Raman scattering in liquids. *Chem. Phys.*, *34*:11–16 (1978).

577. J. Schroeder, V. H. Schiemann, and J. Jonas. Raman study of molecular reorientation in liquid chloroform and chloroform-d under high pressure. *J. Chem. Phys.*, *69*:5479–5488 (1978).

578. M. Schwartz and C. H. Wang. Temperature dependent study of Rayleigh wing scattering in liquid acetonitrile. *Chem. Phys. Lett.*, *29*:383–388 (1974).

579. A. V. Sechkarev and P. T. Nikolaenko. Investigation of intermolecular dynamics in the condensed states of matter by the method of vibrational spectroscopy, Part 1: Intensity distribution and intermolecular light scattering spectra in the neighborhood of the Rayleigh line. *Izv. Vuz. Fiz.* (*U.S.S.R.*), *4*:104–110 (1969).

580. H. K. Shin. Collision induced light scattering in liquids. *J. Chem. Phys.*, *56*:2617–2622 (1972).

581. B. Simic-Glavaski, D. A. Jackson, and J. G. Powles. Low frequency Raman lines in the depolarized wing of liquid toluene and benzene. *Phys. Lett. A*, *32*:329–330 (1970).

582. B. Simic-Glavaski, D. A. Jackson, and J. G. Powles. Cabannes-Daure effect in liquids. *Phys. Lett. A*, *34*:255–256 (1971).

583. B. Simic-Glavaski and D. A. Jackson. Rayleigh depolarized light scattered from isotropic and anisotropic molecular liquids. *J. Phys.* (*Paris*), *33*:Cl:183–Cl:189 (1972).

584. K. Singer, J. V. L. Singer, and A. J. Taylor. Molecular dynamics of liquids modelled by "2-Lennard-Jones centres" pair potentials: II. Translational and rotational autocorrelation functions. *Molec. Phys.*, *37*:1239–1262 (1979).

585. R. E. Slusher and C. M. Surko. Raman scattering from condensed phases of He^3 and He^4. *Phys. Rev. Lett.*, *27*:1699–1702 (1971).

586. W. A. Steele. "Molecular Reorientation in Dense Systems." In J. van Kranendonk (ed.), *Intermolecular Spectroscopy and Dynamical Properties of Dense Systems—Proceedings of the International School of Physics "Enrico Fermi," Course LXXV*, North-Holland, Amsterdam, 1980, pp. 325–374.

587. W. A. Steele and H. A. Posch. "Liquids and Liquid State Interactions: A Workshop Report." In G. Birnbaum (ed.), *Phenomena Induced by Intermolecular Interactions*, Plenum Press, New York, 1985, pp. 549–556.

588. W. A. Steele and R. Vallauri. Computer simulations of pair dynamics in molecular fluids. *Molec. Phys.*, *61*:1019–1030 (1987).

589. M. J. Stephen. Raman scattering in liquid helium. *Phys. Rev.*, *187*:279–285 (1969).

590. B. P. Stoicheff. "Brillouin and Raman Spectroscopy with Lasers." In M. S. Feld and M. A. Kurnit, (eds.), *Fundamentals of Laser Physics*, Wiley, New York, 1974, pp. 573–611.

591. H. L. Strauss. "The Use of Quasi-elastic Light Scattering for the Determination of the Collective Properties of Molecules." In C. B. Moore (ed.), *Chemical and Biochemical Applications of Lasers*, Academic Press, New York, 1974, pp. 281–307.

592. W. B. Streelt and D. J. Tildesley. Computer simulation of polyatomic molecules: II. Molecular dynamics study of diatomic liquids with atom-atom and quadrupole-quadrupole potentials. *Proc. Roy. Soc. (London), A, 355*:239–266 (1977).

593. R. A. Stuckart, C. J. Montrose, and T. A. Litovitz. "Comparison of Interaction Induced Light Scattering and Infrared Absorption in Liquids." In *Faraday Symposium of Chemical Society: Newer Aspects of Molecular Relaxation Processes, No. 11*, Chemical Society, London, 1977, pp. 94–105.

594. C. M. Surko and R. E. Slusher. Two-roton Raman scattering in He^3–He^4 solutions. *Phys. Rev. Lett., 30*:1111–1114 (1973).

595. G. C. Tabisz, W. R. Wall, and D. P. Shelton. Collision induced light scattering from liquid CCl_4 and C_6H_{12}. *Chem. Phys. Lett., 15*:387–391 (1972).

596. G. C. Tabisz, W. R. Wall, D. P. Shelton, and J. Ho. "The Spectral Profile of Collision Induced Light Scattering from some Molecular Liquids." In *Advances in Raman Spectroscopy*, Heyden & Son, London, 1972, pp. 466–471.

597. H. N. V. Temperley. "Liquid State Physics." In J. van Kranendonk (ed.), *Intramolecular Spectroscopy and Dynamical Properties of Dense Systems—Proceedings of the International School of Physics "Enrico Fermi," Course LXXV*, North-Holland, Amsterdam, 1980, pp. 393–398.

598. D. J. Tildesley and P. A. Madden. Time correlation functions for a model of liquid carbon disulfide. *Molec. Phys., 48*:129–152 (1983).

599. U. M. Titulaer and J. M. Deutch. Dielectric model of roton interactions in superfluid helium. *Phys. Rev. A, 10*:1345–1354 (1974).

600. V. Tognetti and R. Vallauri. Molecular dynamics studies and memory-function approach to collision-induced light scattering in the low-density limit. *Phys. Rev. A, 20*:2634–2637 (1979).

601. J. H. Topalian, J. F. Maguire, and J. P. McTague. Liquid Cl_2 dynamics studies by depolarized Raman and Rayleigh scattering. *J. Chem. Phys., 71*:1884–1888 (1979).

602. R. Vallauri, Balucani U, and U. Tognetti. Lineshape theory and molecular dynamics in collision induced light scattering. *Phys. Rev. A, 19*:177–186 (1979).

603. R. Vallauri. "Molecular Dynamics Studies of Interaction Induced Absorption and Light Scattering in Diatomic Systems." In G. Birnbaum (ed.), *Phenomena Induced by Intermolecular Interactions*, Plenum Press, New York, 1985, pp. 457–473.

604. J. van der Elsken and R. A. Huijts. Density dependence of the depolarized light scattering spectrum of xenon. *J. Chem. Phys., 88*:3007–3015 (1988).

605. P. van Konynenburg and W. A. Steele. Angular time-correlation functions from spectra for some molecular liquids. *J. Chem. Phys., 56*:4776–4787 (1972).

606. P. van Konynenburg and W. A. Steele. Molecular rotation in some simple fluids. *J. Chem. Phys., 62*:2301–2311 (1975).

607. D. Varshneya, S. F. Shirron, T. A. Litovitz, M. Zoppi, and F. Barocchi. Collision induced light scattering: Integrated intensity of argon. *Phys. Rev. A, 23*:77–86 (1981).

608. A. A. van Well, I. M. de Schepper, P. Verkerk, and R. A. Huijts. Depolarized light scattering and neutron scattering spectra for noble gas fluids. *J. Chem. Phys., 87*:687–696 (1987).

609. J. Vermesse, D. Levesque, and J. J. Weis. Influence of the potential on the intensity of depolarized light scattering in liquid argon. *Chem. Phys. Lett., 85*:120–122 (1982).

610. H. Versmold and U. Zimmermann. Depolarized Rayleigh scattering: Temperature dependence of molecular reorientation of CO_2 at constant density. *Molec. Phys.*, *50*:65–75 (1983).

611. H. Versmold and U. Zimmermann. Density dependence of interaction-induced scattering: Contributions to the depolarized Rayleigh band of ethane. *J. Chem. Soc. Faraday Trans. 2*, *83*:1815–1824 (1987).

612. J. L. Viovy, G. M. Searby, F. Fried, M. J. Vellutini, and M. J. Sixou. Comportement non lorentzien de la raie Rayleigh deépolarisée dans quelques liquides. Discussion des théories "á trois vaiables." *Molec. Phys.*, *38*:1275–1299 (1979).

613. V. Volterra, J. A. Bucaro, and T. A. Litovitz. Molecular motion and light scattering in liquids. *Ber. Bunsen Ges. Phys. Chem.*, *75*:309–315 (1971).

614. C. H. Wang and R. B. Wright. Is an exponential spectral shape a criterion for collision induced light scattering in hydrogen bonded molecular fluids? *Chem. Phys. Lett.*, *11*:277–280 (1971).

615. G. H. Watson, Jr. Raman Scattering from Solid and Fluid Helium at High Pressure. Ph.D. Dissertation, Physics Department, University of Delaware, Newark, Del., 1984.

616. J. J. Weis and B. J. Alder. Effect of multiple scattering on the intensity of depolarized light. *Chem. Phys. Lett.*, *81*:113–114 (1981).

617. R. B. Wright, M. Schwartz, and C. H. Wang. Temperature dependent Raman study of molecular motions and interactions of CH_3I in the liquid phase. *J. Chem. Phys.*, *58*:5125–5134 (1973).

618. R. B. Wright and C. H. Wang. Effect of density on the Raman scattering of molecular fluids: II. Study of intermolecular interaction in CO_2. *J. Chem. Phys.*, *61*:2707–2710 (1974).

619. A. Zawadowski, J. Ruvalds, and J. Solana. Bound roton pairs in superfluid helium. *Phys. Rev. A*, *5*:399–421 (1972).

620. T. W. Zerda, S. Perry, and J. Jonas. Pressure and temperature Raman study of the rotational motion of propyne. *Chem. Phys. Lett.*, *83*:600–604 (1981).

621. T. Zerda, J. Schroeder, and J. Jonas. Raman band shape and dynamics of molecular motion of SF_6 in the supercritical dense fluid region. *J. Chem. Phys.*, *75*:1612–1622 (1981).

622. T. W. Zerda, X. Song, and J. Jonas. Temperature and density study of the Rayleigh line shape of fluid N_2O. *J. Phys. Chem.*, *90*:771–774 (1986).

623. T. W. Zerda, X. Song, J. Jonas, B. Ladanyi, and L. Geiger. Experimental and molecular dynamics study of depolarized Rayleigh scattering by O_2. *J. Chem. Phys.*, *87*:840–851 (1987).

624. M. Zoppi and G. Spinelli. Interaction induced translational Raman scattering of liquid argon: The spectral moments. *Phys. Rev. A*, *33*:939–945 (1986).

II.3. *Solids*

625. B. J. Alder, H. L. Strauss, J. J. Weis, J. P. Hansen, and M. L. Klein. A molecular dynamics study of the intensity and band shape of depolarized light scattered from rare-gas crystals. *Physica B*, *83*:249–258 (1976).

626. P. J. Berkhout and I. F. Silvera. Mixing of rotational states, breakdown of the independent polarizability approximation and renormalized interactions in the solid hydrogens under pressure. *Communic. Phys.*, *2*:109–114 (1977).

627. G. Briganti, V. Mazzacurati, G. Signorelli, and M. Nardone. Interaction induced light scattering in orientationally disordered crystals: The translational phonon region. *Molec. Phys.*, *43*:1347–1356 (1981).

628. E. Cazzanelli, A. Fontana, G. Mariotto, V. Mazzacurati, G. Ruocco, and G. Signorelli. Analysis of the Raman spectral shape in α–AgI. *Phys. Rev. B*, *28*:7269–7276 (1983).

629. T. I. Cox and P. A. Madden. The forbidden Raman ν_2, π_u transition in crystalline CS_2. *Chem. Phys. Lett.*, *77*:511–513 (1981).

630. R. K. Crawford, D. G. Bruns, D. A. Gallagher, and M. V. Klein. Raman scattering from condensed argon. *Phys. Rev. B*, *17*:4871–4883 (1978).

631. S. Cunsolo and G. Signorelli. "Some Considerations on Spectra Induced by Intermolecular interactions in molecular solids and amorphous systems: A workshop report." In G. Birnbaum (ed.), *Phenomena Induced by Intermolecular Interactions*, Plenum Press, New York, 1985, pp. 609–611.

632. P. Depondt, M. Debeau, and R. M. Pick. Rotational motion in plastic neopentane by Raman spectroscopy. *J. Chem. Phys.*, *77*:2779–2785 (1982).

633. P. A. Fleury and J. P. McTague. Short-wavelength collective excitation in liquid and solid hydrogen. *Phys. Rev. Lett.*, *31*:914–918 (1973).

634. P. A. Fleury, J. M. Worlock, and H. L. Carter. Molecular dynamics by light scattering in the condensed phases of Ar, Kr, Xe. *Phys. Rev. Lett.*, *30*:591–594 (1973).

635. P. A. Fleury and J. P. McTague. Molecular interactions in the condensed phases of ortho-para hydrogen mixtures. *Phys. Rev. A*, *12*:317–326 (1975).

636. P. A. Fleury. "Spectroscopy of Collective Pair Excitations." In G. K. Horton and A. A. Maradudin (ed.) *Dynamical Properties of Solids*, North-Holland, Amsterdam, 1980, pp. 197–244.

637. R. Folland, D. A. Jackson, and S. Rajagopal. Light scattering in plastic crystals: II. Raman and Rayleigh scattering studies of molecular reorientation in norbornylene. *Molec. Phys.*, *30*:1063–1071 (1975).

638. W. N. Hardy, I. F. Silvera, and J. P. McTague. Raman scattering in oriented crystals of paradeuterium and orthohydrogen. *Phys. Rev. B*, *12*:753–789 (1975).

639. D. D. Klug and E. Whalley. The relation between simple analytic models for optical spectra of disordered solids. *J. Chem. Phys.*, *71*:2903–2910 (1979).

640. K. B. Lyons, P. A. Fleury, and H. L. Carter. Two-phonon difference scattering in solid xenon. *Phys. Rev.*, *21*:1653–1657 (1980).

641. P. A. Madden and J. A. Board. Light scattering by liquid and solid sodium chloride. *J. Chem. Soc. Faraday Trans. 2*, *83*:1891–1908 (1987).

642. M. Marchi, J. S. Tse, and M. L. Klein. Infrared and Raman spectra of hexagonal ice in the lattice-mode region. *J. Chem. Soc. Faraday Trans. 2*, *83*:1867–1874 (1987).

643. G. Mariotto, A. Fontana, E. Cazzanelli, and M. P. Fontana. Temperature dependence of Raman scattering and partial disorder in AgI crystals. *Phys. Stat. Solidi B*, *101*:341–351 (1981).

644. G. Mariotto, A. Fontana, E. Cazzanelli, F. Rocca, M. P. Fontana, V. Mazzacurati, and G. Signorelli. Temperature dependence of the Raman depolarization ratio in α–AgI. *Phys. Rev. B*, *23*:4782–4783 (1981).

645. V. Mazzacurati, M. Nardone, and G. Signorelli. Light scattering from disordered media. *Molec. Phys.*, *38*:1379–1391 (1979).

646. V. Mazzacurati, C. Pona, G. Signorelli, G. Briganti, M. A. Ricci, E. Mazzega, M. Nardone, A. De Santis, and M. Sampoli. Interaction induced light scattering: the translational spectra of ice I_h single crystals. *Molec. Phys.*, *44*:1163–1175 (1981).

647. V. Mazzacurati, G. Ruocco, G. Signorelli, E. Cazzanelli, A. Fontana, and G. Mariotto.

Theoretical model for the temperature dependence of Raman scattering in α-AgI. *Phys. Rev. B*, *26*:2216–2223 (1982).

648. V. Mazzacurati, G. Ruocco, and G. Signorelli. "Induced Light Scattering in Disordered Solids." In G. Birnbaum (ed.), *Phenomena Induced by Intermolecular Interactions*, Plenum Press, New York, 1985, pp. 567–588.
649. J. P. McTague, M. J. Mandell, and A. Rahman. Raman spectrum of a Lennard-Jones glass. *J. Chem. Phys.*, *68*:1876–1878 (1978).
650. C. Raptis. Evidence of temperature defect induced first-order Raman scattering in pure NaCl crystals. *Phys. Rev. B*, *33*:1350–1352 (1986).
651. G. Signorelli, V. Mazzacurati, M. Nardone, and C. Pona. "Depolarized Raman Scattering from Disordered Systems." In J. van Kranendonk (ed.), *Intermolecular Spectroscopy and Dynamical Properties of Dense Systems—Proceedings of the International School of Physics "Enrico Fermi," Course LXXV*, North-Holland, Amsterdam, 1980, pp. 294–306.
652. I. F. Silvera, W. N. Hardy, and J. P. McTague. Raman active phonons in the hexagonal phases of solid H_2, D_2, and HD. *Phys. Rev. B*, *5*:1578–1586 (1972).
653. I. F. Silvera. "Infrared and Raman Scattering in Molecular Solids, Mainly Hydrogen." In J. van Kranendonk (ed.), *Intermolecular Spectroscopy and Dynamical Properties of Dense Systems—Proceedings of the International School of Physics "Enrico Fermi," Course LXXV*, North-Holland, Amsterdam, 1980, pp. 399–429.
654. R. E. Slusher and C. M. Surko. Raman scattering from condensed phases of helium: I Optic phonons in solid helium. *Phys. Rev. B*, *13*:1086–1094 (1976).
655. R. E. Slusher and C. M. Surko. Raman scattering from condensed phases of helium: II. Excitation of more than one phonon. *Phys. Rev. B*, *13*:1095–1104 (1976).
656. C. M. Surko and R. E. Slusher. "Raman Scattering from Condensed Phases of ^{3}He and ^{4}He." In *Low Temperature Physics—LT13*, Plenum Press, New York, 1972, pp. 100–104.
657. C. M. Surko and R. E. Slusher. "The Raman Spectra of Solid and Liquid Helium at Large Frequency Shifts." In *Low-Temperature Physics—LT14*, American Elsevier, New York, 1975, p. 487.
658. C. M. Surko and R. E. Slusher. Raman spectra from condensed phases of helium, II. Excitation of more than one phonon. *Phys. Rev. B*, *13*:1095–1104 (1976).
659. J. van Kranendonk. *Solid Hydrogen*. Plenum Press, New York, 1983.
660. G. H. Watson, Jr. and W. B. Daniels. Raman scattering from solid argon at high pressures. *Phys. Rev. B*, *37*:2669–2673 (1988).
661. N. R. Werthamer. Light scattering from solid helium. *Phys. Rev.*, *185*:348–355 (1969).
662. N. R. Werthamer, R. L. Gray, and T. R. Koehler. Computation of Raman scattering cross sections in rare-gas crystal, I. Neon and argon. *Phys. Rev. B*, *2*:4199–4201 (1970).
663. N. R. Werthamer, R. L. Gray, and T. R. Koehler. Computation of Raman scattering cross sections in rare-gas crystals, II. Helium. *Phys. Rev. B*, *4*:1324–1327 (1971).
664. E. Whalley and J. E. Bertie. Optical spectra of orientationally disordered crystals: I. Theory for translational lattice vibrations. *J. Chem. Phys.*, *46*:1264–1270 (1966).

Appendix

665. G. S. Agarwal and J. Cooper. Effective two-level description of pressure-induced resonances in four-wave mixing. *Phys. Rev. A*, *26*:2761–2767 (1982).
666. J. R. Andrews, R. M. Hochstrasser, and H. P. Trommsdorff. Vibrational transitions in excited states of molecules using coherent Stokes Raman spectroscopy: application to ferrocytochrome-c. *Chem. Phys.*, *62*:87–101 (1981).

667. H. F. Arnoldus, T. F. George, K. S. Lam, J. F. Scipione, P. L. DeVries, and J. M. Yuan. "Recent Progress in the Theory of Laser-assisted Collisions." In D. K. Evans (ed.), *Laser Applications in Physical Chemistry*, Marcel Dekker, New York, 1987.

668. F. G. Baglin, U. Zimmerman, and H. Versmold. Higher-order collision induced Raman scattering from the dipole forbidden ν_9 mode of ethane. *Molec. Phys.*, *52*:877–890 (1984).

669. T. Bancewicz and S. Kielich. Intermolecular interaction effect on the line shape of hyper-Rayleigh light scattering by molecular liquids. *Molec. Phys.*, *31*:615–627 (1976).

670. W. E. Baylis, E. Walentinowicz, E. Phaneuf, and R. A. Krause. Rotational effects in collisionally induced fine-structure transitions. *Phys. Rev. Lett.*, *31*:741–744 (1973).

671. B. J. Berne and R. Pecora. Light scattering as a probe of fast-reaction kinetics: The depolarized spectrum of Rayleigh scattered light from a chemically reacting medium. *J. Chem. Phys.*, *50*:783–791 (1969); see also Erratum, *J. Chem. Phys.*, *51*:475–476 (1969).

672. G. Birnbaum. "Comments on Hyper-Rayleigh Scattering." In G. Birnbaum (ed.), *Phenomena Induced by Intermolecular Interactions*, Plenum Press, New York, 1985, pp. 773–774.

673. A. R. Bogdan, Y. Prior, and N. Bloembergen. Pressure-induced degenerate frequency resonance in four-wave light mixing. *Opt. Lett.*, *6*:82–83 (1981).

674. A. R. Bogdan, M. W. Downer, and N. Bloembergen. Quantitative characteristics of pressure induced degenerate frequency resonance in four-wave mixing with continuous wave laser beams. *Opt. Lett.*, *6*:348–353 (1981).

675. A. R. Bogdan, M. W. Downer, and N. Bloembergen. Quantitative characteristics of pressure-induced four-wave mixing signals observed with cw laser beams. *Phys. Rev. A*, *24*:623–626 (1981).

676. J. Borysow, L. Frommhold, and J. W. Keto. The third-order Raman spectrum of argon pairs. *Molec. Phys.*, to appear in 1989.

677. K. Burnett, J. Cooper, R. J. Ballagh, and E. W. Smith. Collisional redistribution of radiation: I. The density matrix. *Phys. Rev. A*, *22*:2005–2026 (1980).

678. K. Burnett and J. Cooper. Collisional redistribution of radiation: II. The effects of degeneracy on the equations of motion for the density matrix. *Phys. Rev. A*, *22*:2027–2043 (1980).

679. K. Burnett and J. Cooper. Collisional redistribution of radiation: III. The equation of motion for the correlation function and the scattered spectrum. *Phys. Rev. A*, *22*:2044–2060 (1980).

680. J. L. Carlsten and A. Szöke. Spectral resolution of near-resonant Rayleigh scattering and collision induced resonance fluorescence. *Phys. Rev. Lett.*, *36*:667–671 (1976).

681. J. L. Carlsten and A. Szöke. Near-resonant Rayleigh scattering and collision induced fluorescence. *Optics Communic.*, *18*:138–139 (1976).

682. J. L. Carlsten and A. Szöke. Collisional redistribution of near-resonant scattered light in Sr vapor. *J. Phys. B*, *9*:L231–L235 (1976).

683. J. L. Carlsten, A. Szöke, and M. G. Raymer. Collisional redistribution and saturation of near-resonance scattered light. *Phys. Rev. A*, *15*:1029–1045 (1977).

684. J. H. R. Clarke, G. J. Hills, C. J. Oliver, and J. M. Vaughan. Rayleigh light scattering from ionic solutions and ionic association reactions. *J. Chem. Phys.*, *61*:2810–2813 (1974).

685. S. L. Cyvin, J. E. Rauch, and J. C. Decius. Theory of hyper-Raman effects (nonlinear elastic light scattering): Selection rules and depolarization ratios for the second-order polarizability. *J. Chem. Phys.*, *43*:4083–4095 (1965).

686. B. D. Fainberg. On the theory of collision-induced lines forbidden in Raman scattering. *Zh. Eksp. Teor. Fiz.* (*U.S.S.R.*), *69*:1935–1942 (1975) [*Sov. Phys. JETP*, *42*:982–985 (1976)].

687. S. R. Federman and L. Frommhold. Recombination of hydrogen atoms via free-to-bound Raman transitions. *Phys. Rev. A*, *25*: 2012–2016 (1982).

688. W. Finkelnburg. *Kontinuierliche Spektren*. Springer, Berlin, 1938.

689. W. Finkelnburg and T. Peters. "Kontinuierliche Spektren." In S. Flügge (ed.), *Handbook of Physics—Spectroscopy II*, Springer, Berlin, 1957, pp. 79–204.

690. J. Fiutak and J. van Kranendonk. The effect of collisions on resonance fluorescence and Rayleigh scattering at high intensities. *J. Phys. B*, *13*:2869–2884 (1980).

691. W. M. Gelbart. Second harmonic generation by atomic fluids. *Chem. Phys. Lett.*, *23*:53–55 (1973).

692. J. I. Gersten. Collision-induced fluorescent light scattering in atomic gases. *Phys. Rev. Lett.*, *31*:73–76 (1973).

693. B. I. Greene, P. A. Fleury, H. L. Carter, and R. C. Farrow. Microscopic dynamics in simple liquids by subpicosecond birefringence. *Phys. Rev. A*, *29*:271–274 (1984).

694. P. S. Julienne. "Collision Induced Radiative Transitions at Optical Frequencies." In G. Birnbaum (ed.), *Phenomena Induced by Intermolecular Interactions*, Plenum, New York, 1985, pp. 749–771.

695. S. Kielich. On three-photon light scattering in atomic fluids. *Acta Phys. Polonica*, *32*:297–300 (1967).

696. S. Kielich. Second harmonic light scattering by dense, isotropic media. *Acta Phys. Polonica*, *33*:89–104 (1968).

697. S. Kielich. On the relation between nonlinear refractive index and molecular light scattering in liquids. *Chem. Phys. Lett.*, *2*:112–115 (1968).

698. S. Kielich. DC electric field-induced optical second harmonic generation by interacting multipolar molecules. *Chem. Phys. Lett.*, *2*:569–572 (1968).

699. S. Kielich. Molecular interactions in optically induced nonlinearities. *IEEE J. Quantum Electronics*, *QE-4*:744–752 (1968).

700. S. Kielich, J. R. Lalanne, and F. B. Martin. Double photon elastic light scattering by liquids having centrosymmetric molecules. *Phys. Rev. Lett.*, *26*:1295–1298 (1971).

701. S. Kielich and S. Woźniak. Influence of statistical fluctuational processes on higher-order nonlinear refractive index of simpler fluids. *Acta Phys. Polonica, A*, *39*:233–235 (1971).

702. S. Kielich, J. R. Lalanne, and F. B. Martin. second harmonic light scattering induced in liquids by fluctuational electric fields of quadrupolar molecules. *Acta Phys. Polonica, A*, *41*:479–482 (1972).

703. S. Kielich, J. R. Lalanne, and F. B. Martin. Cooperative second harmonic laser light scattering in liquid cyclohexane, benzene and carbon disulfide. *J. Raman Spectrosc.,*, *1*:119–139 (1973).

704. S. Kielich. "The Determination of Molecular Electric Multipoles and Their Polarizabilities by Methods of Nonlinear Intermolecular Spectroscopy of Scattered Laser Light." In J. van Kranendonk (ed.), *Intermolecular Spectroscopy and Dynamical Properties of Dense Systems—Proceedings of the International School of Physics "Enrico Fermi," Course LXXV*, North-Holland, Amsterdam, 1980, pp. 146–155.

705. S. Kielich. Nonlinear refractive index and light scattering due to fluctuations of molecular multipole electric fields. *Optics Communic.*, *34*:367–374 (1980).

706. S. Kielich. *Nonlinear Molecular Optics*. Nauka, Moscow, 1981.

707. S. Kielich. Coherent light scattering by interacting anisotropic molecules with variable dipolar polarizability. *J. Phys. (Paris)*, *43*:1749–1757 (1982).

708. S. Kielich. Multi-photon scattering molecular spectroscopy. *Progr. Opt.*, *20*:155–261 (1983).

709. H. Kildal and S. R. J. Brueck. Orientational and electronic contributions to the third-order susceptibilities of cryogenic liquids. *J. Chem. Phys.*, *73*:4951–4958 (1980).

710. A. M. F. Lau. "On Laser-induced Inelastic Collisions." In N. K. Rahman and C. Guidotti (ed.), *Photon-Assisted Collisions and Related Topics*, Harwood Academic, New York, 1982, pp. 55–92.

711. B. F. Levine. "Studies of Molecular Characteristics and Interactions Using Hyperpolarizabilities as a Probe." In M. Davies (ed.), *A Specialist Periodical Report—Dielectric and Related Molecular Processes*, Chemical Society, London, 1977, pp. 73–107.

712. F. B. Martin and J. R. Lalanne. Agreement between depolarized Rayleigh scattering and optical Kerr effect induced by *Q*-switched laser waves in some liquids. *Phys. Rev. A*, *4*:1275–1278 (1971).

713. J. P. McTague, C. H. Lin, T. K. Gustafson, and R. Y. Chiao. The observation of filaments in liquid argon. *Phys. Lett. A*, *32*:82–83 (1970).

714. Y. Prior, A. R. Bogdan, M. Dagenais, and N. Bloembergen. Pressure-induced extra resonances in four-wave mixing. *Phys. Rev. Lett.*, *46*:111–114 (1981).

715. Y. Prior. "Collision Induced Coherent Phenomena." In N. K. Rahman and C. Guidotti (ed.) *Collisions and Half-Collisions with Lasers*, Harwood Academic, New York, 1984, pp. 295–306.

716. N. K. Rahman and C. Guidotti (eds.). *Photon-assisted Collisions and Related Topics*. Harwood Academic, New York, 1982.

717. N. K. Rahman and C. Guidotti (eds.). *Collisions and Half-Collisions with Lasers*. Harwood Academic, New York, 1984.

718. R. Samson and R. A. Pasmanter. Multi-particle effects in second harmonic generation. *Chem. Phys. Lett.*, *25*:405–408 (1974).

719. S. I. Yakovlenko. Laser induced radiative collisions. *Kvantovaya Elektron. (Moscow)*, *8*:259 (1978) [*Sov. J. Quantum Electron*, *8*:151 (1978)].

720. Y. Yeh and R. N. Keeler. Experimental study of reaction kinetics by light scattering, Part 1: Polarized Rayleigh component. *J. Chem. Phys.*, *51*:1120–1127 (1969).

721. Y. Yeh. Experimental study of reaction kinetics by light scattering, Part 2: Helix-coil transition of the copolymer deoxyadenylate-deoxythymidylate (DAT). *J. Chem. Phys.*, *52*:6218–6224 (1970).

Author Index*

* The numbers refer to the references in the preceding reference list.

GENERAL CONNECTIONS AMONG NUCLEAR ELECTROMAGNETIC SHIELDINGS AND POLARIZABILITIES

PAOLO LAZZERETTI

Dipartimento di Chimica dell'Università degli Studi di Modena
Via Campi 183
41100 Modena, Italy

CONTENTS

I. INTRODUCTION

When a molecule is immersed in an external electromagnetic field, the electron distribution gets polarized and electronic current densities arise. The relative phenomenology can be investigated within the framework of response theory in both the classical and quantum case [1–3].

Accordingly, the behavior of the perturbed system is rationalized in terms of generalized susceptibilities, which are intrinsic properties of the molecule, represented by response tensors.

A comprehensive theory of electric polarizabilities and magnetic susceptibilities has been developed [4–6], which accounts for the electromagnetic moments induced in the electron cloud by the switching on of an external time-dependent perturbation.

This theory has revealed a powerful tool for the interpretation of a wide class of phenomena, including the optical rotatory power, the refractive index and the stopping power of gases, the Stark effect, the long-range intermolecular forces, and so on.

A different class of phenomena can be related to the interaction of the electronic polarization density and induced current with the nuclei. Thus the chemical shifts in nuclear magnetic resonance (NMR) spectroscopy are interpreted in terms of magnetic shielding of the electrons, perturbed by a static magnetic field, at those nuclei possessing an intrinsic magnetic moment [7].

When the molecule is perturbed by a static electric field, nuclear electric shieldings can be defined [8–11], which are related to IR intensities [12–14].

In the case of an external time-dependent magnetic field, the idea of nuclear magnetic shielding can be extended, introducing dynamic tensors, and a further generalization leads to the definition of a wider group of electromagnetic shielding tensors relative to the nuclei of a molecule perturbed by external radiation [15].

These new molecular tensors seem very promising, as they are related to other spectroscopic parameters. A periodic magnetic field induces an electric field at the nuclei, which can be described in terms of an electromagnetic shielding tensor [15]. This is related to the intensity of absorption bands in vibrational circular dichroism (VCD) [16, 20]. Also, in the presence of a periodic electric field, the nuclei are acted upon by an induced magnetic field. This leads to the definition of magnetoelectric shielding [15].

This chapter is concerned with the presentation of the interesting properties that characterize the nuclear electromagnetic shielding tensors: They satisfy translational and rotational sum rules that are very general quantum-mechanical relations, that is, Thomas–Reiche–Kuhn sum rules, gauge-invariance conditions, commutation relations, hypervirial theorems, and constraints expressing the conservation of the current density. All of these are different but deeply interrelated aspects of one and the same physical background and can be reduced to the same unitary perspective. The shielding tensors are related to each other and are connected with the electric polarizability, the optical activity and the paramagnetic susceptibility via simple equations, which, in a sense, constitute some sort of connective tissue in the body of second-order properties.

II. MULTIPOLE EXPANSION, BLOCH POTENTIALS, AND MOLECULAR HAMILTONIAN

Let us consider an isolated molecule perturbed by an electromagnetic field. According to the semiclassical approach, the external radiation is described as a plane monochromatic wave traveling with velocity c and obeying the Maxwell equations [21] (i.e., the fields are not quantized).

The scalar potential ϕ and the vector potential $\mathbf{A}$ of the wave in vacuo satisfy the Lorentz condition

$$\frac{1}{c}\frac{\partial\phi}{\partial t}+\nabla\cdot\mathbf{A}=0. \tag{1}$$

In addition, the "radiation" (or, as it is sometimes called, "solenoidal") gauge is assumed for the free wave

$$\phi=0=\nabla\cdot\mathbf{A}. \tag{2}$$

The explicit form of the vector potential is [21]

$$\mathbf{A}=\Re[\mathbf{A}_0\exp(i\psi)],\qquad \psi=\mathbf{k}\cdot\mathbf{r}-\omega t, \tag{3}$$

where ω is the angular frequency, ψ is the phase, and

$$\mathbf{k}=\frac{\omega}{c}\mathbf{n} \tag{4}$$

is the wave vector. The unit vector $\mathbf{n}$ lies along the direction of propagation. The electric and magnetic vectors are expressed in terms of the potentials:

$$\mathbf{E}=-\nabla\phi-\frac{1}{c}\frac{\partial\mathbf{A}}{\partial t},\qquad \mathbf{B}=\nabla\times\mathbf{A} \tag{5}$$

[according to Eqn. (2), $\nabla\phi=\mathbf{0}$ for the wave in vacuo].

It is sometimes convenient to work with complex quantities, omitting the $\Re$ sign for the real part in Eqn. (3), whenever linear operations are performed, that is,

$$\mathbf{A}=\mathbf{A}_0\exp(i\psi). \tag{3$'$}$$

For a plane monochromatic wave the complex fields are

$$\mathbf{E}=ik\mathbf{A},\qquad \mathbf{B}=i\mathbf{k}\times\mathbf{A}, \tag{5$'$}$$

and the actual fields are obtained by taking the real part of Eqn. (5′):

$$\mathbf{E}=\mathbf{E}_0\cos\omega\left(t-\frac{\mathbf{n}\cdot\mathbf{r}}{c}\right),\qquad \mathbf{B}=\mathbf{n}\times\mathbf{E}. \tag{5$''$}$$

In the absence of perturbation the stationary electronic states of a molecule with n electrons and N nuclei are described in terms of the eigenfunctions $|j\rangle$ to the unperturbed Born–Oppenheimer electronic Hamiltonian

$$H_0 = \sum_{i=1}^{n}\left[\frac{p_i^2}{2m} - \sum_{I=1}^{N}\frac{Z_I e^2}{|\mathbf{r}_i - \mathbf{R}_I|} + \frac{1}{2}\sum_{j\neq i}^{n}\frac{e^2}{|\mathbf{r}_i - \mathbf{r}_j|}\right] + \frac{1}{2}\sum_{I}^{N}\sum_{J\neq I}^{N}\frac{Z_I Z_J e^2}{|\mathbf{R}_I - \mathbf{R}_J|}, \tag{6}$$

where $-e$, m, $\mathbf{r}_i$, and $\mathbf{p}_i(Z_I e, M_I, \mathbf{R}_I, \mathbf{P}_I)$ are charge, mass, coordinates, and canonical momentum of the ith electron (Ith nucleus).

The external fields induce forced oscillations in the electron cloud. The interaction is described in terms of the time-dependent Hamiltonian $H^{(1)}$ within the framework of propagator methods [3] or, equivalently, introducing time-dependent perturbation theory [22, 23]. Relaxing condition (2) for the free wave, the general form of the Hamiltonian becomes, neglecting electron spin,

$$H = H_0 + V, \qquad V = H^{(1)} + H^{(2)},$$

$$V = \sum_{i=1}^{n}\left[\frac{e}{2mc}\mathbf{A}_i\cdot\mathbf{p}_i + \frac{e}{2mc}\mathbf{p}_i\cdot\mathbf{A}_i + \frac{e^2}{2mc^2}A_i^2 - e\phi_i\right],$$

$$\mathbf{A}_i \equiv \mathbf{A}(\mathbf{r}_i), \qquad \phi_i \equiv \phi(\mathbf{r}_i). \tag{7}$$

It is expedient to represent this interaction Hamiltonian as a series of terms in which the electromagnetic multipoles are coupled with the external fields and their derivatives taken at the origin of the coordinate system.

This can be accomplished in two ways: (i) via a canonical transformation of the Hamiltonian (7) and its (perturbed) eigenstates, assuming the radiation gauge (2) [24–26],

$$H \rightarrow H' = \exp\left(\frac{i}{\hbar}G\right)H\exp\left(-\frac{i}{\hbar}G\right) - \dot{G}, \tag{8}$$

$$\Psi_j \rightarrow \Psi_j' = \exp\left(\frac{i}{\hbar}G\right)\Psi_j, \tag{9}$$

where G is a suitable generating function; (ii) by a judicious choice of gauge from the very beginning, which makes (7) *equal* to the multipole Hamiltonian [27–29].

This distinction is largely formal, owing to the substantial identity of the unitary time-dependent transformation (8)–(9) with the gauge transformations of the Hamiltonian and its eigenfunctions [21–22]. However, alternative (ii) presents some advantages, as it offers a more direct derivation of the multipole interaction Hamiltonian, in which definitions of Hermitian electric and magnetic multipole operators, consistent with classical radiation theory [30], are explicitly displayed.

According to Bloch [27], whose notation is retained here, the real potentials acting on a particle with coordinates **r** are expressed as Taylor series:

$$\phi(r_1, r_2, r_3, t) = \phi(0, 0, 0, t) + \sum_{k=1}^{\infty} \frac{1}{k!} \frac{\partial^k \phi}{\partial r_{\alpha_1} \partial r_{\alpha_2} \cdots \partial r_{\alpha_k}} r_{\alpha_1} r_{\alpha_2} \cdots r_{\alpha_k}, \tag{10}$$

$$A_\alpha(r_1, r_2, r_3, t) = A_\alpha(0, 0, 0, t) + \sum_{k=1}^{\infty} \frac{1}{k!} \frac{\partial^k A_\alpha}{\partial r_{\alpha_1} \partial r_{\alpha_2} \cdots \partial r_{\alpha_k}} r_{\alpha_1} r_{\alpha_2} \cdots r_{\alpha_k}, \tag{11}$$

where the x, y, and z components of vector **r** are rewritten r_α, $\alpha = 1, 2, 3$, the partial derivatives are taken at $\mathbf{r} = \mathbf{0}$, and sum over repeated Greek indices is implied. These potentials are not constrained to satisfy condition (2). Customarily the term $\phi_0 \equiv \phi(0, 0, 0, t)$ is not retained [27, 28]. In any case, the corresponding energy will vanish for a neutral collection of charges.

The Taylor series for the potentials (10) and (11) must be consistent with the corresponding ones for the electric and magnetic fields. To this end let us consider the gauge transformation

$$\phi^{\mathscr{B}} = \phi - \frac{1}{c}\dot{f}^{\mathscr{B}}, \tag{12a}$$

$$\mathbf{A}^{\mathscr{B}} = \mathbf{A} + \nabla f^{\mathscr{B}}, \tag{12b}$$

$$f^{\mathscr{B}}(r_1, r_2, r_3, t) = -\sum_{k=1}^{\infty} \frac{1}{k!} \frac{\partial^{k-1} A_{\alpha_1}}{\partial r_{\alpha_2} \partial r_{\alpha_3} \cdots \partial r_{\alpha_k}} r_{\alpha_1} r_{\alpha_2} \cdots r_{\alpha_k}, \tag{12c}$$

which, using Eqn. (5), leads to the definition of the Bloch potentials [27]

$$\phi^{\mathscr{B}} = -\sum_{k=0}^{\infty} \frac{1}{(k+1)!} \left(\mathbf{r} \cdot \frac{\partial^k \mathbf{E}}{\partial r_{\alpha_1} \partial r_{\alpha_2} \cdots \partial r_{\alpha_k}} \right) r_{\alpha_1} r_{\alpha_2} \cdots r_{\alpha_k}, \tag{13}$$

$$\mathbf{A}^{\mathscr{B}} = -\sum_{k=0}^{\infty} \frac{k+1}{(k+2)!} \left(\mathbf{r} \times \frac{\partial^k \mathbf{B}}{\partial r_{\alpha_1} \partial r_{\alpha_2} \cdots \partial r_{\alpha_k}} \right) r_{\alpha_1} r_{\alpha_2} \cdots r_{\alpha_k}, \tag{14}$$

and to the Bloch gauge for the vector potential

$$\nabla \cdot \mathbf{A}^{\mathscr{B}} = \sum_{k=0}^{\infty} \frac{(k+1)(k+2)}{(k+3)!} \left(\mathbf{r} \cdot \frac{\partial^k (\nabla \times \mathbf{B})}{\partial r_{\alpha_1} \partial r_{\alpha_2} \cdots \partial r_{\alpha_k}} \right) r_{\alpha_1} r_{\alpha_2} \cdots r_{\alpha_k}. \tag{2'}$$

The first-order Hamiltonian of a charge distribution is then given the form of a series

$$\begin{aligned} H^{(1)} &= \int d\tau\, \hat{\rho}\phi^{\mathscr{B}} - \frac{1}{c}\int d\tau\,(\hat{\mathbf{J}}\cdot\mathbf{A}^{\mathscr{B}}) \\ &= -\sum_{k=0}^{\infty}\Bigg[\left(\frac{\partial^k E_\alpha}{\partial r_{\alpha_1}\partial r_{\alpha_2}\cdots\partial r_{\alpha_k}}\right)_{r=0}\mu_{\alpha\alpha_1\cdots\alpha_k} \\ &\quad + \left(\frac{\partial^k B_\alpha}{\partial r_{\alpha_1}\partial r_{\alpha_2}\cdots\partial r_{\alpha_k}}\right)_{r=0} m_{\alpha\alpha_1\cdots\alpha_k}\Bigg], \end{aligned} \tag{15}$$

where the electric and magnetic moments are defined as tensors of rank $k+1$:

$$\mu_{\alpha\alpha_1\cdots\alpha_k} = \frac{1}{(k+1)!}\int d\tau\,\hat{\rho} r_\alpha r_{\alpha_1}\cdots r_{\alpha_k}, \tag{16}$$

$$m_{\alpha\alpha_1\cdots\alpha_k} = \frac{k+1}{(k+2)!}\frac{1}{c}\int d\tau\,(\mathbf{r}\times\hat{\mathbf{J}})_\alpha r_{\alpha_1} r_{\alpha_2}\cdots r_{\alpha_k}. \tag{17}$$

In these formulas $\hat{\rho}(\mathbf{r})$ is the charge density operator and $\hat{\mathbf{J}}(\mathbf{r})$ is the current density operator. For a discrete n-electron charge distribution

$$\hat{\rho}(\mathbf{r}) = -e\sum_{i=1}^{n}\delta(\mathbf{r}-\mathbf{r}_i), \tag{18}$$

$$\hat{\mathbf{J}}(\mathbf{r}) = -\frac{e}{2m}\sum_{i=1}^{n}[\mathbf{p}_i\delta(\mathbf{r}-\mathbf{r}_i) + \delta(\mathbf{r}-\mathbf{r}_i)\mathbf{p}_i], \tag{19}$$

and

$$\mu_{\alpha\alpha_1\cdots\alpha_k} = -\frac{1}{(k+1)!}e\sum_{i=1}^{n}[r_\alpha r_{\alpha_1}\cdots r_{\alpha_k}]_i, \tag{20}$$

$$m_{\alpha\alpha_1\cdots\alpha_k} = -\frac{k+1}{(k+2)!}\frac{e}{2mc}\sum_{i=1}^{n}[l_\alpha r_{\alpha_1} r_{\alpha_2}\cdots r_{\alpha_k} + r_{\alpha_1} r_{\alpha_2}\cdots r_{\alpha_k} l_\alpha]_i. \tag{21}$$

If Eqs. (20) and (21) are multiplied by the "normalization" factor $(k+1)!$, one obtains the same definitions as in the Raab paper [30]. The first terms of the series are, using Raab's normalization,

$$\mu_\alpha = -e\sum_{i=1}^{n} r_{i\alpha}, \tag{22}$$

$$\mu_{\alpha\beta} = -e \sum_{i=1}^{n} (r_\alpha r_\beta)_i, \tag{23}$$

$$\mu_{\alpha\beta\gamma} = -e \sum_{i=1}^{n} (r_\alpha r_\beta r_\gamma)_i, \tag{24}$$

$$m_\alpha = -\frac{e}{2mc} \sum_{i=1}^{n} l_{i\alpha}, \qquad \mathbf{l}_i = \mathbf{r}_i \times \mathbf{p}_i, \tag{25}$$

$$m_{\alpha\beta} = -\frac{e}{3mc} \sum_{i=1}^{n} (l_\alpha r_\beta + r_\beta l_\alpha)_i, \tag{26}$$

$$m_{\alpha\beta\gamma} = -\frac{3e}{8mc} \sum_{i=1}^{n} (l_\alpha r_\beta r_\gamma + r_\beta r_\gamma l_\alpha)_i. \tag{27}$$

The electronic magnetic multipoles (25)–(27) are *unperturbed*, or *permanent*, moment operators. In the presence of a vector potential $\mathbf{A}(\mathbf{r}, t)$ (we simplify the notation, omitting the $\mathscr{B}$ index), the canonical momentum is replaced by the mechanical momentum

$$\mathbf{\Pi} = \sum_{i=1}^{n} \boldsymbol{\pi}_i, \qquad \boldsymbol{\pi}_i = \mathbf{p}_i + \frac{e}{c}\mathbf{A}_i, \tag{28}$$

and the angular momentum becomes

$$\mathbf{L}' = \mathbf{L} + \frac{e}{c} \sum_{i=1}^{n} \mathbf{r}_i \times \mathbf{A}_i, \qquad \mathbf{L} = \sum_{i=1}^{n} \mathbf{l}_i. \tag{29}$$

Within the Bloch gauge for the vector potential, the perturbed magnetic moments become, according to Eqn. (29),

$$m'_\alpha = m_\alpha + \frac{e^2}{4mc^2} \sum_{i=1}^{n} (r_\alpha r_\beta - r^2 \delta_{\alpha\beta})_i B(0)_\beta + \frac{e^2}{6mc^2} \sum_{i=1}^{n} [(r_\alpha r_\beta - r^2 \delta_{\alpha\beta}) r_\gamma]_i B(0)_{\gamma\beta} + \cdots, \tag{30}$$

$$m'_{\alpha\beta} = m_{\alpha\beta} + \frac{e^2}{3mc^2} \sum_{i=1}^{n} [(r_\alpha r_\gamma - r^2 \delta_{\alpha\gamma}) r_\beta]_i B(0)_\gamma + \cdots, \tag{31}$$

and so on, in which $B(0)_\alpha$ is the magnetic field at the origin and $B(0)_{\alpha\beta} \equiv (\nabla_\alpha B_\beta)_{r=0}$, and so on. These definitions, allowing for the Raab conventions, fulfill the equations

$$\mu_\alpha = -\frac{\partial V}{\partial E(0)_\alpha}, \qquad \mu_{\alpha\beta} = -2\frac{\partial V}{\partial E(0)_{\beta\alpha}}, \tag{32}$$

$$m'_\alpha = -\frac{\partial V}{\partial B(0)_\alpha}, \qquad m'_{\alpha\beta} = -2\frac{\partial V}{\partial B(0)_{\beta\alpha}}, \tag{33}$$

and so on, in which each partial derivative is taken at constant values of all the other fields and field gradients. When expansion (15) is truncated, one must be careful to consider all the terms of the same order of magnitude in the Hamiltonian to avoid inconsistencies. Thus, within the "dipole approximation" the vector potential [and therefore the electric field (5)] is assumed to be spatially uniform over the molecular domain and only the electric dipole (22) is retained in the Hamiltonian. To the next higher "quadrupole approximation," the electric field gradient and the magnetic field are assumed to be uniform: Both electric quadrupole and magnetic dipole must be considered in Eqn. (15).

The electric multipoles are sometimes redefined to be traceless in any two suffixes [6, 21]. For instance, the electronic contribution to the traceless electric quadrupole operator is

$$\Theta_{\alpha\beta} = \tfrac{1}{2}(3\mu_{\alpha\beta} - \mu_{\gamma\gamma}\delta_{\alpha\beta}) = -\frac{e}{2}\sum_{i=1}^{n} [3r_\alpha r_\beta - r^2\delta_{\alpha\beta}]_i. \tag{34}$$

Raab [30] has shown the inadequacies of this procedure, which, in the case of dynamic fields, cannot be extended beyond the quadrupole terms and may lead to inconsistencies with the interaction Hamiltonian (7). Accordingly, we retain here definition (34) within the limits of the quadrupole approximation. To higher order only the set of nontraceless moments (22)–(27) can be properly used.

Within the quadrupole approximation, the terms entering the first-order Hamiltonian (15) are

$$\mathscr{H}^{E} = h^{E_\alpha} E_\alpha(0, t), \tag{35}$$

$$\mathscr{H}^{\nabla E} = h^{E_{\alpha\beta}} E_{\alpha\beta}(0, t), \tag{36}$$

$$\mathscr{H}^{B} = h^{B_\alpha} B_\alpha(0, t), \tag{37}$$

where

$$h^{E_\alpha} = -\mu_\alpha = eR_\alpha, \tag{38}$$

$$\mathbf{R} = \sum_{i=1}^{n} \mathbf{r}_i, \tag{39}$$

$$h^{\nabla_\alpha E_\beta} \equiv h^{E_{\alpha\beta}} = -\tfrac{1}{2}\mu_{\beta\alpha}, \tag{40}$$

$$h^{B_\alpha} = -m_\alpha = \frac{e}{2mc} L_\alpha. \tag{41}$$

If the traceless form (34) of quadrupole operator is retained, then the corresponding Hamiltonian becomes

$$\mathscr{H}^{\nabla E} = -\tfrac{1}{3}\Theta_{\beta\alpha}E_{\alpha\beta}(0,t). \tag{36$'$}$$

It is sometimes advantageous [31, 32] to introduce alternative forms of the Hamiltonians (35)–(37). Classically it can be argued that one is always free to change the Lagrangian

$$L \rightarrow L + \frac{dG}{dt},$$

where G is an arbitrary function of time, coordinates and external fields, without affecting the equations of motion. Then the Hamiltonian undergoes a canonical transformation [33, 34].

In quantum mechanics the unitary transformation (8) leads to different gauges for the Hamiltonian, which are sometimes also referred to as different "formalisms." An alternative first-order Hamiltonian can be defined from (8) as

$$H^{(1)\prime} = H^{(1)} - \frac{d}{dt}G, \tag{42}$$

$$\frac{dG}{dt} = \frac{i}{\hbar}[H_0, G] + \frac{\partial G}{\partial t}, \tag{43}$$

and G is chosen so that

$$H^{(1)} = \frac{\partial G}{\partial t}, \qquad H^{(1)\prime} = -\frac{i}{\hbar}[H_0, G]. \tag{44}$$

Introducing the generating functions [31, 32]

$$G^A = -\frac{e}{c}\,\mathbf{A}(0,t)\cdot\mathbf{R}, \tag{45}$$

$$G^Z = -\frac{e}{mc^2}\,\mathbf{Z}(0,t)\cdot\mathbf{P}, \tag{45$'$}$$

$$G^{\nabla \times Z} = -\frac{e}{2mc^2}[\nabla \times \mathbf{Z}(\mathbf{r}, t)]_0 \cdot \mathbf{L}, \tag{45''}$$

the dipole velocity first-order Hamiltonian becomes (in the radiation gauge)

$$\mathscr{H}^A = \mathbf{A}(0, t) \cdot \mathbf{h}^A, \qquad \mathbf{h}^A = \frac{e}{mc}\mathbf{P}, \tag{35'}$$

$$\mathbf{P} = \sum_{i=1}^{n} \mathbf{p}_i; \tag{46}$$

(we recall that the inverse canonical transformation has been introduced by Goeppert-Mayer [24] to obtain the dipole length Hamiltonian (35) from (35′), i.e., from (7) within the radiation gauge). The dipole acceleration (or force) first-order Hamiltonian is

$$\mathscr{H}^Z = \mathbf{Z}(0, t) \cdot \mathbf{h}^Z, \qquad \mathbf{h}^Z = \frac{e}{mc^2}\mathbf{F}_n^N, \tag{35''}$$

$$\mathbf{F}_n^N = \sum_{I=1}^{N} \mathbf{F}_n^I = \sum_{i=1}^{n} \mathbf{F}_i^N, \qquad \mathbf{F}_i^I = -e^2 Z_I \frac{\mathbf{r}_i - \mathbf{R}_I}{|\mathbf{r}_i - \mathbf{R}_I|^3}, \tag{47}$$

$$\mathbf{F}_n^N = -e\mathbf{E}_n^N = -e \sum_{I=1}^{N} \sum_{i=1}^{n} \mathbf{E}_i^I, \tag{48}$$

$$\mathbf{E}_i^I = Z_I \mathbf{E}_I^i, \tag{49}$$

$$\mathbf{E}_I^i = e\frac{\mathbf{r}_i - \mathbf{R}_I}{|\mathbf{r}_i - \mathbf{R}_I|^3}, \tag{50}$$

and the torque Hamiltonian is

$$\mathscr{H}^{\nabla \times Z} = (\nabla \times \mathbf{Z})_0 \cdot \mathbf{h}^{\nabla \times Z}, \qquad \mathbf{h}^{\nabla \times Z} = \frac{e}{2mc^2}\mathbf{K}_n^N, \tag{37'}$$

$$\mathbf{K}_n^N = \sum_{I=1}^{N} \mathbf{K}_n^I = \sum_{i=1}^{n} \mathbf{K}_i^N, \qquad \mathbf{K}_i^I = e^2 Z_I \frac{\mathbf{r}_i - \mathbf{R}_I}{|\mathbf{r}_i - \mathbf{R}_I|^3} \times \mathbf{R}_I. \tag{51}$$

In these formulas $\mathbf{Z}$ is the electric Hertz–Righi polarization potential [35, 36, 21]. According to Nisbet's choice [35] of gauge for $\mathbf{Z}$,

$$\mathbf{A}(\mathbf{r}, t) = -\frac{1}{c}\frac{\partial}{\partial t}\mathbf{Z}(\mathbf{r}, t), \tag{52}$$

$$\phi(\mathbf{r}, t) = \nabla \cdot \mathbf{Z}(\mathbf{r}, t) = 0. \tag{53}$$

In deriving (35′), (35″), and (37″), we have used, according to recipe (44),

$$\frac{\partial}{\partial t} G^{A} = \mathscr{H}^{E},$$

$$\frac{\partial}{\partial t} G^{Z} = \mathscr{H}^{A},$$

$$\frac{\partial}{\partial t} G^{\nabla \times Z} = \mathscr{H}^{B}.$$

III. OBSERVABLES AND THEIR EQUATION OF MOTION

The time-dependent Schrödinger equation for the perturbed state $|\Phi_a(t)\rangle$ is

$$(H_0 + H^{(1)})|\Phi_a\rangle = i\hbar \frac{\partial}{\partial t}|\Phi_a\rangle, \tag{54}$$

and the wavefunction is written as a linear combination of stationary states

$$|\Phi_a(t)\rangle = \sum_j c_{ja}(t)|\Psi_j(t)\rangle, \tag{55}$$

$$|\Psi_j(t)\rangle = |j\rangle \exp\left(-\frac{i}{\hbar} E_j t\right), \qquad |j\rangle \equiv |\psi_j\rangle,$$

$$H_0|\psi_j\rangle = E_j|\psi_j\rangle, \tag{56}$$

where H_0 is the time-independent Born–Oppenheimer Hamiltonian. For a monochromatic wave the perturbing operators (35)–(37), (35′), (35″), and (37′) will have the form

$$H^{(1)} = \hat{H}^{(1)} \cos \omega t,$$

where $\hat{H}^{(1)}$ is independent of time and the time-dependent mixing coefficients correct through first order are

$$c_{ja}(t) = -\frac{1}{\hbar(\omega_{ja}^2 - \omega^2)}\left[\langle j|H^{(1)}|a\rangle \omega_{ja} + i\left\langle j \left| \frac{\partial H^{(1)}}{\partial t} \right| a \right\rangle\right] \exp(i\omega_{ja} t), \tag{57}$$

where

$$\omega_{ja} = \frac{1}{\hbar}(E_j - E_a). \tag{58}$$

We shall be interested in the average values, correct to first order in the perturbing radiation, of the multipoles (22)–(27), which may themselves involve the perturbation. Thus, in obvious notation,

$$T = T_0 + T_1 + \cdots,$$

and the observable is

$$\langle \Phi_a | T | \Phi_a \rangle \equiv \langle T \rangle_a = \langle a | T_0 | a \rangle + \langle a | T_1 | a \rangle + 2\Re \left[\sum_{j \neq a} \langle a | T_0 | j \rangle \exp(-i\omega_{ja} t) c_{ja}(t) \right], \tag{59}$$

where T_0 does not depend explicitly on time and T_1 may be explicitly time dependent.

Within the quadrupole approximation and adopting (35)–(37),

$$\begin{aligned} c_{ja}(t) = \frac{\exp(i\omega_{ja} t)}{\hbar(\omega_{ja}^2 - \omega^2)} [&\langle j | \mu_\alpha | a \rangle (\omega_{ja} E(0)_\alpha + i\dot{E}(0)_\alpha) \\ &+ \tfrac{1}{2} \langle j | \mu_{\alpha\beta} | a \rangle (\omega_{ja} E(0)_{\beta\alpha} + i\dot{E}(0)_{\beta\alpha}) \\ &+ \langle j | m_\alpha | a \rangle (\omega_{ja} B(0)_\alpha + i\dot{B}(0)_\alpha)], \end{aligned} \tag{60}$$

and

$$\begin{aligned} \langle T \rangle_a = \langle a | T_0 | a \rangle + \langle a | T_1 | a \rangle + \frac{1}{\hbar} \Re \Bigg\{ &\sum_{j \neq a} \frac{2}{\omega_{ja}^2 - \omega^2} \langle a | T_0 | j \rangle \\ &\times [\langle j | \mu_\alpha | a \rangle (\omega_{ja} E(0)_\alpha + i\dot{E}(0)_\alpha) + \tfrac{1}{2} \langle j | \mu_{\alpha\beta} | a \rangle (\omega_{ja} E(0)_{\beta\alpha} + i\dot{E}(0)_{\beta\alpha}) \\ &+ \langle j | m_\alpha | a \rangle (\omega_{ja} B(0)_\alpha + i\dot{B}(0)_\alpha)] \Bigg\}. \end{aligned} \tag{61}$$

From the Schrödinger equation for the stationary unperturbed state

$$H_0 \Psi_j = i\hbar \frac{\partial}{\partial t} \Psi_j, \tag{62}$$

one finds the equation of motion for $\langle a | T_0 | a \rangle$,

$$\frac{d}{dt} \langle a | T_0 | a \rangle = \frac{d}{dt} \langle \Psi_a | T_0 | \Psi_a \rangle \equiv \left\langle a \left| \frac{dT_0}{dt} \right| a \right\rangle, \tag{63}$$

(which is not very informative, however, if $|a\rangle$ is an eigenfunction to H_0, since it is just $0 = 0$) and the equation for the total time derivative of operator T_0 is [22]

$$\frac{dT_0}{dt} = \frac{i}{\hbar}[H_0, T_0]. \tag{64}$$

Allowing for (62), this result can be generalized for the dynamic observable (59). From (57) and (59), we obtain, to first order,

$$\frac{d}{dt}\langle T\rangle_a = \left\langle \frac{i}{\hbar}[H_0, T]\right\rangle_a + \left\langle a\left|\frac{\partial}{\partial t}T_1\right|a\right\rangle + \left\langle a\left|\frac{i}{\hbar}[H^{(1)}, T_0]\right|a\right\rangle \equiv \left\langle\frac{dT}{dt}\right\rangle_a, \tag{65}$$

where, according to (59),

$$\left\langle \frac{i}{\hbar}[H_0, T]\right\rangle_a = \left\langle a\left|\frac{i}{\hbar}[H_0, T_0]\right|a\right\rangle + \left\langle a\left|\frac{i}{\hbar}[H_0, T_1]\right|a\right\rangle + 2\Re\left[\sum_{j\neq a}\left\langle a\left|\frac{i}{\hbar}[H_0, T_0]\right|j\right\rangle \exp(-i\omega_{ja}t)c_{ja}(t)\right] \tag{66}$$

(at any rate, the first line on the right side is vanishing if $|a\rangle$ is an eigenfunction to H_0).

In the case of monochromatic radiation and within the quadrupole approximation, Eqn. (65) can be recast in the form

$$\frac{d}{dt}\langle T\rangle_a = \left\langle \frac{i}{\hbar}[H_0, T]\right\rangle_a + \left\langle a\left|\frac{\partial}{\partial t}T_1\right|a\right\rangle + \frac{i}{\hbar}\langle a|[T_0, \mu_\beta E(0)_\beta + m_\beta B(0)_\beta + \tfrac{1}{2}\mu_{\beta\gamma}E(0)_{\gamma\beta}]|a\rangle, \tag{67}$$

which is our final formula for the equation of motion of dynamic observables in the presence of monochromatic radiation.

There are two points to note about Eqs. (61) and (67). First, as they hold for any T, Eqs. (65) and (67) provide a general restatement of the Ehrenfest relations [34], that is, the quantum analog of the equations of motion of classical mechanics. Second, they embed a series of constraints among different molecular properties. It will be shown that these can be given the form of *sum rules.*

We consider now some explicit forms of T. According to Eqs. (22), (25), (28), (47), (50), and (51), using the traceless quadrupole (34) and the Hamiltonian (36′) and denoting by

$$\Delta\langle T\rangle_a = \langle T\rangle_a - \langle a|T_0|a\rangle \tag{68}$$

the contribution to the electronic observable induced by the electromagnetic radiation, which adds to the permanent value $\langle a|T_0|a\rangle$, we find (the index a is omitted to avoid cumbersome notation)

$$\begin{aligned}\Delta\langle \mu_\alpha\rangle &= \alpha_{\alpha\beta}E(0)_\beta + \kappa_{\alpha\beta}B(0)_\beta + \hat{\alpha}_{\alpha\beta}\dot{E}(0)_\beta + \hat{\kappa}_{\alpha\beta}\dot{B}(0)_\beta \\ &\quad + \tfrac{1}{3}A_{\alpha,\beta\gamma}E(0)_{\beta\gamma} + \tfrac{1}{3}\hat{A}_{\alpha,\beta\gamma}\dot{E}(0)_{\beta\gamma},\end{aligned} \tag{69}$$

$$\begin{aligned}\Delta\langle m'_\alpha\rangle &= \kappa_{\beta\alpha}E(0)_\beta + \chi_{\alpha\beta}B(0)_\beta - \hat{\kappa}_{\beta\alpha}\dot{E}(0)_\beta + \hat{\chi}^{\mathrm{p}}_{\alpha\beta}\dot{B}(0)_\beta \\ &\quad + \tfrac{1}{3}D_{\alpha,\beta\gamma}E(0)_{\beta\gamma} + \tfrac{1}{3}\hat{D}_{\alpha,\beta\gamma}\dot{E}(0)_{\beta\gamma},\end{aligned} \tag{70}$$

$$\begin{aligned}\Delta\langle \Theta_{\alpha\beta}\rangle &= A_{\gamma,\alpha\beta}E(0)_\gamma + D_{\gamma,\alpha\beta}B(0)_\gamma - \hat{A}_{\gamma,\alpha\beta}\dot{E}(0)_\gamma - \hat{D}_{\gamma,\alpha\beta}\dot{B}(0)_\gamma \\ &\quad + C_{\alpha\beta,\gamma\delta}E(0)_{\gamma\delta} + \hat{C}_{\alpha\beta,\gamma\delta}\dot{E}(0)_{\gamma\delta},\end{aligned} \tag{71}$$

$$\begin{aligned}\Delta\langle \Pi_\alpha\rangle &= o_{\alpha\beta}E(0)_\beta + \upsilon_{\alpha\beta}B(0)_\beta + \hat{o}_{\alpha\beta}\dot{E}(0)_\beta + \hat{\upsilon}^{\mathrm{p}}_{\alpha\beta}\dot{B}(0)_\beta \\ &\quad + \tfrac{1}{3}U_{\alpha,\beta\gamma}E(0)_{\beta\gamma} + \tfrac{1}{3}\hat{U}_{\alpha,\beta\gamma}\dot{E}(0)_{\beta\gamma},\end{aligned} \tag{72}$$

$$\begin{aligned}\Delta\langle E^n_{I\alpha}\rangle &= -\gamma^I_{\alpha\beta}E(0)_\beta + \xi^I_{\alpha\beta}B(0)_\beta - \hat{\gamma}^I_{\alpha\beta}\dot{E}(0)_\beta + \hat{\xi}^I_{\alpha\beta}\dot{B}(0)_\beta \\ &\quad + \tfrac{1}{3}\nu^I_{\alpha,\beta\gamma}E(0)_{\beta\gamma} + \tfrac{1}{3}\hat{\nu}^I_{\alpha,\beta\gamma}\dot{E}(0)_{\beta\gamma},\end{aligned} \tag{73}$$

$$\begin{aligned}\Delta\langle B^{n'}_{I\alpha}\rangle &= \lambda^I_{\alpha\beta}E(0)_\beta - \sigma^I_{\alpha\beta}B(0)_\beta + \hat{\lambda}^I_{\alpha\beta}\dot{E}(0)_\beta - \hat{\sigma}^{\mathrm{p}I}_{\alpha\beta}\dot{B}(0)_\beta \\ &\quad + \tfrac{1}{3}\tau^I_{\alpha,\beta\gamma}E(0)_{\beta\gamma} + \tfrac{1}{3}\hat{\tau}^I_{\alpha,\beta\gamma}\dot{E}(0)_{\beta\gamma},\end{aligned} \tag{74}$$

$$\Delta\langle F^N_{n\alpha}\rangle = -\sum_{I=1}^{N} Z_I e\Delta\langle E^n_{I\alpha}\rangle, \tag{75}$$

$$\begin{aligned}\Delta\langle K^N_{n\alpha}\rangle &= \varepsilon_{\alpha\beta\gamma}\sum_{I=1}^{N} Z_I e\Delta\langle E^n_{I\beta}\rangle R_{I\gamma} \\ &= \iota_{\alpha\beta}E(0)_\beta + \eta_{\alpha\beta}B(0)_\beta + \hat{\iota}_{\alpha\beta}\dot{E}(0)_\beta + \hat{\eta}_{\alpha\beta}\dot{B}(0)_\beta \\ &\quad + \tfrac{1}{3}V_{\alpha,\beta\gamma}E(0)_{\beta\gamma} + \tfrac{1}{3}\hat{V}_{\alpha,\beta\gamma}\dot{E}(0)_{\beta\gamma},\end{aligned} \tag{76}$$

where the operator representing the magnetic field of electrons on nucleus I in the presence of an external magnetic field is

$$\mathbf{B}_I^{n'} = \mathbf{B}_I^n + \frac{e}{2mc^2} \sum_{i=1}^{n} [(\mathbf{B} \cdot E_I^i)\mathbf{r}_i - (\mathbf{E}_I^i \cdot \mathbf{r}_i)\mathbf{B}], \tag{77}$$

$$\mathbf{B}_I^n = -\frac{e}{cm} \mathbf{M}_I^n = \mathbf{B}_{I0}^n, \qquad \mathbf{M}_I^n = \sum_{i=1}^{n} \frac{\mathbf{l}_i(\mathbf{R}_I)}{|\mathbf{r}_i - \mathbf{R}_I|^3}. \tag{78}$$

Equations (75) and (76) yield the nuclear contribution to the average force and torque on the electron distribution in the presence of external radiation. If the hypervirial theorems for **R**, **P**, and **L**, (i.e., the momentum, force, and torque theorems) are satisfied, then

$$\langle a|\mathbf{P}|a\rangle = \langle a|\mathbf{F}_n^N|a\rangle = \langle a|\mathbf{K}_n^N|a\rangle = \mathbf{0},$$

and

$$\Delta\langle P_\alpha\rangle = \langle P_\alpha\rangle, \qquad \Delta\langle F_{n\alpha}^N\rangle = \langle F_{n\alpha}^N\rangle, \qquad \Delta\langle K_{n\alpha}^N\rangle = \langle K_{n\alpha}^N\rangle. \tag{79}$$

Obviously enough, the external radiation can only induce forced oscillations of the electron distribution, whose mean position is fixed in space. Thus, for instance, for the electric perturbation, averaging the terms depending on $\dot{\mathbf{E}}$ over a period $\mathscr{T}$, one finds

$$\int_0^{\mathscr{T}} dt\, \Delta\langle \Pi_\alpha\rangle = -\frac{m}{e}\alpha_{\alpha\beta}[E_\beta(\mathscr{T}) - E_\beta(0)] = 0. \qquad -\frac{m}{e}\boldsymbol{\alpha} = \hat{\mathbf{0}}.$$

From Eqn. (72) and

$$\Delta\langle\Pi_\alpha\rangle = -\frac{m}{e}\int d\tau\, J_\alpha(\mathbf{r}),$$

the following expressions are found for the induced n-electron current densities:

$$J_\alpha^{E_\beta} = -\frac{e}{m\hbar}\sum_{j\neq a}\frac{2\omega_{ja}}{\omega_{ja}^2 - \omega^2}\Re[\langle j|\mu_\beta|a\rangle\psi_a^* P_\alpha\psi_j], \tag{80}$$

$$J_\alpha^{\dot{E}_\beta} = \frac{e}{m\hbar}\sum_{j\neq a}\frac{2}{\omega_{ja}^2 - \omega^2}\Im[\langle j|\mu_\beta|a\rangle\psi_a^* P_\alpha\psi_j], \tag{81}$$

$$J_\alpha^{B_\beta} = -\frac{e}{m\hbar}\sum_{j\neq a}\frac{2\omega_{ja}}{\omega_{ja}^2 - \omega^2}\Re[\langle j|m_\beta|a\rangle\psi_a^* P_\alpha\psi_j]$$

$$+\frac{e}{2mc}\varepsilon_{\alpha\beta\gamma}\psi_a^*\mu_\gamma\psi_a, \tag{82}$$

$$J_\alpha^{\dot{B}_\beta} = \frac{e}{m\hbar} \sum_{j \neq a} \frac{2}{\omega_{ja}^2 - \omega^2} \mathfrak{F}[\langle j|m_\beta|a\rangle \psi_a^* P_\alpha \psi_j], \tag{83}$$

$$J_\alpha^{E_{\beta\gamma}} = -\frac{e}{3m\hbar} \sum_{j \neq a} \frac{2\omega_{ja}}{\omega_{ja}^2 - \omega^2} \mathfrak{R}[\langle j|\Theta_{\beta\gamma}|a\rangle \psi_a^* P_\alpha \psi_j], \tag{84}$$

$$J_\alpha^{\dot{E}_{\beta\gamma}} = \frac{e}{3m\hbar} \sum_{j \neq a} \frac{2}{\omega_{ja}^2 - \omega^2} \mathfrak{F}[\langle j|\Theta_\beta|a\rangle \psi_a^* P_\alpha \psi_j]. \tag{85}$$

The spatial 1-electron current density $J_\alpha(\mathbf{r})$ is found integrating over n electronic spin variables s_i and $n-1$ electronic coordinates $\mathbf{r}_i$ the function $J_\alpha \equiv J_\alpha(\mathbf{x}_1, \ldots \mathbf{x}_n)$, where $\mathbf{x}_i = \mathbf{r}_i s_i$, defined as

$$\begin{aligned} J_\alpha = {} & J_\alpha^{E_\beta} E(0)_\beta + J_\alpha^{B_\beta} B(0)_\beta + J_\alpha^{\dot{E}_\beta} \dot{E}(0)_\beta + J_\alpha^{\dot{B}_\beta} \dot{B}(0)_\beta \\ & + J_\alpha^{E_{\beta\gamma}} E(0)_{\beta\gamma} + J_\alpha^{\dot{E}_{\beta\gamma}} \dot{E}(0)_{\beta\gamma}. \end{aligned} \tag{86}$$

IV. POLARIZABILITIES AND SUSCEPTIBILITIES

In Eqs. (69)–(76) for the electronic observables, the quantities multiplying the electromagnetic fields at the origin are tensors of rank 2, 3 and 4, describing the linear response of the electron distribution. From Eqn. (61) one immediately obtains the equations for those quantities, which are also referred to as second-order properties according to the terminology of perturbation theory. One finds the second-rank dipole tensors:

$$\boldsymbol{\alpha}(\omega) = \frac{e^2}{\hbar} \sum_{j \neq a} \frac{2\omega_{ja}}{\omega_{ja}^2 - \omega^2} \mathfrak{R}(\langle a|\mathbf{R}|j\rangle \langle j|\mathbf{R}|a\rangle), \tag{87}$$

$$\hat{\boldsymbol{\alpha}}(\omega) = -\frac{e^2}{\hbar} \sum_{j \neq a} \frac{2}{\omega_{ja}^2 - \omega^2} \mathfrak{F}(\langle a|\mathbf{R}|j\rangle \langle j|\mathbf{R}|a\rangle), \tag{88}$$

$$\boldsymbol{\kappa}(\omega) = \frac{e^2}{2cm\hbar} \sum_{j \neq a} \frac{2\omega_{ja}}{\omega_{ja}^2 - \omega^2} \mathfrak{R}(\langle a|\mathbf{R}|j\rangle \langle j|\mathbf{L}|a\rangle), \tag{89}$$

$$\hat{\boldsymbol{\kappa}}(\omega) = -\frac{e^2}{2cm\hbar} \sum_{j \neq a} \frac{2}{\omega_{ja}^2 - \omega^2} \mathfrak{F}(\langle a|\mathbf{R}|j\rangle \langle j|\mathbf{L}|a\rangle), \tag{90}$$

$$\boldsymbol{\chi}^{\mathrm{p}}(\omega) = \frac{e^2}{4c^2m^2\hbar} \sum_{j \neq a} \frac{2\omega_{ja}}{\omega_{ja}^2 - \omega^2} \mathfrak{R}(\langle a|\mathbf{L}|j\rangle \langle j|\mathbf{L}|a\rangle), \tag{91}$$

$$\boldsymbol{\chi}^{\mathrm{d}} = -\frac{e^2}{4mc^2} \left\langle a \left| \sum_{i=1}^{n} (r_i^2 \mathbf{1} - \mathbf{r}_i \mathbf{r}_i) \right| a \right\rangle, \tag{92}$$

$$\boldsymbol{\chi} = \boldsymbol{\chi}^{\mathrm{p}} + \boldsymbol{\chi}^{\mathrm{d}}, \tag{93}$$

$$\hat{\boldsymbol{\chi}}^{\mathrm{p}}(\omega) = -\frac{e^2}{4c^2m^2\hbar}\sum_{j\neq a}\frac{2}{\omega_{ja}^2-\omega^2}\mathfrak{F}(\langle a|\mathbf{L}|j\rangle\langle j|\mathbf{L}|a\rangle), \tag{94}$$

$$\mathbf{o}(\omega) = -\frac{e}{\hbar}\sum_{j\neq a}\frac{2\omega_{ja}}{\omega_{ja}^2-\omega^2}\mathfrak{R}(\langle a|\mathbf{P}|j\rangle\langle j|\mathbf{R}|a\rangle), \tag{95}$$

$$\hat{\mathbf{o}}(\omega) = \frac{e}{\hbar}\sum_{j\neq a}\frac{2}{\omega_{ja}^2-\omega^2}\mathfrak{F}(\langle a|\mathbf{P}|j\rangle\langle j|\mathbf{R}|a\rangle), \tag{96}$$

$$\mathfrak{v}^{\mathrm{p}}(\omega) = -\frac{e}{2mc\hbar}\sum_{j\neq a}\frac{2\omega_{ja}}{\omega_{ja}^2-\omega^2}\mathfrak{R}(\langle a|\mathbf{P}|j\rangle\langle j|\mathbf{L}|a\rangle), \tag{97}$$

$$\upsilon_{\alpha\beta}^{d} = \frac{e}{2c}\varepsilon_{\alpha\beta\gamma}\langle a|R_\gamma|a\rangle, \tag{98}$$

$$\mathfrak{v} = \mathfrak{v}^{\mathrm{p}} + \mathfrak{v}^{\mathrm{d}}, \tag{99}$$

$$\hat{\mathfrak{v}}^{\mathrm{p}}(\omega) = \frac{e}{2mc\hbar}\sum_{j\neq a}\frac{2}{\omega_{ja}^2-\omega^2}\mathfrak{F}(\langle a|\mathbf{P}|j\rangle\langle j|\mathbf{L}|a\rangle), \tag{100}$$

$$\imath(\omega) = -\frac{e}{\hbar}\sum_{j\neq a}\frac{2\omega_{ja}}{\omega_{ja}^2-\omega^2}\mathfrak{R}(\langle a|\mathbf{K}_n^N|j\rangle\langle j|\mathbf{R}|a\rangle), \tag{101}$$

$$\hat{\imath}(\omega) = \frac{e}{\hbar}\sum_{j\neq a}\frac{2}{\omega_{ja}^2-\omega^2}\mathfrak{F}(\langle a|\mathbf{K}_n^N|j\rangle\langle j|\mathbf{R}|a\rangle), \tag{102}$$

$$\boldsymbol{\eta}(\omega) = -\frac{e}{2mc\hbar}\sum_{j\neq a}\frac{2\omega_{ja}}{\omega_{ja}^2-\omega^2}\mathfrak{R}(\langle a|\mathbf{K}_n^N|j\rangle\langle j|\mathbf{L}|a\rangle), \tag{103}$$

$$\hat{\boldsymbol{\eta}}(\omega) = \frac{e}{2mc\hbar}\sum_{j\neq a}\frac{2}{\omega_{ja}^2-\omega^2}\mathfrak{F}(\langle a|\mathbf{K}_n^N|j\rangle\langle j|\mathbf{L}|a\rangle). \tag{104}$$

The third-rank tensors are mixed dipole–quadrupole polarizabilities:

$$A(\omega)_{\alpha,\beta\gamma} = \frac{1}{\hbar}\sum_{j\neq a}\frac{2\omega_{ja}}{\omega_{ja}^2-\omega^2}\mathfrak{R}(\langle a|\mu_\alpha|j\rangle\langle j|\Theta_{\beta\gamma}|a), \tag{105}$$

$$\hat{A}(\omega)_{\alpha,\beta\gamma} = -\frac{1}{\hbar}\sum_{j\neq a}\frac{2}{\omega_{ja}^2-\omega^2}\mathfrak{F}(\langle a|\mu_\alpha|j\rangle\langle j|\Theta_{\beta\gamma}|a\rangle), \tag{106}$$

$$D(\omega)_{\alpha,\beta\gamma} = \frac{1}{\hbar}\sum_{j\neq a}\frac{2\omega_{ja}}{\omega_{ja}^2-\omega^2}\mathfrak{R}(\langle a|m_\alpha|j\rangle\langle j|\Theta_{\beta\gamma}|a\rangle), \tag{107}$$

$$\hat{D}(\omega)_{\alpha,\beta\gamma} = -\frac{1}{\hbar}\sum_{j\neq a}\frac{2}{\omega_{ja}^2-\omega^2}\mathfrak{F}(\langle a|m_\alpha|j\rangle\langle j|\Theta_{\beta\gamma}|a\rangle), \tag{108}$$

$$U(\omega)_{\alpha,\beta\gamma} = \frac{1}{\hbar}\sum_{j\neq a}\frac{2\omega_{ja}}{\omega_{ja}^2-\omega^2}\mathfrak{R}(\langle a|P_\alpha|j\rangle\langle j|\Theta_{\beta\gamma}|a\rangle), \tag{109}$$

$$\hat{U}(\omega)_{\alpha,\beta\gamma} = -\frac{1}{\hbar}\sum_{j\neq a}\frac{2}{\omega_{ja}^2-\omega^2}\mathfrak{I}(\langle a|P_\alpha|j\rangle\langle j|\Theta_{\beta\gamma}|a\rangle), \tag{110}$$

$$V(\omega)_{\alpha,\beta\gamma} = \frac{1}{\hbar}\sum_{j\neq a}\frac{2\omega_{ja}}{\omega_{ja}^2-\omega^2}\mathfrak{R}(\langle a|K_{n\alpha}^N|j\rangle\langle j|\Theta_{\beta\gamma}|a\rangle), \tag{111}$$

$$\hat{V}(\omega)_{\alpha,\beta\gamma} = -\frac{1}{\hbar}\sum_{j\neq a}\frac{2}{\omega_{ja}^2-\omega^2}\mathfrak{I}(\langle a|K_{n\alpha}^N|j\rangle\langle j|\Theta_{\beta\gamma}|a), \tag{112}$$

The quadrupole polarizabilities are fourth-rank tensors:

$$C(\omega)_{\alpha\beta,\gamma\delta} = \frac{1}{3\hbar}\sum_{j\neq a}\frac{2\omega_{ja}}{\omega_{ja}^2-\omega^2}\mathfrak{R}(\langle a|\Theta_{\alpha\beta}|j\rangle\langle j|\Theta_{\gamma\delta}|a\rangle), \tag{113}$$

$$\hat{C}(\omega)_{\alpha\beta,\gamma\delta} = -\frac{1}{3\hbar}\sum_{j\neq a}\frac{2}{\omega_{ja}^2-\omega^2}\mathfrak{I}(\langle a|\Theta_{\alpha\beta}|j\rangle\langle j|\Theta_{\gamma\delta}|a\rangle), \tag{114}$$

Some of those have long been known: $\boldsymbol{\alpha}$ is the electric dipole polarizability (a symmetric polar tensor of dimension l^3) in length formalism, $\hat{\boldsymbol{\kappa}}$ (an asymmetric axial tensor of dimension $l^3 t$) is related to the optical activity, **A** is the mixed dipole–quadrupole polarizability, $\boldsymbol{\chi}$ is the magnetic susceptibility (or magnetizability), written as a sum of diamagnetic and paramagnetic components.

We recall that the partition into diamagnetic and paramagnetic contributions to the molecular properties is not unique and, to some extent, "unphysical." For instance, in the case of a static uniform magnetic field, the resolutions of Eqn. (93) into (92) and (91) and of (99) into (97) and (98) stem from the choice

$$\mathbf{A}(\mathbf{r}-\mathbf{r}_0) = \tfrac{1}{2}\mathbf{B}\times(\mathbf{r}-\mathbf{r}_0)$$

for the vector potential. For simplicity, let us assume that **B** lies along the z direction, so that $A_x = -\frac{1}{2}By$, $A_y = \frac{1}{2}Bx$, and $A_z = 0$. If one alternatively chooses [21, 22, 27]

$$\mathbf{A}' = \mathbf{A} + \nabla\Lambda, \qquad \Lambda = \tfrac{1}{2}Bxy,$$

via a gauge transformation, then $A'_x = 0$, $A'_y = Bx$, and

$$\chi_{zz}^{\mathrm{p}'}(0) = \frac{e^2}{\hbar m^2 c^2}\sum_{j\neq a}\frac{2}{\omega_{ja}}\left\langle a\left|\sum_{i=1}^{n}(xp_y)_i\right|j\right\rangle\left\langle j\left|\sum_{i=1}^{n}(xp_y)_i\right|a\right\rangle,$$

$$\chi_{zz}^{\mathrm{d}'} = -\frac{e^2}{mc^2}\left\langle a\left|\sum_{i=1}^{n}x_i^2\right|a\right\rangle.$$

At any rate, if $\mathbf{r}_0$ is the center of mass, it turns out that the paramagnetic part of the susceptibility (91) is related to the rotational c factor [7c] and hence is separately observable, which perhaps justifies the wide use of the traditional definition (91).

Change of origin of the reference system also induces a gauge transformation and, consequently, leads to alternative partitions of tensors (93) and (99). They will be examined in some detail in Section VI.

We observe that the molecular properties defined via Eqs. (88), (89), (94), (95), (100), (102), (103), (106), (107), (109), and (112) are c tensors in the terminology of Birss [38], that is, they change sign under time reversal. Accordingly, they are identically vanishing for diamagnetic molecules. Also, in the absence of an external magnetic field, the eigenfunctions $|j\rangle$ may be chosen to be real if $|a\rangle$ is non-degenerate. When an additional static magnetic field is present, $|j\rangle$ and $|a\rangle$ are complex time-independent perturbed states and, as well as ω_{ja}, are functions of the additional field. In this case the effect of the c tensors is that of adding higher order contributions to the induced moments and fields.

Equations (87)–(112) for the molecular polarizabilities and susceptibilities have been obtained assuming the length and angular momentum gauges; see Eqs. (35)–(37), (60), and (61). One could alternatively use the first-order Hamiltonians (35′), (35″), and (37′) in (57) and (59), defining molecular tensors in different formalisms. This amounts to introducing the off-diagonal hypervirial relations, which are consistent with Eqn. (64):

$$\langle a|\mathbf{R}|j\rangle = \frac{i}{m}\omega_{ja}^{-1}\langle a|\mathbf{P}|j\rangle = -\frac{1}{m}\omega_{ja}^{-2}\langle a|\mathbf{F}_n^N|j\rangle$$

$$= \frac{e}{m}\omega_{ja}^{-2}\sum_{I=1}^{N} Z_I\langle a|\mathbf{E}_I^n|j\rangle, \tag{115}$$

$$\langle a|\mathbf{L}|j\rangle = i\omega_{ja}^{-1}\langle a|\mathbf{K}_n^N|j\rangle. \tag{116}$$

If the hypervirial theorem for the relative operators is satisfied, allowing for the eigenstates $|j\rangle$, then the physically equivalent definitions corresponding to different gauges yield exactly the same numerical estimates. In approximate calculations based on the algebraic approximation, the results provided by the different formalisms give an idea of the completeness of a basis set with respect to a given operator. In any case, it is interesting to observe that the use of the force and torque Hamiltonians leads to a resolution of the molecular properties into atomic terms. Thus, allowing for the identity

$$\frac{\omega_{ja}}{\omega_{ja}^2 - \omega^2} = \frac{1}{\omega_{ja}} + \frac{\omega^2}{\omega_{ja}(\omega_{ja}^2 - \omega^2)}, \tag{117}$$

one finds for the electric polarizability in mixed acceleration–length formalism

$$\boldsymbol{\alpha}(\omega) = \sum_{I=1}^{N} \boldsymbol{\alpha}^I(\omega), \tag{118}$$

$$\boldsymbol{\alpha}^I(\omega) = \frac{e^3}{m\hbar} Z_I \sum_{j \neq a} \frac{2}{\omega_{ja}(\omega_{ja}^2 - \omega^2)} \Re(\langle a|\mathbf{E}_I^n|j\rangle\langle j|\mathbf{R}|a\rangle). \tag{119}$$

Note that (119) is not symmetric, and one could alternatively introduce the definition

$$\tilde{\boldsymbol{\alpha}}^I(\omega) = \frac{e^3}{m\hbar} Z_I \sum_{j \neq a} \frac{2}{\omega_{ja}(\omega_{ja}^2 - \omega^2)} \Re(\langle a|\mathbf{R}|j\rangle\langle j|\mathbf{E}_I^n|a\rangle). \tag{120}$$

Equation (118) yields a partition of the molecular polarizability into *gross* atomic contributions. One could also introduce a different additive scheme defining *net* atomic contributions $\boldsymbol{\alpha}^{II}$ and *bond*, or *pair*, contributions $\boldsymbol{\alpha}^{IJ}$:

$$\boldsymbol{\alpha}^{IJ}(\omega) = \frac{e^4}{m^2\hbar} Z_I Z_J \sum_{j \neq a} \frac{2}{\omega_{ja}^3(\omega_{ja}^2 - \omega^2)} \Re(\langle a|\mathbf{E}_I^n|j\rangle\langle j|\mathbf{E}_J^n|a\rangle). \tag{121}$$

For the optical rotary power, via the identity

$$\frac{1}{\omega_{ja}^2 - \omega^2} = \frac{1}{\omega_{ja}^2} + \frac{\omega^2}{\omega_{ja}^2(\omega_{ja}^2 - \omega^2)}, \tag{122}$$

one similarly finds

$$\hat{\boldsymbol{\kappa}}(\omega) = \sum_{I=1}^{N} \hat{\boldsymbol{\kappa}}^I(\omega), \tag{123}$$

defining two basically different schemes:

$$\hat{\boldsymbol{\kappa}}^I(\omega) = \frac{e^2}{2mc\hbar} \sum_{j \neq a} \frac{2}{\omega_{ja}(\omega_{ja}^2 - \omega^2)} \Re(\langle a|\mathbf{R}|j\rangle\langle j|\mathbf{K}_n^I|a\rangle), \tag{124}$$

$$\hat{\boldsymbol{\kappa}}^I(\omega) = -\frac{e^3}{2cm^2\hbar} Z_I \sum_{j \neq a} \frac{2}{\omega_{ja}^2(\omega_{ja}^2 - \omega^2)} \Im(\langle a|\mathbf{E}_I^n|j\rangle\langle j|\mathbf{L}|a\rangle). \tag{125}$$

Analogously, for the paramagnetic susceptibility

$$\boldsymbol{\chi}^{\mathrm{p}I}(\omega) = -\frac{e^2}{4c^2m^2\hbar} \sum_{j \neq a} \frac{2}{\omega_{ja}^2 - \omega^2} \Im(\langle a|\mathbf{K}_n^I|j\rangle\langle j|\mathbf{L}|a\rangle),$$

$$\boldsymbol{\chi}^{\mathrm{p}IJ}(\omega) = \frac{e^2}{4c^2m^2\hbar} \sum_{j \neq a} \frac{2}{\omega_{ja}(\omega_{ja}^2 - \omega^2)} \Re(\langle a|\mathbf{K}_n^I|j\rangle\langle j|\mathbf{K}_n^J|a\rangle). \tag{126}$$

These results may be used to rationalize the experimental additivity schemes for the molecular tensors. Owing to the $(r - R_I)^{-3}$ factor of force and torque operators, the molecular wavefunction is essentially weighted in the environment of nucleus I, which could imply transferability of *atomic* terms from molecule to molecule in a series of structurally and chemically related homologues.

In some instances it is convenient to introduce a complex representation. This is particularly useful in order to account for absorption and stimulated emission [6]. Allowing for the complex fields (5′) and electric field gradient

$$\mathbf{E} = \mathbf{E}_0 \exp\left[i\omega\left(t - \frac{\mathbf{n}\cdot\mathbf{r}}{c} \right) \right], \qquad \frac{\partial \mathbf{E}}{\partial t} = i\omega \mathbf{E}, \tag{127}$$

$$\mathbf{B} = \mathbf{B}_0 \exp\left[i\omega\left(t - \frac{\mathbf{n}\cdot\mathbf{r}}{c} \right) \right] = \mathbf{n} \times \mathbf{E}, \qquad \frac{\partial \mathbf{B}}{\partial t} = i\omega \mathbf{B}, \tag{128}$$

$$E_{\alpha\beta} = -\frac{i\omega n_\alpha}{c} E_{0\beta} \exp\left[i\omega\left(t - \frac{\mathbf{n}\cdot\mathbf{r}}{c} \right) \right], \tag{129}$$

the new complex properties $\mathbf{X}$ defined hereafter are equal to the sum $\mathbf{X} + i\omega\hat{\mathbf{X}}$, involving the quantities previously introduced in (87)–(112). The complex induced moments become

$$\Delta\langle \boldsymbol{\mu} \rangle = \boldsymbol{\alpha}\cdot\mathbf{E} + \boldsymbol{\kappa}\cdot\mathbf{B} + \tfrac{1}{3}\mathbf{A}:\nabla\mathbf{E}, \tag{130}$$

$$\Delta\langle \mathbf{m}' \rangle = \mathbf{E}\cdot\boldsymbol{\kappa}^* + (\boldsymbol{\chi}^{\mathrm{p}} + \boldsymbol{\chi}^{\mathrm{d}})\cdot\mathbf{B} + \tfrac{1}{3}\mathbf{D}:\nabla\mathbf{E}. \tag{131}$$

The complex tensors are defined as

$$\begin{aligned} \boldsymbol{\alpha}(\omega) = \frac{e^2}{\hbar} \sum_{j \neq a} \frac{2}{\omega_{ja}^2 - \omega^2} [\omega_{ja} \Re(\langle a|\mathbf{R}|j\rangle \langle j|\mathbf{R}|a\rangle). \\ - i\omega \Im(\langle a|\mathbf{R}|j\rangle \langle j|\mathbf{R}|a\rangle)] = \boldsymbol{\alpha}^\dagger, \end{aligned} \tag{132}$$

$$\begin{aligned} \boldsymbol{\kappa}(\omega) = \frac{e^2}{2cm\hbar} \sum_{j \neq a} \frac{2}{\omega_{ja}^2 - \omega^2} [\omega_{ja} \Re(\langle a|\mathbf{R}|j\rangle \langle j|\mathbf{L}|a\rangle) \\ - i\omega \Im(\langle a|\mathbf{R}|j\rangle \langle j|\mathbf{L}|a\rangle)], \end{aligned} \tag{133}$$

$$\begin{aligned} \boldsymbol{\chi}^{\mathrm{p}}(\omega) = \frac{e^2}{4c^2m^2\hbar} \sum_{j \neq a} \frac{2}{\omega_{ja}^2 - \omega^2} [\omega_{ja} \Re(\langle a|\mathbf{L}|j\rangle \langle j|\mathbf{L}|a\rangle) \\ - i\omega \Im(\langle a|\mathbf{L}|j\rangle \langle j|\mathbf{L}|a\rangle)] = \boldsymbol{\chi}^{\mathrm{p}\dagger}, \end{aligned} \tag{134}$$

and similar formulas can be given for the other complex quantities. The actual moments are the real parts of Eqs. (130) and (131).

V. NUCLEAR ELECTROMAGNETIC SHIELDINGS

The physical meaning of the response tensors appearing in formulas (73) and (74) is immediately grasped. The dipole properties are

$$\boldsymbol{\gamma}^I(\omega) = \frac{e}{\hbar}\sum_{j\neq a}\frac{2\omega_{ja}}{\omega_{ja}^2-\omega^2}\Re(\langle a|\mathbf{E}_I^n|j\rangle\langle j|\mathbf{R}|a\rangle), \tag{135}$$

$$\hat{\boldsymbol{\gamma}}^I(\omega) = -\frac{e}{\hbar}\sum_{j\neq a}\frac{2}{\omega_{ja}^2-\omega^2}\Im(\langle a|\mathbf{E}_I^n|j\rangle\langle j|\mathbf{R}|a\rangle), \tag{136}$$

$$\boldsymbol{\xi}^I(\omega) = -\frac{e}{2cm\hbar}\sum_{j\neq a}\frac{2\omega_{ja}}{\omega_{ja}^2-\omega^2}\Re(\langle a|\mathbf{E}_I^n|j\rangle\langle j|\mathbf{L}|a\rangle), \tag{137}$$

$$\hat{\boldsymbol{\xi}}^I(\omega) = \frac{e}{2cm\hbar}\sum_{j\neq a}\frac{2}{\omega_{ja}^2-\omega^2}\Im(\langle a|\mathbf{E}_I^n|j\rangle\langle j|\mathbf{L}|a\rangle), \tag{138}$$

$$\boldsymbol{\sigma}^{\mathrm{p}I}(\omega) = -\frac{e^2}{2c^2m^2\hbar}\sum_{j\neq a}\frac{2\omega_{ja}}{\omega_{ja}^2-\omega^2}\Re(\langle a|\mathbf{M}_I^n|j\rangle\langle j|\mathbf{L}|a\rangle), \tag{139}$$

$$\boldsymbol{\sigma}^{\mathrm{d}I} = \frac{e}{2c^2m}\left\langle a\left|\sum_{i=1}^{n}(\mathbf{r}_i\cdot\mathbf{E}_I^i\mathbf{1}-\mathbf{r}_i\mathbf{E}_I^i)\right|a\right\rangle, \tag{140}$$

$$\boldsymbol{\sigma}^I = \boldsymbol{\sigma}^{\mathrm{p}I}+\boldsymbol{\sigma}^{\mathrm{d}I}, \tag{141}$$

$$\hat{\boldsymbol{\sigma}}^{\mathrm{p}I}(\omega) = \frac{e^2}{2c^2m^2\hbar}\sum_{j\neq a}\frac{2}{\omega_{ja}^2-\omega^2}\Im(\langle a|\mathbf{M}_I^n|j\rangle\langle j|\mathbf{L}|a\rangle), \tag{142}$$

$$\boldsymbol{\lambda}^I(\omega) = \frac{e^2}{cm\hbar}\sum_{j\neq a}\frac{2\omega_{ja}}{\omega_{ja}^2-\omega^2}\Re(\langle a|\mathbf{M}_I^n|j\rangle\langle j|\mathbf{R}|a\rangle), \tag{143}$$

$$\hat{\boldsymbol{\lambda}}^I(\omega) = -\frac{e^2}{cm\hbar}\sum_{j\neq a}\frac{2}{\omega_{ja}^2-\omega^2}\Im(\langle a|\mathbf{M}_I^n|j\rangle\langle j|\mathbf{R}|a\rangle). \tag{144}$$

By taking the scalar dyadic product with the external electric field, Eqn. (73) yields the electric field induced at nucleus I by the perturbed electron via a feedback effect. The total effective field experienced by the nucleus is hence

$$\mathbf{E}_{\mathrm{eff}}(\mathbf{R}^I;\omega) = [\mathbf{1}-\boldsymbol{\gamma}^I(\omega)]\cdot\mathbf{E}. \tag{145}$$

Accordingly, we may call nuclear electric shielding the dimensionless tensor (135). In much the same way, Eqs. (138), (141), and (144), define dynamic electromagnetic, magnetic, and magnetoelectric nuclear shieldings. Equation (139) is a generalization of the Ramsey definition [7] for the static magnetic shielding.

The mixed dipole–quadrupole nuclear shieldings are

$$\nu^I(\omega)_{\alpha,\beta\gamma} = \frac{1}{\hbar}\sum_{j\neq a}\frac{2\omega_{ja}}{\omega_{ja}^2-\omega^2}\Re(\langle a|E^n_{I_\alpha}|j\rangle\langle j|\Theta_{\beta\gamma}|a\rangle), \tag{146}$$

$$\hat{\nu}^I(\omega)_{\alpha,\beta\gamma} = -\frac{1}{\hbar}\sum_{j\neq a}\frac{2}{\omega_{ja}^2-\omega^2}\Im(\langle a|E^n_{I_\alpha}|j\rangle\langle j|\Theta_{\beta\gamma}|a\rangle), \tag{147}$$

$$\tau^I(\omega)_{\alpha,\beta\gamma} = \frac{1}{\hbar}\sum_{j\neq a}\frac{2\omega_{ja}}{\omega_{ja}^2-\omega^2}\Re(\langle a|B^n_{I_\alpha}|j\rangle\langle j|\Theta_{\beta\gamma}|a\rangle), \tag{148}$$

$$\hat{\tau}^I(\omega)_{\alpha,\beta\gamma} = -\frac{1}{\hbar}\sum_{j\neq a}\frac{2}{\omega_{ja}^2-\omega^2}\Im(\langle a|B^n_{I_\alpha}|j\rangle\langle j|\Theta_{\beta\gamma}|a\rangle), \tag{149}$$

and describe the effects of electric field gradient. The c tensors (136), (137), (142), and (143) are identically zero for diamagnetic molecules.

Allowing for the hypervirial relations (115) and (116) and for the identities (117) and (122), one can partition the nuclear shieldings into *atomic* terms, in much the same way as for polarizabilities and susceptibilities. Thus

$$\hat{\boldsymbol{\xi}}^I(\omega) = \sum_{J=1}^{N}\hat{\boldsymbol{\xi}}^{IJ}(\omega), \tag{150}$$

$$\hat{\boldsymbol{\xi}}^{IJ}(\omega) = -\frac{e}{2cm\hbar}\sum_{j\neq a}\frac{2}{\omega_{ja}(\omega_{ja}^2-\omega^2)}\Re(\langle a|\mathbf{E}^n_I|j\rangle\langle j|\mathbf{K}^J_n|a\rangle), \tag{151}$$

$$\boldsymbol{\gamma}^I(\omega) = \sum_{J=1}^{N}\boldsymbol{\gamma}^{IJ}(\omega), \tag{152}$$

$$\boldsymbol{\gamma}^{IJ}(\omega) = \frac{e^2}{m\hbar}Z_J\sum_{j\neq a}\frac{2}{\omega_{ja}(\omega_{ja}^2-\omega^2)}\Re(\langle a|\mathbf{E}^n_I|j\rangle\langle j|\mathbf{E}^n_J|a\rangle), \tag{153}$$

$$Z_I\gamma^{IJ}_{\alpha\beta} = Z_J\gamma^{JI}_{\beta\alpha}, \tag{154}$$

$$\hat{\boldsymbol{\gamma}}^{IJ}(\omega) = -\frac{e^2}{m\hbar}Z_J\sum_{j\neq a}\frac{2}{\omega_{ja}^2(\omega_{ja}^2-\omega^2)}\Im(\langle a|\mathbf{E}^n_I|j\rangle\langle j|\mathbf{E}^n_J|a\rangle), \tag{155}$$

$$Z_I\hat{\gamma}^{IJ}_{\alpha\beta} = -Z_J\hat{\gamma}^{JI}_{\beta\alpha}, \tag{156}$$

$$\boldsymbol{\sigma}^{pIJ}(\omega) = -\frac{e^2}{2c^2m^2\hbar}\sum_{j\neq a}\frac{2}{\omega_{ja}^2-\omega^2}\Im(\langle a|\mathbf{M}^n_I|j\rangle\langle j|\mathbf{K}^J_n|a\rangle), \tag{157}$$

$$\hat{\boldsymbol{\sigma}}^{pIJ}(\omega) = -\frac{e^2}{2c^2m^2\hbar}\sum_{j\neq a}\frac{2}{\omega_{ja}(\omega_{ja}^2-\omega^2)}\Re(\langle a|\mathbf{M}^n_I|j\rangle\langle j|\mathbf{K}^J_n|a\rangle), \tag{158}$$

$$\boldsymbol{\lambda}^{IJ}(\omega) = \frac{e^3}{cm^2\hbar}Z_J\sum_{j\neq a}\frac{2}{\omega_{ja}(\omega_{ja}^2-\omega^2)}\Re(\langle a|\mathbf{M}^n_I|j\rangle\langle j|\mathbf{E}^n_J|a\rangle), \tag{159}$$

$$\hat{\boldsymbol{\lambda}}^{IJ}(\omega) = -\frac{e^3}{cm^2\hbar} Z_J \sum_{j\neq a} \frac{2}{\omega_{ja}^2(\omega_{ja}^2 - \omega^2)} \mathfrak{F}(\langle a|\mathbf{M}_I^n|j\rangle\langle j|\mathbf{E}_J^n|a\rangle). \tag{160}$$

Within the complex representation (127)–(129) the induced fields at the nuclei become

$$\Delta\langle\mathbf{E}_I^n\rangle = -\boldsymbol{\gamma}^I\cdot\mathbf{E} + \boldsymbol{\xi}^I\cdot\mathbf{B} + \tfrac{1}{3}\mathbf{v}^I:\nabla\mathbf{E} \tag{161}$$

$$\Delta\langle\mathbf{B}_I^{n'}\rangle = \boldsymbol{\lambda}^I\cdot\mathbf{E} - (\boldsymbol{\sigma}^{\mathrm{p}I} + \boldsymbol{\sigma}^{\mathrm{d}I})\cdot\mathbf{B} + \tfrac{1}{3}\boldsymbol{\tau}^I:\nabla\mathbf{E}, \tag{162}$$

and the complex shieldings are defined analogously to the complex polarizabilities, for instance,

$$\begin{aligned}\boldsymbol{\gamma}^I(\omega) = \frac{e}{\hbar}\sum_{j\neq a}\frac{2}{\omega_{ja}^2-\omega^2}[\omega_{ja}\mathfrak{R}(\langle a|\mathbf{E}_I^n|j\rangle\langle j|\mathbf{R}|a\rangle)\\ - i\omega\mathfrak{F}(\langle a|\mathbf{E}_I^n|j\rangle\langle j|\mathbf{R}|a\rangle)],\end{aligned} \tag{163}$$

$$\begin{aligned}\boldsymbol{\xi}^I(\omega) = -\frac{e}{2cm\hbar}\sum_{j\neq a}\frac{2}{\omega_{ja}^2-\omega^2}[\omega_{ja}\mathfrak{R}(\langle a|\mathbf{E}_I^n|j\rangle\langle j|\mathbf{L}|a\rangle)\\ - i\omega\mathfrak{F}(\langle a|\mathbf{E}_I^n|j\rangle\langle j|\mathbf{L}|a\rangle)],\end{aligned} \tag{164}$$

$$\begin{aligned}\boldsymbol{\sigma}^{\mathrm{p}I}(\omega) = -\frac{e^2}{2c^2m^2\hbar}\sum_{j\neq a}\frac{2}{\omega_{ja}^2-\omega^2}[\omega_{ja}\mathfrak{R}(\langle a|\mathbf{M}_I^n|j\rangle\langle j|\mathbf{L}|a\rangle)\\ - i\omega\mathfrak{F}(\langle a|\mathbf{M}_I^n|j\rangle\langle j|\mathbf{L}|a\rangle)],\end{aligned} \tag{165}$$

$$\begin{aligned}\boldsymbol{\lambda}^I(\omega) = \frac{e^2}{cm\hbar}\sum_{j\neq a}\frac{2}{\omega_{ja}^2-\omega^2}[\omega_{ja}\mathfrak{R}(\langle a|\mathbf{M}_I^n|j\rangle\langle j|\mathbf{R}|a\rangle)\\ - i\omega\mathfrak{F}(\langle a|\mathbf{M}_I^n|j\rangle\langle j|\mathbf{R}|a\rangle)].\end{aligned} \tag{166}$$

Some of the nuclear shielding tensors are related to spectroscopic parameters. The static electric shielding can be experimentally obtained via infrared (IR) intensities [12], and the static electromagnetic shielding is available from the intensities of vibrational circular dichroism spectra [16–20].

For the ith fundamental band, the integrated absorption coefficient is proportional to squares of derivatives, with respect to normal coordinates, of the dipole moment in the molecular ground state ψ_0, that is,

$$\begin{aligned}A_i^{\mathrm{IR}} \propto \sum_\alpha\left(\frac{\partial}{\partial Q_i}\mathcal{M}_\alpha\right)^2,\\ \mathcal{M}_\alpha = \left\langle\psi_0\middle|\mu_\alpha + e\sum_{I=1}^N Z_I R_{I\alpha}\middle|\psi_0\right\rangle.\end{aligned} \tag{167}$$

Derivatives of the dipole moment with respect to Q_i can be expressed within a Cartesian reference frame via a similarity transformation, introducing atomic polar tensors (APTs) [13, 14]. The connection between the latter and the electric shielding is obtained by means of the Hellmann–Feynman theorem. Within the Born–Oppenheimer approximation and allowing for the dipole length formalism, the perturbed Hamiltonian in the presence of a static external electric field **E** is given by Eqs. (6) and (35).

The molecular (real) wavefunction is expanded in powers of **E**:

$$\psi = \psi_0 + E_\alpha \psi^{E_\alpha} + \ldots, \tag{168}$$

where

$$\psi^{E_\alpha} = -\frac{e}{\hbar} \sum_{j \neq 0} |\psi_j\rangle \langle \psi_j | R_\alpha | \psi_0 \rangle \omega_{j0}^{-1} \tag{169}$$

is the first-order perturbed function. The operator representing the total force on nucleus I is

$$\mathbf{F}_I = -\nabla_I H = \mathbf{F}_I^n + \mathbf{F}_I^E + \mathbf{F}_I^{N-I}, \tag{170}$$

where

$$\mathbf{F}_I^E = Z_I e \mathbf{E} \tag{171}$$

is the electric Lorentz force and the last term represents the repulsive force on nucleus I exerted by the other nuclei:

$$\mathbf{F}_I^{N-I} = Z_I e^2 \sum_{J \neq I} Z_J \frac{\mathbf{R}_I - \mathbf{R}_J}{|\mathbf{R}_I - \mathbf{R}_J|^3}. \tag{172}$$

Summing over I, these internuclear forces cancel out by action equals reaction.

For an electrically neutral molecule

$$\sum_{I=1}^{N} \langle \psi | \mathbf{F}_I | \psi \rangle = \mathbf{0}. \tag{173}$$

To first order in E

$$e \sum_{I=1}^{N} Z_I [\delta_{\alpha\beta} + 2 \langle \psi_0 | E_{I\alpha}^n | \psi^{E_\beta} \rangle] E_\beta = 0. \tag{174}$$

From Eqs. (135) and (169) and

$$\gamma^I(0)_{\alpha\beta} = -2\langle\psi_0|E^n_{I\alpha}|\psi^{E_\beta}\rangle, \tag{175}$$

the equilibrium condition is rewritten as

$$\sum_{I=1}^{N} Z_I[\delta_{\alpha\beta} - \gamma^I(0)_{\alpha\beta}] = 0. \tag{176}$$

For wavefunctions obeying the Hellman–Feynman theorem

$$-\sum_{I=1}^{N} \nabla_I\langle\psi|H|\psi\rangle = -\sum_{I=1}^{N}\langle\psi|\nabla_I H|\psi\rangle = \sum_{I=1}^{N}\langle\psi|\mathbf{F}_I|\psi\rangle. \tag{177}$$

Hence, to first order in E, we find

$$\nabla_{I\alpha}\mathscr{M}_\beta = Z_I e[\delta_{\alpha\beta} - \gamma^I(0)_{\alpha\beta}], \tag{178}$$

which relates the nuclear electric shielding tensor to the atomic polar tensor and to the IR intensities.

The results we have obtained provide a reinterpretation of IR absorption along the following lines: the IR radiation is a dynamic electric field which causes oscillations in the electronic cloud. The perturbed electrons, in turn, induce an additional dynamic electric field at the nuclei via a feedback effect. The latter are hence acted upon by the effective electric field (145), that is, by a frequency-dependent Lorentz force that is responsible for changes of nuclear vibrational motion. Accordingly, the electron distribution of a molecule plays a fundamental role in determining the general features of nuclear vibrations and the magnitude of IR parameters.

As the derivative of the electric dipole moment can also be written

$$\nabla_{I\alpha}\mathscr{M}_\beta = Z_I e\left[\delta_{\alpha\beta} - \frac{2}{Z_I}\langle\psi_0|R_\beta|\nabla_{I\alpha}\psi_0\rangle\right] \tag{179}$$

from the interchange theorem

$$\langle\psi_0|E^n_{I\alpha}|\psi^{E_\alpha}\rangle = -\frac{1}{Z_I}\langle\psi_0|R_\beta|\nabla_{I\alpha}\psi_0\rangle \tag{180}$$

one finds a formula for the Hellmann—Feynman analytic gradient of a molecular wavefunction evaluated at the equilibrium geometry

$$|\nabla_I\psi_0\rangle = -\frac{1}{\hbar}\sum_{j\neq 0}|\psi_j\rangle\langle\psi_j|\mathbf{F}^I_n|\psi_0\rangle\omega_{j0}^{-1}. \tag{181}$$

Hence, the static nuclear electric shielding can be given the alternative general form

$$\gamma^I(0) = \frac{2}{Z_I}\Re(\langle \nabla_I \psi_0 | \mathbf{R} | \psi_0 \rangle). \tag{182}$$

We observe that (181) is the first-order perturbed function related to the acceleration Hamiltonian (35″) in the limit of a static Hertz–Righi potential. Using Eqn. (181), the Stephens overlap integral [17] $I^I_{\alpha\beta}$, appearing in the expression for the intensity of VCD bands, can be immediately rewritten in terms of the electromagnetic shielding [20]:

$$\begin{aligned} I^I_{\alpha\beta} &= -\sum_{j\neq a}\left\langle a\left|\left[\frac{\partial}{\partial R_{I\alpha}}H_0\right]\right|j\right\rangle\langle j|m_\beta|a\rangle(E_a - E_j)^{-2} \\ &= -\frac{iZ_I e}{2\hbar}\hat{\xi}^I(0)_{\alpha\beta}. \end{aligned} \tag{183}$$

VI. SUM RULES

The molecular response tensors are characterized by peculiar properties and satisfy a series of very general quantum-mechanical relations. First we observe that the dynamic properties can be rewritten as a sum of the corresponding static property and a function multiplying the square of the angular frequency. Thus, for instance, in the case of dipole electric polarizability, using Eqn. (117),

$$\boldsymbol{\alpha}(\omega) = \boldsymbol{\alpha}(0) + \frac{e^2}{\hbar}\sum_{j\neq a}\frac{2\omega^2}{\omega_{ja}(\omega_{ja}^2 - \omega^2)}\Re(\langle a|\mathbf{R}|j\rangle\langle j|\mathbf{R}|a\rangle). \tag{184}$$

Analogously, for the optical rotatory power, using Eqn. (122),

$$\hat{\boldsymbol{\kappa}}(\omega) = \hat{\boldsymbol{\kappa}}(0) - \frac{e^2}{2cm\hbar}\sum_{j\neq a}\frac{2\omega^2}{\omega_{ja}^2(\omega_{ja}^2 - \omega^2)}\Im(\langle a|\mathbf{R}|j\rangle\langle j|\mathbf{L}|a\rangle). \tag{185}$$

In some cases the new function has a precise physical meaning, that is, it describes another molecular property. One can easily find, for instance, that

$$\hat{\boldsymbol{\kappa}}(\omega) = \frac{e}{m}\omega^{-2}[\boldsymbol{\upsilon}^{\mathrm{p}}(\omega) - \boldsymbol{\upsilon}^{\mathrm{p}}(0)], \tag{186}$$

which relates the optical rotatory power to the polarizability (97). It can also be proven that

$$\hat{\mathbf{o}}(\omega) = -\frac{m}{e}\boldsymbol{\alpha}(\omega), \quad \mathbf{o}(\omega) = \frac{m}{e}\omega^2\hat{\boldsymbol{\alpha}}(\omega), \quad \mathbf{o}(0) = \mathbf{0},$$

$$\hat{\mathbf{U}}(\omega) = -\frac{m}{e}\mathbf{A}(\omega), \ \mathbf{U}(\omega) = \frac{m}{e}\omega^2\hat{\mathbf{A}}(\omega), \ \mathbf{U}(0) = \mathbf{0},$$

$$\hat{\mathfrak{v}}^{\mathrm{P}}(\omega) = -\frac{m}{e}\boldsymbol{\kappa}(\omega), \quad \mathfrak{v}(0) = \mathbf{0}. \tag{187}$$

These relations are a simple consequence of the equations [9]

$$\frac{\partial}{\partial t}\langle \mu_\alpha \rangle = -\frac{e}{m}\langle \Pi_\alpha \rangle, \quad \frac{\partial^2}{\partial t^2}\mathbf{E} = -\omega^2\mathbf{E},$$

and their physical meaning is immediately gathered from (72). For instance, they restate the velocity theorem in the presence of a static magnetic field [9a] and the fact that a static electric field cannot induce current densities.

The response tensors appearing in (76) are immediately rewritten in terms of the electric and electromagnetic shieldings:

$$\iota(\omega)_{\alpha\beta} = -\sum_{I=1}^{N} Z_I e \varepsilon_{\alpha\lambda\gamma} R_{I\lambda} \gamma^I(\omega)_{\gamma\beta}, \tag{188}$$

$$\hat{\iota}(\omega)_{\alpha\beta} = \sum_{I=1}^{N} Z_I e \varepsilon_{\alpha\lambda\gamma} R_{I\lambda} \hat{\gamma}^I(\omega)_{\gamma\beta}, \tag{189}$$

$$\eta(\omega)_{\alpha\beta} = -\sum_{I=1}^{N} Z_I e \varepsilon_{\alpha\lambda\gamma} R_{I\lambda} \xi^I(\omega)_{\gamma\beta}, \tag{190}$$

$$\hat{\eta}(\omega)_{\alpha\beta} = \sum_{I=1}^{N} Z_I e \varepsilon_{\alpha\lambda\gamma} R_{I\lambda} \hat{\xi}^I(\omega)_{\gamma\beta}, \tag{191}$$

$$V_{\alpha,\beta\gamma} = -\sum_{I=1}^{N} Z_I e \varepsilon_{\alpha\delta\varepsilon} R_{I\delta} v^I(\omega)_{\varepsilon,\beta\gamma}, \tag{192}$$

$$\hat{V}_{\alpha,\beta\gamma} = \sum_{I=1}^{N} Z_I e \varepsilon_{\alpha\delta\varepsilon} R_{I\delta} \hat{v}^I(\omega)_{\varepsilon,\beta\gamma}. \tag{193}$$

These results show that the second-order properties are mutually related. A series of more general sum rules can also be found. To this end, let us introduce the auxiliary tensors

$$(\mathbf{P}, \mathbf{L})_{-1} = \frac{1}{\hbar} \sum_{j \neq a} \frac{1}{\omega_{ja}} \Re(\langle a|\mathbf{P}|j\rangle \langle j|\mathbf{L}|a\rangle), \tag{194}$$

$$(\mathbf{P}, \mathbf{P})_{-1} = \frac{1}{\hbar} \sum_{j \neq a} \frac{2}{\omega_{ja}} \Re(\langle a|\mathbf{P}|j\rangle \langle j|\mathbf{P}|a\rangle), \tag{195}$$

$$(\mathbf{K}_n^N, \mathbf{L})_{-2} = -\frac{1}{\hbar} \sum_{j \neq a} \frac{2}{\omega_{ja}^2} \Im(\langle a|\mathbf{K}_n^N|j\rangle \langle j|\mathbf{L}|a\rangle), \tag{196}$$

$$(\mathbf{K}_n^N, \mathbf{F}_n^N)_{-3} = \frac{1}{\hbar} \sum_{j \neq a} \frac{2}{\omega_{ja}^3} \Re(\langle a|\mathbf{K}_n^N|j\rangle \langle j|\mathbf{F}_n^N|a\rangle), \tag{197}$$

$$(\mathbf{K}_n^N, \mathbf{K}_n^N)_{-3} = \frac{1}{\hbar} \sum_{j \neq a} \frac{2}{\omega_{ja}^3} \Re(\langle a|\mathbf{K}_n^N|j\rangle \langle j|\mathbf{K}_n^N|a\rangle), \tag{198}$$

$$(\mathbf{M}_I^n, \mathbf{P})_{-1} = \frac{1}{\hbar} \sum_{j \neq a} \frac{2}{\omega_{ja}} \Re(\langle a|\mathbf{M}_I^n|j\rangle \langle j|\mathbf{P}|a\rangle), \tag{199}$$

$$(\mathbf{M}_I^n, \mathbf{F}_n^N)_{-2} = \frac{1}{\hbar} \sum_{j \neq a} \frac{2}{\omega_{ja}^2} \Im(\langle a|\mathbf{M}_I^n|j\rangle \langle j|\mathbf{F}_n^N|a\rangle), \tag{200}$$

$$(\mathbf{F}_n^N, \mathbf{R})_{-1} \doteq -\frac{m}{\hbar} \sum_{j \neq a} 2\omega_{ja}^{-1} \Re(\langle a|\mathbf{F}_n^N|j\rangle \langle j|\mathbf{R}|a\rangle), \tag{201}$$

$$(\mathbf{K}_n^N, \mathbf{R})_{-1} = -\frac{m}{\hbar} \sum_{j \neq a} 2\omega_{ja}^{-1} \Re(\langle a|\mathbf{K}_n^N|j\rangle \langle j|\mathbf{R}|a\rangle), \tag{202}$$

$$(\mathbf{L}, \mathbf{F}_n^N)_{-2} = \frac{1}{\hbar} \sum_{j \neq a} 2\omega_{ja}^{-2} \Im(\langle a|\mathbf{L}|j\rangle \langle j|\mathbf{F}_n^N|a\rangle). \tag{203}$$

By means of these formulas and the hypervirial relations, we can prove that

$$\sum_{I=1}^{N} Z_I \gamma^I(0) = n\mathbf{1}, \tag{204}$$

which is the Thomas–Reiche–Kuhn sum rule [34], written in the mixed acceleration–length formalism. The corresponding equation for (136) gives

$$\sum_{I=1}^{N} Z_I \hat{\gamma}^I(0) = -\frac{2m}{\hbar}[\Im(\langle a|\mathbf{RR}|a\rangle) - \Im(\langle a|\mathbf{R}|a\rangle \langle a|\mathbf{R}|a\rangle)] = \mathbf{0}, \tag{205}$$

as the diagonal matrix elements of Hermitian operators are real. Analogously,

$$\sum_{I=1}^{N} Z_I v^I(0)_{\alpha,\beta\gamma} = \langle a|R_\alpha|a\rangle\delta_{\beta\gamma} - \tfrac{3}{2}\langle a|R_\beta|a\rangle\delta_{\alpha\gamma} - \tfrac{3}{2}\langle a|R_\gamma|a\rangle\delta_{\alpha\beta}, \tag{206}$$

$$\sum_{I=1}^{N} Z_I \hat{v}^I(0)_{\alpha,\beta\gamma} = 0. \tag{207}$$

The sum rule

$$\sum_{I=1}^{N} Z_I \xi^I(0)_{\alpha\beta} = \frac{1}{2mc}\langle a|P_\gamma|a\rangle\varepsilon_{\alpha\beta\gamma}, \tag{208}$$

is trivially satisfied in the case of real eigenfunctions. (In fact, it is vanishing, owing to the momentum theorem [9a]. If the states $|j\rangle$ have been determined in the presence of a static magnetic field, however, the velocity theorem [9a] holds and Eqn. (208) may be different from zero).

We observe that Eqn. (204) is the same as the equilibrium condition (176). It can also be easily proven that it states the translational invariance of the dipole moment of a neutral molecule. Introducing the operator

$$\mathscr{T}(\mathbf{d}) = \exp\left(\frac{i}{\hbar}\sum_{I=1}^{N} \mathbf{d}\cdot\mathbf{P}_I\right), \tag{209}$$

which generates a rigid translation of $\mathbf{d}$ of the nuclear skeleton, and owing to the definition (181) of the gradient function, one finds immediately, to first order in $\mathbf{d}$,

$$\nabla_{I\alpha}\langle a|R_\beta|a\rangle = Z_I\gamma^I(0)_{\alpha\beta}, \tag{210}$$

$$\nabla_{I\alpha}\langle a|L_\beta|a\rangle = -2cmZ_I\xi^I(0)_{\alpha\beta}, \tag{211}$$

$$\nabla_{I\alpha}\langle a|\Theta_{\beta\gamma}|a\rangle = Z_I e v^I(0)_{\alpha,\beta\gamma}, \tag{212}$$

$$\frac{i}{\hbar}\sum_{I=1}^{N} P_{I\alpha}\langle a|R_\beta|a\rangle = n\delta_{\alpha\beta} = \sum_{I=1}^{N} Z_I\gamma^I(0)_{\alpha\beta}, \tag{213}$$

$$\frac{i}{\hbar}\sum_{I=1}^{N} P_{I\alpha}\langle a|L_\beta|a\rangle = -\varepsilon_{\alpha\beta\gamma}\langle a|P_\gamma|a\rangle = -2cm\sum_{I=1}^{N} Z_I\xi^I(0)_{\alpha\beta}. \tag{214}$$

Hence Eqs. (213) and (214), [i.e., (204) and (208)] are constraints for translational invariance of expectation values. If the momentum theorem is satisfied, then, according to (214), the average angular momentum is independent of origin.

A relation between the paramagnetic susceptibility and the electromagnetic nuclear shieldings is easily established via Eqs. (196) and (203):

$$\sum_{I=1}^{N} Z_I \varepsilon_{\alpha\beta\gamma} R_{I\beta} \hat{\xi}^I(0)_{\gamma\delta} = \frac{1}{2cm} \sum_{I=1}^{N} \varepsilon_{\alpha\beta\gamma} R_{I\beta} (F^I_{n\gamma}, L_\delta)_{-2}$$

$$= \frac{1}{2cm} (K^N_{n\alpha}, L_\delta)_{-2} = \frac{2mc}{e^2} \chi^{\mathrm{p}}(0)_{\alpha\delta}. \tag{215}$$

This is another way of writing the partition into Pascalian terms (126):

$$\chi^{\mathrm{p}I}(0)_{\alpha\delta} = \frac{e^2}{4m^2c^2} (K^I_{n\alpha}, L_\delta)_{-2} = \frac{e^2}{2mc} Z_I \varepsilon_{\alpha\beta\gamma} R_{I\beta} \hat{\xi}^I(0)_{\gamma\delta}. \tag{215$'$}$$

The identity between the first and the last sides of (215) holds for any frequency, that is,

$$\sum_{I=1}^{N} Z_I \varepsilon_{\alpha\beta\gamma} R_{I\beta} \hat{\xi}^I(\omega)_{\gamma\delta} = \frac{2mc}{e^2} \chi^{\mathrm{p}}(\omega)_{\alpha\delta}. \tag{216}$$

The same formulas can be rewritten within the torque formalism:

$$\sum_{I=1}^{N} \varepsilon_{\alpha\beta\gamma} R_{I\beta} (K^N_{n\delta}, F^I_{n\gamma})_{-3} = (K^N_{n\delta}, K^N_{n\alpha})_{-3} = \frac{4m^2c^2}{e^2} \chi^{\mathrm{p}}(0)_{\delta\alpha}. \tag{215$''$}$$

Similar equations can be found for the nuclear shieldings (136), (137), (146), and (147):

$$\sum_{I=1}^{N} Z_I \varepsilon_{\alpha\beta\gamma} R_{I\beta} \xi^I(0)_{\gamma\delta} = \frac{1}{2cm} \varepsilon_{\alpha\delta\gamma} \langle a | L_\gamma | a \rangle, \tag{217}$$

$$\sum_{I=1}^{N} Z_I \varepsilon_{\alpha\beta\gamma} R_{I\beta} \hat{\gamma}^I(\omega)_{\gamma\delta} = -\frac{2cm}{e^2} \kappa(\omega)_{\delta\alpha}. \tag{217$'$}$$

$$\sum_{I=1}^{N} Z_I \varepsilon_{\alpha\beta\gamma} R_{I\beta} \nu^I(0)_{\gamma,\varepsilon\delta} = \frac{1}{e} (\langle a | \Theta_{\beta\delta} | a \rangle \varepsilon_{\alpha\beta\varepsilon} + \langle a | \Theta_{\beta\varepsilon} | a \rangle \varepsilon_{\alpha\beta\delta}), \tag{218}$$

$$\sum_{I=1}^{N} Z_I \varepsilon_{\alpha\beta\gamma} R_{I\beta} \hat{\nu}^I(\omega)_{\gamma,\varepsilon\delta} = \frac{2cm}{e^2} D(\omega)_{\alpha,\varepsilon\delta}. \tag{218$'$}$$

Using Eqn. (200), one also finds

$$\sum_{J=1}^{N} \varepsilon_{\alpha\beta\gamma} R_{J\beta} (F^J_{n\gamma}, M^n_{I\delta})_{-2} = (K^N_{n\alpha}, M^n_{I\delta})_{-2} = -\frac{2m^2c^2}{e^2} \sigma^{\mathrm{p}I}(0)_{\delta\alpha}. \tag{219}$$

Some of these equations are sum rules arising from constraints for rotational invariance and are also determined introducing the operator

$$\mathscr{R}(\boldsymbol{\theta}) = \exp\left(\frac{i}{\hbar}\sum_{I=1}^{N} \boldsymbol{\theta}\cdot\mathbf{L}_I\right), \qquad \boldsymbol{\theta} = \theta\mathbf{n}_\theta, \qquad \mathbf{L}_I = \mathbf{R}_I \times \mathbf{P}_I, \tag{220}$$

which generates a rigid rotation of the nuclear framework through an angle θ about the $\mathbf{n}_\theta$ axis. Thus, for instance, to first order in $\boldsymbol{\theta}$,

$$\frac{i}{\hbar}\sum_{I=1}^{N} L_{I\alpha}\langle a|R_\delta|a\rangle = \frac{i}{\hbar}\langle a|[L_\alpha, R_\delta]|a\rangle$$

$$= \varepsilon_{\alpha\beta\delta}\langle a|R_\beta|a\rangle = \sum_{I=1}^{N} Z_I\varepsilon_{\alpha\beta\gamma}R_{I\beta}\gamma^I(0)_{\gamma\delta}, \tag{221}$$

$$\frac{i}{\hbar}\sum_{I=1}^{N} L_{I\alpha}\langle a|L_\delta|a\rangle = \varepsilon_{\alpha\beta\delta}\langle a|L_\beta|a\rangle = -2cm\sum_{I=1}^{N} Z_I\varepsilon_{\alpha\beta\gamma}R_{I\beta}\xi^I(0)_{\gamma\delta}. \tag{222}$$

The translational rule for the electromagnetic shieldings is the same as the rotational rule for the electric shielding:

$$\sum_{I=1}^{N} \varepsilon_{\alpha\beta\gamma}R_{I\beta}(F^I_{n\gamma}, R_\delta)_{-1} = (K^N_{n\alpha}, R_\delta)_{-1} = m\sum_{I=1}^{N} Z_I\varepsilon_{\alpha\beta\gamma}R_{I\beta}\gamma^I(0)_{\gamma\delta}$$

$$= m\varepsilon_{\alpha\beta\delta}\langle a|R_\beta|a\rangle = (L_\alpha, F^N_{n\delta})_{-2}$$

$$= 2cm\sum_{I=1}^{N} Z_I\hat{\xi}^I(0)_{\delta\alpha}, \tag{223}$$

and a general relation holds for any frequency:

$$\sum_{I=1}^{N} Z_I\varepsilon_{\alpha\beta\gamma}R_{I\beta}\gamma^I(\omega)_{\gamma\delta} = 2c\sum_{I=1}^{N} Z_I\hat{\xi}^I(\omega)_{\delta\alpha}. \tag{224}$$

Sum rule (223) has a deep physical meaning and synthesizes various aspects. Owing to the results of ref. 31, it restates the translational gauge invariance of magnetizability (93), the commutation relation

$$[L_\alpha, R_\beta] = i\hbar\varepsilon_{\alpha\beta\gamma}R_\gamma,$$

and embodies the conservation theorem for the current density induced by a static magnetic field. From the discussion following Eqs. (178) and (183), sum rule (223) is also a connection between IR and VCD intensities.

Another set of relations connecting different response tensors can be determined allowing for the hypervirial relations (115) and (116) and the identities (117)–(122). For instance, from the definition of the mixed acceleration–length polarizability, one has

$$\boldsymbol{\alpha}(\omega) = \frac{e^2}{m}\omega^{-2}\sum_{I=1}^{N} Z_I[\boldsymbol{\gamma}^I(\omega) - \boldsymbol{\gamma}^I(0)]. \tag{225}$$

A similar partitioning holds for the individual atomic contributions:

$$\tilde{\boldsymbol{\alpha}}^I(\omega) = \frac{e^2}{m}\omega^{-2}\sum_{J=1}^{N} Z_J[\boldsymbol{\gamma}^{JI}(\omega) - \boldsymbol{\gamma}^{JI}(0)]. \tag{226}$$

Similarly, for the imaginary polarizability

$$\hat{\alpha}^I(\omega)_{\beta\alpha} = -\frac{e^2}{m}\omega^{-2}\sum_{J=1}^{N} Z_J[\hat{\gamma}^{JI}(\omega)_{\alpha\beta} - \hat{\gamma}^{JI}(0)_{\alpha\beta}], \tag{226'}$$

$$\hat{\boldsymbol{\alpha}}(\omega) = \frac{e^2}{m}\omega^{-2}\sum_{I=1}^{N} Z_I[\hat{\boldsymbol{\gamma}}^I(\omega) - \hat{\boldsymbol{\gamma}}^I(0)]; \tag{227}$$

[see, however, Eqn. (205)]. The optical rotatory power can be expressed in terms of the nuclear electric and electromagnetic shieldings:

$$\hat{\kappa}(\omega)_{\delta\alpha} = \frac{e^2}{2cm}\omega^{-2}\sum_{I=1}^{N} Z_I \varepsilon_{\alpha\beta\gamma} R_{I\gamma}[\gamma^I(\omega)_{\beta\delta} - \gamma^I(0)_{\beta\delta}], \tag{228}$$

$$\hat{\boldsymbol{\kappa}}(\omega) = -\frac{e^2}{m}\omega^{-2}\sum_{I=1}^{N} Z_I[\hat{\boldsymbol{\xi}}^I(\omega) - \hat{\boldsymbol{\xi}}^I(0)]. \tag{229}$$

The mixed dipole–quadrupole polarizability satisfies a similar equation involving the electric mixed dipole–quadrupole shieldings

$$A(\omega)_{\alpha,\beta\gamma} = -\frac{e^2}{m\hbar}\sum_{I=1}^{N} Z_I \sum_{j\neq a} \frac{2}{\omega_{ja}(\omega_{ja}^2 - \omega^2)} \Re(\langle a|E_{I\alpha}^n|j\rangle\langle j|\Theta_{\beta\gamma}|a\rangle), \tag{230}$$

$$\mathbf{A}(\omega) = -\frac{e^2}{m}\omega^{-2}\sum_{I=1}^{N} Z_I[\mathbf{v}^I(\omega) - \mathbf{v}^I(0)], \tag{231}$$

and

$$\begin{aligned}\hat{D}(\omega)_{\alpha,\delta\varepsilon} &= -\frac{e^2}{2mc}\omega^{-2}\sum_{I=1}^{N} Z_I \varepsilon_{\alpha\beta\gamma} R_{I\beta}[v^I(\omega)_{\gamma,\delta\varepsilon} - v^I(0)_{\gamma,\delta\varepsilon}] \\ &= \frac{1}{2c}\sum_{I=1}^{N} \varepsilon_{\alpha\beta\gamma} R_{I\beta} A^I(\omega)_{\gamma,\delta\varepsilon}.\end{aligned} \tag{232}$$

For the atomic contributions to the electric and electromagnetic shielding one finds

$$\varepsilon_{\beta\gamma\delta} R_{J\gamma} \gamma^{IJ}(\omega)_{\alpha\beta} = -2c\hat{\xi}^{IJ}(\omega)_{\alpha\delta}, \tag{233}$$

$$\sum_{J=1}^{N} \varepsilon_{\beta\gamma\delta} R_{J\gamma} \gamma^{IJ}(\omega)_{\alpha\beta} = -2c\hat{\xi}^{I}(\omega)_{\alpha\delta}, \tag{234}$$

$$\varepsilon_{\alpha\beta\gamma} \hat{\gamma}^{IJ}(\omega)_{\delta\beta} R_{J\gamma} = 2c\omega^{-2}[\xi^{IJ}(\omega)_{\delta\alpha} - \xi^{IJ}(0)_{\delta\alpha}],$$

$$\sum_{J=1}^{N} \varepsilon_{\alpha\beta\gamma} \hat{\gamma}^{IJ}(\omega)_{\delta\beta} R_{J\gamma} = 2c\omega^{-2}[\xi^{I}(\omega)_{\delta\alpha} - \xi^{I}(0)_{\delta\alpha}], \tag{235}$$

$$\sum_{I=1}^{N} \varepsilon_{\alpha\beta\gamma} Z_I \hat{\gamma}^{I}(\omega)_{\beta\delta} R_{I\gamma} = -2c\omega^{-2} \sum_{I=1}^{N} Z_I[\xi^{I}(\omega)_{\delta\alpha} - \xi^{I}(0)_{\delta\alpha}], \tag{236}$$

$$\varepsilon_{\lambda\mu\delta} R_{J\mu} \hat{\lambda}^{IJ}(\omega)_{\gamma\delta} = -2c\omega^{-2}[\sigma^{pIJ}(\omega)_{\gamma\lambda} - \sigma^{pIJ}(0)_{\gamma\lambda}], \tag{237}$$

$$\varepsilon_{\lambda\mu\delta} R_{J\mu} \lambda^{IJ}(\omega)_{\gamma\delta} = -2c\hat{\sigma}^{pIJ}(\omega)_{\gamma\lambda}, \tag{237'}$$

$$\sum_{J=1}^{N} \varepsilon_{\lambda\mu\delta} R_{J\mu} \hat{\lambda}^{IJ}(\omega)_{\gamma\delta} = -2c\omega^{-2}[\sigma^{pI}(\omega)_{\gamma\lambda} - \sigma^{pI}(0)_{\gamma\lambda}], \tag{238}$$

$$\sum_{J=1}^{N} \varepsilon_{\lambda\mu\delta} R_{J\mu} \lambda^{IJ}(\omega)_{\gamma\delta} = -2c\hat{\sigma}^{pI}(\omega)_{\gamma\lambda}, \tag{238'}$$

and the atomic contributions to the molecular polarizability are related to the atomic contributions to the rotatory power tensor:

$$\varepsilon_{\alpha\beta\gamma} \alpha^{I}(\omega)_{\beta\delta} R_{I\gamma} = 2c\hat{\kappa}^{I}(\omega)_{\delta\alpha}, \tag{239}$$

$$\sum_{I=1}^{N} \varepsilon_{\alpha\beta\gamma} \alpha^{I}(\omega)_{\beta\delta} R_{I\gamma} = 2c\hat{\kappa}(\omega)_{\delta\alpha}, \tag{240}$$

$$\sum_{I=1}^{N} \varepsilon_{\alpha\beta\gamma} \tilde{\alpha}^{I}(\omega)_{\delta\beta} R_{I\gamma} = 2c\hat{\kappa}(\omega)_{\delta\alpha}. \tag{240'}$$

The additivity scheme for the optical rotatory power and the pair partitioning schemes for the polarizabilities and the susceptibilities are also correlated:

$$\varepsilon_{\delta\gamma\alpha} \varepsilon_{\lambda\mu\beta} R_{I\gamma} R_{J\mu} \alpha^{IJ}(\omega)_{\alpha\beta} = 4c^2\omega^{-2}[\chi^{pIJ}(\omega)_{\delta\lambda} - \chi^{pIJ}(0)_{\delta\lambda}],$$

$$\varepsilon_{\delta\gamma\alpha} R_{I\gamma} \hat{\kappa}^{I}(\omega)_{\alpha\lambda} = -2c\omega^{-2}[\chi^{pI}(\omega)_{\delta\lambda} - \chi^{pI}(0)_{\delta\lambda}],$$

$$\sum_{I,J=1}^{N} \varepsilon_{\delta\gamma\alpha} \varepsilon_{\lambda\mu\beta} R_{I\gamma} R_{J\mu} \alpha^{IJ}(\omega)_{\alpha\beta} = -2c \sum_{I=1}^{N} \varepsilon_{\delta\gamma\alpha} R_{I\gamma} \hat{\kappa}^{I}(\omega)_{\alpha\lambda}$$
$$= 4c^2\omega^{-2}[\chi^{p}(\omega)_{\delta\lambda} - \chi^{p}(0)_{\delta\lambda}]. \tag{241}$$

The Hessian of the unperturbed molecular energy gives the force constants [39]

$$\nabla_I \nabla_J \langle \psi_0 | H_0 | \psi_0 \rangle = \Omega_{IJ} - \tfrac{1}{2} \langle \psi_0 | (\nabla_I \mathbf{F}_J^n + \nabla_J \mathbf{F}_I^n) | \psi_0 \rangle$$
$$- 2 \sum_{j \neq 0} \langle \psi_0 | \mathbf{F}_n^I | \psi_j \rangle \langle \psi_j | \mathbf{F}_n^J | \psi_0 \rangle (E_j - E_0)^{-1}. \quad (242)$$

The first term on the right side contains the derivatives of the nuclear repulsions; the third is the electronic relaxation term and, for $\omega = 0$, it can be rewritten as $Z_I Z_J \boldsymbol{\phi}^{IJ}(0)$, where

$$\boldsymbol{\phi}^{IJ}(\omega) = -\frac{e^2}{\hbar} \sum_{j \neq 0} \frac{2\omega_{j0}}{\omega_{j0}^2 - \omega^2} \langle \psi_0 | \mathbf{E}_I^n | \psi_j \rangle \langle \psi_j | \mathbf{E}_J^n | a \rangle. \quad (243)$$

Then we find a connection between the electric shielding and the relaxation term:

$$\boldsymbol{\gamma}^I(\omega) = -m^{-1} \omega^{-2} \sum_{J=1}^{N} Z_J [\boldsymbol{\phi}^{IJ}(\omega) - \boldsymbol{\phi}^{IJ}(0)]. \quad (244)$$

Equations (186), (225), and (229) can be interpreted as relations involving generalized susceptibilities in ω space. In configuration space they are replaced by the Ehrenfest relations, written in terms of the observables (69)–(76). From (67) and

$$\frac{i}{\hbar} \langle a | [\mu_\alpha, m_\beta] | a \rangle B(0)_\beta = -\frac{e^2}{mc} \langle a | A_\alpha | a \rangle,$$

where $A_\alpha = \varepsilon_{\alpha\beta\gamma} B_\beta R_\gamma$, one finds that the equation of motion

$$\frac{d}{dt} \langle \mu_\alpha \rangle = -\frac{e}{m} \langle \Pi_\alpha \rangle \quad (245)$$

is satisfied. Accordingly, the static velocity theorem can be written in terms of Eqn. (99) [see also Eqn. (187)]

$$\langle \Pi_\alpha \rangle = v(0)_{\alpha\beta} B_\beta = 0,$$

$$(P_\alpha, L_\beta)_{-1} = -\frac{2mc}{e} v^{\mathrm{p}}(0)_{\alpha\beta} = m \varepsilon_{\alpha\beta\gamma} \langle a | R_\gamma | a \rangle. \quad (246)$$

Within the quadrupole approximation, the magnetic field is irrotational and the Hermitian operator representing the Lorentz magnetic force on the electrons is

$$F^{B}_{n\alpha} = -\frac{e}{mc}\,\varepsilon_{\alpha\beta\gamma}B(0)_{\gamma}P_{\beta}; \tag{247}$$

the operator for the electric Lorentz force is

$$F^{E}_{n\alpha} = -enE(0)_{\alpha} - eR_{\gamma}E(0)_{\gamma\alpha}, \tag{248}$$

so that

$$\langle a|F^{L}_{n\alpha}|a\rangle = \langle a|F^{E}_{n\alpha}|a\rangle + \langle a|F^{B}_{n\alpha}|a\rangle.$$

From the results

$$\frac{i}{\hbar}[P_{\alpha}, m_{\beta}] = \frac{e}{2mc}\,\varepsilon_{\alpha\beta\gamma}P_{\gamma},$$

$$\frac{i}{\hbar}[P_{\alpha}, \Theta_{\beta\gamma}] = eR_{\alpha}\delta_{\beta\gamma} - \tfrac{3}{2}e(R_{\gamma}\delta_{\alpha\beta} + R_{\beta}\delta_{\alpha\gamma}),$$

and the Maxwell equation

$$\varepsilon_{\alpha\beta\gamma}E(0)_{\beta\gamma} = -\frac{1}{c}\dot{B}(0)_{\alpha},$$

we find from Eqn. (67) the equation of motion

$$\frac{d}{dt}\langle \Pi_{\alpha}\rangle = \langle F^{N}_{n\alpha}\rangle + \langle a|F^{L}_{n\alpha}|a\rangle. \tag{249}$$

The operator representing the torque on the electrons due to the external electric field (Lorentz electric torque) is

$$K^{E}_{n\alpha} = \varepsilon_{\alpha\beta\gamma}[\mu_{\beta}E(0)_{\gamma} + \mu_{\beta\delta}E(0)_{\delta\gamma}] \tag{250}$$

within the quadrupole approximation. The operator representing the torque on the electrons arising from the external magnetic field (Lorentz magnetic torque) can be written

$$K^{B}_{n\alpha} = -\frac{e}{2mc}\,B_{\beta}(0)\sum_{i=1}^{n}[r_{\beta}p_{\alpha} + p_{\alpha}r_{\beta} - (r_{\gamma}p_{\gamma} + p_{\gamma}r_{\gamma})\delta_{\alpha\beta}]_{i}. \tag{251}$$

The first-order contribution to the oscillating magnetic dipole moment is

$$m_\alpha^{(1)} = -\frac{e^2}{2mc^2}\varepsilon_{\alpha\beta\gamma}\sum_{i=1}^{n}(r_\beta A_\gamma)_i,$$

with $A_{i\gamma} = \frac{1}{2}\varepsilon_{\gamma\delta\varepsilon}B_\delta r_{i\varepsilon}$ in the Bloch gauge (14). Then, using the Maxwell equation

$$\dot{B}_\alpha = -c\varepsilon_{\alpha\beta\gamma}E_{\beta\gamma},$$

one has, to first order in the field,

$$\frac{\partial m_\alpha^{(1)}}{\partial t} = \frac{e^2}{4mc}\varepsilon_{\alpha\beta\gamma}[E(0)_{\delta\gamma} - E(0)_{\gamma\delta}]\sum_{i=1}^{n}(r_\beta r_\delta)_i,$$

and using the results

$$\frac{i}{\hbar}[m_\alpha, \mu_\beta]E(0)_\beta = -\frac{e}{2mc}\varepsilon_{\alpha\beta\gamma}\mu_\beta E(0)_\gamma, \tag{252}$$

$$\frac{i}{\hbar}[m_\alpha, m_\beta]B(0)_\beta = -\frac{e}{2mc}\varepsilon_{\alpha\beta\gamma}m_\beta B(0)_\gamma, \tag{253}$$

$$\begin{aligned}\frac{i}{3\hbar}[m_\alpha, \Theta_{\beta\gamma}]E(0)_{\beta\gamma} &= -\frac{e}{6mc}(\varepsilon_{\alpha\lambda\beta}\Theta_{\lambda\gamma} + \varepsilon_{\alpha\lambda\gamma}\Theta_{\lambda\beta})E(0)_{\beta\gamma}\\ &= \frac{e^2}{4mc}\varepsilon_{\alpha\beta\gamma}[E(0)_{\gamma\delta} + E(0)_{\delta\gamma}]\sum_{i=1}^{n}(r_\beta r_\delta)_i,\end{aligned} \tag{254}$$

one finds from (67)

$$\frac{d}{dt}\langle m'_\alpha\rangle = -\frac{e}{2mc}\langle K_{n\alpha}^N\rangle - \frac{e}{2mc}\langle a|K_{n\alpha}^L|a\rangle, \tag{255}$$

where

$$K_{n\alpha}^L = K_{n\alpha}^E + K_{n\alpha}^B. \tag{256}$$

Some molecular tensors (electric dipole polarizability, electric and magnetoelectric shielding) are origin independent, as can be immediately found by inspection of definitions (87)–(112). Other tensors depend on the origin assumed for the multipole expansion. For instance, in a change of origin

$$\mathbf{r}'' = \mathbf{r}' + \mathbf{d}, \tag{257}$$

$$\hat{\kappa}_{\alpha\beta}(\mathbf{r}'') = \hat{\kappa}_{\alpha\beta}(\mathbf{r}') - \frac{1}{2c}\varepsilon_{\beta\gamma\delta} d_\delta \alpha_{\alpha\gamma}, \qquad \mathrm{Tr}\{\hat{\boldsymbol{\kappa}}(\mathbf{r}'')\} = \mathrm{Tr}\{\hat{\boldsymbol{\kappa}}(\mathbf{r}')\}, \tag{258}$$

$$A_{\alpha,\beta\gamma}(\mathbf{r}'') = A_{\alpha,\beta\gamma}(\mathbf{r}') - \tfrac{3}{2} d_\beta \alpha_{\gamma\alpha} - \tfrac{3}{2} d_\gamma \alpha_{\alpha\beta} + d_\delta \alpha_{\delta\alpha}\delta_{\beta\gamma}, \tag{259}$$

$$\chi^{\mathrm{p}}_{\alpha\beta}(\mathbf{r}'') + \chi^{\mathrm{d}}_{\alpha\beta}(\mathbf{r}'') = \chi^{\mathrm{p}}_{\alpha\beta}(\mathbf{r}') + \chi^{\mathrm{d}}_{\alpha\beta}(\mathbf{r}') + \frac{1}{4c^2}\omega^2 \varepsilon_{\alpha\gamma\delta}\varepsilon_{\beta\lambda\mu} d_\delta d_\mu \alpha_{\gamma\lambda}$$

$$- \frac{1}{2c}\omega^2(\varepsilon_{\alpha\gamma\delta} d_\delta \hat{\kappa}_{\gamma\beta} + \varepsilon_{\beta\gamma\delta} d_\delta \hat{\kappa}_{\gamma\alpha}), \tag{260}$$

$$\hat{D}_{\alpha,\beta\gamma}(\mathbf{r}'') = \hat{D}_{\alpha,\beta\gamma}(\mathbf{r}') + \tfrac{3}{2} d_\beta \hat{\kappa}_{\gamma\alpha} + \tfrac{3}{2} d_\gamma \hat{\kappa}_{\beta\alpha} - d_\lambda \hat{\kappa}_{\lambda\alpha}\delta_{\beta\gamma}$$

$$- \frac{1}{2c}\varepsilon_{\alpha\lambda\mu} d_\lambda A_{\mu,\beta\gamma}(\mathbf{r}''), \tag{261}$$

$$\sigma^{\mathrm{p}I}_{\alpha\beta}(\mathbf{r}'') + \sigma^{\mathrm{d}I}_{\alpha\beta}(\mathbf{r}'') = \sigma^{\mathrm{p}I}_{\alpha\beta}(\mathbf{r}') + \sigma^{\mathrm{d}I}_{\alpha\beta}(\mathbf{r}') - \frac{\omega^2}{2c}\varepsilon_{\beta\gamma\delta} d_\delta \hat{\lambda}^I_{\alpha\gamma}, \tag{262}$$

$$\hat{\tau}^I_{\alpha,\beta\gamma}(\mathbf{r}'') = \hat{\tau}^I_{\alpha,\beta\gamma}(\mathbf{r}') - \tfrac{3}{2} d_\beta \hat{\lambda}^I_{\alpha\gamma} - \tfrac{3}{2} d_\gamma \hat{\lambda}^I_{\alpha\beta} + d_\delta \hat{\lambda}^I_{\alpha\delta}\delta_{\beta\gamma}, \tag{263}$$

$$\hat{\xi}^I_{\alpha\beta}(\mathbf{r}'') = \hat{\xi}^I_{\alpha\beta}(\mathbf{r}') + \frac{1}{2c}\varepsilon_{\beta\gamma\delta}\gamma^I_{\alpha\gamma} d_\delta, \tag{264}$$

$$\nu^I_{\alpha,\beta\gamma}(\mathbf{r}'') = \nu^I_{\alpha,\beta\gamma}(\mathbf{r}') + \tfrac{3}{2} d_\beta \gamma^I_{\alpha\gamma} + \tfrac{3}{2} d_\gamma \gamma^I_{\alpha\beta} - d_\delta \gamma^I_{\alpha\delta}\delta_{\beta\gamma}, \tag{265}$$

and similar equations are found for the other properties. Equations (258), (262), and (264) are valid if the hypervirial relations (115) are exactly satisfied. For instance, to fulfill this requirement within the coupled Hartree–Fock method and random-phase approximation [10, 12], a complete basis set must be used. In any numerical calculation adopting a truncated basis, the equations describing the origin dependence of molecular tensors must be properly interpreted. Thus, for instance, if the optical activity tensor is defined via Eqn. (90), then in Eqn. (258) the electric polarizability in mixed length–velocity gauge must be used. Also, if the electromagnetic shielding (138) is dealt with in Eqn. (264), then the electric shielding must be expressed in the velocity formalism.

To some purposes, one can define new molecular tensors that are independent of the origin [40]. At any rate, it can be easily proven that the induced moments (69)–(72) and fields (73)–(74) are, order by order, independent of the origin chosen for the multipole expansion, provided that all the terms of the same order of magnitude are retained. Thus, within the quadrupole approximation, both the magnetic field and the electric field gradient must be taken into account. Then, exploiting again the Maxwell equations and observing that the electric field at the new origin is

$$E_\beta(\mathbf{r}'', t) = E_\beta(\mathbf{r}', t) + d_\gamma E_{\gamma\beta}(0, t), \tag{266}$$

one obtains, for instance,

$$\begin{aligned} \Delta\langle \mu_\alpha(\mathbf{r}'')\rangle &= \Delta\langle \mu_\alpha(\mathbf{r}')\rangle, \\ \Delta\langle m_\alpha(\mathbf{r}'')\rangle &= \Delta\langle m_\alpha(\mathbf{r}')\rangle, \\ \Delta\langle E^n_{I\alpha}(\mathbf{r}'')\rangle &= \Delta\langle E^n_{I\alpha}(\mathbf{r}')\rangle. \end{aligned} \tag{267}$$

In a static magnetic field, the sum of the diamagnetic and paramagnetic contributions to susceptibility and magnetic shielding is origin independent, as can be immediately observed from Eqs. (93) and (141), but (91) and (92) and (139) and (140) vary, depending on the coordinate system. The change of origin can be described as a gauge transformation, the diamagnetic and the paramagnetic contributions in the new system being related to the old ones:

$$\begin{aligned} \chi^{\mathrm{d}}_{\alpha\beta}(\mathbf{r}'') = \chi^{\mathrm{d}}_{\alpha\beta}(\mathbf{r}') + \frac{e^2}{4mc^2}\{&2\langle R_\gamma\rangle d_\gamma \delta_{\alpha\beta} - \langle R_\alpha\rangle d_\beta \\ &- \langle R_\beta\rangle d_\alpha - n d_\gamma d_\gamma \delta_{\alpha\beta} + n d_\alpha d_\beta\}, \end{aligned} \tag{268}$$

$$\begin{aligned} \chi^{\mathrm{p}}_{\alpha\beta}(\mathbf{r}'') = \chi^{\mathrm{p}}_{\alpha\beta}(\mathbf{r}') + \frac{e^2}{4m^2c^2}\{&d_\delta[\varepsilon_{\alpha\gamma\delta}(P_\gamma, L_\beta)_{-1} + \varepsilon_{\beta\gamma\delta}(P_\gamma, L_\alpha)_{-1}] \\ &+ \varepsilon_{\alpha\gamma\delta}\varepsilon_{\beta\lambda\mu} d_\delta d_\mu (P_\gamma, P_\lambda)_{-1}\}, \end{aligned} \tag{269}$$

$$\sigma^{\mathrm{p}I}_{\alpha\beta}(\mathbf{r}'') = \sigma^{\mathrm{p}I}_{\alpha\beta}(\mathbf{r}') - \frac{e^2}{2m^2c^2}\varepsilon_{\beta\gamma\delta} d_\delta (M^n_{I\alpha}, P_\gamma)_{-1}, \tag{270}$$

$$\sigma^{\mathrm{d}I}_{\alpha\beta}(\mathbf{r}'') = \sigma^{\mathrm{d}I}_{\alpha\beta}(\mathbf{r}') - \frac{e}{2mc^2}(d_\gamma\langle E^n_{I\gamma}\rangle\delta_{\alpha\beta} - d_\alpha\langle E^n_{I\beta}\rangle), \tag{271}$$

where, for example, $\langle R_\alpha\rangle$, is the shorthand notation for $\langle a|R_\alpha|a\rangle$.

From these equations one finds the constraints for translational invariance of magnetizability and magnetic shielding [31]:

$$(P_\gamma, L_\alpha)_{-1} = -(P_\alpha, L_\gamma)_{-1} = m\langle a|R_\beta|a\rangle, \tag{272}$$

$$(P_\alpha, P_\beta)_{-1} = mn\delta_{\alpha\beta}, \tag{273}$$

$$(P_\gamma, M^n_{I\alpha})_{-1} = -(P_\alpha, M^n_{I\gamma})_{-1} = \frac{m}{e}\langle a|E^n_{I\beta}|a\rangle, \tag{274}$$

where α, β, γ are a cyclic permutation of x, y, z. These conditions also express the conservation of the current density [31] and are related to Eqs. (204), (223), and (246). They can also be directly obtained from definitions (194), (195) and (199) using the hypervirial relations (115) and the commutation rules

$$[L_\alpha, R_\beta] = i\hbar\varepsilon_{\alpha\beta\gamma}R_\gamma,$$

$$[P_\alpha, R_\beta] = -in\hbar\delta_{\alpha\beta},$$

$$[M^n_{I\alpha}, R_\beta] = \frac{i\hbar}{e}\varepsilon_{\alpha\beta\gamma}E^n_{I\gamma},$$

Another set of sum rules for the static electromagnetic shieldings is provided by the virial theorem [9a, 41, 42], that is, the hypervirial theorem

$$\left\langle a\left|\frac{i}{\hbar}[H, \mathscr{V}]\right|a\right\rangle = \left\langle a\left|\sum_{i=1}^{n}\left(\frac{\partial H}{\partial \mathbf{p}_i}\cdot\mathbf{p}_i - \mathbf{r}_i\cdot\frac{\partial H}{\partial \mathbf{r}_i}\right)\right|a\right\rangle = 0,$$

for the virial operator

$$\mathscr{V} = \sum_{i=1}^{n}\mathbf{r}_i\cdot\mathbf{p}_i - \tfrac{3}{2}i\hbar n.$$

In the absence of perturbation, the virial relation is [9a]

$$\langle H_0\rangle + \langle T_n^{(0)}\rangle = \sum_{I=1}^{N} R_{I\alpha}\langle F_{I\alpha}^{(0)}\rangle, \tag{275}$$

where H_0 is the unperturbed Hamiltonian (6), $T_n^{(0)}$ is the kinetic energy operator for the electrons, and

$$F_{I\alpha}^{(0)} = F_{I\alpha}^{n} + F_{I\alpha}^{N-I} = -\frac{\partial H_0}{\partial R_{I\alpha}}$$

is the force on nucleus I in the absence of perturbation for a given molecular geometry. Let us now introduce a nonuniform static electric field

$$E_\alpha(\mathbf{r}) = E_\alpha(0) + r_\beta E_{\beta\alpha},$$

where, according to the quadrupole approximation, $E_\alpha(0)$ is the field at the origin and the gradient $E_{\alpha\beta}$ is uniform over the molecular dimensions. Therefore we add to the molecular Hamiltonian (6) the electric dipole terms

$$-\mu_\alpha E_\alpha - e\sum_{I=1}^{N} Z_I R_{I\alpha}E_\alpha$$

and the electric quadrupole terms

$$-\tfrac{1}{3}\Theta_{\alpha\beta}E_{\alpha\beta} - \tfrac{1}{6}e\sum_{I=1}^{N} Z_I(3R_{I\alpha}R_{I\beta} - R_I^2\delta_{\alpha\beta})E_{\alpha\beta}.$$

Since the dipole term is homogeneous of degree 1 and the quadrupole term is homogeneous of degree 2 in the $\mathbf{r}_i$ and $\mathbf{R}_I$, from Euler's theorem on homogeneous functions we have [9a, 41]

$$\langle H_0\rangle + \langle T_n^{(0)}\rangle + \mathscr{M}_\alpha E_\alpha + \mathscr{M}_{\alpha\beta}E_{\alpha\beta} = \sum_{I=1}^{N} R_{I\alpha}\langle F_{I\alpha}\rangle, \tag{276}$$

where $\mathscr{M}_\alpha$ has been previously defined in Eqn. (167) and

$$\mathscr{M}_{\alpha\beta} = -e\left\langle \sum_{i=1}^{n} (r_\alpha r_\beta)_i \right\rangle + e\sum_{I=1}^{N} Z_I R_{I\alpha}R_{I\beta}. \tag{277}$$

The expectation values are taken over

$$|a\rangle + |a^{E_\alpha}\rangle E_\alpha(0) + |a^{E_{\alpha\beta}}\rangle E_{\alpha\beta},$$

and the operator for the force on nucleus I in the presence of the nonuniform electric field is [see also Eqn. (170)]

$$F_{I\alpha} = F_{I\alpha}^{n} + F_{I\alpha}^{N-I} + Z_I e E_\alpha(0) + Z_I e R_{I\beta}E_{\beta\alpha}. \tag{278}$$

The first-order functions are

$$|a^{E_\alpha}\rangle = \frac{1}{\hbar}\sum_{j\neq a}\omega_{ja}^{-1}|j\rangle\langle j|\mu_\alpha|a\rangle, \tag{279}$$

$$|a^{E_{\alpha\beta}}\rangle = \frac{1}{3\hbar}\sum_{j\neq a}\omega_{ja}^{-1}|j\rangle\langle j|\Theta_{\alpha\beta}|a\rangle. \tag{280}$$

Hence, to first order in E_α, Eqn. (276) gives [8, 9a]

$$\sum_{I=1}^{N} Z_I R_{I\alpha}\gamma^I(0)_{\alpha\beta} = \langle a|R_\beta|a\rangle + \frac{1}{\hbar}\sum_{j\neq a}\frac{2}{\omega_{ja}}\Re(\langle a|T_n^{(0)}|j\rangle\langle j|R_\beta|a\rangle), \tag{281}$$

and, using Eqn. (178),

$$\sum_{I=1}^{N} R_{I\alpha}\nabla_{I\alpha}\mathscr{M}_\beta = \mathscr{M}_\beta + \frac{1}{\hbar}\sum_{j\neq a}\frac{2}{\omega_{ja}}\Re(\langle a|T_n^{(0)}|j\rangle\langle j|\mu_\beta|a\rangle). \tag{281$'$}$$

Owing to the translational sum rule (176) for dipole derivatives, (281′) is origin independent for electrically neutral molecules. To first order in $E_{\alpha\beta}$, Eqn. (276) gives

$$\sum_{I=1}^{N} Z_I e R_{I\alpha} v^I(0)_{\alpha,\beta\gamma} = 2\langle a|\Theta_{\beta\gamma}|a\rangle + \frac{1}{\hbar}\sum_{j\neq a}\frac{2}{\omega_{ja}}\Re(\langle a|T_n^{(0)}|j\rangle\langle j|\Theta_{\beta\gamma}|a\rangle), \tag{282}$$

$$\sum_{I=1}^{N} R_{I\alpha}\nabla_{I\alpha}\mathcal{Q}_{\beta\gamma} = 2\mathcal{Q}_{\beta\gamma} + \frac{1}{\hbar}\sum_{j\neq a}\frac{2}{\omega_{ja}}\Re(\langle a|T_n^{(0)}|j\rangle\langle j|\Theta_{\beta\gamma}|a\rangle), \tag{282'}$$

where

$$\mathcal{Q}_{\alpha\beta} = \langle a|\Theta_{\alpha\beta}|a\rangle + \frac{e}{2}\sum_{I=1}^{N} Z_I(3R_{I\alpha}R_{I\beta} - R_I^2\delta_{\alpha\beta}) \tag{283}$$

is the molecular quadrupole moment and its derivative is

$$\nabla_{I\alpha}\mathcal{Q}_{\beta\gamma} = Z_I e[\tfrac{3}{2}(R_{I\gamma}\delta_{\alpha\beta} + R_{I\beta}\delta_{\alpha\gamma}) - R_{I\alpha}\delta_{\beta\gamma} + v^I(0)_{\alpha,\beta\gamma}]. \tag{284}$$

Acknowledgments

It is a pleasure to acknowledge a long correspondence, important suggestions, and useful comments raised by Professor S. T. Epstein. I also wish to thank my friend Dr. R. Zanasi for several discussions on this work.

References

1. R. Kubo, *J. Phys. Soc. Japan 12*:570 (1957).
2. S. V. Tyablikov, *Methods in the Quantum Theory of Magnetism*, Plenum, New York, 1967.
3. D. N. Zubarev, *Nonequilibrium Statistical Thermodynamics*, Consultants Bureau, New York, 1974.
4. J. H. Van Vleck, *The Theory of the Electric and Magnetic Susceptibilities*, Oxford University Press, 1932.
5. D. W. Davies, *The Theory of the Electric and Magnetic Properties of Molecules*, Wiley, London, 1967.
6. A. D. Buckingham, *Adv. Chem. Phys. 12*:107 (1967).
7. N. F. Ramsey, *Phys. Rev. 78*:699 (1950); *86*:243 (1952); *Molecular Beams*, Oxford University Press, London, 1956.
8. H. Sambe, *J. Chem. Phys. 58*:4779 (1973).
9. S. T. Epstein, *The Variation Method in Quantum Chemistry*, Academic, New York, 1974; *Theor. Chim. Acta 61*:303 (1982); S. T. Epstein and R. E. Johnson, *J. Chem. Phys. 51*:188 (1969),

10. P. Lazzeretti and R. Zanasi, *Phys. Rev. A 24*:1696 (1981); *Phys. Rev. A 27*:1301 (1983); *J. Chem. Phys. 79*:889 (1983) and references therein.
11. P. W. Fowler and A. D. Buckingham, *Chem. Phys. 98*:167 (1985).
12. P. Lazzeretti and R. Zanasi, *Chem. Phys. Lett. 112*:103 (1984); *109*:89 (1984); *114*:79 (1985); *J. Chem. Phys. 83*:1218 (1985); *84*:3916 (1986).
13. J. F. Biarge, J. Herranz, and J. H. Morcillo, *An. R. Soc. Esp. Fis. Quim. Ser. A 57*:81 (1961); J. H. Morcillo, L. J. Zamorano, and J. M. V. Heredia, *Spectrochim. Acta 22*:1969 (1966).
14. W. B. Person and G. Zerbi (eds.), *Vibrational Intensities in Infrared and Raman Spectroscopy*, Elsevier, Amsterdam, 1982; W. B. Person and J. H. Newton, *J. Chem. Phys. 61*:1040 (1974); *64*:3036 (1976); B. A. Zilles and W. B. Person, *J. Chem. Phys. 79*:65 (1983).
15. P. Lazzeretti and R. Zanasi, *Phys. Rev. A 33*:3727 (1986).
16. A. D. Buckingham, P. W. Fowler, and P. A. Galwas, *Chem. Phys. 112*:1 (1987).
17. P. J. Stephens, *J. Phys. Chem. 89*:748 (1985).
18. M. A. Lowe and P. J. Stephens, *Annu. Rev. Phys. Chem. 36*:231 (1985).
19. M. A. Lowe, G. A. Segal, and P. J. Stephens, *J. Am. Chem. Soc. 108*:248 (1986).
20. P. Lazzeretti, R. Zanasi, and P. J. Stephens, *J. Phys. Chem. 90*:6761 (1986).
21. L. Landau and E. Lifshitz, *Théorie du Champ*, Mir, Moscow, 1966.
22. L. Landau and E. Lifshitz, *Mécanique Quantique*, Mir, Moscow, 1966.
23. P. W. Langhoff, S. T. Epstein, and M. Karplus, *Rev. Mod. Phys. 44*:602 (1972).
24. M. Goeppert-Mayer, *Ann. Phys. 9*:273 (1931).
25. P. I. Richards, *Phys. Rev. 73*:254 (1948).
26. J. Fiutak, *Can. J. Phys. 41*:12 (1963).
27. F. Bloch, in *W. Heisenberg und die Physik unserer Zeit*, Friedr. Vieweg & Sohn, Braunschweig, 1961, p. 93. This reference has been brought to the attention of the author by S. T. Epstein.
28. L. D. Barron and C. G. Gray, *J. Phys. A: Math. Nucl. Gen. 6*:59 (1973).
29. R. G. Woolley, *J. Phys. B: Atom. Mol. Phys. 6*:L97 (1973).
30. R. E. Raab, *Molec. Phys. 29*:1323 (1975).
31. P. Lazzeretti and R. Zanasi, *Phys. Rev. A 32*:2607 (1985).
32. P. Lazzeretti and R. Zanasi, *Chem. Phys. Lett. 118*:217 (1985).
33. H. Goldstein, *Classical Mechanics*, 2nd ed., Addison-Wesley, Reading, Mass., 1980.
34. E. Merzbacher, *Quantum Mechanics*, 2nd ed., Wiley, New York, 1983.
35. A. Nisbet, *Proc. Roy. Soc. A 231*:250 (1955) and references therein.
36. W. McCrea, *Proc. Roy. Soc. A 240*:447 (1957).
37. P. Pascal, *Ann. Phys. Chim. 19*:5 (1910); *25*:289 (1912); *29*:218 (1913); K. G. Denbigh, *Trans. Faraday Soc. 36*:936 (1940); J. Hoarau, *Ann. Chim. 13*:544 (1956).
38. R. R. Birss, *Rep. Progr. Phys. 26*:307 (1963).
39. J. Gerratt and I. M. Mills, *J. Chem. Phys. 49*:1719 (1963).
40. A. D. Buckingham and M. B. Dunn, *J. Chem. Soc. A* 1988 (1971).
41. L. Landau and E. Lifshitz, *Mécanique*, Mir, Moscow, 1966.
42. G. Marc and W. G. McMillan, *Adv. Chem. Phys. 58*:209 (1985).

AUTHOR INDEX

Numbers in parentheses are reference numbers and indicate that the author's work is referred to although his name is not mentioned in the text. Numbers in *italics* show the pages on which the complete references are listed.

SUBJECT INDEX